THE METHOD OF UNDETERMINED COEFFICIENTS

The following table lists trial solutions for the DE $P(D)y = F(x)$, where $P(D)$ is a polynomial differential operator.

$F(x)$	Usual trial solution	Modified trial solution
$cx^k e^{ax}$	If $P(a) \neq 0$: $y_p(x) = e^{ax}(A_0 + A_1 x + \cdots$ $\qquad + A_k x^k)$.	If a is a root of $P(r) = 0$ of multiplicity m: $y_p(x) = x^m e^{ax}(A_0 + A_1 x + \cdots + A_k x^k)$.
$cx^k e^{ax}\cos bx$ or $cx^k e^{ax}\sin bx$	If $P(a + ib) \neq 0$: $y_p(x) = e^{ax}[A_0\cos bx + B_0\sin bx$ $\quad + x(A_1\cos bx + B_1\sin bx)$ $\quad + \cdots + x^k(A_k\cos bx + B_k\sin bx)]$.	If $a + ib$ is a root of $P(r) = 0$ of multiplicity m: $y_p(x) = x^m e^{ax}[A_0\cos bx + B_0\sin bx$ $\quad + x(A_1\cos bx + B_1\sin bx)$ $\quad + \cdots + x^k(A_k\cos bx + B_k\sin bx)]$

If $F(x)$ is the sum of functions of the preceding form then the appropriate trial solution is the corresponding sum.

BASIC INTEGRALS

Function $F(x)$	Integral $\int F(x)\, dx$
$x^n,\ n \neq -1$	$\dfrac{1}{n + 1} x^{n+1} + c$
x^{-1}	$\ln |x| + c$
$e^{ax},\ a \neq 0$	$\dfrac{1}{a} e^{ax} + c$
$\sin x$	$-\cos x + c$
$\cos x$	$\sin x + c$
$\tan x$	$\ln |\sec x| + c$
$\sec x$	$\ln |\sec x + \tan x| + c$
$\csc x$	$\ln |\csc x - \cot x| + c$
$e^{ax}\sin bx$	$\dfrac{1}{a^2 + b^2} e^{ax}(a\sin bx - b\cos bx) + c$
$e^{ax}\cos bx$	$\dfrac{1}{a^2 + b^2} e^{ax}(a\cos bx + b\sin bx) + c$
$\ln x$	$x \ln x - x + c$
$\dfrac{1}{a^2 + x^2}$	$\dfrac{1}{a} \tan^{-1}(x/a) + c$
$\dfrac{1}{\sqrt{a^2 - x^2}},\ a > 0$	$\sin^{-1}(x/a) + c$
$\dfrac{1}{\sqrt{a^2 + x^2}}$	$\ln| x + \sqrt{a^2 + x^2} | + c$
$\dfrac{f'(x)}{f(x)}$	$\ln |f(x)| + c$
$e^{u(x)} \dfrac{du}{dx}$	$e^{u(x)} + c$

Differential Equations and Linear Algebra

Second Edition

Differential Equations and Linear Algebra

STEPHEN W. GOODE

California State University, Fullerton

Prentice Hall, Upper Saddle River, New Jersey 07458

Library of Congress Cataloging-in-Publication Data

Goode, Stephen W.
    An introduction to differential equations and linear algebra /
Stephen W. Goode. — 2nd ed.
      p.   cm.
    Includes bibliographical references and index.
    ISBN 0-13-263757-X
    1. Differential equations.   2. Algebras, Linear.   I. Title.
QA371.G644   2000
515'.35—dc21   99-36734
                                              CIP

Acquisitions Editor: George Lobell
Editor-in-Chief: Jerome Grant
Editorial Assistant: Gale Epps
Editorial/Production Supervision: Bob Walters
Senior Managing Editor: Linda Mihatov Behrens
Executive Managing Editor: Kathleen Schiaparelli
Assistant Vice President Production and Manufacturing: David W. Riccardi
Manufacturing Buyer: Alan Fischer
Manufacturing Manager: Trudy Pisciotti
Marketing Manager: Melody Marcus
Art Director: Jayne Conte
Cover Designer: Bruce Kenselaar
Cover Photo: Clark County Library, Las Vegas, Nevada, exterior circulation—
        Michael Graves & Associates/Bob Freund

Printed in the United States of America
10 9 8 7 6 5 4 3 2 1

ISBN 0-13-263757-X

Prentice-Hall International (UK) Limited, *London*
Prentice-Hall of Australia Pty. Limited, *Sydney*
Prentice-Hall Canada Inc., *Toronto*
Prentice-Hall Hispanoamericana, S.A., *Mexico*
Prentice-Hall of India Private Limited, *New Delhi*
Prentice-Hall of Japan, Inc., *Tokyo*
Prentice-Hall (Singapore) Pte. Ltd.., *Singapore*
Editora Prentice-Hall do Brasil, Ltda., *Rio de Janeiro*

To the memory of my mother

Barbara Goode

Contents

5   Vector Spaces   271

6   Linear Transformations and the Eigenvalue/Eigenvector Problem   355

7   Linear Differential Equations of Order n 422

8   Systems of Differential Equations   448

9   The Laplace Transform and Some Elementary Applications   533

Preface

In *Differential Equations and Linear Algebra*, second edition, the material on linear algebra and differential equations required in many sophomore courses for mathematics, science, and engineering majors is introduced. In writing this text I have endeavored to develop an appreciation for the power of the general vector space framework in formulating and solving linear problems. My aim has been to present the material in a manner that is accessible to the student who has successfully completed three semesters of calculus, and it is definitely the intention that the student read the text, not just the examples. Almost all results are proved in detail. However, it is certainly possible to by-pass many of the proofs and use the text in a more problem solving based setting.

In this second edition of the text, there have been many changes (detailed below) that reflect both my own experience in regularly teaching the material over the past fifteen years, and also the current trends in the teaching of differential equations and linear algebra, including the use of technology in the classroom. Indeed, there are many instances in which the power of technology is illustrated, and a large majority of the exercise sets have problems that require some form of technology (computer algebra system (CAS) or graphing calculator) for their solution. These problems are designated with a ◆. The second edition has been written with maximum flexibility in mind to help accommodate the different emphases that can be placed in a combined differential equations and linear algebra course, the varying backgrounds of students who enroll in this type of course, and also the fact that different institutions have different credit values for such a course. The whole text can be covered in a five credit-hour course. For courses with a lower credit-hour value, some selectivity will have to be exercised. For example, in the differential equations parts of the text, it is possible to omit much of Chapter 1, Sections 2.5 – 2.7, 2.9 (if series solutions are not to be discussed), and some, or all, of Sections 8.8 – 8.12. The core material in linear algebra is given in Sections 3.1 – 3.6, 4.4 (for instructors who wish to de-emphasize the determinant), 5.1 – 5.7, 6.1, 6.3, 6.5 – 6.6. At California State University, Fullerton we have a four credit-hour course for sophomores which is based around the material in Chapters 1–8.

The CAS Maple has been used quite extensively both within the text, and also in constructing the exercises sets. Furthermore, most of the exercises have been checked using Maple.

Major Changes in the Second Edition

(1)   Applications of first-order differential equations are now incorporated into Chapter 1. They are first used to motivate the study of differential equations, and then they are used to illustrate the applicability of the results. In courses where the students have already had an introduction to the classical techniques for solving first-order differential equations, Chapter 1 could be used as a review.

(2)   A discussion of slope and direction fields, numerical solutions to first-order differential equations, and a general introduction to the phase plane have been included in Chapter 1. This allows more emphasis to be placed on the geometry of differential equations.

(3)   One of the most significant changes in the text is the insertion of Chapter 2, which gives an introduction to second-order linear differential equations and the standard techniques for their solution in the constant coefficient case. The primary motivation for this change is that it enables some of the key ideas from linear algebra (linear combinations, linear dependence, and linear independence) to be introduced in a concrete setting. Instructors of students who have already seen the basic solution techniques in a previous course can focus on the linear structure of the solutions, and then quickly move into the linear algebra part of the text.

(4)   Chapters 3 and 4 from the first edition have been combined, and new material has been added on elementary matrices and the LU factorization.

(5)   The chapter on determinants has been rearranged and streamlined. In response to the belief of many instructors that determinants should be de-emphasized, a summary section has been included that can be used as a stand-alone introduction to determinants.

(6)   The chapter on vector spaces has been almost totally rewritten. More emphasis is placed on $\mathbf{R}^n$ and visualization. Furthermore, since we already have the basic results from Chapter 2 on the structure of solutions to second-order linear differential equations, these can be used to motivate and illustrate the more general vector space concepts. A new section on the row space and column space of a matrix has been added. This aids in the discussion of linear transformations in the following chapter. There is also an optional section on the rank-nullity theorem which enables students to see the complete analogy between the structure of the solution set to a linear second-order differential equation, and the solution set to a linear system of algebraic equations.

(7)   The material on linear transformations has been rewritten with more emphasis on $\mathbf{R}^n$. A section on transformations of $\mathbf{R}^2$ has been included to help students to visualize the effects of a linear transformation. The discussion of the algebraic eigenvalue/eigenvector problem now appears within the linear transformation chapter, and its study is motivated from the solution of a system of linear differential equations. Solving a system of differential equations is also used as the motivation for studying the diagonalization problem. A brief introduction to quadratic forms has been included as an application of the orthogonal diagonalization problem.

(8)   Chapter 7 deals with the theory and solution techniques for nth order differential equations. The major emphasis here is on the application of the vector space techniques to derive the general theory. The discussion of solution techniques is fairly brief, since they are an extension of the techniques already presented in Chapter 2. The method of undetermined coefficients for differential equations of order higher than two is first introduced via guessing the appropriate trial solution, and then annihilators are introduced

to justify the method. I have found that introducing the technique in this manner is easier for students to understand, compared to launching straight into annihilators. Indeed, some instructors may choose not to cover the annihilator material.

(9)   The major change in the chapter on systems of differential equations (now Chapter 8) is the inclusion of a section dealing with the phase plane for linear autonomous systems, and a section on nonlinear systems. Technology is used in both sections to generate the appropriate phase portraits.

(10) The material on Laplace transforms and series solutions of differential equations is largely unaltered.

Acknowledgments

I would like to acknowledge the thoughtful input from the following external reviewers for the second edition:

Paul Hernandez, Palo Alto Community College

Johnny Henderson, Auburn University

I. Gary Rosen, University of Southern California

Donald Hartig, California Polytechnic State University, San Luis Obispo

Jeffrey Stopple, University of California, Santa Barbara

Martin Forrest, Louisiana State University

William Stout, Salve Regina University

Edith Mooers, University of California, Los Angeles

All of their comments were considered carefully in the final preparation of the text. Ernie Solheid at CSUF provided detailed comments on the manuscript, and also checked the final version thoroughly for mathematical accuracy.

I would also like to express my deep debt of gratitude to my wife, Christina, and my daughters, Megan and Tobi, who have provided an unbounded level of support, encouragement, and understanding throughout the development of this project.

Stephen W. Goode (sgoode@fullerton.edu)

1

First-Order Differential Equations

1.1 HOW DIFFERENTIAL EQUATIONS ARISE

In this section we will introduce the idea of a differential equation through the mathematical formulation of a variety of problems. These problems are then used throughout the chapter to illustrate the applicability of the techniques that will be introduced.

1. NEWTON'S SECOND LAW OF MOTION

Newton's second law of motion states that the rate of change of momentum of an object is equal to the sum of the applied forces that are acting on the object. If we consider an object of mass m moving in one dimension under the influence of a force F, then the mathematical statement of this law is

$$\frac{d}{dt}(mv) = F,$$

where $v(t)$ denotes the velocity of the object at time t. If the mass is constant, then this can be written in the familiar form

$$m\frac{dv}{dt} = F. \tag{1.1.1}$$

We let $y(t)$ denote the displacement of the object at time t. Then, using the fact that velocity and displacement are related via

$$v = \frac{dy}{dt},$$

1

it follows that (1.1.1) can be written as

$$m\frac{d^2y}{dt^2} = F. \tag{1.1.2}$$

This is an example of a **differential equation** (DE), so called because it involves *derivatives* of the unknown function $y(t)$.

As a specific example, consider the case of an object falling freely under the influence of gravity (see Figure 1.1.1). In this case the only force acting on the object is

Positive y-direction

mg

Figure 1.1.1   Particle falling under the influence of gravity.

$F = mg$, where g denotes the (constant) acceleration due to gravity. Choosing the positive y-direction as downward, it follows from equation (1.1.2) that the motion of the object is governed by the DE

$$m\frac{d^2y}{dt^2} = mg, \tag{1.1.3}$$

or equivalently,

$$\frac{d^2y}{dt^2} = g.$$

Since g is a constant, we can integrate this equation to determine $y(t)$. Performing one integration yields

$$\frac{dy}{dt} = gt + c_1,$$

where c_1 is an arbitrary integration constant. Integrating once more with respect to t we obtain

$$y(t) = \frac{1}{2}gt^2 + c_1 t + c_2, \tag{1.1.4}$$

where c_2 is a second integration constant. We see that the DE has an infinite number of solutions parametrized by the constants c_1 and c_2. In order to uniquely specify the motion, we must augment the DE with *initial conditions* that specify the initial position and initial velocity of the object. For example, if the object is released at $t = 0$ from $y = y_0$ with a velocity v_0, then, in addition to the DE, we have the initial conditions

$$y(0) = y_0, \quad \frac{dy}{dt}(0) = v_0. \tag{1.1.5}$$

These conditions must be imposed on the solution (equation (1.1.4)) in order to determine the values of c_1 and c_2 that correspond to the particular problem under investigation. Setting $t = 0$ in equation (1.1.4) and using the first initial condition from (1.1.5) we find that

$$y_0 = c_2.$$

Substituting this into equation (1.1.4), we get

$$y(t) = \frac{1}{2} gt^2 + c_1 t + y_0. \qquad (1.1.6)$$

In order to impose the second initial condition from (1.1.5), we first differentiate equation (1.1.6) to obtain

$$\frac{dy}{dt} = gt + c_1.$$

Consequently the second initial condition in (1.1.5) requires

$$c_1 = v_0.$$

From equation (1.1.6) it follows that the position of the object at time t is

$$y(t) = \frac{1}{2} gt^2 + v_0 t + y_0.$$

The DE (1.1.3) together with the initial conditions (1.1.5) is an example of an *initial-value problem*.

As a second application of Newton's law of motion, consider the spring-mass system depicted in Figure 1.1.2, where for simplicity we are neglecting frictional and external forces. In this case the only force acting on the mass is the restoring force (or

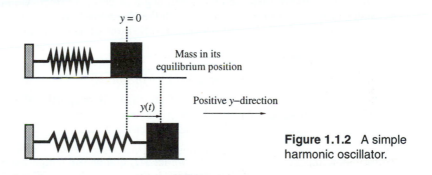

Figure 1.1.2   A simple harmonic oscillator.

spring force), F_s, due to the displacement of the spring from its equilibrium (unstretched) position. We use Hooke's law to model this force:

Hooke's Law

The restoring force of a spring is directly proportional to the displacement of the spring from its equilibrium position and is directed towards the equilibrium position.

If $y(t)$ denotes the displacement of the spring from its equilibrium position at time t (see Figure 1.1.2), then according to Hooke's law, the restoring force is

$$F_s = -ky,$$

where k is a positive constant called the **spring constant**. Consequently, Newton's law implies that the motion of the spring-mass system is governed by the DE

$$m\frac{d^2y}{dt^2} = -ky,$$

which we write in the equivalent form

$$\frac{d^2y}{dt^2} + \omega^2 y = 0, \tag{1.1.7}$$

where $\omega = (k/m)^{1/2}$. At present we cannot solve this DE. However, we leave it as an exercise to verify by direct substitution that

$$y(t) = A\cos(\omega t - \phi)$$

is a solution to the DE (1.1.7), where A and ϕ are constants (determined from the initial conditions for the problem). We see that the resulting motion is periodic with amplitude A. This is consistent with what we might expect physically, since no frictional forces or external forces are acting on the system. This type of motion is referred to as **simple harmonic motion**, and the physical system is called a **simple harmonic oscillator**.

2. NEWTON'S LAW OF COOLING

We now build a mathematical model describing the cooling (or heating) of an object. Suppose that we bring an object into a room. If the temperature of the object is hotter than that of the room, then the object will begin to cool. Further, we might expect that the major factor governing the rate at which the object cools is the temperature difference between it and the room. Indeed, according to Newton's law of cooling:

Newton's Law of Cooling

The rate of change of temperature of an object is proportional to the temperature difference between the object and its surrounding medium.

To formulate this law mathematically, we let $T(t)$ denote the temperature of the object at time t, and let $T_m(t)$ denote the temperature of the surrounding medium. Newton's law of cooling can then be expressed as the differential equation

$$\frac{dT}{dt} = -k(T - T_m), \tag{1.1.8}$$

where k is a constant. The minus sign in front of the constant k is traditional. It ensures that k will always be positive[1]. Once we have studied Section 1.4 it will be easy to show that, when T_m is constant, the solution to this DE is (see also Problem 6 at the end of this section)

$$T(t) = T_m + ce^{-kt}, \tag{1.1.9}$$

where c is a constant. Newton's law of cooling therefore predicts that as t approaches

---

[1]If $T > T_m$, then the object will cool, so that $dT/dt < 0$. Hence from equation (1.1.8) k must be positive. Similarly, if $T < T_m$, then $dT/dt > 0$, and once more equation (1.1.8) implies that k must be positive.

infinity ($t \to \infty$) the temperature of the object approaches that of the surrounding medium ($T \to T_m$). This is certainly consistent with our everyday experience (see Figure 1.1.3).

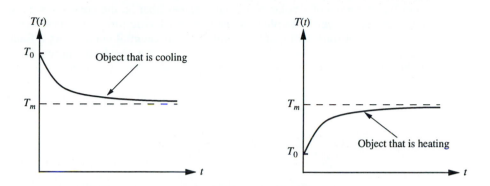

Figure 1.1.3 According to Newton's law of cooling the temperature of an object approaches room temperature exponentially.

3. THE ORTHOGONAL TRAJECTORY PROBLEM

Next we consider a geometric problem that has many interesting and important applications. Suppose

$$F(x, y, c) = 0 \qquad\qquad (1.1.10)$$

defines a family of curves in the xy-plane, where the constant c labels the different curves. We assume that every curve has a well-defined tangent at each point. Associated with this family is a second family of curves, say,

$$G(x, y, k) = 0, \qquad\qquad (1.1.11)$$

with the property that whenever a curve from the family (1.1.10) intersects a curve from the family (1.1.11) it does so at right angles.[1] We say that the curves in the family (1.1.11) are **orthogonal trajectories** of the family (1.1.10), and vice versa. For example, from elementary geometry, it follows that the lines $y = kx$ are orthogonal trajectories of the family of concentric circles $x^2 + y^2 = c^2$. (See Figure 1.1.4.)

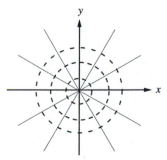

Figure 1.1.4 The family of curves $x^2 + y^2 = c^2$ and the orthogonal trajectories $y = kx$.

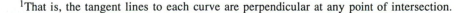

[1]That is, the tangent lines to each curve are perpendicular at any point of intersection.

Orthogonal trajectories arise in various applications. For example, a family of curves and its orthogonal trajectories can be used to define an orthogonal coordinate system in the xy-plane. In Figure 1.1.4 the families $x^2 + y^2 = c^2$ and $y = kx$ are the coordinate curves of a polar coordinate system (that is, the curves $r =$ constant and $\theta =$ constant respectively). In physics, the lines of electric force of a static configuration are the orthogonal trajectories of the family of equipotential curves. As a final example, if we consider a two-dimensional heated plate, then the heat energy flows along the orthogonal trajectories to the constant temperature curves (isotherms).

Statement of the Problem: Given the equation of a family of curves, find the equation of the family of orthogonal trajectories.

Mathematical Formulation: We recall that curves that intersect at right angles satisfy the following:[1]

The product of the slopes at the point of intersection $= -1$.

Thus if the given family has slope $m_1 = f(x, y)$ at the point (x, y), then the slope of the family of orthogonal trajectories is $m_2 = -1/f(x, y)$, and therefore the DE that determines the orthogonal trajectories is

$$\frac{dy}{dx} = -\frac{1}{f(x, y)}.$$

Example 1.1.1   Determine the equation of the family of orthogonal trajectories to the curves with equation

$$y^2 = cx. \tag{1.1.12}$$

Solution   According to the preceding discussion, the DE determining the orthogonal trajectories is

$$\frac{dy}{dx} = -\frac{1}{f(x, y)},$$

where $f(x, y)$ denotes the slope of the given family at the point (x, y). To determine $f(x, y)$, we differentiate equation (1.1.12) implicitly with respect to x to obtain

$$2y\frac{dy}{dx} = 2c. \tag{1.1.13}$$

We must now eliminate c from the previous equation to obtain an expression that gives the slope at the point (x, y). From equation (1.1.12) we have

$$c = \frac{y^2}{x},$$

which, when substituted into equation (1.1.13), yields

$$\frac{dy}{dx} = \frac{y}{2x}.$$

Consequently, the slope of the given family at the point (x, y) is

---

[1]By the slope of a curve at a given point we mean the slope of the tangent line to the curve at that point.

$$f(x, y) = \frac{y}{2x},$$

so that the orthogonal trajectories are obtained by solving the DE

$$\frac{dy}{dx} = -\frac{2x}{y}.$$

A key point to notice is that we cannot solve this DE by simply integrating with respect to x, since the function on the right-hand side of the DE depends on *both* x and y. However, multiplying by y we see that

$$y\frac{dy}{dx} = -2x,$$

or equivalently,

$$\frac{d}{dx}\left(\frac{1}{2}y^2\right) = -2x.$$

Since the right-hand side of this equation depends only on x whereas the term on the left-hand side is a derivative with respect to x, we can integrate both sides of the equation with respect to x to obtain

$$\frac{1}{2}y^2 = -x^2 + c_1,$$

which we write as

$$2x^2 + y^2 = k,$$

where $k = 2c_1$. We see that the curves in the given family (1.1.12) are parabolas, and the orthogonal trajectories (1.1.14) are a family of ellipses. This is illustrated in Figure 1.1.5.

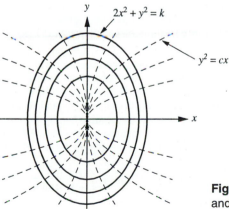

Figure 1.1.5 The family of curves $y^2 = cx$ and its orthogonal trajectories $2x^2 + y^2 = k$.

EXERCISES 1.1

1. An object is released from rest at a height of 100m above the ground. Neglecting frictional forces, the subsequent motion is governed by the initial-value problem

$$\frac{d^2y}{dt^2} = g, \quad y(0) = 0, \frac{dy}{dt}(0) = 0,$$

where $y(t)$ denotes the displacement of the object from its initial position at time t. Solve this initial-value problem and use your solution to determine the time when the object hits the ground.

2. An object that is initially thrown vertically upward with a speed of 2 m/s from a height of h meters takes 10 seconds to reach the ground. Set up and solve the initial-value problem that governs the motion of the object, and determine h.

3. An object that is released from a height h meters above the ground with a vertical velocity v_0 meters/second hits the ground after t_0 seconds. Neglecting frictional forces, set up and solve the initial-value problem governing the motion, and use your solution to show that

$$v_0 = \frac{1}{2t_0}(2h - gt_0{}^2).$$

4. Verify that $y(t) = A\cos(\omega t - \phi)$ where A and ϕ are constants, is a solution to the DE (1.1.7). Determine the constants A and ϕ in the particular case when the initial conditions are $y(0) = a$, $dy/dt(0) = 0$.

5. Verify that $y(t) = c_1\cos \omega t + c_2\sin \omega t$ is a solution to the DE (1.1.7). Show that the amplitude of the motion is

$$A = \sqrt{c_1{}^2 + c_2{}^2}.$$

6. By writing equation (1.1.8) in the form

$$\frac{1}{T - T_m}\frac{dT}{dt} = -k$$

and using $u^{-1}du/dt = d(\ln u)/dt$, derive formula (1.1.9).

For problems 7−14 the following formulas will be needed in order to obtain a DE that can be integrated:

$$y^k\frac{dy}{dx} = \frac{1}{k+1}\frac{d}{dx}(y^{k+1}), \text{ where } k \neq -1,$$

$$y^{-1}\frac{dy}{dx} = \frac{d}{dx}(\ln y).$$

For problems 7−15, find the equation of the orthogonal trajectories to the given family of curves. In each case sketch some curves from each family.

7. $x^2 + 4y^2 = c.$

8. $y = c/x.$

9. $y = cx^2.$

10. $y^2 = 2x + c.$

11. $y = ce^x.$

For problems 12−15, m (not equal to 0) is a fixed constant and c is the constant labeling the different curves in the given family. In each case find the equation of the orthogonal trajectories.

12. $y = mx + c.$

13. $y = cx^m.$

14. $y^2 + mx^2 = c.$

15. $y^2 = mx + c.$

16. A coordinate system (u, v) is called orthogonal if its coordinate curves (the two families of curves $u = $ constant and $v = $ constant) are orthogonal trajectories (for example, a Cartesian coordinate system or a polar coordinate system). Let (u, v) be orthogonal coordinates where $u = x^2 + 2y^2$, and x and y are Cartesian coordinates. Find the Cartesian equation of the v−coordinate curves, and sketch the (u, v) coordinate system.

17. Any curve with the property that whenever it intersects a curve of a given family it does so at an angle $a \neq \pi/2$ is called an **oblique trajectory** of the given family. Let m_1 (equal to $\tan a_1$) denote the slope of the required family at the point (x, y), and let m_2 (equal to $\tan a_2$) denote the slope of the given family. (See Figure 1.1.6.) Show that

$$m_1 = \frac{m_2 - \tan a}{1 + m_2\tan a}.$$

(Hint: From Figure 1.1.6, $\tan a_1 = \tan(a_2 - a)$.) Thus the equation of the family of oblique trajectories is obtained by solving

$$\frac{dy}{dx} = \frac{m_2 - \tan a}{1 + m_2\tan a}.$$

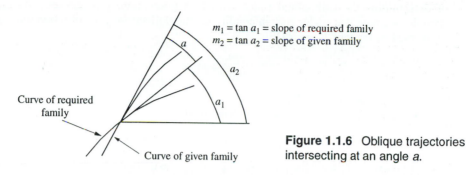

$m_1 = \tan a_1 =$ slope of required family
$m_2 = \tan a_2 =$ slope of given family

Curve of required family

a_2

a_1

Curve of given family

Figure 1.1.6 Oblique trajectories intersecting at an angle a.

1.2   BASIC IDEAS AND TERMINOLOGY

In the previous section we have used some applied problems to illustrate how differential equations arise. We now turn our attention to formalizing mathematically several of the ideas that were introduced through these examples. We begin with a very general definition of a differential equation.

> ***Definition 1.2.1:***   A DE is an equation involving one or more derivatives of an unknown function.

Example 1.2.1   The following are all DE:[1]

(a) $\dfrac{dy}{dx} + y = x^2$,   (b) $\dfrac{d^2y}{dx^2} = -k^2y$,   (c) $\dfrac{d^3y}{dx^3} + \left(\dfrac{d^2y}{dx^2}\right)^5 + \cos x = 0$,

(d) $\sin(dy/dx) + \tan^{-1}y = 1$,   (e) $\phi_{xx} + \phi_{yy} - \phi_x = e^x + x \sin y$.   ❏

The DE occurring in (a)–(d) are called **ordinary** DE since the unknown function $y(x)$ depends only on one variable x. In (e), the unknown function $\phi(x, y)$ depends on more than one variable, and hence, the equation involves partial derivatives. Such a DE is called a **partial** DE. In this text we consider only ordinary DE. We begin by introducing some definitions and terminology.

> ***Definition 1.2.2:***   The order of the highest derivative occurring in a DE is called the **order** of the DE.

In Example 1.2.1, (a) has order 1, (b) has order 2, (c) has order 3, and (d) has order 1. If we look back at the examples from the previous section, we see that problems formulated using Newton's second law of motion will always be governed by a second-order DE (for the position of the object). Indeed, second-order DE play a very fundamental role in applied problems. However, DE of other orders also arise in applications. For example, the DE obtained from Newton's law of cooling is a first-order DE, as is the DE

---

[1]Throughout the text the abbreviation DE will be used for *differential equation* and *differential equations*.

for determining the orthogonal trajectories to a given family of curves. As another example, we note that under certain conditions, the deflection, $y(x)$, of a horizontal beam is governed by the *fourth-order* DE

$$\frac{d^4y}{dx^4} = F(x)$$

for an appropriate function $F(x)$.

Any DE of order n can be written in the form

$$G(x, y, y', y'', \ldots, y^{(n)}) = 0, \tag{1.2.1}$$

where we have introduced the prime notation to denote derivatives, and $y^{(n)}$ denotes the nth derivative of y with respect to x (not y to the power of n). Of particular interest to us throughout the text will be linear DE. These arise as the special case of equation (1.2.1) when $y, y', \ldots, y^{(n)}$ occur to the first degree only, and not as products or arguments of other functions. The general form for such a DE is given in the next definition.

Definition 1.2.3:   A DE that can be written in the form

$$a_0(x)y^{(n)} + a_1(x)y^{(n-1)} + \cdots + a_n(x)y = F(x)$$

where $a_0, a_1, \ldots, a_n$ and F are functions of x only, is called a **linear DE of order n.** Such a DE is linear in $y, y', y'', \ldots, y^{(n)}$.

A DE that does not satisfy this definition is called a **nonlinear** DE.

Example 1.2.2

$$y'' + x^2y' + (\sin x)y = e^x \quad \text{and} \quad xy''' + 4x^2y' - \frac{2}{1 + x^2}y = 0$$

are linear DE of order 2 and order 3, respectively, whereas the DE

$$y'' + x\sin(y') - xy = x^2 \quad \text{and} \quad y'' - x^2y' + y^2 = 0$$

are nonlinear. In the first case the nonlinearity arises from the $\sin(y')$ term, whereas in the second DE the nonlinearity is due to the y^2 term.

Example 1.2.3   The general forms for first- and second-order linear DE are

$$a_0(x)\frac{dy}{dx} + a_1(x)y = F(x)$$

and

$$a_0(x)\frac{d^2y}{dx^2} + a_1(x)\frac{dy}{dx} + a_2(x)y = F(x),$$

respectively.   ❑

If we consider the examples from the previous section we see that the DE governing the simple harmonic oscillator is a second order *linear* DE. In this case the linearity was imposed in the modeling process when we assumed that the restoring force was directly proportional to the displacement from equilibrium (Hooke's law). Not all springs satisfy this relationship. For example Duffing's equation

$$m\frac{d^2y}{dt^2} + k_1 y + k_2 y^3 = 0$$

gives a mathematical model of a nonlinear spring-mass system. If $k_2 = 0$, this reduces to the simple harmonic oscillator equation.

Newton's law of cooling assumes a linear relationship between the rate of change of temperature of an object and the temperature difference between the object and that of the surrounding medium. Hence, the resulting DE is a linear DE. This can be seen explicitly by writing equation (1.1.8) as

$$\frac{dT}{dt} + kT = kT_m,$$

which is a first-order *linear* DE. Finally, the DE for determining the orthogonal trajectories of a given family of curves will in general be nonlinear.

SOLUTIONS OF DIFFERENTIAL EQUATIONS

We now define precisely what is meant by a solution to a DE.

Definition 1.2.4:   A function $y = f(x)$ that is (at least) n times differentiable on an interval I is called a **solution** to the DE (1.2.1) on I if the substitution

$$y = f(x),\ y' = f'(x),\ \ldots,\ y^{(n)} = f^{(n)}(x)$$

reduces DE (1.2.1) to an identity valid for all x in I. In this case we say that $y = f(x)$ satisfies the DE.

Example 1.2.4   Verify that $y(x) = c_1\sin x + c_2\cos x$, where c_1 and c_2 are constants, is a solution to the linear DE $y'' + y = 0$ for x in the interval $(-\infty, \infty)$.

Solution   The function $y(x)$ is certainly twice differentiable for all real x. Furthermore,

$$y'(x) = c_1\cos x - c_2\sin x$$

and

$$y''(x) = -(c_1\sin x + c_2\cos x).$$

Consequently

$$y'' + y = -(c_1\sin x + c_2\cos x) + c_1\sin x - c_2\cos x = 0,$$

so that $y'' + y = 0$. It follows from the preceding definition that the given function is a solution to the DE on $(-\infty, \infty)$.   ❑

In the preceding example x could assume all real values. Often, however, the independent variable will be restricted in some manner. For example, the DE

$$\frac{dy}{dx} = \frac{1}{2\sqrt{x}}(y - 1)$$

is undefined when $x \leq 0$ and so any solution would be defined only for $x > 0$. In fact this linear DE has solution

$$y(x) = ce^{\sqrt{x}} + 1, \quad x > 0,$$

where c is a constant. (You can check this by plugging in to the given DE, as was done in Example 1.2.4. In Section 1.4 we will introduce a technique that will enable you to derive this solution.)

We now distinguish two different ways in which solutions to a DE can be expressed. Often, as in Example 1.2.4, we will be able to obtain a solution to a DE in the **explicit** form

$$y = f(x),$$

for some function f. However, when dealing with nonlinear DE, we usually have to be content with a solution written in the **implicit** form

$$F(x, y) = 0,$$

where the function F defines the solution, $y(x)$, implicitly as a function of x. This is illustrated in Example 1.2.5.

Example 1.2.5   Verify that the relation $x^2 + y^2 - 4 = 0$ defines an implicit solution to the nonlinear DE

$$\frac{dy}{dx} = -\frac{x}{y}.$$

Solution   We regard the given relation as defining y as a function of x. Differentiating this relation with respect to x yields[1]

$$2x + 2y \frac{dy}{dx} = 0.$$

That is,

$$\frac{dy}{dx} = -\frac{x}{y},$$

as required. In this example we can obtain y explicitly in terms of x since $x^2 + y^2 - 4 = 0$ implies that

$$y = \pm\sqrt{4 - x^2}.$$

The implicit relation therefore contains the *two* explicit solutions

$$y(x) = \sqrt{4 - x^2}, \quad y(x) = -\sqrt{4 - x^2},$$

which correspond graphically to the two semi-circles sketched in Figure 1.2.1. Since $x = \pm 2$ correspond to $y = 0$ in both of these equations, whereas the DE is only defined for $y \neq 0$, we must omit $x = \pm 2$ from the domains of the solutions. Consequently, both of the foregoing solutions to the DE are valid for $-2 < x < 2$.

---

[1]Note that we have used implicit differentiation in obtaining $d(y^2)/dx = 2y \, dy/dx$.

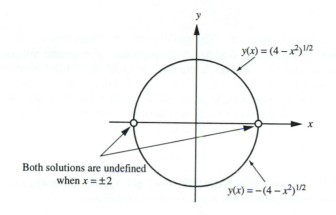

Figure 1.2.1 Two solutions to the DE $y' = -x/y$.

In the previous example the solutions to the DE are more simply expressed in implicit form although, as we have shown, it is quite easy to obtain the corresponding explicit solutions. In the following example the solution must be expressed in implicit form, since it is impossible to solve the implicit relation (analytically) for y as a function of x.

Example 1.2.6   Show that the relation $\sin(xy) + y^2 - x = 0$ defines a solution to

$$\frac{dy}{dx} = \frac{1 - y\cos(xy)}{x\cos(xy) + 2y}.$$

Solution   Differentiating the given relationship implicitly with respect to x yields

$$\cos(xy)\left(y + x\frac{dy}{dx}\right) + 2y\frac{dy}{dx} - 1 = 0.$$

That is,

$$\frac{dy}{dx}[x\cos(xy) + 2y] = 1 - y\cos(xy),$$

which implies that

$$\frac{dy}{dx} = \frac{1 - y\cos(xy)}{x\cos(xy) + 2y}$$

as required.

Now consider the simple DE

$$\frac{d^2y}{dx^2} = 12x.$$

From elementary calculus we know that all functions whose second derivative is $12x$ can be obtained by performing two integrations. Integrating the given DE once yields

$$\frac{dy}{dx} = 6x^2 + c_1,$$

where c_1 is an arbitrary constant. Integrating again we obtain

$$y(x) = 2x^3 + c_1 x + c_2, \tag{1.2.2}$$

where c_2 is another arbitrary constant. The point to notice about this solution is that it contains *two* arbitrary constants. Further, by assigning appropriate values to these constants, we can determine all solutions to the DE. We call (1.2.2) the *general solution* to the DE. In this example the given DE was of second order, and the general solution contained two arbitrary constants, which arose due to the fact that two integrations were required to solve the DE. In the case of an *n*th-order DE we might suspect that the most general *form* of solution that can arise would contain *n* arbitrary constants. This is indeed the case and motivates the following definition:

> *Definition 1.2.5:*   A solution to an *n*th order DE on an interval *I* is called the **general solution** on *I* if it satisfies the following conditions:
>
> 1 . The solution contains *n* constants $c_1, c_2, \ldots, c_n$.
>
> 2 . All solutions to the DE can be obtained by assigning appropriate values to the constants.

REMARK   Not all DE have a general solution. For example, consider

$$(y')^2 + (y-1)^2 = 0.$$

The only solution to this DE is $y(x) = 1$, and hence the DE does not have a solution containing an arbitrary constant.

Example 1.2.7   We have shown in Example 1.2.4 that $y(x) = c_1 \sin x + c_2 \cos x$ is a solution to the DE $y'' + y = 0$. In Chapter 2 we will establish that *all* solutions to $y'' + y = 0$ are of this form, a fact that is by no means obvious at this stage.

Example 1.2.8   Find the general solution to the DE $y'' = e^{-x}$.

Solution   Integrating the given DE with respect to *x* yields

$$y' = -e^{-x} + c_1,$$

where c_1 is an integration constant. Integrating this equation we obtain

$$y(x) = e^{-x} + c_1 x + c_2 \tag{1.2.3}$$

where c_2 is another integration constant. Consequently, all solutions to $y'' = e^{-x}$ are of the form (1.2.3), and therefore, according to Definition 1.2.5, this is the general solution to $y'' = e^{-x}$ on any interval.   ◻

As the previous example illustrates, we can, in principle, always find the general solution to a DE of the form

$$\frac{d^n y}{dx^n} = f(x) \tag{1.2.4}$$

by performing *n* integrations. However, if the function on the right-hand side of the DE is not a function of *x* only, this procedure cannot be used. Indeed, one of the major aims of this text is to determine solution techniques for DE that are more complicated than equation (1.2.4).

A solution to a DE is called a **particular solution** if it does not contain any arbitrary constants. One way in which particular solutions arise is by assigning specific values to the arbitrary constants occurring in the general solution to a DE. For example, from (1.2.3),

$$y(x) = e^{-x} + x$$

is a particular solution to the DE $d^2y/dx^2 = e^{-x}$ (the solution corresponding to $c_1 = 1$, $c_2 = 0$).

INITIAL-VALUE PROBLEMS

As discussed in the previous section, the unique specification of an applied problem requires more than just a DE. We must also give appropriate auxiliary conditions that characterize the problem under investigation. Of particular interest to us is the case of the initial-value problem defined for an nth-order DE as follows:

Definition 1.2.6:   An nth-order DE together with n auxiliary conditions of the form

$$y(x_0) = y_0, \ y'(x_0) = y_1, \ \ldots, \ y^{(n-1)}(x_0) = y_{n-1},$$

where $y_0, y_1, \ldots, y_{n-1}$ are constants, is called an **initial-value problem** (denoted IVP).

Example 1.2.9   Solve the IVP

$$y'' = e^{-x}, \tag{1.2.5}$$
$$y(0) = 1, \ y'(0) = 4. \tag{1.2.6}$$

Solution   From Example 1.2.8, the general solution to equation (1.2.5) is

$$y(x) = e^{-x} + c_1 x + c_2. \tag{1.2.7}$$

We now impose the auxiliary conditions (1.2.6). Setting $x = 0$ in (1.2.7) we see that

$$y(0) = 1 \text{ if and only if } 1 = 1 + c_2. \text{ So } c_2 = 0.$$

Setting $c_2 = 0$ in (1.2.7) and differentiating the result yields

$$y'(x) = -e^{-x} + c_1.$$

Consequently $y'(0) = 4$ if and only if $4 = -1 + c_1$, and hence $c_1 = 5$. Thus the given auxiliary conditions pick out the particular solution to the DE with $c_1 = 5$, and $c_2 = 0$, so that the IVP has the unique solution

$$y(x) = e^{-x} + 5x. \qquad \square$$

IVP play a fundamental role in the theory and applications of DE. In the previous example, the IVP had a unique solution. More generally, suppose we have a DE that can be written in the **normal** form

$$y^{(n)} = f(x, y, y', \ldots, y^{(n-1)}).$$

According to Definition 1.2.6, the IVP for such an nth-order DE is

solve

$$y^{(n)} = f(x, y, y', \ldots, y^{(n-1)})$$

subject to

$$y(x_0) = y_0, \ y'(x_0) = y_1, \ \ldots, \ y^{(n-1)}(x_0) = y_{n-1}, \qquad (1.2.8)$$

where $y_0, y_1, \ldots, y_{n-1}$ are constants.

It can be shown that this IVP always has a unique solution provided f and its partial derivatives with respect to $y, y', \ldots, y^{(n-1)}$, are continuous in an appropriate region. This is a fundamental result in DE theory whose power will be illustrated in Chapters 2 and 7. The special case when $n = 1$ is stated precisely in the next section.

For the remainder of this chapter, we will focus our attention primarily on *first-order* DE and some of their elementary applications. We will investigate such DE qualitatively, analytically, and numerically.

EXERCISES 1.2

For problems 1–5, determine the order of the given DE. For each case state whether the DE is linear or nonlinear.

1. $\dfrac{d^2y}{dx^2} + e^{xy}\dfrac{dy}{dx} = x^2$.

2. $\dfrac{d^3y}{dx^3} + 4\dfrac{d^2y}{dx^2} + \sin x \dfrac{dy}{dx} = xy + \tan x$.

3. $y'' + 3x(y')^3 - y = 1 + 3x$.

4. $\sin x \ e^{y''} + y' - \tan y = \cos x$.

5. $\dfrac{d^4y}{dx^4} + 3\dfrac{d^2y}{dx^2} = x$.

For problems 6–16, verify that the given function is a solution to the given DE (c_1 and c_2 are arbitrary constants). For each case, state the maximum interval over which the solution is valid.

6. $y(x) = c_1 e^x + c_2 e^{-2x}$, $y'' + y' - 2y = 0$.

7. $y(x) = \dfrac{1}{x + 4}$, $y' = -y^2$.

8. $y(x) = c_1 x^{1/2}$, $y' = \dfrac{y}{2x}$.

9. $y(x) = e^{-x}\sin 2x$, $y'' + 2y' + 5y = 0$.

10. $y(x) = c_1\cosh 3x + c_2\sinh 3x$, $y'' - 9y = 0$.

11. $y(x) = c_1 x^{-3} + c_2 x^{-1}$, $x^2 y'' + 5xy' + 3y = 0$.

12. $y(x) = c_1 x^{1/2} + 3x^2$, $2x^2 y'' - xy' + y = 9x^2$.

13. $y(x) = c_1 x^2 + c_2 x^3 - x^2\sin x$, $x^2 y'' - 4xy' + 6y = x^4\sin x$.

14. $y(x) = c_1 e^{ax} + c_2 e^{bx}$, $y'' - (a + b)y' + aby = 0$, where a and b are constants and $a \neq b$.

15. $y(x) = e^{ax}(c_1 + c_2 x)$, $y'' - 2ay' + a^2 y = 0$, where a is a constant.

16. $y(x) = e^{ax}(c_1\cos bx + c_2\sin bx)$, $y'' - 2ay' + (a^2 + b^2)y = 0$, where a and b are constants.

For problems 17–20, determine all values of the constant r such that the given function solves the given DE.

17. $y(x) = e^{rx}$, $y'' + 2y' - 3y = 0$.

18. $y(x) = e^{rx}$, $y'' - 8y' + 16y = 0$.

19. $y(x) = x^r$, $x^2 y'' + xy' - y = 0$.

20. $y(x) = x^r$, $x^2 y'' + 5xy' + 4y = 0$.

21. When N is a positive integer, the **Legendre** equation

$$(1 - x^2)y'' - 2xy' + N(N + 1)y = 0, \ -1 < x < 1$$

has a solution that is a polynomial of degree N. Show by substitution into the DE that in the case $N = 3$ such a solution is

$$y(x) = \frac{1}{2}x(5x^2 - 3).$$

22. Determine a solution to the DE

$$(1 - x^2)y'' - xy' + 4y = 0$$

of the form $y(x) = a_0 + a_1 x + a_2 x^2$ satisfying the normalization condition $y(1) = 1$.

For problems 23–27, show that the given relation defines an implicit solution to the given DE (c is an arbitrary constant).

23. $x \sin y - e^x = c, \ y' = \dfrac{e^x - \sin y}{x \cos y}$.

24. $xy^2 + 2y - x = c, \ y' = \dfrac{1 - y^2}{2(1 + xy)}$.

25. $e^{xy} - x = c, \ y' = \dfrac{1 - ye^{xy}}{xe^{xy}}$. Determine the solution satisfying $y(1) = 0$.

26. $e^{y/x} + xy^2 - x = c, \ y' = \dfrac{x^2(1 - y^2) + ye^{y/x}}{x(e^{y/x} + 2x^2y)}$.

27. $x^2y^2 - \sin x = c, \ y' = \dfrac{\cos x - 2xy^2}{2x^2y}$. Determine the corresponding explicit solution that satisfies $y(\pi) = 1/\pi$.

For problems 28–31, find the general solution to the given DE and the maximum interval on which the solution is valid.

28. $y' = \sin x$.

29. $y' = x^{-1/2}$.

30. $y'' = xe^x$.

31. $y'' = x^n, \ n$ an integer.

For problems 32–34, solve the given IVP.

32. $y' = \ln x, \ y(1) = 2$.

33. $y'' = \cos x, \ y(0) = 2, \ y'(0) = 1$.

34. $y''' = 6x, \ y(0) = 1, \ y'(0) = -1, \ y''(0) = 4$.

A second order DE together with two auxiliary conditions imposed at different values of the independent variable is called a **boundary-value problem**. For problems 35 and 36, solve the given boundary-value problem.

35. $y'' = e^{-x}, \ y(0) = 1, \ y(1) = 0$.

36. $y'' = -2(3 + 2 \ln x), \ y(1) = y(e) = 0$.

37. The DE $y'' + y = 0$ has the general solution $y(x) = c_1 \cos x + c_2 \sin x$.

(a) Show that the boundary-value problem $y'' + y = 0, \ y(0) = 0, \ y(\pi) = 1$ has no solutions.

(b) Show that the boundary-value problem $y'' + y = 0, \ y(0) = 0, \ y(\pi) = 0$ has an infinite number of solutions.

For problems 38–43, verify that the given function is a solution to the given DE. In these problems c_1 and c_2 are arbitrary constants.

◆ **38.** $y(x) = c_1 e^{2x} + c_2 e^{-3x}, \ y'' + y' - 6y = 0$.

◆ **39** $y(x) = c_1 x^4 + c_2 x^{-2}, \ x^2 y'' - xy' - 8y = 0, \ x > 0$.

◆ **40.** $y(x) = c_1 x^2 + c_2 x^2 \ln x + \dfrac{1}{6} x^2 (\ln x)^3, \ x^2 y'' - 3xy' + 4y = x^2 \ln x, \ x > 0$.

◆ **41.** $y(x) = x^a [c_1 \cos(b \ln x) + c_2 \sin(b \ln x)], \ x^2 y'' + (1 - 2a)xy' + (a^2 + b^2)y = 0, \ x > 0$, where a and b are arbitrary constants.

◆ **42.** $y(x) = c_1 e^x + c_2 e^{-x}(1 + 2x + 2x^2), \ xy'' - 2y' + (2 - x)y = 0, \ x > 0$.

◆ **43** $y(x) = \displaystyle\sum_{k=0}^{10} \frac{1}{k!} x^k, \ xy'' - (x + 10)y' + 10y = 0, \ x > 0$.

◆ **44. (a)** Derive the polynomial of degree five that satisfies both the Legendre equation

$$(1 - x^2)y'' - 2xy' + 30y = 0$$

and the normalization condition $y(1) = 1$.

(b) Sketch your solution from (a) and determine approximations to all zeros and local maxima and local minima on the interval $(-1, 1)$.

◆ **45.** One solution to the Bessel equation of (nonnegative) integer order N

$$x^2 y'' + xy' + (x^2 - N^2)y = 0$$

is

$$y(x) = J_N(x) = \sum_{k=0}^{\infty} \frac{(-1)^k}{k!(N + k)!} \left(\frac{x}{2}\right)^{2k+N}.$$

(a) Write the first three terms of $J_0(x)$.

(b) Let $J(0, x, m)$ denote the mth partial sum

$$J(0, x, m) = \sum_{k=0}^{m} \frac{(-1)^k}{(k!)^2} \left(\frac{x}{2}\right)^{2k}.$$

Plot $J(0, x, 4)$ and use your plot to approximate the first positive zero of $J_0(x)$. Compare your value against a tabulated value or one generated by a computer algebra system.

(c)   plot $J_0(x)$ and $J(0, x, 4)$ on the same axes over the interval $[0, 2]$. How well do they compare?

(d) If your system has built-in Bessel functions, plot $J_0(x)$ and $J(0, x, m)$ on the same axes over the interval $[0, 10]$ for various values of m. What is the smallest value of m that gives an accurate approximation to the first *three* positive zeros of $J_0(x)$?

1.3   THE GEOMETRY OF FIRST-ORDER DE

The primary aim of this chapter is to study the first-order DE

$$\frac{dy}{dx} = f(x, y), \tag{1.3.1}$$

where $f(x, y)$ is a given function of x and y. In this section we focus our attention mainly on the geometric aspects of the DE and its solutions.

The graph of any solution to the DE (1.3.1) is called a **solution curve**. If we recall the geometric interpretation of the derivative dy/dx as giving the slope of the tangent line at any point on the curve with equation $y = y(x)$, we see that the function $f(x, y)$ in (1.3.1) gives the slope of the tangent line to the solution curve passing through the point (x, y). Consequently when we solve equation (1.3.1), we are finding all curves whose slope at the point (x, y) is given by the function $f(x, y)$. According to our definition in the previous section, the general solution to the DE[1] will involve one arbitrary constant, and therefore, geometrically, the general solution gives a family of solution curves in the xy-plane, one solution curve corresponding to each value of the arbitrary constant.

Example 1.3.1   Find the general solution to the DE $dy/dx = 2x$, and sketch the corresponding solution curves.

Solution   The DE can be integrated directly to obtain $y(x) = x^2 + c$. Consequently the solution curves are a family of parabolas in the xy-plane. This is illustrated in Figure 1.3.1.

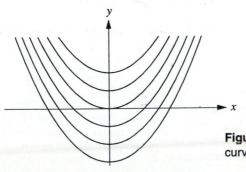

Figure 1.3.1 Some solution curves for the DE $dy/dx = 2x$.

---

[1]Assuming the DE has a general solution.

Figure 1.3.2 gives a Mathematica plot of some solution curves to the DE

$$\frac{dy}{dx} = y - x^2.$$

This illustrates that generally the solution curves of a DE are quite complicated. Upon completion of the material in this section you will be able to obtain Figure 1.3.2 without the necessity of a computer algebra system.

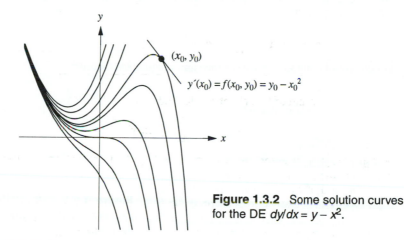

Figure 1.3.2 Some solution curves for the DE $dy/dx = y - x^2$.

EXISTENCE AND UNIQUENESS OF SOLUTIONS

It is useful for the further analysis of the DE (1.3.1) to at this point give a brief discussion of the existence and uniqueness of solutions to the corresponding IVP

$$\frac{dy}{dx} = f(x, y), \quad y(x_0) = y_0. \tag{1.3.2}$$

Geometrically, we are interested in finding the particular solution curve to the DE that passes through the point in the xy-plane with coordinates (x_0, y_0). The following questions arise regarding the IVP:

1. *Existence:* Does the IVP have *any* solutions?

2. *Uniqueness:* If the answer to (1) is yes, does the IVP have *only one* solution?

Certainly in the case of an applied problem we would be interested only in IVP that have precisely one solution. The following theorem establishes conditions on f that guarantee the existence and uniqueness of a solution to the IVP (1.3.2).

Theorem 1.3.1 (Existence and Uniqueness Theorem):   Let $f(x, y)$ be a function that is continuous on the rectangle

$$R = \{(x, y): \ a \leq x \leq b, \ c \leq y \leq d\}.$$

Suppose further that $\dfrac{\partial f}{\partial y}$ is continuous in R. Then for any interior point (x_0, y_0) in the rectangle R, there exists an interval I containing x_0 such that the IVP (1.3.2) has a unique solution for x in I.

PROOF   A full discussion of this theorem is given in Appendix 4. For a complete proof see, for example, G.F. Simmons, *Differential Equations*, McGraw–Hill, 1972. Figure 1.3.3 gives a geometric illustration of the result.   ∎

REMARK   From a geometric viewpoint, if $f(x, y)$ satisfies the hypotheses of the existence and uniqueness theorem in a region R of the xy-plane then throughout that region the solution curves of the DE $dy/dx = f(x, y)$ cannot intersect. For if two solution curves did intersect at (x_0, y_0) in R, then that would imply that there was more than one solution to the IVP

$$\frac{dy}{dx} = f(x, y), \quad y(x_0) = y_0,$$

which would contradict the existence and uniqueness theorem.

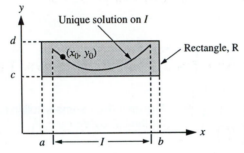
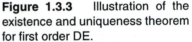

Figure 1.3.3   Illustration of the existence and uniqueness theorem for first order DE.

The following example illustrates how the preceding theorem can be used to establish the existence of a unique solution to a DE, even though at present we do not know how to determine the solution.

Example 1.3.2   Prove that the IVP

$$\frac{dy}{dx} = 3xy^{1/3}, \quad y(0) = a$$

has a unique solution whenever $a \neq 0$.

Solution   In this case the initial point is $x_0 = 0$, $y_0 = a$. If we let $f(x, y) = 3xy^{1/3}$, then $\partial f/\partial y = xy^{-2/3}$. We see that f is continuous at all points in the xy-plane, whereas $\partial f/\partial y$ is continuous at all points not lying on the x-axis ($y \neq 0$). Provided $a \neq 0$, we can certainly draw a rectangle containing $(0, a)$ that does not intersect the x-axis. (See Figure 1.3.4.) In any such rectangle the hypotheses of the existence and uniqueness theorem are satisfied, and therefore the IVP does indeed have a unique solution.

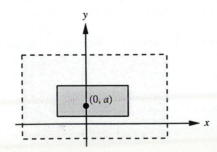

Figure 1.3.4   The IVP in Example 1.3.2 satisfies the hypotheses of the existence and uniqueness theorem in the small rectangle, but not in the large rectangle.

Example 1.3.3   Discuss the existence and uniqueness of solutions to the IVP

$$\frac{dy}{dx} = 3xy^{1/3}, \quad y(0) = 0.$$

Solution   The DE is the same as in the previous example, but the initial condition is imposed on the x-axis. Since $\partial f/\partial y = x/y^{2/3}$ is not continuous along the x-axis there is no rectangle containing (0, 0) in which the hypotheses of the existence and uniqueness theorem are satisfied. We can therefore draw no conclusion from the theorem itself. We leave it as an exercise to verify by direct substitution that the given IVP does in fact have the following two solutions

(a) $y(x) = 0$;   (b) $y(x) = x^3$.

Consequently in this case the IVP does not have a *unique* solution.

SLOPE FIELDS

We now return to our discussion of the geometry of solutions to the DE

$$\frac{dy}{dx} = f(x, y).$$

The fact that the function $f(x, y)$ gives the slope of the tangent line to the solution curves of this DE leads to a simple and important idea for determining the overall shape of the solution curves. We compute the value of $f(x, y)$ at several points and draw through each of the corresponding points in the xy-plane small line segments having $f(x, y)$ as their slopes. The resulting sketch is called the **slope field** for the DE. The key point is that each solution curve must be tangent to the line segments that we have drawn, and therefore by studying the slope field we can obtain the general shape of the solution curves.

Example 1.3.4   Sketch the slope field for the DE $\frac{dy}{dx} = 2x^2$.

Solution   The slope of the solution curves to the DE at each point in the xy-plane depends on x only. Consequently, the slopes of the solution curves will be the same at every point on any line parallel to the y-axis (x = constant). Table 1.3.1 contains the values of the slope of the solution curves at various points in the interval [−1, 1]. Using

TABLE   1.3.1

x	Slope $= 2x^2$
0	0
±0.2	0.08
±0.4	0.32
±0.6	0.72
±0.8	1.28
±1.0	2

this information we obtain the slope field shown in Figure 1.3.5. In this example, we can integrate the DE to obtain the general solution

$$y(x) = \frac{2}{3} x^3 + c.$$

Some solution curves and their relation to the slope field are also shown in Figure 1.3.5.

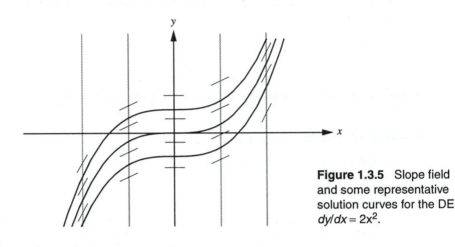

Figure 1.3.5 Slope field and some representative solution curves for the DE $dy/dx = 2x^2$.

In the previous example, the slope field could be obtained fairly easily due to the fact that the slope of the solution curves to the DE were constant on lines parallel to the y-axis. For more complicated DE, further analysis is generally required if we wish to obtain an accurate plot of the slope field and the behavior of the corresponding solution curves. Below we have listed three useful procedures.

1. *Isoclines:* For the DE

$$\frac{dy}{dx} = f(x, y), \tag{1.3.3}$$

the function $f(x, y)$ determines the regions in the xy-plane where the slope of the solution curves is positive, as well as those regions where it is negative. Furthermore, each solution curve will have the same slope k along the family of curves

$$f(x, y) = k.$$

These curves are called the **isoclines** of the DE, and can be very useful in determining slope fields. In general when sketching a slope field we start by drawing several isoclines and the corresponding line segments with slope k at various points along them.

2. *Equilibrium Solutions:*   Any solution to the DE (1.3.3) of the form $y(x) = y_0$ where y_0 is a constant is called an **equilibrium solution** to the DE. The corresponding solution curve is a line parallel to the x-axis. From equation (1.3.3), equilibrium solutions are given by any *constant* values of y for which $f(x, y) = 0$, and therefore can often be obtained by inspection. For example, the DE

$$\frac{dy}{dx} = (y - x)(y + 1)$$

has the equilibrium solution $y(x) = -1$. One of the main reasons that equilibrium solutions are useful in sketching slope fields and determining the general behavior of the full family of solution curves is that, from the existence and uniqueness theorem, we know that no other solution curves can intersect the solution curve corresponding to an equilibrium solution. Consequently, equilibrium solutions serve to divide the xy-plane into different regions.

3. *Concavity Changes:*   By differentiating equation (1.3.3) (implicitly) with respect to x we can obtain an expression for d^2y/dx^2 in terms of x and y. This can be useful in determining the behavior of the concavity of the solution curves to the DE (1.3.3).

The remaining examples illustrate the application of the foregoing procedures.

Example 1.3.5   Sketch the slope field for the DE

$$\frac{dy}{dx} = y - x. \qquad (1.3.4)$$

Solution   By inspection we see that the DE has no equilibrium solutions. The isoclines of the DE are the family of straight lines $y - x = k$. Thus each solution curve of the DE has slope k at all points along the line $y - x = k$. Table 1.3.2 contains several

TABLE   1.3.2

Slope of Solution Curves	Equation of Isocline
$k = -2$	$y = x - 2$
$k = -1$	$y = x - 1$
$k = 0$	$y = x$
$k = 1$	$y = x + 1$
$k = 2$	$y = x + 2$

values for the slopes of the solution curves, and the equations of the corresponding isoclines. We note that the slope at all points along the isocline $y = x + 1$ is unity, which, from Table 1.3.2, coincides with the slope of any solution curve that meets it. This implies that the isocline must in fact coincide with a solution curve. Hence, one solution to the DE (1.3.4) is $y(x) = x + 1$ and, by the existence and uniqueness theorem, no other solution curve can intersect this one. In order to determine the behavior of the concavity of the solution curves, we differentiate the given DE implicitly with respect to x to obtain

$$\frac{d^2y}{dx^2} = \frac{dy}{dx} - 1 = y - x - 1,$$

where we have used (1.3.4) to substitute for dy/dx in the second step. We see that the solution curves are concave up ($y'' > 0$) at all points above the line

$$y = x + 1 \qquad (1.3.5)$$

and concave down ($y'' < 0$) at all points beneath this line. We also note that equation (1.3.5) coincides with the particular solution already identified. Putting all of this information together we obtain the slope field sketched in Figure 1.3.6.

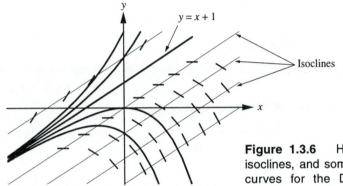

Figure 1.3.6 Hand-drawn slope field, isoclines, and some approximate solution curves for the DE in Example 1.3.5.

❐

GENERATING SLOPE FIELDS USING TECHNOLOGY

Many computer algebra systems (CAS) and graphing calculators have built-in programs to generate slope fields. As an example, in the CAS Maple the command

$$\text{diffeq} := \text{diff}(y(x), x) = y(x) - x;$$

assigns the name diffeq to the DE considered in the previous example. The further command

$$\text{DEplot(diffeq, } y(x), x = -3..3, y = -3..3, \text{arrows=line);}$$

then produces a sketch of the slope field for the DE on the square $-3 \le x \le 3, -3 \le y \le 3$. Initial conditions such as $y(0) = 0, y(0) = 1, y(0) = 2, y(0) = -1$ can be specified using the command

$$\text{IC} := \{[0, 0], [0, 1], [0, 2], [0, -1]\};$$

Then the command

$$\text{DEplot(diffeq, } y(x), x = -3..3, \text{IC, } y = -3..3, \text{arrows=line);}$$

not only plots the slope field, but also gives a *numerical approximation* to each of the solution curves satisfying the specified initial conditions. Some of the methods that can be used to generate such numerical approximations will be discussed in Section 1.11. The preceding sequence of Maple commands was used to generate the Maple plot given in

Figure 1.3.7. Clearly the generation of slope fields and approximate solution curves is one area where technology can be extremely helpful.

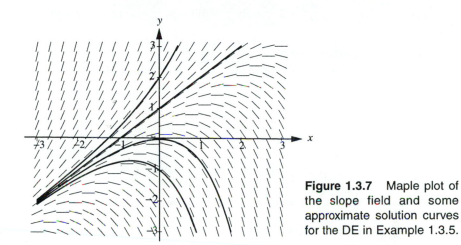

Figure 1.3.7 Maple plot of the slope field and some approximate solution curves for the DE in Example 1.3.5.

Example 1.3.6   Sketch the slope field and some approximate solution curves for the DE

$$\frac{dy}{dx} = y(2 - y). \tag{1.3.6}$$

Solution   We first note that the given DE has the two equilibrium solutions

$$y(x) = 0, \text{ and } y(x) = 2.$$

Consequently, from the existence and uniqueness theorem, the xy-plane can be divided into the three distinct regions $y < 0$, $0 < y < 2$, and $y > 2$. From equation (1.3.6) the behavior of the sign of the slope of the solution curves in each of these regions is given in the following schematic.

sign of slope: $----|++++|----$

y interval: $0 \quad\ \ 2$

The isoclines are determined from

$$y(2 - y) = k.$$

That is,

$$y^2 - 2y + k = 0,$$

so that the solution curves have slope k at all points of intersection with the *horizontal lines*

$$y = 1 \pm \sqrt{1 - k}. \tag{1.3.7}$$

Table 1.3.3 contains some of the isocline equations. Note from equation (1.3.7) that the largest possible positive slope is $k = 1$. We see that the slope of the solution curves quickly become very large and negative for y outside the interval [0, 2]. Finally, differentiating equation (1.3.6) implicitly with respect to x yields

$$\frac{d^2y}{dx^2} = 2\frac{dy}{dx} - 2y\frac{dy}{dx} = 2(1-y)\frac{dy}{dx} = 2y(1-y)(2-y).$$

TABLE 1.3.3

slope of solution curve	Equation of isocline
$k = 1$	$y = 1$
$k = 0$	$y = 2$ and $y = 0$
$k = -1$	$y = 1 \pm \sqrt{2}$
$k = -2$	$y = 1 \pm \sqrt{3}$
$k = -3$	$y = 3$ and $y = -1$
$k = -n,\ n \geq 1$	$y = 1 \pm \sqrt{n + 1}$

The sign of d^2y/dx^2 is given in the following schematic.

sign of y´´: $----|++++|----|++++$
y-interval: $0\quad 1\quad 2$

Using this information leads to the slope field sketched in Figure 1.3.8. We have also

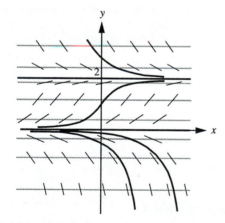

Figure 1.3.8   Hand–drawn slope field, isoclines, and some solution curves for the DE $dy/dx = y(2 - y)$.

included some approximate solution curves. We see from the slope field that for any initial condition $y(x_0) = y_0$, with $0 \leq y_0 \leq 2$, the corresponding unique solution to the DE will be bounded. In contrast, if $y_0 > 2$, the slope field suggests that all corresponding solutions approach $y = 2$ as $x \rightarrow \infty$, whereas if $y_0 < 0$, then all corresponding solutions approach $y = 0$ as $x \rightarrow -\infty$. Furthermore, the behavior of the slope field also suggests that the solution curves that do not lie in the region $0 < y < 2$ may diverge at finite values of x. We leave it as an exercise to verify (by substitution into equation (1.3.6)) that for all values of the constant c,

$$y(x) = \frac{2ce^{2x}}{ce^{2x} - 1}$$

is a solution to the given DE. We see that any initial condition that yields a positive value for c will indeed lead to a solution that has a vertical asymptote at $x = \frac{1}{2}\ln(1/c)$. ☐

The tools that we have introduced in this section enable us to analyze the solution behavior of many first-order DE. However, for complicated functions $f(x, y)$ in equation (1.3.3), performing these computations by hand can be a tedious task. Fortunately, as we have illustrated, there are many computer programs available for drawing slope fields and generating solution curves (numerically). Furthermore, several graphing calculators also have these capabilities. Towards the end of this chapter we will show how the geometric techniques introduced in this section can be used to analyze the solution properties of certain second-order DE.

EXERCISES 1.3

For problems 1–7, determine the DE giving the slope of the tangent line at the point (x, y) for the given family of curves.

1 . $y = c/x$.

2 . $y = cx^2$.

3 . $x^2 + y^2 = 2cx$.

4 . $y^2 = cx$.

5 . $2cy = x^2 - c^2$.

6 . $y^2 - x^2 = c$.

7 . $(x - c)^2 + (y - c)^2 = 2c^2$.

For problems 8–11, verify that the given function (or relation) defines a solution to the given DE and sketch some of the solution curves. If an initial condition is given, label the solution curve corresponding to the resulting unique solution. (In these problems c denotes an arbitrary constant.)

8 . $x^2 + y^2 = c$, $y' = -x/y$.

9 . $y = cx^3$, $y' = 3y/x$, $y(2) = 8$.

10. $y^2 = cx$, $2x\,dy - y\,dx = 0$, $y(1) = 2$.

11. $(x - c)^2 + y^2 = c^2$, $y' = \dfrac{y^2 - x^2}{2xy}$, $y(2) = 2$.

12. Prove that the IVP

$$y' = x\sin(x + y), \quad y(0) = 1$$

has a unique solution.

13. Use the existence and uniqueness theorem to prove that $y(x) = 3$ is the only solution to the IVP

$$y' = \frac{x}{x^2 + 1}\,(y^2 - 9), \quad y(0) = 3.$$

14. Do you think that the IVP

$$y' = xy^{1/2}, \quad y(0) = 0$$

has a unique solution? Justify your answer.

15. Even simple looking DE can have complicated solution curves. For this problem, we investigate the solution curves of the DE

$$y' = -2xy^2. \qquad (15.1)$$

(a) Verify that the hypotheses of the existence and uniqueness theorem are satisfied for the IVP

$$y' = -2xy^2, \quad y(x_0) = y_0$$

for any x_0, y_0. This establishes that the IVP always has a unique solution on some interval containing x_0.

(b) Verify that for all values of the constant c,

$$y(x) = \frac{1}{x^2 + c}$$ is a solution to equation (15.1).

(c) Use the solution to equation (15.1) given in (b) to solve the following IVP. For each case, sketch the corresponding solution curve, and state the maximum interval on which your solution is valid.

(i) $y' = -2xy^2$, $y(0) = 1$.

(ii) $y' = -2xy^2$, $y(1) = 1$.

(iii) $y' = -2xy^2$, $y(0) = -1$.

(d) What is the unique solution to the IVP

$$y' = -2xy^2, \quad y(0) = 0?$$

16. Consider the IVP:

$$y' = y(y - 1), \quad y(x_0) = y_0.$$

(a) Verify that the hypotheses of the existence and uniqueness theorem are satisfied for this IVP for any x_0, y_0. This establishes that the IVP always has a unique solution on some interval containing x_0.

(b) By inspection, determine all equilibrium solutions to the DE.

(c) Determine the regions in the xy-plane where the solution curves are concave up, and determine those regions where they are concave down.

(d) Sketch the slope field for the DE, and determine all values of y_0 for which the IVP has bounded solutions. On your slope field, sketch representative solution curves in the three cases $y_0 < 0$, $0 < y_0 < 1$, and $y_0 > 1$.

For problems 17–24, sketch the slope field and some representative solution curves for the given DE.

17. $y' = 4x$.

18. $y' = 1/x$.

19. $y' = x + y$.

20. $y' = x/y$.

21. $y' = -4x/y$.

22. $y' = x^2 y$.

23. $y' = x^2 \cos y$.

24. $y' = x^2 + y^2$.

25. According to Newton's law of cooling (see Section 1.1), the temperature of an object at time t is governed by the DE

$$\frac{dT}{dt} = -k(T - T_m),$$

where T_m is the temperature of the surrounding medium, and k is a constant. Consider the case when $T = 70$ and $k = 1/80$. Sketch the corresponding slope field and some representative solution curves. What happens to the temperature of the object as $t \to \infty$? Note that this result is independent of the initial temperature of the object.

For problems 26–31, determine the slope field and some representative solution curves for the given DE.

◆ 26. $y' = -2xy$.

◆ 27. $y' = \dfrac{x \sin x}{1 + y^2}$.

◆ 28. $y' = 3x - y$.

◆ 29. $y' = 2x^2 \sin y$.

◆ 30. $y' = \dfrac{2 + y^2}{3 + 0.5x^2}$.

◆ 31. $y' = \dfrac{1 - y^2}{2 + 0.5x^2}$.

◆ 32 (a) Determine the slope field for the DE

$$y' = x^{-1}(3 \sin x - y)$$

on the interval $(0, 10]$.

(b) Plot the solution curves corresponding to each of the following initial conditions:

$$y(0.5) = 0; \quad y(1) = -1; \quad y(1) = 2; \quad y(3) = 0.$$

What do you conclude about the behavior as $x \to 0^+$ of solutions to the DE?

(c) Plot the solution curve corresponding to the initial condition $y(\pi/2) = 6/\pi$. How does this fit in with your answer to part (b)?

(d) Describe the behavior of the solution curves for large positive x.

◆ 33. Consider the family of curves $y = kx^2$, where k is a constant.

(a) Show that the DE of the family of orthogonal trajectories is

$$\frac{dy}{dx} = -\frac{x}{2y}.$$

(b) On the same axes sketch the slope field for the preceding DE and several members of the given family of curves. Describe the family of orthogonal trajectories.

◆ 34 Consider the DE

$$\frac{di}{dt} + ai = b,$$

where a and b are constants. By drawing the slope fields corresponding to various values of a and b, formulate a conjecture regarding the value of $\lim_{t \to \infty} i(t)$.

1.4   SEPARABLE   DE

In the previous section we analyzed first-order DE using qualitative techniques. We now begin an analytical study of these DE by developing some solution techniques that enable us to determine the exact solution to certain types of DE. The simplest DE for which a solution technique can be obtained are the so-called separable equations, which are defined as follows:

Definition 1.4.1:   A first-order DE is called **separable** if it can be written in the form

$$p(y)\frac{dy}{dx} = q(x). \qquad (1.4.1)$$

The solution technique for a separable DE is given in Theorem 1.4.1.

Theorem 1.4.1:   If $p(y)$ and $q(x)$ are continuous, then equation (1.4.1) has the general solution

$$\int p(y)\, dy = \int q(x)\, dx + c, \qquad (1.4.2)$$

where c is an arbitrary constant.

PROOF   Integrating both sides of equation (1.4.2) with respect to x yields

$$\int p(y)\frac{dy}{dx}\, dx = \int q(x)\, dx + c.$$

That is,

$$\int p(y)\, dy = \int q(x)\, dx + c. \qquad \blacksquare$$

REMARK   In *differential form*, equation (1.4.1) can be written as

$$p(y)\, dy = q(x)\, dx,$$

and the general solution (1.4.2) is obtained by integrating the left-hand side with respect to y and the right-hand side with respect to x. This is the general procedure for solving separable equations.

Example 1.4.1   Solve $(1 + y^2)\dfrac{dy}{dx} = x\cos x$.

Solution   By inspection we see that the DE is separable. Integrating both sides of the DE yields

$$\int (1 + y^2)\, dy = \int x\cos x\, dx + c.$$

Using integration by parts to evaluate the integral on the right-hand side we obtain

$$y + \frac{1}{3}y^3 = x\sin x + \cos x + c,$$

or equivalently

$$y^3 + 3y = 3(x \sin x + \cos x) + c_1,$$

where $c_1 = 3c$. As often happens with separable DE, the solution is given in implicit form. ❐

In general, the DE $\dfrac{dy}{dx} = f(x)g(y)$ is separable, since it can be written as

$$\frac{1}{g(y)} \frac{dy}{dx} = f(x),$$

which is of the form of equation (1.4.1) with $p(y) = 1/g(y)$. It is important to note, however, that in writing the given DE in this way, we have assumed that $g(y) \neq 0$. Thus the general solution to the resulting DE may not include solutions of the original equation corresponding to any values of y for which $g(y) = 0$. (These are the equilibrium solutions for the original DE.) We will illustrate with an example.

Example 1.4.2   Find all solutions to

$$y' = -2y^2 x. \tag{1.4.3}$$

Solution   Separating the variables yields

$$y^{-2} \, dy = -2x \, dx. \tag{1.4.4}$$

Integrating both sides we obtain

$$-y^{-1} = -x^2 + c$$

so that

$$y(x) = \frac{1}{x^2 - c}. \tag{1.4.5}$$

This is the general solution to *equation (1.4.4)*. It is not the general solution to equation (1.4.3), since there is no value of the constant c for which $y(x) = 0$, whereas by

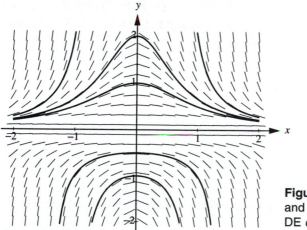

Figure 1.4.1 The slope field and some solution curves for the DE $dy/dx = -2xy^2$.

inspection, we see $y(x) = 0$ is a solution to equation (1.4.3). This solution is not contained in (1.4.5), since in separating the variables, we divided by y and hence assumed implicitly that $y \neq 0$. Thus the solutions to equation (1.4.3) are

$$y(x) = \frac{1}{x^2 - c} \quad \text{and} \quad y(x) = 0.$$

The slope field for the given DE is depicted in Figure 1.4.1, together with some representative solution curves.   ❑

Many of the difficulties that students encounter with first-order DE arise not from the solution techniques themselves, but in the algebraic simplifications that are used to obtain a simple form for the resulting solution. We will explicitly illustrate some of the standard simplifications using the DE

$$\frac{dy}{dx} = -2xy.$$

First notice that $y(x) = 0$ is an equilibrium solution to the DE. Consequently, no other solution curves can cross the x-axis. For $y \neq 0$ we can separate the variables to obtain

$$\frac{1}{y}\, dy = -2x\, dx. \tag{1.4.6}$$

Integrating this equation yields

$$\ln |y| = -x^2 + c.$$

Exponentiating both sides of this solution gives

$$|y| = e^{(-x^2 + c)},$$

or equivalently

$$|y| = e^c e^{-x^2}.$$

We now introduce a new constant c_1 defined by $c_1 = e^c$. Then the preceding expression for $|y|$ reduces to

$$|y| = c_1 e^{-x^2}. \tag{1.4.7}$$

Notice that c_1 is a *positive* constant. This is a perfectly acceptable form for the solution. However, a redefinition of the integration constant can be used to eliminate the absolute value bars as follows. According to (1.4.7), the solution to the DE is

$$y(x) = \begin{cases} c_1 e^{-x^2}, & \text{if } y > 0, \\ -c_1 e^{-x^2}, & \text{if } y < 0. \end{cases} \tag{1.4.8}$$

We can now define a new constant c_2, by

$$c_2 = \begin{cases} c_1, & \text{if } y > 0, \\ -c_1, & \text{if } y < 0, \end{cases}$$

in terms of which the solutions given in (1.4.8) can be combined into the single formula

$$y(x) = c_2 e^{-x^2}. \tag{1.4.9}$$

The appropriate sign for c_2 will be determined from the initial conditions. For example, the initial condition $y(0) = 1$ would require that

$$1 = c_2,$$

with corresponding unique solution

$$y(x) = e^{-x^2}.$$

Similarly the initial condition $y(0) = -1$ leads to

$$c_2 = -1,$$

so that

$$y(x) = -e^{-x^2}.$$

We make one further point about the solution (1.4.9). In obtaining the separable form (1.4.6), we divided the given DE by y, and so, the derivation of the solution obtained assumes that $y \neq 0$. However, as we have already noted, $y(x) = 0$ is indeed a solution to this DE. Formally this solution is the special case $c_2 = 0$ in (1.4.9), and corresponds to the initial condition $y(0) = 0$. Thus (1.4.9) does give the general solution to the DE, provided we allow c_2 to assume the value zero. The slope field for the DE, together with some particular solution curves, is shown in Figure 1.4.2.

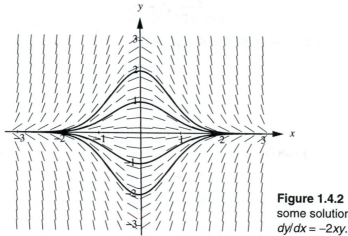

Figure 1.4.2 Slope field and some solution curves for the DE $dy/dx = -2xy$.

Example 1.4.3   An object of mass m falls from rest, starting at a point near the earth's surface. Assuming that the air resistance is proportional to the velocity of the object, determine the subsequent motion.

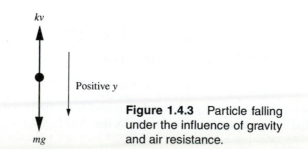

Figure 1.4.3 Particle falling under the influence of gravity and air resistance.

Solution   Let $y(t)$ be the distance traveled by the object at time t from the point it was released, and let the positive y direction be downward. Then, $y(0) = 0$, and the velocity of the object, $v(t)$, is $v = dy/dt$. The forces acting on the object are those due to gravity, $F_g = mg$, and the force due to air resistance, $F_r = -kv$, where k is a positive constant. According to Newton's second law, the DE describing the motion of the object is

$$m\frac{dv}{dt} = F_g + F_r = mg - kv.$$

We are also given the initial condition $v(0) = 0$. Thus the IVP governing the behavior of v is

$$\begin{cases} m\dfrac{dv}{dt} = mg - kv, \\ v(0) = 0. \end{cases} \tag{1.4.10}$$

Separating the variables in equation (1.4.10) yields

$$\frac{m}{mg - kv}\, dv = dt,$$

which can be integrated directly to obtain

$$-\frac{m}{k}\ln|mg - kv| = t + c.$$

Multiplying both sides of this equation by $-k/m$ and exponentiating the result yields

$$|mg - kv| = c_1 e^{-(k/m)t},$$

where $c_1 = e^{-ck/m}$. By redefining the constant c_1, we can write this in the equivalent form

$$mg - kv = c_2 e^{-(k/m)t}.$$

Hence

$$v(t) = \frac{mg}{k} - c_3 e^{-(k/m)t}, \tag{1.4.11}$$

where $c_3 = c_2/k$. Imposing the initial condition $v(0) = 0$ yields

$$c_3 = \frac{mg}{k}.$$

So the solution to the IVP (1.4.10) is

$$v(t) = \frac{mg}{k}[1 - e^{-(k/m)t}]. \tag{1.4.12}$$

Notice that the velocity does not increase indefinitely, but approaches a so-called *limiting velocity* v_L defined by

$$v_L = \lim_{t \to \infty} v(t) = \lim_{t \to \infty} \frac{mg}{k}[1 - e^{-(k/m)t}] = \frac{mg}{k}.$$

The behavior of the velocity as a function of time is shown in Fig. 1.4.4. Due to the negative exponent in (1.4.11), we see that this result is independent of the value of the initial velocity.

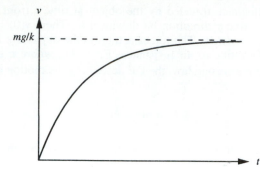

Figure 1.4.4 The behavior of the velocity of the object in Example 1.4.3.

Since $dy/dt = v$, it follows from (1.4.12) that the position of the object at time t can be determined by solving the IVP

$$\frac{dy}{dt} = \frac{mg}{k}[1 - e^{-(k/m)t}], \quad y(0) = 0.$$

The DE can be integrated directly to obtain

$$y(t) = \frac{mg}{k}\left[t + \frac{m}{k}e^{-(k/m)t}\right] + c.$$

Imposing the initial condition $y(0) = 0$ yields

$$c = -\frac{m^2 g}{k^2},$$

so that

$$y(t) = \frac{mg}{k}\left[t + \frac{m}{k}(e^{-(k/m)t} - 1)\right].$$

Example 1.4.4   A hot metal bar whose temperature is $350°$ F is placed in a room whose temperature is constant at $70°$ F. After two min, the temperature of the bar is $210°$ F. Using *Newton's law of cooling*, determine the temperature of the bar after four min, and the time required for the bar to cool to $100°$ F.

Solution   According to Newton's law of cooling (see Section 1.1), the temperature of the object at time t is governed by the DE

$$\frac{dT}{dt} = -k(T - T_m), \tag{1.4.13}$$

where, from the statement of the problem,

$$T_m = 70° \text{ F}; \quad T(0) = 350° \text{ F}; \quad T(2) = 210° \text{ F}.$$

We are required to find the following:

(a) $T(4)$.   (b) The time when $T = 100°$ F.

Substituting for T_m in equation (1.4.13), we have the separable equation

$$\frac{dT}{dt} = -k(T - 70).$$

Separating the variables yields

$$\frac{1}{T-70}\,dT=-k\,dt,$$

which we can integrate immediately to obtain

$$\ln|T-70|=-kt+c.$$

Exponentiating both sides and solving for T yields

$$T(t)=70+c_1e^{-kt}, \qquad\qquad (1.4.14)$$

where we have redefined the integration constant. The two constants c_1 and k can be determined from the given auxiliary conditions as follows.

$T(0)=350°$ F requires that $350=70+c_1$. Hence,

$$c_1=280.$$

Substituting this value for c_1 into (1.4.14) yields

$$T(t)=70(1+4e^{-kt}). \qquad\qquad (1.4.15)$$

Consequently, $T(2)=210°$ F if and only if

$$210=70(1+4e^{-2k}),$$

so that

$$e^{-2k}=\frac{1}{2}.$$

Hence,

$$k=\frac{1}{2}\ln 2,$$

and so, from (1.4.15),

$$T(t)=70\bigl[1+4\,e^{-(t/2)\ln 2}\bigr]. \qquad\qquad (1.4.16)$$

We can now answer (a) and (b).

(a) $T(4)=70(1+4\,e^{-2\ln 2})=70(1+4\,\frac{1}{2^2})=140°$ F.

(b)   From (1.4.16), $T(t)=100°$ F when

$$100=70\bigl[1+4\,e^{-(t/2)\ln 2}\bigr].$$

That is, when

$$e^{-(t/2)\ln 2}=\frac{3}{28}.$$

Taking the natural logarithm of both sides and solving for t yields

$$t=\frac{2\ln(28/3)}{\ln 2}\approx 6.4\ \text{mins.}$$

EXERCISES 1.4

For problems 1–11, solve the given DE.

1. $\dfrac{dy}{dx} = 2xy$.

2. $\dfrac{dy}{dx} = \dfrac{y^2}{x^2 + 1}$.

3. $e^{x+y}\,dy - dx = 0$.

4. $\dfrac{dy}{dx} = \dfrac{y}{x \ln\ x}$.

5. $y\,dx - (x - 2)\,dy = 0$.

6. $\dfrac{dy}{dx} = \dfrac{2x(y - 1)}{x^2 + 3}$.

7. $y - x\dfrac{dy}{dx} = 3 - 2x^2\dfrac{dy}{dx}$.

8. $\dfrac{dy}{dx} = \dfrac{\cos(x - y)}{\sin\ x\ \sin\ y} - 1$.

9. $\dfrac{dy}{dx} = \dfrac{x(y^2 - 1)}{2(x - 2)(x - 1)}$.

10. $\dfrac{dy}{dx} = \dfrac{x^2 y - 32}{16 - x^2} + 2$.

11. $(x - a)(x - b)y' - (y - c) = 0$, a, b, c constants.

In problems 12–15 solve the given IVP.

12. $(x^2 + 1)y' + y^2 = -1$, $y(0) = 1$.

13. $(1 - x^2)y' + xy = ax$, $y(0) = 2a$, where a is a constant.

14. $\dfrac{dy}{dx} = 1 - \dfrac{\sin(x + y)}{\sin\ y\ \cos\ x}$, $y(\pi/4) = \pi/4$.

15. $y' = y^3\sin\ x$, $y(0) = 0$.

16. One solution to the IVP

$$\frac{dy}{dx} = \frac{2}{3}(y - 1)^{1/2}, \quad y(1) = 1$$

is $y(x) = 1$. Determine another solution to this IVP. Does this contradict the existence and uniqueness theorem? Explain.

17. An object of mass m falls from rest, starting at a point near the earth's surface. Assuming that the air resistance varies as the square of the velocity of the object, a simple application of Newton's second law yields the IVP for the velocity, $v(t)$, of the object at time t:

$$m\frac{dv}{dt} = mg - kv^2, \quad v(0) = 0,$$

where k, m, g are positive constants.

(a) Solve the foregoing IVP for v in terms of t.

(b) Does the velocity of the object increase indefinitely? Justify.

(c) Determine the position of the object at time t.

18. Find the equation of the curve that passes through the point $(0, 1/2)$ and whose slope at each point (x, y) is $-x/(4y)$.

19. At time t the velocity, $v(t)$, of an object moving in a straight line satisfies

$$\frac{dv}{dt} = -(1 + v^2). \qquad (19.1)$$

(a) Show that

$$\tan^{-1}(v) = \tan^{-1}(v_0) - t,$$

where v_0 (greater than 0) denotes the velocity of the object at $t = 0$. Hence prove that the object comes to rest after a finite time $\tan^{-1}(v_0)$. Does the object remain at rest?

(b) Use the chain rule to show that equation (19.1) can be written as $v\dfrac{dv}{dx} = -(1 + v^2)$, where $x(t)$ denotes the distance traveled by the object, at time t, from its position at $t = 0$. Determine the distance traveled by the object when it first comes to rest.

20. The DE governing the velocity of an object is

$$\frac{dv}{dt} = -kv^n,$$

where k (greater than 0) and n are constants. At $t = 0$ the object is set in motion with velocity v_0.

(a) Show that the object comes to rest in a finite time if and only if $n < 1$, and determine the maximum distance traveled by the object in this case.

(b) If $1 \le n < 2$, show that the maximum distance traveled by the object in a finite time is less than

$$\frac{v_0^{2-n}}{(2 - n)k}.$$

(c) If $n \ge 2$, show that there is no limit to the distance that the object can travel.

21. The pressure, p, and density, ρ, of the atmosphere at a height y above the earth's surface are related by

$$dp = -g\rho\,dy.$$

Assuming that p and ρ satisfy the adiabatic equation of state $p = p_0 \left(\dfrac{\rho}{\rho_0} \right)^{\gamma}$, where $\gamma \,(\neq 1)$ is a constant and p_0 and ρ_0 denote the pressure and density at the earth's surface, respectively, show that

$$ p = p_0 \left[1 - \frac{(\gamma - 1)}{\gamma} \frac{\rho_0 g\, y}{p_0} \right]^{\gamma/(\gamma - 1)} $$

22. An object whose temperature is 615° F is placed in a room whose temperature is 75° F. At 4 p.m. ($t = 0$) the temperature of the object is 135° F, whereas an hour later its temperature is 95° F. At what time was the object placed in the room ?

23. An inflammable substance whose initial temperature is 50° F is inadvertently placed in a hot oven whose temperature is 450° F. After 20 min, the temperature of the substance is 150° F. Find the temperature of the substance after 40 min. If the substance ignites when its temperature reaches 350° F, find the time of combustion.

24. At 2 p.m. ($t = 0$) on a cool (34° F) afternoon in March, Sherlock Holmes measured the temperature of a dead body to be 38° F. One hour later the temperature was 36° F. After a quick calculation using Newton's law of cooling, and taking the normal temperature of a living body to be 98° F, Holmes concluded that the time of death was 10 a.m. Was Holmes right ?

25. At 4 p.m. a hot coal was pulled out of a furnace and allowed to cool at room temperature (75° F). If after 10 min the temperature of the coal was 415° F, and after 20 min its temperature was 347° F. Find the following:

(a) The temperature of the furnace.

(b) The time when the temperature of the coal was 100° F .

26. A hot object is placed in a room whose temperature is 72° F. After 1 minute, the temperature of the object is 150° F and its rate of change of temperature is 20° F per minute. Find the initial temperature of the object and the rate at which its temperature is changing after 10 min.

27. Consider the situation depicted in Figure 1.4.5 in which an object is placed in a room whose temperature T_1 is itself changing according to Newton's law of cooling. In this case, we must solve the system of differential equations

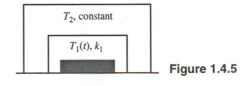

Figure 1.4.5

$$ \frac{dT_1}{dt} = -k_1(T_1 - T_2) $$

and

$$ \frac{dT}{dt} = -k(T - T_1) $$

where k, k_1 and T_2 are constants. Show that, for $k \neq k_1$, the temperature of the object at time t is given by

$$ T(t) = T_2 + \frac{c_1 k}{(k - k_1)} e^{-k_1 t} + c_2 e^{-kt}, $$

where c_1 and c_2 are constants.

28. An object is placed in an ice chest whose temperature is 35° F. The external temperature is constant at 80° F. After 5 min, the temperature of the object is 60° F and is decreasing at a rate of 1° F per minute. After 10 min, the temperature in the ice chest is 40° F. Determine:

(a) The initial temperature of the object.

(b) The time when the temperature of the object begins to increase.

1.5   SOME SIMPLE POPULATION MODELS

In this section we consider two important models of population growth whose mathematical formulation leads to separable DE.

1 . MALTHUSIAN GROWTH

The simplest mathematical model of population growth is obtained by assuming that the rate of increase of the population at any time is proportional to the size of the population at that time. If we let $P(t)$ denote the population at time t, then

$$\frac{dP}{dt} = kP,$$

where k is a *positive* constant. Separating the variables and integrating yields

$$P(t) = P_0 e^{kt}, \qquad (1.5.1)$$

where P_0 denotes the population at $t = 0$. This law predicts an exponential increase in the population with time which gives a reasonably accurate description of the growth of certain algae, bacteria, and cell cultures. It is called the **Malthusian growth model**. The time taken for such a culture to double in size is called the **doubling time**. This is the time, t_d, when $P(t_d) = 2P_0$. Substituting into (1.5.1) yields

$$2P_0 = P_0 e^{kt_d}.$$

Dividing both sides by P_0 and taking logarithms, we find

$$kt_d = \ln 2,$$

so that the doubling time is

$$t_d = \frac{1}{k} \ln 2.$$

Example 1.5.1   The number of bacteria in a certain culture grows at a rate that is proportional to the number present. If the number increased from 500 to 2000 in 2 h, determine the number present after 12 h, and also find the doubling time.

Solution   The behavior of the system is governed by the DE

$$\frac{dP}{dt} = kP,$$

so that

$$P(t) = P_0 e^{kt}.$$

Taking $t = 0$ as the time when the population was 500, we have $P_0 = 500$. Thus,

$$P(t) = 500 e^{kt}.$$

Further, $P(2) = 2000$ implies that

$$2000 = 500 e^{2k},$$

so that

$$k = \frac{1}{2} \ln 4 = \ln 2.$$

Consequently,

$$P(t) = 500 \, e^{t \ln 2}.$$

The number of bacteria present after twelve hours is therefore

$$P(12) = 500 \, e^{12 \ln 2} = 500 \, (2^{12}) = 2,048,000.$$

The doubling time of the system is

$$t_d = \frac{1}{k} \ln 2 = 1 \text{ h.}$$

2. LOGISTIC POPULATION MODEL

The Malthusian growth law (1.5.1) does not provide an accurate model for the growth of a population over a long time period. To obtain a more realistic model we need to take account of the fact that as the population increases several factors will begin to have an effect on the growth rate. For example, there will be increased competition for the limited resources that are available, increases in disease, and overcrowding of the limited available space, all of which would serve to slow the growth rate. In order to model this situation mathematically, we modify the DE leading to the simple exponential growth law by adding in a term that slows the growth down as the population increases. If we consider a closed environment (neglecting factors such as immigration and emigration), then the rate of change of population can be modeled by the DE

$$\frac{dP}{dt} = [B(t) - D(t)]P,$$

where $B(t)$ and $D(t)$ denote the birth rate and death rate per individual respectively. The simple exponential law corresponds to the case when $B(t) = k$ and $D(t) = 0$. In the more general situation of interest now, the increased competition as the population grows would result in a corresponding increase in the death rate per individual. Perhaps the simplest way to take account of this is to assume that the death rate per individual is directly proportional to the instantaneous population, and that the birth rate per individual remains constant. The resulting IVP governing the population growth can then be written as

$$\frac{dP}{dt} = (B_0 - D_0 P)P, \quad P(0) = P_0,$$

where B_0 and D_0 are *positive* constants. It is useful to write the DE in the equivalent form

$$\frac{dP}{dt} = r\left(1 - \frac{P}{C}\right)P, \tag{1.5.2}$$

where $r = B_0$, and $C = B_0/D_0$. This DE is separable, and can be solved without difficulty. Before doing that, however, we give a qualitative analysis of the DE. The constant C in equation (1.5.2) is called the **carrying capacity** of the population. We see from this equation that if $P < C$, then $dP/dt > 0$ and the population increases, whereas if $P > C$, then $dP/dt < 0$ and the population decreases. We can therefore interpret C as representing the maximum population that the environment can sustain. We note that $P(t) = C$ is an equilibrium solution to the DE, as is $P(t) = 0$. The isoclines for equation (1.5.2) are determined from

$$r\left(1 - \frac{P}{C}\right)P = k,$$

where k is a constant. This can be written as

$$P^2 - CP + \frac{kC}{r} = 0,$$

so that the isoclines are the lines

$$P = \frac{1}{2}\left[C \pm \left(C^2 - \frac{4kC}{r} \right)^{1/2} \right].$$

This tells us that the slopes of the solution curves satisfy

$$C^2 - \frac{4kC}{r} \geq 0,$$

so that

$$k \leq rC/4.$$

Furthermore, the largest value that the slope can assume is $k = rC/4$, which corresponds to $P = C/2$. We also note that the slope approaches zero as the solution curves approach the equilibrium solutions $P = 0, C$. Differentiating equation (1.5.2) yields

$$\frac{d^2P}{dt^2} = r\left[\left(1 - \frac{P}{C}\right)\frac{dP}{dt} - \frac{P}{C}\frac{dP}{dt} \right] = r\left(1 - 2\frac{P}{C}\right)\frac{dP}{dt} = \frac{r^2}{C^2}(C - 2P)(C - P)P,$$

where we have substituted for dP/dt from (1.5.2) and simplified the result. Since $P = C$ and $P = 0$ are solutions to the DE (1.5.2), the only points of inflection occur along the line $P = C/2$. The behavior of the concavity is therefore given by the following schematic:

$$\textbf{Sign of } \ddot{P}: \quad | +\!+\!+\!+ \;\; | -\!-\!-\!- \;\; | +\!+\!+\!+$$
$$\textbf{P-interval:} \quad 0 \qquad C/2 \qquad C$$

This information determines the general behavior of the solution curves to the DE (1.5.2). Figure 1.5.1 gives a Maple plot of the slope field and some representative solution curves. Of course, such a figure could have been constructed by hand using the

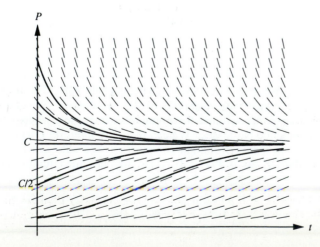

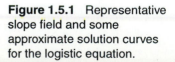

Figure 1.5.1 Representative slope field and some approximate solution curves for the logistic equation.

information we have obtained. From Figure 1.5.1, we see that if the initial population is less than the carrying capacity, then the population increases monotonically towards the carrying capacity. Similarly, if the initial population is bigger than the carrying capacity (not really relevant physically), then the population monotonically decreases towards the carrying capacity. Once more this illustrates the power of the qualitative techniques that have been introduced for analyzing first-order DE.

We turn now to obtaining an analytical solution to the DE (1.5.2). Separating the variables in equation (1.5.2) and integrating yields

$$\int \frac{C}{P(C-P)}\, dP = rt + c_1,$$

where c_1 is an integration constant. Using a partial fraction decomposition on the right-hand side, we find

$$\int \left(\frac{1}{P} + \frac{1}{C-P}\right) dP = rt + c_1,$$

which upon integration gives

$$\ln \left| \frac{P}{C-P} \right| = rt + c_1.$$

Exponentiating, and redefining the integration constant yields

$$\frac{P}{C-P} = c_2 e^{rt},$$

which can be solved algebraically for P to obtain

$$P(t) = \frac{c_2 C e^{rt}}{1 + c_2 e^{rt}}.$$

or equivalently,

$$P(t) = \frac{c_2 C}{c_2 + e^{-rt}}.$$

Imposing the initial condition $P(0) = P_0$ we find that $c_2 = P_0/(C-P_0)$. Inserting this value of c_2 into the preceding expression for $P(t)$ yields

$$P(t) = \frac{C P_0}{P_0 + (C - P_0)e^{-rt}}. \tag{1.5.3}$$

We make two comments regarding this formula. Firstly, we see that due to the negative exponent of the exponential term in the denominator, as $t \to \infty$, the population does indeed tend to the carrying capacity C independently of the initial population P_0. Secondly, by writing (1.5.3) in the equivalent form

$$P(t) = \frac{P_0}{P_0/C + (1 - P_0/C)e^{-rt}},$$

it follows that if P_0 is very small compared to the carrying capacity then for small t the terms involving P_0 in the denominator can be neglected leading to the approximation

$$P(t) \approx P_0 e^{rt}.$$

Consequently, in this case, the Malthusian population model does approximate the logistic model for small time intervals.

Although we now have a formula for the solution to the logistic population model, the qualitative analysis is certainly very enlightening as far as the general overall properties of the solution are concerned. Of course if we want to investigate specific details of a particular model, then we would use the corresponding exact solution (1.5.3).

Example 1.5.2   The initial population (measured in thousands) of a city is 20. After 10 years, this has increased to 50.87, whereas after 15 years the population is 78.68. Use the logistic model to predict the population after 30 years.

Solution   In this problem, we have $P_0 = P(0) = 20$, $P(10) = 50.87$, $P(15) = 78.68$, and we wish to find $P(30)$. Substituting for P_0 into equation (1.5.3) yields

$$P(t) = \frac{20C}{20 + (C - 20)e^{-rt}}. \tag{1.5.4}$$

Imposing the two remaining auxiliary conditions leads to the following pair of equations for determining r and C

$$50.87 = \frac{20C}{20 + (C - 20)e^{-10r}},$$

$$78.68 = \frac{20C}{20 + (C - 20)e^{-15r}}.$$

This is a pair of nonlinear algebraic equations that are tedious to solve by hand. We therefore turn to technology. Using the algebraic capabilities of Maple, we find that

$$r \approx 0.1, \quad C \approx 500.37.$$

Substituting these values of r and C in equation (1.5.4) yields

$$P(t) = \frac{10007.4}{20 + 480.37e^{-0.1t}}.$$

Accordingly, the predicted value of the population after 30 years is

$$P(30) = \frac{10007.4}{20 + 480.37e^{-3}} = 227.87.$$

A sketch of $P(t)$ is given in Figure 1.5.2.

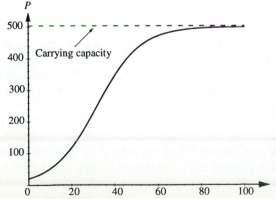

Figure 1.5.2   Solution curve corresponding to the population model in Example 1.5.2. The population is measured in thousands of people.

EXERCISES 1.5

1. The number of bacteria in a culture grows at a rate that is proportional to the number present. Initially there were 10 bacteria in the culture. If the doubling time of the culture is 3 hours, find the number of bacteria that were present after 24 hours.

2. The number of bacteria in a culture grows at a rate that is proportional to the number present. After 10 hours there were 5000 bacteria present, and after 12 hours there were 6000 bacteria present. Determine the initial size of the culture and the doubling time of the population.

3. A certain cell culture has a doubling time of 4 hours. Initially there were 2000 cells present. Assuming an exponential growth law, determine the time it takes for the culture to contain 10^6 cells.

4. At time t, the population, $P(t)$, of a certain city is increasing at a rate that is proportional to the number of residents in the city at that time. In January 1990 ($t = 0$) the population of the city was 10000 and by 1995 it had risen to 20000.

(a) What will the population of the city be at the beginning of the year 2010?

(b) In what year will the population reach 1 million?

In the logistic population model (1.5.3), if $P(t_1) = P_1$ and $P(2t_1) = P_2$ then it can be shown (through some algebra to derive by hand, although easy on a computer algebra system) that

$$r = \frac{1}{t_1} \ln\left[\frac{P_2(P_1 - P_0)}{P_0(P_2 - P_1)}\right], \qquad (1.5.5)$$

$$C = \frac{P_1[P_1(P_0 + P_2) - 2P_0P_2]}{P_1^2 - P_0P_2}. \qquad (1.5.6)$$

These formulas should be used in problems 5–7.

5. The initial population in a small village is 500. After 5 years this has grown to 800, while after 10 years the population is 1000. Using the logistic population model, determine the population after 15 years.

6. An animal sanctuary had an initial population of 50 animals. After 2 years the population was 62, while after 4 years it was 76. Using the logistic population model, determine the carrying capacity and the number of animals in the sanctuary after 20 years.

7. **(a)** Use equations (1.5.5) and (1.5.6) and the fact that r and C are positive to derive two inequalities that P_0, P_1, P_2 must satisfy in order for there to be a solution to the logistic equation satisfying the conditions $P(0) = P_0$, $P(t_1) = P_1$, $P(2t_1) = P_2$.

(b) The initial population in a town is 10,000. After 5 years this has grown to 12,000, while after 10 years the population is 18,000. Is there a solution to the logistic equation that fits this data?

8. Of the 1500 passengers, crew, and staff that board a cruise ship, 5 have the flu. After one day of sailing the number of infected people has risen to 10. Assuming that the rate at which the flu virus spreads is proportional to the product of the number of infected individuals times the number not yet infected, determine how many people will have the flu at the end of the 14 day cruise. Would you like to be a member of the customer relations department for the cruise line the day after the ship docks?

9. Consider the population model

$$\frac{dP}{dt} = r(P - T)P, \quad P(0) = P_0, \qquad (9.1)$$

where r, T and P_0 are positive constants.

(a) Perform a qualitative analysis of the DE in the IVP (9.1) following the steps used in the text for the logistic equation. Identify the equilibrium solutions, the isoclines, and the behavior of the slope and concavity of the solution curves.

(b) Using the information obtained in (a), sketch the slope field for the DE and include representative solution curves.

(c) What predictions can you make regarding the behavior of the population? Consider the cases $P_0 < D$ and $P_0 > D$. The constant T is called the **threshold level**. Based on your predictions, why is this an appropriate term to use for T?

10. In the previous problem a qualitative analysis of the DE in (9.1) was carried out. In this problem we determine the exact solution to the DE and verify the predictions from the qualitative analysis.

(a) Solve the IVP (9.1).

(b) Using your solution from (a), verify that if $P_0 < D$, then $\lim_{t \to \infty} P(t) = 0$. What does this mean for the population?

(c) Using your solution from (a), verify that if $P_0 > D$, then each solution curve has a vertical asymptote at $t = t_e$, where

$$t_e = \frac{1}{rD} \ln\left(\frac{P_0}{P_0 - D}\right).$$

How do you interpret this result in terms of population growth? Note that this was not obvious from the qualitative analysis performed in the previous problem.

11. As a modification to the population model considered in the previous two problems, suppose that $P(t)$ satisfies the IVP

$$\frac{dP}{dt} = r(C - P)(T - P)P, \quad P(0) = P_0,$$

where r, C, T, P_0 are constants, and $0 < T < C$. Perform a qualitative analysis of this model. Sketch the slope field, and some representative solution curves in the three cases $0 < P_0 < T$, $T < P_0 < C$, and $P_0 > C$. Describe the behavior of the corresponding solutions.

The next two problems consider the **Gompertz** population model which is governed by the IVP

$$\frac{dP}{dt} = rP(\ln C - \ln P), \quad P(0) = P_0, \quad (1.5.7)$$

where r, C, and P_0 are positive constants.

12. Determine all equilibrium solutions for the DE in (1.5.7), and the behavior of the slope and concavity of the solution curves. Use this information to sketch the slope field and some representative solution curves.

13. Solve the IVP (1.5.7) and verify that all solutions satisfy $\lim_{t \to \infty} P(t) = C$.

◆ **14.** Use some form of technology to solve the pair of equations

$$P_1 = \frac{CP_0}{P_0 + (C - P_0)e^{-rt_1}},$$

$$P_2 = \frac{CP_0}{P_0 + (C - P_0)e^{-2rt_1}},$$

for r and C, and thereby derive the expressions given in equations (1.5.5) and (1.5.6).

◆ **15.** According to data from the U.S. Bureau of the Census, the population (measured in millions of people) of the U.S. in 1950, 1960, and 1970 was, respectively, 151.3, 179.4, 203.3.

(a) Using the 1950 and 1960 population figures, solve the corresponding Malthusian population model.

(b) Determine the logistic model corresponding to the given data.

(c) On the same set of axes, plot the solution curves obtained in (a) and (b). From your plots, determine the values the different models would have predicted for the population in 1980 and 1990, and compare these predictions to the actual values of 226.54 and 248.71, respectively.

◆ **16.** In a period of 5 years the population of a city doubles from its initial size of 50 (measured in thousands of people). After 10 more years the population has reached 250. Determine the logistic model corresponding to this data. Sketch the solution curve and use your plot to estimate the time it will take for the population to reach 95% of the carrying capacity.

1.6   FIRST-ORDER LINEAR DE

In this section we derive a technique for determining the general solution to any first-order *linear* DE. This is the most important technique in the chapter.

Definition 1.6.1:   A DE that can be written in the form

$$a(x)\frac{dy}{dx} + b(x)y = r(x) \tag{1.6.1}$$

where $a(x)$, $b(x)$, and $r(x)$ are functions defined on an interval (α, β) is called a **first-order linear** DE.

We assume that $a(x) \neq 0$ on (α, β) and divide both sides of (1.6.1) by $a(x)$ to obtain the **standard form**

$$\frac{dy}{dx} + p(x)y = q(x) \tag{1.6.2}$$

where $p(x) = b(x)/a(x)$ and $q(x) = r(x)/a(x)$. The idea behind the solution technique for (1.6.2) is to rewrite the DE in the form

$$\frac{d}{dx}[g(x, y)] = F(x) \tag{1.6.3}$$

for an appropriate function $g(x, y)$. The general solution to the DE can then be obtained by an integration. First consider an example.

Example 1.6.1   Solve the DE

$$\frac{dy}{dx} + \frac{1}{x}y = e^x, \ \ x > 0. \tag{1.6.4}$$

Solution   If we multiply (1.6.4) by x we obtain

$$x\frac{dy}{dx} + y = xe^x.$$

But, from the product rule for differentiation, the left-hand side of this equation is just the expanded form of $\frac{d}{dx}(xy)$. Thus (1.6.4) can be written in the equivalent form

$$\frac{d}{dx}(xy) = xe^x.$$

Integrating both sides of this equation with respect to x yields

$$xy = xe^x - e^x + c$$

so that the general solution (1.6.4) is

$$y(x) = x^{-1}[e^x(x - 1) + c],$$

where c is an arbitrary constant.   ❑

In the previous example we multiplied the given DE by the function $I(x) = x$. This had the effect of reducing the left-hand side of the resulting DE to the integrable form

$$\frac{d}{dx}(xy).$$

Motivated by this example, we now consider the possibility of multiplying the general linear DE

$$\frac{dy}{dx} + p(x)y = q(x) \tag{1.6.5}$$

by a nonzero function $I(x)$, chosen in such a way that the left-hand side of the resulting DE is

$$\frac{d}{dx}[I(x)y].$$

Henceforth we will assume that the functions p and q are continuous on (α, β). Multiplying the DE (1.6.5) by $I(x)$ yields

$$I \frac{dy}{dx} + p(x) \, Iy = Iq(x). \tag{1.6.6}$$

Furthermore, from the product rule for derivatives, we know that

$$\frac{d}{dx}(Iy) = I \frac{dy}{dx} + \frac{dI}{dx} \, y. \tag{1.6.7}$$

Comparing equation (1.6.6) and equation (1.6.7) we see that equation (1.6.6) can indeed be written in the integrable form

$$\frac{d}{dx}(Iy) = Iq(x),$$

provided the function $I(x)$ is a solution to[1]

$$I \frac{dy}{dx} + p(x)Iy = I \frac{dy}{dx} + \frac{dI}{dx} \, y.$$

This will hold whenever $I(x)$ satisfies the separable DE

$$\frac{dI}{dx} = p(x) \, I. \tag{1.6.8}$$

Separating the variables and integrating yields

$$\ln |I| = \int p(x) \, dx + c,$$

so that

$$I(x) = c_1 e^{\int p(x) \, dx},$$

where c_1 is an arbitrary constant. Since we only require one solution to equation (1.6.8) we set $c_1 = 1$, in which case

$$I(x) = e^{\int p(x) \, dx}. \tag{1.6.9}$$

We can therefore draw the following conclusion.

Multiplying the linear DE

$$\frac{dy}{dx} + p(x)y = q(x) \tag{1.6.10}$$

by $e^{\int p(x) \, dx}$ reduces it to the integrable form

$$\frac{d}{dx}\left[e^{\int p(x) \, dx} y \right] = q(x) \, e^{\int p(x) \, dx}. \tag{1.6.11}$$

The general solution to (1.6.10) can now be obtained from (1.6.11) by integration. Formally we have

---

[1]This is obtained by equating the left-hand side of equation (1.6.6) to the right-hand side of equation (1.6.7).

$$y(x) = e^{-\int p(x)\,dx} \left[\int q(x)\, e^{\int p(x)\,dx}\, dx + c \right].$$ (1.6.12)

REMARKS

1. The function $I = e^{\int p(x)\,dx}$ is called an **integrating factor** for the DE (1.6.10) since it enables us to reduce the DE to a form that is directly integrable.

2. You should not memorize (1.6.12). In a specific problem we first evaluate the integrating factor $e^{\int p(x)\,dx}$ and then use (1.6.11).

Example 1.6.2   Solve the IVP

$$\frac{dy}{dx} + xy = xe^{x^2/2}, \quad y(0) = 1.$$

Solution   An appropriate integrating factor in this case is

$$I(x) = e^{\int x\,dx} = e^{x^2/2}.$$

Multiplying the given DE by I and using (1.6.11) yields

$$\frac{d}{dx}(e^{x^2/2}\, y) = xe^{x^2}.$$

Integrating both sides with respect to x, we obtain

$$e^{x^2/2}\, y = \int xe^{x^2}dx + c.$$

Hence,

$$y(x) = e^{-x^2/2}\left(\tfrac{1}{2}\, e^{x^2} + c\right).$$

Imposing the initial condition $y(0) = 1$ yields

$$1 = \frac{1}{2} + c,$$

so that $c = \dfrac{1}{2}$. Thus the required particular solution is

$$y(x) = \frac{1}{2}\, e^{-x^2/2}\,(e^{x^2} + 1) = \frac{1}{2}\,(e^{x^2/2} + e^{-x^2/2}) = \cosh(x^2/2).$$

Example 1.6.3   Solve $x\dfrac{dy}{dx} + 2y = \cos x, \ x > 0$.

Solution   We first write the given DE in standard form. Dividing by x yields

$$\frac{dy}{dx} + 2x^{-1}\, y = x^{-1}\cos x.$$ (1.6.13)

An integrating factor is

$$I(x) = e^{\int 2x^{-1}dx} = e^{2\ln x} = x^2,$$

so that upon multiplying equation (1.6.13) by I, we obtain

$$\frac{d}{dx}(x^2 y) = x \cos x.$$

Integrating and rearranging gives

$$y(x) = x^{-2}(x \sin x + \cos x + c),$$

where we have used integration by parts on the right-hand side.

Example 1.6.4   Solve the IVP

$$y' - y = f(x), \quad y(0) = 0,$$

where $f(x) = \begin{cases} 1, & \text{if } x < 1, \\ 2 - x, & \text{if } x \geq 1. \end{cases}$

Solution   We have sketched $f(x)$ in Figure 1.6.1. An integrating factor for the DE is

$$I(x) = e^{-x}.$$

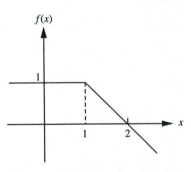

Figure 1.6.1 A sketch of the function $f(x)$ from Example 1.6.4.

Upon multiplication by the integrating factor, the DE reduces to

$$\frac{d}{dx}(e^{-x} y) = e^{-x} f(x).$$

We now integrate this DE over the interval $[0, x]$. To do so we need to use a dummy integration variable which we denote by w. We therefore obtain

$$e^{-w} y(w) \Big]_0^x = \int_0^x e^{-w} f(w) \, dw,$$

or equivalently,

$$e^{-x} y(x) - y(0) = \int_0^x e^{-w} f(w) \, dw.$$

Multiplying by e^x and substituting for $y(0) = 0$ yields

$$y(x) = e^x \int_0^x e^{-w} f(w) \, dw. \tag{1.6.14}$$

Due to the form of $f(x)$, the value of the integral on the right-hand side will depend on whether $x < 1$ or $x \geq 1$.

If $x < 1$, then $f(w) = 1$ and so (1.6.14) can be written as

$$y(x) = e^x \int_0^x e^{-w}dw = e^x(1 - e^{-x}), \qquad (1.6.15)$$

so that

$$y(x) = e^x - 1, \quad x < 1.$$

If $x \geq 1$, then the interval of integration $[0, x]$ must be split into two parts. From (1.6.14) we have

$$y(x) = e^x \left[\int_0^1 e^{-w}dw + \int_1^x (2 - w)e^{-w}dw \right].$$

A straightforward integration leads to

$$y(x) = e^x \left\{ (1 - e^{-1}) + (-2e^{-w} + we^{-w} + e^{-w}) \big]_1^x \right\},$$

which simplifies to

$$y(x) = e^x(1 - e^{-1}) + x - 1.$$

The solution to the IVP can therefore be written as

$$y(x) = \begin{cases} e^x - 1, & \text{if } x < 1, \\ e^x(1 - e^{-1}) + x - 1, & \text{if } x \geq 1. \end{cases}$$

A sketch of the corresponding solution curve is given in Figure 1.6.2. Differentiating

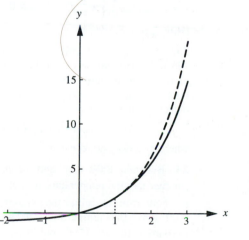

Figure 1.6.2 The solution curve for the IVP in Example 1.6.4. Dashed curve is the continuation of $y(x) = e^x - 1$ for $x > 1$.

both branches of this function we find

$$y'(x) = \begin{cases} e^x, & \text{if } x < 1, \\ e^x(1 - e^{-1}) + 1, & \text{if } x \ge 1. \end{cases} \qquad y''(x) = \begin{cases} e^x, & \text{if } x < 1, \\ e^x(1 - e^{-1}), & \text{if } x \ge 1. \end{cases}$$

We see that even though the function f in the original DE was not differentiable at $x = 1$, the solution to the IVP has a continuous derivative at that point. The discontinuity in the derivative of the driving term does show up in the second derivative of the solution as indeed it must.

EXERCISES 1.6

For problems 1 – 14, solve the given DE.

1. $\dfrac{dy}{dx} - y = e^{2x}$.

2. $x^2 y' - 4xy = x^7 \sin x$, $x > 0$.

3. $y' + 2xy = 2x^3$.

4. $\dfrac{dy}{dx} + \dfrac{2x}{(1 - x^2)} y = 4x$, $-1 < x < 1$.

5. $\dfrac{dy}{dx} + \dfrac{2x}{(1 + x^2)} y = \dfrac{4}{(1 + x^2)^2}$.

6. $2 \cos^2 x \, y' + y \sin 2x = 4 \cos^4 x$, $0 \le x < \pi/2$.

7. $y' + \dfrac{1}{x \ln x} y = 9x^2$.

8. $y' - y \tan x = 8 \sin^3 x$.

9. $t \dfrac{dx}{dt} + 2x = 4e^t$, $t > 0$.

10. $y' = \sin x \, (y \sec x - 2)$.

11. $(1 - y \sin x) \, dx - \cos x \, dy = 0$.

12. $y' - x^{-1} y = 2x^2 \ln x$.

13. $y' + \alpha y = e^{\beta x}$, α, β constants.

14. $y' + m x^{-1} y = \ln x$, m constant.

For problems 15–20, solve the given IVP.

15. $y' + 2x^{-1} y = 4x$, $y(1) = 2$.

16. $(\sin x) \, y' - y \cos x = \sin 2x$, $y(\pi/2) = 2$.

17. $\dfrac{dx}{dt} + \dfrac{2}{(4 - t)} x = 5$, $x(0) = 4$.

18. $(y - e^x) \, dx + dy = 0$, $y(0) = 1$.

19. $y' + y = f(x)$, $y(0) = 3$, where

$$f(x) = \begin{cases} 1, & \text{if } x \le 1, \\ 0, & \text{if } x > 1. \end{cases}$$

20. $y' - 2y = f(x)$, $y(0) = 1$, where

$$f(x) = \begin{cases} 1 - x, & x < 1, \\ 0, & x \ge 1. \end{cases}$$

21. Solve the IVP in Example 1.6.4 as follows. First determine the general solution to the DE on each interval separately. Then use the given initial condition to find the appropriate integration constant for the interval $(-\infty, 1)$. To determine the integration constant on the interval $[1, \infty)$ use the fact that the solution must be continuous at $x = 1$.

22. Find the general solution to the second-order DE

$$\frac{d^2 y}{dx^2} + \frac{1}{x} \frac{dy}{dx} = 9x, \quad x > 0.$$

(Hint: Let $u = dy/dx$.)

23. Suppose that an object is placed in a medium whose temperature is increasing at a constant rate of $\alpha °F$ per minute. Show that, according to Newton's law of cooling, the temperature of the object at time t is given by:

$$T(t) = \alpha(t - k^{-1}) + c_1 + c_2 e^{-kt},$$

where c_1 and c_2 are constants.

24. Between 8:00 a.m. and 12 p.m. on a hot summer day the temperature rose at a rate of $10°F$ per hour from an initial temperature of $65°F$. At 9:00 a.m. the temperature of an object was measured to be $35°F$ and was, at that time, increasing at a rate of $5°F$ per hour. Show that the temperature of the object, at time t, was

$$T(t) = 10t - 15 + 40 \, e^{(1 - t)/8}, \quad 0 \le t \le 4.$$

25. It is known that a certain object has constant of proportionality $k = 1/40$ in Newton's law of cooling. When the temperature of this object is 0°F, it is placed in a medium whose temperature is changing in time according to

$$T_m(t) = 80 \, e^{-t/20}.$$

(a) Using Newton's law of cooling, show that the temperature of the object at time t is

$$T(t) = 80(e^{-t/40} - e^{-t/20}).$$

(b) What happens to the temperature of the object as $t \to +\infty$? Is this reasonable ?

(c) Determine the time, t_{max}, when the temperature of the object is a maximum. Find $T(t_{max})$ and $T_m(t_{max})$.

(d) Make a sketch to depict the behavior of $T(t)$, and $T_m(t)$.

26. The DE

$$\frac{dT}{dt} = -k_1[T - T_m(t)] + A_0 \qquad (26.1)$$

where k_1 and A_0 are positive constants, can be used to model the temperature variation $T(t)$ in a building. In this equation the first term on the right-hand side gives the contribution due to the variation in the outside temperature, and the second term on the right-hand side gives the contribution due to the heating effect from internal sources such as machinery, lighting, people, etc. Consider the case when

$$T_m(t) = A - B\cos \omega t, \quad \omega = \pi/12, \qquad (26.2)$$

where A and B are constants, and t is measured in hours.

(a) Make a sketch of $T_m(t)$. Taking $t = 0$ to correspond to midnight, describe the variation of the external temperature over a 24-hour period.

(b) With T_m given in (26.2), solve (26.1) subject to the initial condition $T(0) = T_0$.

27. This problem demonstrates the **variation-of-parameters method** for first-order linear DE. Consider the first-order linear DE

$$y' + p(x) \, y = q(x). \qquad (27.1)$$

(a) Show that the general solution to the associated *homogeneous* equation

$$y' + p(x) \, y = 0$$

is

$$y_H(x) = c_1 e^{-\int p(x)\,dx}.$$

(b) Determine the function $u(x)$ such that

$$y(x) = u(x) \, e^{-\int p(x)\,dx}$$

is a solution to (27.1), and hence derive the general solution to (27.1).

For problems 28–31, use the technique derived in the previous problem to solve the given DE.

28. $y' + y = e^{-2x}$.

29. $y' + x^{-1}y = \cos x, \; x > 0$.

30. $xy' - y = x^2 \ln x$.

31. $y' + y \cot x = 2 \cos x, \; 0 < x < \pi$.

◆ **33.** Use a DE solver to determine the solution to each of the IVP from ExercIses 15–18, and sketch the corresponding solution curve.

◆ **34.** Use a DE solver to determine the solution to each of the IVP from ExercIses 19 and 20. Sketch the corresponding solution curve.

1.7   TWO MODELING PROBLEMS GOVERNED BY FIRST-ORDER LINEAR DE

There are many examples of applied problems whose mathematical formulation leads to a first-order linear DE. In this section we analyze two in detail.

1. MIXING PROBLEMS

Statement of the Problem: Consider the situation depicted in Figure 1.7.1. A tank initially contains V_0 liters of a solution in which is dissolved A_0 grams of a certain chemical. A solution containing c_1 grams/liter of the same chemical flows into the tank at a constant rate of r_1 liters/minute, and the mixture flows out at a constant rate of r_2 liters/minute. We assume that the mixture is kept uniform by stirring. Then at any time t the concentration of chemical in the tank, $c_2(t)$, is the same throughout the tank and is given by

$$c_2 = \frac{A(t)}{V(t)}, \tag{1.7.1}$$

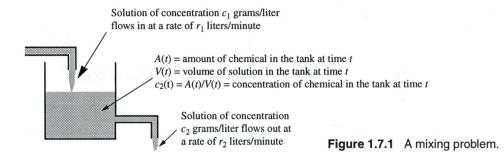

Solution of concentration c_1 grams/liter
flows in at a rate of r_1 liters/minute

$A(t)$ = amount of chemical in the tank at time t
$V(t)$ = volume of solution in the tank at time t
$c_2(t) = A(t)/V(t)$ = concentration of chemical in the tank at time t

Solution of concentration
c_2 grams/liter flows out at
a rate of r_2 liters/minute

Figure 1.7.1 A mixing problem.

where $V(t)$ denotes the volume of solution in the tank at time t and $A(t)$ denotes the amount of chemical in the tank at time t.

Mathematical Formulation: The two variables in the problem are $V(t)$ and $A(t)$. In order to determine how they change with time, we first consider their change during a short time interval, Δt minutes.

In time Δt, $r_1 \Delta t$ liters of solution flow into the tank, whereas $r_2 \Delta t$ liters flow out. Thus during the time interval Δt, the change in the volume of solution in the tank is

$$\Delta V = r_1 \Delta t - r_2 \Delta t = (r_1 - r_2)\Delta t. \tag{1.7.2}$$

Since the concentration of chemical in the inflow is c_1 grams/liter (assumed constant), it follows that in the time interval Δt the amount of chemical that flows into the tank is $c_1 r_1 \Delta t$. Similarly, the amount of chemical that flows out in this same time interval is approximately[1] $c_2 r_2 \Delta t$. Thus, the total change in the amount of chemical in the tank during the time interval Δt, denoted by ΔA, is approximately

$$\Delta A \approx c_1 r_1 \Delta t - c_2 r_2 \Delta t = (c_1 r_1 - c_2 r_2)\Delta t. \tag{1.7.3}$$

Dividing equations (1.7.2) and (1.7.3) by Δt yields

$$\frac{\Delta V}{\Delta t} = r_1 - r_2,$$

$$\frac{\Delta A}{\Delta t} \approx c_1 r_1 - c_2 r_2,$$

---

[1]This is only an approximation, since c_2 is *not* constant over the time interval Δt. The approximation will become more accurate as $\Delta t \to 0$.

respectively. These equations describe the rates of change of V and A over the short, but finite, time interval Δt. In order to determine the instantaneous rates of change of V and A, we take the limit as $\Delta t \to 0$ to obtain

$$\frac{dV}{dt} = r_1 - r_2, \tag{1.7.4}$$

$$\frac{dA}{dt} = c_1 r_1 - \frac{A}{V} r_2, \tag{1.7.5}$$

where we have substituted for c_2 from equation (1.7.1). Since r_1 and r_2 are constants, we can integrate equation (1.7.4) directly, to obtain

$$V(t) = (r_1 - r_2)t + V_0,$$

where V_0 is an integration constant. Substituting for V into equation (1.7.5) and rearranging terms yields the *linear* equation for $A(t)$

$$\frac{dA}{dt} + \frac{r_2}{(r_1 - r_2)t + V_0} A = c_1 r_1. \tag{1.7.6}$$

This DE can be solved, subject to the initial condition $A(0) = A_0$, to determine the behavior of $A(t)$.

REMARK Do *not* memorize equation (1.7.6). You will be expected to derive it for each specific example.

Example 1.7.1   A tank contains 8 L of water in which is dissolved 32 g of chemical. A solution containing 2 g/L of the chemical flows into the tank at a rate of 4 L/min, and the well–stirred mixture flows out at a rate of 2 L/min. Determine the amount of chemical in the tank after 20 mins. What is the concentration of chemical in the tank at that time?

Solution   We are given

$$r_1 = 4 \text{ L/min}, \quad r_2 = 2 \text{ L/min}, \quad c_1 = 2 \text{ g/L}, \quad V(0) = 8, \quad \text{and} \quad A(0) = 32 \text{ g}.$$

We must find $A(20)$ and $A(20)/V(20)$.

Now,

$$\Delta V = r_1\,\Delta t - r_2\,\Delta t$$

implies that

$$\frac{dV}{dt} = 2.$$

Integrating this equation and imposing the initial condition that $V(0) = 8$ yields

$$V(t) = 2(t + 4). \tag{1.7.7}$$

Further,

$$\Delta A \approx c_1 r_1 \Delta t - c_2 r_2 \Delta t$$

implies that

$$\frac{dA}{dt} = 8 - 2c_2.$$

That is, since $c_2 = A/V$,

$$\frac{dA}{dt} = 8 - 2\frac{A}{V}.$$

Substituting for V from (1.7.7), we must solve

$$\frac{dA}{dt} + \frac{1}{t + 4}A = 8. \qquad (1.7.8)$$

This first-order linear equation has integrating factor

$$I = e^{\int 1/(t+4)\, dt} = t + 4.$$

Consequently (1.7.8) can be written in the integrable form

$$\frac{d}{dt}[(t + 4)A] = 8(t + 4)$$

which can be integrated directly to obtain

$$(t + 4)A = 4(t + 4)^2 + c.$$

Hence

$$A(t) = \frac{1}{t + 4}[4(t + 4)^2 + c].$$

Imposing the given initial condition $A(0) = 32$ g implies that $c = 64$. Consequently

$$A(t) = \frac{4}{t + 4}[(t + 4)^2 + 16].$$

Setting $t = 20$ gives

$$A(20) = \frac{1}{6}[(24)^2 + 16] = \frac{296}{3} \text{ g}.$$

Furthermore, using (1.7.7),

$$\frac{A(20)}{V(20)} = \frac{1}{48} \cdot \frac{296}{3} = \frac{37}{18} \text{ g/L}.$$

2. ELECTRIC CIRCUITS

An important application of DE arises from the analysis of simple electric circuits. The most basic electric circuit is obtained by connecting the ends of a wire to the terminals of a battery or generator. This causes a flow of charge, $q(t)$, through the wire thereby producing a current, $i(t)$, defined to be the rate of change of charge. Thus,

$$i(t) = \frac{dq}{dt}. \qquad (1.7.9)$$

In practice a circuit will contain several components which oppose the flow of charge. As current passes through these components work has to be done and so there is a loss of energy which is described by the resulting voltage drop across each component. For the circuits that we will consider, the behavior of the current in the circuit is governed by Kirchoff's second law which can be stated as follows.

Kirchoff's Second Law:

The sum of the voltage drops around a closed circuit is zero.

In order to apply this law we need to know the relationship between the current passing through each component in the circuit and the resulting voltage drop. The components of interest to us are resistors, capacitors and inductors. We briefly describe each of these next.

1. *Resistors:* As its name suggests, a resistor is a component that, due to it's constituency, directly resists the flow of charge through it. According to Ohm's law, the voltage drop, ΔV_R, between the ends of a resistor is directly proportional to the current that is passing through it. This is expressed mathematically as

$$\Delta V_R = iR \qquad (1.7.10)$$

where the constant of proportionality, R, is called the **resistance** of the resistor. The units of resistance are ohms (Ω).

2. *Capacitors:* A capacitor can be thought of as a component that stores charge and thereby opposes the passage of current. If $q(t)$ denotes the charge on the capacitor at time t, then the drop in voltage, ΔV_C, as current passes through it is directly proportional to $q(t)$. It is usual to express this law in the form

$$\Delta V_C = \frac{1}{C} q, \qquad (1.7.11)$$

where the constant C is called the **capacitance** of the capacitor. The units of capacitance are farads (F).

3. *Inductors:* The third component that is of interest to us is an inductor. This can be considered as a component that opposes any *change* in the current flowing through it. The drop in voltage as current passes through an inductor is directly proportional to the rate at which the current is changing. We write this as

$$\Delta V_L = L \frac{di}{dt}, \qquad (1.7.12)$$

where the constant L is called the **inductance** of the inductor measured in units of henrys (H).

4. *EMF:* The final component in our circuits will be a source of voltage that produces an electromotive force (EMF). We can think of this as providing the force that drives the charge through the circuit. As current passes through the voltage source, there is a voltage gain, which we denote by $E(t)$ volts (that is, a voltage drop of $-E(t)$ volts).

A circuit containing all of these components is shown in Figure 1.7.2. According to Kirchoff's second law, the sum of the voltage drops at any instant must be zero. Applying this to the RLC circuit in Figure 1.7.2 we obtain

$$\Delta V_R + \Delta V_C + \Delta V_L - E(t) = 0. \qquad (1.7.13)$$

Substituting into equation (1.7.13) from (1.7.10)–(1.7.12) and rearranging yields the basic DE for an RLC circuit, namely,

$$L \frac{di}{dt} + iR + \frac{q}{C} = E(t). \qquad (1.7.14)$$

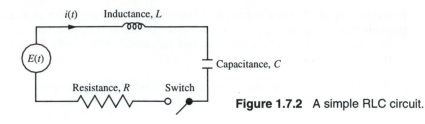

Figure 1.7.2 A simple RLC circuit.

There are three cases that are of importance in applications, two of which are governed by first-order linear DE.

CASE 1: AN RL CIRCUIT   In the case when there is no capacitor present, we have what is referred to as an RL circuit. The DE (1.7.14) then reduces to

$$\frac{di}{dt} + \frac{R}{L} i = \frac{1}{L} E(t). \tag{1.7.15}$$

This is a first-order linear DE for the current in the circuit at any time t.

CASE 2: AN RC CIRCUIT   Now consider the case when there is no inductor present in the circuit. Setting $L = 0$ in equation (1.7.14) yields

$$i + \frac{1}{RC} q = \frac{E}{R}.$$

In this equation we have two unknowns, namely $q(t)$ and $i(t)$. Substituting from (1.7.9) for $i(t) = dq/dt$ we obtain the following DE for $q(t)$:

$$\frac{dq}{dt} + \frac{1}{RC} q = \frac{E}{R}. \tag{1.7.16}$$

In this case, the first-order linear DE (1.7.16) can be solved for the charge $q(t)$ on the plates of the capacitor. The current in the circuit can then be obtained from

$$i(t) = \frac{dq}{dt}$$

by differentiation.

CASE 3: AN RLC CIRCUIT   In the general case, we must consider all three components to be present in the circuit. Substituting from (1.7.9) into equation (1.7.14) yields the following DE for determining the charge on the capacitor:

$$\frac{d^2q}{dt^2} + \frac{R}{L} \frac{dq}{dt} + \frac{1}{LC} q = \frac{1}{L} E(t). \tag{1.7.17}$$

We will develop techniques in Chapter 2 that enable us to solve this DE without difficulty. For the remainder of this section we restrict our attention to RL and RC circuits. Since these are both first-order linear DE, we can solve them using the technique derived in the previous section once the applied EMF, $E(t)$, has been specified. The two most important forms for $E(t)$ are

$$E(t) = E_0 \text{ and } E(t) = E_0 \cos \omega t,$$

where E_0 and ω are constants. The first of these corresponds to a source of EMF such as a battery. The resulting current is called a **direct current** (DC). The second form of EMF oscillates between $\pm E_0$ and is called an **alternating current** (AC).

Example 1.7.2   Determine the current in an RL circuit if the applied EMF is $E(t) = E_0 \cos \omega t$, where E_0 and ω are constants.

Solution   Substituting into equation (1.7.15) for $E(t)$ yields the DE

$$\frac{di}{dt} + \frac{R}{L} i = \frac{E_0}{L} \cos \omega t,$$

which we write as

$$\frac{di}{dt} + ai = \frac{E_0}{L} \cos \omega t, \tag{1.7.18}$$

where $a = \dfrac{R}{L}$. An integrating factor for the latter equation is $I(t) = e^{at}$, so that the equation can be written in the equivalent form

$$\frac{d}{dt}(e^{at}i) = \frac{E_0}{L} e^{at} \cos \omega t.$$

Integrating this equation using the standard integral

$$\int e^{at} \cos \omega t \, dt = \frac{1}{a^2 + \omega^2} e^{at}(a \cos \omega t + \omega \sin \omega t) + c,$$

we obtain

$$e^{at}i = \frac{E_0}{L(a^2 + \omega^2)} e^{at}(a \cos \omega t + \omega \sin \omega t) + c,$$

where c is an integration constant. Consequently,

$$i(t) = \frac{E_0}{L(a^2 + \omega^2)} (a \cos \omega t + \omega \sin \omega t) + c\, e^{-at}.$$

Imposing the initial condition $i(0) = 0$, we find

$$c = -\frac{E_0 a}{L(a^2 + \omega^2)},$$

so that

$$i(t) = \frac{E_0}{L(a^2 + \omega^2)} (a \cos \omega t + \omega \sin \omega t - a e^{-at}). \tag{1.7.19}$$

This solution can be written in the form

$$i(t) = i_S(t) + i_T(t),$$

where

$$i_S(t) = \frac{E_0}{L(a^2 + \omega^2)} (a \cos \omega t + \omega \sin \omega t), \qquad i_T(t) = -\frac{a E_0}{L(a^2 + \omega^2)} e^{-at}.$$

The term i_T decays exponentially with time and is referred to as the **transient part** of the solution. As $t \to +\infty$ the solution (1.7.19) approaches the **steady-state solution** i_S. The steady-state solution can be written in a more illuminating form as follows. If we construct the right-angled triangle (see Figure 1.7.3) with sides a and ω, then the hypotenuse of the triangle is $\sqrt{a^2 + \omega^2}$. Consequently, there exists a unique angle $\phi \in (0, \pi/2)$, such that

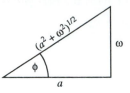

Figure 1.7.3 Defining the phase angle for an RL circuit.

$$\cos \phi = \frac{a}{\sqrt{a^2 + \omega^2}}, \quad \sin \phi = \frac{\omega}{\sqrt{a^2 + \omega^2}}.$$

Equivalently,

$$a = \sqrt{a^2 + \omega^2} \cos \phi, \quad \omega = \sqrt{a^2 + \omega^2} \sin \phi.$$

Substituting for a and ω into the expression for i_S yields

$$i_S(t) = \frac{E_0}{L\sqrt{a^2 + \omega^2}} (\cos \omega t \cos \phi + \sin \omega t \sin \phi),$$

which can be written, using an appropriate trigonometric identity, as

$$i_S(t) = \frac{E_0}{L\sqrt{a^2 + \omega^2}} \cos(\omega t - \phi).$$

This is referred to as the **phase-amplitude** form of the solution. Comparing this with the original driving term, $E_0 \cos \omega t$, we see that the system has responded with a steady-state solution having the same periodic behavior, but with a **phase shift** of ϕ radians. Furthermore the amplitude of the response is

$$A = \frac{E_0}{L\sqrt{a^2 + \omega^2}} = \frac{E_0}{\sqrt{R^2 + \omega^2 L^2}}, \tag{1.7.20}$$

where we have substituted for $a = R/L$. This is illustrated in Figure 1.7.4. The general

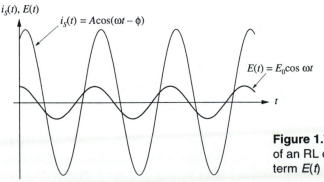

Figure 1.7.4 The response of an RL circuit to the driving term $E(t) = E_0 \cos \omega t$.

picture that we have, therefore, is that the transient part of the solution affects $i(t)$ for a short period after which the current settles into a steady state, which, in the case when the driving EMF $E(t) = E_0\cos \omega t$, is a phase shift of this driving EMF with an amplitude given in equation (1.7.20). This general behavior is illustrated in Figure 1.7.5.

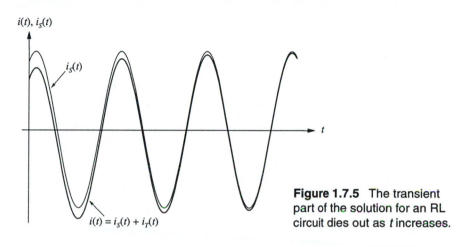

Figure 1.7.5 The transient part of the solution for an RL circuit dies out as t increases.

Our next example illustrates the procedure for solving the DE (1.7.16) governing the behavior of an RC circuit.

Example 1.7.3   Consider the RC circuit in which $R = 0.5 \ \Omega$, $C = 0.1$ F, and $E_0 = 20$ V. Given that the capacitor has zero initial charge, determine the current in the circuit after 0.25 s.

Solution   In this case we first solve equation (1.7.16) for $q(t)$ and then determine the current in the circuit by differentiating the result. Substituting for R, C and E into equation (1.7.16) yields

$$\frac{dq}{dt} + 20q = 40,$$

which has general solution

$$q(t) = 2 + ce^{-20t},$$

where c is an integration constant. Imposing the initial condition $q(0) = 0$ yields $c = -2$, so that

$$q(t) = 2(1 - e^{-20t}).$$

Differentiating this expression for q gives the current in the circuit

$$i(t) = \frac{dq}{dt} = 40e^{-20t}.$$

Consequently,

$$i(0.25) = 40 \ e^{-5} \approx 0.27 \text{ A}.$$

EXERCISES 1.7

1. A container initially contains 10 L of water in which is dissolved 20 g of salt. A solution containing 4 g/L of salt is pumped into the container at a rate of 2 L/min, and the well–stirred mixture runs out at a rate of 1 L/min. How much salt is in the tank after 40 min ?

2. A tank initially contains 600 L of solution in which is dissolved 1500 g of chemical. A solution containing 5 g/L of the chemical flows into the tank at a rate of 6 L/min, and the well–stirred mixture flows out at a rate of 3 L/min. Determine the concentration of chemical in the tank after one hour.

3. A tank whose volume is 40 L initially contains 20 L of water. A solution containing 10 g/L of salt is pumped into the tank at a rate of 4 L/min, and the well–stirred mixture flows out at a rate of 2 L/min. How much salt is in the tank just before the solution overflows?

4. A tank whose volume is 200 L is initially half full of a solution that contains 100 g of chemical. A solution containing 0.5 g/L of the same chemical flows into the tank at a rate of 6 L/min, and the well–stirred mixture flows out at a rate of 4 L/min. Determine the concentration of chemical in the tank just before the solution overflows.

5. A container initially contains 10 L of a salt solution. Water flows into the container at a rate of 3 L/min, and the well–stirred mixture flows out at a rate of 2 L/min. After 5 min the concentration of salt in the container is 0.2 g/L. Find:

(a) The amount of salt in the container initially.

(b) The volume of solution in the container when the concentration of salt is 0.1 g/L.

6. A tank initially contains 20 L of water. A solution containing 1 g/L of chemical flows into the tank at a rate of 3 L/min, and the mixture flows out at a rate of 2 L/min.

(a) Set up and solve the IVP for $A(t)$, the amount of chemical in the tank at time t.

(b) Determine the time when the concentration of chemical in the tank reaches 0.5 g/L.

7. A tank initially contains w liters of a solution in which is dissolved A_0 grams of chemical. A solution containing k grams/liter of the same chemical flows into the tank at a rate of r liters/minute, and the mixture flows out at the same rate.

(a) Show that the amount of chemical, $A(t)$, in the tank at time t is

$$A(t) = e^{-(rt)/w}\left[kw(e^{(rt)/w} - 1) + A_0\right].$$

(b) Show that the concentration of chemical in the tank eventually approaches k grams/liter. Is this result reasonable?

8. Consider the double mixing problem depicted in Figure 1.7.6.

(a) Show that the DE for determining $A_1(t)$ and $A_2(t)$ are

$$\frac{dA_1}{dt} + \frac{r_2}{(r_1 - r_2)t + V_1}A_1 = c_1r_1,$$

$$\frac{dA_2}{dt} + \frac{r_3}{(r_2 - r_3)t + V_2}A_2 = \frac{r_2A_1}{(r_1 - r_2)t + V_1},$$

where V_1 and V_2 are constants.

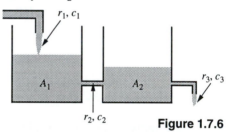

Figure 1.7.6

(b) Consider the case in which $r_1 = 6$ L/min, $r_2 = 4$ L/min, $r_3 = 3$ L/min, and $c_1 = 0.5$ g/L. If the first tank initially holds 40 L of water in which 4 g of chemical is dissolved, whereas the second tank initially contains 20 g of chemical dissolved in 20 L of water, determine the amount of chemical in the second tank after 10 min.

9. Consider the RL circuit in which $R = 4\ \Omega$, $L = 0.1$ H, and $E(t) = 20$ V. If there is no current flowing initially, determine the current in the circuit for $t \geq 0$.

10. Consider the RC circuit which has $R = 5\ \Omega$, $C = \frac{1}{50}$ F, and $E(t) = 100$ V. If the capacitor is uncharged initially, determine the current in the circuit for $t \geq 0$.

11. An RL circuit has EMF $E(t) = 10 \sin 4t$ volts. If $R = 2\ \Omega$, $L = \frac{2}{3}$ H, and there is no current flowing initially, determine the current for $t \geq 0$.

12. Consider the RC circuit with $R = 2\ \Omega$, $C = 1/8$ F, and $E(t) = 10 \cos 3t$ volts. If $q(0) = 1$ C, determine the current in the circuit for $t \geq 0$.

13. Consider the general RC circuit with $E(t) = 0$. Suppose that $q(0) = 5$ C. Determine the charge on the capacitor for $t > 0$. What happens as $t \to \infty$? Is this reasonable?

14. Determine the current in an RC circuit if the capacitor has zero charge initially and the driving EMF is $E = E_0$, where E_0 is a constant. Make a sketch depicting the change in the charge $q(t)$ on the capacitor with time and show that $q(t)$ approaches a constant value as t increases. What happens to the current in the circuit as $t \to \infty$?

15. Determine the current flowing in an RL circuit if the applied EMF is $E(t) = E_0 \sin \omega t$, where E_0 and ω are constants. Identify the transient part of the solution and the steady-state solution.

16. Determine the current flowing in an RL circuit if the applied EMF is constant and the initial current is zero.

17. Determine the current flowing in an RC circuit if the capacitor is initially uncharged and the driving EMF is given by $E(t) = E_0 e^{-at}$, where E_0 and a are constants.

18. Consider the special case of the RLC circuit in which the resistance is negligible and the driving EMF is zero. In this case the DE governing the charge on the capacitor is

$$\frac{d^2q}{dt^2} + \frac{1}{LC} q = 0.$$

If the capacitor has an initial charge of q_0 coulombs, and there is no current flowing initially, determine the charge on the capacitor for $t > 0$, and the corresponding current in the circuit. [Hint: Let $u = dq/dt$ and use the chain rule to show that this implies $du/dt = u(du/dq)$.]

19. Repeat question 18 in the case when the driving EMF is $E(t) = E_0$, a constant.

1.8   CHANGE OF VARIABLES

So far we have introduced techniques for solving separable and first-order linear DE. Clearly, most first-order DE are not of these two types. In this section, we consider two further types of DE that can be solved by using a change of variables to reduce them to one of the types we know how to solve. The key point to grasp in this section, however, is not the specific changes of variables that we discuss, but the general idea of changing variables in a DE. Further examples are considered in the exercises.

FIRST-ORDER HOMOGENEOUS DE

We first require a preliminary definition.

> **Definition 1.8.1:** A function $f(x, y)$ is said to be homogeneous of degree zero[1] if
>
> $$f(tx, ty) = f(x, y)$$
>
> for all positive values of t for which (tx, ty) is in the domain of f.

REMARK   Equivalently we can say that f is homogeneous of degree zero if it is invariant under a rescaling of the variables x and y.

Example 1.8.1   The simplest nonconstant functions that are homogeneous of degree zero are $f(x, y) = \dfrac{y}{x}$, and $f(x, y) = \dfrac{x}{y}$.

---

[1] More generally $f(x, y)$ is said to be homogeneous of degree m if $f(tx, ty) = t^m f(x, y)$.

Example 1.8.2   If $f(x, y) = \dfrac{x^2 - y^2}{2xy + y^2}$, then $f(tx, ty) = \dfrac{t^2(x^2 - y^2)}{t^2(2xy + y^2)} = f(x, y)$, so that f is homogeneous of degree zero.   ❏

In the previous example, if we factor an x^2 term from the numerator and denominator, then the function f can be written in the form

$$f(x, y) = \frac{x^2[1 - (y/x)^2]}{x^2[2y/x + (y/x)^2]}.$$

That is,

$$f(x, y) = \frac{1 - (y/x)^2}{2y/x + (y/x)^2}.$$

Thus f can be considered to depend on the single variable $V = y/x$. The following Theorem establishes that this is a basic property of all functions that are homogeneous of degree zero.

Theorem 1.8.1: A function $f(x, y)$ is homogeneous of degree zero if and only if it depends on y/x only.

PROOF   Suppose that f is homogeneous of degree zero. We must consider two cases separately.

(a)   If $x > 0$, we can take $t = 1/x$ in Definition 1.8.1 to obtain

$$f(x, y) = f(1, y/x),$$

which is a function of $V = y/x$ only.

(b)   If $x < 0$, then we can take $t = -1/x$ in Definition 1.8.1. In this case we obtain

$$f(x, y) = f(-1, -y/x),$$

which once more depends on y/x only.
    Conversely, suppose that $f(x, y)$ depends only on y/x. If we replace x by tx and y by ty then f is unaltered and hence is homogeneous of degree zero.   ■

REMARK   Do not memorize the formulas in the preceding theorem. Just remember that a function $f(x, y)$ that is homogeneous of degree zero depends only on the combination y/x and hence can be considered as a function of a single variable, say, $F(V)$, where $V = y/x$.

We now consider solving DE that satisfy the following definition:

Definition 1.8.2:   If $f(x, y)$ is homogeneous of degree zero, then the DE

$$\frac{dy}{dx} = f(x, y)$$

is called a **homogeneous first-order DE.**

In general, if

$$\frac{dy}{dx} = f(x, y)$$

is a homogeneous first-order DE, then we cannot solve it directly. However, our preceding discussion implies that such a DE can be written in the equivalent form

$$\frac{dy}{dx} = F(y/x), \tag{1.8.1}$$

for an appropriate function F. This suggests that instead of using the variables x and y, we should use the variables x and V, where $V = y/x$, or equivalently,

$$y = xV(x). \tag{1.8.2}$$

Substitution of (1.8.2) into the right-hand side of equation (1.8.1) has the effect of reducing it to a function of V only. We must also determine how the derivative term dy/dx transforms. Differentiating (1.8.2) with respect to x using the product rule yields the following relationship between $\frac{dy}{dx}$ and $\frac{dV}{dx}$

$$\frac{dy}{dx} = x\frac{dV}{dx} + V. \tag{1.8.3}$$

Substituting into equation (1.8.1) we therefore obtain

$$x\frac{dV}{dx} + V = F(V)$$

or, equivalently,

$$x\frac{dV}{dx} = F(V) - V.$$

The variables can now be separated to yield

$$\frac{1}{F(V) - V}\,dV = \frac{1}{x}\,dx,$$

which can be solved directly by integration. We have therefore established the next theorem.

Theorem 1.8.2: The change of variables $y = xV(x)$ reduces a homogeneous first-order DE $dy/dx = f(x, y)$ to the separable equation

$$\frac{1}{F(V) - V}\,dV = \frac{1}{x}\,dx.$$

REMARK The separable equation that results in the previous technique can be integrated to obtain a relationship between V and x. We then obtain the solution to the given DE by substituting y/x for V in this relationship.

Example 1.8.3 Find the general solution to

$$\frac{dy}{dx} = \frac{4x + y}{x - 4y}. \tag{1.8.4}$$

Solution The function on the right-hand side of equation (1.8.4) is homogeneous of degree zero, so that we have a first-order homogeneous DE. Substituting $y = xV$ into the equation yields

$$\frac{d}{dx}(xV) = \frac{4 + V}{1 - 4V}.$$

That is,

$$x\frac{dV}{dx} + V = \frac{4 + V}{1 - 4V},$$

or equivalently,

$$x\frac{dV}{dx} = \frac{4(1 + V^2)}{1 - 4V}.$$

Separating the variables gives

$$\frac{1 - 4V}{4(1 + V^2)}\,dV = \frac{1}{x}\,dx.$$

We write this as

$$\left[\frac{1}{4(1 + V^2)} - \frac{V}{1 + V^2}\right]dV = \frac{1}{x}\,dx,$$

which can be integrated directly to obtain

$$\frac{1}{4}\tan^{-1}V - \frac{1}{2}\ln(1 + V^2) = \ln|x| + c.$$

Substituting $V = y/x$ and multiplying through by 2 yields

$$\frac{1}{2}\tan^{-1}(y/x) - \ln\,[(x^2 + y^2)/x^2] = \ln x^2 + c_1,$$

which simplifies to

$$\frac{1}{2}\tan^{-1}(y/x) - \ln(x^2 + y^2) = c_1. \tag{1.8.5}$$

This solution is more easily expressed in terms of polar coordinates

$$x = r\cos\theta,\ y = r\sin\theta \ \Leftrightarrow\ r = (x^2 + y^2)^{1/2},\ \theta = \tan^{-1}(y/x).$$

Substituting into equation (1.8.5) yields

$$\frac{1}{2}\theta - \ln r^2 = c_1,$$

or equivalently,

$$\ln r = \frac{1}{4}\theta + c_2.$$

Exponentiating both sides of this equation gives

$$r = c_3 e^{\theta/4}.$$

For each value of c_3, this is the equation of a logarithmic spiral. The particular spiral with equation $r = \frac{1}{2}e^{\theta/4}$ is shown in Figure 1.8.1.

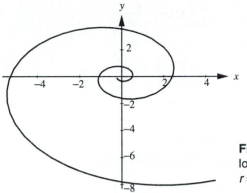

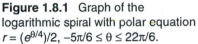

Figure 1.8.1 Graph of the logarithmic spiral with polar equation $r = (e^{\theta/4})/2$, $-5\pi/6 \leq \theta \leq 22\pi/6$.

Example 1.8.4   Find the equation of the orthogonal trajectories to the family

$$x^2 + y^2 - 2cx = 0. \tag{1.8.6}$$

(Completing the square in x, we obtain $(x - c)^2 + y^2 = c^2$, which represents the family of circles centered at $(c, 0)$, with radius c.)

Solution   First we need an expression for the slope of the given family at the point (x, y). Differentiating equation (1.8.6) implicitly with respect to x yields

$$2x + 2y \frac{dy}{dx} - 2c = 0,$$

which simplifies to

$$\frac{dy}{dx} = \frac{c - x}{y}. \tag{1.8.7}$$

This is not the DE of the given family, since it still contains the constant c, and hence is dependent on the individual curves in the family. Therefore, we must eliminate c to obtain an expression for the slope of the family that is independent of any particular curve in the family. From equation (1.8.6) we have

$$c = \frac{x^2 + y^2}{2x}.$$

Substituting this expression for c into equation (1.8.7) and simplifying gives

$$\frac{dy}{dx} = \frac{y^2 - x^2}{2xy}.$$

Therefore, the DE of the orthogonal trajectories is

$$\frac{dy}{dx} = -\frac{2xy}{y^2 - x^2}. \tag{1.8.8}$$

This DE is first-order homogeneous. Substituting $y = xV(x)$ into equation (1.8.8) yields

$$\frac{d}{dx}(xV) = \frac{2V}{1 - V^2},$$

so that

$$x \frac{dV}{dx} + V = \frac{2V}{1 - V^2}.$$

Hence

$$x \frac{dV}{dx} = \frac{V + V^3}{1 - V^2},$$

or in separated form,

$$\frac{1 - V^2}{V(1 + V^2)} dV = \frac{1}{x} dx.$$

Decomposing the left-hand side into partial fractions yields

$$\left(\frac{1}{V} - \frac{2V}{1 + V^2} \right) dV = \frac{1}{x} dx,$$

which can be integrated directly to obtain

$$\ln |V| - \ln(1 + V^2) = \ln |x| + c,$$

or equivalently,

$$\ln \left(\frac{|V|}{1 + V^2} \right) = \ln |x| + c.$$

Exponentiating both sides and redefining the constant yields

$$\frac{V}{1 + V^2} = c_1 x.$$

Substituting back for $V = y/x$, we obtain

$$\frac{xy}{x^2 + y^2} = c_1 x.$$

That is,

$$x^2 + y^2 = c_2 y,$$

where $c_2 = 1/c_1$. Completing the square in y yields

$$x^2 + (y - k)^2 = k^2, \qquad (1.8.9)$$

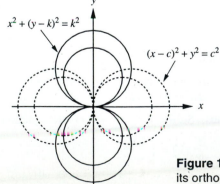

Figure 1.8.2 The family $(x - c)^2 + y^2 = c^2$ and its orthogonal trajectories $x^2 + (y - k)^2 = k^2$.

where $k = c_2/2$. Equation (1.8.9) is the equation of the family of orthogonal trajectories. This is the family of circles centered at $(0, k)$ with radius k (circles along the y-axis). (See Figure 1.8.2.)

BERNOULLI EQUATIONS

We now consider a nonlinear DE that can be reduced to a linear equation by a change of variables.

Definition 1.8.3:   A DE that can be written in the form

$$\frac{dy}{dx} + p(x)y = q(x)y^n, \tag{1.8.10}$$

where n is a constant, is called a **Bernoulli equation**.

If $n \neq 0$ and $n \neq 1$, then a Bernoulli equation is nonlinear, but can be reduced to a linear equation as follows. We first divide equation (1.8.10) by y^n to obtain

$$y^{-n}\frac{dy}{dx} + y^{1-n}p(x) = q(x). \tag{1.8.11}$$

We now make the change of variables

$$u(x) = y^{1-n}, \tag{1.8.12}$$

which implies that

$$\frac{du}{dx} = (1-n)\, y^{-n}\frac{dy}{dx}.$$

That is,

$$y^{-n}\frac{dy}{dx} = \frac{1}{(1-n)}\frac{du}{dx}.$$

Substituting into equation (1.8.11) for y^{1-n} and $y^n\frac{dy}{dx}$ yields the *linear* DE

$$\frac{1}{(1-n)}\frac{du}{dx} + p(x)u = q(x),$$

or in standard form,

$$\frac{du}{dx} + (1-n)p(x)u = (1-n)\, q(x). \tag{1.8.13}$$

The linear equation (1.8.13) can now be solved for u as a function of x. The solution to the original equation is then obtained from (1.8.12).

Example 1.8.5   Solve $\dfrac{dy}{dx} + \dfrac{3}{x}\, y = \dfrac{12y^{2/3}}{(1+x^2)^{1/2}},\ \ x > 0.$

Solution   The DE is a Bernoulli equation. Dividing both sides of the DE by $y^{2/3}$ yields

$$y^{-2/3}\frac{dy}{dx} + \frac{3}{x}y^{1/3} = \frac{12}{(1+x^2)^{1/2}}. \tag{1.8.14}$$

We now let

$$u = y^{1/3}, \tag{1.8.15}$$

which implies that

$$\frac{du}{dx} = \frac{1}{3}y^{-2/3}\frac{dy}{dx}.$$

Substituting into equation (1.8.14) yields

$$3\frac{du}{dx} + \frac{3}{x}u = \frac{12}{(1+x^2)^{1/2}},$$

or in standard form,

$$\frac{du}{dx} + \frac{1}{x}u = \frac{4}{(1+x^2)^{1/2}}. \tag{1.8.16}$$

An integrating factor for this linear equation is

$$I(x) = e^{\int(1/x)dx} = e^{\ln x} = x,$$

so that equation (1.8.16) can be written as

$$\frac{d}{dx}(xu) = \frac{4x}{(1+x^2)^{1/2}}.$$

Integrating, we obtain

$$u(x) = x^{-1}[4(1+x^2)^{1/2} + c],$$

and so, from (1.8.15), the solution to the original DE is

$$y^{1/3} = x^{-1}[4(1+x^2)^{1/2} + c].$$

EXERCISES 1.8

For problems 1–8, determine whether the given function is homogeneous of degree zero. Rewrite those that are as functions of the single variable $V = y/x$.

1. $f(x, y) = \dfrac{x^2 - y^2}{xy}$.

2. $f(x, y) = x - y$.

3. $f(x, y) = \dfrac{x\sin(x/y) - y\cos(y/x)}{y}$.

4. $f(x, y) = \dfrac{\sqrt{x^2 + y^2}}{x - y}$, $x > 0$.

5. $f(x, y) = \dfrac{y}{x - 1}$.

6. $f(x, y) = \dfrac{x - 3}{y} + \dfrac{5y + 9}{3y}$.

7. $f(x, y) = \dfrac{\sqrt{x^2 + y^2}}{x}$, $x < 0$.

8. $f(x, y) = \dfrac{\sqrt{x^2 + 4y^2} - x + y}{x + 3y}$, $x \neq 0, y \neq 0$.

For problems 9–22, solve the given DE.

9. $(3x - 2y)\dfrac{dy}{dx} = 3y$.

10. $y' = \dfrac{(x + y)^2}{2x^2}$.

11. $\sin(y/x)\,(xy' - y) = x\cos(y/x)$.

12. $xy' = (16x^2 - y^2)^{1/2} + y, \ x > 0.$

13. $xy' - y = (9x^2 + y^2)^{1/2}, \ x > 0.$

14. $y(x^2 - y^2) \, dx - x(x^2 + y^2) \, dy = 0.$

15. $xy' + y \ln x = y \ln y.$

16. $\dfrac{dy}{dx} = \dfrac{y^2 + 2xy - 2x^2}{x^2 - xy + y^2}.$

17. $2xy \, dy - (x^2 e^{-y^2/x^2} + 2y^2) \, dx = 0.$

18. $x^2 \dfrac{dy}{dx} = y^2 + 3xy + x^2.$

19. $yy' = (x^2 + y^2)^{1/2} - x, \ x > 0.$

20. $2x(y + 2x)y' = y(4x - y).$

21. $x \dfrac{dy}{dx} = x \tan(y/x) + y.$

22. $\dfrac{dy}{dx} = \dfrac{x\sqrt{x^2 + y^2} + y^2}{xy}, \ x > 0.$

23. Solve the DE in Example 1.8.3 by first transforming it into polar coordinates. (Hint: Write the DE in differential form and then express dx and dy in terms of r and θ.)

For problems 24–26, solve the given IVP.

24. $\dfrac{dy}{dx} = \dfrac{2(2y - x)}{x + y}, \ y(0) = 2.$

25. $\dfrac{dy}{dx} = \dfrac{2x - y}{x + 4y}, \ y(1) = 1.$

26. $\dfrac{dy}{dx} = \dfrac{y - \sqrt{x^2 + y^2}}{x}, \ y(3) = 4.$

27. Find *all* solutions to
$$x \dfrac{dy}{dx} - y = (4x^2 - y^2)^{1/2}, \ x > 0.$$

28. (a) Show that the general solution to the DE
$$\dfrac{dy}{dx} = \dfrac{x - ay}{ax - y}$$
can be written in polar form as $r = ke^{a\theta}$.

(b) For the particular case when $a = 1/2$, determine the solution satisfying the initial condition $y(1) = 1$, and find the maximum x–interval on which this solution is valid. (Hint: When does the solution curve have a vertical tangent?)

◆ (c) On the same set of axes sketch the spiral corresponding to your solution in (b), and the line $y = x/2$. Thereby verify graphically the x–interval obtained in (b).

For problems 29 and 30, determine the orthogonal trajectories to the given family of curves. Sketch some curves from each family.

29. $x^2 + y^2 = 2cy.$

30. $(x - c)^2 + (y - c)^2 = 2c^2.$

31. Let S_1 denote the family of circles, centered on the line $y = mx$, each member of which passes through the origin.

(a) Show that the equation of S_1 can be written in the form
$$(x - a)^2 + (y - ma)^2 = a^2(m^2 + 1),$$
where a is a constant that labels particular members of the family.

(b) Determine the equation of the family of orthogonal trajectories to S_1, and show that it consists of the family of circles centered on the line $x = -my$ that pass through the origin.

◆ (c) Sketch some curves from both families when $m = \sqrt{3}/3$.

Let F_1 and F_2 be two families of curves with the property that whenever a curve from the family F_1 intersects one from the family F_2, it does so at an angle $\alpha \neq \pi/2$. If we know the equation of F_2, then it can be shown (see Problem 17 in Section 1.1) that the DE for determining F_1 is

$$\dfrac{dy}{dx} = \dfrac{m_2 - \tan \alpha}{1 + m_2 \tan \alpha}, \qquad (1.8.17)$$

where m_2 denotes the slope of the family F_2 at the point (x, y).

For problems 32–34, use equation (1.8.17) to determine the equation of the family of curves that cuts the given family at an angle $\alpha = \pi/4$.

32. $x^2 + y^2 = c.$

33. $y = cx^6.$

34. $x^2 + y^2 = 2cx.$

35. (a) Use equation (1.8.17) to find the equation of the family of curves that intersects the family of hyperbolas $y = c/x$ at an angle $\alpha = \alpha_0$.

◆ (b) When $\alpha_0 = \pi/4$, sketch several curves from each family.

36. (a) Use equation (1.8.17) to show that the family of curves that intersects the family of concentric circles $x^2 + y^2 = c$ at an angle $\alpha = \tan^{-1} m$ has polar equation $r = ke^{m\theta}$.

◆ (b) When $\alpha = \pi/6$, sketch several curves from each family.

For problems 37–49, solve the given DE.

37. $y' - x^{-1}y = 4x^2 y^{-1} \cos x, \ x > 0.$

38. $\dfrac{dy}{dx} + \dfrac{1}{2}(\tan x)\, y = 2y^3\sin x.$

39. $\dfrac{dy}{dx} - \dfrac{3}{2x}\, y = 6y^{1/3}x^2\ln x.$

40. $y' + 2x^{-1}y = 6(1 + x^2)^{1/2}\sqrt{y}\,,\ x > 0.$

41. $y' + 2x^{-1}y = 6y^2x^4.$

42. $2x(y' + y^3x^2) + y = 0.$

43. $(x - a)(x - b)(y' - y^{1/2}) = 2(b - a)y,\ a,\ b$ constants.

44. $y' + 6x^{-1}y = 3x^{-1}y^{2/3}\cos x\,,\ x > 0.$

45. $y' + 4xy = 4x^3y^{1/2}.$

46. $\dfrac{dy}{dx} - \dfrac{1}{2x\ln x}\, y = 2xy^3.$

47. $\dfrac{dy}{dx} - \dfrac{1}{(\pi - 1)x}\, y = \dfrac{3}{(1 - \pi)}\, xy^{\pi}.$

48. $2y' + y\cot x = 8y^{-1}\cos^3 x.$

49. $(1 - \sqrt{3})\, y' + y\sec x = y^{\sqrt{3}}\sec x.$

For problems 50 and 51, solve the given IVP.

50. $\dfrac{dy}{dx} + \dfrac{2x}{1 + x^2}\, y = xy^2,\ y(0) = 1.$

51. $y' + y\cot x = y^3\sin^3 x,\ y(\pi/2) = 1.$

52. Consider the DE

$$y' = F(ax + by + c), \qquad (52.1)$$

where $a, b\ (\neq 0)$ and c are constants. Show that the change of variables from x and y to x and V, where

$$V = ax + by + c$$

reduces equation (52.1) to the separable form

$$\frac{1}{bF(V) + a}\, dV = dx.$$

For problems 53–55 use the result from the previous problem to solve the given DE. For problem 53, impose the given initial condition.

53. $y' = (9x - y)^2,\ y(0) = 0.$

54. $y' = (4x + y + 2)^2.$

55. $y' = \sin^2(3x - 3y + 1).$

56. Show that the change of variables $V = xy$ transforms the DE

$$\frac{dy}{dx} = \frac{y}{x}\, F(xy)$$

into the separable DE

$$\frac{1}{V[F(V) + 1]}\frac{dV}{dx} = \frac{1}{x}.$$

57. Use the result from the previous problem to solve

$$\frac{dy}{dx} = \frac{y}{x}[\ln(xy) - 1].$$

58. Consider the DE

$$\frac{dy}{dx} = \frac{x + 2y - 1}{2x - y + 3}. \qquad (58.1)$$

(a) Show that the change of variables defined by

$$x = u - 1,\ y = v + 1$$

transforms equation (58.1) into the homogeneous equation

$$\frac{dv}{du} = \frac{u + 2v}{2u - v}. \qquad (58.2)$$

(b) Find the general solution to equation (58.2), and hence, solve equation (58.1).

59. A DE of the form

$$y' + p(x)y + q(x)y^2 = r(x) \qquad (59.1)$$

is called a **Riccati** equation.

(a) If $y = Y(x)$ is a known solution to equation (59.1), show that the substitution

$$y = Y(x) + v^{-1}(x)$$

reduces it to the linear equation

$$v' - [p(x) + 2Y(x)q(x)]v = q(x).$$

(b) Find the general solution to the Riccati equation $x^2y' - xy - x^2y^2 = 1,\ x > 0$, given that $y = -x^{-1}$ is a solution.

60. Consider the Riccati equation

$$y' + 2x^{-1}y - y^2 = -2x^{-1},\ x > 0. \qquad (60.1)$$

(a) Determine the values of the constants a and r such that $y(x) = ax^r$ is a solution to equation (60.1)

(b) Use the result from problem 59(a) to determine the general solution to equation (60.1).

61. (a) Show that the change of variables $y = x^{-1} + w(x)$ transforms the Riccati DE

$$y' + 7x^{-1}y - 3y^2 = 3x^{-2} \qquad (61.1)$$

into the Bernoulli equation

$$w' + x^{-1}w = 3w^2. \qquad (61.2)$$

(b) Solve equation (61.2), and hence determine the general solution to (61.1).

62. Consider the DE

$$y^{-1}y' + p(x)\ln y = q(x), \qquad (62.1)$$

where $p(x)$ and $q(x)$ are continuous functions on some interval (a, b). Show that the change of variables $u = \ln y$ reduces equation (62.1) to the linear DE

$$u' + p(x)u = q(x),$$

and hence show that the general solution to equation (62.1) is

$$y(x) = \exp\left\{ I^{-1}\left[\int I(x)\, q(x)\, dx + c \right] \right\},$$

where

$$I = e^{\int p(x)\, dx} \qquad (62.2)$$

and c is an arbitrary constant.

63. Use the technique derived in the previous problem to solve the IVP

$$y^{-1}y' - 2x^{-1}\ln y = x^{-1}(1 - 2\ln x), \quad y(1) = e.$$

64. Consider the DE

$$f'(y)\frac{dy}{dx} + p(x)f(y) = q(x), \qquad (64.1)$$

where p and q are continuous functions on some interval (a, b), f is an invertible function, and a prime denotes differentiation with respect to y. Show that equation (64.1) can be written as

$$\frac{du}{dx} + p(x)u = q(x),$$

where $u = f(y)$ and hence show that the general solution to equation (64.1) is

$$y(x) = f^{-1}\left\{ I^{-1}\left[\left(\int I(x)\, q(x)\, dx + c \right) \right] \right\}$$

where I is given in (62.2), f^{-1} is the inverse of f, and c is an arbitrary constant.

65. Solve

$$\sec^2 y\, \frac{dy}{dx} + \frac{1}{2\sqrt{1 + x}} \tan y = \frac{1}{2\sqrt{1 + x}}.$$

1.9   EXACT DE

For the next technique it is best to consider first-order DE written in differential form

$$M(x, y)\, dx + N(x, y)\, dy = 0, \qquad (1.9.1)$$

where M and N are given functions, assumed to be sufficiently smooth. The method that we will consider is based on the idea of a *differential*.

You should recall from a previous calculus course that if $\phi = \phi(x, y)$ is a function of two variables, x and y, then the differential of ϕ, denoted $d\phi$, is defined by

$$d\phi = \frac{\partial \phi}{\partial x}\, dx + \frac{\partial \phi}{\partial y}\, dy. \qquad (1.9.2)$$

The right-hand side of (1.9.2) is similar to the expression in equation (1.9.1). This observation can sometimes be used to solve a DE.

Example 1.9.1   Solve

$$2x \sin y\, dx + x^2\cos y\, dy = 0. \qquad (1.9.3)$$

Solution   This equation is separable, however we will use a different technique to solve it. By inspection, we notice that

$$2x \sin y\, dx + x^2\cos y\, dy = d(x^2\sin y).$$

Consequently, equation (1.9.3) can be written as

$$d(x^2\sin y) = 0.$$

This implies that

$$x^2\sin y = \text{constant},$$

and hence, the general solution to equation (1.9.3) is

$$\sin y = \frac{c}{x^2},$$

where c is an arbitrary constant.   ◻

In the foregoing example we were able to write the given DE in the form $d\phi(x, y) = 0$, and hence obtain its solution. However, we cannot always do this. Indeed we see by comparing equation (1.9.1) with (1.9.2) that the DE $M(x, y)\, dx + N(x, y)\, dy = 0$ can be written as $d\phi = 0$ if and only if

$$M = \frac{\partial \phi}{\partial x}, \quad \text{and} \quad N = \frac{\partial \phi}{\partial y},$$

for some function ϕ. This motivates the following definition:

Definition 1.9.1:   The DE $M(x, y)\, dx + N(x, y)\, dy = 0$ is said to be **exact** in a region R of the xy-plane if there exists a function $\phi(x, y)$ such that

$$\frac{\partial \phi}{\partial x} = M, \quad \frac{\partial \phi}{\partial y} = N, \qquad (1.9.4)$$

for all (x, y) in R.

Any function ϕ satisfying (1.9.4) is called a **potential function** for the DE $M(x, y)\, dx + N(x, y)\, dy = 0$. We emphasize that if such a function exists, then the DE $M(x, y)\, dx + N(x, y)\, dy = 0$ can be written as

$$d\phi = 0.$$

This is why such a DE is called an *exact* DE. From the previous example, a potential function for the DE

$$2x \sin y \, dx + x^2 \cos y \, dy = 0$$

is

$$\phi(x, y) = x^2 \sin y.$$

We now show that if a DE is exact then, *provided we can find a potential function* ϕ, its solution can be written down immediately.

Theorem 1.9.1:   The general solution to an exact equation

$$M(x, y)\, dx + N(x, y)\, dy = 0$$

is defined implicitly by

$$\phi(x, y) = c,$$

where ϕ satisfies (1.9.4) and c is an arbitrary constant.

PROOF   We rewrite the DE in the form

$$M(x, y) + N(x, y) \frac{dy}{dx} = 0.$$

That is, from (1.9.4) (assuming exactness),

$$\frac{\partial \phi}{\partial x} + \frac{\partial \phi}{\partial y} \frac{dy}{dx} = 0.$$

But this is just $\dfrac{d\phi}{dx} = 0$, which implies that $\phi(x, y) = c$, where c is a constant. ∎

REMARKS

1. The potential function ϕ is a function of two variables x and y, and we interpret the relationship $\phi(x, y) = c$ as defining y implicitly as a function of x. The preceding theorem states that this relationship defines the general solution to the DE for which ϕ is a potential function.

2. Geometrically, Theorem 1.9.1 says that the solution curves of an exact DE are the family of curves $\phi(x, y) = $ constant. These are called the **level curves** of the function $\phi(x, y)$.

The following two questions now arise:

1. How can we tell whether a given DE is exact?

2. If we have an exact equation, how do we find a potential function?

The answers are given in the next theorem.

*Theorem 1.9.2 (**Test for Exactness**):* Let M, N, and their first partial derivatives M_y and N_x, be continuous in a (simply connected[1]) region R of the xy-plane. Then the DE $M(x, y)\, dx + N(x, y)\, dy = 0$ is exact for all x, y in R if and only if

$$\frac{\partial M}{\partial y} = \frac{\partial N}{\partial x}. \tag{1.9.5}$$

PROOF We first prove that exactness implies the validity of equation (1.9.5). If the DE is exact, then by definition there exists a potential function $\phi(x, y)$ such that $\phi_x = M$ and $\phi_y = N$. Thus, taking partial derivatives, $\phi_{xy} = M_y$ and $\phi_{yx} = N_x$. Since M_y and N_x are continuous in R, it follows that ϕ_{xy} and ϕ_{yx} are continuous in R. But this implies that $\phi_{xy} = \phi_{yx}$ and hence that $M_y = N_x$.

We now prove the converse. Thus we assume that equation (1.9.5) holds and must prove that there exists a potential function ϕ such that

$$\frac{\partial \phi}{\partial x} = M \tag{1.9.6}$$

and

$$\frac{\partial \phi}{\partial y} = N. \tag{1.9.7}$$

The proof is constructional. That is we actually find a potential function ϕ. We begin by integrating equation (1.9.6) with respect to x, holding y fixed (this is a partial integration) to obtain

$$\phi(x, y) = \int^{x} M(s, y)\, ds + h(y), \tag{1.9.8}$$

---

[1] Roughly speaking, simply connected means that the interior of any closed curve drawn in the region also lies in the region. For example the interior of a circle is a simply connected region, although the region between two concentric circles is not.

where $h(y)$ is an arbitrary function of y (this is the integration "constant" that we must allow to depend on y, since we held y fixed in performing the integration)[1]. We now show how to determine $h(y)$ so that the function ϕ defined in (1.9.8) also satisfies equation (1.9.7). Differentiating (1.9.8) partially with respect to y yields

$$\frac{\partial \phi}{\partial y} = \frac{\partial}{\partial y} \int^x M(s, y)\, ds + \frac{dh}{dy}.$$

In order that ϕ satisfy equation (1.9.7) we must choose $h(y)$ to satisfy

$$\frac{\partial}{\partial y} \int^x M(s, y)\, ds + \frac{dh}{dy} = N(x, y).$$

That is,

$$\frac{dh}{dy} = N(x, y) - \frac{\partial}{\partial y} \int^x M(s, y)\, ds. \tag{1.9.9}$$

Since the left-hand side of this expression is a function of y only, we must show, for consistency, that the right-hand side also depends only on y. Taking the derivative of the right-hand side with respect to x yields

$$\frac{\partial}{\partial x}\left(N - \frac{\partial}{\partial y} \int^x M(s, y)\, ds \right) = \frac{\partial N}{\partial x} - \frac{\partial^2}{\partial x \partial y} \int^x M(s, y)\, ds$$

$$= \frac{\partial N}{\partial x} - \frac{\partial}{\partial y}\left(\frac{\partial}{\partial x} \int^x M(s, y)\, ds \right)$$

$$= \frac{\partial N}{\partial x} - \frac{\partial M}{\partial y}.$$

Thus, using (1.9.5), we have

$$\frac{\partial}{\partial x}\left(N - \frac{\partial}{\partial y} \int^x M(s, y)\, ds \right) = 0,$$

so that the right-hand side of equation (1.9.9) does just depend on y. It follows that (1.9.9) is a consistent equation, and hence, we can integrate both sides with respect to y to obtain

$$h(y) = \int^y N(x, t)\, dt - \int^y \frac{\partial}{\partial t}\left(\int^x M(s, t)\, ds \right) dt.$$

Finally, substituting into (1.9.8) yields the potential function

$$\phi(x, y) = \int^x M(s, y)\, ds + \int^y N(x, t)\, dt - \int^y \frac{\partial}{\partial t}\left(\int^x M(s, t)\, ds \right) dt. \qquad \blacksquare$$

---

[1]Throughout the text $\int^x f(t)\, dt$ means evaluate the indefinite integral $\int f(t)\, dt$ and replace t with x in the result.

REMARK   You should *not* attempt to memorize the final result for ϕ. For each particular problem you will construct an appropriate potential function from first principles. This is illustrated in Examples 1.9.3 and 1.9.4.

Example 1.9.2   Determine whether the given DE is exact.

(a) $[1 + \ln(xy)]\, dx + (x/y)\, dy = 0$.

(b) $x^2 y\, dx - (xy^2 + y^3)\, dy = 0$.

Solution

(a)   In this case, $M = 1 + \ln(xy)$ and $N = x/y$, so that $M_y = 1/y = N_x$. It follows from the previous theorem that the DE is exact.

(b)   In this case, we have $M = x^2 y$, $N = -(xy^2 + y^3)$, so that $M_y = x^2$, whereas $N_x = -y^2$. Since $M_y \ne N_x$, the DE is not exact.

Example 1.9.3   Find the general solution to $2xe^y\, dx + (x^2 e^y + \cos y)\, dy = 0$.

Solution   We have

$$M(x, y) = 2xe^y, \quad N(x, y) = x^2 e^y + \cos y,$$

so that

$$M_y = 2xe^y = N_x.$$

Hence the given DE is exact, and so there exists a potential function ϕ such that (see Definition 1.9.1)

$$\frac{\partial \phi}{\partial x} = 2xe^y, \tag{1.9.10}$$

$$\frac{\partial \phi}{\partial y} = x^2 e^y + \cos y. \tag{1.9.11}$$

Integrating equation (1.9.10) with respect to x, holding y fixed, yields

$$\phi(x, y) = x^2 e^y + h(y) \tag{1.9.12}$$

where h is an arbitrary function of y. We now determine $h(y)$ such that (1.9.12) also satisfies equation (1.9.11). Taking the derivative of (1.9.12) with respect to y yields

$$\frac{\partial \phi}{\partial y} = x^2 e^y + \frac{dh}{dy}. \tag{1.9.13}$$

Equations (1.9.11) and (1.9.13) give two expressions for $\frac{\partial \phi}{\partial y}$. This allows us to determine h. Subtracting equation (1.9.11) from equation (1.9.13) gives the consistency requirement

$$\frac{dh}{dy} = \cos y,$$

which implies, upon integration, that

$$h(y) = \sin y,$$

where we have set the integration constant equal to zero without loss of generality since we only require one potential function. Substitution into (1.9.12) yields the potential function

$$\phi(x, y) = x^2 e^y + \sin y.$$

Consequently, the given DE can be written as

$$d(x^2 e^y + \sin y) = 0,$$

and so, from Theorem 1.9.1, the general solution is

$$x^2 e^y + \sin y = c.$$ □

Notice that the solution obtained in the preceding example is an implicit solution. Due to the nature of the way in which the potential function for an exact equation is obtained, this is usually the case.

Example 1.9.4   Find the general solution to

$$[\sin(xy) + xy \cos(xy) + 2x] \, dx + [x^2 \cos(xy) + 2y] \, dy = 0.$$

Solution   We have

$$M(x, y) = \sin(xy) + xy \cos(xy) + 2x \quad \text{and} \quad N(x, y) = x^2 \cos(xy) + 2y.$$

Thus,

$$M_y = 2x \cos(xy) - x^2 y \sin(xy) = N_x,$$

and so the DE is exact. Hence there exists a potential function $\phi(x, y)$ such that

$$\frac{\partial \phi}{\partial x} = \sin(xy) + xy \cos(xy) + 2x, \tag{1.9.14}$$

$$\frac{\partial \phi}{\partial y} = x^2 \cos(xy) + 2y. \tag{1.9.15}$$

In this case, equation (1.9.15) is the simpler equation, and so we integrate it with respect to y, holding x fixed, to obtain

$$\phi(x, y) = x \sin(xy) + y^2 + g(x), \tag{1.9.16}$$

where $g(x)$ is an arbitrary function of x. We now determine $g(x)$, and hence ϕ, from (1.9.14) and (1.9.16). Differentiating (1.9.16) partially with respect to x yields

$$\frac{\partial \phi}{\partial x} = \sin(xy) + xy \cos(xy) + \frac{dg}{dx}. \tag{1.9.17}$$

Equations (1.9.14) and (1.9.17) are consistent if and only if

$$\frac{dg}{dx} = 2x.$$

Hence, upon integrating,

$$g(x) = x^2,$$

where we have once more set the integration constant to zero without loss of generality, since we only require one potential function. Substituting into (1.9.16) gives the potential function

$$\phi(x, y) = x \sin xy + x^2 + y^2.$$

The original DE can therefore be written as

$$d(x \sin xy + x^2 + y^2) = 0,$$

and hence the general solution is

$$x \sin xy + x^2 + y^2 = c. \qquad \square$$

REMARK   At first sight the above procedure appears to be quite complicated. However with a little bit of practice you will see that the steps are, in fact, fairly straightforward. As we have shown in Theorem 1.9.2, the method works in general, *provided you start with an exact DE.*

INTEGRATING FACTORS

Usually a given DE will not be exact. However, sometimes it is possible to multiply the DE by a nonzero function to obtain an exact equation that can then be solved using the technique of the previous section. Notice that the solution to the resulting exact equation will be the same as that of the original equation, since we multiply by a *nonzero* function.

Definition 1.9.2:   A nonzero function $I(x, y)$ is called an **integrating factor** for $M(x, y)\, dx + N(x, y)\, dy = 0$ if the DE

$$I(x, y)\, M(x, y)\, dx + I(x, y)\, N(x, y)\, dy = 0$$

is exact.

Example 1.9.5   Show that $I = x^2 y$ is an integrating factor for the DE

$$(3y^2 + 5x^2 y)\, dx + (3xy + 2x^3)\, dy = 0. \qquad (1.9.18)$$

Solution   Multiplying the given DE (which is not exact) by $x^2 y$ yields

$$(3x^2 y^3 + 5x^4 y^2)\, dx + (3x^3 y^2 + 2x^5 y)\, dy = 0. \qquad (1.9.19)$$

Thus,

$$M_y = 9x^2 y^2 + 10x^4 y = N_x,$$

so that the DE (1.9.19) is exact, and hence $I = x^2 y$ is an integrating factor for equation (1.9.18). Indeed we leave it as an exercise to verify that (1.9.19) can be written as

$$d(x^3 y^3 + x^5 y^2) = 0,$$

so that the general solution to equation (1.9.19) (and hence the general solution to equation (1.9.18)) is defined implicitly by

$$x^3 y^3 + x^5 y^2 = c.$$

That is,

$$x^3 y^2 (y + x^2) = c. \qquad \square$$

As shown in the next theorem, using the test for exactness it is straightforward to determine the conditions that a function $I(x, y)$ must satisfy in order to be an integrating factor for the DE $M(x, y)\, dx + N(x, y)\, dy = 0$.

Theorem 1.9.3: $I(x, y)$ is an integrating factor for

$$M(x, y)\, dx + N(x, y)\, dy = 0 \qquad\qquad (1.9.20)$$

if and only if it is a solution to the partial differential equation

$$N \frac{\partial I}{\partial x} - M \frac{\partial I}{\partial y} = \left(\frac{\partial M}{\partial y} - \frac{\partial N}{\partial x} \right) I. \qquad\qquad (1.9.21)$$

PROOF   Multiplying equation (1.9.20) by I yields

$$IM\, dx + IN\, dy = 0.$$

This equation is exact if and only if

$$\frac{\partial}{\partial y}(IM) = \frac{\partial}{\partial x}(IN),$$

that is, if and only if

$$\frac{\partial I}{\partial y} M + I \frac{\partial M}{\partial y} = \frac{\partial I}{\partial x} N + I \frac{\partial N}{\partial x}.$$

Rearranging the terms in this equation yields equation (1.9.21). $\qquad\blacksquare$

The previous theorem is not too useful in general, since it is usually no easier to solve the partial DE (1.9.21) to find I than it is to solve the original equation (1.9.20). However, it sometimes happens that an integrating factor exists that depends only on one variable. We now show that Theorem 1.9.3 can be used to determine when such an integrating factor exists and also to actually find a corresponding integrating factor.

Theorem 1.9.4:   Consider the DE $M(x, y)\, dx + N(x, y)\, dy = 0$.

1. There exists an integrating factor that depends only on x if and only if $(M_y - N_x)/N = f(x)$, a function of x only. In such a case, an integrating factor is

$$I(x) = e^{\int f(x)\, dx}.$$

2. There exists an integrating factor that depends only on y if and only if $(M_y - N_x)/M = g(y)$, a function of y only. In such a case, an integrating factor is

$$I(y) = e^{-\int g(y)\, dy}.$$

PROOF   We will prove only (1). Suppose first that $I = I(x)$ is an integrating factor for $M(x, y)\, dx + N(x, y)\, dy = 0$. Then $\dfrac{\partial I}{\partial y} = 0$, and so, from (1.9.21), I is a solution to

$$\frac{dI}{dx} N = (M_y - N_x)I.$$

That is,

$$\frac{1}{I} \frac{dI}{dx} = \frac{M_y - N_x}{N}.$$

Since, by assumption, I is a function of x only, it follows that the left-hand side of this expression depends only on x and hence also the right-hand side.

Conversely, suppose that $M_y - N_x/N = f(x)$, a function of x only. Then, dividing (1.9.21) by N, it follows that I is an integrating factor for $M(x, y) \, dx + N(x, y) \, dy = 0$ if and only if it is a solution to

$$\frac{\partial I}{\partial x} - \frac{M}{N} \frac{\partial I}{\partial y} = I f(x). \tag{1.9.22}$$

We must show that this DE has a solution I that depends on x only. We do this by explicitly integrating the DE under the assumption that $I = I(x)$. Indeed, if $I = I(x)$, then equation (1.9.22) reduces to

$$\frac{dI}{dx} = I f(x),$$

which is a separable equation with solution

$$I(x) = e^{\int f(x) \, dx}.$$

The proof of (2) is similar, and so we leave it as an exercise.   ■

Example 1.9.6   Solve

$$(2x - y^2) \, dx + xy \, dy = 0, \quad x > 0. \tag{1.9.23}$$

Solution   The equation is not exact ($M_y \neq N_x$). However,

$$\frac{M_y - N_x}{N} = \frac{-2y - y}{xy} = -\frac{3}{x},$$

which is a function of x only. It follows from (1) of the preceding theorem that an integrating factor for equation (1.9.23) is

$$I(x) = e^{-\int (3/x) \, dx} = e^{-3\ln x} = x^{-3}.$$

Multiplying equation (1.9.23) by I yields the exact equation

$$(2x^{-2} - x^{-3}y^2) \, dx + x^{-2}y \, dy = 0. \tag{1.9.24}$$

(You should check that this *is* exact ($M_y = N_x$), although it must be by the previous theorem.) We leave it as an exercise to verify that a potential function for equation (1.9.24) is

$$\phi(x, y) = \frac{1}{2}x^{-2}y^2 - 2x^{-1}$$

and hence the general solution to (1.9.23) is given implicitly by

$$\frac{1}{2}x^{-2}y^2 - 2x^{-1} = c,$$

or equivalently,

$$y^2 - 4x = c_1 x^2.$$

EXERCISES 1.9

For problems 1–3, determine whether the given DE is exact.

1. $(y + 3x^2)\, dx + x\, dy = 0.$

2. $[\cos(xy) - xy\sin(xy)]\, dx - x^2\sin(xy)\, dy = 0.$

3. $y\, e^{xy}\, dx + (2y - xe^{xy})\, dy = 0.$

For problems 4–12, solve the given DE.

4. $2xy\, dx + (x^2 + 1)\, dy = 0.$

5. $(y^2 + \cos x)\, dx + (2xy + \sin y)\, dy = 0.$

6. $x^{-1}(xy - 1)\, dx + y^{-1}(xy + 1)\, dy = 0.$

7. $(4e^{2x} + 2xy - y^2)\, dx + (x - y)^2\, dy = 0.$

8. $(y^2 - 2x)\, dx + 2xy\, dy = 0.$

9. $\left(\dfrac{1}{x} - \dfrac{y}{x^2 + y^2}\right)dx + \dfrac{x}{x^2 + y^2}\, dy = 0.$

10. $[1 + \ln(xy)\,]\, dx + xy^{-1}\, dy = 0.$

11. $[y\cos(xy) - \sin x]\, dx + x\cos(xy)\, dy = 0.$

12. $(2xy + \cos y)\, dx + (x^2 - x\sin y - 2y)\, dy = 0.$

For problems 13–15, solve the given IVP.

13. $(3x^2 \ln x + x^2 - y)\, dx - x\, dy = 0,\ y(1) = 5.$

14. $2x^2 y' + 4xy = 3\sin x,\ y(2\pi) = 0.$

15. $(ye^{xy} + \cos x)\, dx + xe^{xy}\, dy = 0,\ y(\pi/2) = 0.$

16. Show that if $\phi(x, y)$ is a potential function for $M(x, y)\, dx + N(x, y)\, dy = 0$, then so is $\phi(x, y) + c$, where c is an arbitrary constant. This shows that potential functions are only defined up to an additive constant.

For problems 17–19, determine whether the given function is an integrating factor for the given DE.

17. $I(x, y) = \cos(xy)$,
    $[\tan(xy) + xy]\, dx + x^2\, dy = 0.$

18. $I(x) = \sec x$,
    $[2x - (x^2 + y^2)\tan x]\, dx + 2y\, dy = 0.$

19. $I(x, y) = y^{-2} e^{-x/y},\ y(x^2 - 2xy)\, dx - x^3\, dy = 0.$

For problems 20–26, determine an integrating factor for the given DE, and hence find the general solution.

20. $(xy - 1)\, dx + x^2\, dy = 0.$

21. $y\, dx - (2x + y^4)\, dy = 0.$

22. $x^2 y\, dx + y(x^3 + e^{-3y}\sin y)\, dy = 0.$

23. $(y - x^2)\, dx + 2x\, dy = 0,\ x > 0.$

24. $xy[2\ln(xy) + 1]\, dx + x^2\, dy = 0,\ x > 0.$

25. $\dfrac{dy}{dx} + \dfrac{2x}{1 + x^2}\, y = \dfrac{1}{(1 + x^2)^2}.$

26. $(3xy - 2y^{-1})\, dx + x(x + y^{-2})\, dy = 0.$

For problems 27–29, determine the values of the constants r and s such that $I(x, y) = x^r y^s$ is an integrating factor for the given DE.

27. $(y^{-1} - x^{-1})\, dx + (xy^{-2} - 2y^{-1})\, dy = 0.$

28. $y(5xy^2 + 4)\, dx + x(xy^2 - 1)\, dy = 0.$

29. $2y(y + 2x^2)\, dx + x(4y + 3x^2)\, dy = 0.$

30. Prove that if $(M_y - N_x)/M = g(y)$, a function of y only, then an integrating factor for

$$M(x, y)\, dx + N(x, y)\, dy = 0$$

is $I(y) = e^{-\int g(y)\, dy}$.

31. Consider the general first-order *linear* DE

$$\frac{dy}{dx} + p(x)y = q(x), \qquad (31.1)$$

where $p(x)$ and $q(x)$ are continuous functions on some interval (a, b).

(a) Rewrite equation (31.1) in differential form, and show that an integrating factor for the resulting equation is

$$I(x) = e^{\int p(x)\, dx}. \qquad (31.2)$$

(b) Show that the general solution to equation (31.1) can be written in the form

$$y(x) = I^{-1}\left\{ \int^x I(t)\, q(t)\, dt + c \right\},$$

where I is given in equation (31.2), and c is an arbitrary constant.

1.10   SUMMARY OF TECHNIQUES

The techniques that we have derived for solving first-order DE will always work, provided they are applied to the correct type of equation. The difficulty that is often encountered at this stage is not in actually applying the techniques, but rather in determining which technique is appropriate. In Table 1.10.1 we have summarized the five basic types of DE that we have developed techniques for solving.

TABLE 1.10.1   A SUMMARY OF THE BASIC SOLUTION TECHNIQUES FOR $y' = f(x, y)$

Type	Standard Form	Technique
Separable	$p(y)\, y' = q(x)$	Separate the variables and integrate directly.
First-order linear	$y' + p(x)y = q(x)$	Rewrite as $\dfrac{d}{dx}\left[ye^{\int p(x)\,dx}\right] = q(x)e^{\int p(x)\,dx}$, and integrate with respect to x.
First-order homogeneous	$y' = f(x, y)$ with f homogeneous of degree zero $[f(tx, ty) = f(x, y)]$	Change variables: $y = xV(x)$ and reduce to a separable equation.
Bernoulli equation	$y' + p(x)y = q(x)y^n$	Divide by y^n and make the change of variables $u = y^{1-n}$. This reduces the DE to a linear equation.
Exact	$M(x, y)\, dx + N(x, y)\, dy = 0$, with $M_y = N_x$.	The solution is $\phi(x, y) = c$, where ϕ is determined by integrating $\phi_x = M, \phi_y = N$.

If a given DE cannot be written in one of these forms then the next step is to try to determine an integrating factor. If that fails then we might try to find a change of variables that would reduce the DE to one of the above types.

Example 1.10.1   Determine which of the above types, if any, the following DE falls into

$$\frac{dy}{dx} = -\frac{(8x^5 + 3y^4)}{4xy^3}.$$

Solution   Since the given DE is written in the form $dy/dx = f(x, y)$, we first check whether it is separable or homogeneous. By inspection, we see that it is neither of these. We next check to see whether it is a linear or a Bernoulli equation. We therefore rewrite the equation in the equivalent form

$$\frac{dy}{dx} + \frac{3}{4x}\, y = -2x^4 y^{-3}, \tag{1.10.1}$$

which we recognize as a Bernoulli equation with $n = -3$. We could therefore solve the equation using the appropriate technique. Due to the y^{-3} term in equation (1.10.1) it follows that the equation is *not* a linear equation. Finally, we check for exactness. The natural differential form to try for the given DE is

$$(8x^5 + 3y^4)\, dx + 4xy^3\, dy = 0. \tag{1.10.2}$$

In this form, we have

$$M_y = 12y^3, \quad N_x = 4y^3,$$

so that the equation is *not* exact. However, we see that

$$(M_y - N_x)/N = 2x^{-1},$$

so that according to Theorem 1.9.4, $I(x) = x^2$ is an integrating factor. Therefore, we could multiply equation (1.10.2) by x^2 and then solve it as an exact equation.

EXERCISES 1.10

For problems 1–20, determine which of the five types of DE we have studied the given equation falls into, and use an appropriate technique to find the general solution.

1. $\dfrac{dy}{dx} = \dfrac{2 \ln x}{x y}$.

2. $xy' - 2y = 2x^2 \ln x$.

3. $\dfrac{dy}{dx} = -\dfrac{2xy}{x^2 + 2y}$.

4. $(y^2 + 3xy - x^2)\, dx - x^2 dy = 0$.

5. $y' + y\,(\tan x + y \sin x) = 0$.

6. $\dfrac{dy}{dx} + \dfrac{2e^{2x}}{1 + e^{2x}}\, y = \dfrac{1}{e^{2x} - 1}$.

7. $y' - x^{-1}y = x^{-1}\sqrt{x^2 - y^2}$.

8. $\dfrac{dy}{dx} = \dfrac{\sin y + y \cos x + 1}{1 - x \cos y - \sin x}$.

9. $\dfrac{dy}{dx} + \dfrac{1}{x}\, y = \dfrac{25x^2 \ln x}{2y}$.

10. $e^{2x+y}\, dy - e^{x-y}\, dx = 0$.

11. $y' + y \cot x = \sec x$.

12. $\dfrac{dy}{dx} + \dfrac{2 e^x}{1 + e^x}\, y = 2\, y^{1/2}\, e^{-x}$.

13. $y[\ln(y/x) + 1]\, dx - x\, dy = 0$.

14. $(1 + 2xe^y)\, dx - (e^y + x)\, dy = 0$.

15. $y' + y \sin x = \sin x$.

16. $(3y^2 + x^2)\, dx - 2xy\, dy = 0$.

17. $2x(\ln x\,)y' - y = -9x^3y^3 \ln x$.

18. $(1 + x)\, y' = y(2 + x)$.

19. $(x^2 - 1)(y' - 1) + 2y = 0$.

20. $x \sec^2(xy)\, dy = -[y \sec^2(xy) + 2x]\, dx$.

21. Determine all values of the constants m and n, if there are any, for which the DE

$$(x^5 + y^m)\, dx - x^n y^3\, dy = 0$$

is each of the following.

(a) Exact. (b) Separable. (c) Homogeneous. (d) Linear. (e) Bernoulli equation.

1.11   NUMERICAL SOLUTION TO FIRST-ORDER DE

So far in this chapter we have investigated first-order DE geometrically via slope fields, and analytically, by trying to construct exact solutions to certain types of DE. Certainly, for most first-order DE, it simply is not possible to find analytic solutions, since they will not fall into the few classes for which solution techniques are available. Our final approach to analyzing first-order DE is to look at the possibility of constructing a numerical approximation to the unique solution to the IVP

$$y' = f(x, y), \; y(x_0) = y_0. \tag{1.11.1}$$

We consider three techniques that give varying levels of accuracy. In each case, we generate a sequence of approximations $y_1, y_2, \ldots$ to the value of the exact solution at the points $x_1, x_2, \ldots$, where $x_{n+1} = x_n + h$, $n = 0, 1, \ldots$, and h is a real number. We emphasize that numerical methods do *not* generate a formula for the solution to the DE. Rather they generate a sequence of approximations to the value of the solution at specified points. Furthermore, if we use a sufficient number of points then by plotting the points (x_i, y_i) and joining them with straight line segments we are able to obtain an overall approximation to the *solution curve* corresponding to the solution of the given IVP. This is how the approximate solution curves were generated in the preceding sections via the CAS Maple. There are many subtle ideas associated with constructing numerical solutions to IVP that are beyond the scope of this text. Indeed, a full discussion of the application of numerical methods to DE is best left for a future course in numerical analysis.

EULER'S METHOD

Suppose we wish to approximate the solution to the IVP (1.11.1) at $x = x_1 = x_0 + h$, where h is small. The idea behind *Euler's Method* is to use the tangent line to the solution curve through (x_0, y_0) to obtain such an approximation. (See Figure 1.11.1.) The equation of the tangent line through (x_0, y_0) is

$$y(x) = y_0 + y'(x_0, y_0)(x - x_0).$$

That is, since from (1.11.1) $y'(x_0, y_0) = f(x_0, y_0)$,

$$y(x) = y_0 + f(x_0, y_0)(x - x_0).$$

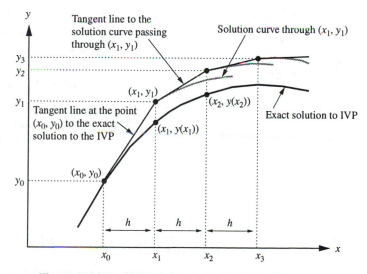

Figure 1.11.1 Euler's method for approximating the solution to the IVP $dy/dx = f(x, y)$, $y(x_0) = y_0$.

Setting $x = x_1$ in this equation yields the Euler approximation to the exact solution through x_1, namely,

$$y_1 = y_0 + f(x_0, y_0)(x_1 - x_0),$$

which we write as

$$y_1 = y_0 + hf(x_0, y_0).$$

Now suppose we wish to obtain an approximation to the exact solution to the IVP (1.11.1) at $x_2 = x_1 + h$. We can use the same idea, except we now use the tangent line to the solution curve through (x_1, y_1). From (1.11.1), the slope of this tangent line is $f(x_1, y_1)$, so that the equation of the required tangent line is

$$y(x) = y_1 + f(x_1, y_1)(x - x_1).$$

Setting $x = x_2$ yields the approximation

$$y_2 = y_1 + hf(x_1, y_1)$$

where we have substituted for $x_2 - x_1 = h$, to the solution to the IVP at $x = x_2$. Continuing in this manner, we determine the sequence of approximations

$$y_{n+1} = y_n + hf(x_n, y_n), \quad n = 0, 1, \ldots$$

to the solution to the IVP (1.11.1) at the points $x_{n+1} = x_n + h$.

Euler's Method for approximating the solution to the IVP

$$y' = f(x, y), \quad y(x_0) = y_0$$

at the points $x_{n+1} = x_0 + nh$ ($n = 0, 1, \ldots$) is

$$y_{n+1} = y_n + hf(x_n, y_n), \quad n = 0, 1, \ldots \qquad (1.11.2)$$

Example 1.11.1   Consider the IVP

$$y' = y - x, \quad y(0) = 1/2.$$

Use Euler's method with (a) $h = 0.1$ and (b) $h = 0.05$ to obtain an approximation to $y(1)$. Given that the exact solution to the IVP is $y(x) = x + 1 - \frac{1}{2}e^x$, compare the errors in the two approximations to $y(1)$.

Solution   In this problem we have

$$f(x, y) = y - x, \quad x_0 = 0, \quad y_0 = 1/2.$$

(a)   Setting $h = 0.1$ in (1.11.2) yields

$$y_{n+1} = y_n + 0.1(y_n - x_n).$$

Hence,

$$y_1 = y_0 + 0.1(y_0 - x_0) = 0.5 + 0.1(0.5 - 0) = 0.55,$$

$$y_2 = y_1 + 0.1(y_1 - x_1) = 0.55 + 0.1(0.55 - 0.1) = 0.595.$$

Continuing in this manner, we generate the approximations listed in Table 1.11.1, where we have rounded the calculations to six decimal places. We have also listed the values of the exact solution and the absolute value of the error. In this case, the approximation to $y(1)$ is

$$y_{10} = 0.703129,$$

with an absolute error of

$$|y(1) - y_{10}| = 0.062270. \qquad (1.11.3)$$

TABLE 1.11.1   The results of applying Euler's method with $h = 0.1$ to the IVP in Example 1.11.1

n	x_n	y_n	Exact Solution	Absolute Error
1	0.1	0.55	0.547414	0.002585
2	0.2	0.595	0.589299	0.005701
3	0.3	0.6345	0.625070	0.009430
4	0.4	0.66795	0.654088	0.013862
5	0.5	0.694745	0.675639	0.019106
6	0.6	0.714219	0.688941	0.025278
7	0.7	0.725641	0.693124	0.032518
8	0.8	0.728205	0.687229	0.040976
9	0.9	0.721026	0.670198	0.050828
10	1.0	0.703129	0.640859	0.062270

(b)   When $h = 0.05$, Euler's method gives

$$y_{n+1} = y_n + 0.05(y_n - x_n), \qquad n = 0, 1, \ldots, 19,$$

which generates the approximations given in Table 1.11.2, where we have only listed every other intermediate approximation. We see that the approximation to $y(1)$ is

$$y_{20} = 0.686525$$

and that the absolute error in this approximation is

$$|y(1) - y_{20}| = 0.032492.$$

TABLE 1.11.2   The results of applying Euler's method with $h = 0.05$ to the IVP in Example 1.11.1.

n	x_n	y_n	Exact Solution	Absolute Error
2	0.1	0.54875	0.547414	0.001335
4	0.2	0.592247	0.589299	0.002948
6	0.3	0.629952	0.625070	0.004881
8	0.4	0.661272	0.654088	0.007185
10	0.5	0.685553	0.675639	0.009913
12	0.6	0.702072	0.688941	0.013131
14	0.7	0.710034	0.693124	0.016910
16	0.8	0.708563	0.687229	0.021333
18	0.9	0.696690	0.670198	0.026492
20	1.0	0.686525	0.640859	0.032492

Comparing this with (1.11.3), we see that the smaller step size has led to a better approximation. In fact, it has almost halved the error at $y(1)$. In Figure 1.11.2 we have plotted the exact solution and the Euler approximations just obtained. ❑

In the preceding example we saw that halving the step size had the effect of essentially halving the error. However, even then the accuracy was not as good as we probably would have liked. Of course we could just keep decreasing the step size (provided we do not take h to be so small that round-off errors start to play a role) to increase the accuracy, but then the number of steps we would have to take would make the calculations very cumbersome. A better approach is to derive methods that have a higher order of accuracy. We will consider two such methods.

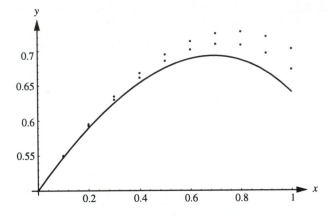

Figure 1.11.2 The exact solution to the IVP considered in Example 1.11.1 and the two approximations obtained using Euler's method.

MODIFIED EULER METHOD (HEUN'S METHOD)

The next method that we consider is an example of what is called a **predictor–corrector** method. The idea is to use the formula from Euler's method to obtain a first approximation to the solution $y(x_{n+1})$. We denote this approximation by y^*_{n+1}, so that

$$y^*_{n+1} = y_n + hf(x_n, y_n).$$

We now improve (or "correct") this approximation by once more applying Euler's method. But this time, we use the average of the slopes of the solution curves through (x_n, y_n) and (x_{n+1}, y^*_{n+1}). This gives

$$y_{n+1} = y_n + \frac{1}{2} h[f(x_n, y_n) + f(x_{n+1}, y^*_{n+1})].$$

As illustrated in Figure 1.11.3 for the case $n = 1$, we can interpret the modified Euler approximations as arising from first stepping to the point $P(x_n + h/2, y_n + hf(x_n, y_n)/2)$ along the tangent line to the solution curve through (x_n, y_n) and then stepping from P to (x_{n+1}, y_{n+1}) along the line through P whose slope is $f(x_n, y^*_n)$.

The Modified Euler Method for approximating the solution to the IVP

$$y' = f(x, y), \ \ y(x_0) = y_0$$

at the points $x_{n+1} = x_0 + nh$ $(n = 0, 1, \ldots)$ is

$$y_{n+1} = y_n + \frac{1}{2} h[f(x_n, y_n) + f(x_{n+1}, y^*_{n+1})],$$

where

$$y^*_{n+1} = y_n + hf(x_n, y_n), \ \ \ \ n = 0, 1, \ldots.$$

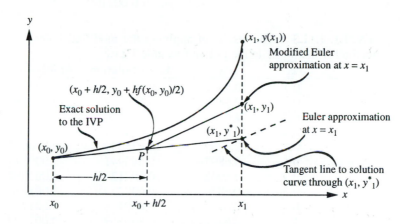

Figure 1.11.3 Derivation of the first step in the modified Euler method.

Example 1.11.2   Apply the modified Euler method with $h = 0.1$ to determine an approximation to the solution to the IVP

$$y' = y - x, \quad y(0) = 0.5$$

at $x = 1$.

Solution   Taking $h = 0.1$, and $f(x, y) = y - x$ in the modified Euler method yields

$$y^*_{n+1} = y_n + 0.1(y_n - x_n),$$

$$y_{n+1} = y_n + 0.05(y_n - x_n + y^*_{n+1} - x_{n+1}).$$

Hence,

$$y_{n+1} = y_n + 0.05\{y_n - x_n + [y_n + 0.1(y_n - x_n)] - x_{n+1}\}.$$

That is,

$$y_{n+1} = y_n + 0.05(2.1y_n - 1.1x_n - x_{n+1}), \quad n = 0, 1, \ldots, 9.$$

When $n = 0$, $y_1 = y_0 + 0.05(2.1y_0 - 1.1x_0 - x_1) = 0.5475$.

and when $n = 1$, $y_2 = y_1 + 0.05(2.1y_1 - 1.1x_1 - x_2) = 0.5894875$.

Continuing in this manner, we generate the results displayed in Table 1.11.3. From this table, we see that the approximation to $y(1)$ according to the modified Euler method is

$$y_{10} = 0.642960.$$

As seen in the previous example, the value of the exact solution at $x = 1$ is

$$y(1) = 0.640859.$$

Consequently, the absolute error in the approximation at $x = 1$ using the modified Euler approximation with $h = 0.1$ is

$$|y(1) - y_{10}| = 0.002100.$$

TABLE 1.11.3   The results of applying the modified Euler
method with $h = 0.1$ to the IVP in Example 1.11.2

n	x_n	y_n	Exact Solution	Absolute Error
1	0.1	0.5475	0.547414	0.000085
2	0.2	0.589487	0.589299	0.000189
3	0.3	0.625384	0.625070	0.000313
4	0.4	0.654549	0.654088	0.000461
5	0.5	0.676277	0.675639	0.000637
6	0.6	0.689786	0.688941	0.000845
7	0.7	0.694213	0.693124	0.001089
8	0.8	0.688605	0.687229	0.001376
9	0.9	0.671909	0.670198	0.001711
10	1.0	0.642959	0.640859	0.002100

Comparing this with the results of the previous example we see that the modified Euler
method has picked up approximately one decimal place of accuracy when using a step size
$h = 0.1$. This is indicative of the general result that the error in the modified Euler
method behaves as order h^2 as compared to the order h behavior of the Euler method. In
Figure 1.11.4 we have sketched the exact solution to the DE and the modified Euler
approximation with $h = 0.1$.

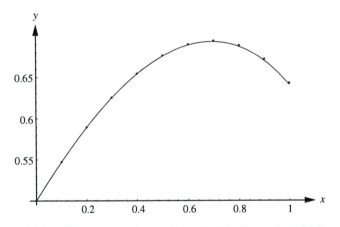

Figure 1.11.4   The exact solution to the IVP in Example 1.11.2 and the
approximations obtained using the modified Euler method with $h = 0.1$.

RUNGE–KUTTA METHOD OF ORDER FOUR

The final method that we consider is somewhat more tedious to use in hand
calculations, but is very easily programmed into a calculator or computer. It is a fourth–
order method, which, in the case of a DE of the form $y' = f(x)$, reduces to Simpson's Rule
(which you have probably studied in a previous calculus course) for numerically
evaluating definite integrals. Without justification, we state the algorithm.

> **Fourth-Order Runge–Kutta Method (RK4)** for approximating the solution to the IVP
>
> $$y' = f(x, y), \ y(x_0) = y_0$$
>
> at the points $x_{n+1} = x_0 + nh \ (n = 0, 1, \dots)$ is
>
> $$y_{n+1} = y_n + \frac{1}{6}(k_1 + 2k_2 + 2k_3 + k_4),$$
>
> where
>
> $$k_1 = hf(x_n, y_n), \quad k_2 = hf(x_n + \tfrac{1}{2}h, y_n + \tfrac{1}{2}k_1), \quad k_3 = hf(x_n + \tfrac{1}{2}h, y_n + \tfrac{1}{2}k_2),$$
>
> $$k_4 = hf(x_{n+1}, y_n + k_3),$$
>
> $n = 0, 1, \dots$

REMARK   In the previous sections, we used Maple to generate slope fields and approximate solution curves for first-order DE. The solution curves were in fact generated using a Runge–Kutta approximation.

Example 1.11.3   Apply RK4 with $h = 0.1$ to determine an approximation to the solution to the IVP

$$y' = y - x, \ y(0) = 0.5$$

at $x = 1$.

Solution   We take $h = 0.1$, and $f(x, y) = y - x$ in RK4, and need to determine y_{10}. First we determine k_1, k_2, k_3, k_4.

$k_1 = 0.1 f(x_n, y_n) = 0.1(y_n - x_n)$.

$k_2 = 0.1 f(x_n + 0.05, y_n + 0.5k_1) = 0.1(y_n + 0.5k_1 - x_n - 0.05)$

$k_3 = 0.1 f(x_n + 0.05, y_n + 0.5k_2) = 0.1(y_n + 0.5k_2 - x_n - 0.05)$

$k_4 = 0.1 f(x_{n+1}, y_n + k_3) = 0.1(y_n + k_3 - x_{n+1})$.

When $n = 0$,

$$k_1 = 0.1(0.5) = 0.05,$$

$$k_2 = 0.1[0.5 + (0.5)(0.05) - 0.05] = 0.0475,$$

$$k_3 = 0.1[0.5 + (0.5)(0.0475) - 0.05] = 0.047375,$$

$$k_4 = 0.1(0.5 + 0.047375 - 0.1) = 0.0447375,$$

so that

$$y_1 = y_0 + \frac{1}{6}(k_1 + 2k_2 + 2k_3 + k_4) = 0.5 + \frac{1}{6}(0.2844875) = 0.54741458,$$

rounded to eight decimal places. Continuing in this manner, we obtain the results displayed in Table 1.11.4. In particular, we see that the RK4 approximation to $y(1)$ is

$$y_{10} = 0.64086013,$$

so that

$$|y(1) - y_{10}| = 0.00000104.$$

Clearly this is an excellent approximation. If we increase the step size to $h = 0.2$, the corresponding approximation to $y(1)$ becomes

$$y_5 = 0.640874,$$

with absolute error

$$|y(1) - y_5| = 0.000015,$$

which is still very impressive.

TABLE 1.11.4   The results of applying RK4 to the IVP with $h = 0.1$ in Example 1.11.3.

n	x_n	y_n	Exact Solution	Absolute Error
1	0.1	0.54741458	0.54741454	0.00000004
2	0.2	0.58929871	0.58929862	0.00000009
3	0.3	0.62507075	0.62507060	0.00000015
4	0.4	0.65408788	0.65408765	0.00000022
5	0.5	0.67563968	0.67563936	0.00000032
6	0.6	0.68894102	0.68894060	0.00000042
7	0.7	0.69312419	0.69312365	0.00000054
8	0.8	0.68723022	0.68722954	0.00000068
9	0.9	0.67019929	0.67019844	0.00000085
10	1.0	0.64086013	0.64085909	0.00000104

EXERCISES 1.11

For problems 1–5, use Euler's method with the specified step size to determine the solution to the given IVP at the specified point.

1 . $y' = 4y - 1$, $y(0) = 1$, $h = 0.05$, $y(0.5)$.

2 . $y' = -2xy/(1 + x^2)$, $y(0) = 1$, $h = 0.1$, $y(1)$.

3 . $y' = x - y^2$, $y(0) = 2$, $h = 0.05$, $y(0.5)$.

4 . $y' = -x^2y$, $y(0) = 1$, $h = 0.2$, $y(1)$.

5 . $y' = 2xy^2$, $y(0) = 1$, $h = 0.1$, $y(1)$.

For problems 6–10, use the modified Euler method with the specified step size to determine the solution of the given IVP at the specified point. In each case compare your answer to that obtained using Euler's method.

6 . The IVP in problem 1.

7 . The IVP in problem 2.

8 . The IVP in problem 3.

9 . The IVP in problem 4.

10. The IVP in problem 5.

For problems 11–15, use RK4 with the specified step size to determine the solution of the given IVP at the specified point. In each case compare your answer to that obtained using Euler's method.

11. The IVP in problem 1.

12. The IVP in problem 2.

13. The IVP in problem 3.

14. The IVP in problem 4.

15. The IVP in problem 5.

◆ 16. Use RK4 with $h = 0.5$ to approximate the solution to the IVP

$$y' + \frac{1}{10}y = e^{-x/10}\cos x, \quad y(0) = 0$$

at the points $x = 0.5, 1.0, \ldots, 25$. Plot these points and describe the behavior of the corresponding solution.

1.12   SOME HIGHER ORDER DE

So far we have developed analytical techniques only for solving special types of first-order DE. The methods that we have discussed do not apply directly to higher order DE and so the solution to such equations usually requires the derivation of new techniques. One approach is to replace a higher order DE by an equivalent *system* of first-order equations. (This will be developed further in Chapter 8.) For example, any second order DE that can be written in the form

$$\frac{d^2y}{dx^2} = F\left(x, \ y \ , \ \frac{dy}{dx}\right), \tag{1.12.1}$$

where F is a known function, can be replaced by an equivalent pair of first-order DE as follows.

We let $v = dy/dx$. Then $d^2y/dx^2 = dv/dx$, and so solving equation (1.12.1) is equivalent to solving the following *two* first-order DE

$$\frac{dy}{dx} = v, \tag{1.12.2}$$

$$\frac{dv}{dx} = F(x, y, v). \tag{1.12.3}$$

In general the DE (1.12.3) cannot be solved directly, since it involves *three* variables, namely, $x, y,$ and v. However, for certain forms of the function F, equation (1.12.3) will involve only two variables and then can sometimes be solved for v using one of our previous techniques. Having obtained v, we can then substitute into equation (1.12.2) to obtain a first-order DE for y. We now discuss two forms of F for which this is certainly the case.

CASE 1   SECOND-ORDER EQUATIONS WITH THE DEPENDENT VARIABLE MISSING

If y does not occur explicitly in the function F, then equation (1.12.1) assumes the form

$$\frac{d^2y}{dx^2} = F\left(x, \frac{dy}{dx}\right). \tag{1.12.4}$$

Substituting $v = dy/dx$ and $dv/dx = d^2y/dx^2$ into this equation allows us to replace it with the two first-order equations

$$\frac{dy}{dx} = v, \tag{1.12.5}$$

$$\frac{dv}{dx} = F(x, v). \tag{1.12.6}$$

Thus, to solve equation (1.12.4), we first solve equation (1.12.6) for v in terms of x and then solve equation (1.12.5) for y as a function of x.

Example 1.12.1   Find the general solution to

$$\frac{d^2y}{dx^2} = \frac{1}{x}\left(\frac{dy}{dx} + x^2\cos x\right), \quad x > 0. \tag{1.12.7}$$

Solution   In equation (1.12.7), the dependent variable is missing, and so we let $v = dy/dx$ which implies that $d^2y/dx^2 = dv/dx$. Substituting into equation (1.12.7) yields the following equivalent first-order system

$$\frac{dy}{dx} = v, \tag{1.12.8}$$

$$\frac{dv}{dx} = \frac{1}{x}(v + x^2\cos x). \tag{1.12.9}$$

Equation (1.12.9) is a first-order linear DE with standard form

$$\frac{dv}{dx} - x^{-1}v = x\cos x. \tag{1.12.10}$$

An appropriate integrating factor is

$$I(x) = e^{-\int x^{-1}\,dx} = e^{-\ln x} = x^{-1}.$$

Multiplying equation (1.12.10) by x^{-1} reduces it to

$$\frac{d}{dx}(x^{-1}v) = \cos x,$$

which can be integrated directly to obtain

$$x^{-1}v = \sin x + c_1.$$

Thus,

$$v(x) = x\sin x + c_1 x. \tag{1.12.11}$$

Substituting this expression for v into equation (1.12.8) gives

$$\frac{dy}{dx} = x\sin x + c_1 x$$

which we can integrate to obtain

$$y(x) = -x\cos x + \sin x + c_1 x^2 + c_2,$$

where we have absorbed a factor of $1/2$ into c_1.

CASE 2   SECOND-ORDER EQUATIONS WITH THE INDEPENDENT VARIABLE MISSING

If x does not occur explicitly in the function F in equation (1.12.1), then we must solve a DE of the form

$$\frac{d^2y}{dx^2} = F\left(y, \frac{dy}{dx}\right). \tag{1.12.12}$$

In this case, we still let

$$v = \frac{dy}{dx},$$

as previously, but now we use the chain rule to express d^2y/dx^2 in terms of dv/dy. Specifically, we have

$$\frac{d^2y}{dx^2} = \frac{dv}{dx} \quad \underset{\underset{\text{via chain rule}}{\uparrow}}{=} \quad \frac{dv}{dy}\frac{dy}{dx} = v\frac{dv}{dy}.$$

Substituting for $\dfrac{dy}{dx}$ and $\dfrac{d^2y}{dx^2}$ into equation (1.12.12) reduces the second-order equation to the equivalent first-order system

$$\frac{dy}{dx} = v, \qquad (1.12.13)$$

$$v\frac{dv}{dy} = F(y, v). \qquad (1.12.14)$$

In this case, we first solve equation (1.12.14) for v as a function of y and then solve equation (1.12.13) for y as a function of x.

Example 1.12.2   Find the general solution to

$$\frac{d^2y}{dx^2} = -\frac{2}{(1 - y)}\left(\frac{dy}{dx}\right)^2. \qquad (1.12.15)$$

Solution   In this DE, the independent variable does not occur explicitly. Therefore, we let $v = dy/dx$ and use the chain rule to obtain

$$\frac{d^2y}{dx^2} = \frac{dv}{dx} = \frac{dv}{dy}\frac{dy}{dx} = v\frac{dv}{dy}.$$

Substituting into equation (1.12.15) results in the equivalent system

$$\frac{dy}{dx} = v, \qquad (1.12.16)$$

$$v\frac{dv}{dy} = -\frac{2}{1 - y}v^2. \qquad (1.12.17)$$

Separating the variables in the DE (1.12.17) gives

$$\frac{1}{v}\,dv = -\frac{2}{1 - y}\,dy, \qquad (1.12.18)$$

which can be integrated to obtain

$$\ln |v| = 2\ln |1 - y| + c.$$

Combining the logarithm terms and exponentiating yields

$$v(y) = c_1(1 - y)^2, \qquad (1.12.19)$$

where we have redefined $c_1 = \pm e^c$. Notice that in solving equation (1.12.17), we implicitly assumed that $v \neq 0$, since we divided by it to obtain equation (1.12.18). However the general form (1.12.19) does include the solution $v = 0$, provided we allow c_1 to equal zero. Substituting for v into equation (1.12.16) yields

$$\frac{dy}{dx} = c_1(1 - y)^2.$$

Separating the variables and integrating we obtain

$$(1 - y)^{-1} = c_1 x + d_1.$$

That is,

$$1 - y = \frac{1}{c_1 x + d_1}.$$

Solving for y gives

$$y(x) = \frac{c_1 x + (d_1 - 1)}{c_1 x + d_1},$$ (1.12.20)

which can be written in the simpler form

$$y(x) = \frac{x + a}{x + b},$$ (1.12.21)

where the constants a and b are defined by $a = (d_1 - 1)/c_1$, and $b = d_1/c_1$. Notice that the form (1.12.21) does not include the solution $y = $ constant, which *is* contained in (1.12.20) (set $c_1 = 0$). This is because in dividing by c_1, we implicitly assumed that $c_1 \neq 0$. Thus in specifying the solution in the form (1.12.21), we should also include the statement that $y = $ constant is a solution.

Example 1.12.3   Determine the displacement at time t of a simple harmonic oscillator that is extended a distance a units from its equilibrium position and released from rest at $t = 0$.

Solution   According to the derivation in Section 1.1, the motion of the simple harmonic oscillator is governed by the IVP

$$\frac{d^2 y}{dt^2} = -\omega^2 y,$$ (1.12.22)

$$y(0) = a, \quad \frac{dy}{dt}(0) = 0,$$ (1.12.23)

where ω is a positive constant. The DE (1.12.22) has the independent variable (in this case, t) missing. We therefore let $v = dy/dt$ and use the chain rule to write

$$\frac{d^2 y}{dt^2} = v \frac{dv}{dy}.$$

It then follows that equation (1.12.22) can be replaced by the equivalent first-order system

$$\frac{dy}{dt} = v,$$ (1.12.24)

$$v \frac{dv}{dy} = -\omega^2 y.$$ (1.12.25)

Separating the variables and integrating equation (1.12.25) yields

$$\frac{v^2}{2} = -\frac{\omega^2 y^2}{2} + c,$$

which implies that

$$v(y) = \pm \sqrt{c_1 - \omega^2 y^2},$$

where $c_1 = 2c$. Substituting for v into equation (1.12.24) yields

$$\frac{dy}{dt} = \pm \sqrt{c_1 - \omega^2 y^2}.$$ (1.12.26)

Setting $t = 0$ in this equation and using the initial conditions (1.12.23), we find that $c_1 = \omega^2 a^2$. Equation (1.12.26) therefore gives

$$\frac{dy}{dt} = \pm \omega \sqrt{a^2 - y^2}\ .$$

By separating the variables and integrating, we obtain

$$\sin^{-1}(y/a) = \pm \omega t + b$$

where b is an integration constant. Thus,

$$y(t) = a\ \sin(b \pm \omega t).$$

The initial condition $y(0) = a$ implies that $\sin(b) = 1$, and so we can choose $b = \pi/2$. We therefore have

$$y(t) = a\ \sin(\pi/2 \pm \omega t).$$

That is,

$$y(t) = a\ \cos \omega t.$$

Consequently the predicted motion is that the mass oscillates between $\pm a$ for all t. This solution makes sense physically since the simple harmonic oscillator does not include dissipative forces that would slow the motion.   □

REMARK In Chapter 2 you will learn how to solve the IVP (1.12.22), (1.12.23) in just a few lines of work without requiring any integration!

EXERCISES 1.12

For problems 1–13, solve the given DE.

1. $y'' = 2x^{-1}y' + 4x^2$.

2. $(x - 1)(x - 2)y'' = y' - 1$.

3. $y'' + 2y^{-1}(y')^2 = y'$.

4. $y'' = (y')^2 \tan y$.

5. $y'' + y' \tan x = (y')^2$.

6. $\ddot{x} = (\dot{x})^2 + 2\dot{x}$, where $\cdot \equiv d/dt$.

7. $y'' - 2x^{-1}y' = 6x^4$.

8. $t\ddot{x} = 2(t + \dot{x})$, where $\cdot \equiv d/dt$.

9. $y'' - \alpha(y')^2 - \beta y' = 0$, where α and β are nonzero constants.

10. $y'' - 2x^{-1}y' = 18x^4$.

11. $(1 + x^2)y'' = -2xy'$.

12. $y'' + y^{-1}(y')^2 = ye^{-y}(y')^3$.

13. $y'' - y' \tan x = 1$, $0 \le x < \pi/2$.

In problems 14 and 15 solve the given IVP.

14. $yy'' = 2y'^2 + y^2$, $y(0) = 1$, $y'(0) = 0$.

15. $y'' = \omega^2 y$, $y(0) = a$, $y'(0) = 0$, ω, a positive constants.

16. The following IVP arises in the analysis of a cable suspended between two fixed points

$$y'' = \frac{1}{a}\ [1 + (y')^2]^{1/2},\ y(0) = a,\ y'(0) = 0,$$

where a is a nonzero constant. Solve this IVP for $y(x)$. The corresponding solution curve is called a *catenary*.

17. Consider the general second order linear DE with dependent variable missing

$$y'' + p(x)y' = q(x).$$

Replace this DE with an equivalent pair of first-order equations and express the solution in terms of integrals.

18. Consider the general third-order DE of the form

$$y''' = F(x, y'').\qquad (18.1)$$

(a) Show that equation (18.1) can be replaced by the equivalent first-order system

$$\frac{du_1}{dx} = u_2, \quad \frac{du_2}{dx} = u_3, \quad \frac{du_3}{dx} = F(x, u_3),$$

where the variables u_1, u_2, u_3 are defined by

$$u_1 = y, \quad u_2 = y', \quad u_3 = y''.$$

(b) Solve $y''' = x^{-1}(y'' - 1)$.

19. A simple pendulum consists of a particle of mass m supported by a piece of string of length L. Assuming that the pendulum is displaced through an angle θ_0 radians from the vertical and then released from rest, the resulting motion is described by the IVP

$$\frac{d^2\theta}{dt^2} + \frac{g}{L}\sin\theta = 0, \ \theta(0) = \theta_0, \ \frac{d\theta}{dt}(0) = 0. \quad (19.1)$$

(a) For small oscillations, $\theta \ll 1$, we can use the approximation $\sin\theta \approx \theta$ in equation (19.1) to obtain the linear equation

$$\frac{d^2\theta}{dt^2} + \frac{g}{L}\theta = 0, \ \theta(0) = \theta_0, \ \frac{d\theta}{dt}(0) = 0.$$

Solve this IVP for θ as a function of t. Is the predicted motion reasonable?

(b) Obtain the following first integral of (19.1):

$$\frac{d\theta}{dt} = \pm [(2g/L)(\cos\theta - \cos\theta_0)]^{1/2}. \quad (19.2)$$

(c) Show from equation (19.2) that the time T (equal to the period of motion / 4) required for θ to change from 0 to θ_0 is given by the *elliptic integral of the first kind*

$$T = (L/2g)^{1/2} \int_0^{\theta_0} \frac{1}{(\cos\theta - \cos\theta_0)^{1/2}} \, d\theta. \quad (19.3)$$

(d) Show that (19.3) can be written as

$$T = (L/g)^{1/2} \int_0^{\pi/2} \frac{1}{(1 - k^2\sin^2 u)^{1/2}} \, du,$$

where $k = \sin(\theta_0/2)$.

(Hint: First express $\cos\theta$ and $\cos\theta_0$ in terms of $\sin^2(\theta/2)$ and $\sin^2(\theta_0/2)$.)

1.13*   THE PHASE PLANE

Just as slope fields are useful in determining the overall solution properties of a first-order DE, qualitative techniques can be used to analyze certain second order DE. At present we will restrict our attention to DE that can be written in the form

$$\ddot{y} = F(y, \dot{y}), \qquad (1.13.1)$$

where an overdot denotes differentiation with respect to t. It is useful to interpret the DE (1.13.1) as describing the motion of an object moving in one dimension under the action of a force per unit mass $F(y, \dot{y})$, so that $y(t)$ gives the displacement of the object at time t. If we introduce new variables u and v defined by

$$u = y, \quad v = \dot{y}, \qquad (1.13.2)$$

then equation (1.13.1) can be replaced by the equivalent pair of equations

$$\dot{u} = v, \qquad (1.13.3)$$

$$\dot{v} = F(u, v). \qquad (1.13.4)$$

In this formulation, u and v denote the position and velocity of the object at time t. A solution to the system of equations (1.13.3) and (1.13.4) consists of a pair of functions $u(t), v(t)$ that solves both equations simultaneously. In general, it is not possible to find

---

* This section can be omitted without loss of continuity. The discussion of the phase plane for linear systems in Chapter 8 does not depend on the material in this section.

such solutions. However, combining equations (1.13.3) and (1.13.4) yields the single first-order DE

$$\frac{dv}{du} = \frac{F(u, v)}{v},\tag{1.13.5}$$

which, if solvable, determines the relationship between u and v. Plotting the general solution to equation (1.13.5) in the uv-plane yields the **phase portrait** for the system (or DE (1.13.1)). Individual solution curves to (1.13.5) are called **phase paths**, **trajectories**, or **orbits** of the system. The uv-plane itself is called the **phase plane**. A given pair of values for $u(t_0)$ and $v(t_0)$ serves as initial conditions for equation (1.13.1). Hence, by following the particular trajectory in the phase plane that passes through the point $(u(t_0), v(t_0))$, we are viewing how the physical system evolves in time. From equation (1.13.3), we see that $du/dt > 0$ when $v > 0$, so that trajectories in the upper half plane are traversed from left to right as time increases, whereas if $v < 0$, then $du/dt < 0$, so that trajectories in the lower half plane are traversed from right to left as time increases.

Before considering examples, we make some general comments about the system of equations (1.13.3) and (1.13.4)

(**1**)   As we shall see later in the text, a special role is played by constant solutions to the system. Physically, for any such solution, the particle is at rest for all time and therefore these solutions are represented by points in the phase plane called **equilibrium points**, or **rest points**. The equilibrium points are given by any values of u and v that satisfy $du/dt = 0$, $dv/dt = 0$, and therefore, from equations (1.13.3) and (1.13.4), are any solutions to

$$v = 0, \quad F(u, 0) = 0.$$

Consequently, equilibrium points for the system of equations (1.13.3) and (1.13.4) must lie on the u-axis.

(**2**)   There is a unique trajectory corresponding to each set of initial conditions $(u(t_0), v(t_0))$. Consequently, trajectories cannot intersect one another. In particular, since equilibrium points correspond to (constant) solutions of the system, no trajectories can enter or leave an equilibrium point.

(**3**)   Phase paths that are closed curves correspond to periodic solutions.

(**4**)   If a trajectory crosses the u-axis, then, from (2), the point of intersection cannot be an equilibrium point. Consequently we have $v = 0$, and $F(u, 0) \neq 0$ at such an intersection point. It follows from equation (1.13.5) that the phase curve has a vertical tangent line at the point of intersection and therefore is perpendicular to the u-axis.

(**5**)   Even if we cannot solve equation (1.13.5) to determine an explicit formula for the trajectories, we can obtain an idea of the behavior of the trajectories by sketching the slope field for the equation. In this case, however, it is useful to append an arrow head to each line segment in the slope field to indicate the direction that the trajectory through a particular point is traversed. The resulting slope field is then called a **direction field**.

Example 1.13.1   Determine the phase portrait for $\ddot{y} - 4y = 0$.

Solution   We let $u = y$, $v = \dot{y}$, so that the DE is replaced by the equivalent system

$$\dot{u} = v, \quad \dot{v} = 4u.$$

The only equilibrium point is the origin (0, 0), and the trajectories are determined by solving

$$\frac{dv}{du} = 4\frac{u}{v}. \tag{1.13.6}$$

This is a separable DE with general solution

$$4u^2 - v^2 = c,$$

where c is an integration constant. Consequently, each trajectory is a hyperbola. The asymptotes of this family of hyperbolas are the lines $v = \pm 2u$. We therefore obtain the phase portrait shown in Figure 1.13.1. Also included in this figure is the direction field for the DE (1.13.6). Since (0, 0) is an equilibrium point, the asymptotes $v = \pm 2u$ actually correspond to the four different trajectories $v = \pm 2u$, $u > 0$ and $v = \pm 2u$, $u < 0$. In this case, the equilibrium point is called a **saddle point**. This equilibrium point is said to be **unstable**, since not every trajectory that approaches the equilibrium point remains close to the equilibrium point for all later t.

   The general behavior of solutions to the given system can be read off from the phase portrait. The first point to note is that there are no periodic solutions since none of the trajectories are closed. Now suppose, for example, that we are interested in the solution corresponding to the initial conditions $y(0) = 1$, $dy/dt(0) = 0$. In the phase plane variables these initial conditions are $u = 1$ and $v = 0$ at $t = 0$, which corresponds to the point (1, 0) in the phase plane. By drawing the corresponding trajectory through this point, we can follow the evolution of u and v with time. We see from the phase portrait that both u and v increase with t along this trajectory.

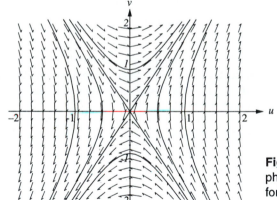

Figure 1.13.1 Maple generated phase portrait and direction field for the system in Example 1.13.1.

Example 1.13.2 Determine the phase portrait for $\ddot{y} + \omega^2 y = 0$, where ω is a positive constant.

Solution   The corresponding system in the phase plane is

$$\dot{u} = v, \quad \dot{v} = -\omega^2 u.$$

We see that the only equilibrium point in the phase plane is (0, 0), which corresponds to the solution $y(t) = 0$ for all t. To determine the trajectories, we must solve

$$\frac{dv}{du} = -\omega^2\frac{u}{v},$$

which is a separable DE with general solution

$$\omega^2 u^2 + v^2 = c,$$

where c is an integration constant. Consequently, each trajectory is an ellipse, and all solutions to the given DE are therefore periodic. As was shown in Section 1.1, the given DE describes the motion of a spring-mass system in the absence of friction or external driving forces. Hence the periodic behavior of the solution. The phase portrait is given in Figure 1.13.2. The equilibrium point $(0, 0)$ is called a **center**. It is a **stable** equilibrium point, since all trajectories that are close to $(0, 0)$ remain close to $(0, 0)$ for all later time.

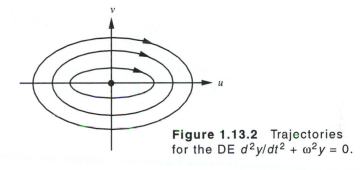

Figure 1.13.2   Trajectories for the DE $d^2y/dt^2 + \omega^2 y = 0$.

The following example is somewhat more complicated than the previous two. Even though we can determine the equation of the trajectories, a direction field analysis is also needed in order to completely understand the phase portrait.

Example 1.13.3   Determine the phase portrait for $\ddot{y} = \dot{y}\,(y - \dot{y})$.

Solution   Introducing the variables $u = y$, $v = \dot{y}$, yields the corresponding system

$$\dot{u} = v, \qquad \dot{v} = v(u - v).$$

In this case we see that each point on the u-axis is an equilibrium point. Hence no trajectory can cross the u-axis. To determine the trajectories we must solve

$$\frac{dv}{du} = u - v. \tag{1.13.7}$$

Rearranging yields the linear DE

$$\frac{dv}{du} + v = u$$

which has general solution

$$v = u - 1 + ce^{-u}, \tag{1.13.8}$$

where c is an integration constant. One particular solution is

$$v = u - 1, \tag{1.13.9}$$

which corresponds to $c = 0$. This is the equation of a line, but actually corresponds to two trajectories, since we must delete the point $(1, 0)$, which is an equilibrium point. Determining particular trajectories can be useful in constructing the overall phase portrait, since we now know that no other trajectories can intersect the line. Furthermore, from equation (1.13.8) we see that as $u \to +\infty$ the trajectories approach the line (1.13.9). From equation (1.13.7) we see that $dv/du = 0$ at all points along the curve $v = u$ and that

$dv/du > 0$ if $v < u$, whereas $dv/du < 0$ if $v > u$. Finally, differentiating equation (1.13.7) with respect to u yields

$$\frac{d^2v}{du^2} = 1 - \frac{dv}{du} = 1 - u + v.$$

Consequently, there is a change in concavity at all points along the line $1 - u + v = 0$, which coincides with the trajectory (1.13.9). Putting all of this information together leads to the phase portrait and direction field shown in Figure 1.13.3.

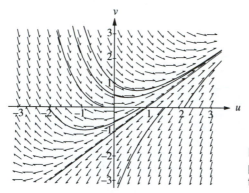

Figure 1.13.3 Maple generated phase portrait and direction field for the system in Example 1.13.3.

CONSERVATIVE SYSTEMS

We next consider the more restricted, but still physically important, case of DE that can be written in the form

$$\ddot{y} + F(y) = 0. \tag{1.13.10}$$

We can interpret equation (1.13.10) as the DE governing the motion of an object of unit mass moving under the influence of a force $-F(y)$, where $y(t)$ denotes the position of the object at time t. If we introduce the phase plane variables $u = y$, $v = \dot{y}$, then the DE can be replaced by the system

$$\dot{u} = v, \qquad \dot{v} = -F(u). \tag{1.13.11}$$

Using the chain rule, the second of these equations can be written as

$$v\frac{dv}{du} = -F(u),$$

which we can formally integrate to obtain

$$\frac{1}{2}v^2 + \int_0^u F(w)dw = E, \tag{1.13.12}$$

where E is an integration constant. The first term in (1.13.12) represents *kinetic energy*. Further, the *potential energy* associated with the given physical system, $V(u)$, is defined by

$$V(u) = \int_0^u F(w)dw, \tag{1.13.13}$$

so that (1.13.12) can be written as

$$\frac{1}{2}v^2 + V(u) = E,$$

which is the statement that the total energy is conserved along any solution curve to equation (1.13.10).   For this reason, a DE of the form (1.13.10) is called a **conservative** DE. From equation (1.13.12), it follows that the trajectories for the system are given by

$$v = \pm\sqrt{2[E - V(u)]}. \tag{1.13.14}$$

The key to sketching the phase portrait for the system is to note from equation (1.13.14) that E and u can assume only those values that maintain

$$E - V(u) \geq 0. \tag{1.13.15}$$

Therefore, if we sketch the potential energy function V, we can determine the possible values of E and the corresponding interval(s) for u such that inequality (1.13.15) holds. We can then sketch the trajectories using equation (1.13.14). The following two points are important.

(1)   Corresponding to each u for which the inequality (1.13.15) is valid, (1.13.14) determines two points in the phase plane symmetrically placed with respect to the u-axis. Consequently, the trajectories are symmetric with respect to the u-axis and crossing the u-axis corresponds to changing signs in equation (1.13.14).

(2)   From equation (1.13.13) we have $F(u) = V'(u)$. Consequently, the equilibrium points for the system (1.13.11) occur on the u-axis at any values of u for which $V'(u) = 0$. From elementary calculus, these are *critical points* of V, and therefore they correspond to local maxima, local minima, or horizontal points of inflection for V.

There are four basic types of behavior for the trajectories depending on the nature of the critical points of $V(u)$.

1. Equilibrium point corresponds to a local minimum of V

A representative potential energy function is sketched in Figure 1.13.4. If $E = E_0$, then the only value of u for which inequality (1.13.15) can hold is $u = u_0$, which corresponds to the equilibrium point.   For $E = E_1$, the inequality will hold for u in the interval $[u_1, u_2]$ shown in the figure.   For each u in $[u_1, u_2]$ we plot the resulting pairs of points in the phase plane obtained from equation (1.13.14), thereby generating a closed curve, such as that shown in the figure.   Consequently, the solutions in the neighborhood of a local minimum of V are periodic.   The equilibrium point in this case is a stable **center**.

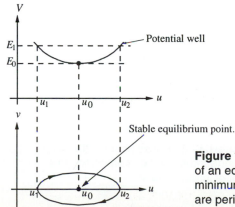

Figure 1.13.4   Phase plane in the neighborhood of an equilibrium point corresponding to a local minimum of V. All solutions in the neighborhood are periodic

2.   Equilibrium point corresponds to a local maximum of V

A representative potential energy function is shown in Figure 1.13.5. For $E = E_1$, as u increases towards the critical point of V, $E - V$ decreases to zero and then increases from zero after the critical point. Consequently, there are four corresponding trajectories that approach the equilibrium point. (Remember that trajectories cannot pass through an equilibrium point.) For $E = E_0$, there are two different intervals of u for which the inequality (1.13.15) holds. These give rise to the two phase curves shown in the Figure 1.13.5 that intersect the u-axis. Finally, for $E = E_2$ we again get two phase curves. These curves are closest to the u-axis when $u = u_0$, but do not intersect the u-axis.

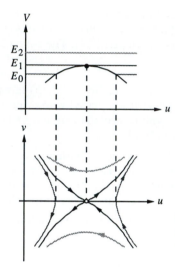

Figure 1.13.5 Phase plane in the neighborhood of an equilibrium point corresponding to a local maximum of V.

3.   Equilibrium point corresponds to a horizontal point of inflection of V

Similar arguments to those used in (1) and (2) lead to Figure 1.13.6.

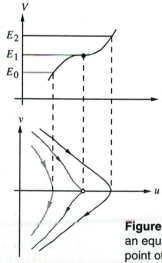

Figure 1.13.6 Phase plane in the neighborhood of an equilibrium point corresponding to a horizontal point of inflection of V.

We illustrate the preceding ideas with an example.

Example 1.13.4   A simple pendulum consists of a mass m attached to the end of a light rod of length L, whose other end is fixed. (See Figure 1.13.7.) If we let y denote the angle the rod is displaced from its vertical position at time t, then the equation of motion of the pendulum is

$$\ddot{y} + \omega^2 \sin y = 0,$$

where $\omega = \sqrt{g/L}$. Determine the phase plane associated with this DE in the case when $\omega = 1$.

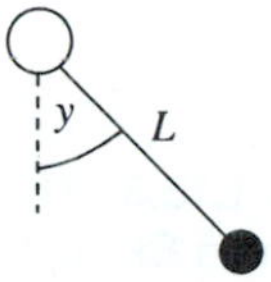

Figure 1.13.7   The simple pendulum

Solution   The DE of interest is

$$\ddot{y} + \sin y = 0,$$

which is a conservative DE with $F(y) = \sin y$. Consequently, the potential energy is

$$V(y) = \int_0^y \sin w \, dw = 1 - \cos y,$$

which we have sketched in Figure 1.13.8. Introducing the phase plane variables $u = y$, $v = dy/dt$ yields the system

$$\dot{u} = v, \quad \dot{v} = -\sin u.$$

We see that there are equilibrium points at $(n\pi, 0)$, $n = 0, \pm 1, \pm 2, \ldots$. The equation for the trajectories is

$$v = \pm\sqrt{2[E - V(u)]} = \pm\sqrt{2[E - (1 - \cos u)]}.$$

Four different types of behavior arise, depending on the value of the constant E.

When $E = 0$:   This value of E corresponds to the equilibrium points at $(\pm 2n\pi, 0)$. Physically, the pendulum hangs vertically downward in stable equilibrium.

When $0 < E < 2$: As shown in Figure 1.13.8, any value of E in this interval corresponds to a periodic solution. Consequently, the equilibrium points at $(\pm 2n\pi, 0)$ are centers.

When $E = 2$:   The equilibrium points at $(\pm(2n + 1)\pi, 0)$ have $E = 2$, as do the two trajectories joining each such pair of equilibrium points. The corresponding solutions are *not* periodic, since the two trajectories cannot pass through the equilibrium points and therefore do not join to make a closed curve. These trajectories are called **separatrices**. They are the boundary curves between regions in the phase plain containing periodic solutions (shaded in the figure) and those that do not. Physically, the equilibrium points correspond to the pendulum being balanced vertically [$y = \pm(2n + 1)\pi$] which is an unstable equilibrium. The separatrices correspond to initial conditions that just impart enough energy into the system for the pendulum to swing exactly once to a position of unstable equilibrium.

When $E > 2$:   In this case the trajectories are not closed, and the solutions are not periodic. Physically, the initial conditions impart so much energy into the system that it continues to rotate. This is often referred to as a whirring motion.

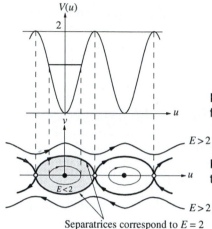

Figure 1.13.8 The potential energy function in Example 1.13.4.

Figure 1.13.9 The phase plane for the pendulum in Example 1.13.4.

Separatrices correspond to $E = 2$

EXERCISES 1.13

1. Sketch the phase portrait for the DE
$$\ddot{y} + 16y = 0.$$
Label the trajectory corresponding to each of the following initial conditions

(a) $y(0) = 0,\ \dot{y}(0) = 1.$ (b) $y(0) = 1,\ \dot{y}(0) = 1.$

2. Sketch the phase portrait for the DE
$$\ddot{y} - 9y = 0.$$
Label the trajectory corresponding to each of the following sets initial conditions. For each set of initial conditions, determine the behavior of the solution as $t \to \infty$.

(a) $y(0) = 1,\ \dot{y}(0) = 1.$

(b) $y(0) = 0,\ \dot{y}(0) = 1.$

(c) $y(0) = -1,\ \dot{y}(0) = 1.$

(d) $y(0) = 0,\ \dot{y}(0) = -1.$

3. Sketch the phase portrait for the DE
$$\ddot{y} = 3\dot{y}y^2.$$
Label the trajectory corresponding to each of the following sets initial conditions.

(a) $y(0) = 1,\ \dot{y}(0) = 0.$ (b) $y(0) = 1,\ \dot{y}(0) = 1.$

(c) $y(0) = -1,\ \dot{y}(0) = -1.$

4. Sketch the phase portrait for the DE
$$\ddot{y} = 2\dot{y}y.$$
Label the trajectory corresponding to each of the following sets initial conditions.

(a) $y(0) = 1,\ \dot{y}(0) = 0.$ (b) $y(0) = 0,\ \dot{y}(0) = 1.$

(c) $y(0) = 1,\ \dot{y}(0) = 0,$ (d) $y(0) = -1,\ \dot{y}(0) = 0.$

(e) $y(0) = 0,\ \dot{y}(0) = -1.$

5. By solving the IVP
$$\ddot{y} = 2\dot{y}y,\ y(0) = 1,\ \dot{y}(0) = 0,$$

verify that the corresponding trajectory in the phase plane is traversed in a finite time interval.

6. By solving the IVP

$$\ddot{y} = 2\dot{y}y, \quad y(0) = 0, \quad \dot{y}(0) = -1,$$

verify that the corresponding trajectory in the phase plane satisfies

$$\lim_{t \to \infty} u(t) = -1, \quad \lim_{t \to \infty} u(t) = 1,$$

$$\lim_{t \to -\infty} v(t) = 0, \quad \lim_{t \to -\infty} v(t) = 0,$$

7. Consider the DE

$$\ddot{y} = \dot{y}(2\dot{y} - y).$$

(a) Show that in the phase plane variables $u = y$, $v = \dot{y}$, the trajectories are obtained by solving

$$\frac{dv}{du} = 2v - u,$$

and sketch the direction field and representative trajectories for this DE.

8. Consider the DE

$$\ddot{y} = \dot{y}(ay + b\dot{y}),$$

where a and b are constants, and $b > 0$.

(a) Show that in the phase plane variables $u = y$, $v = \dot{y}$, the trajectories are obtained by solving

$$\frac{dv}{du} = au + bv. \tag{8.1}$$

and sketch the direction field and representative trajectories for this DE.

(b) Solve (8.1) and show that one solution is

$$v = -\frac{a}{b}u - \frac{a}{b^2}.$$

(c) Determine where the concavity of the solution curves to (8.1) changes.

(d) Sketch the phase portrait (you will need to consider the cases $a > 0$, $a < 0$, $a = 0$ separately.)

9. **(a)** Sketch the potential energy function, and determine the phase portrait for the DE

$$\ddot{y} + y + 0.5y^2 = 0.$$

(b) Use your phase portrait to determine the values of the positive constant α for which the IVP

$$\ddot{y} + y + 0.5y^2 = 0, \quad y(0) = \alpha, \quad \dot{y}(0) = 0$$

has periodic solutions.

10. **(a)** Sketch the potential energy function, and determine the phase portrait for the DE

$$\ddot{y} + y - 0.5y^2 = 0.$$

(b) Use your phase portrait to determine the values of the positive constant α for which the IVP

$$\ddot{y} + y + 0.5y^2 = 0, \quad y(0) = \alpha, \quad \dot{y}(0) = 0$$

has periodic solutions.

11. **(a)** Sketch the potential energy function, and determine the phase portrait for the DE

$$\ddot{y} + y - 0.5y^3 = 0.$$

(b) Use your phase portrait to establish that the solution to the IVP

$$\ddot{y} + y - 0.5y^3 = 0, \quad y(0) = 1, \quad \dot{y}(0) = 0$$

is periodic.

12. The motion of a bead that is free to slide on a circular wire (radius a) which is being rotated at a constant angular velocity ω is governed by the IVP

$$\frac{d^2\theta}{dt^2} - \sin\theta (\cos\theta - \varepsilon) = 0, \quad \theta(0) = \theta_0, \quad \frac{d\theta}{dt}(0) = 0,$$

where $\varepsilon = g/(a\omega^2)$. (See Figure 1.13.10.)

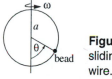

Figure 1.13.10 A bead sliding on a rotating circular wire.

(a) What physical situation does $\varepsilon \ll 1$ correspond to?

(b) Using the representative potential function $V(\theta)$ given in Figure 1.13.11, determine the phase portrait for the DE and describe the physical motion corresponding to each separatrix. Indicate all periodic regimes in your diagram.

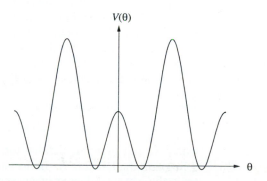

Figure 1.13.11 Representative potential function for Problem 1.13.12.

2

Second-Order Linear Differential Equations

We now begin our study of linear mathematics by considering second-order *linear* DE. Our approach will be somewhat different to that taken in Chapter 1. The first task will be to develop the full theory for these DE. We will begin by establishing that every solution to any DE of the form

$$y'' + a_1(x)y' + a_2(x)y = 0 \tag{2.0.1}$$

can be written as

$$y(x) = c_1 y_1(x) + c_2 y_2(x), \tag{2.0.2}$$

where y_1 and y_2 are any two solutions to the DE that are nonproportional ($y_2/y_1 \neq$ constant). The power of this type of result is that if we can find two such solutions by any method whatsoever then we know that by forming the *linear* combination (2.0.2) we have determined all solutions to the DE.

The next step in the theoretical development is to demonstrate that for the general linear DE

$$y'' + a_1(x)y' + a_2(x)y = F(x) \tag{2.0.3}$$

every solution can be written as

$$y(x) = c_1 y_1(x) + c_2 y_2(x) + y_p(x),$$

where y_1 and y_2 are as defined for the previous situation, and $y_p(x)$ is any *particular* solution to the DE (2.0.3). In the later chapters, we will see that these theoretical results are really just special cases of a general framework that holds for all linear problems.

According to the preceding discussion, our theory will tell us that in order to solve equation (2.0.3) we need two nonproportional solutions to equation (2.0.1) and one particular solution to equation (2.0.3). The remainder of the chapter is then concerned with developing techniques for obtaining these solutions. Essentially, the only situation where general techniques are available is when a_1 and a_2 are constants. In this case, we will be able to determine appropriate nonproportional solutions y_1 and y_2 with no more work than solving a quadratic algebraic equation. In particular, unlike the techniques developed in Chapter 1 for solving first-order DE, there will be no integration. For many forms of the function $F(x)$ in equation (2.0.3) we will also be able to determine y_p algebraically, and for general $F(x)$, we will have a more sophisticated technique that will always, in principle, enable y_p to be determined.

As an application of the techniques that are derived in this chapter, we give a full analysis of a linear spring-mass system. Such a system and its generalization to the case of more than one spring and mass (considered in Chapter 8) can be used to model an extensive variety of applied problems, such as an automobile suspension system, the swaying of a tall building due to a high wind or earthquake, or the displacement of atoms in a crystal, to mention just a few. We also show that the DE governing the charge on the capacitor in an RLC circuit, derived in Section 1.7, (see equation (1.7.17)) is mathematically equivalent to that of a corresponding spring-mass system.

2.1   BASIC THEORETICAL RESULTS

A linear second-order DE is any DE that can be written in the standard form

$$y'' + a_1(x)y' + a_2(x)y = F(x), \qquad (2.1.1)$$

where a_1, a_2 and F are sufficiently smooth functions of x on the interval of interest I. From a theoretical viewpoint, the following existence and uniqueness theorem is the cornerstone result.

Theorem 2.1.1:   Let a_1, a_2, and F be continuous functions on an interval I. Then, for each $x_0 \in I$, the IVP

$$y'' + a_1(x)y' + a_2(x)y = F(x), \quad y(x_0) = y_0, \ y'(x_0) = y_1$$

has a unique solution on I.

PROOF   Omitted. See, for example, E.A. Coddington, *An Introduction to Ordinary Differential Equations*, Dover, 1961. ■

The geometric interpretation of this theorem is a little different to that of the existence and uniqueness theorem for first-order DE discussed in Section 1.3. In particular, solution curves of second-order linear DE can intersect. What the theorem tells us, however, is that if at any such point of intersection, the solution curves also have the same slope, then they are identical. Physically, if we take the independent variable as time, t, (rather than x), and let $y(t)$ denote the position of an object at time t, then the theorem can be interpreted as saying that specification of the initial position and initial velocity of an object governed by a second-order linear DE uniquely specifies its motion.

If $F(x)$ is identically zero in equation (2.1.1), the DE is called **homogeneous**. Hence, the general homogeneous second-order linear DE is

$$y'' + a_1(x)y' + a_2(x)y = 0. \qquad (2.1.2)$$

We note by inspection that this always has the solution

$$y(x) = 0,$$

for all $x \in I$. Furthermore, applying the existence and uniqueness theorem, this solution is the unique solution to the IVP

$$y'' + a_1(x)y' + a_2(x)y = 0, \;\; y(x_0) = 0, \; y'(x_0) = 0.$$

If the function $F(x)$ in equation (2.1.1) is not identically zero on I, then the DE (2.1.1) is called **nonhomogeneous**. Unlike the homogeneous problem, in this case the unique solution to the IVP

$$y'' + a_1(x)y' + a_2(x)y = F(x), \;\; y(x_0) = 0, \; y'(x_0) = 0$$

is a nonzero function. Therefore, we can consider the resulting solution as being driven, or forced by $F(x)$. Consequently, $F(x)$ is often referred to as the **forcing term** or **driving term** for the DE.

We first focus our attention on the homogeneous DE (2.1.2). The next theorem introduces the two fundamental properties of solutions to such a DE. We will see in the later chapters how these properties are characteristic of solutions to any linear problem.

Theorem 2.1.2: If $y = y_1(x)$ and $y = y_2(x)$ are two solutions to the homogeneous DE (2.1.2), then

(**a**) $y(x) = y_1(x) + y_2(x)$ is a solution to equation (2.1.2)

and

(**b**) $y(x) = cy_1(x)$ is a solution to equation (2.1.2) for any constant c.

PROOF   We are given that both

$$y_1'' + a_1(x)y_1' + a_2(x)y_1 = 0 \quad \text{and} \quad y_2'' + a_1(x)y_2' + a_2(x)y_2 = 0. \qquad (2.1.3)$$

(**a**) If $y(x) = y_1(x) + y_2(x)$, then

$$
\begin{aligned}
y'' + a_1(x)y' + a_2(x)y &= (y_1 + y_2)'' + a_1(x)(y_1 + y_2)' + a_2(x)(y_1 + y_2) \\
&= [y_1'' + a_1(x)y_1' + a_2(x)y_1] + [y_2'' + a_1(x)y_2' + a_2(x)y_2] \\
&= 0 + 0 = 0,
\end{aligned}
$$

where we have used (2.1.3). Consequently, $y(x) = y_1(x) + y_2(x)$ is a solution to the DE (2.1.2).

(**b**) If $y(x) = cy_1(x)$, then

$$
\begin{aligned}
y'' + a_1 y' + a_2 y &= (cy_1)'' + a_1(cy_1)' + a_2(cy_1) \\
&= c(y_1'' + a_1 y_1' + a_2 y_1) = 0,
\end{aligned}
$$

where we have once more used (2.1.3). This establishes that $y(x) = cy_1(x)$ is a solution to equation (2.1.2). ∎

The two properties illustrated in the foregoing theorem are called the **linearity properties**. If we combine them, we see that for any constants c_1 and c_2, the function

$$y(x) = c_1 y_1(x) + c_2 y_2(x)$$

is a solution to the DE (2.1.2) whenever $y = y_1(x)$ and $y = y_2(x)$ are solutions to the DE. This is called the **principle of superposition** or the **linearity principle**. An expression of the form

$$c_1 y_1(x) + c_2 y_2(x)$$

is called a **linear combination** of y_1 and y_2.

Before stating the fundamental theorem for homogeneous linear second-order DE, we need to discuss linear dependence and linear independence of functions. These concepts are defined, for two functions, in Definition 2.1.1.

> **Definition 2.1.1:**   Two functions $y_1(x)$, $y_2(x)$ that are proportional ($y_1 = cy_2$ or $y_2 = dy_1$ where c and d constants) on an interval I are said to be **linearly dependent** on I. Functions that are not linearly dependent on I are said to be **linearly independent** on I.

Example 2.1.1   The functions $y_1(x) = x^2$ and $y_2(x) = 7x^2$ are linearly dependent on any interval, since $y_2 = 7y_1$. The functions $y_1(x) = x$, $y_2(x) = x^2$ are linearly independent on any interval, since neither is a *constant* multiple of the other.   □

Although it is generally straightforward to check by inspection whether a pair of functions is linearly dependent on a given interval, when discussing solutions of a homogeneous linear DE it is useful to have an analytic condition that characterizes linear dependence. The appropriate result is given in the next theorem, whose proof relies on the existence and uniqueness theorem.

> **Theorem 2.1.3:**   Let $y = y_1(x)$ and $y = y_2(x)$ be solutions to
>
> $$y'' + a_1(x)y' + a_2(x)y = 0 \qquad\qquad (2.1.4)$$
>
> on an interval I. Then y_1 and y_2 are linearly dependent on I if and only if
>
> $$y_1 y_2' - y_1' y_2 = 0, \qquad\qquad (2.1.5)$$
>
> for all $x \in I$.

PROOF   Suppose first that y_1 and y_2 are linearly dependent on I. If either of these functions is identically zero on I, then clearly equation (2.1.5) holds for all $x \in I$. Now consider the case when neither function is identically zero on I. Then, from the definition of linear dependence, it follows that $y_2 = cy_1$ for some constant c. Hence, $y_2' = cy_1'$. Eliminating c between these two equations yields the condition (2.1.5).

Conversely suppose that equation (2.1.5) holds for all $x \in I$. We must establish that y_2 and y_1 are proportional. If y_2 were identically zero on I, then proportionality would follow from $y_2 = 0y_1$. Assume, therefore, that y_2 is not identically zero on I. Then, since y_2 is continuous on I, there is some subinterval, say, J, of I on which y_2 is never zero. Dividing equation (2.1.5) by y_2^2 we obtain

$$\frac{y_1 y_2{}' - y_1{}' y_2}{y_2{}^2} = 0.$$

From the quotient rule for differentiation, this latter equation is just the expanded form of

$$-\left(\frac{y_1}{y_2}\right)' = 0.$$

Hence,

$$\frac{y_1}{y_2} = c$$

on the interval J. Consequently,

$$y_1 = cy_2$$

on J, from which it follows that $y_1{}' = cy_2{}'$ on J. Since J is a subinterval of I, the existence and uniqueness theorem can now be applied to conclude that $y_1(x) = cy_2(x)$ for all x in I. ∎

The combination of functions $y_1 y_2{}' - y_1{}' y_2$ plays a fundamental role in linear DE theory, and so we give it a special name and notation.

Definition 2.1.2:   Let y_1 and y_2 be differentiable functions on the interval I. The **Wronskian** of the these functions, denoted $W[y_1, y_2]$, is defined by

$$W[y_1, y_2] = y_1 y_2{}' - y_1{}' y_2.$$

We now state the basic theorem about solutions to the DE (2.1.2).

Theorem 2.1.4:   Let $y(x) = y_1(x)$ and $y(x) = y_2(x)$ be two solutions to

$$y'' + a_1(x)y' + a_2(x)y = 0 \tag{2.1.6}$$

that are linearly independent on the interval of interest. Then every solution to this DE is of the form

$$y(x) = c_1 y_1(x) + c_2 y_2(x), \tag{2.1.7}$$

where c_1, c_2 are arbitrary constants.

PROOF   In order to maintain the flow of the discussion, the proof of this theorem has been relegated to the end of the section. ∎

The importance of Theorem 2.1.4 stems from the fact that if, by any means whatsoever, we can find two solutions to the DE (2.1.6) that are linearly independent on the interval of interest, then by forming a linear combination of those solutions, we have determined all solutions to the DE. Since every solution to the DE can be obtained by assigning appropriate values to the constants in the solution (2.1.7), this is the **general solution** to the DE (2.1.6).

Example 2.1.2   Determine all solutions to the DE $y'' + y' - 6y = 0$ of the form $y(x) = e^{rx}$, where r is a constant. Use your solutions to determine the general solution to the DE.

Solution   Substituting $y(x) = e^{rx}$ into the given DE yields

$$e^{rx}(r^2 + r - 6) = 0,$$

or equivalently,

$$(r + 3)(r - 2) = 0.$$

Hence, two solutions to the DE are

$$y_1(x) = e^{2x}, \ y_2(x) = e^{-3x}.$$

Furthermore, the Wronskian of these two solutions is

$$W[y_1, y_2](x) = e^{2x}(-3e^{-3x}) - (2e^{2x})(e^{-3x}) = -5e^{-x} \neq 0,$$

so that the solutions are linearly independent on any interval (of course the linear independence could have been seen by inspection.) It follows directly from Theorem 2.1.4 that the general solution to the DE is

$$y(x) = c_1 e^{2x} + c_2 e^{-3x}. \hspace{3cm} \square$$

We now turn our attention to the linear nonhomogeneous DE

$$y'' + a_1(x)y' + a_2(x)y = F(x). \hspace{2cm} (2.1.8)$$

The general solution to the associated homogeneous DE is called the **complementary function** for equation (2.1.8) and is denoted $y_c(x)$. Thus, if $y_1(x)$ and $y_2(x)$ denote two linearly independent solutions to the associated homogeneous DE, then

$$y_c(x) = c_1 y_1(x) + c_2 y_2(x).$$

Our next theorem contains the major result regarding the solution to equation (2.1.8).

Theorem 2.1.5:   Consider the nonhomogeneous linear DE (2.1.8). Let $y_c(x)$ denote the complementary function, and let $y_p(x)$ denote one particular solution to equation (2.1.8). Then all solutions to equation (2.1.8) are of the form

$$y(x) = y_c(x) + y_p(x).$$

PROOF   We are given that

$$y_p'' + a_1(x)y_p' + a_2(x)y_p = F(x).$$

If $y(x)$ is any solution to equation (2.1.8) we then also have

$$y'' + a_1(x)y' + a_2(x)y = F(x).$$

Subtracting these two equations yields

$$(y - y_p)'' + a_1(x)(y - y_p)' + a_2(x)(y - y_p) = 0.$$

Consequently, the function $y - y_p$ is a solution to the associated homogeneous equation. Applying Theorem 2.1.4, we can therefore conclude that

$$y(x) - y_p(x) = c_1 y_1(x) + c_2 y_2(x),$$

where y_1, y_2 are any two linearly independent solutions to the associated homogeneous DE. Rearranging this latter equation yields

$$y(x) = c_1 y_1(x) + c_2 y_2(x) + y_p(x) = y_c(x) + y_p(x). \qquad \blacksquare$$

Example 2.1.3   Verify that $y_p(x) = \dfrac{1}{3} e^{5x}$ is a particular solution to the DE

$$y'' + y' - 6y = 8e^{5x},$$

and determine the general solution.

Solution   For the given function, we have

$$y_p'' + y_p' - 6y_p = \frac{25}{3} e^{5x} + \frac{5}{3} e^{5x} - 2e^{5x} = 8e^{5x}.$$

Hence, $y_p(x) = \dfrac{1}{3} e^{5x}$ is a solution to the given DE. We have seen in the previous example that the general solution to the associated homogeneous DE is

$$y_c(x) = c_1 e^{2x} + c_2 e^{-3x}.$$

So, Theorem 2.1.5 tells us that the general solution to the given DE is

$$y(x) = c_1 e^{2x} + c_2 e^{-3x} + \frac{1}{3} e^{5x}. \qquad \square$$

Our final result regarding nonhomogeneous DE once more stresses the importance of linearity.

Theorem 2.1.6:   If $y = y_{p_1}(x)$ and $y = y_{p_2}(x)$ are particular solutions to the DE

$$y'' + a_1(x)y + a_2(x)y = F_1(x), \quad y'' + a_1(x)y + a_2(x)y = F_2(x),$$

respectively, then a particular solution to the DE

$$y'' + a_1(x)y + a_2(x)y = F_1(x) + F_2(x)$$

is

$$y(x) = y_{p_1}(x) + y_{p_2}(x).$$

PROOF   If $y(x) = y_{p_1}(x) + y_{p_2}(x)$, then

$$y'' + a_1(x)y + a_2(x)y = (y_{p_1} + y_{p_2})'' + a_1(x)(y_{p_1} + y_{p_2})' + a_2(x)(y_{p_1} + y_{p_2})$$

$$= [y_{p_1}'' + a_1(x)y_{p_1}' + a_2(x)y_{p_1}] + [y_{p_2}'' + a_1(x)y_{p_2}' + a_2(x)y_{p_2}]$$

$$= F_1(x) + F_2(x). \qquad \blacksquare$$

Finally in this section, we establish Theorem 2.1.4. We begin with a result about the Wronskian of solutions to the homogeneous DE (2.1.2).

Theorem 2.1.7:   Let y_1, y_2 be two solutions to the DE (2.1.2) on an interval I, and let $W = W[y_1, y_2]$ denote their Wronskian. Then W is either identically zero on I, or W is never zero on I.

PROOF   Differentiating the expression for the Wronskian yields

$$W' = y_1 y_2'' + y_1' y_2' - y_1'' y_2 - y_1' y_2' = y_1 y_2'' - y_1'' y_2.$$

Since y_1 and y_2 are solutions to equation (2.1.2), we have

$$y_1'' + a_1(x)y_1' + a_2(x)y_1 = 0, \tag{2.1.9}$$

$$y_2'' + a_1(x)y_2' + a_2(x)y_2 = 0. \tag{2.1.10}$$

Multiplying equation (2.1.10) by y_1, equation (2.1.9) by y_2, and subtracting the resulting equations yields

$$W' + a_1(x)W = 0.$$

This is a first-order linear DE with integrating factor $e^{\int a_1(x)dx}$. Consequently,

$$W = ce^{-\int a_1(x)dx},$$

where c is a constant. Since the exponential term is never zero, it follows that W is either identically zero on I ($c = 0$) or never zero on I ($c \neq 0$). $\blacksquare$

PROOF OF THEOREM 2.1.4   Let y_1 and y_2 be two linearly independent solutions to the DE

$$y'' + a_1(x)y' + a_2(x)y = 0$$

on the interval I. We must establish that every solution to this DE is of the form

$$y(x) = c_1 y_1(x) + c_2 y_2(x).$$

From the existence and uniqueness theorem, it suffices to show that at every point $x_0 \in I$ the system of equations

$$c_1 y_1(x_0) + c_2 y_2(x_0) = y(x_0) \tag{2.1.11}$$

$$c_1 y_1'(x_0) + c_2 y_2'(x_0) = y'(x_0) \tag{2.1.12}$$

has a solution. Multiplying equation (2.1.11) by $y_2'(x_0)$, equation (2.1.12) by $y_2(x_0)$ and subtracting the resulting equations yields

$$c_1[y_2'(x_0)y_1(x_0) - y_2(x_0)y_1'(x_0)] = y_2'(x_0)\,y(x_0) - y_2(x_0)y'(x_0),$$

or equivalently,

$$c_1 W(x_0) = y_2'(x_0)\,y(x_0) - y_2(x_0)y'(x_0).$$

Since y_1 and y_2 are linearly independent on I, it follows that $W(x_0) \neq 0$, for any x_0 in I. Consequently,

$$c_1 = \frac{y_2'(x_0)\,y(x_0) - y_2(x_0)\,y'(x_0)}{W(x_0)}.$$

A similar computation yields

$$c_2 = \frac{y_1(x_0)\, y'(x_0) - y_1{}'(x_0)\, y(x_0)}{W(x_0)}.$$

Consequently, the system does indeed always have a solution, and the theorem is established.   ∎

EXERCISES 2.1

1. What is the unique solution to the IVP
$$y'' + x^2 y + e^x y = 0, \quad y(0) = 0, \quad y'(0) = 0?$$

For problems 2 and 3, show that the given functions are solutions to the given DE, and use them to determine the general solution of the DE.

2. $y'' - y' - 6y = 0$, $y_1(x) = e^{3x}$, $y_2(x) = e^{-2x}$.

3. $y'' + 4y = 0$, $y_1(x) = \cos 2x$, $y_2(x) = \sin 2x$.

4. Determine all values of the constant r such that $y(x) = e^{rx}$ is a solution to
$$y'' - 4y' + 3y = 0. \tag{4.1}$$
Hence, find the general solution to equation (4.1).

5. Determine all values of the constant r such that $y(x) = x^r$ is a solution to
$$x^2 y'' + 3xy' - 8y = 0, \quad x > 0. \tag{5.1}$$
Hence, find the general solution to equation (5.1) on $(0, \infty)$.

6. Consider the DE
$$y'' + y' - 6y = 18e^{3x}. \tag{6.1}$$

(a) Determine all values of the constant r such that $y(x) = e^{rx}$ is a solution to the associated homogeneous equation on $(-\infty, \infty)$. Hence, determine the complementary function for equation (6.1).

(b) Determine the value of the constant A_0 such that $y_p(x) = A_0 e^{3x}$ is a particular solution to equation (6.1).

(c) Use your results from (a) and (b) to determine the general solution to equation (6.1).

7. Consider the DE
$$y'' + y' - 2y = 4x^2. \tag{7.1}$$

(a) Determine all values of the constant r such that $y(x) = e^{rx}$ is a solution to the associated homogeneous equation. Hence determine the complementary function for equation (7.1).

(b) Determine the values of the constants a_0, a_1, a_2 such that
$$y_p(x) = a_0 + a_1 x + a_2 x^2$$
is a particular solution to equation (7.1).

(c) Use your results from (a) and (b) to determine the general solution to equation (7.1).

2.2   REDUCTION OF ORDER

We now introduce a powerful technique for determining the general solution to any second-order linear DE, assuming that we know just one solution to the associated homogeneous equation. The technique, usually referred to as **reduction of order**, is useful in the general derivation of solution techniques for linear DE.

Consider first the general second-order linear homogeneous DE
$$y'' + a_1(x)y' + a_2(x)y = 0, \tag{2.2.1}$$

where we assume that the functions a_1 and a_2 are continuous on an interval I. We know that the general solution to equation (2.2.1) is of the form

$$y(x) = c_1 y_1(x) + c_2 y_2(x),$$

where y_1 and y_2 are linearly independent solutions to equation (2.2.1) on I. Suppose that we have found one solution, say, $y = y_1(x)$. If $y = y_2(x)$ is a second solution to equation (2.2.1), then in order that y_1 and y_2 be linearly independent on I, the ratio y_2/y_1 must be nonconstant. We can therefore write

$$y_2(x) = u(x)y_1(x), \tag{2.2.2}$$

for some (nonconstant) function u. The following derivation establishes that $u(x)$ can, in theory, always be determined from equation (2.2.1). Differentiating (2.2.2) twice with respect to x yields

$$y_2{}' = u'y_1 + uy_1{}',$$

$$y_2{}'' = u''y_1 + 2u'y_1{}' + uy_1{}''.$$

Substituting into equation (2.2.1) gives

$$(u''y_1 + 2u'y_1{}' + uy_1{}'') + a_1(x)(u'y_1 + uy_1{}') + a_2(x)(uy_1) = 0,$$

so that (2.2.2) solves equation (2.2.1) provided u satisfies

$$u[y_1{}'' + a_1(x)y_1{}' + a_2(x)y_1] + u''y_1 + u'[2y_1{}' + a_1(x)y_1] = 0. \tag{2.2.3}$$

Since $y = y_1(x)$ is a solution to equation (2.2.1) the coefficient of u in this expression vanishes. Consequently equation (2.2.3) reduces to

$$u''y_1 + u'[2y_1{}' + a_1(x)y_1] = 0,$$

which is a first-order separable (and linear) DE for u'. We have therefore reduced the order of the DE, hence the name of the technique. Separating the variables in the preceding DE yields

$$\frac{u''}{u'} = -\left(\frac{2y_1{}'}{y_1} + a_1\right).$$

The functions appearing on the right-hand side are all known, and so we can formally integrate this equation to obtain

$$\ln |u'| = -2\ln |y_1| - \int^x a_1(s)\, ds + c,$$

which upon exponentiation gives

$$u'(x) = cy_1^{-2}(x) \exp\left(-\int^x a_1(s)\,ds\right), \tag{2.2.4}$$

where we have redefined the (nonzero) constant c. One more integration yields

$$u(x) = c\int^x y_1^{-2}(t) \exp\left(-\int^t a_1(s)\,ds\right)dt, \tag{2.2.5}$$

where we have set the integration constant to zero without loss of generality. We have therefore shown that

$$y_2(x) = y_1(x)u(x)$$

is a solution to equation (2.2.1) provided that u is given by (2.2.5). Further, from (2.2.4) we see that $u' \neq 0$, so that the y_1 and y_2 are linearly independent on I. We therefore have the following theorem.

Theorem 2.2.1:   If $y = y_1(x)$ is a solution to

$$y'' + a_1(x)y' + a_2(x)y = 0 \tag{2.2.6}$$

on an interval I, then a second linearly independent solution to equation (2.2.6) on I is

$$y_2(x) = cy_1 \int^x y_1^{-2}(t) \, \exp\left(-\int^t a_1(s)ds\right) dt$$

where c is a constant that can be chosen to have any convenient nonzero value.

REMARK   Do *not* memorize the preceding formula for y_2. Just remember the idea behind the technique, namely, that if $y = y_1(x)$ is a solution to equation (2.2.5) then a second linearly independent solution of the form $y_2(x) = y_1(x)u(x)$ can be determined by substitution into the DE.

Example 2.2.1   Find the general solution to

$$xy'' - 2y' + (2 - x)y = 0, \ x > 0, \tag{2.2.7}$$

given that one solution is $y_1(x) = e^x$.

Solution   We know that there is a second linearly independent solution of the form

$$y_2(x) = y_1(x)u(x) = e^x u(x). \tag{2.2.8}$$

Differentiating y_2 twice with respect to x yields

$$y_2' = e^x(u' + u),$$
$$y_2'' = e^x(u'' + 2u' + u).$$

Substituting into equation (2.2.7) we find that u must satisfy

$$x(u'' + 2u' + u) - 2(u' + u) + (2 - x)u = 0,$$

which simplifies to

$$xu'' + 2u'(x - 1) = 0.$$

Separating the variables yields

$$\frac{u''}{u'} = 2(x^{-1} - 1).$$

By integrating, we obtain

$$\ln |u'| = 2(\ln x - x) + c,$$

which can be written as

$$u' = c_1 x^2 e^{-2x}.$$

Integrating once more and setting the resulting integration constant to zero, gives

$$u(x) = -\frac{1}{4} c_1 e^{-2x}(1 + 2x + 2x^2).$$

Substituting into (2.2.8) yields the second linearly independent solution

$$y_2(x) = e^{-x}(1 + 2x + 2x^2),$$

where we have set $c_1 = -4$ without loss of generality. Consequently, the general solution to equation (2.2.7) is

$$y(x) = c_1 e^x + c_2 e^{-x}(1 + 2x + 2x^2). \qquad \square$$

The reduction of order technique can also be used to determine the general solution to a second-order *nonhomogeneous* linear DE, assuming that we know one solution to the associated homogeneous equation. Rather than derive the general result, which we leave as an exercise, we will illustrate with an example.

Example 2.2.2   Find the general solution to

$$x^2 y'' + 3xy' + y = 4 \ln x, \quad x > 0 \qquad (2.2.9)$$

given that one solution to the associated homogeneous equation is $y(x) = x^{-1}$.

Solution   We try for a solution of the form

$$y(x) = x^{-1}u(x), \qquad (2.2.10)$$

where $u(x)$ is to be determined. Differentiating y twice yields

$$y' = x^{-1}u' - x^{-2}u, \quad y'' = x^{-1}u'' - 2x^{-2}u' + 2x^{-3}u.$$

Substituting into equation (2.2.9) and collecting terms we obtain the following DE for u:

$$u'' + x^{-1}u' = 4x^{-1}\ln x. \qquad (2.2.11)$$

To facilitate the integration, we let

$$v = u', \qquad (2.2.12)$$

in which case equation (2.2.11) can be written as the first-order linear DE

$$v' + x^{-1}v = 4x^{-1}\ln x. \qquad (2.2.13)$$

An integrating factor for this linear equation is $I = e^{\int x^{-1}dx} = x$, so that equation (2.2.13) can be written in the equivalent form

$$\frac{d}{dx}(xv) = 4 \ln x.$$

Integrating both sides with respect to x yields

$$xv = 4x(\ln x - 1) + c_1,$$

where c_1 is a constant. Thus,

$$v(x) = 4(\ln x - 1) + c_1 x^{-1}.$$

Substituting into (2.2.12), we have

$$u'(x) = 4(\ln x - 1) + c_1 x^{-1},$$

which can be integrated directly to obtain

$$u(x) = 4x(\ln x - 2) + c_1 \ln x + c_2,$$

where c_2 is another integration constant. Inserting this expression for u into (2.2.10) yields

$$y(x) = 4(\ln x - 2) + c_1 x^{-1} \ln x + c_2 x^{-1}.$$

Is this the general solution to equation (2.2.9)? The answer is *yes*. The complementary function is

$$y_c(x) = c_1 x^{-1} \ln x + c_2 x^{-1},$$

whereas

$$y_p(x) = 4(\ln x - 2)$$

is a particular solution.

EXERCISES 2.2

For problems 1–6, y_1 is a solution to the given DE. Use the method of reduction of order to determine a second linearly independent solution.

1. $x^2 y'' - 3xy' + 4y = 0$, $x > 0$, $y_1(x) = x^2$.

2. $x^2 y'' - 2xy' + (x^2 + 2)y = 0$, $x > 0$, $y_1(x) = x \sin x$.

3. $xy'' + (1 - 2x)y' + (x - 1)y = 0$, $x > 0$, $y_1(x) = e^x$.

4. $y'' - x^{-1}y' + 4x^2 y = 0$, $x > 0$, $y_1(x) = \sin(x^2)$.

5. $(1 - x^2)y'' - 2xy' + 2y = 0$, $-1 < x < 1$, $y_1(x) = x$,

6. $4x^2 y'' + 4xy' + (4x^2 - 1)y = 0$, $x > 0$, $y_1(x) = x^{-1/2} \sin x$.

7. Consider the *Cauchy–Euler equation*

$$x^2 y'' - (2m - 1)xy' + m^2 y = 0, \quad x > 0, \quad (7.1)$$

where m is a constant.

(a) Determine a particular solution to equation (7.1) of the form $y_1(x) = x^r$.

(b) Use your solution from (a) and the method of reduction of order to obtain a second linearly independent solution.

8. Determine the values of the constants a_0, a_1, and a_2 such that

$$y(x) = a_0 + a_1 x + a_2 x^2$$

is a solution to

$$(4 + x^2)y'' - 2y = 0,$$

and use the reduction of order technique to find a second linearly independent solution.

9. Consider the DE

$$xy'' - (\alpha x + \beta)y' + \alpha\beta y = 0, \; x > 0, \quad (9.1)$$

where α and β are constants.

(a) Show that $y_1(x) = e^{\alpha x}$ is a solution to equation (9.1).

(b) Use reduction of order to derive the second linearly independent solution

$$y_2(x) = e^{\alpha x} \int x^\beta e^{-\alpha x}\, dx.$$

(c) In the particular case when $\alpha = 1$ and β is a nonnegative integer, show that a second linearly independent solution to equation (9.1) is

$$y_2(x) = 1 + x + \frac{1}{2!}x^2 + \cdots + \frac{1}{\beta!}x^\beta.$$

For problems 10–15, y_1 is a solution to the associated homogeneous equation. Use the method of reduction of order to determine the general solution to the given DE.

10. $y'' - 6y' + 9y = 15 e^{3x} x^{1/2}$, $x > 0$, $y_1(x) = e^{3x}$.

11. $y'' - 4y' + 4y = 4e^{2x} \ln x$, $x > 0$, $y_1(x) = e^{2x}$.

12. $4x^2 y'' + y = x^{1/2} \ln x$, $x > 0$, $y_1(x) = x^{1/2}$.

13. $y'' + y = \csc x$, $0 < x < \pi$, $y_1(x) = \sin x$.

14. $xy'' - (2x + 1)y' + 2y = 8x^2 e^{2x}$, $x > 0$, $y_1(x) = e^{2x}$.

15. $x^2 y'' - 3xy' + 4y = 8x^4$, $x > 0$, $y_1(x) = x^2$.

16. Consider the DE

$$y'' + p(x)y' + q(x)y = r(x), \quad (16.1)$$

where p, q, and r are continuous on an interval I. If $y = y_1(x)$ is a solution to the associated homogeneous equation, show that $y_2 = u(x)y_1(x)$ is a solution to equation (16.1) provided $v = u'$ is a solution to the linear DE

$$v' + \left(2\frac{y_1'}{y_1} + p \right) v = \frac{r}{y_1}.$$

Express the solution to equation (16.1) in terms of integrals. Identify two linearly independent solutions to the associated homogeneous equation and a particular solution to equation (16.1).

2.3   SECOND-ORDER HOMOGENEOUS CONSTANT COEFFICIENT LINEAR DE

In general, it is not possible to determine solutions to linear DE in terms of elementary functions, although representation of solutions in terms of some type of convergent infinite series is often possible. (See Chapter 10.) In the next few sections, we will restrict our attention to the simple, but important, case of DE of the form

$$y'' + a_1 y' + a_2 y = 0, \tag{2.3.1}$$

where a_1 and a_2 are constants. Naturally associated with the DE (2.3.1) is the real polynomial

$$P(r) = r^2 + a_1 r + a_2.$$

We refer to $P(r)$ as the **auxiliary polynomial**, and the quadratic equation $P(r) = 0$ is called the **auxiliary equation**. Before determining the solution to the DE (2.3.1), we recall two properties of the complex exponential function. (See Appendix 1 for a fuller discussion of complex exponential functions.)

1. Euler's formula:

$$e^{(a+ib)x} = e^{ax}(\cos bx + i\sin bx). \tag{2.3.2}$$

2. If $r = a + ib$, then

$$\frac{d}{dx}(e^{rx}) = re^{rx}. \tag{2.3.3}$$

The basic result that enables us to solve equation (2.3.1) is given in the following theorem:

Theorem 2.3.1: $y(x) = e^{rx}$ is a solution to the DE (2.3.1) whenever r is a real or complex root of the auxiliary equation $P(r) = 0$.

PROOF   Substituting $y(x) = e^{rx}$ into equation (2.3.1) using equation (2.3.3) yields

$$y'' + a_1 y' + a_2 y = r^2 e^{rx} + a_1 r e^{rx} + a_2 e^{rx} = e^{rx}(r^2 + a_1 r + a_2) = e^{rx}P(r).$$

So, if $P(r) = 0$, then $y(x) = e^{rx}$ solves the DE (2.3.1).   ■

Using Theorem 2.3.1 we can now find the general solution to any DE of the form

$$y'' + a_1 y' + a_2 y = 0, \tag{2.3.4}$$

where a_1 and a_2 are constants. The auxiliary equation is $r^2 + a_1 r + a_2 = 0$, with roots

$$r_1 = \frac{-a_1 + \sqrt{a_1{}^2 - 4a_2}}{2}, \qquad r_2 = \frac{-a_1 - \sqrt{a_1{}^2 - 4a_2}}{2}. \tag{2.3.5}$$

The next three subcases arise depending on the nature of these roots.

1.   Real and Distinct Roots $(a_1{}^2 - 4a_2 > 0)$

If r_1 and r_2 are real and distinct $(r_1 \neq r_2)$, it follows directly from Theorem 2.3.1 that two solutions to equation (2.3.4) are $y_1(x) = e^{r_1 x}$ and $y_2(x) = e^{r_2 x}$. Since these functions are not proportional on any interval we can conclude (Theorem 2.1.4) that the general solution to the DE in this case is

$$y(x) = c_1 e^{r_1 x} + c_2 e^{r_2 x}.$$

2.   Repeated Real Roots $(a_1{}^2 - 4a_2 = 0)$

The auxiliary polynomial can be factored as

$$P(r) = (r - r_1)^2 = r^2 - 2r_1 r + r_1{}^2,$$

where, from (2.3.5),

$$r_1 = -\frac{1}{2} a_1. \tag{2.3.6}$$

Consequently, one solution to the DE (2.3.4) is

$$y_1(x) = e^{r_1 x}.$$

To obtain the general solution to the DE, we need a second linearly independent solution. This is where the reduction of order method can be applied. If we let y_2 denote a second such solution, then we can write

$$y_2(x) = u(x) y_1(x) = u(x) e^{r_1 x}$$

for some nonconstant function $u(x)$. Differentiating y_2 yields

$$y_2{}'(x) = e^{r_1 x}(u' + r_1 u), \quad y_2{}''(x) = e^{r_1 x}(u'' + 2r_1 u' + r_1{}^2 u),$$

which we substitute into equation (2.3.4) to obtain

$$e^{r_1 x}[(u'' + 2r_1 u' + r_1{}^2 u) + a_1(u' + r_1 u) + a_2 u] = 0,$$

or equivalently,

$$u'' + u'(2r_1 + a_1) + u(r_1{}^2 + a_1 r_1 + a_2) = 0.$$

Since r_1 is a zero of the auxiliary polynomial, the terms multiplying u in the previous equation vanish. Furthermore, using (2.3.6), the terms multiplying u' vanish also. Consequently, we have

$$u'' = 0,$$

which can be integrated directly to obtain

$$u(x) = Ax + B,$$

where A and B are arbitrary constants. The simplest choice to make is $A = 1$, $B = 0$, in which case we have

$$y_2(x) = xe^{r_1 x}.$$

Since y_1 and y_2 are not proportional on any interval, we can conclude (Theorem 2.1.4) that the general solution to the DE is

$$y(x) = c_1 e^{r_1 x} + c_2 x e^{r_1 x}.$$

3.   Complex Conjugate Roots $(a_1^2 - 4a_2 < 0)$

The roots of the auxiliary equation can be written as $r = a \pm ib$, where $b \neq 0$. Hence two complex-valued solutions to the DE (2.3.4) are

$$w_1(x) = e^{(a+ib)x} = e^{ax}(\cos bx + i\sin bx),$$

$$w_2(x) = e^{(a-ib)x} = e^{ax}(\cos bx - i\sin bx),$$

where we have used Euler's formula (2.3.2). We require two linearly independent real-valued solutions to the DE. According to the principle of superposition, any linear combination of w_1 and w_2 will also solve the DE. In particular, defining $y_1(x)$ and $y_2(x)$ by

$$y_1(x) = \frac{1}{2}[w_1(x) + w_2(x)] = e^{ax}\cos bx,$$

$$y_2(x) = \frac{1}{2i}[w_1(x) - w_2(x)] = e^{ax}\sin bx,$$

respectively, yields an appropriate pair of *linearly independent* solutions. Consequently, the general solution to equation (2.3.4) in this case is

$$y(x) = c_1 e^{ax}\cos bx + c_2 e^{ax}\sin bx.$$

We write this solution in the equivalent form

$$y(x) = e^{ax}(c_1 \cos bx + c_2 \sin bx).$$

The preceding discussion is summarized in Table 2.3.1.

TABLE   2.3.1

Summary

Linearly independent solutions to
$$y'' + a_1 y' + a_2 y = 0$$
are determined as follows:

Roots of the Auxiliary Equation	Linearly Independent Solutions to DE
Real distinct: $r_1 \neq r_2$	$y_1(x) = e^{r_1 x}, y_2(x) = e^{r_2 x}$.
Real repeated: $r_1 = r_2$	$y_1(x) = e^{r_1 x}, y_2(x) = xe^{r_1 x}$.
Complex conjugate: $r_1 = a + ib, r_2 = a - ib$	$y_1(x) = e^{ax}\cos bx, y_2(x) = e^{ax}\sin bx$.

Example 2.3.1   Determine the general solution to $y'' - y' - 2y = 0$.

Solution   The auxiliary polynomial is $P(r) = r^2 - r - 2 = (r - 2)(r + 1)$. Therefore, the auxiliary equation has roots $r_1 = 2$, $r_2 = -1$, so that two linearly independent solutions to the given DE are

$$y_1(x) = e^{2x}, \quad y_2(x) = e^{-x}.$$

Hence, the general solution to the DE is

$$y(x) = c_1 e^{2x} + c_2 e^{-x}.$$

Some representative solution curves are sketched in Figure 2.3.1.

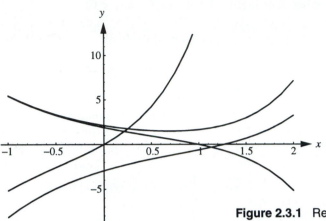

Figure 2.3.1 Representative solution curves for the DE in Example 2.3.1.

Example 2.3.2   Find the general solution to $y'' + 6y' + 25y = 0$.

Solution   The auxiliary equation is $r^2 + 6r + 25 = 0$, with roots $r = -3 \pm 4i$. Consequently, two linearly independent real-valued solutions to the DE are

$$y_1(x) = e^{-3x}\cos 4x, \quad y_2(x) = e^{-3x}\sin 4x,$$

and the general solution to the DE is

$$y(x) = e^{-3x}(c_1\cos 4x + c_2\sin 4x).$$

We see that due to the presence of the trigonometric functions the solutions are oscillatory. The *negative* exponential term implies that the amplitude of the oscillations decays as x increases. Some representative solution curves are given in Figure 2.3.2.

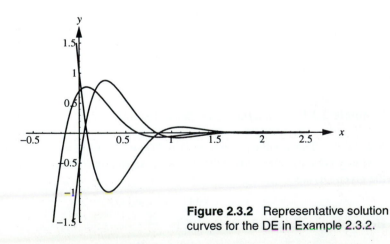

Figure 2.3.2 Representative solution curves for the DE in Example 2.3.2.

Example 2.3.3 Solve the IVP

$$y'' + 4y' + 4y = 0, \quad y(0) = 1, \ y'(0) = 4.$$

Solution The auxiliary polynomial is $P(r) = r^2 + 4r + 4 = (r + 2)^2$. Thus, $r = -2$ is a repeated root of the auxiliary equation, and therefore two linearly independent solutions to the given DE are

$$y_1(x) = e^{-2x}, \quad y_2(x) = xe^{-2x}.$$

Consequently, the general solution is

$$y(x) = e^{-2x}(c_1 + c_2 x).$$

Due to the presence of the negative exponential term, it follows that all solutions approach zero as $x \to \infty$. The initial condition $y(0) = 1$ implies that $c_1 = 1$. Thus,

$$y(x) = e^{-2x}(1 + c_2 x).$$

Differentiating this expression yields

$$y'(x) = -2e^{-2x}(1 + c_2 x) + c_2 e^{-2x},$$

so that the second initial condition requires $c_2 = 6$. Hence, the unique solution to the given IVP is

$$y(x) = e^{-2x}(1 + 6x).$$

Some solution curves are sketched in Figure 2.3.3. Which one corresponds to the given initial conditions?

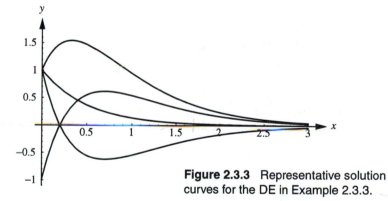

Figure 2.3.3 Representative solution curves for the DE in Example 2.3.3.

EXERCISES 2.3

For problems 1–12, find the general solution to the given DE.

1. $y'' - y' - 2y = 0.$

2. $y'' - 6y' + 9y = 0.$

3. $y'' + 6y' + 25y = 0.$

4. $y'' - 4y' - 5y = 0.$

5. $y'' + 4y' + 4y = 0.$

6. $y'' - 6y' + 34y = 0.$

7. $y'' + 10y' + 25y = 0.$

8. $y'' - 2y = 0.$

9. $y'' + 8y' + 20y = 0.$

10. $y'' + 2y' + 2y = 0.$

11. $y'' - 2y' - 8y = 0$.

12. $y'' - 14y' + 58y = 0$.

For problems 13–15, solve the given IVP.

13. $y'' + y' - 6y = 0$, $y(0) = 3$, $y'(0) = 1$.

14. $y'' - 8y' + 16y = 0$, $y(0) = 2$, $y'(0) = 7$.

15. $y'' - 4y' + 5y = 0$, $y(0) = 3$, $y'(0) = 5$.

16. Solve the IVP

$$y'' - 2my' + (m^2 + k^2)y = 0, \quad y(0) = 0, \quad y'(0) = k,$$

where m and k are positive constants.

17. Find the general solution to

$$y'' - 2my' + (m^2 - k^2)y = 0,$$

where m and k are positive constants. Show that the solution can be written in the form:

$$y(x) = e^{mx}(c_1 \cosh kx + c_2 \sinh kx).$$

18. An object of mass m is attached to one end of a spring, and the other end of the unstretched spring is attached to a fixed wall. (See Figure 2.3.4.) The object is pulled to the right a distance y_0 and released from rest. Assuming that there is a damping force that is proportional to the velocity of the object, an application of Hooke's law and Newton's second law of motion yields an IVP that can be written in the form (using appropriate units)

$$\frac{d^2y}{dt^2} + 2c\frac{dy}{dt} + k^2y = 0, \quad y(0) = y_0, \quad \frac{dy}{dt}(0) = 0,$$

where $y(t)$ denotes the displacement of the spring from its equilibrium position at time t, and c and k are positive constants.

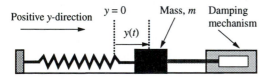

Figure 2.3.4 The spring-mass system considered in problem 18.

(a) Assuming that $c^2 < k^2$ solve the preceding IVP to obtain

$$y(t) = \left(\frac{y_0}{\omega}\right)e^{-ct}(\omega \cos \omega t + c \sin \omega t),$$

where $\omega = \sqrt{k^2 - c^2}$.

(b) Show that the solution in (a) can be written in the form

$$y(t) = \left(\frac{ky_0}{\omega}\right)e^{-ct} \sin(\omega t + \phi), \text{ where } \phi = \tan^{-1}(\omega/c),$$

and then sketch the graph of y against t. Is the predicted motion reasonable?

19. Consider the *partial differential equation* (Laplace's equation)

$$\frac{\partial^2 u}{\partial x^2} + \frac{\partial^2 u}{\partial y^2} = 0. \tag{19.1}$$

(a) Show that the substitution

$$u(x, y) = e^{x/\alpha}f(\xi),$$

where $\xi = \beta x - \alpha y$, (and α and β are positive constants) reduces equation (19.1) to the DE

$$\frac{d^2f}{d\xi^2} + 2p\frac{df}{d\xi} + \frac{q}{\alpha^2}f = 0, \tag{19.2}$$

where

$$p = \frac{\beta}{\alpha(\alpha^2 + \beta^2)}, \quad q = \frac{1}{(\alpha^2 + \beta^2)}. \tag{19.3}$$

(HINT: Use the chain rule, for example

$$\frac{\partial f}{\partial x} = \frac{df}{d\xi}\frac{\partial \xi}{\partial x}.)$$

(b) Solve equation (19.2), and hence, find the corresponding solution to (19.1). (HINT: In solving equation (19.2), you will need to use (19.3) in order to obtain a simple form of solution.)

20. Consider the DE

$$y'' + a_1 y' + a_2 y = 0, \tag{20.1}$$

where a_1 and a_2 are real constants.

(a) If the auxiliary equation has real roots r_1 and r_2, what conditions on these roots would guarantee that every solution to equation (20.1) satisfies

$$\lim_{x \to +\infty} y(x) = 0?$$

(b) If the auxiliary equation has complex conjugate roots $r = a \pm ib$, what conditions on these roots would guarantee that every solution to equation (20.1) satisfies

$$\lim_{x \to +\infty} y(x) = 0?$$

(c) If a_1 and a_2 are positive, prove that $\lim_{x \to +\infty} y(x) = 0$, for every solution to equation (20.1).

(d) If $a_1 > 0$ and $a_2 = 0$, prove that all solutions to equation (20.1) approach a constant value as $x \to +\infty$.

(e) If $a_1 = 0$ and $a_2 > 0$, prove that all solutions   to equation (20.1) remain bounded as $x \to +\infty$.

2.4   THE METHOD OF UNDETERMINED COEFFICIENTS

In Section 2.3, we have seen how to determine the general solution to any second-order linear homogeneous constant coefficient DE. We now turn our attention to the corresponding nonhomogeneous DE

$$y'' + a_1 y' + a_2 y = F(x) \tag{2.4.1}$$

where a_1 and a_2 are constants. Since we can always determine the complementary function for this DE, Theorem 2.1.5 tells us that we need then find only one particular solution to equation (2.4.1) in order to obtain its general solution. Although it is possible to derive a general method for obtaining such a particular solution (see Section 2.8), in this section we will consider a restricted method that is easier to apply and is effective in many situations of physical interest. The method will work for DE (2.4.1), provided the nonhomogeneous term $F(x)$ has one of the following forms

(1) Ae^{ax}; (2) $A \cos bx + B \sin bx$; (3) Ax^k; (4) Sums or products of (1)–(3).

The idea behind the technique is to guess the general form of an appropriate particular solution up to some arbitrary constants (coefficients) and then substitute the proposed function into the DE to determine the values of the constants that do indeed yield a solution. We will introduce the technique with an example.

Example 2.4.1   Determine the general solution to

$$y'' - y = 16e^{3x}. \tag{2.4.2}$$

Solution   We first obtain the complementary function. The auxiliary polynomial is

$$P(r) = r^2 - 1 = (r - 1)(r + 1),$$

so that

$$y_c(x) = c_1 e^x + c_2 e^{-x}.$$

We now need to determine a particular solution. Since taking derivatives of e^{3x} gives back multiples of e^{3x}, we might suspect that there is a solution to equation (2.4.2) of the form

$$y_p(x) = A_0 e^{3x}, \tag{2.4.3}$$

for an appropriately chosen value of the constant A_0. We call (2.4.3) a **trial solution** for the DE (2.4.2). We now substitute the proposed trial solution into the given DE to determine the appropriate value of A_0. Differentiating (2.4.3) twice yields

$$y_p'(x) = 3A_0 e^{3x}, \quad y_p''(x) = 9A_0 e^{3x},$$

so that $y_p(x) = A_0 e^{3x}$ is a solution to equation (2.4.2) if and only if A_0 satisfies

$$(9A_0 - A_0)e^{3x} = 16e^{3x}.$$

Thus,

$$A_0 = 2$$

and a particular solution to equation (2.4.2) is

$$y_p(x) = 2e^{3x}.$$

It follows from Theorem 2.1.5, that the general solution to the DE is

$$y(x) = y_c(x) + y_p(x) = c_1 e^x + c_2 e^{-x} + 2e^{3x}. \qquad \qquad \square$$

More generally, consider the DE

$$y'' + a_1 y' + a_2 y = Ae^{ax}. \tag{2.4.4}$$

Arguing as we did in Example 2.4.1, we might suspect that there is a particular solution to this DE of the form

$$y_p(x) = A_0 e^{ax}, \tag{2.4.5}$$

where the undetermined coefficient A_0 can be determined by substitution into the DE. This is indeed the case, *provided that $r = a$ is not a root of the auxiliary equation*. If $r = a$ is a root of the auxiliary equation, then e^{ax} solves the associated homogeneous equation, and thus, substitution of the proposed trial solution (2.4.5) into the left-hand side of equation (2.4.4) would produce zero, not Ae^{ax}. When this happens, we need to modify the trial solution. The appropriate modification depends on whether $r = a$ is a simple root or a repeated root of the auxiliary equation. The modification is dealt with in the next theorem.

Theorem 2.4.1:   Consider the DE

$$y'' + a_1 y + a_2 y = Ae^{ax}, \tag{2.4.6}$$

where a_1, a_2, A, and a are constants.

1. If $r = a$ is *not* a root of the auxiliary equation, then there exists a particular solution to equation (2.4.6) of the form

$$y_p(x) = A_0 e^{ax}.$$

2. If $r = a$ is a **simple** root of the auxiliary equation, then there exists a particular solution to equation (2.4.6) of the form

$$y_p(x) = A_0 x e^{ax}.$$

3. If $r = a$ is a **repeated** root of the auxiliary equation, then there exists a particular solution to equation (2.4.6) of the form

$$y_p(x) = A_0 x^2 e^{ax}.$$

PROOF

1.   If $y_p(x) = A_0 e^{ax}$, then $y_p'(x) = A_0 a e^{ax}$ and $y_p''(x) = A_0 a^2 e^{ax}$. Substituting into equation (2.4.6) yields

$$A_0(a^2 + a_1 a + a_2) = A.$$

That is,

$$P(a)A_0 = A.$$

Since $P(a) \neq 0$ by assumption, we can divide the preceding equation by $P(a)$ to determine A_0.

2. To establish (2), we need only verify that the proposed particular solution does indeed solve equation (2.4.6) when A_0 is chosen appropriately. If $y_p(x) = A_0 x e^{ax}$, then $y_p'(x) = A_0 e^{ax}(ax + 1)$ and $y_p''(x) = A_0 e^{ax}(a^2 x + 2a)$. Substitution into equation (2.4.6) yields

$$A_0[(a^2 x + 2a) + a_1(ax + 1) + a_2 x] = A,$$

or equivalently,

$$A_0[(a^2 + a_1 a + a_2)x + (a_1 + 2a)] = A. \tag{2.4.7}$$

Since $P(a) = 0$, the coefficient of x vanishes. Furthermore, since a is not a repeated root of the auxiliary equation, it follows that $a \neq -a_1/2$. Consequently, we can solve equation (2.4.7) to obtain

$$A_0 = \frac{1}{a_1 + 2a} A.$$

3. In this case we have $a = -a_1/2$. Consequently, the trial solutions in (1) and (2) cannot yield a solution to the DE (2.4.6). However, if $y_p(x) = A_0 x^2 e^{ax}$, then

$$y_p'(x) = A_0 e^{ax}(ax^2 + 2x), \quad y_p''(x) = A_0 e^{ax}(a^2 x^2 + 4ax + 2)$$

and substitution into (2.3.6) yields, after simplification,

$$A_0[x^2(a^2 + aa_1 + a_2) + 2x(a_1 + 2a) + 2] = A.$$

The two terms in parentheses vanish, since in this case, $r = a$ is a *repeated* root of the auxiliary equation. Consequently, we are left with

$$2A_0 = A,$$

so that

$$A_0 = \frac{1}{2}A. \qquad\blacksquare$$

Example 2.4.2   Find the general solution to

$$y'' - 3y' - 4y = 15e^{4x}. \tag{2.4.8}$$

Solution   The auxiliary polynomial for the given equation is $P(r) = (r - 4)(r + 1)$, so that

$$y_c(x) = c_1 e^{-x} + c_2 e^{4x}.$$

The usual trial solution corresponding to the nonhomogeneous term $F(x) = 15e^{4x}$ is

$$y_p(x) = A_0 e^{4x}.$$

However, since $r = 4$ is a simple root of the auxiliary equation, we must use the modified trial solution

$$y_p(x) = A_0 x e^{4x}.$$

To determine A_0, we substitute y_p into equation (2.4.8). Differentiating y_p twice yields

$$y_p{'}(x) = A_0 e^{4x}(4x + 1), \quad y_p{''}(x) = A_0 e^{4x}(16x + 8).$$

Substitution into equation (2.4.8) gives the following equation for A_0:

$$A_0 e^{4x}[(16x + 8) - 3(4x + 1) - 4x] = 15e^{4x}.$$

That is,

$$5A_0 = 15,$$

so that

$$A_0 = 3.$$

Consequently, a particular solution to equation (2.4.8) is

$$y_p(x) = 3xe^{4x},$$

and hence, the general solution is

$$y(x) = c_1 e^{-x} + c_2 e^{4x} + 3xe^{4x}. \qquad \square$$

Now consider the DE

$$y'' + a_1 y' + a_2 y = A \cos bx + B \sin bx. \qquad (2.4.9)$$

Since derivatives of $\sin bx$ and $\cos bx$ give back multiples of $\sin bx$ and $\cos bx$, it is at least plausible that an appropriate trial solution is

$$y_p(x) = A_0 \cos bx + B_0 \sin bx.$$

Provided that $r = ib$ is not a root of the auxiliary equation, the undetermined coefficients A_0 and B_0 can indeed be obtained by substitution into the DE. If, however, $r = ib$ is a root of the auxiliary equation, then the proposed trial solution coincides with the complementary function and so cannot produce a particular solution to equation (2.4.9). In this case, we must modify the usual trial solution by multiplication by x. These results are summarized in the next theorem.

Theorem 2.4.2:   Consider the DE

$$y'' + a_1 y + a_2 y = A \cos bx + B \sin bx \qquad (2.4.10)$$

where a_1, a_2, A, B, and b are constants.

1. If $r = ib$ is *not* a root of the auxiliary equation, then there exists a particular solution to equation (2.4.10) of the form

$$y_p(x) = A_0 \cos bx + B_0 \sin bx.$$

2. If $r = ib$ is a root of the auxiliary equation, then there exists a particular solution to equation (2.4.10) of the form

$$y_p(x) = x(A_0 \cos bx + B_0 \sin bx).$$

PROOF   The proofs of these results are left as exercises.   ■

Example 2.4.3   Solve the IVP

$$y'' - y' - 2y = 10 \sin x, \qquad (2.4.11)$$
$$y(0) = 0, \quad y'(0) = 1.$$

Solution   The auxiliary polynomial is

$$P(r) = r^2 - r - 2 = (r - 2)(r + 1),$$

so that

$$y_c(x) = c_1 e^{2x} + c_2 e^{-x}.$$

In this case, an appropriate trial solution is

$$y_p(x) = A_0 \sin x + A_1 \cos x.$$

Substituting this trial solution into equation (2.4.11) yields

$$(-A_0 \sin x - A_1 \cos x) - (A_0 \cos x - A_1 \sin x) - 2(A_0 \sin x + A_1 \cos x) = 10 \sin x.$$

That is,

$$(-3A_0 + A_1) \sin x - (A_0 + 3A_1) \cos x = 10 \sin x.$$

This equation is satisfied for all x if and only if

$$-3A_0 + A_1 = 10, \quad A_0 + 3A_1 = 0.$$

The unique solution to this system of equations is

$$A_0 = -3, \quad A_1 = 1,$$

so that a particular solution to equation (2.4.11) is

$$y_p(x) = -3 \sin x + \cos x.$$

Consequently (2.4.11) has general solution

$$y(x) = c_1 e^{2x} + c_2 e^{-x} - 3 \sin x + \cos x. \tag{2.4.12}$$

We now impose the given initial conditions. From (2.4.12), $y(0) = 0$ if and only if

$$c_1 + c_2 = -1, \tag{2.4.13}$$

whereas $y'(0) = 1$ if and only if

$$2c_1 - c_2 = 4. \tag{2.4.14}$$

Solving equations (2.4.13) and (2.4.14) yields

$$c_1 = 1, \quad c_2 = -2,$$

so that, from (2.4.12), the unique solution to the given IVP is

$$y(x) = e^{2x} - 2e^{-x} - 3 \sin x + \cos x. \qquad \square$$

Next consider the DE

$$y'' + a_1 y' + a_2 y = Ax^k$$

where k is a nonnegative integer. Our first guess for a trial solution might be $y_p(x) = A_0 x^k$. However, since taking derivatives of this function reduces the power of x, substitution into the left-hand side of the DE would, in general, not yield a solution. After a little thought, the following trial solution should seem reasonable

$$y_p(x) = A_0 + A_1 x + \cdots + A_k x^k.$$

Thus, in general we must consider all powers of x up to the kth power. However, care is needed if $r = 0$ is a root of the auxiliary equation. In this case, the trial solution has terms in common with the complementary function and once more modification is required to eliminate such coincidence. The appropriate modification is to multiply the usual trial solution by x or x^2 depending on whether $r = 0$ is a simple or repeated root of the auxiliary equation respectively. We therefore have the next theorem.

Theorem 2.4.3:   Consider the DE

$$y'' + a_1 y + a_2 y = Ax^k, \qquad (2.4.15)$$

where a_1, a_2, and A are constants, and k is a nonnegative integer..

1. If $r = 0$ is *not* a root of the auxiliary equation, then there exists a particular solution to equation (2.4.15) of the form

$$y_p(x) = A_0 + A_1 x + \cdots + A_k x^k.$$

2. If $r = 0$ is a *simple* root of the auxiliary equation, then there exists a particular solution to equation (2.4.15) of the form

$$y_p(x) = x(A_0 + A_1 x + \cdots + A_k x^k).$$

3. If $r = 0$ is a *repeated* root of the auxiliary equation, then there exists a particular solution to equation (2.4.15) of the form

$$y_p(x) = x^2(A_0 + A_1 x + \cdots + A_k x^k).$$

PROOF   Once more, verifications of these statements are left as exercises.   ■

Example 2.4.4   Determine the general solution to

$$y'' + 5y' = 4x^2. \qquad (2.4.16)$$

Solution   The auxiliary equation is $r^2 + 5r = 0$, with roots $r = 0, -5$. Consequently,

$$y_c(x) = c_1 + c_2 e^{-5x}.$$

The usual trial solution corresponding to the nonhomogeneous term $F(x) = 4x^2$ is

$$y_p(x) = A_0 + A_1 x + A_2 x^2.$$

However, since $r = 0$ is a simple root of the auxiliary equation, we must use the modified trial solution

$$y_p(x) = x(A_0 + A_1 x + A_2 x^2).$$

Inserting this expression into equation (2.4.16) yields

$$2A_1 + 6A_2 x + 5(A_0 + 2A_1 x + 3A_2 x^2) = 4x^2,$$

which can be written as

$$5A_0 + 2A_1 + (10A_1 + 6A_2)x + 15A_2 x^2 = 4x^2.$$

Consequently, the coefficients must be chosen to satisfy

$$5A_0 + 2A_1 = 0, \quad 10A_1 + 6A_2 = 0, \quad 15A_2 = 4.$$

Solving these equations gives

$$A_0 = \frac{8}{125}, \quad A_1 = -\frac{4}{25}, \quad A_2 = \frac{4}{15},$$

so that a particular solution to equation (2.4.16) is

$$y_p(x) = x(\frac{8}{125} - \frac{4}{25}x + \frac{4}{15}x^2) = \frac{4}{375}x(6 - 15x + 25x^2).$$

The general solution to the DE is therefore

$$y(x) = c_1 + c_2e^{-5x} + \frac{4}{375}x(6 - 15x + 25x^2).$$ ❑

If the nonhomogeneous term in our DE is a product of the forms already discussed then an appropriate trial solution is obtained by taking the corresponding product of trial solutions. The results are summarized in the next theorem.

Theorem 2.4.4:   Consider the DE

$$y'' + a_1y + a_2y = x^ke^{ax}(A \cos bx + B \sin bx) \qquad (2.4.17)$$

where a_1, a_2, a, b, A, and B are constants, and k is a nonnegative integer.

1. If $r = a + ib$ is not a root of the auxiliary equation, then there exists a particular solution to equation (2.4.17) of the form

$$y_p(x) = e^{ax}[(A_0\cos bx + B_0\sin bx) + x(A_1\cos bx + B_1\sin bx) + \cdots$$
$$+ x^k(A_k\cos bx + B_k\sin bx)].$$

2. If $r = a + ib$ is a simple root of the auxiliary equation, then there exists a particular solution to equation (2.4.17) of the form

$$y_p(x) = xe^{ax}[(A_0\cos bx + B_0\sin bx) + x(A_1\cos bx + B_1\sin bx) + \cdots$$
$$+ x^k(A_k\cos bx + B_k\sin bx)].$$

3. If $b = 0$ and $r = a$ is a repeated root of the auxiliary equation, then there exists a particular solution to equation (2.4.17) of the form

$$y_p(x) = x^2e^{ax}(A_0 + A_1x + \cdots + A_kx^k).$$

Finally, we consider the case when the nonhomogeneous term in our DE is a sum of the forms already considered. The following theorem is a direct consequence of Theorem 2.1.6.

Theorem 2.4.5:   Consider the DE

$$y'' + a_1y + a_2y = F(x),$$

where a_1, and a_2 are constants. If $F(x)$ is a sum of terms of the form considered in Theorems 2.4.1–2.4.4, then the appropriate trial solution is obtained by taking the sum of the corresponding trial solutions.

Example 2.4.5   In each case, write an appropriate trial solution. Do not solve for the coefficients in your proposed solution.

(a) $y'' + y = 4e^{3x} + 5x^2$;   (b) $y'' - 4y' + 5y = 3e^{2x}\sin x$;   (c) $y'' - y = 3xe^x$.

Solution

(a)   The auxiliary equation is $r^2 + 1 = 0$, with roots $r = \pm i$. Consequently, the complementary function is

$$y_c(x) = c_1\cos x + c_2\sin x.$$

Therefore, a trial solution corresponding to the nonhomogenous term $F_1(x) = 4e^{3x}$ is

$$y_{p_1}(x) = A_0 e^{3x},$$

whereas a trial solution corresponding to the nonhomogenous term $F_2(x) = 5x^2$ is

$$y_{p_2}(x) = B_0 + B_1 x + B_2 x^2.$$

Hence, an appropriate trial solution for the given DE is

$$y_p(x) = A_0 e^{3x} + B_0 + B_1 x + B_2 x^2.$$

(**b**)   The auxiliary equation is $r^2 - 4r + 5 = 0$, with roots $r = 2 \pm i$.   Hence, the complementary function is

$$y_c(x) = e^{2x}(c_1\cos x + c_2\sin x).$$

The usual trial solution corresponding to $F(x) = 3e^{2x}\sin x$ is

$$y_p(x) = e^{2x}(A_0\cos x + B_0\sin x).$$

However, due to the form of the complementary function we see that this y_p solves the associated homogeneous equation and therefore must be modified in order to yield a solution to the nonhomogeneous DE. In this case we must therefore choose

$$y_p(x) = xe^{2x}(A_0\cos x + B_0\sin x).$$

(**c**)   The complementary function is

$$y_c(x) = c_1 e^x + c_2 e^{-x}.$$

The usual trial solution for the nonhomogeneous term $F(x) = 3xe^x$ is

$$y_p(x) = (A_0 + A_1 x)e^x.$$

Since the first term in y_p coincides with a term in the complementary function, we need to use the modified form

$$y_p(x) = x(A_0 + A_1 x)e^x.$$

EXERCISES 2.4

For problems 1–15, determine the general solution to the given DE.

1. $y'' - 3y' + 2y = 4e^{3x}$.

2. $y'' - 2y' - 3y = 15 e^{4x}$.

3. $y'' - 2y' + y = 3x(x - 4)$.

4. $y'' + 2y' = 49 e^x \sin 2x$.

5. $y'' - 2y' - 3y = 8 e^{3x}$.

6. $y'' + 3y' + y = 6(2e^{2x} + 3e^x)$.

7. $y'' + 16y = 24\cos 4x$.

8. $y'' - y = 3xe^x$.

9. $y'' + 2y' + 2y = 4x^2$.

10. $y'' + 4y' + 4y = 10e^{-2x}$.

11. $y'' + 4y = 16x\cos 2x$.

12. $y'' + 2y' - 3y = 8 e^x - 12 e^{3x}$.

13. $y'' - y' - 2y = 40\sin^2 x$.

14. $y'' + 4y' + 5y = 24\sin x$.

15. $y'' - 4y = 2\cos x + 2e^{2x}$.

For problems 16–19, solve the given IVP.

16. $y'' + 9y = 5\cos 2x$, $y(0) = y'(\pi/2) = 2$.

17. $y'' - y = 9xe^{2x}$, $y(0) = 0$, $y'(0) = 7$.

18. $y'' + y' - 2y = 4\cos x - 2\sin x$, $y(0) = -1$, $y'(0) = 4$.

19. $y'' + y' - 2y = -10 \sin x$, $y(0) = 2$, $y'(0) = 1$.

20. At time t the displacement, $y(t)$, from the equilibrium position of a spring-mass system of mass m is governed by the IVP

$$\frac{d^2y}{dt^2} + \omega^2 y = \frac{F_0}{m} \cos \omega t, \quad y(0) = 1, \quad \frac{dy}{dt}(0) = 0,$$

where F_0 and ω are positive constants. Solve this IVP to determine the motion of the system. What happens as $t \to +\infty$?

21. Consider the electric circuit shown in Figure 2.4.1, where R is the resistance, C is the capacitance, L is the inductance, and E is the electromotive force (EMF).

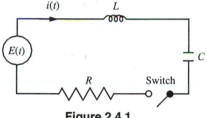

Figure 2.4.1

An application of Kirchoff's law (see Section 1.7 for a derivation) leads to the following DE for the current, i, in the circuit

$$L\frac{di}{dt} + Ri + \frac{1}{C}q = E(t), \tag{22.1}$$

where q is the charge on the capacitor, and we assume that R, L, and C are constants. q and i are related by

$$i = \frac{dq}{dt},$$

so that equation (22.1) can be written as

$$\frac{d^2q}{dt^2} + \frac{R}{L}\frac{dq}{dt} + \frac{1}{LC}q = \frac{1}{L}E(t). \tag{22.2}$$

(a) Find the general solution to equation (22.2) when $E(t) = 0$ and $R^2 = 4L/C$, and hence, find the corresponding current $i(t)$.

(b) Consider the general case when $R = 10 \ \Omega$, $L = 0.5$ H, $C = 1/450$ F, and $E(t) = 600 \cos 30t$ V. Find the general solution to equation (22.2) and the corresponding current $i(t)$. Note that your solution for $i(t)$ consists of an exponential part that dies out quickly (the *transient* current) and an oscillatory part that dominates as $t \to \infty$ (the *steady-state* current).

2.5   COMPLEX-VALUED TRIAL SOLUTIONS

The method of undetermined coefficients for solving

$$y'' + a_1 y' + a_2 y = F(x)$$

can be very tedious to apply if $F(x)$ contains terms of the form $x^k e^{ax}\sin bx$ or $x^k e^{ax}\cos bx$. However, the computations can be reduced significantly from the observation that

$$x^k e^{ax}\cos bx = \text{Re}\{x^k e^{(a+ib)x}\} \quad \text{and} \quad x^k e^{ax}\sin bx = \text{Im}\{x^k e^{(a+ib)x}\},$$

where Re and Im denote the real part and the imaginary part of a complex-valued function, respectively. To see why this observation is useful, we need the next theorem.

Theorem 2.5.1:   If $y(x) = u(x) + iv(x)$ is a complex-valued solution to

$$y'' + a_1 y' + a_2 y = F(x) + iG(x), \tag{2.5.1}$$

then

$$u'' + a_1 u' + a_2 u = F(x) \quad \text{and} \quad v'' + a_1 v' + a_2 v = G(x).$$

PROOF   If $y(x) = u(x) + iv(x)$, then

$$y'' + a_1 y' + a_2 y = [u(x) + iv(x)]'' + a_1[u(x) + iv(x)]' + a_2[u(x) + iv(x)]$$
$$= (u'' + a_1 u' + a_2 u) + i(v'' + a_1 v' + a_2 v).$$

Since y solves equation (2.5.1) we must have

$$(u'' + a_1 u' + a_2 u) + i(v'' + a_1 v' + a_2 v) = F(x) + iG(x).$$

Equating real and imaginary parts on either side of this equation yields the desired result.

∎

Consequently, if we solve the complex equation

$$y'' + a_1 y' + a_2 y = cx^k e^{(a+ib)x}, \qquad (2.5.2)$$

then by taking the real and imaginary parts of the resulting complex-valued solution, we can directly determine solutions to

$$y'' + a_1 y' + a_2 y = cx^k e^{ax}\cos bx \quad \text{and} \quad y'' + a_1 y' + a_2 y = cx^k e^{ax}\sin bx. \quad (2.5.3)$$

The key point is that equation (2.5.2) is a simpler equation to solve than its real counterparts given in (2.5.3). We illustrate the technique with some examples.

Example 2.5.1   Solve

$$y'' + y' - 6y = 4\cos 2x. \qquad (2.5.4)$$

Solution   The complementary function for equation (2.5.4) is

$$y_c(x) = c_1 e^{-3x} + c_2 e^{2x}.$$

In determining a particular solution, we consider the complex DE

$$z'' + z' - 6z = 4e^{2ix}. \qquad (2.5.5)$$

An appropriate complex-valued trial solution for this DE is

$$z_p(x) = A_0 e^{2ix}, \qquad (2.5.6)$$

where A_0 is a complex constant. The first two derivatives of z_p are

$$z_p'(x) = 2iA_0 e^{2ix}, \qquad z_p''(x) = -4A_0 e^{2ix}$$

so that z_p is a solution to equation (2.5.5) if and only if

$$(-4A_0 + 2iA_0 - 6A_0)e^{2ix} = 4e^{2ix},$$

that is, if and only if

$$A_0 = \frac{2}{-5 + i} = -\frac{1}{13}(5 + i).$$

Substituting this value of A_0 into (2.5.6) yields

$$z_p(x) = -\frac{1}{13}(5 + i)\, e^{2ix} = -\frac{1}{13}(5 + i)(\cos 2x + i\sin 2x)$$

$$= \frac{1}{13}(\sin 2x - 5\cos 2x) - \frac{1}{13}i(\cos 2x + 5\sin 2x).$$

Consequently, a particular solution to equation (2.5.4) is

$$y_p(x) = \text{Re}\{z_p\} = \frac{1}{13}(\sin 2x - 5\cos 2x),$$

so that the general solution to equation (2.5.4) is

$$y(x) = y_c(x) + y_p(x) = c_1 e^{-3x} + c_2 e^{2x} + \frac{1}{13}(\sin 2x - 5\cos 2x).$$

Notice that we can also write down the general solution to the DE

$$y'' + y' - 6y = 4\sin 2x,$$

since a particular solution will just be $\text{Im}\{z_p\}$.

Example 2.5.2   Solve

$$y'' - 2y' + 5y = 8e^x \sin 2x. \tag{2.5.7}$$

Solution   The complementary function is

$$y_c(x) = e^x(c_1 \cos 2x + c_2 \sin 2x).$$

In order to determine a particular solution to equation (2.5.7), we consider the complex counterpart

$$z'' - 2z' + 5z = 8e^{(1+2i)x}. \tag{2.5.8}$$

Since $1 + 2i$ is a root of the auxiliary equation, an appropriate trial solution for equation (2.5.8) is

$$z_p(x) = A_0 x e^{(1+2i)x}. \tag{2.5.9}$$

Differentiating with respect to x yields

$$z_p'(x) = A_0 e^{(1+2i)x}[(1 + 2i)x + 1],$$

$$z_p''(x) = A_0 e^{(1+2i)x}[(1 + 2i)^2 x + 2(1 + 2i)] = A_0 e^{(1+2i)x}[(-3 + 4i)x + 2(1 + 2i)].$$

Substituting into equation (2.5.8) leads to the following condition on A_0:

$$A_0[(-3 + 4i)x + 2(1 + 2i) - 2(1 + 2i)x - 2 + 5x] = 8.$$

Hence,

$$A_0 = \frac{2}{i} = -2i.$$

It follows from (2.5.9) that a complex-valued solution to equation (2.5.8) is

$$z_p(x) = -2ixe^{(1+2i)x} = -2ixe^x(\cos 2x + i\sin 2x),$$

and so a particular solution to the original DE is

$$y_p(x) = \text{Im}\{z_p\} = -2xe^x \cos 2x.$$

Consequently, equation (2.5.7) has general solution

$$y(x) = e^x(c_1 \cos 2x + c_2 \sin 2x) - 2xe^x \cos 2x.$$

EXERCISES 2.5

For the following problems, use a complex-valued trial solution to determine a particular solution to the given DE.

1. $y'' + 2y' + y = 50 \sin 3x.$

2. $y'' - y = 10e^{2x} \cos x.$

3. $y'' + 4y' + 4y = 169 \sin 3x.$

4. $y'' - y' - 2y = 40\sin^2 x.$

5. $y'' + y = 3e^x \cos 2x.$

6. $y'' + 2y' + 2y = 2e^{-x}\sin x.$

7. $y'' - 4y = 100xe^x\sin x.$

8. $y'' + 2y' + 5y = 4e^{-x}\cos 2x.$

9. $y'' - 2y' + 10y = 24e^x\cos 3x.$

10. $y'' + 16y = 34e^x + 16\cos 4x - 8\sin 4x.$

11. $\dfrac{d^2y}{dt^2} + \omega_0^2 y = F_0\cos \omega t$, where ω_0 and ω are positive constants, and F_0 is an arbitrary constant. You will need to consider the cases $\omega \neq \omega_0$ and $\omega = \omega_0$ separately.

2.6   OSCILLATIONS OF A MECHANICAL SYSTEM

In this section we analyze in some detail the motion of a mechanical system consisting of a mass attached to a spring. We will see that even such a simple physical system has interesting and varied behavior. We begin by constructing an appropriate mathematical model of the physical situation under consideration.

MATHEMATICAL FORMULATION

Statement of the problem: A mass of m kilograms is attached to the end of a spring whose natural length is l_0 meters. At $t = 0$, the mass is displaced a distance y_0 meters from its equilibrium position and released with a velocity v_0 meters/second. We wish to determine the IVP that governs the resulting motion.

Mathematical formulation of the problem: We assume that the motion takes place vertically and adopt the convention that distances are measured positive in the *downward* direction. In order to formulate the problem mathematically, we need to determine the forces acting on the mass. Consider first the *static equilibrium* position, in which the mass hangs freely from the spring with no motion. (See Figure 2.6.1.) The

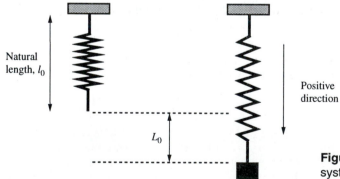

Figure 2.6.1 Spring-mass system in static equilibrium.

forces acting on the mass in this equilibrium position are

1. The force due to gravity

$$F_g = mg.$$

2. The *spring force*, F_s. According to Hooke's law (see Section 1.1)

$$F_s = -kL_0,$$

where k is the spring constant and L_0 is the displacement of the spring from its equilibrium position.

Since the system is in static equilibrium, these forces must exactly balance, so that $F_s + F_g = 0$. Hence,

$$mg = kL_0. \tag{2.6.1}$$

Now consider the situation when the mass has been set in motion. (See Figure 2.6.2.) We let $y(t)$ denote the position of the mass at time t and take $y = 0$ to coincide with the equilibrium position of the system. The equation of motion of the mass can

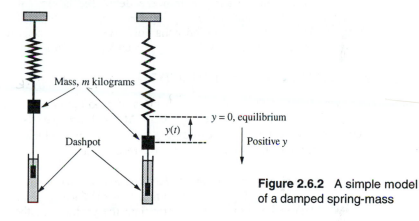

Figure 2.6.2 A simple model of a damped spring-mass

then be obtained from Newton's second law. The forces that now act on the mass are as follows:

1. The force due to gravity F_g. Once more this is

$$F_g = mg. \tag{2.6.2}$$

2. The spring force F_s. At time t the total displacement of the spring from its natural length is $L_0 + y(t)$, so that, according to Hooke's law,

$$F_s = -k[L_0 + y(t)]. \tag{2.6.3}$$

3. A damping force F_d. In general, the motion will be damped due, for example, to air resistance or, as shown in Figure 2.6.2, an external damping system, such as a dashpot. We assume that any damping forces that are present are directly proportional to the velocity of the mass. Under this assumption, we have

$$F_d = -c\frac{dy}{dt}, \tag{2.6.4}$$

where c is a **positive** constant called the **damping constant**. Note that the negative sign is inserted in equation (2.6.4), since F_d always acts in the opposite direction to that of the motion.

4. Any external driving forces, $F(t)$, that are present. For example, the top of the spring or the mass itself may be subjected to an external force.

The total force acting on the system will be the sum of the preceding forces. Thus, using Newton's second law, the DE governing the motion of the mass is

$$m \frac{d^2y}{dt^2} = F_g + F_s + F_d + F(t).$$

Substituting from (2.6.2)–(2.6.4) yields

$$m\frac{d^2y}{dt^2} = mg - k(L_0 + y) - c\frac{dy}{dt} + F(t).$$

That is, using (2.6.1), and rearranging terms,

$$\frac{d^2y}{dt^2} + \frac{c}{m}\frac{dy}{dt} + \frac{k}{m}y = \frac{1}{m}F(t). \tag{2.6.5}$$

In addition, we also have the initial conditions

$$y(0) = y_0, \quad \frac{dy}{dt}(0) = v_0.$$

The motion of the spring-mass system is therefore governed by the IVP

$$\begin{cases} \dfrac{d^2y}{dt^2} + \dfrac{c}{m}\dfrac{dy}{dt} + \dfrac{k}{m}y = \dfrac{1}{m}F(t), \\[2mm] y(0) = y_0, \dfrac{dy}{dt}(0) = v_0. \end{cases} \tag{2.6.6}$$

FREE OSCILLATIONS OF A MECHANICAL SYSTEM

We first consider the case when there are no external forces acting on the system. In the preceding formulation this corresponds to setting $F(t) \equiv 0$, so that the IVP (2.6.6) reduces to

$$\begin{cases} \dfrac{d^2y}{dt^2} + \dfrac{c}{m}\dfrac{dy}{dt} + \dfrac{k}{m}y = 0, \\[2mm] y(0) = y_0, \dfrac{dy}{dt}(0) = v_0. \end{cases}$$

For most of the discussion we will concentrate on the DE alone, since the initial conditions do not significantly affect the behavior of its solutions. We must therefore solve the constant coefficient homogeneous DE

$$\frac{d^2y}{dt^2} + \frac{c}{m}\frac{dy}{dt} + \frac{k}{m}y = 0. \tag{2.6.7}$$

We divide the discussion of the solution to equation (2.6.7) into several subcases.

CASE 1:   NO DAMPING   This is the simplest case that can arise and is of importance for understanding the more general situation. Setting $c = 0$ in equation (2.6.7) yields

$$\frac{d^2y}{dt^2} + \omega_0^2 y = 0, \tag{2.6.8}$$

where

$$\omega_0 = (k/m)^{1/2}. \tag{2.6.9}$$

Equation (2.6.8) has general solution

$$y(t) = c_1 \cos \omega_0 t + c_2 \sin \omega_0 t. \tag{2.6.10}$$

It is instructive to introduce two new constants A_0 and ϕ defined in terms of c_1 and c_2 by (see Figure 2.6.3)

$$A_0 \cos \phi = c_1, \quad A_0 \sin \phi = c_2. \tag{2.6.11}$$

That is,

$$A_0 = \sqrt{c_1{}^2 + c_2{}^2}, \quad \phi = \tan^{-1}\left(\frac{c_2}{c_1}\right).$$

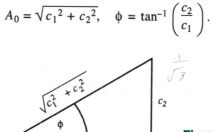
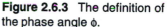

Figure 2.6.3 The definition of the phase angle ϕ.

Substituting from (2.6.11) into (2.6.10) yields

$$y(t) = A_0(\cos \omega_0 t \cos \phi + \sin \omega_0 t \sin \phi).$$

Consequently,

$$y(t) = A_0 \cos(\omega_0 t - \phi). \tag{2.6.12}$$

Clearly, the motion described by (2.6.12) is periodic. We refer to such motion as **simple harmonic motion** (SHM). Figure 2.6.4 depicts this motion for typical values of the constants A_0, ω_0, and ϕ.

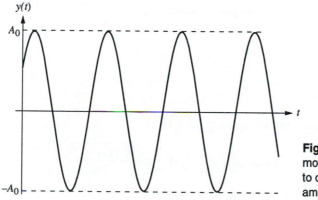

Figure 2.6.4 Simple harmonic motion. The mass continues to oscillate with a constant amplitude A_0.

The standard names for the constants arising in the solution are as follows:

A_0:   the **amplitude** of the motion.

ω_0:   the circular **frequency** of the system.

ϕ: the **phase** of the motion.

The fundamental **period** of oscillation (that is, the time for the system to undergo one complete cycle), T, is

$$T = \frac{2\pi}{\omega_0} = 2\pi(m/k)^{1/2}. \tag{2.6.13}$$

Consequently, the frequency of oscillation (number of oscillations per second), f, is given by

$$f = \frac{1}{T} = \frac{\omega_0}{2\pi} = \frac{1}{2\pi}(k/m)^{1/2}.$$

Notice that this is independent of the initial conditions. It is truly a property of the system.

CASE 2: DAMPING   We now discuss the motion of the spring-mass system when the damping constant, c, is *nonzero*. In this case, the auxiliary polynomial for equation (2.6.7) is

$$P(r) = r^2 + \frac{c}{m}r + \frac{k}{m}$$

with roots

$$r = \frac{-c \pm \sqrt{c^2 - 4km}}{2m}.$$

As we might expect, the behavior of the system is dependent on whether the auxiliary polynomial has distinct real roots, coincident real roots, or complex conjugate roots. These three situations will arise, depending on the magnitude of the (dimensionless) combination of the system variables $c^2/(4km)$. For a given spring and mass, only the damping can be altered, which leads to the following terminology. We say that the system is

(a) *Underdamped* if $c^2/(4km) < 1$	(complex conjugate roots),
(b) *Critically damped* if $c^2/(4km) = 1$	(repeated real root),
(c) *Overdamped* if $c^2/(4km) > 1$	(two distinct real roots).

The corresponding solutions to equation (2.6.7) are

$$\textbf{(a)} \quad y(t) = e^{-ct/(2m)}(c_1 \cos \mu t + c_2 \sin \mu t), \quad \mu = \frac{\sqrt{4km - c^2}}{2m}. \tag{2.6.14}$$

$$\textbf{(b)} \quad y(t) = e^{-ct/(2m)}(c_1 + c_2 t). \tag{2.6.15}$$

$$\textbf{(c)} \quad y(t) = e^{-ct/(2m)}(c_1 e^{\mu t} + c_2 e^{-\mu t}), \quad \mu = \frac{\sqrt{c^2 - 4km}}{2m}. \tag{2.6.16}$$

In all three cases, we have (see the following discussion)

$$\lim_{t \to \infty} y(t) = 0,$$

which implies that the motion dies out for large t. This is certainly consistent with our everyday experience. We will discuss the different cases separately.

CASE 2a: UNDERDAMPING   In this case, the position of the mass at time t is given in (2.6.14), which reduces to SHM when $c = 0$. Once more it is convenient to introduce constants A_0 and ϕ defined by

$$A_0\cos\phi = c_1, \quad A_0\sin\phi = c_2.$$

Now (2.6.14) can be written in the equivalent form

$$y(t) = A_0 e^{-ct/(2m)}\cos(\mu t - \phi). \tag{2.6.17}$$

We see that the mass oscillates between $\pm A_0 e^{-ct/(2m)}$. The corresponding motion is depicted in Figure 2.6.5 for the case when $y(0) > 0$, and $\dfrac{dy}{dt}(0) > 0$.

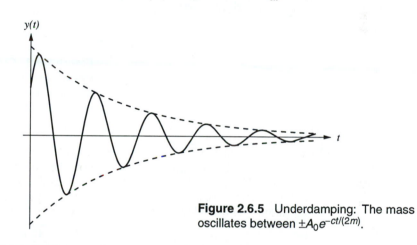

Figure 2.6.5   Underdamping: The mass oscillates between $\pm A_0 e^{-ct/(2m)}$.

In general the motion *is* oscillatory, but it is *not* periodic. The amplitude of the motion dies out exponentially with time, although the time interval, T, between successive maxima (or minima) of $y(t)$ has the constant value (see problem 13)

$$T = \frac{2\pi}{\mu} = \frac{4\pi m}{\sqrt{4km - c^2}}.$$

This is called the **quasiperiod** of the motion.

CASE 2b: CRITICAL DAMPING   This case arises when $c^2/(4km) = 1$. From equation (2.6.7), the motion is governed by the DE

$$\frac{d^2y}{dt^2} + \frac{c}{m}\frac{dy}{dt} + \frac{c^2}{4m^2}y = 0,$$

with general solution

$$y(t) = e^{-ct/(2m)}(c_1 + c_2 t). \tag{2.6.18}$$

Now the damping is so severe that the system can pass through the equilibrium position at most once, and so we do not have oscillatory behavior. If we impose the initial conditions

$$y(0) = y_0, \quad \frac{dy}{dt}(0) = v_0,$$

then it is easily shown (see problem 14) that (2.6.18) can be written in the form

$$y(t) = e^{-ct/(2m)}[y_0 + t(v_0 + \frac{c}{2m}y_0)].$$

Consequently, the system will pass through the equilibrium position, provided y_0 and $v_0 + \frac{c}{2m}y_0$ have opposite signs. A sketch of the motion described by (2.6.18) is given in Figure 2.6.6.

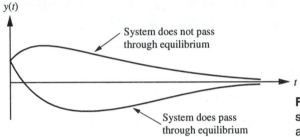

Figure 2.6.6 Critical damping: The system can pass through equilibrium at most once

CASE 2c: OVER DAMPING   In this case we have $c^2/(4km) > 1$. The roots of the auxiliary equation corresponding to equation (2.6.7) are

$$r_1 = \frac{-c + \sqrt{c^2 - 4km}}{2m}, \quad r_2 = \frac{-c - \sqrt{c^2 - 4km}}{2m},$$

so that the general solution to equation (2.6.7) is

$$y(t) = e^{-ct/(2m)}(c_1 e^{\mu t} + c_2 e^{-\mu t}), \quad \mu = \frac{\sqrt{c^2 - 4km}}{2m}.$$

Since c, k, and m are positive, it follows that both of the roots of the auxiliary equation are negative, which implies that both terms in $y(t)$ decay in time. Once more, we do not have oscillatory behavior. The motion is very similar to that of the critically damped case. The system can pass through the equilibrium position at most once. (The graphs given in Figure 2.6.6 are representative of this case also.)

FORCED OSCILLATIONS

We now consider the case when an external force acts on the spring-mass system. As shown at the beginning of the section, the appropriate DE describing the motion of the system is

$$\frac{d^2y}{dt^2} + \frac{c}{m}\frac{dy}{dt} + \frac{k}{m}y = \frac{F(t)}{m}.$$

The situation of most interest arises when the applied force is periodic in time, and we therefore restrict attention to a driving term of the form

$$F(t) = F_0 \cos \omega t,$$

where F_0 and ω are constants. Then the DE governing the motion is

$$\frac{d^2y}{dt^2} + \frac{c}{m}\frac{dy}{dt} + \frac{k}{m}y = \frac{F_0}{m}\cos \omega t. \tag{2.6.19}$$

Once more we will divide our discussion into several cases.

CASE 1: NO DAMPING Setting $c = 0$ in equation (2.6.19) yields

$$\frac{d^2y}{dt^2} + \omega_0^2 y = \frac{F_0}{m}\cos \omega t, \tag{2.6.20}$$

where

$$\omega_0 = (k/m)^{1/2}$$

denotes the circular frequency of the system. The complementary function for equation (2.6.20) is

$$y_c(t) = c_1 \cos \omega_0 t + c_2 \sin \omega_0 t,$$

which can be written in the form

$$y_c(t) = A_0 \cos(\omega_0 t - \phi), \tag{2.6.21}$$

for appropriate constants A_0 and ϕ. We therefore need to find a particular solution to equation (2.6.20). The right-hand side of equation (2.6.20) is of an appropriate form to use the method of undetermined coefficients, although the trial solution will depend on whether $\omega \neq \omega_0$ or $\omega = \omega_0$.

Case 1a $\omega \neq \omega_0$: In this case, the appropriate trial solution is

$$y_p(t) = A \cos \omega t + B \sin \omega t.$$

A straightforward calculation yields the particular solution for equation (2.6.20) (see Problem 25)

$$y_p(t) = \frac{F_0}{m(\omega_0^2 - \omega^2)} \cos \omega t, \tag{2.6.22}$$

so that the general solution to equation (2.6.20) is

$$y(t) = A_0 \cos(\omega_0 t - \phi) + \frac{F_0}{m(\omega_0^2 - \omega^2)} \cos \omega t. \tag{2.6.23}$$

Comparing this with (2.6.12), we see that the resulting motion consists of a superposition of two simple harmonic oscillation modes. One of these modes has the circular frequency, ω_0, of the system, whereas the other mode has the frequency of the driving force. Consequently, the motion is oscillatory and bounded for all time, but, in general, it is *not* periodic. Indeed, it can be shown that the motion is periodic only if the ratio ω/ω_0 is a rational number, say,

$$\frac{\omega}{\omega_0} = \frac{p}{q}, \tag{2.6.24}$$

where p and q are positive integers (see Problem 26). In such a case, the fundamental period of the motion is

$$T = \frac{2\pi q}{\omega_0} = \frac{2\pi p}{\omega} ,$$

where p and q are the smallest integers satisfying equation (2.6.24). A typical (nonperiodic) motion of the form (2.6.23) is sketched in Figure 2.6.7.

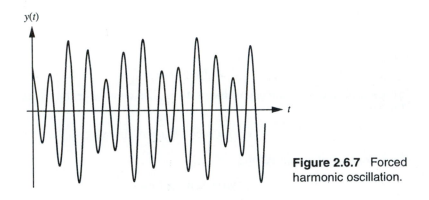

y(t)

t

Figure 2.6.7 Forced harmonic oscillation.

An interesting occurrence arises when the driving frequency ω is close (but not equal to) the natural frequency of the system. To investigate this situation we first impose the zero initial conditions $y(0) = 0$, $dy/dt(0) = 0$ on the general solution (2.6.23). These conditions imply that A_0 and ϕ must satisfy

$$A_0 \cos \phi + \frac{F_0}{m(\omega_0^2 - \omega^2)} = 0, \qquad \omega_0 A_0 \sin \phi = 0.$$

Hence

$$A_0 = -\frac{F_0}{m(\omega_0^2 - \omega^2)}, \qquad \phi = 0.$$

Substituting these values into the general solution (2.6.23) gives

$$y(t) = \frac{F_0}{m(\omega_0^2 - \omega^2)} (\cos \omega t - \cos \omega_0 t).$$

We next use the trigonometric identity $2 \sin A \sin B = \cos(A - B) - \cos(A + B)$ with $A = (\omega_0 - \omega)t/2$ and $B = (\omega_0 + \omega)t/2$ to obtain

$$y(t) = \frac{2F_0}{m(\omega_0^2 - \omega^2)} \sin\left[\left(\frac{\omega_0 - \omega}{2}\right)t\right] \sin\left[\left(\frac{\omega_0 + \omega}{2}\right)t\right].$$

If ω and ω_0 are nearly equal then $\sin[(\omega_0 - \omega)t/2)]$ is slowly varying compared to $\sin[(\omega_0 + \omega)t/2)]$. Thus, $y(t)$ behaves like a rapidly oscillating SHM mode whose amplitude is slowly varying in time. (See Figure 2.6.8.) One of the simplest occurrences of this phenomenon is when two tuning forks whose frequencies are nearly equal are struck simultaneously.

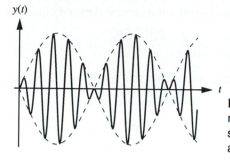

Figure 2.6.8 When $\omega_0 \approx \omega$ the resulting motion can be interpreted as being simple harmonic with a slowly varying amplitude.

Case 1b: $\omega = \omega_0$ *Resonance:* When the frequency of the driving term coincides with the frequency of the system, we must solve

$$\frac{d^2y}{dt^2} + \omega_0{}^2 y = \frac{F_0}{m} \cos \omega_0 t. \tag{2.6.25}$$

The complementary function can be written as

$$y_c(t) = A_0 \cos(\omega_0 t - \phi),$$

and an appropriate trial solution is

$$y_p(t) = t(A \cos \omega_0 t + B \sin \omega_0 t).$$

A straightforward application of the method of undetermined coefficients yields the particular solution (see problem 25)

$$y_p(t) = \frac{F_0}{2m\omega_0} t \sin \omega_0 t, \tag{2.6.26}$$

so that the general solution to equation (2.6.25) is

$$y(t) = A_0 \cos(\omega_0 t - \phi) + \frac{F_0}{2m\omega_0} t \sin \omega_0 t.$$

We see that the motion is oscillatory, but we also see that the amplitude increases without bound as $t \to \infty$. This phenomenon, which occurs when the driving and natural frequencies coincide, is called **resonance**. Its physical consequences cannot be overemphasized. For example, the occurrence of resonance in the present situation would eventually lead to the spring's elastic limit being exceeded, and hence, the system would be destroyed. This situation is depicted in Figure 2.6.9.

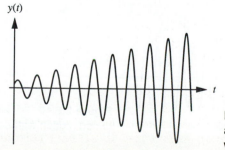

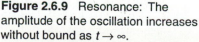

Figure 2.6.9 Resonance: The amplitude of the oscillation increases without bound as $t \to \infty$.

CASE 2:   DAMPING   We now consider the general damped equation

$$\frac{d^2y}{dt^2} + \frac{c}{m}\frac{dy}{dt} + \frac{k}{m}y = \frac{F_0}{m}\cos \omega t, \tag{2.6.27}$$

where $c \neq 0$. An appropriate trial solution for this equation is

$$y_p(t) = A\cos \omega t + B\sin \omega t.$$

A fairly lengthy, but straightforward, computation yields the particular solution (see problem 28)

$$y_p(t) = \frac{F_0}{(k - m\omega^2)^2 + c^2\omega^2}\,[(k - m\omega^2)\cos \omega t + c\omega \sin \omega t], \tag{2.6.28}$$

which can be written in the form

$$y_p(t) = \frac{F_0}{H}\cos(\omega t - \eta), \tag{2.6.29}$$

where

$$\cos \eta = \frac{m(\omega_0^2 - \omega^2)}{H}, \quad \sin \eta = \frac{c\omega}{H}, \quad H = \sqrt{m^2(\omega_0^2 - \omega^2)^2 + c^2\omega^2}\,,$$

and

$$\omega_0 = (k/m)^{1/2}\,.$$

Consider first the case of underdamping. Using the homogeneous solution from (2.6.17) and the foregoing particular solution, it follows that the general solution to equation (2.6.27) in this case is

$$y(t) = A_0 e^{-ct/(2m)}\cos(\mu t - \phi) + \frac{F_0}{H}\cos(\omega t - \eta). \tag{2.6.30}$$

For large t, we see that y_p is dominant. For this reason, we refer to the complementary function as the **transient** part of the solution and y_p is called the **steady-state** solution. We recognize equation (2.6.30) as consisting of a superposition of two harmonic oscillations, one damped and the other undamped. The motion is eventually simple harmonic with a frequency coinciding with that of the driving term.

The cases for critical damping and overdamping are similar, since in both cases the complementary function (transient part of the solution) dies out exponentially and the steady-state solution (2.6.29) dominates. A typical motion of a forced mechanical system with damping is shown in Figure 2.6.10.

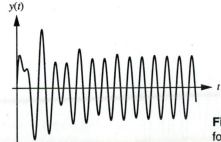

Figure 2.6.10 An example of forced motion with damping.

EXERCISES 2.6

For problems 1 and 2, consider the spring-mass system whose motion is governed by the given IVP. Determine the circular frequency of the system and the amplitude, phase, and period of the motion.

1. $\dfrac{d^2y}{dt^2} + 4y = 0$, $y(0) = 2$, $\dfrac{dy}{dt}(0) = 4$.

2. $\dfrac{d^2y}{dt^2} + \omega_0^2 y = 0$, $y(0) = y_0$, $\dfrac{dy}{dt}(0) = v_0$, where ω_0, y_0, v_0 are constants.

3. A force of 3 N stretches a spring by 1 m.

(a) Find the spring constant k.

(b) A mass of 4 kg is attached to the spring. At $t = 0$ the mass is pulled down a distance 1 m from equilibrium and released with a downward velocity of 0.5 m/s. Assuming that damping is negligible, determine an expression for the position of the mass at time t. Find the circular frequency of the system and the amplitude, phase, and period of the motion.

For problems 4–9, determine the motion of the spring-mass system governed by the given IVP. In each case state whether the motion is underdamped, critically damped or overdamped, and make a sketch depicting the motion.

4. $\dfrac{d^2y}{dt^2} + 2\dfrac{dy}{dt} + 5y = 0$, $y(0) = 1$, $\dfrac{dy}{dt}(0) = 3$.

5. $\dfrac{d^2y}{dt^2} + 3\dfrac{dy}{dt} + 2y = 0$, $y(0) = 1$, $\dfrac{dy}{dt}(0) = 0$.

6. $4\dfrac{d^2y}{dt^2} + 12\dfrac{dy}{dt} + 5y = 0$, $y(0) = 1$, $\dfrac{dy}{dt}(0) = -3$.

7. $\dfrac{d^2y}{dt^2} + 2\dfrac{dy}{dt} + y = 0$, $y(0) = -1$, $\dfrac{dy}{dt}(0) = 2$.

8. $4\dfrac{d^2y}{dt^2} + 4\dfrac{dy}{dt} + y = 0$, $y(0) = 4$, $\dfrac{dy}{dt}(0) = -1$.

9. $\dfrac{d^2y}{dt^2} + 4\dfrac{dy}{dt} + 7y = 0$, $y(0) = 2$, $\dfrac{dy}{dt}(0) = 6$.

10. (a) Determine the motion of the spring-mass system governed by

$$\dfrac{d^2y}{dt^2} + 5\dfrac{dy}{dt} + 6y = 0, \quad y(0) = -1, \dfrac{dy}{dt}(0) = 4.$$

(b) Find the time at which the mass passes through the equilibrium position, and determine the maximum positive displacement of the mass from equilibrium.

(c) Make a sketch depicting the motion.

11. Consider the spring-mass system whose motion is governed by the DE

$$\dfrac{d^2y}{dt^2} + 2\alpha\dfrac{dy}{dt} + y = 0.$$

Determine all values of the (positive) constant α for which the system is (i) underdamped, (ii) critically damped, and (iii) overdamped. In the case of overdamping solve the system completely. If the initial velocity of the system is zero, determine whether the mass passes through equilibrium.

12. Consider the spring-mass system whose motion is governed by the IVP

$$\dfrac{d^2y}{dt^2} + 3\dfrac{dy}{dt} + 2y = 0, \quad y(0) = 1, \dfrac{dy}{dt}(0) = -3.$$

(a) Determine the position of the mass at time t.

(b) Determine the time when the mass passes through the equilibrium position.

(c) Make a sketch depicting the general motion of the system.

13. Consider the general solution for an *underdamped* spring-mass system.

(a) Show that the time between successive maxima (or minima) of $y(t)$ is

$$T = \frac{2\pi}{\mu} = \frac{4\pi m}{\sqrt{4km - c^2}}.$$

(b) Show that if $\dfrac{c^2}{4km} \ll 1$ then

$$T \approx 2\pi(m/k)^{1/2}.$$

Is this result reasonable?

14. Show that the general solution for the motion of a *critically damped* spring-mass system, with initial displacement y_0 and initial velocity v_0, can be written in the form

$$y(t) = e^{-ct/(2m)}[y_0 + t(v_0 + \frac{c}{2m}y_0)]$$

and that the system can pass through the equilibrium position at most once.

15. A cylinder of side L meters lies one quarter submerged and upright in a certain fluid. At $t = 0$ the cylinder is pushed down a distance $L/2$ meters and released from rest. Show that the resulting motion is simple harmonic, and determine the circular frequency and period of the motion.[1]

A simple pendulum consists of a mass, m kilograms, attached to the end of a light rod of length L meters, whose other end is fixed. (See Figure 2.6.11.) If we let θ radians denote the

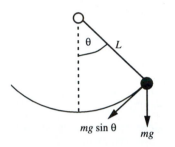

Figure 2.6.11 The simple pendulum

angle the rod is displaced from the vertical at time t, then the component of the velocity in the direction of motion is $v = L d\theta/dt$, so that the component of the acceleration in this direction is $L d^2\theta/dt^2$. Further, the tangential component of the force is $F_T = -mg \sin \theta$, so that, from Newton's second law, the equation of motion of the pendulum is

$$mL \frac{d^2\theta}{dt^2} = -mg \sin \theta.$$

That is,

$$\frac{d^2\theta}{dt^2} + \frac{g}{L} \sin \theta = 0. \qquad (2.6.31)$$

This is a nonlinear DE. However, if we recall the Maclaurin expansion for $\sin \theta$, namely,

$$\sin \theta = \theta - \frac{1}{3!} \theta^3 + \frac{1}{5!} \theta^5 - \cdots,$$

it follows that for small oscillations, we can approximate $\sin \theta$ by θ. Then equation (2.6.31) can be replaced to reasonable accuracy by the simple *linear* DE

$$\frac{d^2\theta}{dt^2} + \frac{g}{L} \theta = 0. \qquad (2.6.32)$$

Problems 16–19 deal with the simple pendulum whose motion is described by equation (2.6.32).

16. A pendulum of length 0.5 m is displaced an angle 0.1 rad from the equilibrium position and released from rest. Determine the resulting motion.

---

[1] According to Archimedes' principle, when an object is partially or wholly immersed in a fluid, it experiences an upward force equal to the weight of fluid displaced.

17. A pendulum of length L meters is displaced an angle α radians from the vertical and released with an angular velocity of β radians/second. Determine the amplitude, phase, and period of the resulting motion.

18. Show that the period of the simple pendulum is $T = 2\pi(L/g)^{1/2}$. Determine the length of a pendulum that takes one second to swing from its extreme position on the right to its extreme position on the left. Take $g = 9.8$ m/s².

19. A clock has a pendulum of length 90 cm. If the clock ticks each time the pendulum swings from its extreme position on the right to its extreme position on the left, determine the number of times the clock ticks in one minute. Take g = 9.8 m/s².

20. An object of mass m is attached to the mid-point of a light elastic string of natural length $6a$. When the ends of the string are fixed at the same level a distance $6a$ apart and the mass is allowed to hang in equilibrium, the length of the stretched string is $10a$. (See Figure 2.6.12.) The mass is pulled down a small vertical distance from

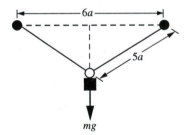

Figure 2.6.12 The static equilibrium position.

equilibrium and released. Show that, *for small oscillations*, the period of the resulting motion is

$$T = \frac{20\pi}{7} (a/g)^{1/2}.$$

21. Repeat the previous problem if the string has natural length $2L_0$ and in equilibrium, the stretched string has length $2L$.

22. Consider the damped spring-mass system whose motion is governed by

$$\frac{d^2y}{dt^2} + 2\frac{dy}{dt} + 5y = 17 \sin 2t, \; y(0) = -2, \frac{dy}{dt}(0) = 0.$$

(a) Determine whether the motion is underdamped, overdamped or critically damped.

(b) Determine the solution to the given IVP and identify the transient and steady-state parts.

23. Consider the spring-mass system whose motion is governed by

$$\frac{d^2y}{dt^2} + \omega_0^2 y = F_0 \sin \omega t, \quad y(0) = 0, \quad \frac{dy}{dt}(0) = 0.$$

Determine the solution if the system is resonating.

24. Consider the spring-mass system whose motion is governed by

$$\frac{d^2y}{dt^2} + 3\frac{dy}{dt} + 2y = 10 \sin t.$$

Determine the steady-state solution, y_p, and express your answer in the form

$$y_p(t) = A_0 \sin(t - \phi),$$

for appropriate constants A_0 and ϕ.

25. Consider the forced undamped spring-mass system whose motion is governed by

$$\frac{d^2y}{dt^2} + \omega_0^2 y = \frac{F_0}{m} \cos \omega t.$$

Derive the particular solutions given in equations (2.6.22) and (2.6.26). (You will need to consider $\omega \neq \omega_0$ and $\omega = \omega_0$ separately.)

26. The general solution to the forced undamped (non-resonating) spring-mass system is

$$y(t) = A_0 \cos(\omega_0 t - \phi) + \frac{F_0}{m(\omega_0^2 - \omega^2)} \cos \omega t.$$

If $\omega/\omega_0 = p/q$, where p and q are integers, show that the motion is periodic with period $T = 2\pi q/\omega_0$.

27. Determine the period of the motion for the spring-mass system governed by the DE

$$\frac{d^2y}{dt^2} + \frac{9}{16} y = 55 \cos 2t.$$

28. Consider the damped forced motion described by

$$\frac{d^2y}{dt^2} + \frac{c}{m}\frac{dy}{dt} + \frac{k}{m} y = \frac{F_0}{m} \cos \omega t.$$

Derive the steady-state solution (2.6.28) given in the text.

29. Consider the damped forced motion described by

$$\frac{d^2y}{dt^2} + \frac{c}{m}\frac{dy}{dt} + \frac{k}{m} y = \frac{F_0}{m} \cos \omega t.$$

We have shown that the steady-state solution can be written in the form

$$y_p(t) = \frac{F_0}{H} \cos(\omega t - \eta),$$

where

$$\cos \eta = \frac{m(\omega_0^2 - \omega^2)}{H}, \quad \sin \eta = \frac{c\omega}{H}, \quad \omega_0 = (k/m)^{1/2}$$

and

$$H = \sqrt{m^2(\omega_0^2 - \omega^2)^2 + c^2\omega^2}.$$

Assuming that $c^2/(2m^2\omega_0^2) < 1$, show that the amplitude of the steady-state solution is a maximum when

$$\omega = [\omega_0^2 - c^2/(2m^2)]^{1/2}.$$

(Hint: The maximum occurs at the value of ω that makes H a minimum. Assume that H is a function of ω, and determine the value of ω that minimizes H.)

30. Consider the damped spring-mass system with $m = 1$, $k = 5$, $c = 2$, and $F(t) = 8 \cos \omega t$.

(a) Determine the transient part of the solution and the steady-state solution.

(b) Determine the value of ω that maximizes the amplitude of the steady-state solution and express the corresponding solution in the form

$$y_p(t) = A_0 \cos(\omega t - \eta),$$

for appropriate constants A_0, ω, and η.

31. Consider the spring-mass system whose motion is governed by the DE

$$\frac{d^2y}{dt^2} + 2\frac{dy}{dt} + 5y = 4e^{-t}\cos 2t.$$

(a) Describe the variation with time of the applied external force.

(b) Determine the motion of the mass. What happens as $t \to \infty$?

32. Consider the spring-mass system whose motion is governed by the DE

$$\frac{d^2y}{dt^2} + 16y = 130 e^{-t}\cos t.$$

Determine the resulting motion, and identify any transient and steady-state parts of your solution.

2.7   RLC CIRCUITS

In Section 1.7, we used Kirchoff's second law to derive the DE

$$\frac{di}{dt} + \frac{R}{L} i + \frac{1}{LC} q = \frac{1}{L} E(t), \tag{2.7.1}$$

which governs the behavior of the RLC circuit shown in Figure 2.7.1. Here, q is the charge on the capacitor at time t, the constants R, L, and C are the resistance, inductance, and capacitance of the circuit elements respectively, and $E(t)$ denotes the driving electromotive force (EMF). The current in the circuit is related to the charge on the capacitor via

$$i(t) = \frac{dq}{dt}. \tag{2.7.2}$$

Substituting this expression for i into equation (2.7.1) yields the second-order constant-coefficient DE

$$\frac{d^2q}{dt^2} + \frac{R}{L} \frac{dq}{dt} + \frac{1}{LC} q = \frac{1}{L} E(t). \tag{2.7.3}$$

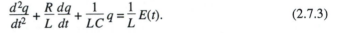
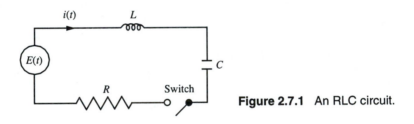

Figure 2.7.1   An RLC circuit.

A comparison of equation (2.7.3) with the basic DE governing the motion of a spring-mass system, namely,

$$\frac{d^2y}{dt^2} + \frac{c}{m} \frac{dy}{dt} + \frac{k}{m} y = \frac{F(t)}{m},$$

reveals that, although the two problems are distinct physically, from a purely mathematical standpoint they are identical. The correspondence between the variables and parameters in an RLC circuit and a spring-mass system is given in Table 2.7.1. It follows that the results derived in the previous section for a spring-mass system can be translated into corresponding results for RLC circuits. Rather than repeating these results, we will make some general observations and then consider one illustrative example. The full investigation of the behavior of an RLC circuit is left for the exercises.

TABLE 2.7.1

RLC circuit	Spring-mass system
$q(t)$	$y(t)$
L	m
R	c
$1/C$	k
$E(t)$	$F(t)$

Consider first the homogeneous DE

$$\frac{d^2q}{dt^2} + \frac{R}{L}\frac{dq}{dt} + \frac{1}{LC}q = 0. \qquad (2.7.4)$$

This has auxiliary equation

$$r^2 + \frac{R}{L}r + \frac{1}{LC} = 0,$$

with roots

$$r = \frac{-R \pm \sqrt{R^2 - 4L/C}}{2L}.$$

Three familiar cases arise. The circuit is said to be

1 .   *Underdamped if $R^2 < 4L/C$.*

2 .   *Critically damped if $R^2 = 4L/C$.*

3 .   *Overdamped if $R^2 > 4L/C$.*

The corresponding solutions to equation (2.7.4) are

1 . $q(t) = e^{-Rt/(2L)}(c_1 \cos \mu t + c_2 \sin \mu t), \quad \mu = \dfrac{\sqrt{4L/C - R^2}}{2L}.$ $\qquad (2.7.5)$

2 . $q(t) = e^{-Rt/(2L)}(c_1 + c_2 t).$

3 . $q(t) = e^{-Rt/(2L)}(c_1 e^{\mu t} + c_2 e^{-\mu t}), \quad \mu = \dfrac{\sqrt{R^2 - 4L/C}}{2L}.$

In all cases with $R \neq 0$,

$$\lim_{t \to \infty} q(t) = 0.$$

Equivalently, we can state that the complementary function for equation (2.7.3), $q = q_c$, satisfies

$$\lim_{t \to \infty} q_c(t) = 0$$

We refer to q_c as the *transient part* of the solution to equation (2.7.3), since it decays exponentially with time. As a specific example, we consider the case of a periodic driving EMF in an underdamped circuit.

Example 2.7.1   Determine the current in the RLC circuit

$$\frac{d^2q}{dt^2} + \frac{R}{L}\frac{dq}{dt} + \frac{1}{LC}q = \frac{E_0}{L}\cos \omega t, \qquad (2.7.6)$$

where E_0 and ω are positive constants and $R^2 < 4L/C$.

Solution   The complementary function given in equation (2.7.5) can be written in phase-amplitude form as

$$q_c(t) = A_0 e^{-Rt/(2L)}\cos(\mu t - \phi),$$

where A_0 and ϕ are defined in the usual manner. A particular solution to equation (2.7.6) can be obtained by using Table 2.7.1 to make the appropriate replacements in the solution (2.6.22) for the corresponding spring-mass system. The result is

$$q_p(t) = \frac{E_0}{H} \cos(\omega t - \eta),$$

where

$$H = \sqrt{L^2(\omega_0{}^2 - \omega^2)^2 + R^2 \omega^2},$$

and

$$\cos \eta = \frac{L(\omega_0{}^2 - \omega^2)}{H}, \quad \sin \eta = \frac{R\omega}{H}, \quad \omega_0 = (1/LC)^{1/2}.$$

Consequently, the charge on the capacitor at time t is

$$q(t) = A_0 e^{-Rt/(2L)} \cos(\mu t - \phi) + \frac{E_0}{H} \cos(\omega t - \eta),$$

and the corresponding current in the circuit can be determined from

$$i(t) = \frac{dq}{dt}.$$

Rather than compute this derivative, we consider the late-time behavior of q and the corresponding late-time behavior of the current. Since q_c tends to zero as $t \to +\infty$, for large t, the particular solution q_p will be the dominant part of $q(t)$. For this reason, we refer to q_p as the **steady-state** solution. The corresponding **steady-state** current in the circuit, denoted i_S, is given by

$$i_S(t) = \frac{dq_p}{dt} = -\frac{\omega E_0}{H} \sin(\omega t - \eta).$$

We see that this is periodic and that the frequency of the oscillation coincides with the frequency of the driving EMF. The amplitude of the oscillation is

$$A = \frac{\omega E_0}{H}.$$

That is, upon substituting for H,

$$A = \frac{\omega E_0}{\sqrt{L^2(\omega_0{}^2 - \omega^2)^2 + R^2 \omega^2}}. \qquad (2.7.7)$$

It is often required to determine the value of ω that maximizes this amplitude. In order to do so, we rewrite (2.7.7) in the equivalent form

$$A = \frac{E_0}{\sqrt{\omega^{-2} L^2(\omega_0{}^2 - \omega^2)^2 + R^2}}.$$

This will be a maximum when the term in parentheses vanishes, which occurs when

$$\omega^2 = \omega_0{}^2.$$

Substituting for $\omega_0{}^2 = 1/LC$, it follows that the amplitude of the steady-state current will be a maximum when $\omega = \omega_{max}$, where

$$\omega_{max} = (LC)^{-1/2}.$$

The corresponding value of A is

$$A_{max} = \frac{E_0}{R} .$$

The behavior of A as a function of ω for typical values of E_0, R, L, and C is shown in Figure 2.7.2.

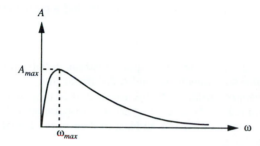

Figure 2.7.2 The behavior of the amplitude of the steady-state current as a function of the driving frequency.

EXERCISES 2.7

1 . Determine the steady-state current in the RLC circuit that has $R = \frac{3}{2}\,\Omega$, $L = \frac{1}{2}\,H$, $C = \frac{2}{3}\,F$, and $E(t) = 13 \cos 3t$ V.

2 . Determine the charge on the capacitor at time t in the RLC circuit that has $R = 4\,\Omega$, $L = 4\,H$, $C = \frac{1}{17}\,F$, and $E = E_0$ V, where E_0 is constant. What happens to the charge on the capacitor as $t \to +\infty$? Describe the behavior of the current in the circuit.

3 . Consider the RLC circuit with $E(t) = E_0 \cos \omega t$ volts, where E_0 and ω are constants. If there is no resistor in the circuit, show that the charge on the capacitor satisfies

$$\lim_{t \to \infty} q(t) = +\infty$$

if and only if $\omega = 1/\sqrt{LC}$. What happens to the current in the circuit as $t \to +\infty$?

4 . Consider the RLC circuit with $R = 16\,\Omega$, $L = 8\,H$, $C = \frac{1}{40}\,F$, and $E(t) = 17 \cos 2t$ V. Determine the current in the circuit for $t > 0$, given that at $t = 0$ the capacitor is uncharged and there is no current flowing.

5 . Consider the RLC circuit with $R = 3\,\Omega$, $L = \frac{1}{2}\,H$, $C = \frac{1}{5}\,F$ and $E(t) = 2 \cos \omega t$ V. Determine the current in the circuit at time t, and find the value of ω that maximizes the amplitude of the steady-state current.

6 . Show that the DE governing the behavior of an RLC circuit can be written directly in terms of the current $i(t)$ as

$$\frac{d^2 i}{dt^2} + \frac{R}{L}\frac{di}{dt} + \frac{1}{LC}\,i = \frac{1}{L}\frac{dE}{dt} .$$

7 . Determine the current in the general RLC circuit with $R^2 < 4L/C$, if $E(t) = E_0 e^{-at}$, where E_0 and a are constants.

8 . Consider the RLC circuit with $R = 2\,\Omega$, $L = \frac{1}{2}\,H$, $C = \frac{2}{5}\,F$. Initially the capacitor is uncharged, and there is no current flowing in the circuit. Determine the current for $t > 0$, if the applied EMF is (see Figure 2.7.3)

$$E(t) = \begin{cases} 50t, \, 0 \le t < \pi, \\ 50\pi, \quad t \ge \pi. \end{cases}$$

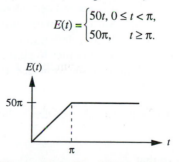

Figure 2.7.3 The EMF from problem 8.

2.8   THE VARIATION-OF-PARAMETERS METHOD

The method of undetermined coefficients has two severe limitations. Firstly, it is only applicable to DE with constant coefficients, and secondly, it can only be applied to DE whose nonhomogeneous terms are of the form described in Section 2.4. For example, we could not use the method of undetermined coefficients to find a particular solution to the DE

$$y'' + 4y' - 6y = x^2\ln x.$$

In this section, we introduce a very powerful technique, called the variation-of-parameters method, for obtaining particular solutions to second-order linear *nonhomogeneous* DE, assuming that we know the *general solution* to the associated homogeneous equation. Unlike the method of undetermined coefficients, the variation-of-parameters method is not restricted to DE with constant coefficients, and, at least in theory, the actual form of the nonhomogeneous term is immaterial.

Consider the linear second-order nonhomogeneous DE

$$y'' + a_1 y' + a_2 y = F, \tag{2.8.1}$$

where we assume that a_1, a_2, and F are continuous on an interval I. Suppose that $y = y_1(x)$ and $y = y_2(x)$ are two *linearly independent* solutions to the associated homogeneous equation

$$y'' + a_1 y' + a_2 y = 0 \tag{2.8.2}$$

on I, so that the general solution to equation (2.8.2) on I is

$$y_c(x) = c_1 y_1(x) + c_2 y_2(x). \tag{2.8.3}$$

The variation-of-parameters method consists of replacing the constants c_1 and c_2 by functions $u_1(x)$ and $u_2(x)$ (that is, we allow the parameters c_1, c_2 to vary) determined in such a way that the resulting function

$$y_p(x) = u_1(x)y_1(x) + u_2(x)y_2(x) \tag{2.8.4}$$

is a particular solution to equation (2.8.1).

Differentiating equation (2.8.4) with respect to x yields

$$y_p' = u_1' y_1 + u_1 y_1' + u_2' y_2 + u_2 y_2'.$$

It is tempting to differentiate this expression once more and then substitute into equation (2.8.1) to determine u_1 and u_2. However, if we did this, the resulting expression for y_p'' would involve second derivatives of u_1 and u_2, and hence, we would have complicated our problem. Since y_p contains *two* unknown functions, whereas equation (2.8.1) gives only one condition for determining them, we have the freedom to impose a further constraint on u_1 and u_2. In order to eliminate second derivatives of u_1 and u_2 arising in y_p'', we try for solutions of the form (2.8.4) satisfying the constraint

$$u_1' y_1 + u_2' y_2 = 0. \tag{2.8.5}$$

The expression for y_p' then reduces to

$$y_p' = u_1 y_1' + u_2 y_2',$$

so that

$$y_p'' = u_1' y_1' + u_1 y_1'' + u_2' y_2' + u_2 y_2''.$$

Substituting into equation (2.8.1) and collecting terms yields

$$u_1(y_1'' + a_1 y_1' + a_2 y_1) + u_2(y_2'' + a_1 y_2' + a_2 y_2) + (u_1' y_1' + u_2' y_2') = F(x).$$

The terms multiplying u_1 and u_2 vanish, since y_1 and y_2 each solve $y'' + a_1 y' + a_2 y = 0$. We therefore require that

$$u_1' y_1' + u_2' y_2' = F. \tag{2.8.6}$$

We may therefore conclude that $y_p(x) = u_1(x) y_1(x) + u_2(x) y_2(x)$ is a solution to equation (2.8.1), provided that u_1 and u_2 satisfy equations (2.8.5) and (2.8.6). That is,

$$y_1 u_1' + y_2 u_2' = 0,$$
$$y_1' u_1' + y_2' u_2' = F.$$

This is a linear algebraic system of equations for the unknowns u_1' and u_2'. Successively eliminating u_1' and u_2' from these equations yields

$$(y_1 y_2' - y_1' y_2) u_1' = -y_2 F, \quad (y_1 y_2' - y_1' y_2) u_2' = y_1 F.$$

We notice that the coefficient of u_1' and u_2' is the Wronskian of y_1 and y_2, which is nonzero, since y_1 and y_2 are linearly independent on I. Consequently, we can divide each expression by the Wronskian to obtain

$$u_1'(x) = -\frac{y_2 F}{W[y_1, y_2]}, \quad u_2'(x) = \frac{y_1 F}{W[y_1, y_2]},$$

so that

$$u_1(x) = -\int \frac{y_2 F}{W[y_1, y_2]}\, dx, \quad u_2(x) = \int \frac{y_1 F}{W[y_1, y_2]}\, dx. \tag{2.8.7}$$

We have therefore established the next theorem.

Theorem 2.8.1 (Variation-of-Parameters Method):   Consider

$$y'' + a_1 y' + a_2 y = F, \tag{2.8.8}$$

where a_1, a_2, and F are assumed to be (at least) continuous on the interval I. Let y_1 and y_2 be linearly independent solutions to the associated homogeneous equation

$$y'' + a_1 y' + a_2 y = 0$$

on I. Then a particular solution to equation (2.8.8) is

$$y_p = u_1 y_1 + u_2 y_2,$$

where u_1 and u_2 satisfy

$$y_1 u_1' + y_2 u_2' = 0,$$
$$y_1' u_1' + y_2' u_2' = F.$$

Example 2.8.1   Solve $y'' + y = \sec x$.

Solution   Two linearly independent solutions to the associated homogeneous equation are $y_1(x) = \cos x$ and $y_2(x) = \sin x$. Thus, a particular solution to the given DE is

$$y_p(x) = u_1 y_1 + u_2 y_2 = u_1 \cos x + u_2 \sin x, \tag{2.8.9}$$

where u_1 and u_2 satisfy

$$\cos x \, u_1' + \sin x \, u_2' = 0,$$
$$-\sin x \, u_1' + \cos x \, u_2' = \sec x.$$

The solution to this system is

$$u_1' = -\sec x \sin x, \quad u_2' = \sec x \cos x.$$

Consequently,

$$u_1(x) = -\int \sec x \sin x \, dx = -\int \frac{\sin x}{\cos x} \, dx = \ln |\cos x|,$$

and

$$u_2(x) = \int \sec x \cos x \, dx = x,$$

where we have set the integration constants to zero, since we only require one particular solution. Substitution into equation (2.8.9) yields

$$y_p(x) = \cos x \ln |\cos x| + x \sin x,$$

so that the general solution to the given DE is

$$y(x) = c_1 \cos x + c_2 \sin x + \cos x \ln |\cos x| + x \sin x.$$

Example 2.8.2   Solve $y'' + 4y' + 4y = e^{-2x} \ln x, \quad x > 0.$

Solution   In this case, two linearly independent solutions to the associated homogeneous equation are $y_1(x) = e^{-2x}$, $y_2(x) = xe^{-2x}$, and hence, a particular solution to the given DE is

$$y_p(x) = u_1 e^{-2x} + u_2 x e^{-2x},$$

where u_1 and u_2 satisfy

$$e^{-2x} u_1' + x e^{-2x} u_2' = 0,$$
$$-2e^{-2x} u_1' + e^{-2x}(1 - 2x) u_2' = e^{-2x} \ln x.$$

The solution to this system is

$$u_1' = -x \ln x, \quad u_2' = \ln x.$$

Integrating both of these expressions by parts (and setting the integration constants to zero), we obtain

$$u_1(x) = \frac{1}{4} x^2 (1 - 2\ln x), \quad u_2(x) = x(\ln x - 1).$$

Thus,

$$y_p(x) = \frac{1}{4} x^2 e^{-2x}(1 - 2\ln x) + x^2 e^{-2x}(\ln x - 1).$$

That is,

$$y_p(x) = \frac{1}{4} x^2 e^{-2x}(2\ln x - 3).$$

Consequently, the general solution to the given DE is

$$y(x) = e^{-2x}\left[c_1 + c_2 x + \frac{1}{4}x^2(2\ln x - 3)\right].$$

❏

Sometimes we can use a combination of the variation-of-parameters method and the method of undetermined coefficients to determine a particular solution to a DE. We illustrate this with an example.

Example 2.8.3   Determine the general solution to the DE

$$y'' + 9y = 6\cot^2 3x + 5e^{2x}, \quad 0 < x < \pi/6. \tag{2.8.10}$$

Solution   The complementary function for the given DE is

$$y_c(x) = c_1 \cos 3x + c_2 \sin 3x.$$

Application of the variation-of-parameters technique directly to the DE (2.8.10) leads to some rather nasty integrals arising from the e^{2x} term in the nonhomogeneous term. However, if we determine a particular solution to each of the DE

$$y'' + 9y = 6\cot^2 3x \tag{2.8.11}$$

and

$$y'' + 9y = 5e^{2x}, \tag{2.8.12}$$

then Theorem 2.1.6 can be applied to conclude that the sum of these two solutions will itself be a particular solution to equation (2.8.10). The key point is that equation (2.8.12) can be solved easily using the method of undetermined coefficients, and therefore, we have alleviated the problem of evaluating the integrals mentioned previously. Consider first equation (2.8.11). According to the variation-of-parameters method, there is a particular solution to this DE of the form

$$y_{p_1}(x) = u_1 \cos 3x + u_2 \sin 3x,$$

where u_1 and u_2 satisfy

$$\cos 3x \, u_1' + \sin 3x \, u_2' = 0,$$

$$-\sin 3x \, u_1' + \cos 3x \, u_2' = 2\cot^2 3x.$$

Solving this system of equations yields

$$u_1' = -2\cot^2 3x \sin 3x, \quad u' = 2\cot^2 3x \cos 3x.$$

Consequently,

$$u_1 = -2\int \cot^2 3x \sin 3x \, dx = -2\int \frac{\cos^2 3x}{\sin 3x} \, dx = -2\int (\csc 3x - \sin 3x) \, dx$$

$$= -\frac{2}{3}[\ln(\csc 3x - \cot 3x) + \cos 3x]$$

and

$$u_2 = 2\int \cot^2 3x \cos 3x \, dx = 2\int \frac{(1 - \sin^2 3x)}{\sin^2 3x} \cos 3x \, dx$$

$$= 2\int \left(\frac{\cos 3x}{\sin^2 3x} - \cos 3x\right) dx = -\frac{2}{3}(\csc 3x + \sin 3x).$$

Therefore,

$$y_{p_1}(x) = -\frac{2}{3} \cos 3x \, [\ln(\csc 3x - \cot 3x) + \cos 3x] - \frac{2}{3}\sin 3x (\csc 3x + \sin 3x),$$

which simplifies to

$$y_{p_1}(x) = -\frac{2}{3} [\cos 3x \, \ln(\csc 3x - \cot 3x) + 2].$$

Next consider equation (2.8.12). An appropriate trial solution for this DE is

$$y_{p_2}(x) = A_0 e^{2x}$$

and substitution into equation (2.8.12) yields $A_0 = 5/13$. Consequently, a particular solution to equation (2.8.12) is

$$y_{p_2}(x) = \frac{5}{13} e^{2x}.$$

It follows directly from Theorem 2.1.6 that a particular solution to equation (2.8.10) is

$$y_p(x) = y_{p_1}(x) + y_{p_2}(x) = -\frac{2}{3} [\cos 3x \, \ln(\csc 3x - \cot 3x) + 2] + \frac{5}{13} e^{2x}.$$

The general solution to equation (2.8.10) is therefore

$$y(x) = c_1 \cos 3x + c_2 \sin 3x - \frac{2}{3} [\cos 3x \, \ln(\csc 3x - \cot 3x) + 2] + \frac{5}{13} e^{2x}.$$

GREEN'S FUNCTIONS

According to Theorem 2.8.1, a particular solution to the DE

$$y'' + a_1 y' + a_2 y = F, \quad a < x < b$$

where a_1, a_2, F are continuous on (a, b) can be written in the form

$$y_p(x) = -y_1(x) \int_{x_0}^{x} \frac{y_2(t)F(t)}{W[y_1, y_2](t)} \, dt + y_2(x) \int_{x_0}^{x} \frac{y_1(t)F(t)}{W[y_1, y_2](t)} \, dt$$

where $x_0 \in (a, b)$ and we have used the expressions in (2.8.7) for u_1 and u_2. Combining the two terms on the right-hand side of the preceding equation yields

$$y_p(x) = \int_{x_0}^{x} \left\{ \frac{y_1(t)y_2(x) - y_2(t)y_1(x)}{W[y_1, y_2](t)} \right\} F(t) \, dt. \tag{2.8.13}$$

which we write as

$$y_p(x) = \int_{x_0}^{x} K(x, t) F(t) \, dt$$

where

$$K(x, t) = \frac{y_1(t)y_2(x) - y_2(t)y_1(x)}{W[y_1, y_2](t)}. \tag{2.8.14}$$

The function $K(x, t)$ is called a **Green's function** for the problem. We see that it depends only on the solutions to the associated homogeneous problem and not on the nonhomogeneous term $F(x)$.

Example 2.8.4 Use a Green's function to determine a particular solution to the DE

$$y'' + 16y = F(x).$$

Solution   Two linearly independent solutions to the associated homogeneous DE are

$$y_1(x) = \cos 4x, \quad y_2(x) = \sin 4x$$

with Wronskian

$$W[y_1, y_2](x) = \begin{vmatrix} \cos 4x & \sin 4x \\ -4\sin 4x & 4\cos 4x \end{vmatrix} = 4.$$

Substitution into (2.8.14) yields

$$K(x, t) = \frac{1}{4}(\cos 4t \sin 4x - \sin 4t \cos 4x) = \frac{1}{4}\sin 4(x - t).$$

Consequently, from (2.8.13),

$$y_p(x) = \frac{1}{4}\int_{x_0}^{x} \sin 4(x - t)\, F(t)\, dt.$$

The general solution to the given DE can therefore be expressed as

$$y(x) = c_1\cos 4x + c_2\sin 4x + \frac{1}{4}\int_{x_0}^{x} \sin 4(x - t)\, F(t)\, dt$$

EXERCISES 2.8

For problems 1–14, use the variation-of-parameters method to find the general solution to the given DE.

1. $y'' + 6y' + 9y = \dfrac{2e^{-3x}}{x^2 + 1}$.

2. $y'' - 4y = \dfrac{8}{e^{2x} + 1}$.

3. $y'' - 4y' + 5y = e^{2x}\tan x,\ 0 < x < \pi/2$.

4. $y'' - 6y' + 9y = 4e^{3x}\ln x,\ x > 0$.

5. $y'' + 4y' + 4y = x^{-2}e^{-2x},\ x > 0$.

6. $y'' + 9y = 18\sec^3(3x),\ |x| < \pi/6$.

7. $y'' - y = 2\tanh x$.

8. $y'' - 2my' + m^2 y = \dfrac{e^{mx}}{1 + x^2},\ m$ constant.

9. $y'' - 2y' + y = 4e^x x^{-3}\ln x,\ x > 0$.

10. $y'' + 2y' + y = \dfrac{e^{-x}}{(4 - x^2)^{1/2}},\ |x| < 2$.

11. $y'' + 2y' + 17y = \dfrac{64e^{-x}}{3 + \sin^2(4x)}$.

12. $y'' + 9y = \dfrac{36}{4 - \cos^2(3x)}$.

13. $y'' - 10y' + 25y = \dfrac{2e^{5x}}{4 + x^2}$.

14. $y'' - 6y' + 13y = 4e^{3x}\sec^2(2x),\ |x| < \pi/4$.

15. $y'' + y = \sec x + 4e^x,\ |x| < \pi/2$.

16. $y'' + y = \operatorname{cosec} x + 2x^2 + 5x + 1,\ 0 < x < \pi$.

17. $y'' + 4y' + 4y = 15e^{-2x}\ln x + 25\cos x,\ x > 0$.

18. $y'' + 4y' + 4y = \dfrac{4e^{-2x}}{1 + x^2} + 2x^2 - 1$.

For problems 19–21, use a Green's function to determine a particular solution to the given DE.

19. $y'' - y = F(x)$.

20. $y'' + y' - 2y = F(x)$.

21. $y'' + 5y' + 4y = F(x)$.

For problems 22 and 23, use a Green's function to solve the given IVP. (Hint: Choose $x_0 = 0$.)

22. $y'' + y = \sec x,\ y(0) = 0,\ y'(0) = 1$.

23. $y'' - 4y' + 4y = 5xe^{2x},\ y(0) = 1,\ y'(0) = 0$.

24. Determine a Green's function for

$$y'' - 2ay' + a^2y = F(x), \qquad (24.1)$$

where a is a constant, and use it to find a particular solution to (24.1) when:

(a) $F(x) = \dfrac{\alpha e^{ax}}{x^2 + \beta^2}$.

(b) $F(x) = \dfrac{\alpha e^{ax}}{(\beta^2 - x^2)^{1/2}}, \; -\beta < x < \beta$.

(c) $F(x) = x^\alpha \ln x, \; x > 0$.

In each case α and β are constants.

25. Consider the DE $y'' + y = F(x)$, where F is continuous on the interval $[a, b]$. If $x_0 \in (a, b)$, show that the solution to the IVP

$$y'' + y = F(x), \; y(x_0) = y_0, \; y'(x_0) = y_1$$

is

$$y(x) = y_0 \cos(x - x_0) + y_1 \sin(x - x_0)$$
$$+ \int_{x_0}^{x} F(t) \sin(x - t) \, dt.$$

2.9   A DIFFERENTIAL EQUATION WITH NONCONSTANT COEFFICIENTS

We end this chapter on second-order linear DE by considering a particular type of DE that has *nonconstant* coefficients. The solution to this DE will be useful in Chapter 10 and also will enable us to give a further illustration of the power of the variation-of-parameters technique introduced in the preceding section.

Definition 2.9.1:   A DE of the form

$$x^2 \frac{d^2y}{dx^2} + a_1 x \frac{dy}{dx} + a_2 y = 0,$$

where a_1, and a_2 are constants, is called a **Cauchy–Euler** equation.

Notice that if we replace x by λx, where λ is a constant, then the form of a Cauchy–Euler equation is unaltered. Such a rescaling of x can be interpreted as a dimensional change (for example, inches $\rightarrow$ cm) and so Cauchy–Euler equations are sometimes called **equi-dimensional** equations.

We will restrict attention to the interval $x > 0$ and leave the extension to the interval $(-\infty, 0)$ for the exercises. Thus, consider the DE

$$x^2 y'' + a_1 xy' + a_2 y = 0, \quad x > 0, \qquad (2.9.1)$$

where a_1 and a_2 are constants. The solution technique is based on the observation that if we substitute $y(x) = x^r$ into (2.9.1), then each of the resulting terms on the left-hand side will be multiplied by the same power of x, which suggests that there may be solutions of the form

$$y(x) = x^r \qquad (2.9.2)$$

for an appropriately chosen constant r. In order to investigate this possibility we differentiate (2.9.2) twice to obtain

$$y' = rx^{r-1}, \quad y'' = r(r - 1)x^{r-2}.$$

Substituting these expressions into equation (2.9.1) yields the condition

$$x^r[r(r - 1) + a_1 r + a_2] = 0,$$

so that (2.9.2) is indeed a solution to equation (2.9.1) provided that r satisfies

$$r(r-1) + a_1 r + a_2 = 0.$$

That is

$$r^2 + (a_1 - 1)r + a_2 = 0. \tag{2.9.3}$$

This is referred to as the **indicial equation** associated with equation (2.9.1). The roots of equation (2.9.3) are

$$r_1 = \frac{-(a_1 - 1) + \sqrt{(a_1 - 1)^2 - 4a_2}}{2}, \quad r_2 = \frac{-(a_1 - 1) - \sqrt{(a_1 - 1)^2 - 4a_2}}{2},$$

so that there are three cases to consider.

Case 1 r_1, r_2 real and distinct: In this case two solutions to equation (2.9.1) are

$$y_1(x) = x^{r_1}, \quad y_2(x) = x^{r_2}.$$

It is easily shown that

$$W[y_1, y_2](x) = (r_2 - r_1) \, x^{r_1 + r_2 - 1},$$

so that y_1 and y_2 are linearly independent on $(0, \infty)$. Consequently, the general solution to equation (2.9.1) is

$$y(x) = c_1 x^{r_1} + c_2 x^{r_2}. \tag{2.9.4}$$

Case 2 $r_1 = r_2 = -\dfrac{(a_1 - 1)}{2}$:   In this case, we obtain only one solution to equation (2.9.1), namely

$$y_1(x) = x^{r_1}.$$

A second linearly independent solution to equation (2.9.1) can be obtained using the reduction of order technique. Setting

$$y_2(x) = x^{r_1} u(x)$$

and differentiating with respect to x yields

$$y_2'(x) = x^{r_1} u' + r_1 x^{r_1 - 1} u, \quad y_2''(x) = x^{r_1} u'' + 2 r_1 x^{r_1 - 1} u' + r_1(r_1 - 1) x^{r_1 - 2} u.$$

Substituting these expressions into equation (2.9.1) we obtain the following equation for u:

$$x^2 [x^{r_1} u'' + 2 r_1 x^{r_1 - 1} u' + r_1(r_1 - 1) x^{r_1 - 2} u] + a_1 x (x^{r_1} u' + r_1 x^{r_1 - 1} u) + a_2 x^{r_1} u = 0.$$

Equivalently,

$$x^{r_1 + 2} u'' + (2 r_1 + a_1) x^{r_1 + 1} u' + x^{r_1} [r_1(r_1 - 1) + a_1 r_1 + a_2] u = 0.$$

The last term on the left-hand side vanishes, since r_1 is a root of the indicial equation (2.9.3). Thus, u must satisfy

$$u'' + x^{-1} u' = 0,$$

where we have substituted for $r_1 = -(a_1 - 1)/2$. This separable equation can be written as

$$\frac{u''}{u'} = -\frac{1}{x},$$

which can be integrated directly to obtain

$$\ln |u'| = -\ln x + c.$$

We can therefore choose

$$u' = x^{-1}.$$

Integrating once more yields

$$u(x) = \ln x,$$

where we have set the integration constant to zero, since we only require one solution. Consequently, a second solution to equation (2.9.1) in this case is

$$y_2(x) = x^{r_1} \ln x.$$

We leave it as an exercise to verify that

$$W[y_1, y_2](x) = x^{2r_1 - 1},$$

so that y_1 and y_2 are linearly independent on $(0, \infty)$. The general solution to equation (2.9.1) is therefore given by

$$y(x) = c_1 x^{r_1} + c_2 x^{r_1} \ln x = x^{r_1}(c_1 + c_2 \ln x). \tag{2.9.5}$$

Case 3   Complex conjugate roots, $r_1 = a + ib$, $r_2 = a - ib$, $b \neq 0$:   In this case, two complex-valued solutions to equation (2.9.1) are

$$w_1(x) = x^{a+ib} = e^{(a+ib)\ln x} = x^a[\cos(b \ln x) + i\sin(b \ln x)],$$

$$w_2(x) = x^{a-ib} = e^{(a-ib)\ln x} = x^a[\cos(b \ln x) - i\sin(b \ln x)],$$

where we have used Euler's formula. Two corresponding real-valued solutions are

$$y_1(x) = \frac{1}{2}[w_1(x) + w_2(x)] = x^a \cos(b \ln x),$$

$$y_2(x) = \frac{1}{2i}[w_1(x) - w_2(x)] = x^a \sin(b \ln x).$$

These solutions have Wronskian

$$W[y_1, y_2](x) = bx^{2a-1},$$

which is nonzero, since $b \neq 0$. Consequently, y_1 and y_2 are linearly independent on $(0, \infty)$, and so the general solution to equation (2.9.1) in this case is

$$y(x) = x^a[c_1 \cos(b \ln x) + c_2 \sin(b \ln x)]. \tag{2.9.6}$$

The preceding results are summarized in Table 2.9.1.

TABLE 2.9.1

Roots of indicial equation $r^2 + (a_1 - 1)r + a_2 = 0$	Linearly independent solutions to $x^2 y'' + a_1 xy' + a_2 y = 0$, $x > 0$
Real distinct: $r_1 \neq r_2$	$y_1(x) = x^{r_1}$, $y_2(x) = x^{r_2}$
Real repeated: $r_1 = r_2$	$y_1(x) = x^{r_1}$, $y_2(x) = x^{r_1} \ln x$
Complex conjugates:	
$r_1 = a + ib$, $r_2 = a - ib$	$y_1(x) = x^a \cos(b \ln x)$,
	$y_2(x) = x^a \sin(b \ln x)$.

Example 2.9.1   Solve
$$x^2 y'' - xy' - 8y = 0, \qquad x > 0. \tag{2.9.7}$$

Solution   Inserting $y = x^r$ into equation (2.9.7) yields the indicial equation
$$r(r-1) - r - 8 = 0.$$

That is
$$r^2 - 2r - 8 = (r-4)(r+2) = 0.$$

Hence, two linearly independent solutions to equation (2.9.7) are
$$y_1(x) = x^4, \qquad y_2(x) = x^{-2}.$$

Consequently, equation (2.9.7) has general solution
$$y(x) = c_1 x^4 + c_2 x^{-2}.$$

Example 2.9.2   Solve the IVP
$$x^2 y'' - 3xy' + 13y = 0, \tag{2.9.8}$$
$$y(1) = 2, \ y'(1) = -5.$$

Solution   Substituting $y = x^r$ into equation (2.9.8) yields the indicial equation
$$r^2 - 4r + 13 = 0,$$

which has the complex conjugate roots
$$r = 2 \pm 3i.$$

It follows that two linearly independent solutions to equation (2.9.8) are
$$y_1(x) = x^2 \cos(3\ln x), \qquad y_2(x) = x^2 \sin(3\ln x)$$

so that the general solution is
$$y(x) = c_1 x^2 \cos(3\ln x) + c_2 x^2 \sin(3\ln x),$$

which we write as
$$y(x) = x^2 [c_1 \cos(3\ln x) + c_2 \sin(3\ln x)].$$

The first initial condition, $y(1) = 2$, requires that
$$c_1 \cos 0 + c_2 \sin 0 = 2,$$

so that $c_1 = 2$. Inserting this value of c_1 into the general solution and differentiating with respect to x yields
$$y'(x) = 2x[2\cos(3\ln x) + c_2\sin(3\ln x)] + x^2[-6x^{-1}\sin(3\ln x) + 3x^{-1}c_2\cos(3\ln x)].$$

The second initial condition therefore requires
$$2(2 + 0) + (0 + 3c_2) = -5,$$

so that $c_2 = -3$. Consequently the solution to the IVP is
$$y(x) = x^2 [2\cos(3\ln x) - 3\sin(3\ln x)].$$

A sketch of the corresponding solution curve is given in Figure 2.9.1. Due to the trigonometric terms the solution is oscillatory. The amplitude of the oscillation is growing rapidly with x due to the multiplicative factor x^2. Furthermore, as $x \to 0^+$, the amplitude also approaches zero.

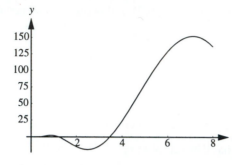

Figure 2.9.1 The solution to the IVP in Example 2.9.2.

Now consider the nonhomogeneous equation

$$x^2 y'' + a_1 x y' + a_2 y = g(x), \tag{2.9.9}$$

where a_1 and a_2 are constants. Since the associated homogeneous equation is a Cauchy–Euler equation, we can determine the complementary function, and hence the variation-of-parameters method can be used to determine a particular solution. We must remember, however, that the formulas derived in the variation-of-parameters technique are based around a DE written in the standard form

$$y'' + a_1(x)y' + a_2(x)y = F(x).$$

Consequently, when applying the method to a DE of the form (2.9.9), the appropriate formulas for determining u_1 and u_2 are

$$
\begin{aligned}
y_1 u_1' + y_2 u_2' &= 0, \\
y_1' u_1' + y_2' u_2' &= x^{-2} g(x).
\end{aligned}
$$

We illustrate with an example.

Example 2.9.3   Find the general solution to

$$x^2 y'' - 3xy' + 4y = x^2 \ln x, \quad x > 0. \tag{2.9.10}$$

Solution   The associated homogeneous equation is the Cauchy–Euler equation

$$x^2 y'' - 3xy' + 4y = 0. \tag{2.9.11}$$

Substituting $y = x^r$ into this equation yields the indicial equation

$$r^2 - 4r + 4 = 0.$$

That is

$$(r - 2)^2 = 0.$$

Hence two linearly independent solutions to equation (2.9.11) are

$$y_1(x) = x^2, \quad y_2(x) = x^2 \ln x.$$

According to the variation-of-parameters technique, a particular solution to equation (2.9.10) is

$$y_p(x) = y_1(x) u_1(x) + y_2(x) u_2(x) = x^2 u_1 + x^2 \ln x \, u_2, \tag{2.9.12}$$

where u_1 and u_2 are determined from

$$x^2 u_1{}' + x^2 \ln x\, u_2{}' = 0,$$

$$2xu_1{}' + (2x\ln x + x)u_2{}' = \ln x.$$

Hence,

$$u_1{}' = -x^{-1}(\ln x)^2, \quad u_2{}' = x^{-1}\ln x,$$

which upon integration gives

$$u_1(x) = -\frac{1}{3}(\ln x)^3, \quad u_2(x) = \frac{1}{2}(\ln x)^2,$$

where we have set the integration constants to zero without loss of generality. Substitution into (2.9.12) yields

$$y_p(x) = -\frac{1}{3}x^2(\ln x)^3 + \frac{1}{2}x^2(\ln x)^3 = \frac{1}{6}x^2(\ln x)^3.$$

Thus, equation (2.9.10) has general solution

$$y(x) = c_1 x^2 + c_2 x^2 \ln x + \frac{1}{6}x^2(\ln x)^3,$$

which can be written as

$$y(x) = \frac{1}{6}x^2[c_1 + c_2\ln x + (\ln x)^3],$$

where we have redefined the constants.

EXERCISES 2.9

For problems 1–8, determine the general solution to the given DE on $(0, \infty)$.

1. $x^2 y'' - xy' + 5y = 0.$

2. $x^2 y'' - 6y = 0.$

3. $x^2 y'' - 3xy' + 4y = 0.$

4. $x^2 y'' - 4xy' + 4y = 0.$

5. $x^2 y'' + 3xy' + y = 0.$

6. $x^2 y'' + 5xy' + 13y = 0.$

7. $x^2 y'' - xy' - 35y = 0.$

8. $x^2 y'' + xy' + 16y = 0.$

For problems 9–11, solve the given Cauchy–Euler equation on the interval $(0, \infty)$. In each case, m and k are positive constants.

9. $x^2 y'' + xy' - m^2 y = 0.$

10. $x^2 y'' - x(2m - 1)y' + m^2 y = 0.$

11. $x^2 y'' - x(2m - 1)y' + (m^2 + k^2)y = 0.$

12. Consider the Cauchy–Euler equation

$$x^2 y'' + xa_1 y' + a_2 y = 0, \quad x > 0. \quad (12.1)$$

(a) Show that the change of independent variable defined by $x = e^z$ transforms equation (12.1) into the constant coefficient equation

$$\frac{d^2 y}{dz^2} + (a_1 - 1)\frac{dy}{dz} + a_2 y = 0. \quad (12.2)$$

(b) Show that if $y_1(z)$, $y_2(z)$ are linearly independent solutions to equation (12.2), then $y_1(\ln x)$, $y_2(\ln x)$ are linearly independent solutions to equation (12.1). (Hint: We already know from (a) that y_1 and y_2 are solutions to equation (12.1). To show that they are linearly independent, verify that

$$W[y_1, y_2](x) = \frac{dz}{dx}W[y_1, y_2](z).)$$

13. Consider the Cauchy–Euler equation

$$x^2 y'' + axy' + by = 0, \quad x < 0. \quad (13.1)$$

Show that the substitution $y = (-x)^r$ yields the indicial equation

$$r^2 + (a - 1)r + b = 0.$$

Thus, linearly independent solutions to equation (13.1) on $(-\infty, 0)$ can be determined by the replacing x with $-x$, in (2.9.4), (2.9.5), and (2.9.6). Consequently, if we replace x by $|x|$ in these solutions we will obtain solutions to equation (2.9.1) that are valid for all $x \neq 0$.

For problems 14–21, solve the given DE on the interval $x > 0$. Remember to write your equation in standard form before applying the variation-of-parameters method.

14. $x^2y'' + 4xy' + 2y = 4\ln x$.

15. $x^2y'' - 4xy' + 6y = x^4\sin x$.

16. $x^2y'' + 6xy' + 6y = 4e^{2x}$.

17. $x^2y'' - 3xy' + 4y = \dfrac{x^2}{\ln x}$.

18. $x^2y'' + 4xy' + 2y = \cos x$.

19. $x^2y'' + xy' + 9y = 9\ln x$.

20. $x^2y'' - xy' + 5y = 8x(\ln x)^2$.

21. $x^2y'' - (2m - 1)xy' + m^2y = x^m(\ln x)^k$, where m and k are constants.

22. (a) Solve the IVP

$$x^2y'' - xy' + 5y = 0,$$

$$y(1) = \sqrt{2}, \quad y'(1) = 3\sqrt{2},$$

and show that your solution can be written in the form

$$y(x) = 2x\cos(2\ln x - \pi/4).$$

(b) Determine all zeros of $y(x)$.

◆ **(c)** Sketch the corresponding solution curve on the interval $[0.001, 16]$, and verify the zeros in the interval $[3, 16]$.

23. The motion of a physical system is governed by the IVP

$$t^2\frac{d^2y}{dt^2} + t\frac{dy}{dt} + 25y = 0$$

$$y(1) = 3\sqrt{3}/2, \quad y'(1) = 15/2.$$

(a) Solve the given IVP, and show that your solution can be written in the form

$$y(t) = 3\cos(5\ln t - \pi/6).$$

(b) Determine all zeros of $y(t)$.

◆ **(c)** Sketch the corresponding solution curve on the interval $[0.01, 2]$.

(d) Is the system performing simple harmonic motion? Justify your answer.

24. We have shown that in the case of complex conjugate roots, $r = a \pm ib$, $b \neq 0$, of the indicial equation the general solution to the Cauchy–Euler equation

$$y'' + a_1y' + a_2y = 0, x > 0$$

is

$$y(x) = x^a[c_1\cos(b\ln x) + c_2\sin(b\ln x)]. \quad (24.1)$$

(a) Show that (24.1) can be written in the form

$$y(x) = A_0x^a\cos(b\ln x - \phi)$$

for appropriate constants A and ϕ.

(b) Determine all zeros of $y(x)$, and the distance between successive zeros. What happens to this distance as $x \to \infty$, and as $x \to 0^+$?

(c) Describe the behavior of the solution as $x \to \infty$ and as $x \to 0^+$ in each of the three cases $a > 0$, $a < 0$, and $a = 0$. In each case, give a general sketch of a generic solution curve.

3

**Matrices and Systems
of Linear Equations**

We will see in the later chapters that most problems in linear algebra can be reduced to questions regarding the solutions of systems of linear algebraic equations. In preparation for this, the next two chapters are concerned with giving a detailed introduction to the theory and solution techniques for such systems. An example of a linear system of algebraic equations in the unknowns x_1, x_2, x_3 is

$$3x_1 + 4x_2 - 7x_3 = 5,$$
$$2x_1 - 3x_2 + 9x_3 = 7,$$
$$7x_1 + 2x_2 - 3x_3 = 4.$$

We see that this system is completely determined by the array of numbers

$$\begin{bmatrix} 3 & 4 & -7 & 5 \\ 2 & -3 & 9 & 7 \\ 7 & 2 & -3 & 4 \end{bmatrix},$$

which contains the coefficients of the unknowns on the left-hand side of the system and the numbers appearing on the right-hand side of the system. Such an array is an example of a matrix. In this chapter, we see that, in general, linear algebraic systems of equations are best represented in terms of matrices and that once such a representation has been made, the solution to the system can be easily determined (providing it has a solution). In the first few sections of this chapter we therefore introduce the basics of matrix algebra. We then apply matrices to solve systems of linear algebraic equations. In Chapter 8 we will see how matrices also give a natural framework for formulating and solving systems of linear differential equations.

3.1   MATRICES: DEFINITIONS AND NOTATION

We begin our discussion of matrices with a definition.

> ***Definition 3.1.1:***   An $m \times n$ **matrix** is a rectangular array of numbers arranged in m horizontal rows and n vertical columns. Matrices are usually denoted by upper case letters, such as A and B. The entries in the matrix are called the **elements** of the matrix.

Example 3.1.1   The following are examples of a 2×3 and a 3×3 matrix

respectively

$$A = \begin{bmatrix} \frac{3}{2} & \frac{5}{4} & \frac{1}{5} \\ 0 & -\frac{3}{7} & \frac{5}{9} \end{bmatrix} \qquad B = \begin{bmatrix} 2 & -1 & 3 \\ 1 & 1 & -1 \\ 0 & 0 & 1 \end{bmatrix}.$$

□

We will use the index notation to denote the elements of a matrix. According to this notation, the element in the ith row and jth column of the matrix A is denoted a_{ij}. Thus, for the matrices in the previous example we have

$$a_{13} = \frac{1}{5}, \quad a_{22} = -\frac{3}{7}, \quad b_{23} = -1, \quad \text{etc.}$$

Using the index notation, a general $m \times n$ matrix A is written

$$A = \begin{bmatrix} a_{11} & a_{12} & \cdots & a_{1n} \\ a_{21} & a_{22} & \cdots & a_{2n} \\ \vdots & \vdots & & \vdots \\ a_{m1} & a_{m2} & \cdots & a_{mn} \end{bmatrix},$$

or, in a more abbreviated form, $A = [a_{ij}]$.

ROW VECTORS AND COLUMN VECTORS

Of special interest to us in the future will be $1 \times n$ and $n \times 1$ matrices. For this reason we give them special names.

> ***Definition 3.1.2:***   A $1 \times n$ matrix is called a **row vector**. An $n \times 1$ matrix is called a **column vector**. The elements of a row or column vector are called the **components** of the vector.

REMARKS

(**1**) If we wish to emphasize the number of components in a row vector or a column vector, we will refer to such vectors as row n-vectors and column n-vectors, respectively.

(**2**) We will see later in the chapter that when a system of linear equations is written using matrices, the basic unknown in the reformulated system is a column vector. A similar formulation will also be given for systems of differential equations.

Example 3.1.2 $\mathbf{a} = [\, 1 \quad -1 \quad 3 \quad 4 \quad 0 \,]$ is a row 5-vector, whereas $\mathbf{b} = \begin{bmatrix} \frac{2}{3} \\ -\frac{1}{5} \\ \frac{4}{7} \end{bmatrix}$ is a

column 3-vector.   ◻

  As indicated in the above example we usually denote a column or row vector by a lowercase letter in **bold** print.

  Associated with any $m \times n$ matrix are m row n-vectors and n column m-vectors which are referred to as the **row vectors** of the matrix and the **column vectors** of the matrix, respectively.

Example 3.1.3   Associated with the matrix $A = \begin{bmatrix} -2 & 1 & 3 & 4 \\ 1 & 2 & 1 & 1 \\ 3 & -1 & 2 & 5 \end{bmatrix}$ are the row

vectors

$$[-2 \ 1 \ 3 \ 4], \ [1 \ 2 \ 1 \ 1], \ \text{and} \ [3 \ -1 \ 2 \ 5];$$

and the column vectors

$$\begin{bmatrix} -2 \\ 1 \\ 3 \end{bmatrix}, \ \begin{bmatrix} 1 \\ 2 \\ -1 \end{bmatrix}, \ \begin{bmatrix} 3 \\ 1 \\ 2 \end{bmatrix}, \ \text{and} \ \begin{bmatrix} 4 \\ 1 \\ 5 \end{bmatrix}.$$ ◻

  Conversely, if $\mathbf{a}_1, \mathbf{a}_2, \ldots, \mathbf{a}_n$ are each column m-vectors, then we let $[\mathbf{a}_1, \mathbf{a}_2, \ldots, \mathbf{a}_n]$ denote the $m \times n$ matrix whose column vectors are $\mathbf{a}_1, \mathbf{a}_2, \ldots, \mathbf{a}_n$. Similarly, if $\mathbf{b}_1, \mathbf{b}_2, \ldots, \mathbf{b}_m$ are each row n-vectors then we write

$$\begin{bmatrix} \mathbf{b}_1 \\ \mathbf{b}_2 \\ \vdots \\ \mathbf{b}_m \end{bmatrix}$$

for the $m \times n$ matrix with row vectors $\mathbf{b}_1, \mathbf{b}_2, \ldots, \mathbf{b}_m$.

Example 3.1.4   If $\mathbf{a}_1 = \begin{bmatrix} \frac{1}{5} \\ 2 \\ 3 \end{bmatrix}$, $\mathbf{a}_2 = \begin{bmatrix} \frac{4}{7} \\ 5 \\ 9 \end{bmatrix}$, and $\mathbf{a}_3 = \begin{bmatrix} -\frac{1}{3} \\ 3 \\ 11 \end{bmatrix}$, then

$$[\mathbf{a}_1, \mathbf{a}_2, \mathbf{a}_3] = \begin{bmatrix} \frac{1}{5} & \frac{4}{7} & -\frac{1}{3} \\ \frac{2}{3} & \frac{5}{9} & \frac{3}{11} \end{bmatrix}.$$

□

If we interchange the row vectors and column vectors in an $m \times n$ matrix, we obtain an $n \times m$ matrix called the **transpose** of A. We denote this matrix by A^T. In index notation, the ijth element of A^T, denoted a^T_{ij}, is given by

$$a^T_{ij} = a_{ji}.$$

Example 3.1.5   If $A = \begin{bmatrix} 1 & 2 & 6 & 2 \\ 0 & 3 & 4 & 7 \end{bmatrix}$, then $A^T = \begin{bmatrix} 1 & 0 \\ 2 & 3 \\ 6 & 4 \\ 2 & 7 \end{bmatrix}$. If $A = \begin{bmatrix} 1 & -3 & 5 \\ 2 & 0 & 7 \\ 3 & 4 & 9 \end{bmatrix}$,

then $A^T = \begin{bmatrix} 1 & 2 & 3 \\ -3 & 0 & 4 \\ 5 & 7 & 9 \end{bmatrix}$.

$n \times n$ MATRICES

An $n \times n$ matrix is called a **square matrix**, since it has the same number of rows as columns. If A is a square matrix, then the elements a_{ii}, $1 \le i \le n$, make up the **main diagonal**, or **leading diagonal**, of the matrix. (See Figure 3.1.1 for the 3×3 case.)

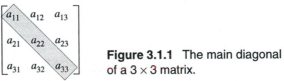

Figure 3.1.1 The main diagonal of a 3×3 matrix.

The sum of the main diagonal elements of an $n \times n$ matrix A is called the **trace** of A and is denoted tr(A). Thus,

$$\text{tr}(A) = a_{11} + a_{22} + \cdots + a_{nn}.$$

An $n \times n$ matrix A is said to be **lower triangular** if $a_{ij} = 0$ whenever $i < j$ (zeros everywhere above the main diagonal), and it is said to be **upper triangular** if $a_{ij} = 0$ whenever $i > j$ (zeros everywhere below the main diagonal). The following are examples of an upper triangular and a lower triangular matrix, respectively:

$$\begin{bmatrix} 1 & 8 & 5 \\ 0 & 3 & 9 \\ 0 & 0 & -1 \end{bmatrix}, \quad \begin{bmatrix} 2 & 0 & 0 \\ 0 & 1 & 0 \\ 6 & 7 & 3 \end{bmatrix}.$$

If every element on the main diagonal of a lower (upper) triangular matrix is a one, the matrix is called a **unit** lower (upper) triangular matrix.

An $n \times n$ matrix $D = [d_{ij}]$ that has all *off–diagonal* elements equal to zero is called a **diagonal** matrix. Such a matrix is completely determined by giving its main diagonal elements. Consequently, we can specify a diagonal matrix in the compact form

$$D = \text{diag}(d_1, d_2, \ldots, d_n),$$

where the d_i denote the diagonal elements.

Example 3.1.6 $D = \text{diag}(1, 2, 0, -3)$ is the 4×4 diagonal matrix

$$D = \begin{bmatrix} 1 & 0 & 0 & 0 \\ 0 & 2 & 0 & 0 \\ 0 & 0 & 0 & 0 \\ 0 & 0 & 0 & -3 \end{bmatrix}.$$

☐

The transpose naturally picks out two important types of $n \times n$ matrices as follows.

Definition   3.1.3:

1. A square matrix A satisfying $A^T = A$ is called a **symmetric matrix**.

2. A square matrix A satisfying $A^T = -A$ is called a **skew-symmetric** (or **anti–symmetric**) **matrix**.

Example 3.1.7 $A = \begin{bmatrix} 1 & -1 & 1 & 5 \\ -1 & 2 & 2 & 6 \\ 1 & 2 & 3 & 4 \\ 5 & 6 & 4 & 9 \end{bmatrix}$ is a symmetric matrix, whereas

$B = \begin{bmatrix} 0 & -1 & -5 & 3 \\ 1 & 0 & 1 & -2 \\ 5 & -1 & 0 & 7 \\ -3 & 2 & -7 & 0 \end{bmatrix}$ is a skew-symmetric matrix.

☐

Notice that the main diagonal elements of the skew-symmetric matrix in the preceding example are all zero. This is true in general, since if A is a skew-symmetric matrix, then $a_{ij} = -a_{ji}$, which implies that when $i = j$, $a_{ii} = -a_{ii}$, so that $a_{ii} = 0$.

EXERCISES 3.1

1. If $A = \begin{bmatrix} 1 & -2 & 3 & 2 \\ 7 & -6 & 5 & -1 \\ 0 & 2 & -3 & -4 \end{bmatrix}$, determine a_{31}, a_{24}, a_{14}, a_{32}, a_{21}, a_{34}.

For problems 2–6, write the matrix with the given elements. In each case specify the dimension of the matrix.

2. $a_{11} = 1$, $a_{21} = -1$, $a_{12} = 5$, $a_{22} = 3$.

3. $a_{11} = 2$, $a_{12} = 1$, $a_{13} = -1$, $a_{21} = 0$, $a_{22} = 4$, $a_{23} = -2$.

4. $a_{11} = -1$, $a_{41} = -5$, $a_{31} = 1$, $a_{21} = 1$.

5. $a_{11} = 1$, $a_{31} = 2$, $a_{42} = -1$, $a_{32} = 7$, $a_{13} = -2$, $a_{23} = 0$, $a_{33} = 4$, $a_{21} = 3$, $a_{41} = -4$, $a_{12} = -3$, $a_{22} = 6$, $a_{43} = 5$.

6. $a_{12} = -1$, $a_{13} = 2$, $a_{23} = 3$, $a_{ji} = -a_{ij}$, $1 \leq i \leq 3$, $1 \leq j \leq 3$.

For problems 7–9, determine $tr(A)$ for the given matrix.

7. $A = \begin{bmatrix} 1 & 0 \\ 2 & 3 \end{bmatrix}$.

8. $A = \begin{bmatrix} 1 & 2 & -1 \\ 3 & 2 & -2 \\ 7 & 5 & -3 \end{bmatrix}$.

9. $A = \begin{bmatrix} 2 & 0 & 1 \\ 3 & 2 & 5 \\ 0 & 1 & -5 \end{bmatrix}$.

For problems 10–12, write the column vectors and row vectors of the given matrix.

10. $A = \begin{bmatrix} 1 & -1 \\ 3 & 5 \end{bmatrix}$.

11. $A = \begin{bmatrix} 1 & 3 & -4 \\ -1 & -2 & 5 \\ 2 & 6 & 7 \end{bmatrix}$.

12. $A = \begin{bmatrix} 2 & 10 & 6 \\ 5 & -1 & 3 \end{bmatrix}$.

13. If $\mathbf{a}_1 = [1 \quad 2]$, $\mathbf{a}_2 = [3 \quad 4]$, and $\mathbf{a}_3 = [5 \quad 1]$, write the matrix

$$A = \begin{bmatrix} \mathbf{a}_1 \\ \mathbf{a}_2 \\ \mathbf{a}_3 \end{bmatrix},$$

and determine the column vectors of A.

14. If

$$\mathbf{b}_1 = \begin{bmatrix} 2 \\ -1 \\ 4 \end{bmatrix}, \quad \mathbf{b}_2 = \begin{bmatrix} 5 \\ 7 \\ -6 \end{bmatrix},$$

$$\mathbf{b}_3 = \begin{bmatrix} 0 \\ 0 \\ 0 \end{bmatrix}, \quad \mathbf{b}_4 = \begin{bmatrix} 1 \\ 2 \\ 3 \end{bmatrix},$$

write the matrix $B = [\mathbf{b}_1, \mathbf{b}_2, \mathbf{b}_3, \mathbf{b}_4]$, and determine the row vectors of B.

15. If $\mathbf{a}_1, \mathbf{a}_2, ..., \mathbf{a}_p$ are each column q-vectors, what are the dimensions of the matrix that has $\mathbf{a}_1$, $\mathbf{a}_2, ..., \mathbf{a}_p$ as its column vectors ?

For problems 16–18, give an example of a matrix of the specified form

16. 3×3 diagonal matrix.

17. 4×4 upper triangular matrix.

18. 4×4 skew-symmetric matrix.

19. 3×3 upper triangular symmetric matrix.

20. 3×3 lower triangular skew-symmetric.

21. Prove that a symmetric upper triangular matrix is diagonal.

22. Determine all elements of the 3×3 skew–symmetric matrix A with $a_{21} = 1$, $a_{31} = 3$, $a_{23} = -1$.

3.2   MATRIX ALGEBRA

In the previous section we introduced the general idea of a matrix. The next step is to develop the algebra of matrices. We first need to define what is meant by equality of matrices.

Definition 3.2.1:   Two matrices A and B are **equal**, written $A = B$, if and only if

(a)   They both have the same dimensions.

(b)   All corresponding elements in the matrices are equal.

According to Definition 3.2.1, even though the matrices

$$A = \begin{bmatrix} 1 & 2 & 3 \\ 4 & 5 & 6 \end{bmatrix} \text{ and } B = \begin{bmatrix} 4 & 2 \\ 3 & 6 \\ 1 & 5 \end{bmatrix}$$

contain the same six numbers, and therefore store the same basic information, they are not equal as matrices.

ADDITION AND SUBTRACTION OF MATRICES AND MULTIPLICATION OF A MATRIX BY A SCALAR

Addition (and subtraction) of matrices is only defined for matrices with the same dimensions.

> **Definition 3.2.2:**   If A and B are both $m \times n$ matrices then we define the **sum** of A and B, denoted by $A + B$, to be the $m \times n$ matrix whose elements are obtained by adding *corresponding* elements of A and B. Thus, if $A = [a_{ij}]$, $B = [b_{ij}]$, and $C = A + B$, then C has elements $c_{ij} = a_{ij} + b_{ij}$.

Example 3.2.1

$$\begin{bmatrix} 2 & -1 & 3 \\ 4 & -5 & 0 \end{bmatrix} + \begin{bmatrix} -1 & 0 & 5 \\ -5 & 2 & 7 \end{bmatrix} = \begin{bmatrix} 1 & -1 & 8 \\ -1 & -3 & 7 \end{bmatrix}.$$
❐

We see directly from Definition 3.2.2 that if A and B are both $m \times n$ matrices, then

$$A + B = B + A \qquad \text{(Matrix addition is commutative)}$$
$$A + (B + C) = (A + B) + C \qquad \text{(Matrix addition is associative)}$$

In order that we can model oscillatory physical phenomena, in much of the later work we will need to use complex[1] as well as real numbers. Throughout the text we will use the term **scalar** to mean a real or complex number.

> **Definition 3.2.3:**   If A is an $m \times n$ matrix and s is a scalar, then we let sA denote the matrix obtained by multiplying every element of A by s. This procedure is called **scalar multiplication**. In index notation, if $A = [a_{ij}]$, then $sA = [sa_{ij}]$.

Example 3.2.2   If $A = \begin{bmatrix} 2 & -1 \\ 4 & 6 \end{bmatrix}$, then $5A = \begin{bmatrix} 10 & -5 \\ 20 & 30 \end{bmatrix}$.

Example 3.2.3   If $A = \begin{bmatrix} 1 + i & i \\ 2 + 3i & 4 \end{bmatrix}$ and $s = 1 - 2i$, where $i = \sqrt{-1}$, find sA.

---

[1] A review of complex numbers is given in Appendix 1. We have already seen in Chapter 2 how complex numbers arise naturally when modeling oscillatory phenomena.

Solution

$$sA = \begin{bmatrix} (1 - 2i)(1 + i) & (1 - 2i)i \\ (1 - 2i)(2 + 3i) & (1 - 2i)4 \end{bmatrix} = \begin{bmatrix} 3 - i & 2 + i \\ 8 - i & 4 - 8i \end{bmatrix}.$$

Definition 3.2.4:   We define **subtraction** of two matrices with the *same dimensions* as

$$A - B = A + (-1)B.$$

That is, we subtract corresponding elements. In index notation $A - B = [a_{ij} - b_{ij}]$.

Further properties satisfied by the operations of matrix addition and multiplication of a matrix by a scalar are as follows:

$$1 \cdot A = A,$$
$$s(A + B) = sA + sB,$$
$$(s + t)A = sA + tA,$$
$$s(tA) = (st)A = (ts)A = t(sA),$$

for any scalars s, t.

The $m \times n$ **zero matrix**, denoted 0, is the $m \times n$ matrix whose elements are all zeros. If we wish to emphasize the dimensions of the zero matrix, then we will use the notation $0_{m \times n}$ or, in the case of the $n \times n$ zero matrix, 0_n. The first property of the zero matrix listed in the next theorem indicates that the zero matrix plays a similar role in matrix addition to that played by the number zero in real-number addition.

Properties of Zero Matrix   For all matrices A,

(a) $A + 0 = A$
(b) $A - A = 0$
(c) $0(A) = 0$.

MULTIPLICATION OF MATRICES

The definition of how to multiply a matrix by a scalar is essentially the only possibility if, in the case when s is a positive integer, we want sA to be the same matrix as the one obtained when A is added to itself s times. We now define how to multiply two matrices together. In this case the multiplication operation is by no means obvious, and so we will build up to the general definition in three stages.

CASE 1: Product of a row n-vector and a column n-vector   We begin by generalizing a concept from elementary calculus. If $\mathbf{a}$ and $\mathbf{b}$ are either row or column n-vectors, with components $a_1, a_2, \ldots, a_n$, and $b_1, b_2, \ldots, b_n$, respectively, then their **dot product**, denoted $\mathbf{a} \cdot \mathbf{b}$, is the *number*

$$\mathbf{a} \cdot \mathbf{b} = a_1b_1 + a_2b_2 + \cdots + a_nb_n.$$

As we will see, this is the key formula in defining the product of two matrices. Now let $\mathbf{a}$ be a **row** n-vector, and let $\mathbf{x}$ be a **column** n-vector. Then their matrix product $\mathbf{ax}$ is the *1×1 matrix* whose single element is obtained by taking the dot product of $\mathbf{a}$ with $\mathbf{x}$. Thus,

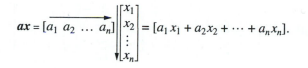

$$ax = [a_1 \ a_2 \ \cdots \ a_n]\begin{bmatrix} x_1 \\ x_2 \\ \vdots \\ x_n \end{bmatrix} = [a_1 x_1 + a_2 x_2 + \cdots + a_n x_n].$$

Example 3.2.4   If $a = [2 \ \ -1 \ \ 3 \ \ 5]$ and $x = \begin{bmatrix} 3 \\ 2 \\ -3 \\ 4 \end{bmatrix}$, then

$$ax = [2 \ \ -1 \ \ 3 \ \ 5]\begin{bmatrix} 3 \\ 2 \\ -3 \\ 4 \end{bmatrix} = [(2)(3) + (-1)(2) + (3)(-3) + (5)(4)] = [15].$$

CASE 2:   Product of an $m \times n$ matrix with a column n-vector   If A is an $m \times n$ matrix and x is a column n-vector, then the product Ax is defined to be the $m \times 1$ matrix whose ith element is obtained by taking the dot product of the ith row vector of A with x. (See Figure. 3.2.1.)

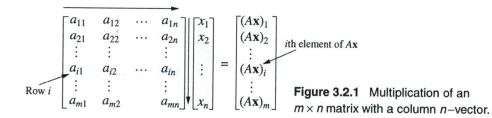

Figure 3.2.1   Multiplication of an $m \times n$ matrix with a column n–vector.

The ith row vector of A, $\mathbf{a}_i$, is

$$\mathbf{a}_i = [a_{i1} \ \ a_{i2} \ \ \cdots \ \ a_{in}],$$

so that Ax has ith element

$$(A\mathbf{x})_i = a_{i1}x_1 + a_{i2}x_2 + \cdots + a_{in}x_n.$$

Consequently the column vector Ax has elements

$$(A\mathbf{x})_i = \sum_{k=1}^{n} a_{ik}x_k, \qquad 1 \le i \le m. \tag{3.2.1}$$

As illustrated in the next example, in practice, we do not use the formula (3.2.1), rather, we explicitly take the dot products of the row vectors of A with the column vector $\mathbf{x}$.

Example 3.2.5   Find $A\mathbf{x}$ if $A = \begin{bmatrix} 2 & 3 & -1 \\ 1 & 4 & -6 \\ 5 & -2 & 0 \end{bmatrix}$ and $\mathbf{x} = \begin{bmatrix} 7 \\ -3 \\ 1 \end{bmatrix}$.

Solution

$$A\mathbf{x} = \begin{bmatrix} 2 & 3 & -1 \\ 1 & 4 & -6 \\ 5 & -2 & 0 \end{bmatrix} \begin{bmatrix} 7 \\ -3 \\ 1 \end{bmatrix} = \begin{bmatrix} 4 \\ -11 \\ 41 \end{bmatrix}.$$

$\square$

The following result regarding multiplication of a column vector by a matrix will be used repeatedly in the later chapters.

Theorem 3.2.1:   If $A = [\mathbf{a}_1, \mathbf{a}_2, \ldots, \mathbf{a}_n]$ is an $m \times n$ matrix and $\mathbf{c} = \begin{bmatrix} c_1 \\ c_2 \\ \vdots \\ c_n \end{bmatrix}$ is a

column n-vector, then

$$A\mathbf{c} = c_1 \mathbf{a}_1 + c_2 \mathbf{a}_2 + \cdots + c_n \mathbf{a}_n. \tag{3.2.2}$$

PROOF   The ikth element of A is the ith component of $\mathbf{a}_k$, so

$$a_{ik} = (\mathbf{a}_k)_i.$$

Applying formula (3.2.1) for multiplication of a column vector by a matrix yields

$$(A\mathbf{c})_i = \sum_{k=1}^{n} a_{ik} c_k = \sum_{k=1}^{n} (\mathbf{a}_k)_i c_k = \sum_{k=1}^{n} (c_k \mathbf{a}_k)_i.$$

Consequently,

$$A\mathbf{c} = \sum_{k=1}^{n} c_k \mathbf{a}_k = c_1 \mathbf{a}_1 + c_2 \mathbf{a}_2 + \cdots + c_n \mathbf{a}_n,$$

as required.   ∎

If $\mathbf{x}_1, \mathbf{x}_2, \ldots, \mathbf{x}_n$ are column m-vectors and $c_1, c_2, \ldots, c_n$ are scalars, then an expression of the form

$$c_1 \mathbf{x}_1 + c_2 \mathbf{x}_2 + \cdots + c_n \mathbf{x}_n$$

is called a **linear combination** of the column vectors. Therefore, from equation (3.2.2), we see that the vector $A\mathbf{c}$ is obtained by taking a linear combination of the column vectors of A. For example, if $A = \begin{bmatrix} 2 & -1 \\ 4 & 3 \end{bmatrix}$ and $\mathbf{c} = \begin{bmatrix} 5 \\ -1 \end{bmatrix}$, then

$$A\mathbf{c} = c_1 \mathbf{a}_1 + c_2 \mathbf{a}_2 = 5 \begin{bmatrix} 2 \\ 4 \end{bmatrix} + (-1) \begin{bmatrix} -1 \\ 3 \end{bmatrix} = \begin{bmatrix} 11 \\ 17 \end{bmatrix}.$$

CASE 3:   Product of an $m \times n$ matrix and an $n \times p$ matrix   The
generalization of matrix multiplication to the case when the second matrix in a product
has more than one column vector is now straightforward. We simply multiply each of
the column vectors of the second matrix by the first matrix, using the preceding rule. We
illustrate with an example.

Example 3.2.6   If $A = \begin{bmatrix} 1 & 4 & 2 \\ 3 & 5 & 7 \end{bmatrix}$ and $B = \begin{bmatrix} 2 & 3 \\ 5 & -2 \\ 8 & 4 \end{bmatrix}$, then

$$AB = \begin{bmatrix} 1 & 4 & 2 \\ 3 & 5 & 7 \end{bmatrix} \begin{bmatrix} 2 & 3 \\ 5 & -2 \\ 8 & 4 \end{bmatrix}$$

$$= \begin{bmatrix} [(1)(2) + (4)(5) + (2)(8)] & [(1)(3) + (4)(-2) + (2)(4)] \\ [(3)(2) + (5)(5) + (7)(8)] & [(3)(3) + (5)(-2) + (7)(4)] \end{bmatrix}$$

$$= \begin{bmatrix} 38 & 3 \\ 87 & 27 \end{bmatrix}$$

◻

The preceding discussion of the matrix product is summarized in the next definition.

Definition 3.2.5:

1. If $A = [a_{ij}]$ is an $m \times n$ matrix and $\mathbf{b} = [b_i]$ is a column n-vector, then $A\mathbf{b}$ is the
column m-vector with components

$$(A\mathbf{b})_i = \sum_{k=1}^{n} a_{ik}b_k \qquad 1 \le i \le m.$$

2. If A is an $m \times n$ matrix and $B = [\mathbf{b}_1, \mathbf{b}_2, \dots, \mathbf{b}_p]$ is an $n \times p$ matrix, then the
product of A and B is the $m \times p$ matrix, denoted AB, defined by

$$AB = [A\mathbf{b}_1, A\mathbf{b}_2, A\mathbf{b}_3 \dots, A\mathbf{b}_p]. \qquad (3.2.3)$$

In order for the product AB to be defined, we see from Definition 3.2.5 that A and B
must satisfy

number of columns of A = number of rows of B.

In such a case, if C represents the product matrix AB, then the relationship between the
dimensions of the matrices is

Example 3.2.7   If $A = \begin{bmatrix} 1 & 3 \\ 2 & 4 \end{bmatrix}$ and $B = \begin{bmatrix} 2 & -2 & 0 \\ 1 & 5 & 3 \end{bmatrix}$, then

$$AB = \begin{bmatrix} 1 & 3 \\ 2 & 4 \end{bmatrix} \begin{bmatrix} 2 & -2 & 0 \\ 1 & 5 & 3 \end{bmatrix} = \begin{bmatrix} 5 & 13 & 9 \\ 8 & 16 & 12 \end{bmatrix}.$$

Example 3.2.8   If $A = [1 \quad 2 \quad -1]$ and $B = \begin{bmatrix} -1 & 1 \\ 0 & 1 \\ 1 & 2 \end{bmatrix}$, then

$$AB = [1 \quad 2 \quad -1] \begin{bmatrix} -1 & 1 \\ 0 & 1 \\ 1 & 2 \end{bmatrix} = [-2 \quad 1].$$

Example 3.2.9   If $A = \begin{bmatrix} 2 \\ -1 \\ 3 \end{bmatrix}$ and $B = [1 \quad 4 \quad -6]$, then

$$AB = \begin{bmatrix} 2 \\ -1 \\ 3 \end{bmatrix} [1 \quad 4 \quad -6] = \begin{bmatrix} 2 & 8 & -12 \\ -1 & -4 & 6 \\ 3 & 12 & -18 \end{bmatrix}.$$

Example 3.2.10   If $A = \begin{bmatrix} 1-i & i \\ 2+i & 1+i \end{bmatrix}$ and $B = \begin{bmatrix} 3+2i & 1+4i \\ i & -1+2i \end{bmatrix}$, then

$$AB = \begin{bmatrix} 1-i & i \\ 2+i & 1+i \end{bmatrix} \begin{bmatrix} 3+2i & 1+4i \\ i & -1+2i \end{bmatrix} = \begin{bmatrix} 4-i & 3+2i \\ 3+8i & -5+10i \end{bmatrix}.$$

REMARK   For an $n \times n$ matrix we use the usual power notation to denote the operation of multiplying A by itself. Thus,

$$A^2 = AA, \quad A^3 = AAA, \quad \text{etc.}$$

Definition 3.2.5 tells us how to evaluate the product of two matrices in practice. However, in order to prove results related to the matrix product, we require an expression for the ijth element of AB in terms of the elements of the matrices A and B. According to our rule for multiplying two matrices, the ijth element of AB is obtained by taking the dot product of the ith row vector of A with the jth column vector of B. But the ith row vector of A is

$$\mathbf{a}_i = [a_{i1} \; a_{i2} \; \dots \; a_{in}],$$

and the jth column vector of B is

$$\mathbf{b}_j = \begin{bmatrix} b_{1j} \\ b_{2j} \\ \vdots \\ b_{nj} \end{bmatrix},$$

so that their dot product is

$$(AB)_{ij} = a_{i1}b_{1j} + a_{i2}b_{2j} + \cdots + a_{in}b_{nj}.$$

Expressing this using the summation notation yields the following important result:

Definition 3.2.6:   If $A = [a_{ij}]$ is an $m \times n$ matrix, $B = [b_{ij}]$ is an $n \times p$ matrix, and $C = AB$, then

$$c_{ij} = \sum_{k=1}^{n} a_{ik}b_{kj} \qquad 1 \le i \le m, \; 1 \le j \le p. \qquad (3.2.4)$$

This is called the **index form** of the matrix product.

The formula (3.2.4) for the ijth element of AB is very important and will often be required in the future. You should memorize it.

We can now establish some basic properties of matrix multiplication.

Theorem 3.2.2:   If A, B and C have appropriate dimensions for the operations to be performed, then

$$A(BC) = (AB)C, \qquad (3.2.5)$$

$$A(B + C) = AB + AC, \qquad (3.2.6)$$

$$(A + B)C = AC + BC. \qquad (3.2.7)$$

PROOF   The idea behind the proof of each of these results is to use the definition of matrix multiplication to show that the ijth element of the matrix on the left-hand side of each equation is equal to the ijth element of the matrix on the right-hand side. We illustrate by proving (3.2.7), but we leave the proofs of the remaining properties as exercises. Suppose that A and B are $m \times n$ matrices and that C is an $n \times p$ matrix. Then, from equation (3.2.4),

$$[(A + B)C]_{ij} = \sum_{k=1}^{n} (a_{ik} + b_{ik})c_{kj} = \sum_{k=1}^{n} a_{ik}c_{kj} + \sum_{k=1}^{n} b_{ik}c_{kj}$$

$$= (AC)_{ij} + (BC)_{ij}$$

$$= (AC + BC)_{ij}, \qquad 1 \le i \le m, \; 1 \le j \le p.$$

Consequently,

$$(A + B)C = AC + BC \qquad \blacksquare$$

Theorem 3.2.2 states that matrix multiplication is associative and distributive (over addition). We now consider the question of commutativity of matrix multiplication. If A is an $m \times n$ matrix and B is an $n \times m$ matrix, we can form both of the products AB and BA. In the first of these, we say that B has been **premultiplied** by A, whereas in the second product, we say that B has been **postmultiplied** by A. If $m \ne n$, then the matrices AB and BA will have different dimensions, and so, they cannot be equal. It is important to realize, however, that even if $m = n$, in general (that is, except for special cases)

$$AB \neq BA.$$

This is the statement that matrix multiplication is *not* commutative. With a little bit of thought this should not be too surprising in view of the fact that the ijth element of AB is obtained by taking the dot product of the ith row vector of A with the jth column vector of B, whereas the ijth element of BA is obtained by taking the dot product of the ith row vector of B with the jth column vector of A. We illustrate with an example.

Example 3.2.11   If $A = \begin{bmatrix} 1 & 2 \\ -1 & 3 \end{bmatrix}$ and $B = \begin{bmatrix} 3 & 1 \\ 2 & -1 \end{bmatrix}$, find AB and BA.

Solution

$$AB = \begin{bmatrix} 1 & 2 \\ -1 & 3 \end{bmatrix}\begin{bmatrix} 3 & 1 \\ 2 & -1 \end{bmatrix} = \begin{bmatrix} 7 & -1 \\ 3 & -4 \end{bmatrix}.$$

$$BA = \begin{bmatrix} 3 & 1 \\ 2 & -1 \end{bmatrix}\begin{bmatrix} 1 & 2 \\ -1 & 3 \end{bmatrix} = \begin{bmatrix} 2 & 9 \\ 3 & 1 \end{bmatrix}.$$

Thus we see that in this example, $AB \neq BA$.   ❏

The **identity matrix**, I_n (or just I if the dimensions are obvious), is the $n \times n$ matrix with ones on the leading diagonal and zeros elsewhere. For example,

$$I_2 = \begin{bmatrix} 1 & 0 \\ 0 & 1 \end{bmatrix}, \quad I_4 = \begin{bmatrix} 1 & 0 & 0 & 0 \\ 0 & 1 & 0 & 0 \\ 0 & 0 & 1 & 0 \\ 0 & 0 & 0 & 1 \end{bmatrix}.$$

The elements of I_n can be represented by the **Kronecker delta symbol**, δ_{ij}, defined by

$$\delta_{ij} = \begin{cases} 1, & \text{if } i = j, \\ 0, & \text{if } i \neq j. \end{cases}$$

Then,

$$\boxed{I_n = [\delta_{ij}].}$$

The following properties of the identity matrix indicate that it plays the same role in matrix multiplication as the number 1 does in the multiplication of real numbers.

Properties of the Identity Matrix

(a) $A_{m \times n} I_n = A_{m \times n}.$

(b) $I_m A_{m \times p} = A_{m \times p}.$

PROOF OF (a)   Using the index form of the matrix product we have

$$(AI)_{ij} = \sum_{k=1}^{n} a_{ik}\, \delta_{kj} = a_{i1}\, \delta_{1j} + a_{i2}\, \delta_{2j} + \cdots + a_{ij}\, \delta_{jj} + \cdots + a_{in}\, \delta_{nj}.$$

But, from the definition of the Kronecker delta symbol, we see that all terms in the summation with $k \neq j$ vanish, so that we are left with

$$(AI)_{ij} = a_{ij}\, \delta_{jj} = a_{ij}, \quad 1 \leq i \leq m, \quad 1 \leq j \leq n. \qquad \blacksquare$$

We leave the proof of (b) as an exercise. The next example illustrates the property.

Example 3.2.12   If $A = \begin{bmatrix} 2 & -1 \\ 3 & 5 \\ 0 & -2 \end{bmatrix}$, then

$$I_3 A = \begin{bmatrix} 1 & 0 & 0 \\ 0 & 1 & 0 \\ 0 & 0 & 1 \end{bmatrix}\begin{bmatrix} 2 & -1 \\ 3 & 5 \\ 0 & -2 \end{bmatrix} = \begin{bmatrix} 2 & -1 \\ 3 & 5 \\ 0 & -2 \end{bmatrix} = A. \qquad \square$$

PROPERTIES OF THE TRANSPOSE

The operation of taking the transpose of a matrix was introduced in the previous section. The next theorem gives three important properties satisfied by the transpose. These should be memorized.

Theorem 3.2.3:   Let A and C be $m \times n$ matrices, and let B be an $n \times p$ matrix. Then

(a) $(A^T)^T = A$.

(b) $(A + C)^T = A^T + C^T$.

(c) $(AB)^T = B^T A^T$.

PROOF   The proofs of (a) and (b) are almost immediate, and hence, they are left as exercises. We prove (c). From the definition of the transpose and the index form of the matrix product we have

$$[(AB)^T]_{ij} \quad = \quad (AB)_{ji} \quad = \quad \sum_{k=1}^{n} a_{jk} b_{ki}$$
$$\uparrow \qquad\qquad\qquad\qquad \uparrow$$

$$\text{Definition of transpose} \qquad \text{Index form of matrix product}$$

$$= \sum_{k=1}^{n} b_{ki} a_{jk} = \sum_{k=1}^{n} b^T_{ik}\, a^T_{kj} = (B^T A^T)_{ij}.$$

Consequently,

$$(AB)^T = B^T A^T. \qquad \blacksquare$$

RESULTS FOR TRIANGULAR MATRICES

Upper and lower triangular matrices play a significant role in the analysis of linear systems of algebraic equations. The following theorem and its corollary will be needed in Section 3.7.

Theorem 3.2.4:   The product of two lower (upper) triangular matrices is a lower (upper) triangular matrix.

PROOF   Suppose that A and B are $n \times n$ lower triangular matrices. Then, $a_{ik} = 0$ whenever $i < k$, and $b_{kj} = 0$ whenever $k < j$. If we let $C = AB$, then we must prove that $c_{ij} = 0$ whenever $i < j$. Using the index form of the matrix product we have

$$c_{ij} = \sum_{k=1}^{n} a_{ik} b_{kj} = \sum_{k=j}^{n} a_{ik} b_{kj} \text{ (since } b_{kj} = 0 \text{ if } k < j). \qquad (3.2.8)$$

We now impose the condition that $i < j$. Then, since $k \geq j$ in (3.2.8), it follows that $k > i$. However, this implies that $a_{ik} = 0$ (since A is lower triangular), and hence, from (3.2.8), that

$$c_{ij} = 0 \text{ whenever } i < j$$

as required.

To establish the result for upper triangular matrices, we use the result just proven for lower triangular matrices and properties of the transpose. If A and B are $n \times n$ upper triangular matrices, then $(AB)^T = B^T A^T$. Consequently, $(AB)^T$ is a lower triangular matrix and therefore AB is upper triangular. ∎

Corollary 3.2.1:   The product of two unit lower triangular matrices is a unit lower triangular matrix.

PROOF   Let A and B be unit lower triangular $n \times n$ matrices. We know from Theorem 3.2.4 that $C = AB$ is a lower triangular matrix. We must establish that $c_{ii} = 1$ for each i. The elements on the main diagonal of C can be obtained by setting $j = i$ in (3.2.8):

$$c_{ii} = \sum_{k=i}^{n} a_{ik} b_{ki}. \qquad (3.2.9)$$

Since $a_{ik} = 0$ whenever $k > i$, the only nonzero term in the summation in (3.2.9) occurs when $k = i$. Consequently,

$$c_{ii} = a_{ii} b_{ii}, \quad i = 1, 2, \dots, n,$$

so that for *unit* lower triangular matrices A and B

$$c_{ii} = 1, \quad i = 1, 2, \dots, n. \qquad ∎$$

EXERCISES 3.2

1. If $A = \begin{bmatrix} 1 & 2 & -1 \\ 3 & 5 & 2 \end{bmatrix}$ and $B = \begin{bmatrix} 2 & -1 & 3 \\ 1 & 4 & 5 \end{bmatrix}$
find $2A$, $-3B$, $A - 2B$, and $3A + 4B$.

2. If

$$A = \begin{bmatrix} 2 & -1 & 0 \\ 3 & 1 & 2 \\ -1 & 1 & 1 \end{bmatrix}, B = \begin{bmatrix} 1 & -1 & 2 \\ 3 & 0 & 1 \\ -1 & 1 & 0 \end{bmatrix}, \text{ and}$$

$$C = \begin{bmatrix} -1 & -1 & 1 \\ 1 & 2 & 3 \\ -1 & 1 & 0 \end{bmatrix},$$

find the matrix D such that $2A + B - 3C + 2D = A + 4C$.

3. Let $A = \begin{bmatrix} 1 & -1 & 2 \\ 3 & 1 & 4 \end{bmatrix}$, $B = \begin{bmatrix} 2 & -1 & 3 \\ 5 & 1 & 2 \\ 4 & 6 & -2 \end{bmatrix}$,

$C = \begin{bmatrix} 1 \\ -1 \\ 2 \end{bmatrix}$, $D = \begin{bmatrix} 2 & -2 & 3 \end{bmatrix}$. Find, if possible,

AB, BC, CA, DC, DB, AD, and CD.

For problems 4–6, determine AB for the given matrices. In these problems, i denotes $\sqrt{-1}$.

4. $A = \begin{bmatrix} 2 - i & 1 + i \\ -i & 2 + 4i \end{bmatrix}$,

$B = \begin{bmatrix} i & 1 - 3i \\ 0 & 4 + i \end{bmatrix}$.

5. $A = \begin{bmatrix} 3 + 2i & 2 - 4i \\ 5 + i & -1 + 3i \end{bmatrix}$,

$B = \begin{bmatrix} -1 + i & 3 + 2i \\ 4 - 3i & 1 + i \end{bmatrix}$.

6. $A = \begin{bmatrix} 3 - 2i & i \\ -i & 1 \end{bmatrix}$,

$B = \begin{bmatrix} -1 + i & 2 - i & 0 \\ 1 + 5i & 0 & 3 - 2i \end{bmatrix}$.

7. Let $A = \begin{bmatrix} 1 & -1 & 2 & 3 \\ -2 & 3 & 4 & 6 \end{bmatrix}$,

$B = \begin{bmatrix} 3 & 2 \\ 1 & 5 \\ 4 & -3 \\ -1 & 6 \end{bmatrix}$, and $C = \begin{bmatrix} -3 & 2 \\ 1 & -4 \end{bmatrix}$. Find ABC and CAB.

8. Let $A = \begin{bmatrix} 1 & -2 \\ 3 & 1 \end{bmatrix}$, $B = \begin{bmatrix} -1 & 2 \\ 5 & 3 \end{bmatrix}$, and

$C = \begin{bmatrix} 3 \\ -1 \end{bmatrix}$. Find $(2A - 3B)C$.

For problems 9–11, determine Ac by computing an appropriate linear combination of the column vectors of A.

9. $A = \begin{bmatrix} 1 & 3 \\ -5 & 4 \end{bmatrix}$, $c = \begin{bmatrix} 6 \\ -2 \end{bmatrix}$.

10. $A = \begin{bmatrix} 3 & -1 & 4 \\ 2 & 1 & 5 \\ 7 & -6 & 3 \end{bmatrix}$, $c = \begin{bmatrix} 2 \\ 3 \\ -4 \end{bmatrix}$.

11. $A = \begin{bmatrix} -1 & 2 \\ 4 & 7 \\ 5 & -4 \end{bmatrix}$, $c = \begin{bmatrix} 5 \\ -1 \end{bmatrix}$.

12. If A is an $m \times n$ matrix and C is an $r \times s$ matrix, what must the dimensions of B be in order for the product ABC to be defined? Write an expression for the ijth element of ABC in terms of the elements of A, B, and C.

13. Find A^2, A^3, and A^4 if

(a) $A = \begin{bmatrix} 1 & -1 \\ 2 & 3 \end{bmatrix}$.

(b) $A = \begin{bmatrix} 0 & 1 & 0 \\ -2 & 0 & 1 \\ 4 & -1 & 0 \end{bmatrix}$.

14. If A and B are $n \times n$ matrices prove that

(a) $(A + B)^2 = A^2 + AB + BA + B^2$.

(b) $(A - B)^2 = A^2 - AB - BA + B^2$.

15. If $A = \begin{bmatrix} 3 & -1 \\ -5 & -1 \end{bmatrix}$, calculate A^2 and verify

that A satisfies $A^2 - 2A - 8I_2 = 0_2$.

16. Find a matrix $A = \begin{bmatrix} 1 & x & z \\ 0 & 1 & y \\ 0 & 0 & 1 \end{bmatrix}$ such that

$$A^2 + \begin{bmatrix} 0 & -1 & 0 \\ 0 & 0 & -1 \\ 0 & 0 & 0 \end{bmatrix} = I_3.$$

17. If $A = \begin{bmatrix} x & 1 \\ -2 & y \end{bmatrix}$ determine all values of x and y for which $A^2 = A$.

18. The Pauli spin matrices σ_1, σ_2, and σ_3 are defined by

$$\sigma_1 = \begin{bmatrix} 0 & 1 \\ 1 & 0 \end{bmatrix}, \quad \sigma_2 = \begin{bmatrix} 0 & -i \\ i & 0 \end{bmatrix}, \quad \text{and}$$

$$\sigma_3 = \begin{bmatrix} 1 & 0 \\ 0 & -1 \end{bmatrix}.$$

Verify that they satisfy

$$\sigma_1\sigma_2 = i\sigma_3, \quad \sigma_2\sigma_3 = i\sigma_1, \quad \text{and} \quad \sigma_3\sigma_1 = i\sigma_2.$$

If A and B are $n \times n$ matrices, we define their **commutator**, denoted $[A, B]$, by

$$[A, B] = AB - BA.$$

Thus $[A, B] = 0$ if and only if A and B commute. That is, $AB = BA$. Problems 19–22 require the commutator.

19. If $A = \begin{bmatrix} 1 & -1 \\ 2 & 1 \end{bmatrix}$ and $B = \begin{bmatrix} 3 & 1 \\ 4 & 2 \end{bmatrix}$, find $[A, B]$.

20. If $A_1 = \begin{bmatrix} 1 & 0 \\ 0 & 1 \end{bmatrix}$, $A_2 = \begin{bmatrix} 0 & 1 \\ 0 & 0 \end{bmatrix}$, and $A_3 = \begin{bmatrix} 0 & 0 \\ 1 & 0 \end{bmatrix}$, compute all of the commutators $[A_i, A_j]$, and determine which of the matrices commute.

21. If $A_1 = \frac{1}{2}\begin{bmatrix} 0 & i \\ i & 0 \end{bmatrix}$, $A_2 = \frac{1}{2}\begin{bmatrix} 0 & -1 \\ 1 & 0 \end{bmatrix}$, and $A_3 = \frac{1}{2}\begin{bmatrix} i & 0 \\ 0 & -i \end{bmatrix}$, verify that

$$[A_1, A_2] = A_3, \quad [A_2, A_3] = A_1, \quad \text{and} \quad [A_3, A_1] = A_2.$$

22. If A, B, and C are any $n \times n$ matrices, find $[A, [B, C]]$ and prove the *Jacobi identity*

$$[A, [B, C]] + [B, [C, A]] + [C, [A, B]] = 0.$$

23. Use the index form of the matrix product to prove properties (3.2.5) and (3.2.6).

24. Prove property (b) of the identity matrix.

25. If A and B are $n \times n$ matrices, prove that $\text{tr}(AB) = \text{tr}(BA)$.

26. Let $A = \begin{bmatrix} 1 & -1 & 1 & 4 \\ 2 & 0 & 2 & -3 \\ 3 & 4 & -1 & 0 \end{bmatrix}$ and

$B = \begin{bmatrix} 0 & 1 \\ -1 & 2 \\ 1 & 1 \\ 2 & 1 \end{bmatrix}$, find A^T, B^T, AA^T, AB, and B^TA^T.

27. Let $A = \begin{bmatrix} 2 & 2 & 1 \\ 2 & 5 & 2 \\ 1 & 2 & 2 \end{bmatrix}$, and let S be the matrix

with column vectors $\mathbf{s}_1 = \begin{bmatrix} -x \\ 0 \\ x \end{bmatrix}$, $\mathbf{s}_2 = \begin{bmatrix} -y \\ y \\ -y \end{bmatrix}$,

and $\mathbf{s}_3 = \begin{bmatrix} z \\ 2z \\ z \end{bmatrix}$, where x, y, z are constants.

(a) Show that $AS = [\mathbf{s}_1, \mathbf{s}_2, 7\mathbf{s}_3]$.

(b) Find all values of x, y, z such that $S^TAS = \text{diag}(1, 1, 7)$.

28. A matrix that is a multiple of I_n is called an $n \times n$ **scalar** matrix. Determine the 4×4 scalar matrix whose trace is 8.

If A is an $n \times n$ matrix, then the matrices S and T defined by

$$S = \frac{1}{2}(A + A^T), \quad T = \frac{1}{2}(A - A^T)$$

are called the symmetric and skew-symmetric parts of A respectively. Problems 29–32 investigate properties of S and T.

29. Use the properties of the transpose to show that S and T are symmetric and skew-symmetric respectively.

30. Find S and T for the matrix

$$A = \begin{bmatrix} 1 & -5 & 3 \\ 3 & 2 & 4 \\ 7 & -2 & 6 \end{bmatrix}.$$

31. If A is an $n \times n$ symmetric matrix show that $T = 0$. What is the corresponding result for skew-symmetric matrices?

32. Show that any $n \times n$ matrix can be written as the sum of a symmetric and a skew-symmetric matrix.

33. Prove that if A is an $n \times p$ matrix and $D = \text{diag}(d_1, d_2, \ldots, d_n)$, then DA is the matrix obtained by multiplying the ith row vector of A by d_i $(1 \le i \le n)$.

34. Use properties of the transpose to prove that

(a) AA^T is a symmetric matrix.

(b) $(ABC)^T = C^T B^T A^T$.

3.3 TERMINOLOGY AND NOTATION FOR SYSTEMS OF LINEAR EQUATIONS

As we mentioned in Section 3.1, one of the main aims of this chapter is to apply matrices to determine the solution properties of any system of linear algebraic equations. We are now in a position to pursue that aim. We begin by introducing some notation and terminology. The general $m \times n$ system of linear equations is of the form

$$
\begin{aligned}
a_{11}x_1 + a_{12}x_2 + \cdots + a_{1n}x_n &= b_1, \\
a_{21}x_1 + a_{22}x_2 + \cdots + a_{2n}x_n &= b_2, \\
&\vdots \\
a_{m1}x_1 + a_{m2}x_2 + \cdots + a_{mn}x_n &= b_m,
\end{aligned}
\tag{3.3.1}
$$

where the **system coefficients** a_{ij} and the **system constants** b_j are given scalars and $x_1, x_2, \ldots, x_n$ denote the unknowns in the system. If $b_i = 0$ for all i then the system is called **homogeneous**, otherwise it is called **nonhomogeneous**.

> **Definition 3.3.1:** By a **solution** to the system (3.3.1) we mean an ordered n–tuple of scalars, say, $(c_1, c_2, \ldots, c_n)$, which, when substituted for $x_1, x_2, \ldots, x_n$ into the left-hand side of system (3.3.1), yield the right-hand side. The set of all solutions to system (3.3.1) is called the **solution set** to the system.

REMARKS

1. Usually the a_{ij} and b_j will be real numbers and we will then be interested in determining only the real solutions to system (3.3.1). However, many of the problems that arise in the later chapters will require the solution to systems with complex coefficients in which case the corresponding solutions will also be complex.

2. If $(c_1, c_2, \ldots, c_n)$ is a solution to the system (3.3.1), we will sometimes specify this solution by writing $x_1 = c_1, x_2 = c_2, \ldots, x_n = c_n$. For example, the ordered pair of numbers $(1, 2)$ is a solution to the system

$$
\begin{aligned}
x_1 + x_2 &= 3, \\
3x_1 - 2x_2 &= -1,
\end{aligned}
$$

and we could express this solution in the equivalent form $x_1 = 1, x_2 = 2$.

The following questions need addressing:

1. Does the system (3.3.1) have a solution?
2. If the answer to (1) is yes, then how many solutions are there?
3. How do we determine all of the solutions?

To obtain an idea of the answer to questions (1) and (2), consider the special case of a system of three equations in three unknowns. The linear system (3.3.1) then reduces to

$$a_{11}x_1 + a_{12}x_2 + a_{13}x_3 = b_1,$$
$$a_{21}x_1 + a_{22}x_2 + a_{23}x_3 = b_2,$$
$$a_{31}x_1 + a_{32}x_2 + a_{33}x_3 = b_3,$$

which can be interpreted as defining three planes in space. An ordered triple (c_1, c_2, c_3) is a solution to this system if and only if it corresponds to the coordinates of a point of intersection of the three planes. There are precisely four possibilities:

1. The planes have no intersection point.
2. The planes intersect in just one point.
3. The planes intersect in a line.
4. The planes are coincident.

In (1), the corresponding system has no solution, whereas in (2), the system has just one solution. Finally, in (3) and (4), every point on the line or plane (respectively) is a solution to the linear system and hence the system has an infinite number of solutions. Cases (1) – (3) are illustrated in Figure 3.3.1. We have therefore proved, geometrically,

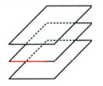

Three parallel planes (no intersection): no solution

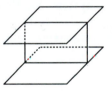

No common intersection: no solution

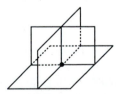

Planes intersect at a point: a unique solution

Planes intersect in a line: an infinite number of solutions

Figure 3.3.1 The geometric interpretation of the solution of three linear algebraic equations in three unknowns.

that there are precisely three possibilities for the solutions of a system of three equations in three unknowns. The system either has no solution, it has just one solution, or it has an infinite number of solutions. In Section 3.5, we will establish that these are the only possibilities for the general $m \times n$ system (3.3.1).

A system of equations that has at least one solution is said to be **consistent**, whereas a system that has no solution is called **inconsistent**. Our problem will be to determine whether a given system is consistent and then, in the case when we do have a consistent system, to find its solution set.

Naturally associated with the system (3.3.1) are the following two matrices

1. The **matrix of coefficients** $A = \begin{bmatrix} a_{11} & a_{12} & \cdots & a_{1n} \\ a_{21} & a_{22} & \cdots & a_{2n} \\ & \vdots & \\ a_{m1} & a_{m2} & \cdots & a_{mn} \end{bmatrix}$.

2. The **augmented matrix** $A^{\#} = \begin{bmatrix} a_{11} & a_{12} & \cdots & a_{1n} & b_1 \\ a_{21} & a_{22} & \cdots & a_{2n} & b_2 \\ & \vdots & & & \vdots \\ a_{m1} & a_{m2} & \cdots & a_{mn} & b_m \end{bmatrix}$.

The augmented matrix completely characterizes a system of equations since it contains all of the system coefficients and system constants. We will see in the following sections that it is the relationship between A and $A^{\#}$ that determines the solution properties of a linear system. Notice that the matrix of coefficients is the submatrix consisting of the first n columns of $A^{\#}$.

Example 3.3.1   Write the system of equations with the following augmented matrix

$$\begin{bmatrix} 1 & 2 & 9 & -1 & 1 \\ 2 & -3 & 7 & 4 & 2 \\ 1 & 3 & 5 & 0 & -1 \end{bmatrix}.$$

Solution

$$\begin{aligned} x_1 + 2x_2 + 9x_3 - x_4 &= 1, \\ 2x_1 - 3x_2 + 7x_3 + 4x_4 &= 2, \\ x_1 + 3x_2 + 5x_3 \quad\quad &= -1. \end{aligned}$$

VECTOR FORMULATION

We next show that the matrix product can be used to write a linear system as a single vector equation. For example, the system

$$\begin{aligned} x_1 + 3x_2 - 4x_3 &= 1, \\ 2x_1 + 5x_2 - x_3 &= 5, \\ x_1 \quad\quad + 6x_3 &= 3, \end{aligned}$$

can be written as the vector equation

$$\begin{bmatrix} 1 & 3 & -4 \\ 2 & 5 & -1 \\ 1 & 0 & 6 \end{bmatrix} \begin{bmatrix} x_1 \\ x_2 \\ x_3 \end{bmatrix} = \begin{bmatrix} 1 \\ 5 \\ 3 \end{bmatrix},$$

since this vector equation is satisfied if and only if

$$\begin{bmatrix} x_1 + 3x_2 - 4x_3 \\ 2x_1 + 5x_2 - x_3 \\ x_1 + 6x_3 \end{bmatrix} = \begin{bmatrix} 1 \\ 5 \\ 3 \end{bmatrix},$$

that is, if and only if each equation of the given system is satisfied. Similarly the general $m \times n$ system of linear equations

$$a_{11}x_1 + a_{12}x_2 + \cdots + a_{1n}x_n = b_1,$$
$$a_{21}x_1 + a_{22}x_2 + \cdots + a_{2n}x_n = b_2,$$
$$\vdots$$
$$a_{m1}x_1 + a_{m2}x_2 + \cdots + a_{mn}x_n = b_m,$$

can be written as the **vector equation**

$$Ax = b, \tag{3.3.2}$$

where A is the matrix of coefficients and

$$x = \begin{bmatrix} x_1 \\ x_2 \\ \vdots \\ x_n \end{bmatrix}, \qquad b = \begin{bmatrix} b_1 \\ b_2 \\ \vdots \\ b_m \end{bmatrix}.$$

We will refer to the n-vector x as the **vector of unknowns**, and the m-vector b will be called the **right-hand side vector**. In this formulation, the unknown in the system is the column vector x.

In Chapter 8, a similar vector formulation for systems of differential equations will be used not only in developing the theory for such systems, but also in deriving solution techniques. As an example of this formulation, consider the system of DE

$$\frac{dx_1}{dt} = 3x_1 + 9x_2 + 6e^t,$$

$$\frac{dx_2}{dt} = 2x_1 - 7x_2 + 3e^t.$$

This system can be written as the vector equation

$$\frac{dx}{dt} = Ax(t) + b(t),$$

where

$$\mathbf{x}(t) = \begin{bmatrix} x_1(t) \\ x_2(t) \end{bmatrix}, \; A = \begin{bmatrix} 3 & 9 \\ 2 & -7 \end{bmatrix}, \; \frac{d\mathbf{x}}{dt} = \begin{bmatrix} \dfrac{dx_1}{dt} \\ \dfrac{dx_2}{dt} \end{bmatrix}, \; \text{and } \mathbf{b}(t) = \begin{bmatrix} 6e^t \\ 3e^t \end{bmatrix}.$$

In this formulation the basic unknown is the column vector $\mathbf{x}(t)$.

EXERCISES 3.3

For problems 1 and 2, show that the given triple of real numbers is a solution to the given system.

1. $(1, -1, 2)$;

$$2x_1 - 3x_2 + 4x_3 = 13,$$
$$x_1 + x_2 - x_3 = -2,$$
$$5x_1 + 4x_2 + x_3 = 3.$$

2. $(2, -3, 1)$;

$$x_1 + x_2 - 2x_3 = -3,$$
$$3x_1 - x_2 - 7x_3 = 2,$$
$$x_1 + x_2 + x_3 = 0,$$
$$2x_1 + 2x_2 - 4x_3 = -6.$$

3. Show that for all values of t, $(1 - t, 2 + 3t, 3 - 2t)$ is a solution to the linear system

$$x_1 + x_2 + x_3 = 6,$$
$$x_1 - x_2 - 2x_3 = -7,$$
$$5x_1 + x_2 - x_3 = 4.$$

4. Show that for all values of s and t, $(s, s - 2t, 2s + 3t, t)$ is a solution to the linear system

$$x_1 + x_2 - x_3 + 5x_4 = 0,$$
$$2x_2 - x_3 + 7x_4 = 0,$$
$$4x_1 + 2x_2 - 3x_3 + 13x_4 = 0.$$

5. By making a sketch in the xy-plane, prove that the following linear system has no solution:

$$2x + 3y = 1,$$
$$2x + 3y = 2.$$

For problems 6–8, determine the coefficient matrix, A, the right-hand side vector, $\mathbf{b}$, and the augmented matrix $A^{\#}$ of the given system.

6.
$$x_1 + 2x_2 - 3x_3 = 1,$$
$$2x_1 + 4x_2 - 5x_3 = 2,$$
$$7x_1 + 2x_2 - x_3 = 3.$$

7.
$$x + y + z - w = 3,$$
$$2x + 4y - 3z + 7w = 2.$$

8.
$$x_1 + 2x_2 - x_3 = 0,$$
$$2x_1 + 3x_2 - 2x_3 = 0,$$
$$5x_1 + 6x_2 - 5x_3 = 0.$$

For problems 9 and 10, write the system of equations with the given coefficient matrix and right-hand side vector.

9. $A = \begin{bmatrix} 1 & -1 & 2 & 3 \\ 1 & 1 & -2 & 6 \\ 3 & 1 & 4 & 2 \end{bmatrix}$, $\mathbf{b} = \begin{bmatrix} 1 \\ -1 \\ 2 \end{bmatrix}$.

10. $A = \begin{bmatrix} 2 & 1 & 3 \\ 4 & -1 & 2 \\ 7 & 6 & 3 \end{bmatrix}$, $\mathbf{b} = \begin{bmatrix} 1 \\ -1 \\ 2 \end{bmatrix}$.

11. Consider the $m \times n$ homogeneous system of linear equations

$$A\mathbf{x} = \mathbf{0}. \qquad (11.1)$$

(a) If

$$\mathbf{x} = [x_1 \; x_2 \; \dots \; x_n]^T \text{ and } \mathbf{y} = [y_1 \; y_2 \; \dots \; y_n]^T$$

are solutions to (11.1), show that

$$\mathbf{z} = \mathbf{x} + \mathbf{y} \text{ and } \mathbf{w} = c\mathbf{x}$$

are also solutions, where c is an arbitrary scalar. This establishes that solutions to a homogeneous linear system of algebraic equations satisfy the same linearity properties as those satisfied by a second-order linear homogeneous DE. (See Theorem 2.1.2.)

(b) Is the result in (a) true when **x** and **y** are solutions to the nonhomogeneous system $A\mathbf{x} = \mathbf{b}$?

3.4   ELEMENTARY ROW OPERATIONS AND ROW-ECHELON MATRICES

In the next section, we will develop methods for solving a system of linear equations. These methods will consist of reducing a given system of equations to a new system that has the same solution set as the given system, but is easier to solve. In this section, we introduce the requisite mathematical results.

ELEMENTARY ROW OPERATIONS

The first step in deriving systematic procedures for solving a linear system is to determine what operations can be performed on such a system *without altering its solution set.*

Example 3.4.1   Consider the system of equations

$$x_1 + 2x_2 + 4x_3 = 2, \tag{3.4.1}$$
$$2x_1 - 5x_2 + 3x_3 = 6, \tag{3.4.2}$$
$$4x_1 + 6x_2 - 7x_3 = 8. \tag{3.4.3}$$

If we permute (interchange), say, equations (3.4.1) and (3.4.2), the resulting system is

$$2x_1 - 5x_2 + 3x_3 = 6,$$
$$x_1 + 2x_2 + 4x_3 = 2,$$
$$4x_1 + 6x_2 - 7x_3 = 8,$$

which certainly has the same solution set as the original system. Further if we multiply, say, equation (3.4.2) by 5, we obtain the system

$$x_1 + 2x_2 + 4x_3 = 2,$$
$$10x_1 - 25x_2 + 15x_3 = 30,$$
$$4x_1 + 6x_2 - 7x_3 = 8,$$

which again has the same solution set as the original system. Finally, if we add, say, twice equation (3.4.1) to equation (3.4.3) we obtain the system

$$x_1 + 2x_2 + 4x_3 = 2, \tag{3.4.4}$$
$$2x_1 - 5x_2 + 3x_3 = 6, \tag{3.4.5}$$
$$(4x_1 + 6x_2 - 7x_3) + 2(x_1 + 2x_2 + 4x_3) = 8 + 2(2). \tag{3.4.6}$$

We see that if (3.4.4)–(3.4.6) are satisfied, then so are (3.4.1)–(3.4.3), and vice versa. It follows that the system of equations (3.4.4)–(3.4.6) has the same solution set as the original system of equations (3.4.1)–(3.4.3).   ❏

More generally, a similar reasoning can be used to show that the following three operations can be performed on any $m \times n$ system of linear equations *without altering the solution set*:

1. Permute equations.

2. Multiply an equation by a nonzero constant.

3. Add a multiple of one equation to another.

Since the operations (1)–(3) involve changes only in the system coefficients and constants (and not changes in the variables), they can be represented by the following operations on the augmented matrix of the system:

1. Permute rows.

2 . Multiply a row by a nonzero constant.

3 . Add a multiple of one row to another row.

These three operations are called **elementary row operations** (ERO) and will be a basic computational tool throughout the text even in cases when the matrix under consideration is not derived from a system of linear equations. The following notation will be used to describe ERO performed on a matrix A.

---

1 . P_{ij}:   **Permute** the ith and jth rows in A.

2 . $M_i(k)$: **Multiply** every element of the ith row of A by a nonzero scalar k.

3 . $A_{ij}(k)$: **Add** to the elements of the jth row of A, a scalar k times the corresponding elements of the ith row of A.

Furthermore, the notation $A \overset{n}{\sim} B$ will mean that matrix B has been obtained from matrix A by a sequence of ERO, with the natural number n referencing the operations used in step n.

Example 3.4.2:   The one step operations performed on the system in Example 3.4.1 can be described as follows using ERO on the augmented matrix of the system:

$$\begin{bmatrix} 1 & 2 & 4 & 2 \\ 2 & -5 & 3 & 6 \\ 4 & 6 & -7 & 8 \end{bmatrix} \overset{1}{\sim} \begin{bmatrix} 2 & -5 & 3 & 6 \\ 1 & 2 & 4 & 2 \\ 4 & 6 & -7 & 8 \end{bmatrix} \quad \text{1. } P_{12}. \text{ Permute (3.4.1) and (3.4.2).}$$

$$\begin{bmatrix} 1 & 2 & 4 & 2 \\ 2 & -5 & 3 & 6 \\ 4 & 6 & -7 & 8 \end{bmatrix} \overset{1}{\sim} \begin{bmatrix} 1 & 2 & 4 & 2 \\ 10 & -25 & 15 & 30 \\ 4 & 6 & -7 & 8 \end{bmatrix} \quad \text{1. } M_2(5). \text{ Multiply (3.4.2) by 5.}$$

$$\begin{bmatrix} 1 & 2 & 4 & 2 \\ 2 & -5 & 3 & 6 \\ 4 & 6 & -7 & 8 \end{bmatrix} \overset{1}{\underset{\sim}{}} \begin{bmatrix} 1 & 2 & 2 & 2 \\ 2 & -5 & 3 & 6 \\ 6 & 10 & 1 & 12 \end{bmatrix}$$ 1. $A_{13}(2)$.   Add 2 times (3.4.1) to (3.4.3).

◻

It is important to realize that each ERO is reversible. Specifically, in terms of the notation just introduced, the inverse operations are determined as follows.

ERO APPLIED TO A: A ~ B	REVERSE ERO APPLIED TO B: B ~ A
P_{ij}	P_{ji}: permute row j and i in B.
$M_i(k)$	$M_i(1/k)$: multiply the ith row of B by $1/k$.
$A_{ij}(k)$	$A_{ij}(-k)$: Add to the elements of the jth row of B a scalar $-k$ times the corresponding elements of the ith row of B.

We introduce a special term for matrices that are related via ERO.

Definition 3.4.1:   Let A be an $m \times n$ matrix. Any matrix obtained from A by a finite sequence of ERO is said to be **row-equivalent** to A.

Thus, all of the matrices in the previous example are row-equivalent (but *not* equal). Since ERO do not alter the solution set of a linear system, we have the next theorem.

Theorem 3.4.1:   Systems of linear equations with row-equivalent augmented matrices have the same solution sets.

ROW-ECHELON MATRICES

Our methods for solving a system of linear equations will consist of using ERO to reduce the augmented matrix of the given system to a simple form. But how simple a form should we aim for? In order to answer this question, consider the system

$$x_1 + x_2 - x_3 = 4, \qquad (3.4.7)$$
$$x_2 - 3x_3 = 5, \qquad (3.4.8)$$
$$x_3 = 2. \qquad (3.4.9)$$

This system can be solved most easily as follows. From equation (3.4.9), $x_3 = 2$. Substituting this value into equation (3.4.8) and solving for x_2 yields $x_2 = 5 + 6 = 11$. Finally, substituting for x_3 and x_2 into equation (3.4.7) and solving for x_1, we obtain $x_1 = -5$. Thus, the solution to the given system of equations is $(-5, 11, 2)$. This technique is called **back substitution** and could be used because the given system has a simple form. The augmented matrix of the system is

$$\begin{bmatrix} 1 & 1 & -1 & 4 \\ 0 & 1 & -3 & 5 \\ 0 & 0 & 1 & 2 \end{bmatrix}.$$

We see that the submatrix (consisting of the first three columns) which corresponds to the matrix of coefficients is a unit upper triangular matrix. The back substitution method will work on any system of linear equations with an augmented matrix of this form. Unfortunately, not all systems of equations have augmented matrices that can be reduced to such a form. However, there is a simple type of matrix, closely related to a unit upper triangular matrix, that any matrix can be reduced to by ERO, and which also represents a system of equations that can be solved (if it has a solution) by back substitution. This type of matrix is called a *row-echelon matrix* and is defined as follows:

Definition 3.4.2:   An $m \times n$ matrix is called a **row-echelon matrix** if it satisfies the following three conditions:

1. If there are any rows consisting entirely of zeros they are grouped together at the bottom of the matrix.

2. The first nonzero element in any nonzero row[1] is a one (This is called a **leading 1**).

3. The leading 1 of any row (except the first) is further to the right than the leading 1 of the row above it.

Example 3.4.3   Examples of row-echelon matrices are

$$\begin{bmatrix} 1 & -2 & 3 & 7 \\ 0 & 1 & 5 & 0 \\ 0 & 0 & 0 & 1 \end{bmatrix}, \quad \begin{bmatrix} 0 & 0 & 1 \\ 0 & 0 & 0 \\ 0 & 0 & 0 \end{bmatrix}, \text{ and } \begin{bmatrix} 1 & -1 & 6 & 5 & 9 \\ 0 & 0 & 1 & 2 & 5 \\ 0 & 0 & 0 & 1 & 0 \\ 0 & 0 & 0 & 0 & 0 \end{bmatrix},$$

whereas

$$\begin{bmatrix} 1 & 0 & -1 \\ 0 & 1 & 2 \\ 0 & 1 & -1 \end{bmatrix} \text{ and } \begin{bmatrix} 1 & 0 & 0 \\ 0 & 0 & 0 \\ 0 & 1 & -1 \\ 0 & 0 & 1 \end{bmatrix}$$

are not row-echelon matrices.   ❐

The basic result that will allow us to determine the solution set to any system of linear equations is stated in the next theorem.

Theorem 3.4.2:   Any matrix is row-equivalent to a row-echelon matrix.

According to this theorem, by applying an appropriate sequence of ERO to any $m \times n$ matrix, we can always reduce it to a row-echelon matrix. When a matrix, A, has been reduced to a row-echelon matrix in this way, we say that it has been reduced to **row-echelon form** (REF) and refer to the resulting matrix as an REF of A. The proof of Theorem 3.4.2 consists of giving an algorithm that will reduce an arbitrary $m \times n$ matrix to a row-echelon matrix after a finite sequence of ERO. Before presenting such an algorithm we first illustrate the result with an example.

---

[1]A *nonzero row* (*nonzero column*) is any row (column) that does not consist entirely of zeros.

Example 3.4.4   Use ERO to reduce $\begin{bmatrix} 2 & 1 & -1 & 3 \\ 1 & -1 & 2 & 1 \\ -4 & 6 & -7 & 1 \\ 2 & 0 & 1 & 3 \end{bmatrix}$ to REF.

Solution   We show each step in detail.

Step 1:   Put a leading 1 in the (11) position.

   This can be most easily accomplished by permuting rows 1 and 2.

$$\begin{bmatrix} 2 & 1 & -1 & 3 \\ 1 & -1 & 2 & 1 \\ -4 & 6 & -7 & 1 \\ 2 & 0 & 1 & 3 \end{bmatrix} \overset{1}{\sim} \begin{bmatrix} 1 & -1 & 2 & 1 \\ 2 & 1 & -1 & 3 \\ -4 & 6 & -7 & 1 \\ 2 & 0 & 1 & 3 \end{bmatrix}$$

Step 2:   Use the leading 1 to put zeros beneath it in column 1.

   This is accomplished by adding appropriate multiples of row 1 to the remaining rows.

$$\overset{2}{\sim} \begin{bmatrix} 1 & -1 & 2 & 1 \\ 0 & 3 & -5 & 1 \\ 0 & 2 & 1 & 5 \\ 0 & 2 & -3 & 1 \end{bmatrix}$$
Step 2 row operations: $\begin{cases} \text{Add } -2*\text{row 1 to row 2} \\ \text{Add } 4*\text{row 1 to row 3} \\ \text{Add } -2*\text{row 1 to row 4} \end{cases}$

Step 3:   Put a leading 1 in the (22) position.

   We could accomplish this by multiplying row 2 by 1/3. However, this would introduce fractions into the matrix and thereby complicate the remaining computations. In hand calculations, fewer algebraic errors result if we avoid the use of fractions. In this case, we can obtain a leading 1 without the use of fractions by subtracting row 3 from row 2.

$$\overset{3}{\sim} \begin{bmatrix} 1 & -1 & 2 & 1 \\ 0 & 1 & -6 & -4 \\ 0 & 2 & 1 & 5 \\ 0 & 2 & -3 & 1 \end{bmatrix}$$

Step 4:   Use the leading 1 in the (22) position to put zeros beneath it in column 2.

   We now add appropriate multiples of row 2 to the rows *beneath* it.

$$\overset{4}{\sim} \begin{bmatrix} 1 & -1 & 2 & 1 \\ 0 & 1 & -6 & -4 \\ 0 & 0 & 13 & 13 \\ 0 & 0 & 9 & 9 \end{bmatrix}$$
Step 4 row operations: $\begin{cases} \text{Add } -2*\text{row 2 to row 3} \\ \text{Add } -2*\text{row 2 to row 4} \end{cases}$

Step 5:   Put a leading 1 in the (33) position.

This can be accomplished by multiplying row 3 by 1/13.

$$5 \atop{\sim} \begin{bmatrix} 1 & -1 & 2 & 1 \\ 0 & 1 & -6 & -4 \\ 0 & 0 & 1 & 1 \\ 0 & 0 & 9 & 9 \end{bmatrix}$$

Step 6:   Use the leading 1 in the (33) position to put zeros *beneath* it in column 3.

The appropriate row operation is to add −9 times row three to row four.

$$6 \atop{\sim} \begin{bmatrix} 1 & -1 & 2 & 1 \\ 0 & 1 & -6 & -4 \\ 0 & 0 & 1 & 1 \\ 0 & 0 & 0 & 0 \end{bmatrix}.$$

This is a row-echelon matrix, and hence, the given matrix has been reduced to REF. The specific operations used at each step are given next using the notation introduced previously in this section. In future examples, the following is how we will summarize the row operations:

---

1. P_{12} **2.** $A_{12}(-2)$, $A_{13}(4)$, $A_{14}(-2)$ **3.** $A_{32}(-1)$
4. $A_{23}(-2)$, $A_{24}(-2)$ **5.** $M_3(1/13)$ **6.** $A_{34}(-9)$

You should study the foregoing example very carefully, since it illustrates the general procedure for reducing an $m \times n$ matrix to REF using ERO. This is a procedure that will be used repeatedly throughout the text. The idea behind reduction to REF is to start at the upper left-hand corner of the matrix and proceed downward and to the right of the matrix. The following algorithm formalizes the steps that reduce any $m \times n$ matrix to REF using a finite number of ERO and thereby provides a proof of Theorem 3.4.2. An illustration of this algorithm is given in Figure 3.4.1.

Algorithm for Reducing an $m \times n$ Matrix A to Row-Echelon Form

(1)   Start with an $m \times n$ matrix A. If $A = 0$, go to (7).

(2)   Determine the leftmost nonzero column (this is called a **pivot column** and the top most position in this column is called a **pivot position**).

(3)   Use ERO to put a 1 in the pivot position.

(4)   Use ERO to put zeros below the pivot position.

(5)   If there are no more nonzero rows below the pivot position go to 7, otherwise go to 6.

(6)   Apply (2)–(5) to the submatrix consisting of the rows that lie below the pivot position.

(7)   The matrix is a row-echelon matrix.

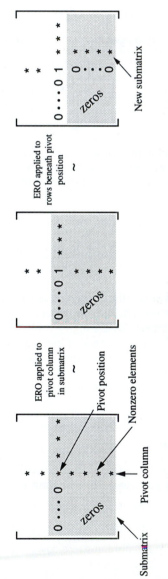

Figure 3.4.1   Illustration of an algorithm for reducing an $m \times n$ matrix to REF.

REMARK   In order to obtain a row-echelon matrix, we put a 1 in each pivot position. However, many algorithms for solving systems of linear equations numerically are based around the preceding algorithm except in step (3) we place a nonzero number (not necessarily a 1) in the pivot position. Of course, the matrix resulting from an application of this algorithm differs from a row-echelon matrix since it will have nonunit elements in the pivot positions.

Example 3.4.5 Reduce $\begin{bmatrix} 3 & 2 & -5 & 2 \\ 1 & 1 & -2 & 1 \\ 1 & 0 & -3 & 4 \end{bmatrix}$ to REF.

Solution   Applying the row-reduction algorithm leads to the following sequence of ERO. The specific row operations used at each step are given at the end of the process.

Pivot position $\longrightarrow \begin{bmatrix} \boxed{3} & 2 & -5 & 2 \\ 1 & 1 & -1 & 1 \\ 1 & 0 & -3 & 4 \end{bmatrix} \overset{1}{\sim} \begin{bmatrix} 1 & 1 & -1 & 1 \\ 3 & 2 & -5 & 2 \\ 1 & 0 & -3 & 4 \end{bmatrix} \overset{2}{\sim} \begin{bmatrix} 1 & 1 & -1 & 1 \\ 0 & -1 & -2 & -1 \\ 0 & -1 & -2 & 3 \end{bmatrix}$

Pivot column

Pivot position, Pivot column, Submatrix

$\overset{3}{\sim} \begin{bmatrix} 1 & 1 & -1 & 1 \\ 0 & 1 & 2 & 1 \\ 0 & -1 & -2 & 3 \end{bmatrix} \overset{4}{\sim} \begin{bmatrix} 1 & 1 & -1 & 1 \\ 0 & 1 & 2 & 1 \\ 0 & 0 & 0 & 4 \end{bmatrix} \overset{5}{\sim} \begin{bmatrix} 1 & 1 & -1 & 1 \\ 0 & 1 & 2 & 1 \\ 0 & 0 & 0 & 1 \end{bmatrix}.$

Submatrix, Pivot column, Pivot position

This is a row-echelon matrix and hence we are done.

1. P_{12} **2.** $A_{12}(-3)$, $A_{13}(-1)$ **3.** $M_2(-1)$ **4.** $A_{23}(1)$ **5.** $M_3(1/4)$.

THE RANK OF A MATRIX

We now derive some further results on row-echelon matrices that will be required in the next section to develop the theory for systems of linear algebraic equations.

If we reduce a matrix A to REF, then the resulting matrix is *not* unique, since we can always add multiples of a row to any row above it while maintaining REF. However, it follows from the row-reduction algorithm that the pivot positions in any REF of A will coincide. We therefore have the following theorem (in Chapter 5 we will see how the proof of this theorem arises naturally from the more sophisticated ideas from linear algebra yet to be introduced).

Theorem 3.4.3:   Let A be an $m \times n$ matrix. All row-echelon matrices that are row-equivalent to A have the same number of nonzero rows.

Theorem 3.4.3 associates a number with any $m \times n$ matrix A namely, the number of nonzero rows in any REF of A. As we will see in the next section, this number is fundamental in determining the solution properties of linear systems, and it indeed plays a central role in linear algebra in general. For this reason, we give it a special name.

Definition 3.4.3:   The number of nonzero rows in any REF of a matrix A is called the **rank** of A, and is denoted rank(A).

Example 3.4.6   Determine rank(A) if $A = \begin{bmatrix} 3 & 1 & 4 \\ 4 & 3 & 5 \\ 2 & -1 & 3 \end{bmatrix}$.

Solution In order to determine rank(A), we must first reduce A to REF.

$$\begin{bmatrix} 3 & 1 & 4 \\ 4 & 3 & 5 \\ 2 & -1 & 3 \end{bmatrix} \overset{1}{\sim} \begin{bmatrix} 1 & 2 & 1 \\ 4 & 3 & 5 \\ 2 & -1 & 3 \end{bmatrix} \overset{2}{\sim} \begin{bmatrix} 1 & 2 & 1 \\ 0 & -5 & 1 \\ 0 & -5 & 1 \end{bmatrix}$$

$$\overset{3}{\sim} \begin{bmatrix} 1 & 2 & 1 \\ 0 & 1 & -\frac{1}{5} \\ 0 & -5 & 1 \end{bmatrix} \overset{4}{\sim} \begin{bmatrix} 1 & 2 & 1 \\ 0 & 1 & -\frac{1}{5} \\ 0 & 0 & 0 \end{bmatrix}.$$

Since there are two nonzero rows in this REF of A it follows from Definition 3.4.3 that

$$\text{rank}(A) = 2.$$

1. $A_{31}(-1)$ **2.** $A_{12}(-4)$, $A_{13}(-2)$ **3.** $M_2(-1/5)$ **4.** $A_{23}(5)$.

❑

In the previous example, the original matrix A had three nonzero rows, whereas any REF of A has only two nonzero rows. We can interpret this geometrically as follows. The three row vectors of A can be considered as vectors in 3-space with components

$$\mathbf{a}_1 = (3, 1, 4), \ \mathbf{a}_2 = (4, 3, 5), \ \mathbf{a}_3 = (2, -1, 3).$$

In performing ERO on A, we are taking combinations of these vectors in the following way:

$$c_1 \mathbf{a}_1 + c_2 \mathbf{a}_2 + c_3 \mathbf{a}_3.$$

That is, we have been combining the vectors *linearly*. The fact that we obtained a row of zeros in the REF means that there are values of the constants c_1, c_2, c_3 such that

$$c_1 \mathbf{a}_1 + c_2 \mathbf{a}_2 + c_3 \mathbf{a}_3 = \mathbf{0},$$

where $\mathbf{0}$ denotes the zero vector $(0, 0, 0)$. Equivalently, one of the vectors can be written in terms of the other two vectors, and therefore, the three vectors lie in a plane. Reducing the matrix to REF has uncovered this dependency. We shall have much more to say about this in Chapter 5.

REDUCED ROW-ECHELON MATRICES

It will be necessary in the future to consider the special row-echelon matrices that arise when 0s are placed above, as well as beneath, each leading 1. Any such matrix is called a reduced row-echelon matrix and is defined precisely as follows.

> **Definition 3.4.4:**   An $m \times n$ matrix is called a **reduced row-echelon matrix** if it satisfies the following conditions.
>
> **1.** It is a row-echelon matrix.
>
> **2.** Any *column* that contains a leading 1 has zeros everywhere else.

Example 3.4.7   The following are examples of reduced row-echelon matrices

$$\begin{bmatrix} 1 & 3 & 0 & 0 \\ 0 & 0 & 1 & 0 \\ 0 & 0 & 0 & 1 \end{bmatrix}, \begin{bmatrix} 1 & -1 & 7 & 0 \\ 0 & 0 & 0 & 1 \end{bmatrix}, \begin{bmatrix} 1 & 0 & 5 & 3 \\ 0 & 1 & 2 & 1 \\ 0 & 0 & 0 & 0 \end{bmatrix}, \text{ and } \begin{bmatrix} 1 & 0 & 0 \\ 0 & 1 & 0 \\ 0 & 0 & 1 \end{bmatrix}.$$

Although an $m \times n$ *matrix A* does not have a unique REF, in reducing A to RREF we are making a particular choice of row-echelon matrix since we arrange that all elements above the leading 1s are 0s. In view of this the following theorem should not be too surprising.

> **Theorem 3.4.4:**   An $m \times n$ matrix is row-equivalent to a *unique* reduced row-echelon matrix.

The unique reduced row-echelon matrix to which a matrix A is row-equivalent will be called the RREF of A. As illustrated in the next example, the row reduction algorithm is easily extended to determine the RREF of A — we just put zeros above and beneath each leading 1.

Example 3.4.8   Determine the RREF of $A = \begin{bmatrix} 3 & -1 & 22 \\ -1 & 5 & 2 \\ 2 & 4 & 24 \end{bmatrix}$.

Solution   We apply the row reduction algorithm, but put 0s above any leading 1.

$$\begin{bmatrix} 3 & -1 & 22 \\ -1 & 5 & 2 \\ 2 & 4 & 24 \end{bmatrix} \overset{1}{\sim} \begin{bmatrix} 1 & 9 & 26 \\ -1 & 5 & 2 \\ 2 & 4 & 24 \end{bmatrix} \overset{2}{\sim} \begin{bmatrix} 1 & 9 & 26 \\ 0 & 14 & 28 \\ 0 & -14 & -28 \end{bmatrix}$$

$$\overset{3}{\sim} \begin{bmatrix} 1 & 9 & 26 \\ 0 & 1 & 2 \\ 0 & -14 & -28 \end{bmatrix} \overset{4}{\sim} \begin{bmatrix} 1 & 0 & 8 \\ 0 & 1 & 2 \\ 0 & 0 & 0 \end{bmatrix}$$

which is the RREF of A.

1. $A_{21}(2)$ **2.** $A_{12}(1)$, $A_{13}(-2)$ **3.** $M_2(1/14)$ **4.** $A_{21}(-9)$, $A_{23}(-14)$.

EXERCISES 3.4

For problems 1–8, determine which of the given matrices are row-echelon or reduced row-echelon matrices.

1. $\begin{bmatrix} 1 & 0 & -1 & 0 \\ 0 & 0 & 1 & 2 \\ 0 & 0 & 0 & 0 \end{bmatrix}$.

2. $\begin{bmatrix} 1 & 0 & 2 & 5 \\ 1 & 0 & 0 & 2 \\ 0 & 1 & 1 & 0 \end{bmatrix}$.

3. $\begin{bmatrix} 1 & 0 & 0 & 0 \\ 0 & 0 & 0 & 1 \end{bmatrix}$.

4. $\begin{bmatrix} 0 & 1 \\ 1 & 0 \end{bmatrix}$.

5. $\begin{bmatrix} 1 & 1 \\ 0 & 0 \end{bmatrix}$.

6. $\begin{bmatrix} 1 & 0 & 1 & 2 \\ 0 & 0 & 1 & 1 \\ 0 & 0 & 0 & 1 \\ 0 & 0 & 0 & 0 \end{bmatrix}$.

7. $\begin{bmatrix} 0 & 0 & 0 & 0 \\ 0 & 0 & 0 & 0 \\ 0 & 0 & 0 & 0 \end{bmatrix}$.

8. $\begin{bmatrix} 0 & 1 & 0 & 0 \\ 0 & 0 & 1 & 0 \\ 0 & 0 & 0 & 1 \\ 0 & 0 & 0 & 0 \end{bmatrix}$.

For problems 9–18 use ERO to reduce the given matrix to REF, and hence determine the rank of each matrix.

9. $\begin{bmatrix} 2 & 1 \\ 1 & -3 \end{bmatrix}$.

10. $\begin{bmatrix} 2 & -4 \\ -4 & 8 \end{bmatrix}$.

11. $\begin{bmatrix} 2 & 1 & 4 \\ 2 & -3 & 4 \\ 3 & -2 & 6 \end{bmatrix}$.

12. $\begin{bmatrix} 0 & 1 & 3 \\ 0 & 1 & 4 \\ 0 & 3 & 5 \end{bmatrix}$.

13. $\begin{bmatrix} 2 & -1 \\ 3 & 2 \\ 2 & 5 \end{bmatrix}$.

14. $\begin{bmatrix} 2 & -1 & 3 \\ 3 & 1 & -2 \\ 2 & -2 & 1 \end{bmatrix}$.

15. $\begin{bmatrix} 2 & -1 & 3 & 4 \\ 1 & -2 & 1 & 3 \\ 1 & -5 & 0 & 5 \end{bmatrix}$.

16. $\begin{bmatrix} 2 & -2 & -1 & 3 \\ 3 & -2 & 3 & 1 \\ 1 & -1 & 1 & 0 \\ 2 & -1 & 2 & 2 \end{bmatrix}$.

17. $\begin{bmatrix} 4 & 7 & 4 & 7 \\ 3 & 5 & 3 & 5 \\ 2 & -2 & 2 & -2 \\ 5 & -2 & 5 & -2 \end{bmatrix}$.

18. $\begin{bmatrix} 2 & 1 & 3 & 4 & 2 \\ 1 & 0 & 2 & 1 & 3 \\ 2 & 3 & 1 & 5 & 7 \end{bmatrix}$.

For problems 19–25, reduce the given matrix to RREF and hence determine the rank of each matrix.

19. $\begin{bmatrix} 3 & 2 \\ 1 & -1 \end{bmatrix}$.

20. $\begin{bmatrix} 3 & 7 & 10 \\ 2 & 3 & -1 \\ 1 & 2 & 1 \end{bmatrix}$.

21. $\begin{bmatrix} 3 & -3 & 6 \\ 2 & -2 & 4 \\ 6 & -6 & 12 \end{bmatrix}$.

22. $\begin{bmatrix} 3 & 5 & -12 \\ 2 & 3 & -7 \\ -2 & -1 & 1 \end{bmatrix}$.

23. $\begin{bmatrix} 1 & -1 & -1 & 2 \\ 3 & -2 & 0 & 7 \\ 2 & -1 & 2 & 4 \\ 4 & -2 & 3 & 8 \end{bmatrix}$.

24. $\begin{bmatrix} 1 & -2 & 1 & 3 \\ 3 & -6 & 2 & 7 \\ 4 & -8 & 3 & 10 \end{bmatrix}$.

25. $\begin{bmatrix} 0 & 1 & 2 & 1 \\ 0 & 3 & 1 & 2 \\ 0 & 2 & 0 & 1 \end{bmatrix}$.

26. The matrix in problem 14.

27. The matrix in problem 15.

28. The matrix in problem 18.

Many forms of technology have commands for performing ERO on a matrix A. For example, in the linear algebra package of Maple, the three ERO are:

swaprow(A, i, j) : permute rows i and j in A.

mulrow(A, i, k):   multiply row i by k.

addrow(A, i, j, k): add $k * $ row i to row j.

◆ For problems 26–28, use some form of technology to determine an REF of the given matrix.

◆ Many forms of technology also have built in functions for directly determining the RREF of a given matrix A. For example in the linear algebra package of Maple the appropriate command is rref(A). In problems 29–31 use technology to determine directly the RREF of the given matrix.

29. The matrix in problem 21.

30. The matrix in problem 24.

31. The matrix in problem 25.

3.5   GAUSSIAN ELIMINATION

We now illustrate how ERO applied to the augmented matrix of a system of linear equations can be used first to determine whether a system is consistent, and second, if the system is consistent, to find all of its solutions. In doing so, we will develop the general theory for linear systems of algebraic equations.

Example 3.5.1   Determine the solution set to

$$3x_1 - 2x_2 + 2x_3 = 9,$$
$$x_1 - 2x_2 + x_3 = 5, \tag{3.5.1}$$
$$2x_1 - x_2 - 2x_3 = -1.$$

Solution   We first use ERO to reduce the augmented matrix of the system to REF.

$$\begin{bmatrix} 3 & -2 & 2 & 9 \\ 1 & -2 & 1 & 5 \\ 2 & -1 & -2 & -1 \end{bmatrix} \overset{1}{\sim} \begin{bmatrix} 1 & -2 & 1 & 5 \\ 3 & -2 & 2 & 9 \\ 2 & -1 & -2 & -1 \end{bmatrix} \overset{2}{\sim} \begin{bmatrix} 1 & -2 & 1 & 5 \\ 0 & 4 & -1 & -6 \\ 0 & 3 & -4 & -11 \end{bmatrix}$$

$$\overset{3}{\sim} \begin{bmatrix} 1 & -2 & 1 & 5 \\ 0 & 1 & 3 & 5 \\ 0 & 3 & -4 & -11 \end{bmatrix} \overset{4}{\sim} \begin{bmatrix} 1 & -2 & 1 & 5 \\ 0 & 1 & 3 & 5 \\ 0 & 0 & -13 & -26 \end{bmatrix} \overset{5}{\sim} \begin{bmatrix} 1 & -2 & 1 & 5 \\ 0 & 1 & 3 & 5 \\ 0 & 0 & 1 & 2 \end{bmatrix}.$$

The system corresponding to this REF of the augmented matrix is

$$x_1 - 2x_2 + x_3 = 5, \tag{3.5.2}$$
$$x_2 + 3x_3 = 5, \tag{3.5.3}$$
$$x_3 = 2, \tag{3.5.4}$$

which can be solved by *back substitution*. From equation (3.5.4), $x_3 = 2$. Substituting into equation (3.5.3) and solving for x_2, we find that $x_2 = -1$. Finally, substituting into equation (3.5.2) for x_3 and x_2 and solving for x_1 yields $x_1 = 1$. Thus, our original system of equations has the unique solution $(1, -1, 2)$, and the solution set to the system is

$$S = \{(1, -1, 2)\}.$$

---

1. P_{12} **2.** $A_{12}(-3), A_{13}(-2)$ **3.** $A_{32}(-1)$ **4.** $A_{23}(-3)$ **5.** $M_3(-1/13)$

$\square$

The process of reducing the augmented matrix to REF and then using back substitution to solve the equivalent system is called **Gaussian Elimination**. The particular case of Gaussian elimination that arises when the augmented matrix is reduced to RREF is called **Gauss–Jordan elimination**.

Example 3.5.2   Use Gauss–Jordan elimination to determine the solution set to

$$x_1 + 2x_2 - x_3 = 1,$$
$$2x_1 + 5x_2 - x_3 = 3,$$
$$x_1 + 3x_2 + 2x_3 = 6.$$

Solution   In this case, we first reduce the augmented matrix of the system to RREF.

$$\begin{bmatrix} 1 & 2 & -1 & 1 \\ 2 & 5 & -1 & 3 \\ 1 & 3 & 2 & 6 \end{bmatrix} \overset{1}{\sim} \begin{bmatrix} 1 & 2 & -1 & 1 \\ 0 & 1 & 1 & 1 \\ 0 & 1 & 3 & 5 \end{bmatrix} \overset{2}{\sim} \begin{bmatrix} 1 & 0 & -3 & -1 \\ 0 & 1 & 1 & 1 \\ 0 & 0 & 2 & 4 \end{bmatrix}$$

$$\overset{3}{\sim} \begin{bmatrix} 1 & 0 & -3 & -1 \\ 0 & 1 & 1 & 1 \\ 0 & 0 & 1 & 2 \end{bmatrix} \overset{4}{\sim} \begin{bmatrix} 1 & 0 & 0 & 5 \\ 0 & 1 & 0 & -1 \\ 0 & 0 & 1 & 2 \end{bmatrix}.$$

The augmented matrix is now in RREF. The equivalent system is

$$x_1 = 5,$$
$$x_2 = -1,$$
$$x_3 = 2.$$

and the solution can be read off directly as $(5, -1, 2)$. Consequently the given system has solution set $\{(5, -1, 2)\}$.

---

1. $A_{12}(-2), A_{13}(-1)$ **2.** $A_{21}(-2), A_{23}(-1)$ **3.** $M_3(1/2)$ **4.** $A_{31}(3), A_{32}(-1)$

$\square$

We see from the previous two examples that the advantage of Gauss–Jordan elimination over Gaussian elimination is that it does not require back substitution. However, the disadvantage is that reducing the augmented matrix to RREF requires more ERO than reduction to REF. It can be shown, in fact, that in general, Gaussian elimination is the more computationally efficient technique. As we will see in the next section, the main reason for introducing the Gauss–Jordan method is its application to the computation of the inverse of an $n \times n$ matrix.

REMARK   The Gaussian elimination method is so systematic that it can be programmed easily on a computer. Indeed many large scale programs for solving $n \times n$ linear systems are based on the row reduction method.

In both of the preceding examples,

$$\text{rank}(A) = \text{rank}(A^\#) = \text{number of unknowns in the system}$$

and the system had a unique solution. More generally, we have the following lemma:

Lemma 3.5.1:   Consider the $m \times n$ linear system $A\mathbf{x} = \mathbf{b}$. Let $A^\#$ denote the augmented matrix of the system. If $\text{rank}(A) = \text{rank}(A^\#) = n$, then the system has a unique solution.

PROOF   If $\text{rank}(A) = \text{rank}(A^\#) = n$, then there are n leading ones in any REF of A, and hence, back substitution gives a unique solution. ∎

There are only two other possibilities for the relationship between $\text{rank}(A)$ and $\text{rank}(A^\#)$, namely, $\text{rank}(A) < \text{rank}(A^\#)$ or $\text{rank}(A) = \text{rank}(A^\#) < n$. We now consider what happens in these cases.

Example 3.5.3   Determine the solution set to

$$\begin{aligned}
x_1 + x_2 - x_3 + x_4 &= 1, \\
2x_1 + 3x_2 + x_3 &= 4, \\
3x_1 + 5x_2 + 3x_3 - x_4 &= 5.
\end{aligned}$$

Solution   We use ERO to reduce the augmented matrix:

$$\begin{bmatrix} 1 & 1 & -1 & 1 & 1 \\ 2 & 3 & 1 & 0 & 4 \\ 3 & 5 & 3 & -1 & 5 \end{bmatrix} \overset{1}{\sim} \begin{bmatrix} 1 & 1 & -1 & 1 & 1 \\ 0 & 1 & 3 & -2 & 2 \\ 0 & 2 & 6 & -4 & 2 \end{bmatrix} \overset{2}{\sim} \begin{bmatrix} 1 & 1 & -1 & 1 & 1 \\ 0 & 1 & 3 & -2 & 2 \\ 0 & 0 & 0 & 0 & -2 \end{bmatrix}$$

The last row tells us that the system of equations has no solution (that is, it is inconsistent), since it requires

$$0x_1 + 0x_2 + 0x_3 + 0x_4 = -2,$$

which is clearly impossible. The solution set to the system is thus the empty set ∅.

---

1. $A_{12}(-2), A_{13}(-3)$ **2.** $A_{23}(-2)$

□

In the previous example $\text{rank}(A) = 2$, whereas $\text{rank}(A^\#) = 3$. Thus, $\text{rank}(A) < \text{rank}(A^\#)$, and the corresponding system has no solution. Next we establish that this result is true in general.

Lemma 3.5.2:   Consider the $m \times n$ linear system $A\mathbf{x} = \mathbf{b}$. Let $A^{\#}$ denote the augmented matrix of the system. If rank$(A) <$ rank$(A^{\#})$, then the system is inconsistent.

PROOF   If rank$(A) <$ rank$(A^{\#})$, then there will be one row in the RREF of the augmented matrix whose first nonzero element arises in the last column. Such a row corresponds to an equation of the form

$$0x_1 + 0x_2 + \cdots + 0x_n = 1,$$

which has no solution. Consequently, the system is inconsistent.   ∎

Now for the final possibility.

Example 3.5.4   Determine the solution set to

$$5x_1 - 6x_2 + x_3 = 4,$$
$$2x_1 - 3x_2 + x_3 = 1,$$
$$4x_1 - 3x_2 - x_3 = 5.$$

Solution   We begin by reducing the augmented matrix of the system.

$$\begin{bmatrix} 5 & -6 & 1 & 4 \\ 2 & -3 & 1 & 1 \\ 4 & -3 & -1 & 5 \end{bmatrix} \overset{1}{\sim} \begin{bmatrix} 1 & -3 & 2 & -1 \\ 2 & -3 & 1 & 1 \\ 4 & -3 & -1 & 5 \end{bmatrix} \overset{2}{\sim} \begin{bmatrix} 1 & -3 & -4 & -1 \\ 0 & 3 & -3 & 3 \\ 0 & 9 & -9 & 9 \end{bmatrix}$$

$$\overset{3}{\sim} \begin{bmatrix} 1 & -3 & -4 & -1 \\ 0 & 1 & -1 & 1 \\ 0 & 9 & -9 & 9 \end{bmatrix} \overset{4}{\sim} \begin{bmatrix} 1 & -3 & -4 & -1 \\ 0 & 1 & -1 & 1 \\ 0 & 0 & 0 & 0 \end{bmatrix}.$$

| **1.** $A_{31}(-1)$ **2.** $A_{12}(-2)$, $A_{13}(-4)$ **3.** $M_2(1/3)$ **4.** $A_{23}(-9)$ |

The augmented matrix is now in REF, and the equivalent system is

$$x_1 - 3x_2 - 4x_3 = -1, \tag{3.5.5}$$
$$x_2 - x_3 = 1. \tag{3.5.6}$$

Since we have three variables, but only two equations relating them, we are free to specify one of the variables arbitrarily. The variable that we choose to specify is called a **free variable**. The remaining variables are then determined by the system of equations and are called **bound variables**. In the foregoing system, we take x_3 as the free variable and set

$$x_3 = t,$$

where t can assume any real value. It follows from (3.5.6) that

$$x_2 = 1 + t.$$

Further, from equation (3.5.5),

$$x_1 = -1 + 3(1 + t) + 4t = 2 + 7t.$$

Thus the solution set to the given system of equations is

$$S = \{(2 + 7t, \ 1 + t, \ t) : t \in \mathbf{R}\}.$$

The system has an infinite number of solutions obtained by allowing the parameter t to assume all real values. For example, two particular solutions of the system are

$$(2, 1, 0) \quad \text{and} \quad (-12, -1, -2),$$

corresponding to $t = 0$ and $t = -2$, respectively.   ❑

REMARK   The geometry of the foregoing solution is as follows. The given equations can be interpreted as defining three planes in 3-space. Any solution to the system gives the coordinates of a point of intersection of the three planes. In the preceding example the planes intersect in a line whose parametric equations are

$$x_1 = 2 + 7t, \qquad x_2 = 1 + t, \qquad x_3 = t.$$

(See Figure. 3.3.1.)

In general, the solution to a consistent $m \times n$ system of linear equations may involve more than one free variable. Indeed, the number of free variables will depend on how many nonzero rows arise in any REF of the augmented matrix, $A^\#$, of the system, that is, on the rank of $A^\#$. More precisely, if rank$(A^\#) = r^\#$, then the equivalent system will have only $r^\#$ relationships between the n variables. Consequently, provided the system is consistent,

$$\text{Number of free variables} = n - r^\#.$$

We therefore have the following lemma.

Lemma 3.5.3:   Consider the $m \times n$ linear system $A\mathbf{x} = \mathbf{b}$. Let $A^\#$ denote the augmented matrix of the system. If rank(A) = rank$(A^\#)$ < n, then the system has an infinite number of solutions.

PROOF   Suppose rank$(A^\#)$ = rank(A) = $r^\#$. If $r^\# < n$, then, as discussed before, any row-echelon equivalent system will have only $r^\#$ equations involving the n variables, and so, there will be $n - r^\#$ free variables. If we assign arbitrary values to these free variables, then the remaining $r^\#$ variables will be uniquely determined, by back substitution, from the system. Since we have one solution for each value of the free variables, in this case there are an infinite number of solutions.   ∎

Example 3.5.5   Use Gaussian elimination to solve

$$\begin{aligned} x_1 - 2x_2 + 2x_3 - x_4 &= 3, \\ 3x_1 + x_2 + 6x_3 + 11x_4 &= 16, \\ 2x_1 - x_2 + 4x_3 + 4x_4 &= 9. \end{aligned}$$

Solution   An REF of the augmented matrix of the system is

$$\begin{bmatrix} 1 & -2 & 2 & -1 & 3 \\ 0 & 1 & 0 & 2 & 1 \\ 0 & 0 & 0 & 0 & 0 \end{bmatrix},$$

so that we have two free variables. The equivalent system is

$$x_1 - 2x_2 + 2x_3 - x_4 = 3, \qquad (3.5.7)$$

$$x_2 \quad\quad + 2x_4 = 1. \qquad (3.5.8)$$

Notice that we cannot choose any two variables freely. For example, from equation (3.5.8), we cannot specify both x_2 and x_4 independently. The bound variables should be taken as those that correspond to leading ones in the REF of $A^\#$, since these are the variables that can always be determined by back substitution.

> Choose as free variables those variables that **do not** correspond to a leading 1 in a REF of $A^\#$.

Applying this rule to the system of equations (3.5.7) and (3.5.8), we choose x_3 and x_4 as free variables and therefore set

$$x_3 = s, \qquad x_4 = t.$$

It then follows from equation (3.5.8) that

$$x_2 = 1 - 2t.$$

Substitution into equation (3.5.7) yields

$$x_1 = 5 - 2s - 3t,$$

so that the solution set to the given system is

$$S = \{(5 - 2s - 3t, 1 - 2t, s, t) : s, t \in \mathbf{R}\}. \qquad \square$$

Lemmas 3.5.1–3.5.3 completely characterize the solution properties of an $m \times n$ linear system. Combining the results of these three lemmas gives the next theorem.

Theorem 3.5.1:   Consider the $m \times n$ linear system $Ax = b$. Let r denote the rank of A, and let $r^\#$ denote the rank of the augmented matrix of the system. Then

1. If $r < r^\#$ the system is inconsistent.

2. If $r = r^\#$ the system is consistent and

    (a) There exists a unique solution if and only if $r^\# = n$.

    (b) There exists an infinite number of solutions if and only if $r^\# < n$.

HOMOGENEOUS LINEAR SYSTEMS

Many of the problems that we will meet in the future will require the solution to a homogeneous system of linear equations. The general form for such a system is

$$a_{11}x_1 + a_{12}x_2 + \cdots + a_{1n}x_n = 0,$$
$$a_{21}x_1 + a_{22}x_2 + \cdots + a_{2n}x_n = 0,$$
$$\vdots \qquad\qquad\qquad (3.5.9)$$
$$a_{m1}x_1 + a_{m2}x_2 + \cdots + a_{mn}x_n = 0,$$

or, in matrix form, $Ax = 0$, where A is the coefficient matrix of the system and 0 denotes the m-vector whose elements are all zeros.

Specializing the results of Theorem 3.5.1 to the case of a homogeneous system enables us to make some stronger statements in this case. Our main result is the following corollary to the previous theorem.

Corollary 3.5.1:   The homogeneous linear system $A\mathbf{x} = \mathbf{0}$ is always cons–istent.

PROOF   The augmented matrix, $A^{\#}$, of a homogeneous linear system only differs from that of the coefficient matrix, A, by the addition of a column of zeros. Consequently, for a homogeneous system, rank($A^{\#}$) = rank(A), and therefore, from Theorem 3.5.1, such a system is necessarily consistent.   ■

REMARKS

1. We can see directly from (3.5.9) that a homogeneous system always has the solution

$$(0, 0, \ldots, 0).$$

This solution is referred to as the **trivial solution**. Consequently, a homogeneous system either has *only* the trivial solution or it has an infinite number of solutions (one of which must be the trivial solution).

2. Once more it is worth mentioning the geometric interpretation of Corollary 3.5.1 in the case of a homogeneous system with three unknowns. We can regard each equation of such a system as defining a plane. Due to the homogeneity, each plane passes through the origin, and hence, the planes intersect at least at the origin.

Often we will be interested in determining whether a given homogeneous system has an infinite number of solutions, and not in actually obtaining the solutions. The following corollary to Theorem 3.5.1 can sometimes be *helpful* in determining by inspection whether a given homogeneous system has nontrivial solutions:

Corollary 3.5.2:   A homogeneous system of m linear equations in n unknowns, *with $m < n$*, has an infinite number of solutions.

PROOF   Let $r^{\#}$ denote the rank of the augmented matrix of the system. Then, $r^{\#} \leq m < n$, and so, by Theorem 3.5.1, the system has an infinite number of solutions. (Note that we have again used rank(A) = rank($A^{\#}$) for a homogeneous system.)   ■

REMARK   If $m \geq n$, then we may or may not have nontrivial solutions, depending on whether the rank, $r^{\#}$, of the augmented matrix of the system satisfies $r^{\#} < n$, or $r^{\#} = n$, respectively.

We end this section with two types of systems that will be of particular interest later in the text.

Example 3.5.6:   Determine the solution set to $A\mathbf{x} = \mathbf{0}$, if $A = \begin{bmatrix} 0 & 2 & 3 \\ 0 & 1 & -1 \\ 0 & 3 & 7 \end{bmatrix}$.

Solution   The augmented matrix of the system is

$$\begin{bmatrix} 0 & 2 & 3 & 0 \\ 0 & 1 & -1 & 0 \\ 0 & 3 & 7 & 0 \end{bmatrix},$$

with RREF

$$\begin{bmatrix} 0 & 1 & 0 & 0 \\ 0 & 0 & 1 & 0 \\ 0 & 0 & 0 & 0 \end{bmatrix}.$$

The equivalent system is

$$x_2 = 0$$
$$x_3 = 0.$$

It is tempting to conclude that the solution to the system is $x_1 = x_2 = x_3 = 0$. However, this is incorrect. Since x_1 does not occur in the system, it is a free variable and therefore *not necessarily* zero. Consequently the correct solution to the foregoing system is $(r, 0, 0)$, where r is a free variable, and the solution set is $\{(r, 0, 0) : r \in \mathbf{R}\}$.   □

The linear systems that we have so far encountered have all had real coefficients, and we have considered corresponding real solutions. The techniques that we have developed for solving linear systems are also applicable to the case when our system has complex coefficients. The corresponding solutions will also be complex.

REMARK In general, the simplest method of putting a leading 1 in a position that contains the complex number $a + ib$ is to multiply the corresponding row by $\frac{1}{a^2 + b^2}(a - ib)$. This is illustrated in steps 1 and 4 in the next example. The student is strongly encouraged to work through each step of this example. If difficulties are encountered, then this is an indication that consultation of Appendix 1 is in order.

Example 3.5.7   Determine the solution set to

$$(1 + 2i)x_1 + \qquad 4x_2 + (3 + i)x_3 = 0,$$
$$(2 - i)x_1 + (1 + i)x_2 + \qquad 3x_3 = 0,$$
$$5ix_1 + (7 - i)x_2 + (3 + 2i)x_3 = 0.$$

Solution   We reduce the augmented matrix of the system.

$$\begin{bmatrix} 1 + 2i & 4 & 3 + i & 0 \\ 2 - i & 1 + i & 3 & 0 \\ 5i & 7 - i & 3 + 2i & 0 \end{bmatrix} \overset{1}{\sim} \begin{bmatrix} 1 & \frac{4}{5}(1 - 2i) & 1 - i & 0 \\ 2 - i & 1 + i & 3 & 0 \\ 5i & 7 - i & 3 + 2i & 0 \end{bmatrix}$$

$$\overset{2}{\sim} \begin{bmatrix} 1 & \frac{4}{5}(1 - 2i) & 1 - i & 0 \\ 0 & (1 + i) - \frac{4}{5}(1 - 2i)(2 - i) & 3 - (1 - i)(2 - i) & 0 \\ 0 & (7 - i) - 4i(1 - 2i) & (3 + 2i) - 5i(1 - i) & 0 \end{bmatrix}$$

$$= \begin{bmatrix} 1 & \frac{4}{5}(1-2i) & 1-i & 0 \\ 0 & 1+5i & 2+3i & 0 \\ 0 & -1-5i & -2-3i & 0 \end{bmatrix} \overset{3}{\sim} \begin{bmatrix} 1 & \frac{4}{5}(1-2i) & 1-i & 0 \\ 0 & 1+5i & 2+3i & 0 \\ 0 & 0 & 0 & 0 \end{bmatrix}$$

$$\overset{4}{\sim} \begin{bmatrix} 1 & \frac{4}{5}(1-2i) & 1-i & 0 \\ 0 & 1 & \frac{1}{26}(17-7i) & 0 \\ 0 & 0 & 0 & 0 \end{bmatrix}.$$

The matrix is now in REF. The equivalent system is

$$x_1 + \tfrac{4}{5}(1-2i)x_2 + (1-i)x_3 = 0,$$
$$x_2 + \tfrac{1}{26}(17-7i)x_3 = 0.$$

There is one free variable, which we take to be $x_3 = r$. Applying back substitution yields

$$x_2 = \tfrac{1}{26}r(-17+7i)$$
$$x_1 = -\tfrac{2}{65}r(1-2i)(-17+7i) - r(1-i) = -\tfrac{1}{65}r(59+17i)$$

so that the solution set to the system is

$$\{(-\tfrac{1}{65}r(59+17i), \tfrac{1}{26}r(-17+7i), r) : r \in \mathbf{C}\}.$$

---

1. $M_1(\tfrac{1}{5}(1-2i))$ 2. $A_{12}(-(2-i))$, $A_{13}(-5i)$
3. $A_{23}(1)$ 4. $M_2(\tfrac{1}{26}(1-5i))$

EXERCISES 3.5

For problems 1–9, use Gaussian elimination to determine the solution set to the given system.

1. $x_1 + 2x_2 + x_3 = 1,$
   $3x_1 + 5x_2 + x_3 = 3,$
   $2x_1 + 6x_2 + 7x_3 = 1.$

2. $3x_1 - x_2 = 1,$
   $2x_1 + x_2 + 5x_3 = 4,$
   $7x_1 - 5x_2 - 8x_3 = -3.$

3. $3x_1 + 5x_2 - x_3 = 14,$
   $x_1 + 2x_2 + x_3 = 3,$
   $2x_1 + 5x_2 + 6x_3 = 2.$

4. $6x_1 - 3x_2 + 3x_3 = 12,$
   $2x_1 - x_2 + x_3 = 4,$
   $-4x_1 + 2x_2 - 2x_3 = -8.$

5. $2x_1 - x_2 + 3x_3 = 14,$
   $3x_1 + x_2 - 2x_3 = -1,$
   $7x_1 + 2x_2 - 3x_3 = 3,$
   $5x_1 - x_2 - 2x_3 = 5.$

6. $2x_1 - x_2 - 4x_3 = 5,$
   $3x_1 + 2x_2 - 5x_3 = 8,$
   $5x_1 + 6x_2 - 6x_3 = 20,$
   $x_1 + x_2 - 3x_3 = -3.$

7.
$$x_1 + 2x_2 - x_3 + x_4 = 1,$$
$$2x_1 + 4x_2 - 2x_3 + 2x_4 = 2,$$
$$5x_1 + 10x_2 - 5x_3 + 5x_4 = 5.$$

8.
$$x_1 + 2x_2 - x_3 + x_4 = 1,$$
$$2x_1 - 3x_2 + x_3 - x_4 = 2,$$
$$x_1 - 5x_2 + 2x_3 - 2x_4 = 1,$$
$$4x_1 + x_2 - x_3 + x_4 = 3.$$

9.
$$x_1 + 2x_2 + x_3 + x_4 - 2x_5 = 3,$$
$$x_3 + 4x_4 - 3x_5 = 2,$$
$$2x_1 + 4x_2 - x_3 - 10x_4 + 5x_5 = 0.$$

For problems 10–15, use Gauss–Jordan elimination to determine the solution set to the given system.

10.
$$2x_1 - x_2 - x_3 = 2,$$
$$4x_1 + 3x_2 - 2x_3 = -1,$$
$$x_1 + 4x_2 + x_3 = 4.$$

11.
$$3x_1 + x_2 + 5x_3 = 2,$$
$$x_1 + x_2 - x_3 = 1,$$
$$2x_1 + x_2 + 2x_3 = 3.$$

12.
$$x_1 \qquad - 2x_3 = -3,$$
$$3x_1 - 2x_2 - 4x_3 = -9,$$
$$x_1 - 4x_2 + 2x_3 = -3.$$

13.
$$2x_1 - x_2 + 3x_3 - x_4 = 3,$$
$$3x_1 + 2x_2 + x_3 - 5x_4 = -6,$$
$$x_1 - 2x_2 + 3x_3 + x_4 = 6.$$

14.
$$x_1 + x_2 + x_3 - x_4 = 4,$$
$$x_1 - x_2 - x_3 - x_4 = 2,$$
$$x_1 + x_2 - x_3 + x_4 = -2,$$
$$x_1 - x_2 + x_3 + x_4 = -8.$$

15.
$$2x_1 - x_2 + 3x_3 + x_4 - x_5 = 11,$$
$$x_1 - 3x_2 - 2x_3 - x_4 - 2x_5 = 2,$$
$$3x_1 + x_2 - 2x_3 - x_4 + x_5 = -2,$$
$$x_1 + 2x_2 + x_3 + 2x_4 + 3x_5 = -3,$$
$$5x_1 - 3x_2 - 3x_3 + x_4 + 2x_5 = 2.$$

For problems 16–20, determine the solution set to the system $Ax = b$ for the given coefficient matrix A and right-hand side vector b.

16. $A = \begin{bmatrix} 1 & -3 & 1 \\ 5 & -4 & 1 \\ 2 & 4 & -3 \end{bmatrix}, b = \begin{bmatrix} 8 \\ 15 \\ -4 \end{bmatrix}.$

17. $A = \begin{bmatrix} 1 & 0 & 5 \\ 3 & -2 & 11 \\ 2 & -2 & 6 \end{bmatrix}, b = \begin{bmatrix} 0 \\ 2 \\ 2 \end{bmatrix}.$

18. $A = \begin{bmatrix} 0 & 1 & -1 \\ 0 & 5 & 1 \\ 0 & 2 & 1 \end{bmatrix}, b = \begin{bmatrix} -2 \\ 8 \\ 5 \end{bmatrix}.$

19. $A = \begin{bmatrix} 1 & -1 & 0 & -1 \\ 2 & 1 & 3 & 7 \\ 3 & -2 & 1 & 0 \end{bmatrix}, b = \begin{bmatrix} 2 \\ 2 \\ 4 \end{bmatrix}.$

20. $A = \begin{bmatrix} 1 & 1 & 0 & 1 \\ 3 & 1 & -2 & 3 \\ 2 & 3 & 1 & 2 \\ -2 & 3 & 5 & -2 \end{bmatrix}, b = \begin{bmatrix} 2 \\ 8 \\ 3 \\ -9 \end{bmatrix}.$

21. Determine all values of the constant k for which the following system has **(a)** no solution, **(b)** an infinite number of solutions, **(c)** a unique solution.

$$x_1 + 2x_2 - x_3 = 3,$$
$$2x_1 + 5x_2 + x_3 = 7,$$
$$x_1 + x_2 - k^2 x_3 = -k.$$

22. Determine all values of the constants a and b for which the following system has **(a)** no solution, **(b)** a unique solution, **(c)** an infinite number of solutions.

$$x_1 + x_2 - 2x_3 = 4,$$
$$3x_1 + 5x_2 - 4x_3 = 16,$$
$$2x_1 + 3x_2 - ax_3 = b.$$

23. Show that the system

$$x_1 + x_2 + x_3 = y_1,$$
$$2x_1 + 3x_2 + x_3 = y_2,$$
$$3x_1 + 5x_2 + x_3 = y_3,$$

has an infinite number of solutions provided (y_1, y_2, y_3) lies on the plane with equation $y_1 - 2y_2 + y_3 = 0$.

24. Consider the system of linear equations

$$a_{11} x_1 + a_{12} x_2 = b_1,$$
$$a_{21} x_1 + a_{22} x_2 = b_2.$$

Define Δ, Δ_1, and Δ_2 by

$$\Delta = a_{11}a_{22} - a_{12}a_{21}, \ \Delta_1 = a_{22}b_1 - a_{12}b_2,$$
$$\Delta_2 = a_{11}b_2 - a_{21}b_1.$$

(a) Show that the given system has a unique solution if and only if $\Delta \neq 0$, and that the unique solution in this case is $x_1 = \Delta_1/\Delta$, $x_2 = \Delta_2/\Delta$.

(b) If $\Delta = 0$ and $a_{11} \neq 0$, determine the conditions on Δ_2 that would guarantee that the system has (i) no solution, (ii) an infinite number of solutions.

(c) Interpret your results in terms of inter-sections of straight lines.

Gaussian elimination with *partial pivoting* uses the following algorithm to reduce the augmented matrix:

(1) Start with augmented matrix $A^{\#}$.
(2) Determine the leftmost nonzero column.
(3) Permute rows to put the element of largest absolute value in the pivot position.
(4) Use ERO to put zeros beneath the pivot position.
(5) If there are no more nonzero rows below the pivot position go to 7, otherwise go to 6.
(6) Apply (2)–(5) to the submatrix consisting of the rows that lie below the pivot position.
(7) The matrix is in reduced form[1].

In problems 25–28 use the preceding algorithm to reduce $A^{\#}$ and then apply back substitution to solve the equivalent system. Technology might be useful in performing the required row operations.

25. The system in problem 1.

26. The system in problem 5.

27. The system in problem 6.

28. The system in problem 10.

29. (a) An $n \times n$ system of linear equations whose matrix of coefficients is a lower triangular matrix is called a lower triangular system. Assuming that $a_{ii} \neq 0$ for each i, devise a method for solving such a system that is analogous to the back substitution method.

(b) Use your method from (a) to solve

$$x_1 \qquad\qquad = 2,$$
$$2x_1 - 3x_2 \qquad = 1,$$
$$3x_1 + x_2 - x_3 = 8.$$

30. Find all solutions to the following nonlinear system of equations

---
[1]Note that this reduced form is *not* a row-echelon matrix.

$$4x_1{}^3 + 2x_2{}^2 + 3x_3 = 12,$$
$$x_1{}^3 - x_2{}^2 + x_3 = 2,$$
$$3x_1{}^3 + x_2{}^2 - x_3 = 2.$$

Does your answer contradict Theorem 3.5.1? Explain.

For problems 31–41, determine the solution set to the given system.

31. $3x_1 + 2x_2 - x_3 = 0,$
$2x_1 + x_2 + x_3 = 0,$
$5x_1 - 4x_2 + x_3 = 0.$

32. $2x_1 + x_2 - x_3 = 0,$
$3x_1 - x_2 + 2x_3 = 0,$
$x_1 - x_2 - x_3 = 0,$
$5x_1 + 2x_2 - 2x_3 = 0.$

33. $2x_1 - x_2 - x_3 = 0,$
$5x_1 - x_2 + 2x_3 = 0,$
$x_1 + x_2 + 4x_3 = 0.$

34. $(1 + 2i)x_1 + (1 - i)x_2 + x_3 = 0,$
$ix_1 + (1 + i)x_2 - ix_3 = 0,$
$2ix_1 + ix_2 + (1 + 3i)x_3 = 0.$

35. $3x_1 + 2x_2 + x_3 = 0,$
$6x_1 - x_2 + 2x_3 = 0,$
$12x_1 + 6x_2 + 4x_3 = 0.$

36. $2x_1 + x_2 - 8x_3 = 0,$
$3x_1 - 2x_2 - 5x_3 = 0,$
$5x_1 - 6x_2 - 3x_3 = 0,$
$3x_1 - 5x_2 + x_3 = 0.$

37. $x_1 + (1 + i)x_2 + (1 - i)x_3 = 0,$
$ix_1 + x_2 + ix_3 = 0,$
$(1 - 2i)x_1 - (1 - i)x_2 + (1 - 3i)x_3 = 0.$

38. $x_1 - x_2 + x_3 = 0,$
$3x_2 + 2x_3 = 0,$
$3x_1 - x_3 = 0,$
$5x_1 + x_2 - x_3 = 0.$

39. $2x_1 - 4x_2 + 6x_3 = 0,$
$3x_1 - 6x_2 + 9x_3 = 0,$
$x_1 - 2x_2 + 3x_3 = 0,$
$5x_1 - 10x_2 + 15x_3 = 0.$

40. $4x_1 - 2x_2 - x_3 - x_4 = 0,$
$3x_1 + x_2 - 2x_3 + 3x_4 = 0,$
$5x_1 - x_2 - 2x_3 + x_4 = 0.$

41. $2x_1 + x_2 - x_3 + x_4 = 0,$
$x_1 + x_2 + x_3 - x_4 = 0,$
$3x_1 - x_2 + x_3 - 2x_4 = 0,$
$4x_1 + 2x_2 - x_3 + x_4 = 0.$

For problems 42–52, determine the solution set to the system $Ax = \mathbf{0}$ for the given matrix A.

42. $A = \begin{bmatrix} 2 & -1 \\ 3 & 4 \end{bmatrix}.$

43. $A = \begin{bmatrix} 1 - i & 2i \\ 1 + i & -2 \end{bmatrix}.$

44. $A = \begin{bmatrix} 1 + i & 1 - 2i \\ -1 + i & 2 + i \end{bmatrix}.$

45. $A = \begin{bmatrix} 1 & 2 & 3 \\ 2 & -1 & 0 \\ 1 & 1 & 1 \end{bmatrix}.$

46. $A = \begin{bmatrix} 1 & 1 & 1 & -1 \\ -1 & 0 & -1 & 2 \\ 1 & 3 & 2 & 2 \end{bmatrix}.$

47. $A = \begin{bmatrix} 2 - 3i & 1 + i & i - 1 \\ 3 + 2i & -1 + i & -1 - i \\ 5 - i & 2i & -2 \end{bmatrix}.$

48. $A = \begin{bmatrix} 1 & 3 & 0 \\ -2 & -3 & 0 \\ 1 & 4 & 0 \end{bmatrix}.$

49. $A = \begin{bmatrix} 1 & 0 & 3 \\ 3 & -1 & 7 \\ 2 & 1 & 8 \\ 1 & 1 & 5 \\ -1 & 1 & -1 \end{bmatrix}.$

50. $A = \begin{bmatrix} 1 & -1 & 0 & 1 \\ 3 & -2 & 0 & 5 \\ -1 & 2 & 0 & 1 \end{bmatrix}.$

51. $A = \begin{bmatrix} 1 & 0 & -3 & 0 \\ 3 & 0 & -9 & 0 \\ -2 & 0 & 6 & 0 \end{bmatrix}.$

52. $A = \begin{bmatrix} 2 + i & i & 3 - 2i \\ i & 1 - i & 4 + 3i \\ 3 - i & 1 + i & 1 + 5i \end{bmatrix}.$

53. Determine all values of the constant k for which the following system has **(a)** a unique solution, **(b)** an infinite number of solutions.

$2x_1 + x_2 - x_3 + x_4 = 0,$
$x_1 + x_2 + x_3 - x_4 = 0,$
$4x_1 + 2x_2 - x_3 + x_4 = 0,$
$3x_1 - x_2 + x_3 + kx_4 = 0.$

3.6 THE INVERSE OF A SQUARE MATRIX

In this section, we investigate the situation when, for a given $n \times n$ matrix A, there exists a matrix B satisfying

$$AB = I_n \quad \text{and} \quad BA = I_n \tag{3.6.1}$$

and derive an efficient method for determining B (when it does exist). As a possible application of the existence of such a matrix B, consider the $n \times n$ linear system

$$Ax = \mathbf{b}. \tag{3.6.2}$$

Premultiplying both sides of (3.6.2) by an $n \times n$ matrix B yields

$$(BA)x = B\mathbf{b}.$$

Assuming that $BA = I_n$, this reduces to

$$\mathbf{x} = B\mathbf{b}. \tag{3.6.3}$$

Thus, we have determined a solution to the system of equations (3.6.2) by a matrix multiplication. It will turn out that this method is not very efficient computationally, and therefore it is generally not used in practice to solve $n \times n$ systems. However, from a theoretical point of view, a formula such as (3.6.3) is very useful. We begin the investigation by establishing that there can be at most one matrix, B, satisfying (3.6.1) for a given $n \times n$ matrix A.

Theorem 3.6.1:   Let A be an $n \times n$ matrix. Suppose B and C are both $n \times n$ matrices satisfying

$$AB = BA = I_n \tag{3.6.4}$$

$$AC = CA = I_n \tag{3.6.5}$$

respectively. Then $B = C$.

PROOF   From (3.6.4), it follows that

$$C = CI_n = C(AB).$$

That is,

$$C = (CA)B = I_n B = B,$$

where we have used (3.6.5) to replace CA by I_n in the second step.   ∎

Since the identity matrix I_n plays the role of the number 1 in the multiplication of matrices, the properties given in (3.6.1) are the analogs for matrices of the properties

$$xx^{-1} = 1, \quad x^{-1}x = 1,$$

which hold for all (nonzero) numbers x. It is therefore natural to denote the matrix B in (3.6.1) by A^{-1} and to call it the inverse of A. The following definition introduces the appropriate terminology.

Definition 3.6.1:   Let A be an $n \times n$ matrix. If there exists a matrix A^{-1} satisfying

$$AA^{-1} = A^{-1}A = I_n,$$

then we call A^{-1} the matrix **inverse** to A, or just the inverse of A. We say that A is **nonsingular** if A^{-1} exists. If A^{-1} does not exist then we say that A is **singular**.

REMARK   It is important to realize that A^{-1} denotes the matrix that satisfies

$$AA^{-1} = A^{-1}A = I_n.$$

It does *not* mean $\frac{1}{A}$, which has no meaning whatsoever.

Example 3.6.1   If $A = \begin{bmatrix} 1 & -1 & 2 \\ 2 & -3 & 3 \\ 1 & -1 & 1 \end{bmatrix}$, verify that $B = \begin{bmatrix} 0 & -1 & 3 \\ 1 & -1 & 1 \\ 1 & 0 & -1 \end{bmatrix}$ is the inverse of A.

Solution   By direct multiplication, we find that

$$AB = \begin{bmatrix} 1 & -1 & 2 \\ 2 & -3 & 3 \\ 1 & -1 & 1 \end{bmatrix} \begin{bmatrix} 0 & -1 & 3 \\ 1 & -1 & 1 \\ 1 & 0 & -1 \end{bmatrix} = \begin{bmatrix} 1 & 0 & 0 \\ 0 & 1 & 0 \\ 0 & 0 & 1 \end{bmatrix} = I_3$$

and

$$BA = \begin{bmatrix} 0 & -1 & 3 \\ 1 & -1 & 1 \\ 1 & 0 & -1 \end{bmatrix} \begin{bmatrix} 1 & -1 & 2 \\ 2 & -3 & 3 \\ 1 & -1 & 1 \end{bmatrix} = \begin{bmatrix} 1 & 0 & 0 \\ 0 & 1 & 0 \\ 0 & 0 & 1 \end{bmatrix} = I_3.$$

Consequently, (3.6.1) is satisfied, and hence, B is indeed the inverse of A. We therefore write

$$A^{-1} = \begin{bmatrix} 0 & -1 & 3 \\ 1 & -1 & 1 \\ 1 & 0 & -1 \end{bmatrix}.$$

We now return to the $n \times n$ system of equations (3.6.2).

Theorem 3.6.2:   If A^{-1} exists, then the $n \times n$ system of linear equations

$$A\mathbf{x} = \mathbf{b}$$

has the *unique* solution

$$\mathbf{x} = A^{-1}\mathbf{b}.$$

PROOF   These results are a direct consequence of the uniqueness of A^{-1} and the calculation leading from (3.6.2) to (3.6.3).   ∎

Our next theorem establishes when A^{-1} exists, and it also uncovers an efficient method for computing A^{-1}.

Theorem 3.6.3:   An $n \times n$ matrix A is nonsingular if and only if rank$(A) = n$.

PROOF   If A^{-1} exists then, as we have just seen, $A\mathbf{x} = \mathbf{b}$ has a unique solution, and so, from Theorem 3.5.1, rank$(A) = n$.
    Conversely, suppose rank$(A) = n$. We must establish that there exists a matrix X satisfying

$$AX = I_n = XA.$$

Let $\mathbf{e}_1, \mathbf{e}_2, \ldots, \mathbf{e}_n$ denote the column vectors of the identity matrix I_n. Then, since rank$(A) = n$, Theorem 3.5.1 implies that each of the linear systems

$$A\mathbf{x}_i = \mathbf{e}_i, \qquad i = 1, 2, \ldots, n \qquad\qquad (3.6.6)$$

has a unique solution.[1]   Consequently if we let $X = [\mathbf{x}_1, \mathbf{x}_2, \ldots, \mathbf{x}_n]$, where $\mathbf{x}_1, \mathbf{x}_2, \ldots, \mathbf{x}_n$ are the unique solutions to the systems in (3.6.6), then

$$A[\mathbf{x}_1, \mathbf{x}_2, \ldots, \mathbf{x}_n] = [A\mathbf{x}_1, A\mathbf{x}_2, \ldots, A\mathbf{x}_n] = [\mathbf{e}_1, \mathbf{e}_2, \ldots, \mathbf{e}_n]$$

---

[1]Notice that for an $n \times n$ system $A\mathbf{x} = \mathbf{b}$, if rank$(A) = n$, then so must rank$(A^\#)$.

that is,

$$AX = I_n. \tag{3.6.7}$$

We must also show that, for the same matrix X,

$$XA = I_n.$$

Postmultiplying both sides of (3.6.7) by A yields

$$(AX)A = A.$$

That is,

$$A(XA - I_n) = 0_n. \tag{3.6.8}$$

Now let $\mathbf{y}_1, \mathbf{y}_2, \ldots, \mathbf{y}_n$ denote the column vectors of the $n \times n$ matrix $XA - I_n$. Equating corresponding column vectors on either side of (3.6.8) implies that

$$A\mathbf{y}_i = \mathbf{0}, \qquad i = 1, 2, \ldots, n. \tag{3.6.9}$$

But, by assumption, rank$(A) = n$, and so each of the systems in (3.6.9) has a unique solution that, since the systems are homogeneous, must be the trivial solution. Consequently,

$$XA - I_n = 0_n,$$

and, therefore,

$$XA = I_n. \tag{3.6.10}$$

Equations (3.6.7) and (3.6.10) imply that $X = A^{-1}$.   ∎

We now have the following converse to Theorem 3.6.2.

Corollary 3.6.1:   Let A be an $n \times n$ matrix. If $A\mathbf{x} = \mathbf{b}$ has a unique solution, then A^{-1} exists.

PROOF   If $A\mathbf{x} = \mathbf{b}$ has a unique solution then, from Theorem 3.5.1, rank$(A) = n$, and so, from the previous theorem, A^{-1} exists.   ∎

Having established when A^{-1} exists, we next develop a method for finding it. Assuming that rank$(A) = n$, let $\mathbf{x}_1, \mathbf{x}_2, \ldots, \mathbf{x}_n$ denote the column vectors of A^{-1}. Then, from (3.6.6), these column vectors can be obtained by solving each of the $n \times n$ systems

$$A\mathbf{x}_i = \mathbf{e}_i, \qquad i = 1, 2, \ldots, n. \tag{3.6.11}$$

As we now show, some computation can be saved if we employ the Gauss–Jordan method in solving these systems. We first illustrate the method when $n = 3$. In this case, from (3.6.11), the column vectors of A^{-1} are determined by solving the three linear systems

$$A\mathbf{x}_1 = \mathbf{e}_1, \qquad A\mathbf{x}_2 = \mathbf{e}_2, \qquad A\mathbf{x}_3 = \mathbf{e}_3.$$

The augmented matrices of these systems can be written as

$$\left[\begin{array}{c|c} A & \begin{matrix} 1 \\ 0 \\ 0 \end{matrix} \end{array} \right], \left[\begin{array}{c|c} A & \begin{matrix} 0 \\ 1 \\ 0 \end{matrix} \end{array} \right], \left[\begin{array}{c|c} A & \begin{matrix} 0 \\ 0 \\ 1 \end{matrix} \end{array} \right],$$

respectively. Furthermore, since rank(A) = 3 by assumption, the RREF of A is I_3. Consequently, using ERO to reduce the augmented matrix of the first system to RREF will yield, schematically,

$$
\left[\begin{array}{cc} A & \begin{matrix} 1 \\ 0 \\ 0 \end{matrix} \end{array}\right] \underset{\sim \cdots \sim}{\overset{\text{ERO}}{}} \left[\begin{array}{ccc|c} 1 & 0 & 0 & a_1 \\ 0 & 1 & 0 & a_2 \\ 0 & 0 & 1 & a_3 \end{array}\right],
$$

which implies that the first column vector of A^{-1} is

$$
\mathbf{x}_1 = \left[\begin{matrix} a_1 \\ a_2 \\ a_3 \end{matrix}\right].
$$

Similarly, for the second system, the reduction

$$
\left[\begin{array}{cc} A & \begin{matrix} 0 \\ 1 \\ 0 \end{matrix} \end{array}\right] \underset{\sim \cdots \sim}{\overset{\text{ERO}}{}} \left[\begin{array}{ccc|c} 1 & 0 & 0 & b_1 \\ 0 & 1 & 0 & b_2 \\ 0 & 0 & 1 & b_3 \end{array}\right]
$$

implies that the second column vector of A^{-1} is

$$
\mathbf{x}_2 = \left[\begin{matrix} b_1 \\ b_2 \\ b_3 \end{matrix}\right].
$$

Finally, for the third system, the reduction

$$
\left[\begin{array}{cc} A & \begin{matrix} 0 \\ 0 \\ 1 \end{matrix} \end{array}\right] \underset{\sim \cdots \sim}{\overset{\text{ERO}}{}} \left[\begin{array}{ccc|c} 1 & 0 & 0 & c_1 \\ 0 & 1 & 0 & c_2 \\ 0 & 0 & 1 & c_3 \end{array}\right]
$$

implies that the third column vector of A^{-1} is

$$
\mathbf{x}_3 = \left[\begin{matrix} c_1 \\ c_2 \\ c_3 \end{matrix}\right].
$$

Consequently,

$$
A^{-1} = [\mathbf{x}_1, \mathbf{x}_2, \mathbf{x}_3] = \left[\begin{matrix} a_1 & b_1 & c_1 \\ a_2 & b_2 & c_2 \\ a_3 & b_3 & c_3 \end{matrix}\right].
$$

The key point to notice is that in solving for $\mathbf{x}_1, \mathbf{x}_2, \mathbf{x}_3$ we use the *same* ERO to reduce A to I_3. We can therefore save a significant amount of work by combining the foregoing operations as follows:

$$
\left[\begin{array}{cc} A & \begin{matrix} 1 & 0 & 0 \\ 0 & 1 & 0 \\ 0 & 0 & 1 \end{matrix} \end{array}\right] \underset{\sim \cdots \sim}{\overset{\text{ERO}}{}} \left[\begin{array}{ccc|ccc} 1 & 0 & 0 & a_1 & b_1 & c_1 \\ 0 & 1 & 0 & a_2 & b_2 & c_2 \\ 0 & 0 & 1 & a_3 & b_3 & c_3 \end{array}\right].
$$

The generalization to the $n \times n$ case is immediate. We form the $n \times 2n$ matrix $[A \quad I_n]$ and reduce A to I_n using ERO. Schematically,

$$[A \quad I_n] \overset{\text{ERO}}{\sim} \cdots \sim [I_n \quad A^{-1}].$$

This method of finding A^{-1} is called the **Gauss–Jordan Technique**.

REMARK   Notice that if we are given an $n \times n$ matrix A, we will not know rank(A), and hence, we will not know whether A^{-1} exists. However, if at any stage in the row reduction of $[A \quad I_n]$ we find that rank$(A) < n$, then it will follow from Theorem 3.6.3 that A is singular and hence that A^{-1} does not exist.

Example 3.6.2   Find A^{-1} if $A = \begin{bmatrix} 1 & 1 & 3 \\ 0 & 1 & 2 \\ 3 & 5 & -1 \end{bmatrix}$.

Solution   Using the Gauss–Jordan technique we proceed as follows.

$$\begin{bmatrix} 1 & 1 & 3 & 1 & 0 & 0 \\ 0 & 1 & 2 & 0 & 1 & 0 \\ 3 & 5 & -1 & 0 & 0 & 1 \end{bmatrix} \overset{1}{\sim} \begin{bmatrix} 1 & 1 & 3 & 1 & 0 & 0 \\ 0 & 1 & 2 & 0 & 1 & 0 \\ 0 & 2 & -10 & -3 & 0 & 1 \end{bmatrix}$$

$$\overset{2}{\sim} \begin{bmatrix} 1 & 0 & 1 & 1 & -1 & 0 \\ 0 & 1 & 2 & 0 & 1 & 0 \\ 0 & 0 & -14 & -3 & -2 & 1 \end{bmatrix} \overset{3}{\sim} \begin{bmatrix} 1 & 0 & 1 & 1 & -1 & 0 \\ 0 & 1 & 2 & 0 & 1 & 0 \\ 0 & 0 & 1 & \frac{3}{14} & \frac{1}{7} & -\frac{1}{14} \end{bmatrix}$$

$$\overset{4}{\sim} \begin{bmatrix} 1 & 0 & 0 & \frac{11}{14} & -\frac{8}{7} & \frac{1}{14} \\ 0 & 1 & 0 & -\frac{3}{7} & \frac{5}{7} & \frac{1}{7} \\ 0 & 0 & 1 & \frac{3}{14} & \frac{1}{7} & -\frac{1}{14} \end{bmatrix}$$

Thus,

$$A^{-1} = \begin{bmatrix} \frac{11}{14} & -\frac{8}{7} & \frac{1}{14} \\ -\frac{3}{7} & \frac{5}{7} & \frac{1}{7} \\ \frac{3}{14} & \frac{1}{7} & -\frac{1}{14} \end{bmatrix}.$$

We leave it as an exercise to check that $AA^{-1} = A^{-1}A = I_n$.

1. $A_{13}(-3)$ **2.** $A_{21}(-1)$, $A_{23}(-2)$ **3.** $M_3(-\frac{1}{14})$ **4.** $A_{31}(-1)$, $A_{32}(-2)$

Example 3.6.3   Use A^{-1} to solve the system

$$\begin{aligned} x_1 + x_2 + 3x_3 &= 2 \\ x_2 + 2x_3 &= 1 \\ 3x_1 + 5x_2 - x_3 &= 4. \end{aligned}$$

Solution   The system can be written as

$$A\mathbf{x} = \mathbf{b},$$

where A is the matrix in the previous example, and $\mathbf{b} = [2 \quad 1 \quad 4]^T$. Since A is nonsingular, the system has a unique solution that can be written as $\mathbf{x} = A^{-1}\mathbf{b}$. That is, from the preceding example,

$$
\mathbf{x} = \begin{bmatrix} \frac{11}{14} & -\frac{8}{7} & \frac{1}{14} \\ -\frac{3}{7} & \frac{5}{7} & \frac{1}{7} \\ \frac{3}{14} & \frac{1}{7} & -\frac{1}{14} \end{bmatrix} \begin{bmatrix} 2 \\ 1 \\ 4 \end{bmatrix} = \begin{bmatrix} \frac{5}{7} \\ \frac{3}{7} \\ \frac{2}{7} \end{bmatrix}.
$$

Thus, $x_1 = \frac{5}{7}$, $x_2 = \frac{3}{7}$, $x_3 = \frac{2}{7}$, so that the solution to the system is $(\frac{5}{7}, \frac{3}{7}, \frac{2}{7})$.   □

REMARK   We now have the following three techniques that can be applied to solve an $n \times n$ linear system with nonsingular coefficient matrix:

1. Solve by Gaussian elimination.
2. Solve by Gauss–Jordan elimination.
3. Find A^{-1} and then obtain the solution directly as $\mathbf{x} = A^{-1}\mathbf{b}$.

Since the first two of these require the row reduction of an $n \times (n + 1)$ augmented matrix, whereas the computation of A^{-1} in (3) requires the row reduction of an $n \times 2n$ matrix, it should be clear that the first two techniques are superior from a computational viewpoint. As mentioned at the beginning of the section, the significance of (3) is more as a theor–etical tool.

PROPERTIES OF THE INVERSE

The inverse of an $n \times n$ matrix satisfies the properties stated in the following theorem, which should be committed to memory:

Theorem 3.6.4:   Let A and B be nonsingular $n \times n$ matrices.   Then

1. A^{-1} is nonsingular and $(A^{-1})^{-1} = A$.
2. AB is nonsingular and $(AB)^{-1} = B^{-1}A^{-1}$.   (3.6.12)
3. A^T is nonsingular and $(A^T)^{-1} = (A^{-1})^T$.   (3.6.13)

PROOF   The proof of each result consists of verifying that the appropriate matrix products yield the identity matrix.

(1)   We must verify that

$$A^{-1}A = I_n \text{ and } AA^{-1} = I_n.$$

Both of these follow directly from Definition 3.6.1.

(2)   We must verify that

$$(AB)(B^{-1}A^{-1}) = I_n \text{ and } (B^{-1}A^{-1})(AB) = I_n.$$ (3.6.14)

We establish the first equality.

$$(AB)(B^{-1}A^{-1}) = A(BB^{-1})A^{-1} = AI_nA^{-1} = AA^{-1} = I_n.$$

Verification of the second equation in (3.6.14) is left as an exercise.

(3) We must verify that

$$A^T(A^{-1})^T = I_n \quad \text{and} \quad (A^{-1})^T A^T = I_n. \tag{3.6.15}$$

First recall the general property of the transpose that $A^T B^T = (BA)^T$. Using this property with $B = A^{-1}$ yields

$$A^T(A^{-1})^T = (A^{-1}A)^T = I_n{}^T = I_n.$$

Verification of the second equation in (3.6.15) is left as an exercise. ∎

SOME FURTHER THEORETICAL RESULTS

Finally in this section, we establish two results that will be required in Section 3.7 and also in one of the proofs that arises in Section 4.2.

Theorem 3.6.5: Let A and B be $n \times n$ matrices. If $AB = I_n$ then both A and B are nonsingular and $B = A^{-1}$.

PROOF Let **b** be an arbitrary column n-vector. Then, since $AB = I_n$, we have

$$A(B\mathbf{b}) = I_n\mathbf{b} = \mathbf{b}.$$

Consequently, for *every* **b**, the system $A\mathbf{x} = \mathbf{b}$ has the solution $\mathbf{x} = B\mathbf{b}$. But this implies that rank(A) = n. To see why, suppose that rank(A) < n, and let A^* denote an REF of A. Choose $\mathbf{b}^*$ to be any column n-vector whose last component is nonzero. Then, since rank(A) < n it follows that the system

$$A^*\mathbf{x} = \mathbf{b}^*$$

is inconsistent. But, applying to the augmented matrix $[A^* \ \mathbf{b}^*]$ the inverse row operations that reduced A to REF, yields $[A \ \mathbf{b}]$ for some **b**. Since $A\mathbf{x} = \mathbf{b}$ has the same solution set as $A^*\mathbf{x} = \mathbf{b}^*$ it follows that $A\mathbf{x} = \mathbf{b}$ is inconsistent. We therefore have a contradiction and so it must be the case that rank(A) = n and therefore A is nonsingular. We must now establish that $A^{-1} = B$. But, since $AB = I_n$,

$$A^{-1} = A^{-1}I_n = A^{-1}(AB) = (A^{-1}A)B = B,$$

as required. It now follows directly from property (1) of the inverse that B is nonsingular with inverse A. ∎

Corollary 3.6.2: Let A and B be $n \times n$ matrices. If AB is nonsingular then both A and B are nonsingular.

PROOF If we let $C = B(AB)^{-1}$, then

$$AC = AB(AB)^{-1} = DD^{-1} = I_n,$$

where $D = AB$. It follows from the preceding theorem that A is nonsingular. Similarly, if we let $C = (AB)^{-1}A$, then

$$CB = (AB)^{-1}AB = I_n.$$

Once more we can apply Theorem 3.6.5 to conclude that B is nonsingular. ∎

EXERCISES 3.6

For problems 1–3, verify, by direct multiplication, that the given matrices are inverses.

1. $A = \begin{bmatrix} 2 & -1 \\ 3 & -1 \end{bmatrix}$, $A^{-1} = \begin{bmatrix} -1 & 1 \\ -3 & 2 \end{bmatrix}$.

2. $A = \begin{bmatrix} 4 & 9 \\ 3 & 7 \end{bmatrix}$, $A^{-1} = \begin{bmatrix} 7 & -9 \\ -3 & 4 \end{bmatrix}$.

3. $A = \begin{bmatrix} 3 & 5 & 1 \\ 1 & 2 & 1 \\ 2 & 6 & 7 \end{bmatrix}$, $A^{-1} = \begin{bmatrix} 8 & -29 & 3 \\ -5 & 19 & -2 \\ 2 & -8 & 1 \end{bmatrix}$.

For problems 4–16, determine A^{-1}, if possible, using the Gauss–Jordan method. If A^{-1} exists, check your answer by verifying that $A A^{-1} = I_n$.

4. $A = \begin{bmatrix} 1 & 2 \\ 1 & 3 \end{bmatrix}$.

5. $A = \begin{bmatrix} 1 & 1+i \\ 1-i & 1 \end{bmatrix}$.

6. $A = \begin{bmatrix} 1 & -i \\ -1+i & 2 \end{bmatrix}$.

7. $A = \begin{bmatrix} 0 & 0 \\ 0 & 0 \end{bmatrix}$.

8. $A = \begin{bmatrix} 1 & -1 & 2 \\ 2 & 1 & 11 \\ 4 & -3 & 10 \end{bmatrix}$.

9. $A = \begin{bmatrix} 3 & 5 & 1 \\ 1 & 2 & 1 \\ 2 & 6 & 7 \end{bmatrix}$.

10. $A = \begin{bmatrix} 0 & 1 & 0 \\ 0 & 0 & 1 \\ 0 & 1 & 2 \end{bmatrix}$.

11. $A = \begin{bmatrix} 4 & 2 & -13 \\ 2 & 1 & -7 \\ 3 & 2 & 4 \end{bmatrix}$.

12. $A = \begin{bmatrix} 1 & 2 & -3 \\ 2 & 6 & -2 \\ -1 & 1 & 4 \end{bmatrix}$.

13. $A = \begin{bmatrix} 1 & i & 2 \\ 1+i & -1 & 2i \\ 2 & 2i & 5 \end{bmatrix}$.

14. $A = \begin{bmatrix} 2 & 1 & 3 \\ 1 & -1 & 2 \\ 3 & 3 & 4 \end{bmatrix}$.

15. $A = \begin{bmatrix} 1 & -1 & 2 & 3 \\ 2 & 0 & 3 & -4 \\ 3 & -1 & 7 & 8 \\ 1 & 0 & 3 & 5 \end{bmatrix}$.

16. $A = \begin{bmatrix} 0 & -2 & -1 & -3 \\ 2 & 0 & 2 & 1 \\ 1 & -2 & 0 & 2 \\ 3 & -1 & -2 & 0 \end{bmatrix}$.

17. Let $A = \begin{bmatrix} 2 & -1 & 4 \\ 5 & 1 & 2 \\ 1 & -1 & 3 \end{bmatrix}$. Find the second column vector of A^{-1} without determining the whole inverse.

For problems 18–22, use A^{-1} to find the solution to the given system.

18. $x_1 + 3x_2 = 1$,
$2x_1 + 5x_2 = 3$.

19. $x_1 + x_2 - 2x_3 = -2$,
$x_2 + x_3 = 3$,
$2x_1 + 4x_2 - 3x_3 = 1$.

20. $x_1 - 2ix_2 = 2$,
$(2-i)x_1 + 4ix_2 = -i$.

21. $3x_1 + 4x_2 + 5x_3 = 1$,
$2x_1 + 10x_2 + x_3 = 1$,
$4x_1 + x_2 + 8x_3 = 1$.

22. $x_1 + x_2 + 2x_3 = 12$,
$x_1 + 2x_2 - x_3 = 24$,
$2x_1 - x_2 + x_3 = -36$.

An $n \times n$ matrix is called **orthogonal** if $A^T = A^{-1}$. For problems 23–26, show that the given matrices are orthogonal.

23. $A = \begin{bmatrix} 0 & 1 \\ -1 & 0 \end{bmatrix}$.

24. $\begin{bmatrix} \sqrt{3}/2 & 1/2 \\ -1/2 & \sqrt{3}/2 \end{bmatrix}$.

25. $\begin{bmatrix} \cos\alpha & \sin\alpha \\ -\sin\alpha & \cos\alpha \end{bmatrix}$.

26. $\dfrac{1}{1+2x^2}\begin{bmatrix} 1 & -2x & 2x^2 \\ 2x & 1-2x^2 & -2x \\ 2x^2 & 2x & 1 \end{bmatrix}$.

27. Complete the proof of Theorem 3.6.4, by verifying the remaining properties in parts (2) and (3).

28. Use properties of the inverse to prove the following:

(a) If A is an $n \times n$ nonsingular *symmetric* matrix, then A^{-1} is symmetric.

(b) If A is an $n \times n$ nonsingular *skew-symmetric* matrix, then A^{-1} is skew-symmetric.

29. Prove that if A, B, C are $n \times n$ matrices satisfying $BA = I_n, AC = I_n$, then $B = C$.

30. Consider the general 2×2 matrix

$$A = \begin{bmatrix} a_{11} & a_{12} \\ a_{21} & a_{22} \end{bmatrix}$$

and let $\Delta = a_{11}a_{22} - a_{12}a_{21}$ with $a_{11} \neq 0$. Show that, provided $\Delta \neq 0$,

$$A^{-1} = \frac{1}{\Delta}\begin{bmatrix} a_{22} & -a_{12} \\ -a_{21} & a_{11} \end{bmatrix}.$$

The quantity Δ defined above is referred to as the determinant of A. We will investigate determinants in more detail in the next chapter.

31. Let A be an $n \times n$ matrix, and suppose that we have to solve the p linear systems

$$A\mathbf{x}_i = \mathbf{b}_i, \qquad i = 1, 2, \dots, p$$

where the $\mathbf{b}_i$ are given. Devise an efficient method for solving these systems.

32. Use your method from the previous problem to solve the three linear systems

$$A\mathbf{x}_i = \mathbf{b}_i, \qquad i = 1, 2, 3$$

if

$$A = \begin{bmatrix} 1 & -1 & 1 \\ 2 & -1 & 4 \\ 1 & 1 & 6 \end{bmatrix}, \mathbf{b}_1 = \begin{bmatrix} 1 \\ 1 \\ -1 \end{bmatrix},$$

$$\mathbf{b}_2 = \begin{bmatrix} -1 \\ 2 \\ 5 \end{bmatrix}, \mathbf{b}_3 = \begin{bmatrix} 2 \\ 3 \\ 2 \end{bmatrix}.$$

33. Let A be an $m \times n$ matrix with $m \leq n$.

(a) If $\text{rank}(A) = m$, prove that there exists a matrix B satisfying $AB = I_m$. Such a matrix is called a **right inverse** of A.

(b) If $A = \begin{bmatrix} 1 & 3 & 1 \\ 2 & 7 & 4 \end{bmatrix}$, determine all right inverses of A.

◆ For problems 34 and 35, reduce the matrix $[A \ I_n]$ to RREF and thereby determine, if possible, the inverse of A.

34. $A = \begin{bmatrix} 5 & 9 & 17 \\ 7 & 21 & 13 \\ 27 & 16 & 8 \end{bmatrix}$.

35. A a randomly generated 4×4 matrix.

◆ For problems 36–38, use built-in functions of some form of technology to determine $\text{rank}(A)$ and, if possible, A^{-1}.

36. $A = \begin{bmatrix} 3 & 5 & -7 \\ 2 & 5 & 9 \\ 13 & -11 & 22 \end{bmatrix}$.

37. $A = \begin{bmatrix} 7 & 13 & 15 & 21 \\ 9 & -2 & 14 & 23 \\ 17 & -27 & 22 & 31 \\ 19 & -42 & 21 & 33 \end{bmatrix}$.

38. A a randomly generated 5×5 matrix.

◆ **39.** For the system in problem 21, determine A^{-1} and use it to solve the system.

◆ **40.** Consider the $n \times n$ *Hilbert* matrix

$$H_n = [1/(i+j)], \ 1 \leq i \leq n, \ 1 \leq j \leq n.$$

(a) Determine H_4, and show that it is non-singular.

(b) Find H_4^{-1} and use it to solve $H_4\mathbf{x} = \mathbf{b}$ if $\mathbf{b} = [2, -1, 3, 5]^T$.

3.7* ELEMENTARY MATRICES AND THE LU FACTORIZATION

We now introduce some matrices that can be used to perform ERO on a matrix. Although they are of limited computational use, they do play a significant role in linear algebra and its applications.

> **Definition 3.7.1:**   Any matrix obtained by performing a single ERO on the identity matrix is called an **elementary matrix.**

In general we will denote elementary matrices by E. If we are describing a specific elementary matrix then, in keeping with the notation introduced previously for ERO, we will use the following notation for the three types of elementary matrices:

Type 1: P_{ij} – permute rows i and j in I_n.

Type 2: $M_i(k)$ – multiply row i of I_n by the nonzero scalar k.

Type 3: $A_{ij}(k)$ – add k times row i of I_n to row j of I_n.

Example 3.7.1   Write all 2×2 elementary matrices.

Solution   From Definition 3.7.1 and using the notation introduced above we have

(1)   Permutation matrix: $P_{12} = \begin{bmatrix} 0 & 1 \\ 1 & 0 \end{bmatrix}$.

(2)   Scaling matrices: $M_1(k) = \begin{bmatrix} k & 0 \\ 0 & 1 \end{bmatrix}$, $M_2(k) = \begin{bmatrix} 1 & 0 \\ 0 & k \end{bmatrix}$.

(3)   Row Combinations: $A_{12}(k) = \begin{bmatrix} 1 & 0 \\ k & 1 \end{bmatrix}$, $A_{21}(k) = \begin{bmatrix} 1 & k \\ 0 & 1 \end{bmatrix}$.   ❏

We leave it as an exercise to verify that the $n \times n$ elementary matrices have the following structure:

P_{ij}:   ones in all positions on the main diagonal except ii and jj, ones in the ij and ji positions, zeros elsewhere.

$M_i(k)$: the diagonal matrix $\text{diag}(1, 1, \ldots, k, \ldots, 1)$ where k appears in the ii position.

$A_{ij}(k)$: ones along the main diagonal, k in the ji position, zeros elsewhere.

One of the key points to note about elementary matrices is the following:

> Pre–multiplying an $n \times p$ matrix A by an $n \times n$ elementary matrix E has the effect of performing the corresponding ERO on A.

---

* The material in this section is used only in one proof in Section 4.2, and in the (optional) Section 6.2.

Rather than proving this statement, which we leave as an exercise, we illustrate with an example.

Example 3.7.2   If $A = \begin{bmatrix} 3 & -1 & 4 \\ 2 & 7 & 5 \end{bmatrix}$ then, for example,

$$M_1(k)A = \begin{bmatrix} k & 0 \\ 0 & 1 \end{bmatrix} \begin{bmatrix} 3 & -1 & 4 \\ 2 & 7 & 5 \end{bmatrix} = \begin{bmatrix} 3k & -k & 4k \\ 2 & 7 & 5 \end{bmatrix}.$$

Similarly,

$$A_{21}(k)A = \begin{bmatrix} 1 & k \\ 0 & 1 \end{bmatrix} \begin{bmatrix} 3 & -1 & 4 \\ 2 & 7 & 5 \end{bmatrix} = \begin{bmatrix} 3 + 2k & -1 + 7k & 4 + 5k \\ 2 & 7 & 5 \end{bmatrix}.$$ □

Since ERO can be performed on a matrix by premultiplication by an appropriate elementary matrix, it follows that any matrix A can be reduced to REF by multiplication by a sequence of elementary matrices. Schematically we can therefore write

$$E_1 E_2 \cdots E_k A = U$$

where U denotes an REF of A.

Example 3.7.3   Determine elementary matrices that reduce $A = \begin{bmatrix} 2 & 3 \\ 1 & 4 \end{bmatrix}$ to REF.

Solution A can be reduced to REF using the following sequence of ERO

$$\begin{bmatrix} 2 & 3 \\ 1 & 4 \end{bmatrix} \underset{\underset{P_{12}}{\uparrow}}{\sim} \begin{bmatrix} 1 & 4 \\ 2 & 3 \end{bmatrix} \underset{\underset{A_{12}(-2)}{\uparrow}}{\sim} \begin{bmatrix} 1 & 4 \\ 0 & -5 \end{bmatrix} \underset{\underset{M_2(-1/5)}{\uparrow}}{\sim} \begin{bmatrix} 1 & 4 \\ 0 & 1 \end{bmatrix}.$$

Consequently

$$M_2(-1/5)\, A_{12}(-2)\, P_{12} A = \begin{bmatrix} 1 & 4 \\ 0 & 1 \end{bmatrix}$$

which we can verify by direct multiplication:

$$M_2(-1/5)\, A_{12}(-2)\, P_{12} A = \begin{bmatrix} 1 & 0 \\ 0 & -1/5 \end{bmatrix} \begin{bmatrix} 1 & 0 \\ -2 & 1 \end{bmatrix} \begin{bmatrix} 0 & 1 \\ 1 & 0 \end{bmatrix} \begin{bmatrix} 2 & 3 \\ 1 & 4 \end{bmatrix}$$

$$= \begin{bmatrix} 1 & 0 \\ 0 & -1/5 \end{bmatrix} \begin{bmatrix} 1 & 0 \\ -2 & 1 \end{bmatrix} \begin{bmatrix} 1 & 4 \\ 2 & 3 \end{bmatrix}$$

$$= \begin{bmatrix} 1 & 0 \\ 0 & -1/5 \end{bmatrix} \begin{bmatrix} 1 & 4 \\ 0 & -5 \end{bmatrix} = \begin{bmatrix} 1 & 4 \\ 0 & 1 \end{bmatrix}.$$ □

Since any ERO is reversible, it follows that each elementary matrix is nonsingular. Indeed in the 2×2 case it is easy to see that

$$M_1(k)^{-1} = \begin{bmatrix} 1/k & 0 \\ 0 & 1 \end{bmatrix}, \quad M_2(k)^{-1} = \begin{bmatrix} 1 & 0 \\ 0 & 1/k \end{bmatrix}, \quad P_{12}^{-1} = P_{12},$$

$$A_{12}(k)^{-1} = \begin{bmatrix} 1 & 0 \\ -k & 1 \end{bmatrix}, \quad A_{21}(k)^{-1} = \begin{bmatrix} 1 & -k \\ 0 & 1 \end{bmatrix}.$$

We leave it as an exercise to verify that in the $n \times n$ case:

$$\boxed{M_i(k)^{-1} = M_i(1/k), \quad P_{ij}^{-1} = P_{ij}, \quad A_{ij}(k)^{-1} = A_{ij}(-k).}$$

Now consider a *nonsingular* $n \times n$ matrix A. Since the unique RREF of such a matrix is the identity matrix I_n it follows from the preceding discussion that there exist elementary matrices $E_1, E_2, \ldots, E_k$ such that

$$E_1 E_2 \cdots E_k A = I_n.$$

But this implies that

$$A^{-1} = E_1 E_2 \cdots E_k$$

or equivalently

$$A^{-1} = E_1 E_2 \cdots E_k I_n.$$

This shows that A^{-1} can be obtained by applying to I_n the same sequence of ERO that reduces A to its unique RREF. This is precisely what we do in computing A^{-1} using the Gauss–Jordan method.

THE LU DECOMPOSITION OF A NONSINGULAR MATRIX

For the remainder of this section we restrict our attention to nonsingular $n \times n$ matrices. In reducing such a matrix to REF we have always placed leading ones on the main diagonal in order that we obtain a row-echelon matrix. We now lift this restriction. As a consequence, the matrix that results from row reduction will be an upper triangular matrix, rather than a row-echelon matrix. Furthermore, reduction to such an upper triangular form can be accomplished without the use of Type 2 row operations.

Example 3.7.4   Reduce the matrix $A = \begin{bmatrix} 2 & 5 & 3 \\ 3 & 1 & -2 \\ -1 & 2 & 1 \end{bmatrix}$ to upper triangular form.

Solution   The given matrix can be reduced to upper triangular form using the following sequence of ERO

$$\begin{bmatrix} 2 & 5 & 3 \\ 3 & 1 & -2 \\ -1 & 2 & 1 \end{bmatrix} \underset{\underset{A_{12}(-3/2),\ A_{13}(1/2)}{\uparrow}}{\sim} \begin{bmatrix} 2 & 5 & 3 \\ 0 & -13/2 & -13/2 \\ 0 & 9/2 & 5/2 \end{bmatrix}$$

$$\underset{\underset{A_{23}(9/13)}{\uparrow}}{\sim} \begin{bmatrix} 2 & 5 & 3 \\ 0 & -13/2 & -13/2 \\ 0 & 0 & -2 \end{bmatrix}.$$

□

When using ERO of Type 3, the multiple of a specific row that is *subtracted* from row i to put a zero in the ij position is called a **multiplier**, and denoted m_{ij}. Thus, in the preceding example there are three multipliers namely

$$m_{21} = 3/2, \quad m_{31} = -1/2, \quad m_{32} = -9/13.$$

The multipliers will be used in the forthcoming discussion.

In Example 3.7.4 we were able to reduce A to upper triangular form using only row operations of Type 3. This is not always the case. For example the matrix $\begin{bmatrix} 0 & 5 \\ 3 & 2 \end{bmatrix}$ requires that rows 1 and 2 be permuted to obtained an upper triangular form. For the moment, however, we will restrict our attention to *nonsingular matrices A for which the reduction to triangular form can be accomplished without permuting rows.* In this case we can therefore reduce A to upper triangular form using row operations of Type 3 only. Furthermore, at the ith step of the reduction, we need only use Type 3 operations that add multiples of row i to rows **beneath** row i. Consequently the corresponding elementary matrices that perform the reduction are *unit lower triangular* matrices. More specifically, in terms of elementary matrices we have

$$E_1 E_2 \cdots E_k A = U$$

where $E_1 E_2 \cdots E_k$ are unit lower triangular Type 3 elementary matrices and U is an upper triangular matrix. Since each elementary matrix is nonsingular we can write the preceding equation as

$$A = E_k^{-1} E_{k-1}^{-1} \cdots E_1^{-1} U. \tag{3.7.1}$$

But, as we have already argued, each of the elementary matrices in (3.7.1) is a unit lower triangular matrix, and we know from Corollary 3.2.1 that the product of two lower triangular matrices is also a lower triangular matrix. Consequently (3.7.1) can be written as

$$A = LU \tag{3.7.2}$$

where

$$L = E_k^{-1} E_{k-1}^{-1} \cdots E_1^{-1} \tag{3.7.3}$$

is a lower triangular matrix (in fact a unit lower triangular matrix) and U is an upper triangular matrix. Equation (3.7.2) is referred to as the **LU factorization of A**. It can be shown (see Exercise 21) that this LU factorization is unique.

Example 3.7.5   Determine the LU factorization of the matrix

$$A = \begin{bmatrix} 2 & 5 & 3 \\ 3 & 1 & -2 \\ -1 & 2 & 1 \end{bmatrix}.$$

Solution   Using the results of Example 3.7.4 we can write

$$E_1 E_2 E_3 A = \begin{bmatrix} 2 & 5 & 3 \\ 0 & -13/2 & -13/2 \\ 0 & 0 & -2 \end{bmatrix}$$

where

$$E_1 = A_{23}(9/13), \quad E_2 = A_{13}(1/2), \quad E_3 = A_{12}(-3/2).$$

Therefore

$$U = \begin{bmatrix} 2 & 5 & 3 \\ 0 & -13/2 & -13/2 \\ 0 & 0 & -2 \end{bmatrix}$$

and, from (3.7.3),

$$L = E_3^{-1} E_2^{-1} E_1^{-1}. \tag{3.7.4}$$

Computing the inverses of the elementary matrices we have

$$E_1^{-1} = A_{23}(-9/13), \quad E_2^{-1} = A_{13}(-1/2), \quad E_3^{-1} = A_{12}(3/2).$$

Substituting these results into (3.7.4) yields

$$L = \begin{bmatrix} 1 & 0 & 0 \\ 3/2 & 1 & 0 \\ 0 & 0 & 1 \end{bmatrix} \begin{bmatrix} 1 & 0 & 0 \\ 0 & 1 & 0 \\ -1/2 & 0 & 1 \end{bmatrix} \begin{bmatrix} 1 & 0 & 0 \\ 0 & 1 & 0 \\ 0 & -9/13 & 1 \end{bmatrix}$$

$$= \begin{bmatrix} 1 & 0 & 0 \\ 3/2 & 1 & 0 \\ -1/2 & -9/13 & 1 \end{bmatrix}.$$

Consequently

$$A = \begin{bmatrix} 1 & 0 & 0 \\ 3/2 & 1 & 0 \\ -1/2 & -9/13 & 1 \end{bmatrix} \begin{bmatrix} 2 & 5 & 3 \\ 0 & -13/2 & -13/2 \\ 0 & 0 & -2 \end{bmatrix}$$

which is easily verified by a matrix multiplication.   ☐

Computing the unit lower triangular matrix L in the LU factorization of A using (3.7.3) can require a significant amount of work. However if we look carefully at the matrix L in Example 3.7.5 we see that the elements beneath the leading diagonal are just the corresponding multipliers. That is,

$$l_{ij} = m_{ij}, \quad i > j. \tag{3.7.5}$$

Furthermore, it can be shown that this relationship holds in general. Consequently we do not need to use (3.7.3) to obtain L. Instead we use row operations of Type 3 to reduce A to upper triangular form and then we can use (3.7.5) to obtain L directly.

Example 3.7.6   Determine the LU decomposition for the matrix

$$A = \begin{bmatrix} 2 & -3 & 1 & 2 \\ 5 & -1 & 2 & 1 \\ 3 & 2 & 6 & -5 \\ -1 & 1 & 3 & 2 \end{bmatrix}.$$

Solution   To determine U we reduce A to upper triangular form using only row operations of Type 3 which add multiples of row i to rows beneath row i.

$$\begin{bmatrix} 2 & -3 & 1 & 2 \\ 5 & -1 & 2 & 1 \\ 3 & 2 & 6 & -5 \\ -1 & 1 & 3 & 2 \end{bmatrix} \underset{\sim}{\overset{1}{}} \begin{bmatrix} 2 & -3 & 1 & 2 \\ 0 & 13/2 & -1/2 & -4 \\ 0 & 13/2 & 9/2 & -8 \\ 0 & -1/2 & 7/2 & 3 \end{bmatrix}$$

$$\underset{\sim}{\overset{2}{}} \begin{bmatrix} 2 & -3 & 1 & 2 \\ 0 & 13/2 & -1/2 & -4 \\ 0 & 0 & 5 & -4 \\ 0 & 0 & 45/13 & 35/13 \end{bmatrix} \underset{\sim}{\overset{3}{}} \begin{bmatrix} 2 & -3 & 1 & 2 \\ 0 & 13/2 & -1/2 & -4 \\ 0 & 0 & 5 & -4 \\ 0 & 0 & 0 & 71/13 \end{bmatrix} = U.$$

Consequently, from (3.7.4),

$$L = \begin{bmatrix} 1 & 0 & 0 & 0 \\ 5/2 & 1 & 0 & 0 \\ 3/2 & 1 & 1 & 0 \\ -1/2 & -1/13 & 9/13 & 1 \end{bmatrix}.$$

We leave it as an exercise to verify that $LU = A$.

Row Operations	*Corresponding Multipliers*
(1) $A_{12}(-5/2)$, $A_{13}(-3/2)$, $A_{14}(1/2)$	$m_{21} = 5/2$, $m_{31} = 3/2$, $m_{41} = -1/2$.
(2) $A_{23}(-1)$, $A_{24}(1/13)$	$m_{32} = 1$, $m_{42} = -1/13$.
(3) $A_{34}(-9/13)$	$m_{43} = 9/13$.

The question that is undoubtedly in the reader's mind is what is the use of the LU decomposition? In order to answer this question, consider the $n \times n$ system of linear equations $Ax = b$, where $A = LU$. If we write the system as

$$LUx = b$$

and let $Ux = y$, then solving $Ax = b$ is equivalent to solving the pair of equations

$$Ly = b,$$

$$Ux = y.$$

Due to the triangular form of each of the coefficient matrices L and U these systems can be solved easily – the first one by "forward" substitution, and the second one by back substitution. In the case when we have a single right-hand side vector $\mathbf{b}$ there is no advantage to using the LU factorization for solving the system over Gaussian elimination. However, if we require the solution of several systems of equations with the same coefficient matrix A, say

$$A\mathbf{x}_i = \mathbf{b}_i, \quad i = 1, 2, \ldots, p$$

then it is more efficient to compute the LU factorization of A once, and then successively solve the triangular systems

$$\left.\begin{array}{l} L\mathbf{y}_i = \mathbf{b}_i, \\ U\mathbf{x}_i = \mathbf{y}_i, \end{array}\right\} \quad i = 1, 2, \ldots, p.$$

Example 3.7.7   Use the LU decomposition of

$$A = \begin{bmatrix} 2 & -3 & 1 & 2 \\ 5 & -1 & 2 & 1 \\ 3 & 2 & 6 & -5 \\ -1 & 1 & 3 & 2 \end{bmatrix}.$$

to solve the system $A\mathbf{x} = \mathbf{b}$ if $\mathbf{b} = [2, -3, 5, 7]$.

Solution   We have shown in the previous example that $A = LU$ where

$$L = \begin{bmatrix} 1 & 0 & 0 & 0 \\ 5/2 & 1 & 0 & 0 \\ 3/2 & 1 & 1 & 0 \\ -1/2 & -1/13 & 9/13 & 1 \end{bmatrix} \text{ and } U = \begin{bmatrix} 2 & -3 & 1 & 2 \\ 0 & 13/2 & -1/2 & -4 \\ 0 & 0 & 5 & -4 \\ 0 & 0 & 0 & 71/13 \end{bmatrix}.$$

We now solve the two triangular systems $L\mathbf{y} = \mathbf{b}$, $U\mathbf{x} = \mathbf{y}$. Using forward substitution on the first of these system we have

$$y_1 = 2, \qquad y_2 = -3 - \frac{5}{2}y_1 = -8,$$

$$y_3 = 5 - \frac{3}{2}y_1 - y_2 = 5 - 3 + 8 = 10,$$

$$y_4 = 7 + \frac{1}{2}y_1 + \frac{1}{13}y_2 - \frac{9}{13}y_3 = 8 - \frac{8}{13} - \frac{90}{13} = \frac{6}{13}.$$

Solving $U\mathbf{x} = \mathbf{y}$ via back substitution yields

$$x_4 = \frac{13}{71}y_4 = \frac{6}{71}, \qquad x_3 = \frac{1}{5}(y_3 + 4x_4) = \frac{1}{5}\left(10 + \frac{24}{71}\right) = \frac{734}{355}.$$

$$x_2 = \frac{2}{13}\left(y_2 + \frac{1}{2}x_3 + 4x_4\right) = \frac{2}{13}\left(-8 + \frac{367}{355} + \frac{24}{71}\right) = -\frac{362}{355}.$$

$$x_1 = \frac{1}{2}(y_1 + 3x_2 - x_3 - 2x_4) = \frac{1}{2}\left(2 - \frac{1086}{355} - \frac{734}{355} - \frac{12}{71}\right) = -\frac{117}{71}.$$

Consequently,

$$\mathbf{x} = \left(-\frac{117}{71}, -\frac{362}{355}, \frac{734}{355}, \frac{6}{71}\right).$$ ❏

In the more general case when row interchanges are required to reduce a nonsingular matrix A to upper triangular form it can be shown that A has a factorization of the form

$$A = PLU \tag{3.7.6}$$

where P is an appropriate product of elementary permutation matrices, L is a unit lower triangular matrix, and U is an upper triangular matrix. From the properties of the elementary permutation matrices it follows (see Exercise 20) that $P^{-1} = P^T$. Using (3.7.6) the linear system $A\mathbf{x} = \mathbf{b}$ can be written as

$$PLU\mathbf{x} = \mathbf{b}$$

or equivalently,

$$LU\mathbf{x} = P^T\mathbf{b}.$$

Consequently to solve $A\mathbf{x} = \mathbf{b}$ in this case we can solve the two triangular systems

$$\begin{cases} L\mathbf{y} = P^T\mathbf{b}, \\ U\mathbf{x} = \mathbf{y}. \end{cases}$$

For a full discussion of this and other factorizations of $n \times n$ matrices, and their applications, the reader is referred to more advanced texts on linear algebra or numerical analysis (for example, B. Noble and J.W. Daniel, *Applied Linear Algebra*, Prentice Hall, 1988; J. Ll. Morris, *Computational Methods in Elementary Numerical Analysis*, Wiley, 1983).

EXERCISES 3.7

1 . Write all 3×3 elementary matrices and their inverses.

For problems 2–5, determine elementary matrices that reduce the given matrix to REF.

2 . $\begin{bmatrix} 3 & 5 \\ 1 & -2 \end{bmatrix}$.

3 . $\begin{bmatrix} 5 & 8 & 2 \\ 1 & 3 & -1 \end{bmatrix}$.

4 . $\begin{bmatrix} 3 & -1 & 4 \\ 2 & 1 & 3 \\ 1 & 3 & 2 \end{bmatrix}$.

5. $\begin{bmatrix} 1 & 2 & 3 & 4 \\ 2 & 3 & 4 & 5 \\ 3 & 4 & 5 & 6 \end{bmatrix}$.

6. Determine elementary matrices $E_1, E_2, \ldots, E_k$ that reduce $A = \begin{bmatrix} 2 & -1 \\ 1 & 3 \end{bmatrix}$ to RREF. Verify by direct multiplication that $E_1 E_2 \cdots E_k A = I_2$.

7. Determine a Type 3 lower triangular elementary matrix E_1, that reduces $A = \begin{bmatrix} 3 & -2 \\ -1 & 5 \end{bmatrix}$ to upper triangular form. Use equation (3.7.3) to determine L and verify equation (3.7.2).

For problems 8–13, determine the LU factorization of the given matrix. Verify your answer by computing the product LU.

8. $A = \begin{bmatrix} 2 & 3 \\ 5 & 1 \end{bmatrix}$.

9. $A = \begin{bmatrix} 3 & 1 \\ 5 & 2 \end{bmatrix}$.

10. $A = \begin{bmatrix} 3 & -1 & 2 \\ 6 & -1 & 1 \\ -3 & 5 & 2 \end{bmatrix}$.

11. $A = \begin{bmatrix} 5 & 2 & 1 \\ -10 & -2 & 3 \\ 15 & 2 & -3 \end{bmatrix}$.

12. $A = \begin{bmatrix} 1 & -1 & 2 & 3 \\ 2 & 0 & 3 & -4 \\ 3 & -1 & 7 & 8 \\ 1 & 3 & 4 & 5 \end{bmatrix}$.

13. $A = \begin{bmatrix} 2 & -3 & 1 & 2 \\ 4 & -1 & 1 & 1 \\ -8 & 2 & 2 & -5 \\ 6 & 1 & 5 & 2 \end{bmatrix}$.

For problems 14–17, use the LU factorization of A to solve the system $Ax = b$.

14. $A = \begin{bmatrix} 1 & 2 \\ 2 & 3 \end{bmatrix}$, $\mathbf{b} = \begin{bmatrix} 3 \\ -1 \end{bmatrix}$.

15. $A = \begin{bmatrix} 1 & -3 & 5 \\ 3 & 2 & 2 \\ 2 & 5 & 2 \end{bmatrix}$, $\mathbf{b} = \begin{bmatrix} 1 \\ 5 \\ -1 \end{bmatrix}$.

16. $A = \begin{bmatrix} 2 & 2 & 1 \\ 6 & 3 & -1 \\ -4 & 2 & 2 \end{bmatrix}$, $\mathbf{b} = \begin{bmatrix} 1 \\ 0 \\ 2 \end{bmatrix}$.

17. $A = \begin{bmatrix} 4 & 3 & 0 & 0 \\ 8 & 1 & 2 & 0 \\ 0 & 5 & 3 & 6 \\ 0 & 0 & -5 & 7 \end{bmatrix}$, $\mathbf{b} = \begin{bmatrix} 2 \\ 3 \\ 0 \\ 5 \end{bmatrix}$.

18. Use the LU factorization of $A = \begin{bmatrix} 2 & -1 \\ -8 & 3 \end{bmatrix}$ to solve each of the systems $Ax_i = b_i$ if

$$\mathbf{b}_1 = \begin{bmatrix} 3 \\ -1 \end{bmatrix}, \mathbf{b}_2 = \begin{bmatrix} 2 \\ 7 \end{bmatrix}, \mathbf{b}_3 = \begin{bmatrix} 5 \\ -9 \end{bmatrix}.$$

19. Use the LU factorization of

$$A = \begin{bmatrix} -1 & 4 & 2 \\ 3 & 1 & 4 \\ 5 & -7 & 1 \end{bmatrix}$$

to solve each of the systems $Ax_i = e_i$ and thereby determine A^{-1}.

20. If $P = P_1 P_2 \cdots P_k$, where each P_i is an elementary permutation matrix, show that $P^{-1} = P^T$.

21. In this problem, we prove that the LU decomposition of a *nonsingular* $n \times n$ matrix is unique in the sense that, if $A = L_1 U_1$ and $A = L_2 U_2$ where L_1 and L_2 are unit lower triangular matrices and U_1 and U_2 are upper triangular matrices, then $L_1 = L_2$ and $U_1 = U_2$.

(a) Apply Corollary 3.6.2 to conclude that L_2 and U_1 are nonsingular and then use the fact that $L_1 U_1 = L_2 U_2$ to establish that $L_2^{-1} L_1 = U_2 U_1^{-1}$.

(b) Use the result from (a) together with Theorem 3.2.4 and Corollary 3.2.1 to prove that $L_2^{-1} L_1 = I_n$ and $U_2 U_1^{-1} = I_n$, from which the required result follows.

22. QR Factorization: It can be shown that any nonsingular $n \times n$ matrix has a factorization of the form

$$A = QR$$

where Q and R are nonsingular and upper triangular with Q satisfying $Q^T Q = I_n$. Determine an algorithm for solving the linear system $A\mathbf{x} = \mathbf{b}$ using this QR factorization.

◆ For problems 23 – 25, use some form of technology to determine the LU factorization of the given matrix. Verify the factorization by computing the product LU.

23. $A = \begin{bmatrix} 3 & 5 & -2 \\ 2 & 7 & 9 \\ -5 & 5 & 11 \end{bmatrix}$.

24. $A = \begin{bmatrix} 27 & -19 & 32 \\ 15 & -16 & 9 \\ 23 & -13 & 51 \end{bmatrix}$.

25. $A = \begin{bmatrix} 34 & 13 & 19 & 22 \\ 53 & 17 & -71 & 29 \\ 21 & 37 & 63 & 59 \\ 81 & 93 & -47 & 39 \end{bmatrix}$.

4

![section divider]

Determinants

In this chapter, we introduce a basic tool in applied mathematics, namely the determinant of an $n \times n$ matrix. The determinant can be considered a number, associated with an $n \times n$ matrix A, whose nonvanishing characterizes when the linear system $A\mathbf{x} = \mathbf{b}$ has a unique solution (or, equivalently, when A^{-1} exists). As we shall see in the later chapters, the determinant also plays a fundamental role in the theory of linear differential equations and linear systems of differential equations. Other areas of applications of the determinant include coordinate geometry and function theory.

Sections 4.1–4.3 give a detailed introduction to determinants and their applications. Alternatively, the summary Section 4.4 can be used to give a non-rigorous and much more abbreviated introduction to the fundamental results required in the remainder of the text.

4.1   THE DEFINITION OF A DETERMINANT

PERMUTATIONS

Before introducing the concept of a determinant, we require some preliminary results about permutations. Consider the first n positive integers $1, 2, 3, \ldots, n$. Any arrangement of these integers in a specific order, say, $(p_1, p_2, \ldots, p_n)$, is called a **permutation**.

Example 4.1.1   There are precisely six distinct permutations of the integers 1,2, and 3:

$$(1, 2, 3), (1, 3, 2), (2, 3, 1), (2, 1, 3), (3, 1, 2), (3, 2, 1). \qquad \square$$

More generally we have the following result:

Theorem 4.1.1:   There are precisely $n!$ distinct permutations of the integers $1, 2, \ldots, n$.

PROOF   The proof is left as an exercise.   ■

The elements in the permutation $(1, 2, 3, \ldots, n)$ are said to be in their natural increasing order. We now introduce a number that describes how far from natural order a given permutation is. The elements p_i and p_j in the permutation $(p_1, p_2, \ldots, p_n)$ are said to be **inverted** if they are out of their natural order — that is, if $p_i > p_j$ with $i < j$. For example, in the permutation $(4, 2, 3, 1)$ the pairs $(4, 2), (4, 3), (4, 1), (2, 1), (3, 1)$ are all out of their natural order, and so, there are a total of five inversions in this permutation. In general we let $N(p_1, p_2, \ldots, p_n)$ denote the total number of inversions in the permutation $(p_1, p_2, \ldots, p_n)$.

Example 4.1.2   Find the number of inversions in the permutations $(1, 3, 2, 4, 5)$ and $(2, 4, 5, 3, 1)$.

Solution   The only pair of elements in the permutation $(1, 3, 2, 4, 5)$ that is out of natural order is $(3, 2)$, so that $N(1, 3, 2, 4, 5) = 1$.

The permutation $(2, 4, 5, 3, 1)$ has the following pairs of elements out of natural order: $(2, 1), (4, 3), (4, 1), (5, 3), (5, 1), (3, 1)$. Thus, $N(2, 4, 5, 3, 1) = 6$. $\square$

It can be shown that the number of inversions gives the minimum number of adjacent interchanges of elements in the permutation that are required to restore the permutation to its natural increasing order. This justifies the claim that the number of inversions describes how far from natural order a given permutation is. For example, $N(3, 2, 1) = 3$ and the permutation $(3, 2, 1)$ can be restored to its natural order by the following sequence of adjacent interchanges:

$$(3, 2, 1) \rightarrow (3, 1, 2) \rightarrow (1, 3, 2) \rightarrow (1, 2, 3).$$

The number of inversions enables us to distinguish two different types of permutations as follows.

Even permutations: If $N(p_1, p_2, \ldots, p_n)$ is an even integer (or zero), we say that $(p_1, p_2, \ldots, p_n)$ is an even permutation. We also say that $(p_1, p_2, \ldots, p_n)$ has **even parity**.

Odd permutations: If $N(p_1, p_2, \ldots, p_n)$ is an odd integer, we say that $(p_1, p_2, \ldots, p_n)$ is an odd permutation. We also say that $(p_1, p_2, \ldots, p_n)$ has **odd parity**.

Example 4.1.3   The permutation $(4, 1, 3, 2)$ has even parity, since $N(4, 1, 3, 2) = 4$, whereas $(3, 2, 1, 4)$ is an odd permutation since $N(3, 2, 1, 4) = 3$. $\square$

We associate a plus or a minus sign with a permutation, depending on whether it has even or odd parity, respectively. Thus, the sign associated with the permutation $(p_1, p_2, \ldots, p_n)$ can be specified by the indicator $\sigma(p_1, p_2, \ldots, p_n)$ defined as follows:

$$\sigma(p_1, p_2, \ldots, p_n) = \begin{cases} +1 & \text{if } (p_1, p_2, \ldots, p_n) \text{ has even parity,} \\ -1 & \text{if } (p_1, p_2, \ldots, p_n) \text{ has odd parity.} \end{cases}$$

This can be expressed directly in terms of the number of inversions as

$$\boxed{\sigma(p_1, p_2, \ldots, p_n) = (-1)^{N(p_1, p_2, \ldots, p_n)}.}$$

Example 4.1.4   It follows from Example 4.1.2 that

$$\sigma(1, 3, 2, 4, 5) = (-1)^1 = -1$$

whereas

$$\sigma(2, 4, 5, 3, 1) = (-1)^6 = +1. \qquad \square$$

The proof of some of our later results will depend upon the next theorem.

Theorem 4.1.2:   If any two elements in a permutation are interchanged then the parity of the resulting permutation is opposite to that of the original permutation.

PROOF   We first show that interchanging two adjacent terms in a permutation changes the parity. Consider an arbitrary permutation $(p_1, p_2, \ldots, p_k, p_{k+1}, \ldots, p_n)$, and suppose we interchange the adjacent elements p_k, and p_{k+1}. Then

1. If $p_k > p_{k+1}$, $N(p_1, p_2, \ldots, p_{k+1}, p_k, \ldots, p_n) = N(p_1, p_2, \ldots, p_k, p_{k+1}, \ldots, p_n) - 1,$

2. If $p_k < p_{k+1}$, $N(p_1, p_2, \ldots, p_{k+1}, p_k, \ldots, p_n) = N(p_1, p_2, \ldots, p_k, p_{k+1}, \ldots, p_n) + 1,$

so that the parity is changed in both cases. Now suppose we interchange the elements p_i and p_k in the permutation $(p_1, p_2, \ldots, p_i, \ldots, p_k, \ldots, p_n)$. We can accomplish this by successively interchanging adjacent elements. In moving p_k to the ith position, we perform $k - i$ interchanges, and the resulting permutation is

$$(p_1, p_2, \ldots, p_k, p_i, \ldots, p_{k-1}, p_{k+1}, \ldots, p_n).$$

We now move p_i to the kth position. This requires $(k - i) - 1$ interchanges. Thus, the total number of interchanges is $2(k - i) - 1$, which is always an odd integer. Since each adjacent interchange changes the parity, the permutation resulting from an odd number of adjacent interchanges has opposite parity to the original permutation. ∎

THE DEFINITION OF A DETERMINANT

We motivate the definition of a determinant by studying the structure of the solution of an $n \times n$ system of linear equations in three cases. These solutions could be derived using Gaussian elimination.

CASE 1: $n = 1$ The system

$$a_{11}x_1 = b_1, \quad \text{with coefficient matrix } A = [a_{11}],$$

has solution

$$x_1 = \frac{b_1}{a_{11}},$$

provided that the 1×1 determinant, $\det(A)$, defined by

$$\boxed{\det(A) = a_{11}}$$

is nonzero.

CASE 2: $n = 2$ The system

$$a_{11}x_1 + a_{12}x_2 = b_1,$$
$$a_{21}x_1 + a_{22}x_2 = b_2,$$

with coefficient matrix $A = \begin{bmatrix} a_{11} & a_{12} \\ a_{21} & a_{22} \end{bmatrix}$, has solution

$$x_1 = -\frac{(a_{12}b_2 - a_{22}b_1)}{a_{11}a_{22} - a_{12}a_{21}}, \qquad x_2 = \frac{(a_{11}b_2 - a_{21}b_1)}{a_{11}a_{22} - a_{12}a_{21}},$$

provided that the 2×2 determinant, $\det(A)$, defined by

$$\boxed{\det(A) = a_{11}a_{22} - a_{12}a_{21}} \tag{4.1.1}$$

is nonzero.

CASE 3: $n = 3$ The system

$$a_{11}x_1 + a_{12}x_2 + a_{13}x_3 = b_1,$$
$$a_{21}x_1 + a_{22}x_2 + a_{23}x_3 = b_2,$$
$$a_{31}x_1 + a_{32}x_2 + a_{33}x_3 = b_3,$$

with coefficient matrix $A = \begin{bmatrix} a_{11} & a_{12} & a_{13} \\ a_{21} & a_{22} & a_{23} \\ a_{31} & a_{32} & a_{33} \end{bmatrix}$, has solution:[1]

$$x_1 = \frac{E_1}{\det(A)}, \qquad x_2 = \frac{E_2}{\det(A)}, \qquad x_3 = \frac{E_3}{\det(A)},$$

provided that the 3×3 determinant, $\det(A)$, is nonzero, where

$$\boxed{\begin{aligned} \det(A) = a_{11}a_{22}a_{33} &+ a_{12}a_{23}a_{31} + a_{13}a_{21}a_{32} \\ &- a_{11}a_{23}a_{32} - a_{12}a_{21}a_{33} - a_{13}a_{22}a_{31}. \end{aligned}}$$

---

[1]Here E_1, E_2, E_3 are certain algebraic expressions involving the a_{ij} and b_i whose exact form need not bother us at the moment.

To generalize the foregoing formulas, we take a closer look at their structure. Each determinant consists of the sum of $n!$ products. Each product term contains precisely one element from each row and one element from each column in the matrix of coefficients of the system. Furthermore, each term is assigned a plus or a minus sign. From the expression for, say, the 3×3 determinant, we see that the row indices of each term have been arranged in their natural order and that the column indices are each a permutation, (p_1, p_2, p_3), of 1, 2, 3. Further, the sign attached to each term coincides with the sign of the permutation of the corresponding column indices, that is, $\sigma(p_1, p_2, p_3)$. These observations motivate the following general definition of the determinant of an $n \times n$ matrix:

Definition 4.1.1: Let $A = [a_{ij}]$ be an $n \times n$ matrix. The **determinant of A**, denoted $\det(A)$, is defined as follows:

$$\det(A) = \sum \sigma(p_1, p_2, \ldots, p_n)\, a_{1p_1} a_{2p_2} a_{3p_3} \cdots a_{np_n} \tag{4.1.2}$$

where the summation is over the $n!$ distinct permutations $(p_1, p_2, \ldots, p_n)$ of the integers 1, 2, 3, ..., n. The determinant of an $n \times n$ matrix is said to have **order n**.

We will also denote $\det(A)$ by

$$\begin{vmatrix} a_{11} & a_{12} & \cdots & a_{1n} \\ a_{21} & a_{22} & \cdots & a_{2n} \\ \vdots & \vdots & & \vdots \\ a_{n1} & a_{n2} & \cdots & a_{nn} \end{vmatrix}.$$

Thus, for example, from (4.1.1),

$$\begin{vmatrix} a_{11} & a_{12} \\ a_{21} & a_{22} \end{vmatrix} = a_{11}a_{22} - a_{12}a_{21}.$$

Example 4.1.5   Use Definition 4.1.1 to derive the expression for the determinant of order 3.

Solution   When $n = 3$, (4.1.2) reduces to

$$\det(A) = \sum \sigma(p_1, p_2, p_3)\, a_{1p_1} a_{2p_2} a_{3p_3},$$

where the summation is over the 3! permutations of 1, 2, 3. It follows that the six terms in this summation are

$$a_{11}a_{22}a_{33}, \quad a_{11}a_{23}a_{32}, \quad a_{12}a_{21}a_{33}, \quad a_{12}a_{23}a_{31}, \quad a_{13}a_{21}a_{32}, \quad a_{13}a_{22}a_{31},$$

so that

$$\det(A) = \sigma(1, 2, 3)a_{11}a_{22}a_{33} + \sigma(1, 3, 2)a_{11}a_{23}a_{32} + \sigma(2, 1, 3)a_{12}a_{21}a_{33}$$

$$+ \sigma(2, 3, 1)a_{12}a_{23}a_{31} + \sigma(3, 1, 2)a_{13}a_{21}a_{32} + \sigma(3, 2, 1)a_{13}a_{22}a_{31}.$$

Using the rules for determining $\sigma(p_1, p_2, p_3)$, we find

$$\sigma(1, 2, 3) = +1, \qquad \sigma(1, 3, 2) = -1, \qquad \sigma(2, 1, 3) = -1,$$
$$\sigma(2, 3, 1) = +1, \qquad \sigma(3, 1, 2) = +1, \qquad \sigma(3, 2, 1) = -1.$$

Hence,

$$\det(A) = \begin{vmatrix} a_{11} & a_{12} & a_{13} \\ a_{21} & a_{22} & a_{23} \\ a_{31} & a_{32} & a_{33} \end{vmatrix} = a_{11}a_{22}a_{33} + a_{12}a_{23}a_{31} + a_{13}a_{21}a_{32}$$
$$- a_{11}a_{23}a_{32} - a_{12}a_{21}a_{33} - a_{13}a_{22}a_{31}. \qquad \square$$

A simple schematic for obtaining the terms in the determinant of order 3 is given in Figure 4.1.1. By taking the product of the elements joined by each arrow and attaching

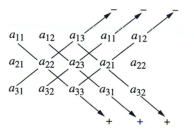

Figure 4.1.1 A schematic for obtaining the determinant of a 3×3 matrix $A = [a_{ij}]$.

the indicated sign to the result, we obtain the six terms in the determinant of the 3×3 matrix $A = [a_{ij}]$. Note that this technique for obtaining the terms in a determinant **does not** generalize to higher order determinants.

Example 4.1.6   Evaluate

$$\text{(a)} \quad |-3|. \qquad \text{(b)} \quad \begin{vmatrix} 3 & -2 \\ 1 & 4 \end{vmatrix}. \qquad \text{(c)} \quad \begin{vmatrix} 1 & 2 & -3 \\ 4 & -1 & 2 \\ 0 & 3 & 1 \end{vmatrix}.$$

Solution

(a) $|-3| = -3.$

(b) $\begin{vmatrix} 3 & -2 \\ 1 & 4 \end{vmatrix} = (3)(4) - (-2)(1) = 14.$

(c)   In this case the schematic in Figure 4.1.1 is

$$\begin{matrix} 1 & 2 & -3 & 1 & 2 \\ 4 & -1 & 2 & 4 & -1 \\ 0 & 3 & 1 & 0 & 3 \end{matrix}$$

so that

$$\begin{vmatrix} 1 & 2 & -3 \\ 4 & -1 & 2 \\ 0 & 3 & 1 \end{vmatrix} = (1)(-1)(1) + (2)(2)(0) + (-3)(4)(3)$$

$$- (0)(-1)(-3) - (3)(2)(1) - (1)(4)(2) = -51$$

GEOMETRIC INTERPRETATION OF THE DETERMINANTS OF ORDERS TWO AND THREE

If $\mathbf{a}$ and $\mathbf{b}$ are two vectors in space, we recall that their dot product is the scalar

$$\mathbf{a} \cdot \mathbf{b} = \| \mathbf{a} \| \, \| \mathbf{b} \| \cos \theta, \tag{4.1.3}$$

whereas the cross product of $\mathbf{a}$ and $\mathbf{b}$ is the vector

$$\mathbf{a} \times \mathbf{b} = \| \mathbf{a} \| \, \| \mathbf{b} \| \sin \theta \, \mathbf{n}. \tag{4.1.4}$$

In these formulas, θ denotes the angle between $\mathbf{a}$ and $\mathbf{b}$, $\| \mathbf{a} \|$ and $\| \mathbf{b} \|$ denote the lengths of $\mathbf{a}$ and $\mathbf{b}$, respectively, and $\mathbf{n}$ denotes a unit vector that is perpendicular to the plane of $\mathbf{a}$ and $\mathbf{b}$ and chosen in such a way that $\{\mathbf{a}, \mathbf{b}, \mathbf{n}\}$ is a right-handed system. If $\mathbf{i}, \mathbf{j}, \mathbf{k}$ denote the unit vectors pointing along the positive x-, y-, and z-axes, respectively, of a rectangular Cartesian coordinate system and $\mathbf{a} = a_1\mathbf{i} + a_2\mathbf{j} + a_3\mathbf{k}$, $\mathbf{b} = b_1\mathbf{i} + b_2\mathbf{j} + b_3\mathbf{k}$, then equation (4.1.4) can be expressed in component form as

$$\mathbf{a} \times \mathbf{b} = (a_2 b_3 - a_3 b_2)\mathbf{i} + (a_3 b_1 - a_1 b_3)\mathbf{j} + (a_1 b_2 - a_2 b_1)\mathbf{k}. \tag{4.1.5}$$

We use these formulas to establish the following results:

(1) The area of a parallelogram with sides determined by the vectors $\mathbf{a} = a_1\mathbf{i} + a_2\mathbf{j}$, and $\mathbf{b} = b_1\mathbf{i} + b_2\mathbf{j}$ is

$$\boxed{\text{Area} = |\det(A)|,}$$

where $A = \begin{bmatrix} a_1 & a_2 \\ b_1 & b_2 \end{bmatrix}$, and the vertical bars denote the absolute value of $\det(A)$.

(2) The volume of the parallelepiped determined by the vectors $\mathbf{a} = a_1\mathbf{i} + a_2\mathbf{j} + a_3\mathbf{k}$, $\mathbf{b} = b_1\mathbf{i} + b_2\mathbf{j} + b_3\mathbf{k}$, $\mathbf{c} = c_1\mathbf{i} + c_2\mathbf{j} + c_3\mathbf{k}$ is

$$\boxed{\text{Volume} = |\det(A)|,}$$

where $A = \begin{bmatrix} a_1 & a_2 & a_3 \\ b_1 & b_2 & b_3 \\ c_1 & c_2 & c_3 \end{bmatrix}$, and once more the vertical bars denote the absolute value of $\det(A)$.

PROOF OF 1   The area of the parallelogram is:

$$\text{Area} = \text{length of base} \times \text{perpendicular height}.$$

From Figure 4.1.2, this can be written as

$$\text{Area} = \| \mathbf{a} \| \, h = \| \mathbf{a} \| \, \| \mathbf{b} \| \, | \sin \theta | = \| \mathbf{a} \times \mathbf{b} \|. \tag{4.1.6}$$

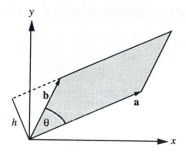

Figure 4.1.2 Determining the area of a parallelogram.

Since the **k** components of **a** and **b** are both zero (the vectors lie in the xy-plane), substitution from equation (4.1.5) yields

$$\text{Area} = \parallel (a_1b_2 - a_2b_1)\mathbf{k} \parallel = \mid (a_1b_2 - a_2b_1) \mid = \mid \det(A) \mid.$$

PROOF OF 2   The volume of the parallelepiped is:

$$\text{Volume} = \text{area of base} \times \text{perpendicular height}.$$

From Figure 4.1.3, we therefore have

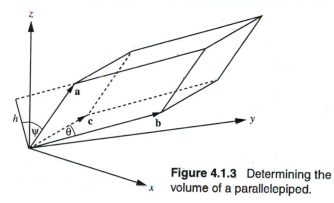

Figure 4.1.3 Determining the volume of a parallelepiped.

$$\text{Volume} = \parallel \mathbf{b} \times \mathbf{c} \parallel \parallel \mathbf{a} \parallel \mid \cos \psi \mid = \parallel \mathbf{b} \times \mathbf{c} \parallel \mid \mathbf{a} \cdot \mathbf{n} \mid,$$

where **n** is a *unit* vector that is perpendicular to the plane containing **b** and **c**, and we have used equation (4.1.3) in the last step. We can now use (4.1.4) and (4.1.5) to obtain

$$\text{Volume} = \mid \mathbf{a} \cdot (\parallel \mathbf{b} \times \mathbf{c} \parallel \mathbf{n}) \mid = \mid \mathbf{a} \cdot (\mathbf{b} \times \mathbf{c}) \mid$$

$$= \mid (a_1\mathbf{i} + a_2\mathbf{j} + a_3\mathbf{k}) \cdot [(b_2c_3 - b_3c_2)\mathbf{i} + (b_3c_1 - b_1c_3)\mathbf{j} + (b_1c_2 - b_2c_1)\mathbf{k}] \mid$$

$$= \mid a_1(b_2c_3 - b_3c_2) + a_2(b_3c_1 - b_1c_3) + a_3(b_1c_2 - b_2c_1) \mid$$

$$= \mid \det(A) \mid$$

as required.

We see from the expression for the volume of a parallelepiped that the condition for three vectors to lie in the same plane (parallelepiped has zero volume) is that $\det(A) = 0$. This will be a useful result in the next chapter.

DETERMINANTS AND LINEAR DIFFERENTIAL EQUATIONS

The main application that we will have for determinants in this text is to linear differential equations and systems of linear differential equations. In Chapter 2, we proved that all solutions to the second-order homogeneous linear DE

$$y'' + a_1(x)y' + a_2(x)y = 0$$

are of the form

$$y(x) = c_1 y_1(x) + c_2 y_2(x),$$

where y_1 and y_2 are any two *linearly independent* solutions to the DE. Furthermore, the analytic criterion for determining whether a given pair of solutions are linearly independent on an interval I is that the Wronskian

$$W[y_1, y_2](x) = y_1 y_2' - y_2 y_1'$$

be nonzero. This criterion can be expressed in terms of determinants as

$$\begin{vmatrix} y_1 & y_2 \\ y_1' & y_2' \end{vmatrix} \neq 0.$$

More generally, we will see later in the text that all solutions to an nth-order linear homogeneous DE can be expressed as

$$y(x) = c_1 y_1(x) + c_2 y_2(x) + \cdots + c_n y_n(x)$$

where $y_1, y_2, \ldots, y_n$ are any n solutions with the property that there are no nonzero values of the constants $d_1, d_2, \ldots, d_n$ such that

$$d_1 y_1 + d_2 y_2 + \cdots + d_n y_n = 0.$$

We will find that the analytic criterion for n solutions to the DE to have this property is that the determinant

$$\begin{vmatrix} y_1(x) & y_2(x) & \cdots & y_n(x) \\ y_1'(x) & y_2'(x) & \cdots & y_n'(x) \\ \vdots & \vdots & & \vdots \\ y_1^{(n-1)}(x) & y_2^{(n-1)}(x) & \cdots & y_n^{(n-1)}(x) \end{vmatrix}$$

be nonzero on the interval of interest. This indicates that determinants play a fundamental role in the solution of linear homogeneous DE.

EXERCISES 4.1

1. Determine the parity of the given permutation.

(**a**) $(2, 1, 3, 4)$.

(**b**) $(1, 3, 2, 4)$.

(**c**) $(1, 4, 3, 5, 2)$.

(**d**) $(5, 4, 3, 2, 1)$.

2. Use the definition of a determinant to derive the general expression for the determinant of A if

$$A = \begin{bmatrix} a_{11} & a_{12} \\ a_{21} & a_{22} \end{bmatrix}.$$

For problems 3–6, evaluate the determinant of the given matrix.

3. $A = \begin{bmatrix} 1 & -1 \\ 2 & 3 \end{bmatrix}$.

4. $A = \begin{bmatrix} 2 & -1 \\ 6 & -3 \end{bmatrix}$.

5. $A = \begin{bmatrix} 1 & -1 & 0 \\ 2 & 3 & 6 \\ 0 & 2 & -1 \end{bmatrix}$.

6. $A = \begin{bmatrix} 2 & 1 & 5 \\ 4 & 2 & 3 \\ 9 & 5 & 1 \end{bmatrix}$.

For problems 7–12, evaluate the given determinant.

7. $\begin{vmatrix} \pi & \pi^2 \\ \sqrt{2} & 2\pi \end{vmatrix}$.

8. $\begin{vmatrix} 2 & 3 & -1 \\ 1 & 4 & 1 \\ 3 & 1 & 6 \end{vmatrix}$.

9. $\begin{vmatrix} 3 & 2 & 6 \\ 2 & 1 & -1 \\ -1 & 1 & 4 \end{vmatrix}$.

10. $\begin{vmatrix} 2 & 3 & 6 \\ 0 & 1 & 2 \\ 1 & 5 & 0 \end{vmatrix}$.

11. $\begin{vmatrix} \sqrt{\pi} & e^2 & e^{-1} \\ \sqrt{67} & 1/30 & 2001 \\ \pi & \pi^2 & \pi^3 \end{vmatrix}$.

12. $\begin{vmatrix} e^{2t} & e^{3t} & e^{-4t} \\ 2e^{2t} & 3e^{3t} & -4e^{-4t} \\ 4e^{2t} & 9e^{3t} & 16e^{-4t} \end{vmatrix}$

13. Verify that $y_1(x) = \cos 2x$, $y_2(x) = \sin 2x$, and $y_3(x) = e^x$ are solutions to the DE

$$y''' - y'' + 4y' - 4y = 0,$$

and show that $\begin{vmatrix} y_1 & y_2 & y_3 \\ y_1' & y_2' & y_3' \\ y_1'' & y_2'' & y_3'' \end{vmatrix}$ is

nonzero on any interval.

14. (**a**) Verify that $y_1(x) = e^x$, $y_2(x) = \cosh x$, $y_3(x) = \sinh x$ are solutions to the DE

$$y''' - y'' - y' + y = 0$$

and show that $\begin{vmatrix} y_1 & y_2 & y_3 \\ y_1' & y_2' & y_3' \\ y_1'' & y_2'' & y_3'' \end{vmatrix}$ is

identically zero.

(**b**) Determine nonzero constants d_1, d_2, d_3 such that

$$d_1 y_1 + d_2 y_2 + d_3 y_3 = 0.$$

15. (**a**) Write all 24 distinct permutations of the integers 1, 2, 3, 4, and determine the parity of each permutation. Hence derive the expression for the determinant of order 4.

(**b**) Use the result from (a) to evaluate

$$\begin{vmatrix} 1 & -1 & 0 & 1 \\ 3 & 0 & 2 & 5 \\ 2 & 1 & 0 & 3 \\ 9 & -1 & 2 & 1 \end{vmatrix}.$$

16. (**a**) If $A = \begin{bmatrix} a_{11} & a_{12} \\ a_{21} & a_{22} \end{bmatrix}$ and c is a constant,

verify that $\det(cA) = c^2 \det(A)$.

(**b**) Use the definition of a determinant to prove that if A is an $n \times n$ matrix and c is a constant, then $\det(cA) = c^n \det(A)$.

17. Determine whether the following are terms in the determinant of order 5. If they are, determine whether the permutation of the column indices has even or odd parity and hence find whether the terms have a plus or a minus sign attached to them:

(**a**) $a_{11} a_{25} a_{33} a_{42} a_{54}$.

(b) $a_{11}a_{23}a_{34}a_{43}a_{52}$.

(c) $a_{11}a_{32}a_{24}a_{43}a_{55}$.

18. Determine the values of the indices p and q such that the following are terms in the determinant of order 4. In each case determine the number of inversions in the permutation of the column indices and hence find the appropriate sign that should be attached to each term.

(a) $a_{13}a_{p4}a_{32}a_{2q}$.

(b) $a_{21}a_{3q}a_{p2}a_{43}$.

(c) $a_{3q}a_{p4}a_{13}a_{42}$.

19. If A is the general $n \times n$ matrix, determine the sign attached to the term

$$a_{1n}a_{2\,n-1}a_{3\,n-2} \cdots a_{n1},$$

which arises in $\det(A)$.

20. The *alternating symbol* ε_{ijk} is defined by

$$\varepsilon_{ijk} = \begin{cases} 1, & \text{if } (ijk) \text{ is an even} \\ & \text{permutation of 1, 2, 3,} \\ -1, & \text{if } (ijk) \text{ is an odd} \\ & \text{permutation of 1, 2, 3,} \\ 0, & \text{otherwise.} \end{cases}$$

(a) Write all nonzero ε_{ijk}, for $1 \le i \le 3$, $1 \le j \le 3$, $1 \le k \le 3$.

(b) If $A = [a_{ij}]$ is a 3×3 matrix, verify that

$$\det(A) = \sum_{i=1}^{3} \sum_{j=1}^{3} \sum_{k=1}^{3} \varepsilon_{ijk} a_{1i} a_{2j} a_{3k}.$$

◆ **21.** Use some form of technology to evaluate the determinants in problems 7–12.

◆ **22.** Let A be an arbitrary 4×4 matrix. By experimenting with various ERO, conjecture how ERO applied to A effect the value of $\det(A)$.

◆ **23.** Verify that $y_1(x) = e^{-2x}\cos 3x$, $y_2(x) = e^{-2x}\sin 3x$, and $y_3(x) = e^{-4x}$ are solutions to the DE

$$y''' + 8y'' + 29y' + 52y = 0,$$

and show that $\begin{vmatrix} y_1 & y_2 & y_3 \\ y_1' & y_2' & y_3' \\ y_1'' & y_2'' & y_3'' \end{vmatrix}$ is nonzero on any interval.

4.2   PROPERTIES OF DETERMINANTS

Evaluating a determinant of order n using the definition given in the previous section is not very practical since the number of terms in the determinant grows as $n!$ (for example, a determinant of order 10 contains 3,628,800 terms). In the next two sections we develop better techniques for evaluating determinants. The following theorem suggests one way to proceed.

Theorem 4.2.1:   If A is an $n \times n$ upper or lower triangular matrix then

$$\det(A) = a_{11}a_{22}a_{33} \cdots a_{nn} = \prod_{i=1}^{n} a_{ii}. \tag{4.2.1}$$

PROOF   We use the definition of a determinant to prove the result in the upper triangular case. From equation (4.1.2),

$$\det(A) = \sum \sigma(p_1, p_2, \ldots, p_n) \, a_{1p_1} a_{2p_2} a_{3p_3} \cdots a_{np_n}. \tag{4.2.2}$$

If A is upper triangular, then $a_{ij} = 0$ whenever $i > j$, and therefore, the only nonzero terms in the preceding summation are those with $p_i \ge i$ for all i. Since all the p_i must be distinct, the only possibility is (start at the right-hand end and work towards the left)

$$p_i = i, \qquad i = 1, 2, \ldots, n,$$

and so equation (4.2.2) reduces to the single term

$$\det(A) = \sigma(1, 2, \ldots, n) a_{11} a_{22} \cdots a_{nn}.$$

Since $\sigma(1, 2, \ldots, n) = 1$, it follows that

$$\det(A) = a_{11} a_{22} \cdots a_{nn}.$$

The proof in the lower triangular case is left as an exercise.   ■

Example 4.2.1   According to the previous theorem,

$$\begin{vmatrix} 2 & 5 & -1 & 3 \\ 0 & -1 & 0 & 4 \\ 0 & 0 & 7 & 8 \\ 0 & 0 & 0 & 5 \end{vmatrix} = (2)(-1)(7)(5) = -70.$$ ❑

We now recall from Chapter 3 that any matrix can be reduced to REF by a sequence of ERO. In the case of an $n \times n$ matrix, any REF will be upper triangular. The preceding theorem suggests, therefore, that we should consider how ERO performed on a matrix, A, alter the value of $\det(A)$.

Elementary Row Operations and Determinants

Let A be an $n \times n$ matrix.

P1.   If B is the matrix obtained by permuting two distinct rows of A, then

$$\det(A) = -\det(B).$$

P2.   If B is the matrix obtained by dividing any row of A by a nonzero scalar k, then

$$\det(A) = k \det(B).$$

P3.   If B is the matrix obtained by adding a multiple of any row of A to a different row of A, then

$$\det(A) = \det(B).$$

REMARK   The main use of P2 is that it enables us to factor a common multiple of any row out of the determinant. For example,

$$\underset{\det(A)}{\underset{\uparrow}{\begin{vmatrix} -5 & 20 \\ 3 & -2 \end{vmatrix}}} = \underset{k}{\underset{\uparrow}{5}} \; \underset{\det(B)}{\underset{\uparrow}{\begin{vmatrix} -1 & 4 \\ 3 & -2 \end{vmatrix}}} = 5(2 - 12) = -50.$$

1. $M_1(\tfrac{1}{5})$.

The proofs of P1 and P3 are given at the end of the section, whereas the proof of P2 is left as an Exercise.

We now illustrate how the foregoing properties together with Theorem 4.2.1 can be used to evaluate a determinant. The basic idea is the same as that for Gaussian elimination. We use ERO to reduce the determinant to upper triangular form and then use Theorem 4.2.1 to evaluate the resulting determinant.

WARNING   When using the properties just given to simplify a determinant, you must remember to take account of any change that arises in the value of the determinant from the operations that you have performed on it.

Example 4.2.2   Evaluate $\begin{vmatrix} 2 & -1 & 3 & 7 \\ 1 & -2 & 4 & 3 \\ 3 & 4 & 2 & -1 \\ 2 & -2 & 8 & -4 \end{vmatrix}$.

Solution

$$\begin{vmatrix} 2 & -1 & 3 & 7 \\ 1 & -2 & 4 & 3 \\ 3 & 4 & 2 & -1 \\ 2 & -2 & 8 & -4 \end{vmatrix} \overset{1}{\underset{= -2}{}} \begin{vmatrix} 1 & -2 & 4 & 3 \\ 2 & -1 & 3 & 7 \\ 3 & 4 & 2 & -1 \\ 1 & -1 & 4 & -2 \end{vmatrix}$$

$$\overset{2}{\underset{= -2}{}} \begin{vmatrix} 1 & -2 & 4 & 3 \\ 0 & 3 & -5 & 1 \\ 0 & 10 & -10 & -10 \\ 0 & 1 & 0 & -5 \end{vmatrix} \overset{3}{\underset{= 20}{}} \begin{vmatrix} 1 & -2 & 4 & 3 \\ 0 & 1 & 0 & -5 \\ 0 & 1 & -1 & -1 \\ 0 & 3 & -5 & 1 \end{vmatrix}$$

$$\overset{4}{\underset{= 20}{}} \begin{vmatrix} 1 & -2 & 4 & 3 \\ 0 & 1 & 0 & -5 \\ 0 & 0 & -1 & 4 \\ 0 & 0 & -5 & 16 \end{vmatrix} \overset{5}{\underset{= 20}{}} \begin{vmatrix} 1 & -2 & 4 & 3 \\ 0 & 1 & 0 & -5 \\ 0 & 0 & -1 & 4 \\ 0 & 0 & 0 & -4 \end{vmatrix} = 80.$$

---

1. P_{12}, $M_4(\frac{1}{2})$ **2.** $A_{12}(-1)$, $A_{13}(-3)$, $A_{14}(-1)$ **3.** P_{24}, $M_3(-\frac{1}{10})$
4. $A_{23}(-1)$, $A_{24}(-3)$ **5.** $A_{34}(-5)$.

THEORETICAL RESULTS FOR $n \times n$ MATRICES AND $n \times n$ LINEAR SYSTEMS

In Chapter 3, we established that an $n \times n$ matrix A was nonsingular if and only if rank$(A) = n$. We now derive a second criterion, based on the determinant of A, for deciding whether A is nonsingular.

Theorem 4.2.2:   An $n \times n$ matrix A is nonsingular if and only if det$(A) \neq 0$.

PROOF   It follows from Theorem 3.6.3 that A is nonsingular if and only if rank$(A) = n$. Thus, if we let A^* denote the RREF of A, then A is nonsingular if and only if det$(A^*) = 1$. But, A^* is obtained from A by performing a sequence of ERO, and so, from properties P1–P3 of determinants, det(A) is just a *nonzero* multiple of det(A^*). That is,

$$\det(A) = k \det(A^*),$$

where k is a *nonzero* constant. Consequently, A is nonsingular if and only if $\det(A) \neq 0$. ∎

Corollary 4.2.1:   The $n \times n$ system of equations $A\mathbf{x} = \mathbf{b}$ has a unique solution if and only if

$$\det(A) \neq 0.$$

PROOF   From Theorem 3.6.2 and Corollary 3.6.1, an $n \times n$ system $A\mathbf{x} = \mathbf{b}$ has a unique solution if and only if A^{-1} exists, that is, from Theorem 4.2.2, if and only if $\det(A) \neq 0$. ∎

If an $n \times n$ system of equations is *nonhomogeneous and $\det(A) = 0$,* then it follows from the results of Chapter 3 and the previous theorem that the system either has an infinite number of solutions or is inconsistent, although we could not tell which without reducing the augmented matrix. However, if the system is *homogeneous*, then it is necessarily consistent, and hence, $\det(A)$ completely characterizes the solution properties of the system.

Corollary 4.2.2:   An $n \times n$ *homogeneous* system of linear equations with coefficient matrix A has an infinite number of solutions if and only if $\det(A) = 0$, and therefore has just the trivial solution if and only if $\det(A) \neq 0$.

PROOF   From Theorem 4.2.2, the system has a unique solution if and only if $\det(A) \neq 0$, and hence, since a homogeneous system is always consistent, there are an infinite number of solutions if and only if $\det(A) = 0$. ∎

REMARK   The preceding corollary is *very* important, since we are often interested only in determining the solution properties of a homogeneous linear system and not actually in finding the solutions themselves. We will refer back to this corollary on many occasions during the remainder of the text.

Example 4.2.3   Verify that the matrix $A = \begin{bmatrix} 1 & -1 & 3 \\ 2 & 4 & -2 \\ 3 & 5 & 7 \end{bmatrix}$ is nonsingular. What can be concluded about the solution to $A\mathbf{x} = \mathbf{0}$?

Solution   It is easily shown that $\det(A) = 52$. Consequently A is nonsingular. It follows from Corollary 4.2.2 that the homogeneous system $A\mathbf{x} = \mathbf{0}$ has only the trivial solution $(0, 0, 0)$.

FURTHER PROPERTIES OF DETERMINANTS

In addition to ERO, the following properties can also be useful in evaluating a determinant.

Let A and B be $n \times n$ matrices.

P4. If A is an $n \times n$ matrix then $\det(A^T) = \det(A)$.

P5. Let $\mathbf{a}_1, \mathbf{a}_2, \ldots, \mathbf{a}_n$ denote the row vectors of A. If the ith row vector of A is the sum of two row vectors, say $\mathbf{a}_i = \mathbf{b}_i + \mathbf{c}_i$, then $\det(A) = \det(B) + \det(C)$, where

$$
B = \begin{bmatrix} \mathbf{a}_1 \\ \vdots \\ \mathbf{a}_{i-1} \\ \mathbf{b}_i \\ \mathbf{a}_{i+1} \\ \vdots \\ \mathbf{a}_n \end{bmatrix}
\quad \text{and} \quad
C = \begin{bmatrix} \mathbf{a}_1 \\ \vdots \\ \mathbf{a}_{i-1} \\ \mathbf{c}_i \\ \mathbf{a}_{i+1} \\ \vdots \\ \mathbf{a}_n \end{bmatrix}.
$$

The corresponding property is also true for columns.

P6. If A has a row (or column) of zeros, then $\det(A) = 0$.

P7. If two rows (or columns) of A are the same, then $\det(A) = 0$.

P8. $\det(AB) = \det(A)\det(B)$.

The proofs of P4, P5, P7, and P8 are given at the end of the section. Property P6 follows directly from the definition of a determinant.

The main importance of P4 is the implication that any results that hold for the rows of a determinant also hold for columns. In particular, "elementary column operations" can also be used to reduce a determinant to upper triangular form. Once more we must be careful to take account of any changes in the value of the determinant that result from the applied operations. We will use the notation

$$\mathrm{CP}_{ij}, \ \mathrm{CM}(k), \ \text{and} \ \mathrm{CA}_{ij}(k)$$

to denote the three types of elementary column operations.

Example 4.2.4 Evaluate
$$
\begin{vmatrix} 6 & 10 & 3 & 4 \\ 3 & 6 & -1 & 2 \\ 9 & 20 & 5 & 4 \\ 15 & 34 & 3 & 8 \end{vmatrix}.
$$

Solution

$$
\begin{vmatrix} 6 & 10 & 3 & 4 \\ 3 & 6 & 1 & 2 \\ 9 & 20 & 5 & 4 \\ 15 & 34 & 3 & 8 \end{vmatrix}
\overset{1}{=} 3 \cdot 2 \cdot 2
\begin{vmatrix} 2 & 5 & 3 & 2 \\ 1 & 3 & -1 & 1 \\ 3 & 10 & 5 & 2 \\ 5 & 17 & 3 & 4 \end{vmatrix}
\overset{2}{=} -12
\begin{vmatrix} 1 & 3 & -1 & 1 \\ 2 & 5 & 3 & 2 \\ 3 & 10 & 5 & 2 \\ 5 & 17 & 3 & 4 \end{vmatrix}
$$

$$
\overset{3}{=} -12
\begin{vmatrix} 1 & 3 & -1 & 1 \\ 0 & -1 & 5 & 0 \\ 0 & 1 & 8 & -1 \\ 0 & 2 & 8 & -1 \end{vmatrix}
\overset{4}{=} -12
\begin{vmatrix} 1 & 3 & -1 & 1 \\ 0 & -1 & 5 & 0 \\ 0 & 0 & 13 & -1 \\ 0 & 0 & 18 & -1 \end{vmatrix}
\overset{5}{=} -12
\begin{vmatrix} 1 & 3 & 17 & 1 \\ 0 & -1 & 5 & 0 \\ 0 & 0 & -5 & -1 \\ 0 & 0 & 0 & -1 \end{vmatrix}
$$

$= -12\,(1)(-1)(-5)(-1) = 60.$

---

1. $CM_1(\tfrac{1}{3})$, $CM_2(\tfrac{1}{2})$, $CM_4(\tfrac{1}{2})$ 2. P_{12} 3. $A_{12}(-2)$, $A_{13}(-3)$, $A_{14}(-5)$
4. $A_{23}(1)$, $A_{24}(2)$ 5. $CA_{43}(18)$.

The property that often gives the most difficulty is P5. We explicitly illustrate its use with an example.

Example 4.2.5   Use property P5 to express

$$\begin{vmatrix} a_1 + b_1 & c_1 + d_1 \\ a_2 + b_2 & c_2 + d_2 \end{vmatrix}$$

as a sum of four determinants.

Solution   Applying P5 to row 1 yields:

$$\begin{vmatrix} a_1 + b_1 & c_1 + d_1 \\ a_2 + b_2 & c_2 + d_2 \end{vmatrix} = \begin{vmatrix} a_1 & c_1 \\ a_2 + b_2 & c_2 + d_2 \end{vmatrix} + \begin{vmatrix} b_1 & d_1 \\ a_2 + b_2 & c_2 + d_2 \end{vmatrix}.$$

We now apply P5 to row 2 in both of the determinants on the right-hand side to obtain

$$\begin{vmatrix} a_1 + b_1 & c_1 + d_1 \\ a_2 + b_2 & c_2 + d_2 \end{vmatrix} = \begin{vmatrix} a_1 & c_1 \\ a_2 & c_2 \end{vmatrix} + \begin{vmatrix} a_1 & c_1 \\ b_2 & d_2 \end{vmatrix} + \begin{vmatrix} b_1 & d_1 \\ a_2 & c_2 \end{vmatrix} + \begin{vmatrix} b_1 & d_1 \\ b_2 & d_2 \end{vmatrix}.$$

Notice that we could also have applied P5 to the columns of the given determinant.

Example 4.2.6   Evaluate

(a) $\begin{vmatrix} 1 & 2 & -3 & 1 \\ -2 & 4 & 6 & 2 \\ -3 & -6 & 9 & 3 \\ 2 & 11 & -6 & 4 \end{vmatrix}.$ (b) $\begin{vmatrix} 2-4x & -4 & 2 \\ 5+3x & 3 & -3 \\ 1-2x & -2 & 1 \end{vmatrix}.$

Solution

(a) $\begin{vmatrix} 1 & 2 & -3 & 1 \\ -2 & 4 & 6 & 2 \\ -3 & -6 & 9 & 3 \\ 2 & 11 & -6 & 4 \end{vmatrix} = -3 \begin{vmatrix} 1 & 2 & 1 & 1 \\ -2 & 4 & -2 & 2 \\ -3 & -6 & -3 & 3 \\ 2 & 11 & 2 & 4 \end{vmatrix} = 0,$

since $C_1 = C_3$.

(b) $\begin{vmatrix} 2-4x & -4 & 2 \\ 5+3x & 3 & -3 \\ 1-2x & -2 & 1 \end{vmatrix} \underset{\substack{\uparrow \\ \text{P5 applied to column 1}}}{=} \begin{vmatrix} 2 & -4 & 2 \\ 5 & 3 & -3 \\ 1 & -2 & 1 \end{vmatrix} + \begin{vmatrix} -4x & -4 & 2 \\ 3x & 3 & -3 \\ -2x & -2 & 1 \end{vmatrix}$

$$= 2 \begin{vmatrix} 1 & -2 & 1 \\ 5 & 3 & -3 \\ 1 & -2 & 1 \end{vmatrix} + x \begin{vmatrix} -4 & -4 & 2 \\ 3 & 3 & -3 \\ -2 & -2 & 1 \end{vmatrix}$$

$$= 0,$$

since $R_1 = R_3$ in the first determinant and $C_1 = C_2$ in the second determinant.

Example 4.2.7   If $A = \begin{bmatrix} \sin\phi & \cos\phi \\ -\cos\phi & \sin\phi \end{bmatrix}$ and $B = \begin{bmatrix} \cos\theta & -\sin\theta \\ \sin\theta & \cos\theta \end{bmatrix}$, show that $\det(AB) = 1$.

Solution   Using P8,

$$\det(AB) = \det(A)\det(B) = (\sin^2\phi + \cos^2\phi)(\cos^2\theta + \sin^2\theta) = 1.$$

Example 4.2.8   Find all x satisfying $\begin{vmatrix} x^2 & x & 1 \\ 1 & 1 & 1 \\ 4 & 2 & 1 \end{vmatrix} = 0.$

Solution   If we expanded the determinant, then we would have a quadratic equation in x. Thus, there are at most two distinct values of x that satisfy the equation. By inspection, the determinant vanishes when $x = 1$ (since then $R_1 = R_2$) and when $x = 2$ (since then $R_1 = R_3$). Consequently, the two values of x satisfying the given equation are $x = 1, 2$.

PROOFS OF THE PROPERTIES OF DETERMINANTS

We now prove several of the properties of determinants.

PROOF OF P1   Let B be the matrix obtained by interchanging row r with row s in A. Then the elements of B are related to those of A as follows

$$b_{ij} = \begin{cases} a_{ij} & \text{if } i \neq r, s, \\ a_{sj} & \text{if } i = r, \\ a_{rj} & \text{if } i = s. \end{cases}$$

Thus, from Definition 4.1.1,

$$\det(B) = \sum \sigma(p_1, p_2, \ldots, p_n)\, b_{1p_1} b_{2p_2} \cdots b_{np_n}$$

$$= \sum \sigma(p_1, p_2, \ldots, p_r, \ldots, p_s, \ldots, p_n)\, a_{1p_1} a_{2p_2} \cdots a_{sp_r} \cdots a_{rp_s} \cdots a_{np_n}.$$

Interchanging p_r and p_s in $\sigma(p_1, p_2, \ldots, p_r, \ldots, p_s, \ldots, p_n)$ and recalling from Theorem 4.1.2 that such an interchange has the effect of changing the parity, we obtain

$$\det(B) = -\sum \sigma(p_1, p_2, \ldots, p_s, \ldots, p_r, \ldots, p_n)\, a_{1p_1} a_{2p_2} \cdots a_{rp_s} \cdots a_{sp_r} \cdots a_{np_n}$$

where we have also rearranged the terms so that the row suffixes are in their natural increasing order. The sum on the right-hand side of this equation is just $\det(A)$, so that

$$\det(B) = -\det(A).\qquad\blacksquare$$

We prove properties P5 and P7 next, since they simplify the proof of P3.

PROOF OF P5   The elements of A are

$$a_{kj} = \begin{cases} a_{kj}, & \text{if } k \neq i, \\ b_{ij} + c_{ij}, & \text{if } k = i. \end{cases}$$

Thus, from Definition 4.1.1,

$$\det(A) = \sum \sigma(p_1, p_2, \ldots, p_n)\, a_{1p_1} a_{2p_2} \cdots a_{np_n}$$

$$= \sum \sigma(p_1, p_2, \ldots, p_n)\, a_{1p_1} a_{2p_2} \cdots a_{i-1p_{i-1}}(b_{ip_i} + c_{ip_i})\, a_{i+1p_{i+1}} \cdots a_{np_n}$$

$$= \sum \sigma(p_1, p_2, \ldots, p_n)\, a_{1p_1} a_{2p_2} \cdots a_{i-1p_{i-1}} b_{ip_i}\, a_{i+1p_{i+1}} \cdots a_{np_n}$$

$$+ \sum \sigma(p_1, p_2, \ldots, p_n) a_{1p_1} a_{2p_2} \cdots a_{i-1p_{i-1}} c_{ip_i}\, a_{i+1p_{i+1}} \cdots a_{np_n}$$

$$= \det(B) + \det(C).\qquad\blacksquare$$

PROOF OF P7   Suppose rows i and j in A are the same. Then if we interchange these rows, the matrix, and hence its determinant, are unaltered. However, according to P1, the determinant of the resulting matrix is $-\det(A)$. Therefore,

$$\det(A) = -\det(A),$$

which implies that

$$\det(A) = 0.\qquad\blacksquare$$

PROOF OF P3   Let $A = [\mathbf{a}_1, \mathbf{a}_2, \ldots, \mathbf{a}_n]^T$, and let B be the matrix obtained from A when k times row j of A is added to row i of A. Then, $B = [\mathbf{a}_1, \mathbf{a}_2, \ldots, \mathbf{a}_i + k\mathbf{a}_j, \ldots, \mathbf{a}_n]^T$ so that, using P5,

$$\det(B) = \det([\mathbf{a}_1, \mathbf{a}_2, \ldots, \mathbf{a}_i + k\mathbf{a}_j, \ldots, \mathbf{a}_n]^T)$$

$$= \det([\mathbf{a}_1, \mathbf{a}_2, \ldots \mathbf{a}_n]^T) + \det([\mathbf{a}_1, \mathbf{a}_2, \ldots, k\mathbf{a}_j, \ldots \mathbf{a}_n]^T)$$

By P2 we can factor out k from row i of the second determinant on the right-hand side. If we do this, it follows that row i and row j of the resulting determinant are the same, and so, from P7, the value of the second determinant is zero. Thus,

$$\det(B) = \det([\mathbf{a}_1, \mathbf{a}_2, \ldots, \mathbf{a}_n]^T) = \det(A)$$

as required. $\blacksquare$

PROOF OF P4   Using Definition 4.1.1, we have

$$\det(A^T) = \sum \sigma(p_1, p_2, \ldots, p_n) a_{p_1 1} a_{p_2 2}\, a_{p_3 3} \cdots a_{p_n n}. \qquad (4.2.3)$$

Since $(p_1, p_2, \ldots, p_n)$ is a permutation of $1, 2, \ldots, n$, it follows that, by rearranging terms,

$$a_{p_1 1} a_{p_2 2} a_{p_3 3} \cdots a_{p_n n} = a_{1q_1} a_{2q_2} a_{3q_3} \cdots a_{nq_n}, \qquad (4.2.4)$$

for an appropriate permutation $(q_1, q_2, \ldots, q_n)$. Furthermore,

$$N(p_1, p_2, ..., p_n) = \text{\# of interchanges in changing } (1, 2, ..., n) \text{ to } (p_1, p_2, ..., p_n)$$

$$= \text{\# of interchanges in changing } (p_1, p_2, ..., p_n) \text{ to } (1, 2, ..., n)$$

$$= \text{\# of interchanges in changing from } (1, 2, ..., n) \text{ to } (q_1, q_2, ..., q_n)$$

$$= N(q_1, q_2, ..., q_n).$$

Thus,

$$\sigma(p_1, p_2, ..., p_n) = \sigma(q_1, q_2, ..., q_n). \tag{4.2.5}$$

Substituting from equations (4.2.4) and (4.2.5) into equation (4.2.3), we have:

$$\det(A^T) = \sum \sigma(q_1, q_2, ..., q_n)\, a_{1q_1} a_{2q_2}\, a_{3q_3} \cdots a_{nq_n}$$

$$= \det(A). \qquad\blacksquare$$

PROOF OF P8   Let E denote an elementary matrix. We leave it as an exercise to verify that

$$\det(E) = \begin{cases} -1 & \text{if } E \text{ permutes rows,} \\ +1 & \text{if } E \text{ adds a multiple of one row to another row,} \\ k & \text{if } E \text{ scales a row by } k. \end{cases}$$

It then follows from properties P1–P3 that in each case

$$\det(EA) = \det(E)\det(A). \tag{4.2.6}$$

Now consider a general product AB. We need to distinguish two cases.

(1) If A is singular, then, from Corollary 3.6.2, so is AB. Consequently, applying Theorem 4.2.2,

$$\det(AB) = 0 = \det(A)\det(B).$$

(2) If A is nonsingular, then from Section 3.7, we know that it can be expressed as the product of elementary matrices, say, $A = E_1 E_2 \cdots E_r$. Hence, repeatedly applying (4.2.6) gives

$$\det(AB) = \det(E_1 E_2 \cdots E_r B) = \det(E_1)\det(E_2 \cdots E_r B)$$

$$= \det(E_1)\det(E_2) \cdots \det(E_r)\det(B)$$

$$= \det(E_1 E_2 \cdots E_r)\det(B) = \det(A)\det(B). \qquad\blacksquare$$

EXERCISES 4.2

For problems 1–12, reduce the given determinant to upper triangular form and then evaluate.

1. $\begin{vmatrix} 1 & 2 & 3 \\ 2 & 6 & 4 \\ 3 & -5 & 2 \end{vmatrix}$.

2. $\begin{vmatrix} 2 & -1 & 4 \\ 3 & 2 & 1 \\ -2 & 1 & 4 \end{vmatrix}$.

3. $\begin{vmatrix} 2 & 1 & 3 \\ -1 & 2 & 6 \\ 4 & 1 & 12 \end{vmatrix}$.

4. $\begin{vmatrix} 0 & 1 & -2 \\ -1 & 0 & 3 \\ 2 & -3 & 0 \end{vmatrix}$.

5. $\begin{vmatrix} 3 & 7 & 1 \\ 5 & 9 & -6 \\ 2 & 1 & 3 \end{vmatrix}$.

6. $\begin{vmatrix} 1 & -1 & 2 & 4 \\ 3 & 1 & 2 & 4 \\ -1 & 1 & 3 & 2 \\ 2 & 1 & 4 & 2 \end{vmatrix}$.

7. $\begin{vmatrix} 2 & 32 & 1 & 4 \\ 26 & 104 & 26 & -13 \\ 2 & 56 & 2 & 7 \\ 1 & 40 & 1 & 5 \end{vmatrix}$.

8. $\begin{vmatrix} 0 & 1 & -1 & 1 \\ -1 & 0 & 1 & 1 \\ 1 & -1 & 0 & 1 \\ -1 & -1 & -1 & 0 \end{vmatrix}$.

9. $\begin{vmatrix} 2 & 1 & 3 & 5 \\ 3 & 0 & 1 & 2 \\ 4 & 1 & 4 & 3 \\ 5 & 2 & 5 & 3 \end{vmatrix}$.

10. $\begin{vmatrix} 2 & -1 & 3 & 4 \\ 7 & 1 & 2 & 3 \\ -2 & 4 & 8 & 6 \\ 6 & -6 & 18 & -24 \end{vmatrix}$.

11. $\begin{vmatrix} 7 & -1 & 3 & 4 \\ 14 & 2 & 4 & 6 \\ 21 & 1 & 3 & 4 \\ -7 & 4 & 5 & 8 \end{vmatrix}$.

12. $\begin{vmatrix} 3 & 7 & 1 & 2 & 3 \\ 1 & 1 & -1 & 0 & 1 \\ 4 & 8 & -1 & 6 & 6 \\ 3 & 7 & 0 & 9 & 4 \\ 8 & 16 & -1 & 8 & 12 \end{vmatrix}$.

For problems 13–19, use Theorem 4.2.2 to determine whether the given matrix is singular or nonsingular.

13. $\begin{bmatrix} 2 & 1 \\ 3 & 2 \end{bmatrix}$.

14. $\begin{bmatrix} -1 & 1 \\ 1 & -1 \end{bmatrix}$.

15. $\begin{bmatrix} 2 & 6 & -1 \\ 3 & 5 & 1 \\ 2 & 0 & 1 \end{bmatrix}$.

16. $\begin{bmatrix} -1 & 2 & 3 \\ 5 & -2 & 1 \\ 8 & -2 & 5 \end{bmatrix}$.

17. $\begin{bmatrix} 1 & 0 & 2 & -1 \\ 3 & -2 & 1 & 4 \\ 2 & 1 & 6 & 2 \\ 1 & -3 & 4 & 0 \end{bmatrix}$.

18. $\begin{bmatrix} 1 & 1 & 1 & 1 \\ -1 & 1 & -1 & 1 \\ 1 & 1 & -1 & -1 \\ -1 & 1 & 1 & -1 \end{bmatrix}$.

19. $\begin{bmatrix} 1 & 2 & -3 & 5 \\ -1 & 2 & -3 & 6 \\ 2 & 3 & -1 & 4 \\ 1 & -2 & 3 & -6 \end{bmatrix}$.

20. Determine all values of the constant k for which the given system has a unique solution

$$x_1 + kx_2 = b_1,$$
$$kx_1 + 4x_2 = b_2.$$

21. Determine all values of the constant k for which the given system has an infinite number of solutions.

$$x_1 + 2x_2 + kx_3 = 0,$$
$$2x_1 - kx_2 + x_3 = 0,$$
$$3x_1 + 6x_2 + x_3 = 0.$$

22. Determine all values of k for which the system

$$x_1 + 2x_2 + x_3 = kx_1,$$
$$2x_1 + x_2 + x_3 = kx_2,$$
$$x_1 + x_2 + 2x_3 = kx_3,$$

has an infinite number of solutions.

23. If $A = \begin{bmatrix} 1 & -1 & 2 \\ 3 & 1 & 4 \\ 0 & 1 & 3 \end{bmatrix}$, find $\det(A)$, and use

properties of determinants to find $\det(A^T)$, $\det(-2A)$.

24. If $A = \begin{bmatrix} 1 & -1 \\ 2 & 3 \end{bmatrix}$ and $B = \begin{bmatrix} 1 & 2 \\ -2 & 4 \end{bmatrix}$,

evaluate $\det(AB)$ and verify P8.

25. If $A = \begin{bmatrix} \cosh x & \sinh x \\ \sinh x & \cosh x \end{bmatrix}$ and

$B = \begin{bmatrix} \cosh y & \sinh y \\ \sinh y & \cosh y \end{bmatrix}$, evaluate $\det(AB)$.

For problems 26–28, use properties of determinants to show that $\det(A) = 0$ for the given matrix A.

26. $A = \begin{bmatrix} 3 & 2 & 1 \\ 6 & 4 & -1 \\ 9 & 6 & 2 \end{bmatrix}$.

27. $A = \begin{bmatrix} 1 & -3 & 1 \\ 2 & -1 & 7 \\ 3 & 1 & 13 \end{bmatrix}$.

28. $A = \begin{bmatrix} 1 + 3a & 1 & 3 \\ 1 + 2a & 1 & 2 \\ 2 & 2 & 0 \end{bmatrix}$.

29. Without expanding the determinant, find all values of x for which $\det(A) = 0$ if

$$A = \begin{bmatrix} 1 & -1 & x \\ 2 & 1 & x^2 \\ 4 & -1 & x^3 \end{bmatrix}.$$

30. Use *only* properties P5, P1, and P2 to show that

$$\begin{vmatrix} \alpha x - \beta y & \beta x - \alpha y \\ \beta x + \alpha y & \alpha x + \beta y \end{vmatrix} = (x^2 + y^2) \begin{vmatrix} \alpha & \beta \\ \beta & \alpha \end{vmatrix}.$$

31. Use *only* properties P5, P1, and P2 to find the value of $\alpha\beta\gamma$ such that

$$\begin{vmatrix} a_1 + \beta b_1 & b_1 + \gamma c_1 & c_1 + \alpha a_1 \\ a_2 + \beta b_2 & b_2 + \gamma c_2 & c_2 + \alpha a_2 \\ a_3 + \beta b_3 & b_3 + \gamma c_3 & c_3 + \alpha a_3 \end{vmatrix} = 0$$

for all values of a_i, b_i, c_i.

32. If A is a nonsingular $n \times n$ matrix, prove that
$$\det(A^{-1}) = \frac{1}{\det(A)}.$$

33. An $n \times n$ matrix A that satisfies $A^T = A^{-1}$ is called an **orthogonal matrix**. Show that if A is an orthogonal matrix, then $\det(A) = \pm 1$.

34. (a) Use the definition of a determinant to prove that if A is an $n \times n$ lower triangular matrix then

$$\det(A) = a_{11}a_{22}a_{33} \cdots a_{nn} = \prod_{i=1}^{n} a_{ii}.$$

(b) Evaluate the following determinant by first reducing it to lower triangular form and then using the result from (a):

$$\begin{vmatrix} 2 & -1 & 3 & 5 \\ 1 & 2 & 2 & 1 \\ 3 & 0 & 1 & 4 \\ 1 & 2 & 0 & 1 \end{vmatrix}.$$

35. If A and S are $n \times n$ matrices with S nonsingular, show that $\det(S^{-1}AS) = \det(A)$.

36. If $\det(A^3) = 0$ is it possible for A to be nonsingular? Justify your answer.

37. Prove property P2 of determinants.

38. Show that $\begin{vmatrix} x & y & 1 \\ x_1 & y_1 & 1 \\ x_2 & y_2 & 1 \end{vmatrix} = 0$ represents

the equation of the straight line through the distinct points (x_1, y_1) and (x_2, y_2),

39. Without expanding the determinant, show that

$$\begin{vmatrix} 1 & x & x^2 \\ 1 & y & y^2 \\ 1 & z & z^2 \end{vmatrix} = (y - z)(z - x)(x - y).$$

40. If A is an $n \times n$ *skew-symmetric* matrix and n is odd, prove that $\det(A) = 0$.

41. Let $A = [\mathbf{a}_1, \mathbf{a}_2, \ldots, \mathbf{a}_n]$ be an $n \times n$ matrix, and let $\mathbf{b} = c_1 \mathbf{a}_1 + c_2 \mathbf{a}_2 + \cdots + c_n \mathbf{a}_n$, where $c_1, c_2, \ldots, c_n$ are constants. If B_k denotes the matrix obtained from A by replacing the kth column vector by $\mathbf{b}$, prove that

$$\det(B_k) = c_k \det(A), \quad k = 1, 2, \ldots, n.$$

◆ **42.** Let A be the general 4×4 matrix.

(a) Verify property P1 of determinants in the case when the first two rows of A are permuted.

(b) Verify property P2 of determinants in the case when row 1 of A is divided by k.

(c) Verify property P3 of determinants in the case when k times row 2 is added to row 1.

◆ **43.** For a randomly generated 5×5 matrix verify that $\det(A^T) = \det(A)$.

◆ **44.** Determine all values of a for which

$$\begin{bmatrix} 1 & 2 & 3 & 4 & a \\ 2 & 1 & 2 & 3 & 4 \\ 3 & 2 & 1 & 2 & 3 \\ 4 & 3 & 2 & 1 & 2 \\ a & 4 & 3 & 2 & 1 \end{bmatrix}$$

is nonsingular.

◆ **45.** If $A = \begin{bmatrix} 1 & 4 & 1 \\ 3 & 2 & 1 \\ 3 & 4 & -1 \end{bmatrix}$, determine all values of the constant k for which the linear system $(A - kI_3)\mathbf{x} = \mathbf{0}$ has an infinite number of solutions, and find the corresponding solutions.

◆ **46.** Use the determinant to show that

$$A = \begin{bmatrix} 1 & 2 & 3 & 4 \\ 2 & 1 & 2 & 3 \\ 3 & 2 & 1 & 2 \\ 4 & 3 & 2 & 1 \end{bmatrix}$$

is nonsingular, and use A^{-1} to solve $A\mathbf{x} = \mathbf{b}$ if $\mathbf{b} = [3, 7, 1, -4]^T$.

4.3   COFACTOR EXPANSIONS

We now obtain an alternative method for evaluating a determinant. The basic idea is that we can reduce a determinant of order n to a sum of determinants of order $n - 1$. Continuing in this manner it is possible to express any determinant as the sum of $(n!)/2$ determinants of order 2. This method is the one most frequently used to evaluate a determinant by hand, although the triangulation procedure introduced in the previous section involves less work in general. When A is nonsingular, the technique we derive leads to formulas for both A^{-1} and the unique solution to $A\mathbf{x} = \mathbf{b}$. We first require two preliminary definitions.

Definition 4.3.1:   Let A be an $n \times n$ matrix. The **minor**, M_{ij}, of the element a_{ij}, is the determinant of the matrix obtained by deleting the ith row vector and jth column vector of A.

REMARK: Notice that M_{ij} is a determinant of order $n - 1$ if A is an $n \times n$ matrix.

Example 4.3.1   If $A = \begin{bmatrix} a_{11} & a_{12} & a_{13} \\ a_{21} & a_{22} & a_{23} \\ a_{31} & a_{32} & a_{33} \end{bmatrix}$, then, for example, $M_{23} = \begin{vmatrix} a_{11} & a_{12} \\ a_{31} & a_{32} \end{vmatrix}$.

Example 4.3.2   Determine the minors $M_{11}, M_{23},$ and M_{31} for

$$A = \begin{bmatrix} 2 & 1 & 3 \\ -1 & 4 & -2 \\ 3 & 1 & 5 \end{bmatrix}.$$

Solution   Using Definition 4.3.1, we have:

$$M_{11} = \begin{vmatrix} 4 & -2 \\ 1 & 5 \end{vmatrix} = 22, \quad M_{23} = \begin{vmatrix} 2 & 1 \\ 3 & 1 \end{vmatrix} = -1, \quad M_{31} = \begin{vmatrix} 1 & 3 \\ 4 & -2 \end{vmatrix} = -14.$$

Definition 4.3.2:   Let A be an $n \times n$ matrix. The **cofactor**, C_{ij}, of the element a_{ij}, is defined by

$$C_{ij} = (-1)^{i+j} M_{ij},$$

where M_{ij} is the minor of a_{ij}.

The appropriate sign in the cofactor C_{ij} is easy to remember, since it alternates in the following manner:

$$\begin{vmatrix} + & - & + & - & + & \cdots \\ - & + & - & + & - & \cdots \\ + & - & + & - & + & \cdots \\ \vdots & \vdots & \vdots & \vdots & \vdots & \end{vmatrix}.$$

Example 4.3.3   Determine the cofactors C_{11}, C_{23}, C_{31} of the matrix

$$A = \begin{bmatrix} 2 & 1 & 3 \\ -1 & 4 & -2 \\ 3 & 1 & 5 \end{bmatrix}.$$

Solution   We have already obtained the minors of A in the previous example. It follows that:

$$C_{11} = +M_{11} = 22, \qquad C_{23} = -M_{23} = 1, \qquad C_{31} = +M_{31} = -14.$$

Example 4.3.4   If $A = \begin{bmatrix} a_{11} & a_{12} \\ a_{21} & a_{22} \end{bmatrix}$, verify that $\det(A) = a_{11}C_{11} + a_{12}C_{12}$.

Solution   In this case,

$$C_{11} = +|\, a_{22}\,| = a_{22}, \qquad C_{12} = -|\, a_{21}\,| = -a_{21},$$

so that

$$a_{11}C_{11} + a_{12}C_{12} = a_{11}a_{22} - a_{12}a_{21} = \det(A). \qquad \square$$

The preceding example is a special case of the following important theorem.

Theorem 4.3.1 (Cofactor Expansion Theorem): Let A be an $n \times n$ matrix. If we multiply the elements in any row (or column) of A by their cofactors, then the sum of the resulting products is $\det(A)$. Thus,

1. If we expand along row i,

$$\det(A) = a_{i1} C_{i1} + a_{i2} C_{i2} + \cdots + a_{in} C_{in} = \sum_{k=1}^{n} a_{ik} C_{ik}.$$

2. If we expand along column j,

$$\det(A) = a_{1j} C_{1j} + a_{2j} C_{2j} + \cdots + a_{nj} C_{nj} = \sum_{k=1}^{n} a_{kj} C_{kj}.$$

PROOF It follows from the definition of a determinant that $\det(A)$ can be written in the form

$$\det(A) = a_{i1} \hat{C}_{i1} + a_{i2} \hat{C}_{i2} + \cdots + a_{in} \hat{C}_{in} \qquad (4.3.1)$$

where the coefficients $\hat{C}_{ij}$ contain no elements from row i or column j. We must show that

$$\hat{C}_{ij} = C_{ij}$$

where C_{ij} is the cofactor of a_{ij}.

Consider first a_{11}. From Definition 4.2.1 the terms in $\det(A)$ that contain a_{11} are given by

$$a_{11} \sum \sigma(1, p_2, p_3, \ldots, p_n)\, a_{2p_2} a_{3p_3} \cdots a_{np_n},$$

where the summation is over the $(n-1)!$ distinct permutations of $2, 3, \ldots, n$. Thus,

$$\hat{C}_{11} = \sum \sigma(1, p_2, p_3, \ldots, p_n)\, a_{2p_2} a_{3p_3} \cdots a_{np_n}.$$

However, this summation is just the minor M_{11}, and since $C_{11} = M_{11}$, we have shown that the coefficient of a_{11} in $\det(A)$ is indeed the cofactor C_{11}. Now consider the element a_{ij}. By successively interchanging adjacent rows and columns in A, we can move a_{ij} into the (11) position *without altering the relative positions of the other rows and columns of A*. We let A' denote the resulting matrix. Obtaining A' from A requires $i - 1$ row interchanges and $j - 1$ column interchanges. Therefore, the total number of interchanges required to obtain A' from A is $i + j - 2$. Consequently,

$$\det(A) = (-1)^{i+j-2} \det(A') = (-1)^{i+j} \det(A').$$

Now for the key point. The coefficient of a_{ij} in $\det(A)$ must be $(-1)^{i+j}$ times the coefficient of a_{ij} in $\det(A')$. But, a_{ij} occurs in the (11) position of A', and so, as we have previously shown, its coefficient in $\det(A')$ is M'_{11}. Since the relative positions of the remaining rows in A have not altered it follows that $M'_{11} = M_{ij}$, and therefore the coefficient of a_{ij} in $\det(A')$ is M_{ij}. Consequently, the coefficient of a_{ij} in $\det(A)$ is $(-1)^{i+j} M_{ij} = C_{ij}$. Applying this result to the elements $a_{i1}, a_{i2}, \ldots, a_{in}$ and comparing with (4.3.1) yields

$$\hat{C}_{ij} = C_{ij}, \quad j = 1, 2, \ldots, n,$$

which establishes the theorem for expansion along a row. The result for expansion along a column follows directly, since $\det(A^T) = \det(A)$. ∎

Example 4.3.5   Evaluate $\begin{vmatrix} 2 & 3 & 4 \\ 1 & -1 & 1 \\ 6 & 3 & 0 \end{vmatrix}$ using the cofactor expansion theorem along (a) row 1, (b) column 3.

Solution

(a) $\begin{vmatrix} 2 & 3 & 4 \\ 1 & -1 & 1 \\ 6 & 3 & 0 \end{vmatrix} = 2 \begin{vmatrix} -1 & 1 \\ 3 & 0 \end{vmatrix} - 3 \begin{vmatrix} 1 & 1 \\ 6 & 0 \end{vmatrix} + 4 \begin{vmatrix} 1 & -1 \\ 6 & 3 \end{vmatrix}$

$$= -6 + 18 + 36 = 48.$$

(b) $\begin{vmatrix} 2 & 3 & 4 \\ 1 & -1 & 1 \\ 6 & 3 & 0 \end{vmatrix} = 4 \begin{vmatrix} 1 & -1 \\ 6 & 3 \end{vmatrix} - 1 \begin{vmatrix} 2 & 3 \\ 6 & 3 \end{vmatrix} + 0 = 36 + 12 + 0 = 48.$ □

Notice that (b) was easier than (a) in the previous example, because of the zero in column 3. Whenever you use the cofactor expansion method to evaluate a determinant, always choose the row or column carefully so as to minimize the amount of work.

Example 4.3.6   Evaluate $\begin{vmatrix} 0 & 3 & -1 & 0 \\ 5 & 0 & 8 & 2 \\ 7 & 2 & 5 & 4 \\ 6 & 1 & 7 & 0 \end{vmatrix}.$

Solution   In this case it is easiest to use either row 1 or column 4. We choose row 1

$\begin{vmatrix} 0 & 3 & -1 & 0 \\ 5 & 0 & 8 & 2 \\ 7 & 2 & 5 & 4 \\ 6 & 1 & 7 & 0 \end{vmatrix} = -3 \begin{vmatrix} 5 & 8 & 2 \\ 7 & 5 & 4 \\ 6 & 7 & 0 \end{vmatrix} + (-1) \begin{vmatrix} 5 & 0 & 2 \\ 7 & 2 & 4 \\ 6 & 1 & 0 \end{vmatrix}$

$$= -3[2(49 - 30) - 4(35 - 48) + 0] - [5(0 - 4) - 0 + 2(7 - 12)] = -240.$$ □

We now have two computational methods for evaluating determinants, namely, the triangulation technique of the previous section and the cofactor expansion theorem. In evaluating a given determinant by hand, it is usually most efficient (and least error prone) to use a combination of the two techniques. More specifically, we use the properties of determinants to set all except one element in a row or column equal to zero and then use the cofactor expansion theorem on the corresponding row or column. We illustrate with an example.

Example 4.3.7   Evaluate $\begin{vmatrix} 2 & 1 & 8 & 6 \\ 1 & 4 & 1 & 3 \\ -1 & 2 & 1 & 4 \\ 1 & 3 & -1 & 2 \end{vmatrix}.$

Solution

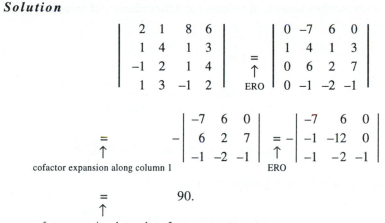

$$\begin{vmatrix} 2 & 1 & 8 & 6 \\ 1 & 4 & 1 & 3 \\ -1 & 2 & 1 & 4 \\ 1 & 3 & -1 & 2 \end{vmatrix} \underset{\underset{\text{ERO}}{\uparrow}}{=} \begin{vmatrix} 0 & -7 & 6 & 0 \\ 1 & 4 & 1 & 3 \\ 0 & 6 & 2 & 7 \\ 0 & -1 & -2 & -1 \end{vmatrix}$$

$$\underset{\underset{\substack{\text{cofactor expansion along column 1}}}{\uparrow}}{=} -\begin{vmatrix} -7 & 6 & 0 \\ 6 & 2 & 7 \\ -1 & -2 & -1 \end{vmatrix} \underset{\underset{\text{ERO}}{\uparrow}}{=} -\begin{vmatrix} -7 & 6 & 0 \\ -1 & -12 & 0 \\ -1 & -2 & -1 \end{vmatrix}$$

$$\underset{\underset{\substack{\text{cofactor expansion along column 3}}}{\uparrow}}{=} \qquad\qquad 90.$$

Example 4.3.8   Determine all k for which the system

$$10x_1 + kx_2 - x_3 = 0,$$
$$kx_1 + x_2 - x_3 = 0,$$
$$2x_1 + x_2 - 3x_3 = 0,$$

has nontrivial solutions.

Solution   The determinant of the matrix of coefficients of the system is

$$\det(A) = \begin{vmatrix} 10 & k & -1 \\ k & 1 & -1 \\ 2 & 1 & -3 \end{vmatrix} = \begin{vmatrix} 10 - 2k & k & -1 + 3k \\ k - 2 & 1 & 2 \\ 0 & 1 & 0 \end{vmatrix}$$

$$= -[2(10 - 2k) - (k - 2)(3k - 1)] = 3(k^2 - k - 6)$$

$$= 3(k - 3)(k + 2).$$

From Corollary 4.2.2, the system has nontrivial solutions if and only if $\det(A) = 0$, that is, if and only if $k = 3$ or $k = -2$.

THE ADJOINT METHOD FOR A^{-1}

We next establish two corollaries to the cofactor expansion theorem that, in the case of a nonsingular matrix A, lead to a method for expressing the elements of A^{-1} in terms of determinants.

Corollary 4.3.1:   If the elements in the ith row (or column) of an $n \times n$ matrix A are multiplied by the cofactors of a different row (or column), then the sum of the resulting products is zero. That is,

1. If we use the elements of row i and the cofactors of row j,

$$\sum_{k=1}^{n} a_{ik}C_{jk} = 0, \quad i \neq j. \tag{4.3.2}$$

2 .   If we use the elements of column i and the cofactors of column j,

$$\sum_{k=1}^{n} a_{ki}C_{kj} = 0, \quad i \neq j. \tag{4.3.3}$$

PROOF   We prove (4.3.2). Adding row i to row j ($i \neq j$) in the matrix A and expanding the determinant of the resulting matrix along row j yields

$$\det(A) = \sum_{k=1}^{n} (a_{jk} + a_{ik})C_{jk} = \sum_{k=1}^{n} a_{jk}C_{jk} + \sum_{k=1}^{n} a_{ik}C_{jk}.$$

That is, using the cofactor expansion theorem,

$$\det(A) = \det(A) + \sum_{k=1}^{n} a_{ik}C_{jk}$$

which implies that

$$\sum_{k=1}^{n} a_{ik}C_{jk} = 0, \quad i \neq j.$$

Equation (4.3.3) can be proved similarly.   ∎

The results of Theorem 4.3.1 and Corollary 4.3.1 can be combined into the following corollary.

Corollary 4.3.2:   Let A be an $n \times n$ matrix. Then,

$$\sum_{k=1}^{n} a_{ik}C_{jk} = \delta_{ij}\det(A), \quad \sum_{k=1}^{n} a_{ki}C_{kj} = \delta_{ij}\det(A), \tag{4.3.4}$$

where δ_{ij} is the Kronecker delta.

The formulas in (4.3.4) should be reminiscent of the index form of the matrix product. Combining this with the fact that the Kronecker delta gives the elements of the identity matrix, we might suspect that (4.3.4) is telling us something about the inverse of A. Before establishing that this suspicion is indeed correct, we need a definition.

Definition 4.3.3:   If every element in an $n \times n$ matrix A is replaced by its cofactor, the resulting matrix is called the **matrix of cofactors** and is denoted M_C. The transpose of the matrix of cofactors is called the **adjoint** of A and is denoted adj(A). Thus, the elements of adj(A) are

$$\text{adj}(A)_{ij} = C_{ji}.$$

Example 4.3.9   Determine adj(A) if $A = \begin{bmatrix} 2 & 0 & -3 \\ -1 & 5 & 4 \\ 3 & -2 & 0 \end{bmatrix}$.

Solution   We first determine the cofactors of A.

$$C_{11} = 8, \ C_{12} = 12. \ C_{13} = -13, \ C_{21} = 6, \ C_{22} = 9, \ C_{23} = 4,$$

$$C_{31} = 15, \ C_{32} = -5, \ C_{33} = 10.$$

Thus,

$$M_C = \begin{bmatrix} 8 & 12 & -13 \\ 6 & 9 & 4 \\ 15 & -5 & 10 \end{bmatrix},$$

so that

$$\text{adj}(A) = (M_C)^T = \begin{bmatrix} 8 & 6 & 15 \\ 12 & 9 & -5 \\ -13 & 4 & 10 \end{bmatrix}.$$

❏

We can now prove the next theorem.

> ***Theorem 4.3.2 (The Adjoint Method for Computing A^{-1}):*** If $\det(A) \neq 0$ then
>
> $$A^{-1} = \frac{1}{\det(A)} \ \text{adj}(A). \tag{4.3.5}$$

PROOF   Let $B = \dfrac{1}{\det(A)} \ \text{adj}(A)$. Then we must establish that

$$AB = I_n = BA.$$

But, using the index form of the matrix product,

$$(AB)_{ij} = \sum_{k=1}^{n} a_{ik} b_{kj} = \sum_{k=1}^{n} a_{ik} \frac{1}{\det(A)} \ \text{adj}(A)_{kj} = \frac{1}{\det(A)} \sum_{k=1}^{n} a_{ik} C_{jk} = \delta_{ij},$$

where we used (4.3.4) in the last step. Consequently, $AB = I_n$. We leave it as an exercise to verify that $BA = I_n$ also.   ■

REMARK   Notice that (4.3.5) gives a formula for the elements of A^{-1} in terms of the elements of $\text{adj}(A)$ as compared with the Gauss–Jordan method for finding A^{-1}, which is a procedure for determining all of the elements of A^{-1}.

Example 4.3.10   For the matrix in the previous example,

$$\det(A) = 55,$$

so that

$$A^{-1} = \frac{1}{55} \begin{bmatrix} 8 & 6 & 15 \\ 12 & 9 & -5 \\ -13 & 4 & 10 \end{bmatrix}.$$

❏

In Chapter 8, we will find that the solution of a system of *differential* equations can be expressed naturally in terms of matrix functions, that is, matrices whose elements are functions defined on some interval. Certain problems will require us to find the inverse of such matrix functions. For 2×2 systems the adjoint method is very quick.

Example 4.3.11   Find A^{-1} if $A = \begin{bmatrix} e^{2t} & e^{-t} \\ 3e^{2t} & 6e^{-t} \end{bmatrix}$.

Solution   In this case, $\det(A) = 3e^t$, and $\mathrm{adj}(A) = \begin{bmatrix} 6e^{-t} & -e^{-t} \\ -3e^{2t} & e^{2t} \end{bmatrix}$, so that

$$A^{-1} = \begin{bmatrix} 2e^{-2t} & -\frac{1}{3}e^{-2t} \\ -e^t & \frac{1}{3}e^t \end{bmatrix}.$$

CRAMER'S RULE

We now derive a technique that enables us, in the case when $\det(A) \neq 0$, to express the unique solution of an $n \times n$ linear system

$$A\mathbf{x} = \mathbf{b}$$

directly in terms of determinants. Let B_k denote the matrix obtained by replacing the kth column vector of A with $\mathbf{b}$. Thus,

$$B_k = \begin{bmatrix} a_{11} & a_{12} & \cdots & b_1 & \cdots & a_{1n} \\ a_{21} & a_{22} & \cdots & b_2 & \cdots & a_{2n} \\ \vdots & \vdots & & \vdots & & \vdots \\ a_{n1} & a_{n2} & \cdots & b_n & \cdots & a_{nn} \end{bmatrix},$$
$$\uparrow$$
$$\text{column } k$$

or, in column vector notation,

$$B_k = \begin{bmatrix} \mathbf{a}_1, & \mathbf{a}_2, & \ldots, & \mathbf{a}_{k-1}, & \mathbf{b}, & \mathbf{a}_{k+1}, & \ldots, & \mathbf{a}_n \end{bmatrix}.$$

The key point to notice is that the cofactors of the elements in the kth column of B_k coincide with the corresponding cofactors of A. Thus, expanding $\det(B_k)$ along the kth column using the cofactor expansion theorem yields

$$\det(B_k) = C_{1k}b_1 + C_{2k}b_2 + \cdots + C_{nk}b_n = \sum_{i=1}^{n} C_{ik}b_i, \quad k = 1, 2, \ldots, n, \quad (4.3.6)$$

where the C_{ij} are the cofactors of A. We can now prove Cramer's rule.

Theorem 4.3.3 (Cramer's Rule):   If $\det(A) \neq 0$, the unique solution to the $n \times n$ system $A\mathbf{x} = \mathbf{b}$ is $(x_1, x_2, \ldots, x_n)$, where

$$x_k = \frac{\det(B_k)}{\det(A)}, \quad k = 1, 2, \ldots, n. \tag{4.3.7}$$

PROOF   If $\det(A) \neq 0$, then the system $A\mathbf{x} = \mathbf{b}$ has the unique solution

$$\mathbf{x} = A^{-1}\mathbf{b}, \tag{4.3.8}$$

where, from Theorem 4.3.2, we can write

$$A^{-1} = \frac{1}{\det(A)} \operatorname{adj}(A). \tag{4.3.9}$$

If we let $\mathbf{x} = [x_1 \ x_2 \ \ldots \ x_n]^T$, $\mathbf{b} = [b_1 \ b_2 \ \ldots \ b_n]^T$, and recall that $\operatorname{adj}(A)_{ij} = C_{ji}$, then substitution from (4.3.9) into (4.3.8) and use of the index form of the matrix product yields

$$x_k = \sum_{i=1}^{n} (A^{-1})_{ki} b_i = \sum_{i=1}^{n} \frac{1}{\det(A)} \operatorname{adj}(A)_{ki} b_i$$

$$= \frac{1}{\det(A)} \sum_{i=1}^{n} C_{ik} b_i, \quad k = 1, 2, \ldots, n.$$

Using (4.3.6), this can be written as

$$x_k = \frac{\det(B_k)}{\det(A)}, \quad k = 1, 2, \ldots, n$$

as required.   ■

REMARK   In general, Cramer's rule requires more work than the Gaussian elimination method, and is also restricted to $n \times n$ systems whose coefficient matrix is nonsingular. However, it is a powerful theoretical tool, since it gives us a formula for the solution of an $n \times n$ system provided $\det(A) \neq 0$.

Example 4.3.12   Solve

$$3x_1 + 2x_2 - x_3 = 4,$$
$$x_1 + x_2 - 5x_3 = -3,$$
$$-2x_1 - x_2 + 4x_3 = 0.$$

Solution   The following determinants are easily evaluated

$$\det(A) = \begin{vmatrix} 3 & 2 & -1 \\ 1 & 1 & -5 \\ -2 & -1 & 4 \end{vmatrix} = 8, \quad \det(B_1) = \begin{vmatrix} 4 & 2 & -1 \\ -3 & 1 & -5 \\ 0 & -1 & 4 \end{vmatrix} = 17,$$

$$\det(B_2) = \begin{vmatrix} 3 & 4 & -1 \\ 1 & -3 & -5 \\ -2 & 0 & 4 \end{vmatrix} = -6, \quad \det(B_3) = \begin{vmatrix} 3 & 2 & 4 \\ 1 & 1 & -3 \\ -2 & -1 & 0 \end{vmatrix} = 7.$$

Inserting these results into (4.3.7) yields $x_1 = 17/8$, $x_2 = -6/8 = -3/4$, $x_3 = 7/8$, so that the solution to the system is $(17/8, -3/4, 7/8)$.

EXERCISES 4.3

For problems 1–3, determine all minors and cofactors of the given matrix.

1. $A = \begin{bmatrix} 1 & -3 \\ 2 & 4 \end{bmatrix}$.

2. $A = \begin{bmatrix} 1 & -1 & 2 \\ 3 & -1 & 4 \\ 2 & 1 & 5 \end{bmatrix}$.

3. $A = \begin{bmatrix} 2 & 10 & 3 \\ 0 & -1 & 0 \\ 4 & 1 & 5 \end{bmatrix}$.

4. If $A = \begin{bmatrix} 1 & 3 & -1 & 2 \\ 3 & 4 & 1 & 2 \\ 7 & 1 & 4 & 6 \\ 5 & 0 & 1 & 2 \end{bmatrix}$, determine the minors $M_{12}, M_{31}, M_{23}, M_{42}$, and the corresponding cofactors.

For problems 5–10, use the cofactor expansion theorem to evaluate the given determinant along the specified row or column.

5. $\begin{vmatrix} 1 & -2 \\ 1 & 3 \end{vmatrix}$, row 1.

6. $\begin{vmatrix} -1 & 2 & 3 \\ 1 & 4 & -2 \\ 3 & 1 & 4 \end{vmatrix}$, column 3.

7. $\begin{vmatrix} 2 & 1 & -4 \\ 7 & 1 & 3 \\ 1 & 5 & -2 \end{vmatrix}$, row 2.

8. $\begin{vmatrix} 3 & 1 & 4 \\ 7 & 1 & 2 \\ 2 & 3 & -5 \end{vmatrix}$, column 1.

9. $\begin{vmatrix} 0 & 2 & -3 \\ -2 & 0 & 5 \\ 3 & -5 & 0 \end{vmatrix}$, row 3.

10. $\begin{vmatrix} 1 & -2 & 3 & 0 \\ 4 & 0 & 7 & -2 \\ 0 & 1 & 3 & 4 \\ 1 & 5 & -2 & 0 \end{vmatrix}$, column 4.

For problems 11–19, evaluate the given determinant using the techniques of this section.

11. $\begin{vmatrix} 1 & 0 & -2 \\ 3 & 1 & -1 \\ 7 & 2 & 5 \end{vmatrix}$.

12. $\begin{vmatrix} -1 & 2 & 3 \\ 0 & 1 & 4 \\ 2 & -1 & 3 \end{vmatrix}$.

13. $\begin{vmatrix} 2 & -1 & 3 \\ 5 & 2 & 1 \\ 3 & -3 & 7 \end{vmatrix}$.

14. $\begin{vmatrix} 0 & -2 & 1 \\ 2 & 0 & -3 \\ -1 & 3 & 0 \end{vmatrix}$.

15. $\begin{vmatrix} 1 & 0 & -1 & 0 \\ 0 & 1 & 0 & -1 \\ -1 & 0 & -1 & 0 \\ 0 & 1 & 0 & 1 \end{vmatrix}$.

16. $\begin{vmatrix} 2 & -1 & 3 & 1 \\ 1 & 4 & -2 & 3 \\ 0 & 2 & -1 & 0 \\ 1 & 3 & -2 & 4 \end{vmatrix}$.

17. $\begin{vmatrix} 3 & 5 & 2 & 6 \\ 2 & 3 & 5 & -5 \\ 7 & 5 & -3 & -16 \\ 9 & -6 & 27 & -12 \end{vmatrix}$.

18.
$$\begin{vmatrix} 2 & -7 & 4 & 3 \\ 5 & 5 & -3 & 7 \\ 6 & 2 & 6 & 3 \\ 4 & 2 & -4 & 5 \end{vmatrix}.$$

19.
$$\begin{vmatrix} 2 & 0 & -1 & 3 & 0 \\ 0 & 3 & 0 & 1 & 2 \\ 0 & 1 & 3 & 0 & 4 \\ 1 & 0 & 1 & -1 & 0 \\ 3 & 0 & 2 & 0 & 5 \end{vmatrix}.$$

20. If $A = \begin{bmatrix} 0 & x & y & z \\ -x & 0 & 1 & -1 \\ -y & -1 & 0 & 1 \\ -z & 1 & -1 & 0 \end{bmatrix}$, show

that $\det(A) = (x + y + z)^2$.

21. (a) Consider the 3×3 *Vandermonde* determinant $V(r_1, r_2, r_3)$ defined by

$$V(r_1, r_2, r_3) = \begin{vmatrix} 1 & 1 & 1 \\ r_1 & r_2 & r_3 \\ r_1^2 & r_2^2 & r_3^2 \end{vmatrix}.$$

Show that $V(r_1, r_2, r_3) = (r_2 - r_1)(r_3 - r_1)(r_3 - r_2)$.

(b) More generally, show that the $n \times n$ Vandermonde determinant

$$V(r_1, r_2, \ldots, r_n) = \begin{vmatrix} 1 & 1 & \cdots & 1 \\ r_1 & r_2 & \cdots & r_n \\ r_1^2 & r_2^2 & \cdots & r_n^2 \\ \vdots & \vdots & & \vdots \\ r_1^{n-1} & r_2^{n-1} & \cdots & r_n^{n-1} \end{vmatrix}$$

has value

$$V(r_1, r_2, \ldots, r_n) = \prod_{1 \le i < m \le n} (r_m - r_i).$$

For problems $22-31$, find, (a) $\det(A)$, (b) the matrix of cofactors, (c) $\text{adj}(A)$, and, if possible, (d) A^{-1}.

22. $A = \begin{bmatrix} 3 & 1 \\ 4 & 5 \end{bmatrix}.$

23. $A = \begin{bmatrix} -1 & -2 \\ 4 & 1 \end{bmatrix}.$

24. $A = \begin{bmatrix} 5 & 2 \\ -15 & -6 \end{bmatrix}.$

25. $A = \begin{bmatrix} 2 & -3 & 0 \\ 2 & 1 & 5 \\ 0 & -1 & 2 \end{bmatrix}.$

26. $A = \begin{bmatrix} -2 & 3 & -1 \\ 2 & 1 & 5 \\ 0 & 2 & 3 \end{bmatrix}.$

27. $A = \begin{bmatrix} 1 & -1 & 2 \\ 3 & -1 & 4 \\ 5 & 1 & 7 \end{bmatrix}.$

28. $A = \begin{bmatrix} 0 & 1 & 2 \\ -1 & -1 & 3 \\ 1 & -2 & 1 \end{bmatrix}.$

29. $A = \begin{bmatrix} 2 & -3 & 5 \\ 1 & 2 & 1 \\ 0 & 7 & -1 \end{bmatrix}.$

30. $A = \begin{bmatrix} 1 & 1 & 1 & 1 \\ -1 & 1 & -1 & 1 \\ 1 & 1 & -1 & -1 \\ -1 & 1 & 1 & -1 \end{bmatrix}.$

31. $A = \begin{bmatrix} 1 & 0 & 3 & 5 \\ -2 & 1 & 1 & 3 \\ 3 & 9 & 0 & 2 \\ 2 & 0 & 3 & -1 \end{bmatrix}.$

32. Let $A = \begin{bmatrix} 1 & -2x & 2x^2 \\ 2x & 1 - 2x^2 & -2x \\ 2x^2 & 2x & 1 \end{bmatrix}.$

(a) Show that $\det(A) = (1 + 2x^2)^3$.

(b) Use the adjoint method to find A^{-1}.

33. Find the element in the second row and third column of A^{-1} if

$$A = \begin{bmatrix} 1 & 0 & 1 & 0 \\ 2 & -1 & 1 & 3 \\ 0 & 1 & -1 & 2 \\ -1 & 1 & 2 & 0 \end{bmatrix}$$

In problems $34-36$ find A^{-1}.

34. $A = \begin{bmatrix} 3e^t & e^{2t} \\ 2e^t & 2e^{2t} \end{bmatrix}.$

35. $A = \begin{bmatrix} e^t \sin 2t & -e^{-t} \cos 2t \\ e^t \cos 2t & e^{-t} \sin 2t \end{bmatrix}$.

36. $A = \begin{bmatrix} e^t & te^t & e^{-2t} \\ e^t & 2te^t & e^{-2t} \\ e^t & te^t & 2e^{-2t} \end{bmatrix}$.

37. If $A = \begin{bmatrix} 1 & 2 & 3 \\ 3 & 4 & 5 \\ 4 & 5 & 6 \end{bmatrix}$, compute the matrix product $A \, \text{adj}(A)$. What can you conclude about $\det(A)$?

For problems 38–41, use Cramer's rule to solve the given linear system:

38. $2x_1 - 3x_2 = 2,$
$\quad x_1 + 2x_2 = 4.$

39. $3x_1 - 2x_2 + x_3 = 4,$
$\quad x_1 + x_2 - x_3 = 2,$
$\quad x_1 \qquad + x_3 = 1.$

40. $x_1 - 3x_2 + x_3 = 0,$
$\quad x_1 + 4x_2 - x_3 = 0,$
$\quad 2x_1 + x_2 - 3x_3 = 0.$

41. $x_1 - 2x_2 + 3x_3 - x_4 = 1,$
$\quad 2x_1 \qquad + x_3 \qquad = 2,$
$\quad x_1 + x_2 \qquad - x_4 = 0,$
$\quad x_2 - 2x_3 + x_4 = 3.$

42. Use Cramer's rule to determine x_1 and x_2 if

$$e^t x_1 + e^{-2t} x_2 = 3 \sin t,$$
$$e^t x_1 - 2e^{-2t} x_2 = 4 \cos t.$$

43. Determine the value of x_2 such that

$$x_1 + 4x_2 - 2x_3 + x_4 = 2,$$
$$2x_1 + 9x_2 - 3x_3 - 2x_4 = 5,$$
$$x_1 + 5x_2 + x_3 - x_4 = 3,$$
$$3x_1 + 14x_2 + 7x_3 - 2x_4 = 6.$$

44. Find all solutions to the system

$$(b+c)x_1 + a(x_2 + x_3) = a,$$
$$(c+a)x_2 + b(x_3 + x_1) = b,$$
$$(a+b)x_3 + c(x_1 + x_2) = c,$$

where a, b, and c are constants. Make sure you consider all cases (that is, those when there is a unique solution, an infinite number of solutions, and no solution).

◆ **45.** Let A be a randomly generated nonsingular 4×4 matrix. Verify the cofactor expansion theorem for expansion along row 1.

◆ **46.** Let A be a randomly generated nonsingular 4×4 matrix. Verify equation (4.3.3) when $i = 2$ and $j = 4$.

◆ **47.** Let A be a randomly generated 5×5 matrix. Determine $\text{adj}(A)$ and compute $A \, \text{adj}(A)$. Use your result to determine $\det(A)$.

◆ **48.** Solve the system of equations

$$1.21x_1 + 3.42x_2 + 2.15x_3 = 3.25,$$
$$5.41x_1 + 2.32x_2 + 7.15x_3 = 4.61,$$
$$21.63x_1 + 3.51x_2 + 9.22x_3 = 9.93.$$

Round your answers to two decimal places.

◆ **49.** Use Cramer's rule to solve the system $A\mathbf{x} = \mathbf{b}$

if $A = \begin{bmatrix} 1 & 2 & 3 & 4 & 4 \\ 2 & 1 & 2 & 3 & 4 \\ 3 & 2 & 1 & 2 & 3 \\ 4 & 3 & 2 & 1 & 2 \\ 4 & 4 & 3 & 2 & 1 \end{bmatrix}$ and $\mathbf{b} = \begin{bmatrix} 68 \\ -72 \\ -87 \\ 79 \\ 43 \end{bmatrix}$.

4.4   SUMMARY OF DETERMINANTS

The primary aim of this section is to serve as a stand-alone introduction to determinants for courses where the determinant is somewhat de-emphasized. It may also be used as a review of the results derived in Sections 4.1–4.3.

DEFINITION OF A DETERMINANT

The determinant of an $n \times n$ matrix A, denoted $\det(A)$, is a scalar whose value can be obtained in the following manner.

(1) If $A = [a_{11}]$, then $\det(A) = a_{11}$.

(2) If $A = \begin{bmatrix} a_{11} & a_{12} \\ a_{21} & a_{22} \end{bmatrix}$, then $\det(A) = a_{11}a_{22} - a_{12}a_{21}$.

(3) For $n > 2$, the determinant of A can be computed using either of the following formulas

$$\det(A) = a_{i1}C_{i1} + a_{i2}C_{i2} + \cdots + a_{in}C_{in}, \tag{4.4.1}$$

$$\det(A) = a_{1j}C_{1j} + a_{2j}C_{2j} + \cdots + a_{nj}C_{nj}, \tag{4.4.2}$$

where $C_{ij} = (-1)^{i+j}M_{ij}$, and M_{ij} is the determinant of the matrix obtained by deleting the ith row and jth column of A. The formulas (4.4.1) and (4.4.2) are referred to as cofactor expansion along the ith row and cofactor expansion along the jth column, respectively. The determinants M_{ij} and C_{ij} are called the **minors** and **cofactors** of A, respectively. We also denote $\det(A)$ by

$$\begin{vmatrix} a_{11} & a_{12} & \cdots & a_{1n} \\ a_{21} & a_{22} & \cdots & a_{2n} \\ \vdots & \vdots & & \vdots \\ a_{n1} & a_{n2} & \cdots & a_{nn} \end{vmatrix}.$$

As an example, consider the general 3×3 matrix $A = \begin{bmatrix} a_{11} & a_{12} & a_{13} \\ a_{21} & a_{22} & a_{23} \\ a_{31} & a_{32} & a_{33} \end{bmatrix}$. Using cofactor expansion along row 1, we have

$$\det(A) = a_{11}C_{11} + a_{12}C_{12} + a_{13}C_{13}. \tag{4.4.3}$$

We next compute the required cofactors.

$$C_{11} = +M_{11} = \begin{vmatrix} a_{22} & a_{23} \\ a_{32} & a_{33} \end{vmatrix} = a_{22}a_{33} - a_{23}a_{32},$$

$$C_{12} = -M_{12} = -\begin{vmatrix} a_{21} & a_{23} \\ a_{31} & a_{33} \end{vmatrix} = -(a_{21}a_{33} - a_{23}a_{31}),$$

$$C_{13} = +M_{13} = \begin{vmatrix} a_{21} & a_{22} \\ a_{31} & a_{32} \end{vmatrix} = a_{21}a_{32} - a_{22}a_{31}.$$

Inserting these expressions for the cofactors into equation (4.4.3) yields

$$\det(A) = a_{11}(a_{22}a_{33} - a_{23}a_{32}) - a_{12}(a_{21}a_{33} - a_{23}a_{31}) + a_{13}(a_{21}a_{32} - a_{22}a_{31}),$$

which can be written as

$$\det(A) = a_{11}a_{22}a_{33} + a_{12}a_{23}a_{31} + a_{13}a_{21}a_{32}$$
$$- a_{11}a_{23}a_{32} - a_{12}a_{21}a_{33} - a_{13}a_{22}a_{31}.$$

Although we chose to use cofactor expansion along the first row to obtain the preceding formula, according to (4.4.1) and (4.4.2), the same result would have been obtained if we had chosen to expand along any row or column of A. A simple schematic for obtaining the terms in the determinant of a 3×3 matrix is given in Figure 4.4.1. By taking the product of the elements joined by each arrow and attaching the indicated sign to the result,

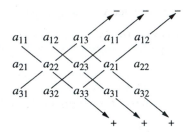

Figure 4.4.1 A schematic for obtaining the determinant of a 3×3 matrix $A = [a_{ij}]$.

we obtain the six terms in the determinant of the 3×3 matrix $A = [a_{ij}]$. Note that this technique for obtaining the terms in a determinant **does not** generalize to larger determinants.

Example 4.4.1   Evaluate $\begin{vmatrix} 2 & -1 & 1 \\ 3 & 4 & 2 \\ 7 & 5 & 8 \end{vmatrix}$.

Solution   In this case, the schematic given in Figure 4.4.1 is

$$\begin{matrix} 2 & -1 & 1 & 2 & -1 \\ 3 & 4 & 2 & 3 & 4 \\ 7 & 5 & 8 & 7 & 5 \end{matrix}$$

so that

$$\begin{vmatrix} 2 & -1 & 1 \\ 3 & 4 & 2 \\ 7 & 5 & 8 \end{vmatrix} = (2)(4)(8) + (-1)(2)(7) + (1)(3)(5)$$

$$- (7)(4)(1) - (5)(2)(2) - (8)(3)(-1)$$

$$= 41.$$

PROPERTIES OF DETERMINANTS

Let A and B be $n \times n$ matrices. The determinant has the following properties:

P1. If B is obtained by permuting two distinct rows (or columns) of A, then

$$\det(A) = -\det(B).$$

P2. If B is obtained by dividing any row (or column) of A by a nonzero scalar k, then
$$\det(A) = k\,\det(B).$$

P3. If B is obtained by adding a multiple of any row (or column) of A to another row (or column) of A, then
$$\det(A) = \det(B).$$

P4. $\det(A^T) = \det(A)$.

P5. Let $\mathbf{a}_1, \mathbf{a}_2, ..., \mathbf{a}_n$ denote the row vectors of A. If the ith row vector of A is the sum of two row vectors, say $\mathbf{a}_i = \mathbf{b}_i + \mathbf{c}_i$, then
$$\det(A) = \det(B) + \det(C),$$
where $B = [\mathbf{a}_1, \mathbf{a}_2, ..., \mathbf{a}_{i-1}, \mathbf{b}_i, \mathbf{a}_{i+1}, ..., \mathbf{a}_n]^T$ and $C = [\mathbf{a}_1, \mathbf{a}_2, ..., \mathbf{a}_{i-1}, \mathbf{c}_i, \mathbf{a}_{i+1}, ..., \mathbf{a}_n]^T$. The corresponding property is also true for columns.

P6. If A has a row (or column) of zeros, then $\det(A) = 0$.

P7. If two rows (or columns) of A are the same, then $\det(A) = 0$.

P8. $\det(AB) = \det(A)\,\det(B)$.

The first three properties tell us how ERO performed on a matrix A alter the value of $\det(A)$. They can be very helpful in reducing the amount of work required to evaluate a determinant, since we can use them to put several zeros in a row or column of A and then use cofactor expansion along the corresponding row or column. We illustrate with an example.

Example 4.4.2   Evaluate
$$\begin{vmatrix} 2 & 1 & 3 & 2 \\ -1 & 1 & -2 & 2 \\ 5 & 1 & -2 & 1 \\ -2 & 3 & 1 & 1 \end{vmatrix}.$$

Solution   Before performing a cofactor expansion, we first use ERO to simplify the determinant.

$$\begin{vmatrix} 2 & 1 & 3 & 2 \\ -1 & 1 & -2 & 2 \\ 5 & 1 & -2 & 1 \\ -2 & 3 & 1 & 1 \end{vmatrix} \overset{1}{=} \begin{vmatrix} 0 & 3 & -1 & 6 \\ -1 & 1 & -2 & 2 \\ 0 & 6 & -12 & 11 \\ 0 & 1 & 5 & -3 \end{vmatrix}$$

$$\underset{\underset{\text{cofactor expansion along column 1}}{\uparrow}}{=} -(-1)\begin{vmatrix} 3 & -1 & 6 \\ 6 & -12 & 11 \\ 1 & 5 & -3 \end{vmatrix}$$

$$\overset{2}{=} \begin{vmatrix} 0 & -16 & 15 \\ 0 & -42 & 29 \\ 1 & 5 & -3 \end{vmatrix} \underset{\underset{\text{cofactor expansion along column 1}}{\uparrow}}{=} \begin{vmatrix} -16 & 15 \\ -42 & 29 \end{vmatrix} = 166.$$

> **1.** $A_{21}(2)$, $A_{23}(5)$, $A_{24}(-2)$ **2.** $A_{31}(-3)$, $A_{32}(-6)$

BASIC THEORETICAL RESULTS

The determinant is a useful theoretical tool in linear algebra. We list next the major results that will be needed for the remainder of the text.

1. The volume of the parallelepiped determined by the vectors $\mathbf{a} = a_1\mathbf{i} + a_2\mathbf{j} + a_3\mathbf{k}$, $\mathbf{b} = b_1\mathbf{i} + b_2\mathbf{j} + b_3\mathbf{k}$, and $\mathbf{c} = c_1\mathbf{i} + c_2\mathbf{j} + c_3\mathbf{k}$ is

$$\boxed{\text{Volume} = |\det(A)|},$$

where $A = \begin{bmatrix} a_1 & a_2 & a_3 \\ b_1 & b_2 & b_3 \\ c_1 & c_2 & c_3 \end{bmatrix}$.

2. An $n \times n$ matrix is nonsingular if and only if $\det(A) \neq 0$.

3. An $n \times n$ linear system $A\mathbf{x} = \mathbf{b}$ has a unique solution if and only if $\det(A) \neq 0$.

4. An $n \times n$ *homogeneous* linear system $A\mathbf{x} = \mathbf{0}$ has an infinite number of solutions if and only if $\det(A) = 0$.

We see, for example, that according to (2), the matrices in the previous two examples are both nonsingular.

If A is an $n \times n$ matrix with $\det(A) \neq 0$, then the following two methods can be derived for obtaining the inverse of A and for finding the unique solution to the linear system $A\mathbf{x} = \mathbf{b}$, respectively.

1. Adjoint Method for A^{-1}:   If A is nonsingular, then

$$A^{-1} = \frac{1}{\det(A)} \text{adj}(A),$$

where $\text{adj}(A)$ denotes the transpose of the matrix obtained by replacing each element in A by its cofactor.

2. Cramer's rule: If $\det(A) \neq 0$, then the unique solution to $A\mathbf{x} = \mathbf{b}$ is $(x_1, x_2, ..., x_n)$, where

$$x_k = \frac{\det(B_k)}{\det(A)}, \quad k = 1, 2, ..., n$$

and B_k denotes the matrix obtained when the kth column vector of A is replaced by $\mathbf{b}$.

Example 4.4.3   Use the adjoint method to determine A^{-1} if $A = \begin{bmatrix} 2 & -1 & 1 \\ 3 & 4 & 2 \\ 7 & 5 & 8 \end{bmatrix}$.

Solution   We have already shown in Example 4.4.1 that $\det(A) = 41$, so that A is nonsingular. Replacing each element in A with its cofactor yields the **matrix of cofactors**

$$M_C = \begin{bmatrix} 22 & -10 & -13 \\ 13 & 9 & -17 \\ -6 & -1 & 11 \end{bmatrix},$$

so that

$$\text{adj}(A) = M_C{}^T = \begin{bmatrix} 22 & 13 & -6 \\ -10 & 9 & -1 \\ -13 & -17 & 11 \end{bmatrix}.$$

Consequently,

$$A^{-1} = \frac{1}{\det(A)} \text{adj}(A) = \begin{bmatrix} \dfrac{22}{41} & \dfrac{13}{41} & -\dfrac{6}{41} \\[2mm] -\dfrac{10}{41} & \dfrac{9}{41} & -\dfrac{1}{41} \\[2mm] -\dfrac{13}{41} & -\dfrac{17}{41} & \dfrac{11}{41} \end{bmatrix}.$$

Example 4.4.4   Use Cramer's rule to solve

$$2x_1 - x_2 + x_3 = 2,$$
$$3x_1 + 4x_2 + 2x_3 = 5,$$
$$7x_1 + 5x_2 + 8x_3 = 3.$$

Solution

The matrix of coefficients is $A = \begin{bmatrix} 2 & -1 & 1 \\ 3 & 4 & 2 \\ 7 & 5 & 8 \end{bmatrix}$, which we have already shown to have a determinant of 41. Consequently, Cramer's rule can indeed be applied. In this problem we have

$$\det(B_1) = \begin{vmatrix} 2 & -1 & 1 \\ 5 & 4 & 2 \\ 3 & 5 & 8 \end{vmatrix} = 91,$$

$$\det(B_2) = \begin{vmatrix} 2 & 2 & 1 \\ 3 & 5 & 2 \\ 7 & 3 & 8 \end{vmatrix} = 22,$$

$$\det(B_3) = \begin{vmatrix} 2 & -1 & 2 \\ 3 & 4 & 5 \\ 7 & 5 & 3 \end{vmatrix} = -78.$$

It therefore follows from Cramer's rule that

$$x_1 = \frac{\det(B_1)}{\det(A)} = \frac{91}{41}, \quad x_2 = \frac{\det(B_2)}{\det(A)} = \frac{22}{41}, \quad x_3 = \frac{\det(B_3)}{\det(A)} = -\frac{78}{41}.$$

EXERCISES 4.4

For problems 1–7, evaluate the given determinant.

1. $\begin{vmatrix} 5 & -1 \\ 3 & 7 \end{vmatrix}$.

2. $\begin{vmatrix} 3 & 5 & 7 \\ -1 & 2 & 4 \\ 6 & 3 & -2 \end{vmatrix}$,

3. $\begin{vmatrix} 5 & 1 & 4 \\ 6 & 1 & 3 \\ 14 & 2 & 7 \end{vmatrix}$.

4. $\begin{vmatrix} 2.3 & 1.5 & 7.9 \\ 4.2 & 3.3 & 5.1 \\ 6.8 & 3.6 & 5.7 \end{vmatrix}$.

5. $\begin{vmatrix} a & b & c \\ b & c & a \\ c & a & b \end{vmatrix}$.

6. $\begin{vmatrix} 3 & 5 & -1 & 2 \\ 2 & 1 & 5 & 2 \\ 3 & 2 & 5 & 7 \\ 1 & -1 & 2 & 1 \end{vmatrix}$.

7. $\begin{vmatrix} 7 & 1 & 2 & 3 \\ 2 & -2 & 4 & 6 \\ 3 & -1 & 5 & 4 \\ 18 & 9 & 27 & 54 \end{vmatrix}$.

For problems 8–12, find det(A). If A is nonsingular, use the adjoint method to find A^{-1}.

8. $A = \begin{bmatrix} 3 & 5 \\ 2 & 7 \end{bmatrix}$.

9. $A = \begin{bmatrix} 1 & 2 & 3 \\ 2 & 3 & 1 \\ 3 & 1 & 2 \end{bmatrix}$.

10. $A = \begin{bmatrix} 3 & 4 & 7 \\ 2 & 6 & 1 \\ 3 & 14 & -1 \end{bmatrix}$.

11. $A = \begin{bmatrix} 2 & 5 & 7 \\ 4 & -3 & 2 \\ 6 & 9 & 11 \end{bmatrix}$.

12. $A = \begin{bmatrix} 5 & -1 & 2 & 1 \\ 3 & -1 & 4 & 5 \\ 1 & -1 & 2 & 1 \\ 5 & 9 & -3 & 2 \end{bmatrix}$.

For problems 13–17, use Cramer's rule to determine the unique solution to the system $A\mathbf{x} = \mathbf{b}$ for the given matrix and vector.

13. $A = \begin{bmatrix} 3 & 5 \\ 6 & 2 \end{bmatrix}, \mathbf{b} = \begin{bmatrix} 4 \\ 9 \end{bmatrix}$.

14. $A = \begin{bmatrix} \cos t & \sin t \\ \sin t & -\cos t \end{bmatrix}, \mathbf{b} = \begin{bmatrix} e^{-t} \\ 3e^{-t} \end{bmatrix}$.

15. $A = \begin{bmatrix} 4 & 1 & 3 \\ 2 & -1 & 5 \\ 2 & 3 & 1 \end{bmatrix}, \mathbf{b} = \begin{bmatrix} 5 \\ 7 \\ 2 \end{bmatrix}$.

16. $A = \begin{bmatrix} 5 & 3 & 6 \\ 2 & 4 & -7 \\ 2 & 5 & 9 \end{bmatrix}, \mathbf{b} = \begin{bmatrix} 3 \\ -1 \\ 4 \end{bmatrix}$.

17. $A = \begin{bmatrix} 3.1 & 3.5 & 7.1 \\ 2.2 & 5.2 & 6.3 \\ 1.4 & 8.1 & 0.9 \end{bmatrix}$,

$\mathbf{b} = \begin{bmatrix} 3.6 \\ 2.5 \\ 9.3 \end{bmatrix}$.

18. If A is a nonsingular $n \times n$ matrix, prove that $\det(A^{-1}) = \dfrac{1}{\det(A)}$.

19. Let A and B be 3×3 matrices with $\det(A) = 3$, $\det(B) = -4$. Determine

$$\det(2A), \quad \det(A^{-1}), \quad \det(A^T B),$$

$$\det(B^5), \quad \det(B^{-1}AB).$$

20. Let A be an $n \times n$ matrix satisfying $AA^T = I_n$. Prove that $\det(A) = \pm 1$.

5

Vector Spaces

The main aim of this text is to study linear mathematics. So far we have considered two different types of linear problems. In Chapter 2, we considered second-order linear DE whereas, in Chapter 3, we studied systems of linear algebraic equations. The theory underlying the solution of both of these problems can be considered as a special case of a general mathematical framework for linear problems. To illustrate this framework, we discuss two examples.

Consider the homogeneous linear algebraic system $A\mathbf{x} = \mathbf{0}$, where

$$A = \begin{bmatrix} 1 & -1 & 2 \\ 2 & -2 & 4 \\ 3 & -3 & 6 \end{bmatrix}.$$

It is straightforward to show that this system has solution set

$$S = \{(r - 2s, r, s) : r, s, \in \mathbf{R}\}.$$

Geometrically we can interpret each solution as defining the coordinates of a point in space or, equivalently, as the geometric vector with components

$$\mathbf{x} = (r - 2s, r, s).$$

Using the standard operations of vector addition and multiplication of a vector by a real number, it follows that $\mathbf{x}$ can be written in the form

$$\mathbf{x} = r(1, 1, 0) + s(-2, 0, 1).$$

We see that every solution to the given linear problem can be expressed as a *linear combination* of the two basic solutions (see Figure 5.0.1):

$$\mathbf{v}_1 = (1, 1, 0), \quad \mathbf{v}_2 = (-2, 0, 1).$$

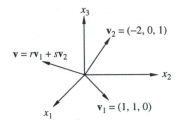

Figure 5.0.1 Two basic solutions to $A\mathbf{x} = \mathbf{0}$ and an example of an arbitrary solution to the system.

This should be reminiscent of the results from Chapter 2, where we established that every solution of the homogeneous *linear* DE

$$y'' + a_1 y' + a_2 y = 0$$

can be written in the form

$$y(x) = c_1 y_1(x) + c_2 y_2(x),$$

where $y_1(x)$ and $y_2(x)$ are any two linearly independent solutions on the interval of interest.

The similarities between the foregoing two problems are quite striking. In both cases we have a set of "vectors," V (in the first problem the vectors are ordered triples of numbers, whereas, in the second problem, they are functions that are at least twice differentiable on an interval I) and a *linear* vector equation defined in V. Further, in both cases, all solutions to the given equation can be expressed as a linear combination of two particular solutions.

In the next two chapters, we develop this way of formulating linear problems in terms of an abstract set of vectors, V, and a linear equation defined on V. We will find that many problems fit into this framework and that the solutions to these problems can be expressed as linear combinations of a certain number (not necessarily two) of basic solutions. The importance of this result cannot be overemphasized. It reduces the search for all solutions to a given problem to that of finding a finite number of solutions. (In the case of a second-order homogeneous linear DE that number was 2.) As specific applications, we will derive the theory underlying linear DE and linear systems of DE as special cases of the general framework.

Before proceeding any further, we give a word of encouragement to the more applied oriented reader. It will probably seem at times that the ideas we are introducing are rather esoteric and that the formalism is pure mathematical abstraction. However, in addition to the inherent mathematical beauty of the formalism, the ideas that it incorporates pervade many areas of applied mathematics, particularly engineering mathematics and mathematical physics, where the problems under investigation are very often linear in nature. Indeed, the linear algebra introduced in the next few chapters should be considered an extremely important addition to your mathematical repertoire, certainly on a par with the ideas of elementary calculus.

5.1   VECTORS IN $\mathbb{R}^n$

In this section, we use some familiar ideas about geometric vectors to motivate the more general and abstract idea of a real vector space. We begin by recalling that a geometric vector can be considered mathematically as a directed line segment (or arrow) that has both a magnitude and a direction attached to it. In calculus courses, we define vector addition according to the parallelogram law (see Figure 5.1.1), namely, the sum of the vectors **x** and **y** is the diagonal of the parallelogram formed by **x** and **y**. We denote the sum by **x** + **y**. It can then be shown geometrically that for all vectors **x**, **y**, **z**,

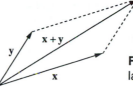

Figure 5.1.1   Parallelogram law of vector addition.

$$\mathbf{x} + \mathbf{y} = \mathbf{y} + \mathbf{x} \tag{5.1.1}$$

and

$$\mathbf{x} + (\mathbf{y} + \mathbf{z}) = (\mathbf{x} + \mathbf{y}) + \mathbf{z}. \tag{5.1.2}$$

These are the statements that the vector addition operation is commutative and associative. The *zero vector*, denoted **0**, is defined as the vector satisfying

$$\mathbf{x} + \mathbf{0} = \mathbf{x}, \tag{5.1.3}$$

for all vectors **x**. We consider the zero vector as having zero magnitude and arbitrary direction. Geometrically we picture the zero vector as corresponding to a point in space. Let –**x** denote the vector that has the same magnitude as **x**, but the opposite direction. Then according to the parallelogram law of addition,

$$\mathbf{x} + (-\mathbf{x}) = \mathbf{0}. \tag{5.1.4}$$

The vector –**x** is called the *additive inverse* of **x**. Properties (5.1.1)–(5.1.4) are the fundamental properties of vector addition. The basic algebra of vectors is completed when we also define the operation of multiplication of a vector by a real number. Geometrically, if **x** is a vector and k is a real number, then $k\mathbf{x}$ is defined to be the vector whose magnitude is $|k|$ times the magnitude of **x** and whose direction is the same as **x** if $k > 0$, and opposite to **x** if $k < 0$. (See Figure 5.1.2.) If $k = 0$, then $k\mathbf{x} = \mathbf{0}$. This **scalar multiplication** operation has several important properties that we now list. Once more, each of these can be established geometrically using only the foregoing definitions of vector addition and scalar multiplication.

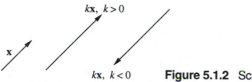

Figure 5.1.2   Scalar multiplication of **x** by k.

For all vectors **x**, **y**, and all real numbers r, s, t,

$$1\mathbf{x} = \mathbf{x}, \tag{5.1.5}$$

$$(st)\mathbf{x} = s(t\mathbf{x}), \tag{5.1.6}$$

$$r(\mathbf{x} + \mathbf{y}) = r\mathbf{x} + r\mathbf{y}, \tag{5.1.7}$$

$$(s + t)\mathbf{x} = s\mathbf{x} + t\mathbf{x}. \tag{5.1.8}$$

We will see in the next section how the concept of a vector space arises as a direct generalization of the ideas associated with geometric vectors. Before performing this abstraction, we want to recall some further features of geometric vectors and give one specific and important extension.

We begin by considering vectors in the plane. Let $\mathbf{R}^2$ denote the set of all ordered pairs of real numbers, thus,

$$\mathbf{R}^2 = \{(x, y) : x \in \mathbf{R}, y \in \mathbf{R}\}.$$

The elements of this set are called *vectors in* $\mathbf{R}^2$, and we use the usual vector notation to denote these elements. Geometrically we identify the vector $\mathbf{x} = (x, y)$ in $\mathbf{R}^2$ with the geometric vector $\mathbf{x}$ from the origin of a Cartesian coordinate system to the point with coordinates (x, y). This identification is illustrated in Figure 5.1.3. The numbers x and y

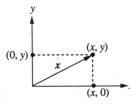

Figure 5.1.3 Identifying vectors in $\mathbf{R}^2$ with geometric vectrs in the plane.

are called the **components** of the geometric vector $\mathbf{x}$. The geometric vector addition and scalar multiplication operations can then be described in $\mathbf{R}^2$ as follows.

If $\mathbf{x} = (x_1, y_1)$, $\mathbf{y} = (x_2, y_2)$, and k is an arbitrary real number, then

$$\mathbf{x} + \mathbf{y} = (x_1, y_1) + (x_2, y_2) = (x_1 + x_2, y_1 + y_2), \tag{5.1.9}$$

$$k\mathbf{x} = k(x_1, y_1) = (kx_1, ky_1). \tag{5.1.10}$$

These are the algebraic statements of the parallelogram law of vector addition and the scalar multiplication law, respectively. (See Figure 5.1.4.) Using the parallelogram law

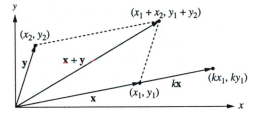

Figure 5.1.4 Vector addition and scalar multiplication in $\mathbf{R}^2$.

of vector addition and equations (5.1.9) and (5.1.10), it follows that any vector $\mathbf{x} = (x, y)$ can be written as

$$\mathbf{x} = x\mathbf{i} + y\mathbf{j} = x(1, 0) + y(0, 1),$$

where $\mathbf{i}$ and $\mathbf{j}$ are the unit vectors pointing along the positive x- and y-coordinate axes, respectively.

The properties (5.1.1)–(5.1.8) are now easily verified for vectors in $\mathbf{R}^2$. In particular, the zero vector in $\mathbf{R}^2$ is the vector

$$\mathbf{0} = (0, 0).$$

Furthermore, equation (5.1.9) implies that

$$(x, y) + (-x, -y) = (0, 0) = \mathbf{0},$$

so that the additive inverse of the general vector $\mathbf{x} = (x, y)$ is $-\mathbf{x} = (-x, -y)$.

It is straightforward to extend these ideas to vectors in space. We define

$$\mathbf{R}^3 = \{(x, y, z) : x \in \mathbf{R}, y \in \mathbf{R}, z \in \mathbf{R}\}$$

and call the elements of this set *vectors in* $\mathbf{R}^3$. As illustrated in Figure 5.1.5, each vector $\mathbf{x} = (x, y, z)$ in $\mathbf{R}^3$ can be identified with the geometric vector $\mathbf{x}$ that joins the origin of a Cartesian coordinate system to the point with coordinates (x, y, z). We call x, y, and z the components of $\mathbf{x}$.

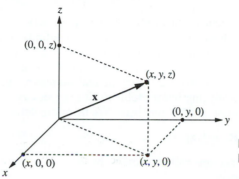

Figure 5.1.5   Identifying vectors in $\mathbf{R}^3$ with geometric vectors in space.

Addition and scalar multiplication in $\mathbf{R}^3$ is defined as follows.

If $\mathbf{x} = (x_1, y_1, z_1)$, $\mathbf{y} = (x_2, y_2, z_2)$, and k is an arbitrary real number, then

$$\mathbf{x} + \mathbf{y} = (x_1, y_1, z_1) + (x_2, y_2, z_2) = (x_1 + x_2, y_1 + y_2, z_1 + z_2), \qquad (5.1.11)$$

$$k\mathbf{x} = k(x_1, y_1, z_1) = (kx_1, ky_1, kz_1). \qquad (5.1.12)$$

Once more, these are, respectively, the component forms of the laws of vector addition and scalar multiplication for geometric vectors. It follows that an arbitrary vector $\mathbf{x} = (x, y, z)$ can be written as

$$\mathbf{x} = x\mathbf{i} + y\mathbf{j} + z\mathbf{k} = x(1, 0, 0) + y(0, 1, 0) + z(0, 0, 1),$$

where $\mathbf{i}$, $\mathbf{j}$, and $\mathbf{k}$ denote the unit vectors which point along the positive x-, y-, and z-coordinate axes, respectively.

We leave it as an exercise to check that the properties (5.1.1) – (5.1.8) are satisfied by vectors in $\mathbf{R}^3$ where

$$\mathbf{0} = (0, 0, 0)$$

and the additive inverse of $\mathbf{x} = (x, y, z)$ is $-\mathbf{x} = (-x, -y, -z)$.

Now let $\mathbf{x} = (x_1, y_1, z_1)$, $\mathbf{y} = (x_2, y_2, z_2)$, and $\mathbf{z} = (x_3, y_3, z_3)$ be three arbitrary vectors in $\mathbf{R}^3$, and consider the corresponding geometric vectors. Using the parallelogram law of vector addition , it follows from Figure 5.1.6 that the sum of these three vectors

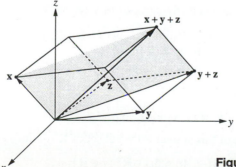

Figure 5.1.6   Adding three vectors in $\mathbf{R}^3$.

is represented geometrically by the diagonal of the parallelepiped whose coterminal sides are determined by the given vectors. Furthermore, as we have seen in Chapter 4, the volume of this parallelepiped is

$$V = |\det([\mathbf{x}, \mathbf{y}, \mathbf{z}])|,$$

so that the vectors are coplanar if and only if $\det([\mathbf{x}, \mathbf{y}, \mathbf{z}]) = 0$.

We now come to our first major abstraction. The definitions of $\mathbf{R}^2$ and $\mathbf{R}^3$ and their associated algebraic operations arise naturally from our experience with Cartesian geometry. However, from an algebraic viewpoint, there is no necessity to restrict attention to ordered pairs or ordered triples of real numbers. Indeed, in developing a general framework for analyzing linear problems, it is necessary to extend our definitions to the case of ordered n-tuples of real numbers. We therefore define

$$\mathbf{R}^n = \{(x_1, x_2, \ldots, x_n) : x_1, x_2, \ldots, x_n \in \mathbf{R}\}. \tag{5.1.13}$$

The elements of this set are called **vectors in $\mathbf{R}^n$**. We maintain the usual vector notation for these elements.

REMARKS

1. We note that $\mathbf{R}^2$ and $\mathbf{R}^3$ are the particular cases $n = 2, 3$, respectively. When $n = 1$, we denote $\mathbf{R}^1$ by $\mathbf{R}$. This is just the set of all real numbers.

2. In Chapter 3, we represented real solutions of a linear system of algebraic equations as an ordered n-tuple of real numbers. Consequently, such a solution defines a vector in $\mathbf{R}^n$, and the set of all real solutions to a linear system in n unknowns is therefore a subset of $\mathbf{R}^n$. Indeed, if A denotes the $m \times n$ coefficient matrix of a linear system, then the set of all real solutions to the system $A\mathbf{x} = \mathbf{b}$ is

$$S = \{\mathbf{x} \in \mathbf{R}^n : A\mathbf{x} = \mathbf{b}\}.$$

Example 5.1.1 $\mathbf{x} = (2, -1/2, 5/9, 2.354, \sqrt{\pi})$ is an example of a vector in $\mathbf{R}^5$.

Example 5.1.2   Express the solution set of the linear system

$$\begin{aligned} x_1 + x_2 + 2x_3 - x_4 &= 0, \\ 3x_1 - 2x_2 + x_3 + 2x_4 &= 0, \\ 5x_1 + 3x_2 + 3x_3 - 2x_4 &= 0, \end{aligned}$$

as a subset of $\mathbf{R}^4$.

Solution   The RREF of the augmented matrix of this system is

$$\begin{bmatrix} 1 & 0 & 0 & 1/5 & 0 \\ 0 & 1 & 0 & -4/5 & 0 \\ 0 & 0 & 1 & -1/5 & 0 \end{bmatrix}.$$

Setting $x_4 = 5t$, it follows that

$$x_3 = t, \quad x_2 = 4t, \quad x_1 = -t.$$

Consequently the solution set of this system is the subset of $\mathbf{R}^4$ defined by

$$S = \{(-t, 4t, t, 5t) \in \mathbf{R}^4 : t \in \mathbf{R}\}. \qquad \square$$

As in the case of $\mathbf{R}^2$ and $\mathbf{R}^3$, the operations of addition and scalar multiplication in $\mathbf{R}^n$ are defined componentwise. Thus, if $\mathbf{x} = (x_1, x_2, \ldots, x_n)$, $\mathbf{y} = (y_1, y_2, \ldots, y_n)$, and k is an arbitrary real number, then

$$\mathbf{x} + \mathbf{y} = (x_1 + y_1, x_2 + y_2, \ldots, x_n + y_n), \tag{5.1.14}$$

$$k\mathbf{x} = (kx_1, kx_2, \ldots, kx_n). \tag{5.1.15}$$

It should be noted that these definitions are direct generalizations of the algebraic operations defined in $\mathbf{R}^2$ and $\mathbf{R}^3$, but that we have no geometric analogy when $n > 3$. It is easily established that these operations satisfy properties (5.1.1)–(5.1.8), where the **zero vector** in $\mathbf{R}^n$ is

$$\mathbf{0} = (0, 0, \ldots, 0)$$

and the additive inverse of the vector $\mathbf{x} = (x_1, x_2, \ldots, x_n)$ is

$$-\mathbf{x} = (-x_1, -x_2, \ldots, -x_n).$$

The verification is left as an exercise.

Example 5.1.3:   If $\mathbf{x} = (1.2, 3.5, 2, 0)$ and $\mathbf{y} = (12.23, 19.65, 23.22, 9.76)$, then

$$\mathbf{x} + \mathbf{y} = (1.2, 3.5, 2, 0) + (12.23, 19.65, 23.22, 9.76) = (13.43, 23.15, 25.22, 9.76)$$

and

$$2.35\mathbf{x} = (2.82, 8.225, 4.7, 0). \qquad \square$$

Notice that we have a natural one-to-one correspondence between vectors in $\mathbf{R}^n$, row n-vectors and column n-vectors defined by:

$$(x_1, x_2, \ldots, x_n) \longleftrightarrow [x_1 \; x_2 \; \ldots \; x_n] \longleftrightarrow \begin{bmatrix} x_1 \\ x_2 \\ \vdots \\ x_n \end{bmatrix}.$$

Further, this correspondence is preserved under the operations of addition and scalar multiplication defined for vectors in $\mathbf{R}^n$ and for row and column vectors. For this reason, we will often treat vectors in $\mathbf{R}^n$, row n-vectors, and column n-vectors as if they are just different representations of the same basic object. We have in fact already adopted this convention in writing the solution to an $m \times n$ linear system $A\mathbf{x} = \mathbf{b}$ in the equivalent forms

$$\mathbf{x} = (x_1, x_2, \ldots, x_n) \quad \text{or} \quad \mathbf{x} = \begin{bmatrix} x_1 \\ x_2 \\ \vdots \\ x_n \end{bmatrix}.$$

EXERCISES 5.1

1. If $\mathbf{x} = (3, 1)$, $\mathbf{y} = (-1, 2)$, determine the vectors $\mathbf{v}_1 = 2\mathbf{x}$, $\mathbf{v}_2 = 3\mathbf{y}$, $\mathbf{v}_3 = 2\mathbf{x} + 3\mathbf{y}$. Sketch the corresponding points in the xy-plane and the equivalent geometric vectors.

2. If $\mathbf{x} = (3, -1, 2, 5)$ and $\mathbf{y} = (-1, 2, 9, -2)$, determine $\mathbf{v} = 5\mathbf{x} - 7\mathbf{y}$ and its additive inverse.

3. Verify the commutative law of addition for vectors in $\mathbf{R}^4$.

4. Verify the associative law of addition for vectors in $\mathbf{R}^4$.

5. Verify properties $(5.1.5) - (5.1.8)$ for vectors in $\mathbf{R}^3$.

5.2   DEFINITION OF A VECTOR SPACE

In the previous section, we showed how the set $\mathbf{R}^n$ of all ordered n-tuples of real numbers, together with the addition and scalar multiplication operations defined on it, has the same algebraic properties as the familiar algebra of geometric vectors. We now push this abstraction one step further and introduce the idea of a vector space. Such an abstraction will enable us to develop a mathematical framework for studying a broad class of linear problems, such as systems of linear algebraic equations, linear DE, and systems of linear differential equations, which have far reaching applications in all areas of applied mathematics, science and engineering.

Let V be a nonempty set. For our purposes, it is useful to call the elements of V vectors and use the usual vector notation $\mathbf{u}$, $\mathbf{v}$, ..., to denote these elements. For example, if V is the set of all 2×2 matrices, then the vectors in V are 2×2 matrices, whereas if V is the set of all positive integers, then the vectors in V are positive integers. We will only be interested in the case when the set V has an addition operation and a scalar multiplication operation defined on its elements in the following senses:

Vector Addition: A rule for combining any two vectors in V. We will use the usual + sign to denote an addition operation, and the result of *adding* the vectors $\mathbf{u}$ and $\mathbf{v}$ will be denoted $\mathbf{u} + \mathbf{v}$.

Real (complex) scalar multiplication: A rule for combining each vector in V with any real (complex) number. We will use the usual notation $k\mathbf{v}$ to denote the result of *scalar multiplying* the vector $\mathbf{v}$ by the real (complex) number k.

To combine the two types of scalar multiplication, we let F denote the set of scalars for which the operation is defined. Thus, for us, F is either the set of all real numbers or the set of all complex numbers. For example, if V is the set of all 2×2 matrices and F denotes the set of all complex numbers, then the usual operation of matrix addition is an addition operation on V, and the usual method of multiplying a matrix by a scalar is a scalar multiplication operation on V. Notice that the result of applying either of these operations is always another vector (2×2 matrix) in V. As a further example, let V be the set of positive integers, and let F be the set of all real numbers, then the usual operations of addition and multiplication within the real numbers define addition and scalar multiplication operations on V. Note in this case, however, that the scalar multiplication operation, in general, will not yield another vector in V, since when we multiply a positive integer by a real number, the result is not, in general, a positive integer.

We are now in a position to give a precise definition of a vector space.

Definition   5.2.1:   Let V be a nonempty set (whose elements we call vectors) on which is defined an addition operation and a scalar multiplication operation with scalars in F. V is called a **vector space over F**, provided the following conditions are satisfied:

A1.   *Closure under addition:* For each pair of vectors $\mathbf{u}$, and $\mathbf{v}$ in V, the sum $\mathbf{u} + \mathbf{v}$ is also in V. We say that V is closed under addition.

A2.   *Closure under scalar multiplication:* For each vector $\mathbf{v}$ in V and each scalar k, the scalar multiple $k\mathbf{v}$ is in V. We say that V is closed under scalar multiplication.

A3.   *Commutativity of addition:* For all $\mathbf{u}, \mathbf{v} \in V$: $\mathbf{u} + \mathbf{v} = \mathbf{v} + \mathbf{u}$.

A4.   *Associativity of Addition:* For all $\mathbf{u}, \mathbf{v}, \mathbf{w} \in V$: $(\mathbf{u} + \mathbf{v}) + \mathbf{w} = \mathbf{u} + (\mathbf{v} + \mathbf{w})$.

A5.   *Existence of a zero vector in V:* In V there is a vector, denoted $\mathbf{0}$, satisfying
$$\mathbf{v} + \mathbf{0} = \mathbf{v}, \quad \text{for all } \mathbf{v} \in V.$$
We call $\mathbf{0}$ the **zero vector** in V.

A6.   *Existence of an additive inverse in V:* For each vector $\mathbf{v} \in V$ there is a vector, denoted $-\mathbf{v}$, in V such that
$$\mathbf{v} + (-\mathbf{v}) = \mathbf{0}.$$
We call $-\mathbf{v}$ the **additive inverse** of $\mathbf{v}$.

A7.   *Unit property:* For all $\mathbf{v} \in V$:
$$1\mathbf{v} = \mathbf{v}.$$

A8.   *Associativity of scalar multiplication:*   For all $\mathbf{v} \in V$ and all scalars $r, s \in F$:
$$(rs)\mathbf{v} = r(s\mathbf{v}).$$

A9.   *Distributive property of scalar multiplication over vector addition:* For all $\mathbf{u}, \mathbf{v} \in V$ and all scalars $r \in F$:
$$r(\mathbf{u} + \mathbf{v}) = r\mathbf{u} + r\mathbf{v}.$$

A10. *Distributive property of scalar multiplication over scalar addition:* For all $\mathbf{v} \in V$ and all scalars $r, s \in F$:
$$(r + s)\mathbf{v} = r\mathbf{v} + s\mathbf{v}.$$

REMARKS

1. A key point to note is that in order to define a vector space, we must start with all of the following:

   (a)   A nonempty set of vectors, V.

   (b)   A set of scalars F (either $\mathbf{R}$ or $\mathbf{C}$).

   (c)   An addition operation defined on V.

   (d)   A scalar multiplication operation defined on V.

   Then we must check that the axioms A1–A10 are satisfied.

2. Terminology: A vector space over the real numbers will be referred to as a **real vector space**, whereas a vector space over the complex numbers will be called a **complex vector space**.

3. As indicated in Definition 5.2.1, we will use bold print to denote vectors in a general vector space. In hand writing it is strongly advised that you denote vectors either as $\vec{v}$ or as $\underset{\sim}{v}$. This will avoid any confusion between vectors in V and scalars in F.

4. When we deal with a familiar vector space, we will use the usual notation for vectors in the space. For example, as shown below, the set of all real-valued functions defined on an interval is a vector space, and we will denote the vectors in this vector space by f, g, ...

EXAMPLES OF VECTOR SPACES

1. The set of all real numbers together with the usual operations of addition and multiplication is a real vector space.

2. The set of all complex numbers is a complex vector space when we use the usual operations of addition and multiplication by a complex number.

3. The set $\mathbf{R}^n$ together with the operations of addition and scalar multiplication defined in (5.1.14), (5.1.15) is a real vector space. As we saw in the previous section, the zero vector in $\mathbf{R}^n$ is the n-tuple of zeros $(0, 0, ..., 0)$, and the additive inverse of the vector $\mathbf{x} = (x_1, x_2, ..., x_n)$ is $-\mathbf{x} = (-x_1, -x_2, ..., -x_n)$.

Example 5.2.1   Let V be the set of all 2×2 matrices with real elements. Show that V, together with the usual operations of matrix addition and multiplication of a matrix by a real number, is a real vector space.

Solution   If A and B are in V (that is, are 2×2 matrices with real entries), then $A + B$ and kA are in V for all real numbers k. Consequently, V is closed under addition and scalar multiplication, and, therefore, Axioms 1 and 2 of the vector space definition hold. We now check the remaining axioms

A3. This axiom is satisfied, since, as we have shown in Chapter 3, matrix addition is commutative.

A4. Once more, the associativity follows directly from our discussion of matrix algebra.

A5. If A is any matrix in V and $0_2 = \begin{bmatrix} 0 & 0 \\ 0 & 0 \end{bmatrix}$, then

$$A + 0_2 = A.$$

Thus, 0_2 is the zero vector in V.

A6. The additive inverse of $A = \begin{bmatrix} a & b \\ c & d \end{bmatrix}$ is $-A = \begin{bmatrix} -a & -b \\ -c & -d \end{bmatrix}$, since

$$A + (-A) = \begin{bmatrix} 0 & 0 \\ 0 & 0 \end{bmatrix} = 0_2.$$

A7. – A10. Each of these are basic properties of matrix algebra.

Thus V, together with the given operations, is a real vector space.   ☐

REMARK   In a similar manner to the previous example, it is easily established that the set of all $m \times n$ matrices with real entries is a real vector space when we use the usual operations of addition of matrices and multiplication of matrices by a real number. Notationally we will denote the vector space of all $n \times n$ matrices with real elements by $M_n(\mathbf{R})$.

Example 5.2.2   Let V be the set of all real-valued functions defined on an interval I. Define addition and scalar multiplication in V as follows. If f and g are in V and k is any real number, then $f + g$ and kf are defined by

$$(f + g)(x) = f(x) + g(x) \text{ for all } x \in I,$$

$$(kf)(x) = kf(x) \text{ for all } x \in I.$$

Show that V, together with the given operations of addition and scalar multiplication, is a real vector space.

Solution   It follows from the given definitions of addition and scalar multiplication that if f and g are in V and k is any real number, then $f + g$ and kf are both real-valued functions on I and are therefore in V. Consequently, the closure axioms A1 and A2 hold. We now check the remaining axioms.

A3. Let f and g be arbitrary functions in V. From the definition of function addition, we have

$$(f + g)(x) = f(x) + g(x) = g(x) + f(x) = (g + f)(x) \text{ for all } x \in I.$$

Consequently, $f + g = g + f$, and so, function addition is commutative.

A4. Let $f, g, h \in V$. Then, for all x in I,

$$\begin{aligned}[(f + g) + h](x) &= (f + g)(x) + h(x) = f(x) + g(x) + h(x)\\ &= f(x) + [g(x) + h(x)] = f(x) + (g + h)(x)\\ &= [f + (g + h)](x).\end{aligned}$$

Consequently, $(f + g) + h = f + (g + h)$, so that function addition is indeed associative.

A5. If we define the zero function, 0, by $0(x) = 0$, for all x in I, then

$$(f + 0)(x) = f(x) + 0(x) = f(x) + 0 = f(x), \text{ for all } f \in V \text{ and all } x \in I,$$

which implies that $f + 0 = f$. Hence, 0 is the zero vector in V. (See Figure 5.2.1.)

A6. If $f \in V$, then $-f$ is defined by $(-f)(x) = -f(x)$ for all x in I, since

$$[f + (-f)](x) = f(x) + (-f)(x) = f(x) - f(x) = 0$$

for all x in I, which implies that $f + (-f) = 0$.

A7. Let $f \in V$. Then, by definition of the scalar multiplication operation, for all $x \in I$,

$$(1f)(x) = 1f(x) = f(x).$$

Consequently, $1f = f$.

A8. Let $f \in V$, and let $r, s \in \mathbf{R}$. Then, for all $x \in I$,

$$[(rs)f](x) = (rs)f(x) = r[sf(x)] = r[(sf)(x)] .$$

Hence, $(rs)f = r(sf)$.

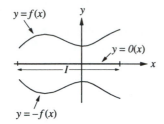

Figure 5.2.1 In the vector space of all functions defined on an interval I, the additive inverse of a given function is obtained by reflecting the graph of the function about the x-axis. The zero vector is the zero function $0(x)$.

A9. Let $f, g \in V$, and let $r \in \mathbf{R}$. Then, for all $x \in I$,

$$[r(f + g)](x) \ = r(f + g)(x) = r[f(x) + g(x)] = rf(x) + rg(x)$$
$$= (rf)(x) + (rg)(x) = (rf + rg)(x).$$

Consequently $r(f + g) = rf + rg$.

A10. Let $f \in V$, and let $r, s \in \mathbf{R}$. Then for all $x \in I$,

$$[(r + s)f](x) = (r + s)f(x) = rf(x) + sf(x) = (rf)(x) + (sf)(x) = (rf + sf)(x).$$

Consequently, $(r + s)f = rf + sf$.

Since all parts of Definition 5.2.1 are satisfied, it follows that V, together with the given operations of addition and scalar multiplication, is a real vector space. ❏

REMARK: As the previous example indicates, a full verification of the vector space definition can be somewhat tedious, although it is usually straightforward. Be careful to not leave out any important steps in such a verification.

THE VECTOR SPACE $\mathbf{C}^n$

We now introduce the most important complex vector space. Let $\mathbf{C}^n$ denote the set of all ordered n-tuples of *complex numbers*. Thus,

$$\mathbf{C}^n = \{(z_1, z_2, \ldots, z_n) : z_1, z_2, \ldots, z_n \in \mathbf{C}\}.$$

We refer to the elements of $\mathbf{C}^n$ as **vectors in $\mathbf{C}^n$**. A typical vector in $\mathbf{C}^n$ is $(z_1, z_2, \ldots, z_n)$, where each z_k is a complex number.

Example 5.2.3   The following are examples of vectors in $\mathbf{C}^2$ and $\mathbf{C}^4$, respectively:

$$\mathbf{u} = (2.1 - 3i, -1.5 + 3.9i), \qquad \mathbf{v} = (5 + 7i, 2 - i, 3 + 4i, -9 - 17i). \qquad ❏$$

In order to obtain a vector space, we must define appropriate operations of addition and multiplication by a scalar in $\mathbf{C}^n$. Based on the corresponding operations in $\mathbf{R}^n$, we define the addition and scalar multiplication operations componentwise. Thus, if $\mathbf{u} = (u_1, u_2, \ldots, u_n)$ and $\mathbf{v} = (v_1, v_2, \ldots, v_n)$ are vectors in $\mathbf{C}^n$ and k is an arbitrary *complex number*, then

$$\mathbf{u} + \mathbf{v} \ = (u_1 + v_1, u_2 + v_2, \ldots, u_n + v_n),$$
$$k\mathbf{u} \ = (ku_1, ku_2, \ldots, ku_n).$$

Example 5.2.4   If $\mathbf{u} = (1 - 3i, 2 + 4i)$, $\mathbf{v} = (-2 + 4i, 5 - 6i)$ and $k = 2 + i$, find $\mathbf{u} + k\mathbf{v}$.

Solution

$$
\begin{aligned}
\mathbf{u} + k\mathbf{v} &= (1 - 3i, 2 + 4i) + (2 + i)(-2 + 4i, 5 - 6i) \\
&= (1 - 3i, 2 + 4i) + (-8 + 6i, 16 - 7i) \\
&= (-7 + 3i, 18 - 3i).
\end{aligned}
$$
                                          ❒

It is straightforward to show that $\mathbf{C}^n$, together with the given operations of addition and scalar multiplication, is a *complex* vector space.

FURTHER PROPERTIES OF VECTOR SPACES

The main reason for formalizing the definition of a vector space is that any results that we can prove based solely on the definition will then apply to all vector spaces. (That is, we do not have to prove separate results for geometric vectors, $n \times n$ matrices, vectors in $\mathbf{R}^n$ or $\mathbf{C}^n$, or real-valued functions etc.) The next theorem lists some results that can be proved using the vector space axioms.

Theorem 5.2.1:   Let V be a vector space.

1. The zero vector is unique, and

  (a) $0\mathbf{u} = \mathbf{0}$, for all $\mathbf{u} \in V$,   (b) $k\mathbf{0} = \mathbf{0}$ for all scalars $k \in F$.

2. The additive inverse of each element in V is unique, and

$$-\mathbf{u} = (-1)\mathbf{u} \text{ for all } \mathbf{u} \in V.$$

PROOF
  Uniqueness of the zero vector: Suppose that there were two zero vectors in V, denoted $\mathbf{0}_1$ and $\mathbf{0}_2$. Then, for any $\mathbf{v} \in V$, we would have

$$\mathbf{v} + \mathbf{0}_1 = \mathbf{v} \qquad\qquad (5.2.1)$$

and

$$\mathbf{v} + \mathbf{0}_2 = \mathbf{v}. \qquad\qquad (5.2.2)$$

We must prove that $\mathbf{0}_1 = \mathbf{0}_2$. But, applying (5.2.1) with $\mathbf{v} = \mathbf{0}_2$, we have

$$
\begin{aligned}
\mathbf{0}_2 &= \mathbf{0}_2 + \mathbf{0}_1 \\
&= \mathbf{0}_1 + \mathbf{0}_2 \quad \text{(Axiom A3)} \\
&= \mathbf{0}_1 \text{ (from (5.2.2) with } \mathbf{v} = \mathbf{0}_1\text{)}.
\end{aligned}
$$

Consequently, $\mathbf{0}_1 = \mathbf{0}_2$, and so, the zero vector is unique in a vector space.

  Proof of property (1a). Let $\mathbf{v}$ be an arbitrary element in a vector space V. Then

$$0 = 0 + 0$$

implies that

$$0\mathbf{v} = (0 + 0)\mathbf{v}.$$

Using Axiom A10, this can be written as

$$0\mathbf{v} = 0\mathbf{v} + 0\mathbf{v}.$$

Adding $-0\mathbf{v}$ to both sides of this equation yields

$$0\mathbf{v} + (-0\mathbf{v}) = (0\mathbf{v} + 0\mathbf{v}) + (-0\mathbf{v}).$$

That is, since addition in a vector space is associative (Axiom A4),

$$0\mathbf{v} + (-0\mathbf{v}) = 0\mathbf{v} + [0\mathbf{v} + (-0\mathbf{v})].$$

Using Axiom A6, this can be written as

$$\mathbf{0} = 0\mathbf{v} + \mathbf{0}.$$

That is, using Axiom A5,

$$\mathbf{0} = 0\mathbf{v}.$$

The remaining proofs are left as exercises.   ∎

We end this section with a list of the most important vector spaces that will be required throughout the remainder of the text. In each case the addition and scalar multiplication operations are the usual ones associated with the set of vectors.

1. $\mathbf{R}^n$, the (real) vector space of all ordered n-tuples of real numbers.

2. $\mathbf{C}^n$, the (complex) vector space of all ordered n-tuples of complex numbers.

3. $M_n(\mathbf{R})$, the (real) vector space of all $n \times n$ matrices with real elements.

4. $C^k(I)$, the vector space of all real-valued functions that are continuous and have (at least) k continuous derivatives on I. We will show that this set of vectors is a (real) vector space in the next section.

5. P_n, the (real) vector space of all real-valued polynomials of degree less than n with real coefficients. That is

$$P_n = \{a_0 + a_1x + a_2x^2 + \cdots + a_{n-1}x^{n-1} : a_0, a_1, \ldots, a_{n-1} \in \mathbf{R}\}.$$

We leave the verification that P_n is a (real) vector space as an exercise.

EXERCISES 5.2

For problems 1–5, determine whether the given set of vectors is closed under addition and closed under scalar multiplication. In each case take the set of scalars to be the set of all real numbers.

1. The set of all rational numbers.

2. The set of all upper triangular matrices with real elements.

3. The set of all solutions to the DE $y'' + 9y = 4x^2$. (Do not solve the DE.)

4. The set of all solutions to the DE $y'' + 9y = 0$ (Do not solve the DE).

5. The set of all solutions to the homogeneous linear system $A\mathbf{x} = \mathbf{0}$.

6. Let $S = \{A \in M_2(\mathbf{R}) : \det(A) = 0\}$.

(a) Is the zero vector from $M_2(\mathbf{R})$ in S?

(b) Give an explicit example illustrating that S is not closed under matrix addition.

(c) Is S closed under scalar multiplication? Justify your answer.

7. Let $\mathbf{N} = \{1, 2, \ldots\}$ denote the set of all positive integers. Give three reasons why $\mathbf{N}$, together with the usual operations of addition and scalar multiplication, is not a real vector space.

8. We have defined the set $\mathbf{R}^2 = \{(x, y) : x \in \mathbf{R}, y \in \mathbf{R}\}$ together with the addition and scalar multiplication operations as follows:

$$\mathbf{x} + \mathbf{y} = (x_1, x_2) + (y_1, y_2) = (x_1 + y_1, x_2 + y_2)$$
$$k\mathbf{x} = k(x_1, x_2) = (kx_1, kx_2).$$

Give a complete verification that each of the vector space axioms is satisfied.

9. Consider the set of all 2×3 matrices, together with the usual operations of matrix addition and scalar multiplication. Determine the zero vector in this set, and the additive inverse of

a general element. (Note that the vector space axioms $1-4$ and $7-10$ follow directly from matrix algebra.)

10. Let P denote the set of all polynomials of degree precisely 2. Is P a vector space?

11. On $\mathbf{R}^2$ define the operations of addition and scalar multiplication as follows:

$$(x_1, x_2) + (y_1, y_2) = (x_1 + y_1, x_2 y_2),$$

$$k(x_1, x_2) = (kx_1, x_2{}^k).$$

Determine whether $\mathbf{R}^2$, together with these algebraic operations, is a vector space.

12. On $\mathbf{R}^2$ define the operations of addition and multiplication by a real number as follows:

$$(x_1, x_2) + (y_1, y_2) = (x_1 - y_1, x_2 - y_2),$$

$$k(x_1, x_2) = (-kx_1, -kx_2).$$

Which of the axioms for a vector space are satisfied by $\mathbf{R}^2$ with these algebraic operations?

13. On $\mathbf{R}^2$ define the operation of addition by

$$(x_1, x_2) + (y_1, y_2) = (x_1 y_1, x_2 y_2).$$

Do axioms A5 and A6 in the definition of a vector space hold ?

14. On $M_2(\mathbf{R})$ define the operation of addition by

$$A + B = AB.$$

Determine which axioms for a vector space are satisfied by $M_2(\mathbf{R})$ with the above addition operation and the usual scalar multiplication operation.

15. In $M_2(\mathbf{R})$ define the operations of addition and multiplication by a real number ($\oplus$ and $\bullet$ respectively) as follows:

$$A \oplus B = -(A + B),$$

$$k \bullet A = -kA,$$

where the operations on the right-hand side are the usual ones associated with $M_2(\mathbf{R})$. Determine which of the axioms for a vector space are satisfied by $M_2(\mathbf{R})$ with the operations $\oplus$ and $\bullet$.

For problems 16 and 17, verify that the given set of objects together with the usual operations of addition and scalar multiplication is a *complex* vector space.

16. $\mathbf{C}^2$.

17. $M_2(\mathbf{C})$, the set of all 2×2 matrices with complex elements.

18. Is $\mathbf{C}^3$ a *real* vector space ?

19. Prove property (1b) of Theorem 5.2.1.

20. Prove property (2) of Theorem 5.2.1.

5.3   SUBSPACES

Let us try to make contact between the abstract vector space idea and the solution of an applied problem. Vector spaces generally arise as the sets containing the unknowns in a given problem. For example, if we are solving a DE, then the basic unknown is a function, and therefore any solution to the DE will be an element of the vector space V of all functions defined on an appropriate interval. Consequently, the solution set of a DE is a subset of V. Similarly, consider the system of linear equations $A\mathbf{x} = \mathbf{b}$, where A is an $m \times n$ matrix with real elements. The basic unknown in this system, $\mathbf{x}$, is a column n–vector, or equivalently a vector in $\mathbf{R}^n$. Consequently, the solution set to the system is a subset of the vector space $\mathbf{R}^n$. As these examples illustrate, the solution set of an applied problem is generally a subset of vectors from an appropriate vector space. The question that we will need to answer in the future is whether this subset of vectors is a vector space in its own right. The following definition introduces the terminology we will use:

> **Definition   5.3.1:**   Let S be a nonempty subset of a vector space V. If S is itself a vector space under the same operations of addition and scalar multiplication as used in V, then we say that S is a **subspace** of V.

In establishing that a given subset of vectors from a vector space V is a subspace of V, it would appear as though we must check that each of the axioms in the vector space definition are satisfied when we restrict our attention to vectors lying only in S. The first and most important theorem of the section tells us that all we need do in fact is check the closure axioms 1 and 2, and if these are satisfied, then the remaining axioms necessarily hold in S. This is a very useful theorem that will be applied on several occasions throughout the remainder of the text.

Theorem 5.3.1:   Let S be a nonempty subset of a vector space V. Then S is a subspace of V if and only if S is closed under the operations of addition and scalar multiplication in V.

PROOF   If S is a subspace of V, then it is a vector space, and hence, it is certainly closed under addition and scalar multiplication. Conversely, assume that S is closed under addition and scalar multiplication. We must prove that axioms A3–A10 of Definition 5.2.1 hold when we restrict to vectors in S. Consider first axioms A3, A4, and A7–A10. These are properties of the addition and scalar multiplication operations, and hence, since we use the same operations in S as in V, axioms A3, A4, and A7–A10 necessarily hold when we restrict to vectors in S.[1] That A5 and A6 also hold for vectors in S can be established as follows. If $\mathbf{u}$ is in S, then, since S is closed under scalar multiplication, both $0\mathbf{u}$, and $(-1)\mathbf{u}$ are in S. Thus, from Theorem 5.2.1, $\mathbf{0}$ and $-\mathbf{u}$ are in S, and therefore, A5 and A6 are satisfied.   ■

In determining whether a subset S of a vector space V is a subspace of V, we must keep clear in our minds what the given vector space is and what conditions on the vectors in V restrict them to lie in the subset S. This is most easily done by expressing S in set notation as follows:

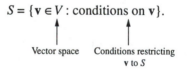

$$S = \{\mathbf{v} \in V : \text{conditions on } \mathbf{v}\}.$$

Vector space   Conditions restricting $\mathbf{v}$ to S

Example 5.3.1   Verify that the set of all real solutions to the following linear system is a subspace of $\mathbf{R}^3$:

$$x_1 + 2x_2 - x_3 = 0,$$
$$2x_1 + 5x_2 - 4x_3 = 0.$$

Solution   The RREF of the augmented matrix of the system is

$$\begin{bmatrix} 1 & 0 & 3 & 0 \\ 0 & 1 & -2 & 0 \end{bmatrix},$$

so that the solution set of the system is

$$S = \{\mathbf{x} \in \mathbf{R}^3 : \mathbf{x} = (-3r, 2r, r), r \in \mathbf{R}\},$$

---

[1]Note that the closure properties are tacitly used here in ensuring that all the vectors appearing in the axioms are indeed vectors in S.

which is a nonempty subset of $\mathbf{R}^3$. If $\mathbf{x} = (-3r, 2r, r)$ and $\mathbf{y} = (-3s, 2s, s)$ are any two vectors in S, then

$$\mathbf{x} + \mathbf{y} = (-3(r + s), 2(r + s), r + s) = (-3t, 2t, t),$$

where $t = r + s$. Consequently, if we add two vectors in S the result is another vector in S. Similarly, if we multiply an arbitrary vector $\mathbf{x} = (-3r, 2r, r)$ in S by a real number k, the resulting vector is

$$k\mathbf{x} = (-3kr, 2kr, kr) = (-3w, 2w, w),$$

where $w = kr$. Hence, whenever a vector in S is multiplied by a real number, the resulting vector also lies in S. We have therefore established that S is closed under both addition and scalar multiplication. Consequently, Theorem 5.3.1 can be applied to conclude that S is a subspace of $\mathbf{R}^3$. Geometrically, the vectors in S lie along the line of intersection of the planes with the given equations. This is the line through the origin in the direction of the vector $\mathbf{v} = (-3, 2, 1)$. (See Figure 5.3.1.)

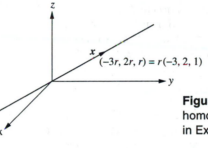

Figure 5.3.1 The solution set to the homogeneous system of linear equations in Example 5.3.1 is a subspace of R^3.

❐

Example 5.3.2   Verify that $S = \{\mathbf{x} \in \mathbf{R}^2 : \mathbf{x} = (r, -3r + 1), r \in \mathbf{R}\}$ is not a subspace of $\mathbf{R}^2$.

Solution   Geometrically, the points in S correspond to those points that lie on the line with Cartesian equation $y = -3x + 1$. Since this line does not pass through the origin, S does not contain the zero vector $(0, 0)$, and therefore, S is not a subspace of $\mathbf{R}^2$. The fact that S is not a subspace of $\mathbf{R}^2$ can also be established by the failure of the closure properties as follows. Let $\mathbf{x}$ and $\mathbf{y}$ be vectors in S. Then

$$\mathbf{x} = (r, -3r + 1), \quad \mathbf{y} = (s, -3s + 1),$$

for appropriate real numbers r and s. Hence,

$$\mathbf{x} + \mathbf{y} = (r, -3r + 1) + (s, -3s + 1)$$
$$= (r + s, -3(r + s) + 2) = (w, -3w + 2),$$

where $w = r + s$. We see that $\mathbf{x} + \mathbf{y}$ is not an element of S. Similarly, for any real number k,

$$k\mathbf{x} = k(r, -3r + 1) = (kr, -3kr + k),$$

which, for $k \neq 1$, is not a vector in S.

Example 5.3.3   Let S denote the set of all real symmetric $n \times n$ matrices. Verify that S is a subspace of $M_n(\mathbf{R})$.

Solution   The subset of interest is

$$S = \{A \in M_n(\mathbf{R}) : A^T = A\}.$$

S is nonempty since, for example, it contains the zero matrix 0_n. We now check for closure. Let A and B be in S. Then

$$A^T = A \quad \text{and} \quad B^T = B.$$

Using these conditions and the properties of the transpose yields

$$(A + B)^T = A^T + B^T = A + B$$

and

$$(kA)^T = kA^T = kA, \quad \text{for all real } k.$$

Consequently $A + B$ and kA are both symmetric matrices, and so, they are elements of S. Hence S is closed under both addition and scalar multiplication and so is indeed a subspace of $M_n(\mathbf{R})$.

Example 5.3.4   Let V be the vector space of all real-valued functions defined on an interval $[a, b]$, and let S denote the set of all functions in V that satisfy $f(a) = 0$. Verify that S is a subspace of V.

Solution   We have

$$S = \{f \in V : f(a) = 0\}$$

which is nonempty, since it contains, for example, the zero function

$$0(x) = 0 \text{ for all } x \text{ in } [a, b].$$

If f and g are in S, then $f(a) = 0$ and $g(a) = 0$. We now check for closure under addition and scalar multiplication. We have

$$(f + g)(a) = f(a) + g(a) = 0 + 0 = 0,$$

which implies that S is closed under addition. Further, if k is any real number,

$$(kf)(a) = kf(a) = k0 = 0,$$

so that S is also closed under scalar multiplication. Theorem 5.3.1 therefore implies that S *is* a subspace of V. Some representative functions from S are sketched in Figure 5.3.2.

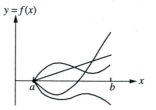

Figure 5.3.2 Some functions in the subspace considered in Example 5.3.4. Each function satisfies $f(a) = 0$.

In the next theorem we establish that the subset $\{\mathbf{0}\}$ of a vector space V is in fact a subspace of V. We call this subspace the **trivial subspace** of V. Any set of vectors from a vector space V that does not consist only of the zero vector will be called a *non–trivial* subset of V.

Theorem 5.3.2:   Let V be a vector space with zero vector $\mathbf{0}$. Then $S = \{\mathbf{0}\}$ is a subspace of V.

PROOF S is certainly nonempty. Further, the closure of S under addition and scalar multiplication follow, respectively, from

$$0 + 0 = 0,$$

$$k0 = 0. \text{ (See Theorem 5.2.1.)}$$

Consequently, S is a subspace of V.   ∎

We now use Theorem 5.3.1 to establish an important result pertaining to homogeneous systems of linear equations that has already been illustrated in Example 5.3.1.

Theorem 5.3.3:   Let A be an $m \times n$ matrix. The solution set of the homog–eneous system of linear equations $Ax = 0$ is a subspace of $\mathbf{C}^n$.

PROOF   Let S denote the solution set of the homogeneous linear system. Then we can write

$$S = \{x \in \mathbf{C}^n : Ax = 0\},$$

a subset of $\mathbf{C}^n$. S is nonempty, since a homogeneous system always admits the trivial solution $x = 0$. If x and y are in S, then

$$Ax = 0 \quad \text{and} \quad Ay = 0.$$

Setting $u = x + y$ and using properties of the matrix product, we have

$$Au = A(x + y) = Ax + Ay = 0 + 0 = 0,$$

so that $x + y$ also solves the system and therefore is in S. Furthermore, if k is any scalar, and $v = kx$, then

$$Av = A(kx) = kAx = k0 = 0,$$

so that kx is also a solution of the system and therefore is in S. Since S is closed under both addition and scalar multiplication, it follows from Theorem 5.3.1 that S is a subspace of $\mathbf{C}^n$.   ∎

The preceding theorem has established that the solution set to any homogeneous linear system of equations is a vector space. Due to the importance of this vector space it is given a special name.

Definition 5.3.2:   Let A be an $m \times n$ matrix. The solution set to the corres–ponding homogeneous linear system $Ax = 0$ is called the **null space of** A and is denoted nullspace(A). Thus,

$$\text{nullspace}(A) = \{x \in \mathbf{C}^n : Ax = 0\}.$$

REMARKS

1. If the matrix A has real elements, then we will consider only the corresponding real solutions to $Ax = 0$. Consequently, in this case,

$$\text{nullspace}(A) = \{x \in \mathbf{R}^n : Ax = 0\},$$

a subspace of $\mathbf{R}^n$.

2. We leave it as an exercise to establish that the previous theorem does not hold for the solution set of a *nonhomogeneous* linear system.

Next we introduce the vector space of primary importance in the study of linear DE. This vector space arises as a subspace of the vector space of all functions that are defined on an interval I.

Example 5.3.5   Let V denote the vector space of all functions that are defined on an interval I, and let $C^k(I)$ denote the set of all functions that are continuous and have (at least) k continuous derivatives on the interval I. Show that $C^k(I)$ is a subspace of V.

Solution   In this case,

$$C^k(I) = \{f \in V : f, f', f'', \dots, f^{(k)} \text{ exist and are continuous on } I\}.$$

It follows from the properties of derivatives, that if we add two functions in $C^k(I)$ the result is a function in $C^k(I)$, and similarly, if we multiply a function in $C^k(I)$ by a scalar, then the result is a function in $C^k(I)$. Thus, Theorem 5.3.1 implies that $C^k(I)$ is a subspace of V.

REMARK   In general, k will be a positive integer. We use the notation $C^\infty(I)$ to denote the vector space of all functions that have continuous derivatives of all orders on I.

Our final result ties together the ideas introduced in this section with those developed in Chapter 2.

Theorem 5.3.4:   The set of all solutions to the homogeneous *linear* DE

$$y'' + a_1 y' + a_2 y = 0 \tag{5.3.1}$$

on an interval I is a vector space.

PROOF   Let S denote the set of all solutions to the given DE. Then S is a nonempty subset of $C^2(I)$ (Why?). We establish that S is in fact a subspace of $C^2(I)$. Let y_1 and y_2 be in S. Then,

$$y_1'' + a_1 y_1' + a_2 y_1 = 0, \text{ and } y_2'' + a_1 y_2' + a_2 y_2 = 0.$$

It follows directly from the linearity properties for second-order linear DE proved in Theorem 2.1.2 that

$$u = y_1 + y_2, \text{ and } v = k y_1$$

also solve the DE, and therefore, S is closed under both addition and scalar multiplication. Consequently, the set of all solutions to equation (5.3.1) is a subspace of $C^2(I)$.   ∎

We will refer to the set of all solutions to a DE of the form (5.3.1) as the **solution space** of the DE.

EXERCISES 5.3

1. Let $S = \{x \in \mathbf{R}^2 : x = (2k, -3k), k \in \mathbf{R}\}$.

(a) Establish that S is a subspace of $\mathbf{R}^2$.

(b) Make a sketch depicting the subspace S in the Cartesian plane.

2. Let $S = \{x \in \mathbf{R}^3 : x = (r - 2s, 3r+ s, s), r, s \in \mathbf{R}\}$.

(a) Establish that S is a subspace of $\mathbf{R}^3$.

(b) Show that the vectors in S lie on the plane with equation $3x - y + 7z = 0$.

For problems 3–19, express S in set notation, and determine whether it is a subspace of the given vector space V.

3. $V = \mathbf{R}^2$, and S is the set of all vectors (x_1, x_2) in V satisfying $3x_1 + 2x_2 = 0$.

4. $V = \mathbf{R}^4$, and S is the set of vectors of the form $(x_1, 0, x_3, 2)$.

5. $V = \mathbf{R}^3$, and S is the set of all vectors (x_1, x_2, x_3) in V satisfying $x_1 + x_2 + x_3 = 1$.

6. $V = \mathbf{R}^n$, and S is the set of all solutions to the nonhomogeneous linear system $A\mathbf{x} = \mathbf{b}$, where A is a fixed $m \times n$ matrix and $\mathbf{b}$ ($\neq \mathbf{0}$) is a fixed vector.

7. $V = \mathbf{R}^2$, and S consists of all vectors (x_1, x_2) satisfying $x_1^2 - x_2^2 = 0$.

8. $V = M_2(\mathbf{R})$, and S is the subset of all 2×2 matrices with $\det(A) = 1$.

9. $V = M_n(\mathbf{R})$, and S is the subset of all $n \times n$ lower triangular matrices.

10. $V = M_n(\mathbf{R})$, and S is the subset of all non-singular matrices.

11. $V = M_2(\mathbf{R})$, and S is the subset of all 2×2 *symmetric* matrices

12. $V = M_2(\mathbf{R})$, and S is the subset of all 2×2 *skew-symmetric* matrices.

13. V is the vector space of all real-valued functions defined on the interval $[a, b]$, and S is the subset of V consisting of those functions satisfying $f(a) = f(b)$.

14. V is the vector space of all real-valued functions defined on the interval $[a, b]$, and S is the subset of V consisting of those functions satisfying $f(a) = 1$.

15. V is the vector space of all real-valued functions defined on the interval $(-\infty, \infty)$, and S is

the subset of V consisting of those functions satisfying $f(-x) = -f(x)$, for all x in $(-\infty, \infty)$.

16. $V = P_3$, and S is the subset of P_3 consisting of all polynomials of the form $p(x) = ax^2 + b$.

17. $V = P_3$, and S is the subset of P_3 consisting of all polynomials of the form $p(x) = ax^2 + 1$.

18. $V = C^2(I)$, and S is the subset of V consisting of those functions satisfying the DE

$$y'' + 2y' - y = 0$$

on I.

19. $V = C^2(I)$, and S is the subset of V consisting of those functions satisfying the DE

$$y'' + 2y' - y = 1$$

on I.

For problems 20–22, determine nullspace(A) for the given matrix.

20. $A = \begin{bmatrix} 1 & -2 & 1 \\ 4 & -7 & -2 \\ -1 & 3 & 4 \end{bmatrix}$.

21. $A = \begin{bmatrix} 1 & 3 & -2 & 1 \\ 3 & 10 & -4 & 6 \\ 2 & 5 & -6 & -1 \end{bmatrix}$.

22. $A = \begin{bmatrix} 1 & i & -2 \\ 3 & 4i & -5 \\ -1 & -3i & i \end{bmatrix}$.

23. Show that the set of all solutions to the nonhomogeneous DE

$$y'' + a_1 y' + a_2 y = F(x),$$

on an interval I, is not a subspace of $C^2(I)$.

24. Let S_1 and S_2 be subspaces of a vector space V. Show that the set theoretic intersection $S_1 \cap S_2$ is a subspace of V.

5.4   SPANNING SETS

The key theoretical result that we established in Chapter 2 regarding the homogeneous linear DE

$$y'' + a_1 y' + a_2 y = 0 \tag{5.4.1}$$

was that *every* solution to the DE can be obtained from the linear combination

$$y(x) = c_1 y_1(x) + c_2 y_2(x), \tag{5.4.2}$$

where y_1, y_2 are any two linearly independent (non–proportional) solutions. The power of this result is impressive: It reduces the search for all solutions to equation (5.4.1) to just finding two linearly independent solutions. In vector space terms, the result can be re–stated as follows:

> Every vector in the solution space to the DE (5.4.1) can be written as a linear combination of any two linearly independent solutions y_1, and y_2.

We say that the solution space is **spanned** by y_1, and y_2. For example, the set of all solutions to the DE

$$y'' + \omega^2 y = 0, \tag{5.4.3}$$

where ω is a positive constant, is spanned by $y_1(x) = \cos \omega x$, $y_2(x) = \sin \omega x$. Physically, as we have seen in Section 2.6, the DE (5.4.3) (with t as independent variable) governs the motion of a free undamped spring-mass system. We can therefore conclude that all possible modes of oscillation of the system can be obtained by taking linear combinations of the basic modes

$$y_1(t) = \cos \omega t, \quad y_2(t) = \sin \omega t.$$

The initial conditions in a specific problem pick out the appropriate linear combination of these basic modes that corresponds to the solution of the problem. We now begin our investigation as to whether this type of idea will work more generally when the solution set to a problem is a vector space. For example, what about the solution set to a homogeneous linear system $A\mathbf{x} = \mathbf{0}$? We might suspect that if there are k free variables defining the vectors in nullspace(A), then every solution to $A\mathbf{x} = \mathbf{0}$ could be expressed as a linear combination of k basic solutions. We will establish that this is indeed the case in Section 5.8. The two key concepts we need to generalize are the idea of a vector space being spanned by a set of vectors, and that of linear independence in a general vector space. These will be addressed in the next two sections respectively. First the spanning concept.

The only algebraic operations that are defined in a vector space are those of addition and scalar multiplication. Consequently, the most general way in which we can combine the vectors $\mathbf{v}_1, \mathbf{v}_2, \ldots, \mathbf{v}_k$ is

$$c_1 \mathbf{v}_1 + c_2 \mathbf{v}_2 + \cdots + c_k \mathbf{v}_k, \tag{5.4.4}$$

where $c_1, c_2, \ldots, c_k$ are scalars. An expression of the form (5.4.4) is called a **linear combination** of $\mathbf{v}_1, \mathbf{v}_2, \ldots, \mathbf{v}_k$. Since V is closed under addition and scalar multiplication, it follows that the foregoing linear combination is itself a vector in V. As just indicated, one of the questions that we wish to answer is whether every vector in a vector space can be obtained by taking linear combinations of a finite set of vectors. The following terminology is used in the case when the answer to this question is affirmative:

> **Definition 5.4.1:** If *every* vector in a vector space V can be written as a linear combination of v_1, v_2, ..., v_k, we say that V is **spanned** or **generated** by v_1, v_2, ..., v_k and call the set of vectors $\{v_1, v_2, ..., v_k\}$ a **spanning set** for V.

Although the terminology is new, as we have already seen, the idea is familiar from our work on DE. As a second illustration, we are all used to representing geometric vectors in terms of their components as (see Section 5.1)

$$v = a\mathbf{i} + b\mathbf{j} + c\mathbf{k},$$

where $\mathbf{i}$, $\mathbf{j}$, and $\mathbf{k}$ denote the unit vectors pointing along the positive x-, y- and z- axes, respectively, of a rectangular Cartesian coordinate system. Using this terminology, we say that v has been expressed as a linear combination of the vectors $\mathbf{i}$, $\mathbf{j}$, and $\mathbf{k}$, and that the vector space of all geometric vectors is spanned by $\mathbf{i}$, $\mathbf{j}$, and $\mathbf{k}$.

We now consider several examples to illustrate the spanning concept in different vector spaces.

Example 5.4.1   Show that $\mathbf{R}^2$ is spanned by the vectors

$$v_1 = (1, 1) \text{ and } v_2 = (2, -1).$$

Solution   We must establish that for every $x = (x_1, x_2)$ in $\mathbf{R}^2$, there exist constants c_1, c_2 such that

$$x = c_1 v_1 + c_2 v_2. \tag{5.4.5}$$

That is, in component form,

$$(x_1, x_2) = c_1(1, 1) + c_2(2, -1).$$

Equating corresponding components in this equation yields the following linear system:

$$c_1 + 2c_2 = x_1,$$
$$c_1 - c_2 = x_2.$$

The determinant of the matrix of coefficients of this system is

$$\begin{vmatrix} 1 & 2 \\ 1 & -1 \end{vmatrix} = -3.$$

Since this is nonzero, the system has a unique solution no matter what the values of x_1, x_2 are. Consequently, equation (5.4.5) can be satisfied for any vector x in $\mathbf{R}^2$, and so, the given vectors do span $\mathbf{R}^2$. Indeed, solving the linear system yields

$$c_1 = \frac{1}{3}(x_1 + 2x_2), \qquad c_2 = \frac{1}{3}(x_1 - x_2).$$

Hence,

$$(x_1, x_2) = \frac{1}{3}(x_1 + 2x_2)v_1 + \frac{1}{3}(x_1 - x_2)v_2.$$

For example, if $x = (2, 1)$, then $c_1 = 4/3$ and $c_2 = 1/3$, so that $x = \frac{4}{3}v_1 + \frac{1}{3}v_2$. This is illustrated in Figure 5.4.1.   □

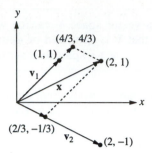
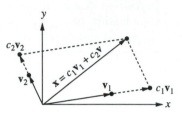
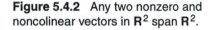

Figure 5.4.1 The vector $\mathbf{x} = (2, 1)$ expressed as a linear combination of $\mathbf{v}_1 = (1, 1)$ and $\mathbf{v}_2 = (2, -1)$.

Figure 5.4.2 Any two nonzero and noncolinear vectors in $\mathbf{R}^2$ span $\mathbf{R}^2$.

More generally, any two nonzero and noncolinear vectors $\mathbf{v}_1$ and $\mathbf{v}_2$ in $\mathbf{R}^2$ span $\mathbf{R}^2$, since, as illustrated geometrically in Figure 5.4.2, every vector in $\mathbf{R}^2$ can be written as a linear combination of $\mathbf{v}_1, \mathbf{v}_2$.

Example 5.4.2   Determine whether the vectors $\mathbf{v}_1 = (1, -1, 4)$, $\mathbf{v}_2 = (-2, 1, 3)$, and $\mathbf{v}_3 = (4, -3, 5)$ span $\mathbf{R}^3$.

Solution   Let $\mathbf{x} = (x_1, x_2, x_3)$ be an arbitrary vector in $\mathbf{R}^3$. We must determine whether there are real numbers c_1, c_2, c_3 such that

$$\mathbf{x} = c_1\mathbf{v}_1 + c_2\mathbf{v}_2 + c_3\mathbf{v}_3 \tag{5.4.6}$$

or, in component form,

$$(x_1, x_2, x_3) = c_1(1, -1, 4) + c_2(-2, 1, 3) + c_3(4, -3, 5).$$

Equating corresponding components on either side of this vector equation yields

$$c_1 - 2c_2 + 4c_3 = x_1,$$
$$-c_1 + c_2 - 3c_3 = x_2,$$
$$4c_1 + 3c_2 + 5c_3 = x_3.$$

Reducing the augmented matrix of this system we obtain

$$\begin{bmatrix} 1 & -2 & 4 & x_1 \\ 0 & 1 & -1 & -x_1 - x_2 \\ 0 & 0 & 0 & 7x_1 + 11x_2 + x_3 \end{bmatrix}.$$

It follows that the system is consistent $[\text{rank}(A) = \text{rank}(A^{\#})]$ if and only if x_1, x_2, x_3 satisfy

$$7x_1 + 11x_2 + x_3 = 0. \tag{5.4.7}$$

Consequently, equation (5.4.6) holds only for those vectors $\mathbf{x} = (x_1, x_2, x_3)$ in $\mathbf{R}^3$ whose components satisfy equation (5.4.7). Hence, $\mathbf{v}_1, \mathbf{v}_2$, and $\mathbf{v}_3$ do *not* span $\mathbf{R}^3$. Geometrically, equation (5.4.7) is the equation of a plane through the origin in space, and so by taking linear combinations of the given vectors, we can obtain only those vectors which lie on this plane. We leave it as an exercise to verify that indeed the three given vectors lie in the plane with equation (5.4.7).   □

The reason that the vectors in the previous example did not span $\mathbf{R}^3$ was because they were coplanar. In general, any three noncoplanar vectors, $\mathbf{v}_1$, $\mathbf{v}_2$, and $\mathbf{v}_3$, in $\mathbf{R}^3$ span $\mathbf{R}^3$, since, as illustrated geometrically in Figure 5.4.3, every vector in $\mathbf{R}^3$ can be written as a linear combination of $\mathbf{v}_1$, $\mathbf{v}_2$, $\mathbf{v}_3$.

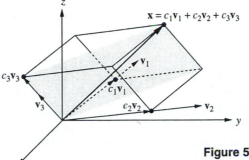

Figure 5.4.3 Any three noncoplanar vectors $\mathbf{v}_1$, $\mathbf{v}_2$, $\mathbf{v}_3$ span $\mathbf{R}^3$.

Example 5.4.3   Verify that $A_1 = \begin{bmatrix} 1 & 0 \\ 0 & 0 \end{bmatrix}$, $A_2 = \begin{bmatrix} 0 & 1 \\ 0 & 0 \end{bmatrix}$, $A_3 = \begin{bmatrix} 0 & 0 \\ 1 & 0 \end{bmatrix}$, $A_4 = \begin{bmatrix} 0 & 0 \\ 0 & 1 \end{bmatrix}$ span $M_2(\mathbf{R})$.

Solution   An arbitrary vector in $M_2(\mathbf{R})$ is of the form $A = \begin{bmatrix} a & b \\ c & d \end{bmatrix}$. This can be written as

$$A = a\begin{bmatrix} 1 & 0 \\ 0 & 0 \end{bmatrix} + b\begin{bmatrix} 0 & 1 \\ 0 & 0 \end{bmatrix} + c\begin{bmatrix} 0 & 0 \\ 1 & 0 \end{bmatrix} + d\begin{bmatrix} 0 & 0 \\ 0 & 1 \end{bmatrix}.$$

That is,

$$A = aA_1 + bA_2 + cA_3 + dA_4.$$

Consequently every vector in $M_2(\mathbf{R})$ can be written as a linear combination of A_1, A_2, A_3, and A_4, and, therefore, these matrices do indeed span $M_2(\mathbf{R})$.

Example 5.4.4   Determine a spanning set for P_3 the vector space of all poly-nomials of degree less than three.

Solution   The general polynomial in P_3 is

$$p(x) = a_0 + a_1 x + a_2 x^2.$$

If we let

$$p_0(x) = 1, \quad p_1(x) = x, \quad p_3(x) = x^2,$$

then

$$p(x) = a_0 p_0(x) + a_1 p_1(x) + a_2 p_2(x).$$

Thus, every vector in P_3 is a linear combination of 1, x, and x^2, and so a spanning set for P_3 is $\{1, x, x^2\}$.

THE LINEAR SPAN OF A SET OF VECTORS

Now let $v_1, v_2, \ldots, v_k$ be vectors in a vector space V. Forming all possible linear combinations of $v_1, v_2, \ldots, v_k$ generates a subset of V called the **linear span** of $\{v_1, v_2, \ldots, v_k\}$, denoted span$\{v_1, v_2, \ldots, v_k\}$. Thus,

$$\text{span}\{v_1, v_2, \ldots, v_k\} = \{v \in V : v = c_1 v_1 + c_2 v_2 + \cdots + c_k v_k, c_1, c_2, \ldots, c_k \in F\}. \quad (5.4.8)$$

For example, suppose $V = C^2(I)$, and let $y_1(x) = \sin x$ and $y_2(x) = \cos x$, then

$$\text{span}\{y_1, y_2\} = \{y \in C^2(I) : y(x) = c_1 \cos x + c_2 \sin x, c_1, c_2 \in \mathbf{R}\}.$$

From our work in Chapter 2, we recognize y_1 and y_2 as being linearly independent solutions to the DE $y'' + y = 0$. Consequently, in this example, the linear span of the given functions coincides with the set of all solutions to the DE $y'' + y = 0$ and therefore is a *subspace* of V. Our next theorem generalizes this result.

Theorem 5.4.1:   Let $v_1, v_2, \ldots, v_k$ be vectors in a vector space V. Then span$\{v_1, v_2, \ldots, v_k\}$ is a subspace of V.

PROOF   Let $S = \text{span}\{v_1, v_2, \ldots, v_k\}$. Then S is certainly nonempty. We must verify closure under addition and closure under scalar multiplication. If u and v are in S, then, from equation (5.4.8),

$$u = a_1 v_1 + a_2 v_2 + \cdots + a_k v_k \quad \text{and} \quad v = b_1 v_1 + b_2 v_2 + \cdots + b_k v_k,$$

for some scalars a_i, b_i. Thus,

$$\begin{aligned}
u + v &= (a_1 v_1 + a_2 v_2 + \cdots + a_k v_k) + (b_1 v_1 + b_2 v_2 + \cdots + b_k v_k) \\
&= (a_1 + b_1) v_1 + (a_2 + b_2) v_2 + \cdots + (a_k + b_k) v_k \\
&= c_1 v_1 + c_2 v_2 + \cdots + c_k v_k,
\end{aligned}$$

where $c_i = a_i + b_i$, $i = 1, 2, \ldots, k$. Consequently, $u + v$ is in S, and so S is closed under addition. Further, if r is any scalar,

$$\begin{aligned}
r u &= r(a_1 v_1 + a_2 v_2 + \cdots + a_k v_k) = (r a_1) v_1 + (r a_2) v_2 + \cdots + (r a_k) v_k \\
&= d_1 v_1 + d_2 v_2 + \cdots + d_k v_k,
\end{aligned}$$

where $d_i = r a_i$, $i = 1, 2, \ldots, k$. Consequently, S is also closed under scalar multiplication. Hence, $S = \text{span}\{v_1, v_2, \ldots, v_k\}$ is a subspace of V.   ∎

REMARK   We will also refer to span$\{v_1, v_2, \ldots, v_k\}$ as the **subspace of V spanned by** $v_1, v_2, \ldots, v_k$.

Example 5.4.5   If $V = \mathbf{R}^2$ and $v_1 = (-1, 1)$, determine span$\{v_1\}$.

Solution   We have

$$\begin{aligned}
\text{span}\{v_1\} &= \{x \in \mathbf{R}^2 : x = c_1 v_1, c_1 \in \mathbf{R}\} = \{x \in \mathbf{R}^2 : x = c_1(-1, 1), c_1 \in \mathbf{R}\} \\
&= \{x \in \mathbf{R}^2 : x = (-c_1, c_1), c_1 \in \mathbf{R}\}.
\end{aligned}$$

Geometrically, this is the line through the origin with parametric equations $x = -c_1$, $y = c_1$, so that the Cartesian equation of the line is $y = -x$. (See Figure 5.4.4.)

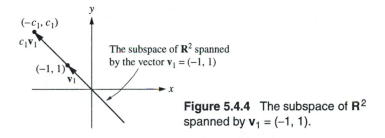

Figure 5.4.4 The subspace of $\mathbf{R}^2$ spanned by $\mathbf{v}_1 = (-1, 1)$.

Example 5.4.6   If $V = \mathbf{R}^3$ and $\mathbf{v}_1 = (1, 0, 1)$ and $\mathbf{v}_2 = (0, 1, 1)$, determine the subspace of $\mathbf{R}^3$ spanned by $\mathbf{v}_1$ and $\mathbf{v}_2$. Does $\mathbf{v} = (1, 1, -1)$ lie in this subspace ?

Solution

$$\begin{aligned}
\text{span}\{\mathbf{v}_1, \mathbf{v}_2\} &= \{\mathbf{x} \in \mathbf{R}^3 : \mathbf{x} = c_1\mathbf{v}_1 + c_2\mathbf{v}_2, c_1, c_2 \in \mathbf{R}\} \\
&= \{\mathbf{x} \in \mathbf{R}^3 : \mathbf{x} = c_1(1, 0, 1) + c_2(0, 1, 1), c_1, c_2 \in \mathbf{R}\} \\
&= \{\mathbf{x} \in \mathbf{R}^3 : \mathbf{x} = (c_1, c_2, c_1 + c_2), c_1, c_2 \in \mathbf{R}\}.
\end{aligned}$$

Since the vector $\mathbf{v} = (1, 1, -1)$ is not of the form $(c_1, c_2, c_1 + c_2)$, it does not lie in span$\{\mathbf{v}_1, \mathbf{v}_2\}$. Geometrically, span$\{\mathbf{v}_1, \mathbf{v}_2\}$ is the plane through the origin determined by the two given vectors $\mathbf{v}_1$ and $\mathbf{v}_2$. It has parametric equations $x = c_1, y = c_2, z = c_1 + c_2$, which implies that its Cartesian equation is $z = x + y$. Thus, the fact that $\mathbf{v}$ is not in span$\{\mathbf{v}_1, \mathbf{v}_2\}$ means that $\mathbf{v}$ does not lie in this plane. The subspace is depicted in Figure 5.4.5.

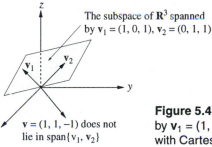

v = (1, 1, -1) does not lie in span$\{\mathbf{v}_1, \mathbf{v}_2\}$

Figure 5.4.5 The subspace of $\mathbf{R}^3$ spanned by $\mathbf{v}_1 = (1, 0, 1)$ and $\mathbf{v}_2 = (0, 1, 1)$ is the plane with Cartesian equation $z = x + y$.

Example 5.4.7   Let $A_1 = \begin{bmatrix} 1 & 0 \\ 0 & 0 \end{bmatrix}$, $A_2 = \begin{bmatrix} 0 & 1 \\ 1 & 0 \end{bmatrix}$, $A_3 = \begin{bmatrix} 0 & 0 \\ 0 & 1 \end{bmatrix}$, in $M_2(\mathbf{R})$.

Determine span$\{A_1, A_2, A_3\}$.

Solution   By definition we have

$$\text{span}\{A_1, A_2, A_3\} = \{A \in M_2(\mathbf{R}) : A = c_1A_1 + c_2A_2 + c_3A_3, c_1, c_2, c_3 \in \mathbf{R}\}$$

$$= \{A \in M_2(\mathbf{R}) : A = c_1 \begin{bmatrix} 1 & 0 \\ 0 & 0 \end{bmatrix} + c_2 \begin{bmatrix} 0 & 1 \\ 1 & 0 \end{bmatrix} + c_3 \begin{bmatrix} 0 & 0 \\ 0 & 1 \end{bmatrix}, c_1, c_2, c_3 \in \mathbf{R}\}$$

$$= \{A \in M_2(\mathbf{R}) : A = \begin{bmatrix} c_1 & c_2 \\ c_2 & c_3 \end{bmatrix}, c_1, c_2, c_3 \in \mathbf{R}\} .$$

This is the set of all real *symmetric* 2×2 matrices.

Example 5.4.8   Determine the subspace of P_3 spanned by

$$p_1(x) = 1 + 3x, \quad p_2(x) = x + x^2.$$

Solution   We have

$$\text{span}\{p_1, p_2\} = \{p \in P_3 : p(x) = c_1 p_1(x) + c_2 p_2(x), c_1, c_2 \in \mathbf{R}\}$$

$$= \{p \in P_3 : p(x) = c_1(1 + 3x) + c_2(x + x^2), c_1, c_2 \in \mathbf{R}\}$$

$$= \{p \in P_3 : p(x) = c_1 + (3c_1 + c_2)x + c_2 x^2, c_1, c_2 \in \mathbf{R}\}.$$

Does $\text{span}\{p_1, p_2\} = P_3$? The answer is no. To establish this, we need only give one example of a polynomial in P_3 that is not in $\text{span}\{p_1, p_2\}$. Consider $p(x) = 1 + x$. If this polynomial were in $\text{span}\{p_1, p_2\}$, then we would be able to find values of c_1, c_2 such that

$$1 + x = c_1 + (3c_1 + c_2)x + c_2 x^2. \tag{5.4.9}$$

Since there is no x^2 term on the left-hand side of this expression, we must set $c_2 = 0$. But then we can only generate polynomials of the form

$$p(x) = c_1(1 + 3x).$$

Consequently there are no values of c_1 and c_2 such that equation (5.4.9) holds, and therefore, $\text{span}\{p_1, p_2\} \neq P_3$.

EXERCISES 5.4

For problems 1–3, determine whether the given set of vectors spans $\mathbf{R}^2$.

1. $\{(1, -1), (2, -2), (2, 3)\}$.

2. $\{(2, 5), (0, 0)\}$.

3. $\{6, -2), (-2, 2/3), (3, -1)\}$.

Recall that three vectors, $\mathbf{v}_1, \mathbf{v}_2, \mathbf{v}_3$ in $\mathbf{R}^3$ are coplanar if and only if $\det([\mathbf{v}_1, \mathbf{v}_2, \mathbf{v}_3]) = 0$. For problems 4–6 use this result to determine whether the given set of vectors spans $\mathbf{R}^3$.

4. $\{(1, -1, 1), (2, 5, 3), (4, -2, 1)\}$.

5. $\{(1, -2, 1), (2, 3, 1), (0, 0, 0), (4, -1, 2)\}$.

6. $\{(2, -1, 4), (3, -3, 5), (1, 1, 3)\}$.

7. Show that the set of vectors $\{(1, 2, 3), (3, 4, 5), (4, 5, 6)\}$ does not span $\mathbf{R}^3$, but that it does span the subspace of $\mathbf{R}^3$ consisting of all vectors lying in the plane with equation $x_1 - 2x_2 + x_3 = 0$.

8. Show that $\mathbf{v}_1 = (2, -1)$, $\mathbf{v}_2 = (3, 2)$ span $\mathbf{R}^2$ and express the vector $\mathbf{v} = (5, -7)$ as a linear combination of $\mathbf{v}_1, \mathbf{v}_2$.

9. Show that $\mathbf{v}_1 = (-1, 3, 2)$, $\mathbf{v}_2 = (1, -2, 1)$, $\mathbf{v}_3 = (2, 1, 1)$ span $\mathbf{R}^3$, and express $\mathbf{x} = (x_1, x_2, x_3)$ as a linear combination of $\mathbf{v}_1, \mathbf{v}_2, \mathbf{v}_3$.

10. Show that $\mathbf{v}_1 = (1, 1)$, $\mathbf{v}_2 = (-1, 2)$, $\mathbf{v}_3 = (1, 4)$ span $\mathbf{R}^2$. Do $\mathbf{v}_1$ and $\mathbf{v}_2$ span $\mathbf{R}^2$ also?

11. Let S be the subspace of $\mathbf{R}^3$ consisting of all vectors of the form $\mathbf{x} = (c_1, c_2, c_2 - 2c_1)$. Show that S is spanned by $\mathbf{v}_1 = (1, 0, -2)$ and $\mathbf{v}_2 = (0, 1, 1)$.

12. Let S be the subspace of $\mathbf{R}^4$ consisting of all vectors of the form $\mathbf{v} = (x_1, x_2, x_2 - x_1, x_1 - 2x_2)$. Determine a set of vectors that spans S.

13. Let S be the subspace of $\mathbf{R}^3$ consisting of all solutions to the linear equation

$$x_1 - 2x_2 - x_3 = 0.$$

Determine a set of vectors that spans S.

For problems 14 and 15, determine a spanning set for nullspace(A).

14. $A = \begin{bmatrix} 1 & 2 & 3 \\ 3 & 4 & 5 \\ 5 & 6 & 7 \end{bmatrix}$.

15. $A = \begin{bmatrix} 1 & 2 & 3 & 5 \\ 1 & 3 & 4 & 2 \\ 2 & 4 & 6 & -1 \end{bmatrix}$.

16. Let S be the subspace of $M_2(\mathbf{R})$ consisting of all *symmetric* 2×2 matrices with real elements. Show that S is spanned by the matrices

$$A_1 = \begin{bmatrix} 1 & 0 \\ 0 & 0 \end{bmatrix}, A_2 = \begin{bmatrix} 0 & 0 \\ 0 & 1 \end{bmatrix}, A_3 = \begin{bmatrix} 0 & 1 \\ 1 & 0 \end{bmatrix}.$$

17. Let S be the subspace of $M_2(\mathbf{R})$ consisting of all *skew-symmetric* 2×2 matrices with real elements. Determine a matrix that spans S.

18. Let S be the subset of $M_2(\mathbf{R})$ consisting of all upper triangular 2×2 matrices.

(a) Verify that S is a subspace of $M_2(\mathbf{R})$.

(b) Determine a set of 2×2 matrices that spans S.

For problems 19 and 20, determine span$\{\mathbf{v}_1, \mathbf{v}_2\}$ for the given vectors in $\mathbf{R}^3$, and describe it geometrically.

19. $\mathbf{v}_1 = (1, -1, 2), \mathbf{v}_2 = (2, -1, 3)$.

20. $\mathbf{v}_1 = (1, 2, -1), \mathbf{v}_2 = (-2, -4, 2)$.

21. Let S be the subspace of $\mathbf{R}^3$ spanned by the vectors $\mathbf{v}_1 = (1, 1, -1), \mathbf{v}_2 = (2, 1, 3), \mathbf{v}_3 = (-2, -2, 2)$. Show that S is also spanned by $\mathbf{v}_1$ and $\mathbf{v}_2$ only.

For problems $22-24$, determine whether the given vector $\mathbf{v}$ lies in span$\{\mathbf{v}_1, \mathbf{v}_2\}$.

22. $\mathbf{v} = (3, 3, 4), \mathbf{v}_1 = (1, -1, 2), \mathbf{v}_2 = (2, 1, 3)$ in $\mathbf{R}^3$.

23. $\mathbf{v} = (5, 3, -6), \mathbf{v}_1 = (-1, 1, 2), \mathbf{v}_2 = (3, 1, -4)$ in $\mathbf{R}^3$.

24. $\mathbf{v} = (1, 1, -2), \mathbf{v}_1 = (3, 1, 2), \mathbf{v}_2 = (-2, -1, 1)$ in $\mathbf{R}^3$.

25. If $p_1(x) = x - 4$, and $p_2(x) = x^2 - x + 3$ in P_3, determine whether $p(x) = 2x^2 - x + 2$ lies in span$\{p_1, p_2\}$.

26. Consider the vectors

$$A_1 = \begin{bmatrix} 1 & -1 \\ 2 & 0 \end{bmatrix}, A_2 = \begin{bmatrix} 0 & 1 \\ -2 & 1 \end{bmatrix}, A_3 = \begin{bmatrix} 3 & 0 \\ 1 & 2 \end{bmatrix}$$

in $M_2(\mathbf{R})$. Determine span$\{A_1, A_2, A_3\}$.

27. Consider the vectors

$$A_1 = \begin{bmatrix} 1 & 2 \\ -1 & 3 \end{bmatrix} \text{ and } A_2 = \begin{bmatrix} -2 & 1 \\ 1 & -1 \end{bmatrix}$$

in $M_2(\mathbf{R})$. Find span$\{A_1, A_2\}$, and determine whether $B = \begin{bmatrix} 3 & 1 \\ -2 & 4 \end{bmatrix}$ lies in this subspace.

28. Let $V = C^\infty(I)$ and let S be the subspace of V spanned by the functions

$$f(x) = \cosh x, \ g(x) = \sinh x.$$

(a) Give an expression for a general vector in S.

(b) Show that S is also spanned by the functions

$$h(x) = e^x, \ j(x) = e^{-x}.$$

For problems $29-32$, give a geometric description of the subspace of $\mathbf{R}^3$ spanned by the given set of vectors.

29. $\{\mathbf{0}\}$

30. $\{\mathbf{v}_1\}$, where $\mathbf{v}_1$ is any nonzero vector in $\mathbf{R}^3$.

31. $\{\mathbf{v}_1, \mathbf{v}_2\}$, where $\mathbf{v}_1, \mathbf{v}_2$ are nonzero and non-colinear vectors in $\mathbf{R}^3$.

32. $\{\mathbf{v}_1, \mathbf{v}_2\}$, where $\mathbf{v}_1, \mathbf{v}_2$ are colinear vectors in $\mathbf{R}^3$.

5.5   LINEAR DEPENDENCE AND LINEAR INDEPENDENCE

As indicated in the previous section, in analyzing a vector space we will be interested in determining a spanning set. However, we don't want to use just any spanning set, we want one that contains the *minimum* number of vectors—a minimal spanning set. But how will we know if we have found a minimal spanning set (assuming one exists)? To help answer this question, we consider a familiar example. From Theorem 2.1.4, we know that the general solution to the DE

$$y'' + y = 0$$

can be obtained by taking all linear combinations of any *two* linearly independent (non–proportional) solutions. The standard choice to use is $y_1(x) = \cos x$, $y_2(x) = \sin x$. However, if we let $y_3(x) = 3 \cos x - 5 \sin x$, for example, then $\{y_1, y_2, y_3\}$ is also a spanning set for the solution space of the DE, since

$$\begin{aligned} \text{span}\{y_1, y_2, y_3\} &= \{c_1 \cos x + c_2 \sin x + c_3(3 \cos x - 2 \sin x) : c_1, c_2, c_3 \in \mathbf{R}\} \\ &= \{(c_1 + 3c_3)\cos x + (c_2 - 2c_3)\sin x : c_1, c_2, c_3 \in \mathbf{R}\} \\ &= \{d_1 \cos x + d_2 \sin x : d_1, d_2 \in \mathbf{R}\} \\ &= \text{span}\{y_1, y_2\}. \end{aligned}$$

The reason that $\{y_1, y_2, y_3\}$ is not a *minimal* spanning set for the solution space is that y_3 is a linear combination of y_1 and y_2, and therefore, as we have just shown, it does not add anything new to the linear span of $\{\cos x, \sin x\}$. Clearly, however, the linear span of just one of the solutions cannot contain all solutions to the DE, since $\{\cos x, \sin x)$ is linearly independent.

As a second example, consider the vectors $\mathbf{v}_1 = (1, 1, 1)$, $\mathbf{v}_2 = (3, -2, 1)$, and $\mathbf{v}_3 = 4\mathbf{v}_1 + \mathbf{v}_2 = (7, 2, 5)$. It is easily verified that $\det([\mathbf{v}_1, \mathbf{v}_2, \mathbf{v}_3]) = 0$. Consequently, the three vectors lie in a plane (see Figure 5.5.1) and therefore, since they are not co–linear, the linear span of these three vectors is the whole of this plane. Furthermore, the

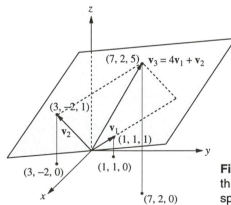

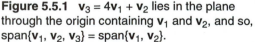

Figure 5.5.1 $\mathbf{v}_3 = 4\mathbf{v}_1 + \mathbf{v}_2$ lies in the plane through the origin containing $\mathbf{v}_1$ and $\mathbf{v}_2$, and so, span$\{\mathbf{v}_1, \mathbf{v}_2, \mathbf{v}_3\}$ = span$\{\mathbf{v}_1, \mathbf{v}_2\}$.

same plane is generated if we consider the linear span of $\mathbf{v}_1$, $\mathbf{v}_2$ alone. As in the DE example, the reason that $\mathbf{v}_3$ does not add anything to the linear span of $\{\mathbf{v}_1, \mathbf{v}_2\}$ is that it is a linear combination of $\mathbf{v}_1$ and $\mathbf{v}_2$. It is not possible, however, to generate all vectors in the plane by taking linear combinations of just one of the given vectors, as we could only generate a line lying in the plane in that case. Consequently, $\{v_1, v_2\}$ is a minimal spanning set for the subspace of $\mathbf{R}^3$ consisting of all points lying on the plane.

   More generally, it is not too difficult to extend the argument used in the preceding DE example to establish the following general result.

   Theorem 5.5.1:   Let $\{v_1, v_2, \ldots, v_k\}$ be a set of at least two vectors in a vector space V. If one of the vectors in the set is a linear combination of the other vectors in the set, then that vector can be deleted from the given set of vectors and the linear span of the resulting set of vectors will be the same subspace of V as the linear span of $\{v_1, v_2, \ldots, v_k\}$.

   PROOF   The proof of this result is left for the exercises   ■

   According to this theorem, if a spanning set $\{v_1, v_2, \ldots, v_k\}$ contains any vector that can be expressed as a linear combination of the other vectors in the set, then it cannot be a minimal spanning set. The problem that we are faced with, therefore, is that of determining when a vector in $\{v_1, v_2, \ldots, v_k\}$ can be expressed as a linear combination of the remaining vectors in the set. The correct formulation for solving this problem comes from generalizing the concepts of linear dependence and linear independence already introduced in Chapter 2 for the case of two functions. First we consider linear dependence.

   Definition 5.5.1:   A finite set of vectors $\{v_1, v_2, \ldots, v_k\}$ in a vector space V is said to be **linearly dependent** if there exist scalars $c_1, c_2, \ldots, c_k$, *not all zero*, such that

$$c_1v_1 + c_2v_2 + \cdots + c_kv_k = 0.$$

A set of vectors that is not linearly dependent (LD) is called linearly independent (LI). This can be stated mathematically as follows:

   Definition 5.5.2:   The set of vectors $\{v_1, v_2, \ldots, v_k\}$ in a vector space V is **linearly independent** if the *only* values of the scalars $c_1, c_2, \ldots, c_k$ for which

$$c_1v_1 + c_2v_2 + \cdots + c_kv_k = 0$$

are $c_1 = c_2 = \cdots = c_k = 0$.

Consider first the simple case of a set containing a single vector v. If $v = 0$, then $\{v\}$ is LD, since for any nonzero scalar c_1

$$c_1 0 = 0.$$

On the other hand, if $v \neq 0$, then the only value of the scalar c_1 for which

$$c_1 v = 0$$

is $c_1 = 0$. Consequently $\{v\}$ is LI. We can therefore state the next theorem.

   Theorem 5.5.2:   A set consisting of a single vector in a vector space V is LD if and only if $v = 0$. Therefore, any set consisting of a single *nonzero* vector is LI.

We next establish that linear dependence of a set containing at least two vectors is equivalent to the property that we are interested in, namely, that at least one vector in the set can be expressed as a linear combination of the remaining vectors in the set.

Theorem 5.5.3:   Let $\{v_1, v_2, \ldots, v_k\}$ be a set of at least two vectors in a vector space V. Then $\{v_1, v_2, \ldots, v_k\}$ is LD if and only if at least one of the vectors in the set can be expressed as a linear combination of the others.

PROOF   If $\{v_1, v_2, \ldots, v_k\}$ satisfies Definition 5.5.1, then there exist scalars $c_1, c_2, \ldots, c_k$, not all zero, such that

$$c_1 v_1 + c_2 v_2 + \cdots + c_k v_k = 0.$$

Suppose that $c_i \neq 0$. Then we can express v_i as a linear combination of the other vectors as follows

$$v_i = \frac{1}{c_i}(c_1 v_1 + c_2 v_2 + \cdots + c_{i-1} v_{i-1} + c_{i+1} v_{i+1} + \cdots + c_k v_k).$$

Conversely, suppose that one of the vectors, say, v_j, can be expressed as a linear combination of the remaining vectors. That is,

$$v_j = c_1 v_1 + c_2 v_2 + \cdots + c_{j-1} v_{j-1} + c_{j+1} v_{j+1} + \cdots + c_k v_k.$$

Adding $(-1)v_j$ to either side of this equation yields

$$c_1 v_1 + c_2 v_2 + \cdots + c_{j-1} v_{j-1} - v_j + c_{j+1} v_{j+1} + \cdots + c_k v_k = 0.$$

Since the coefficient of $v_j = -1 \neq 0$, the set of vectors $\{v_1, v_2, \ldots, v_k\}$ is LD.   ∎

As far as the minimal spanning set idea is concerned, our preceding discussion tells us that a LD spanning set for a (nontrivial) vector space V cannot be a minimal spanning set. The next section is devoted to the study of LI spanning sets and one of the results will indeed establish that any LI spanning set for V is a minimal spanning set for V. For the remainder of this section, however, we focus more on the mechanics of determining whether a given set of vectors is LI or LD. Sometimes this can be done by inspection. For example, from Figure 5.5.2, we see that any set of three vectors in $\mathbf{R}^2$ is LD. As

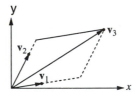

Figure 5.5.2 The set of vectors $\{v_1, v_2, v_3\}$ is LD in $\mathbf{R}^2$ since v_3 is a linear combination of v_1 and v_2.

another example, let V be the vector space of all functions defined on an interval I. If

$$f_1(x) = 1, \quad f_2(x) = 2 \sin^2 x, \quad f_3(x) = -5 \cos^2 x,$$

then $\{f_1, f_2, f_3\}$ is LD in V, since the trigonometric identity $\sin^2 x + \cos^2 x = 1$ implies that for all $x \in I$,

$$f_1(x) = \frac{1}{2} f_2(x) - \frac{1}{5} f_3(x).$$

We can therefore conclude from Theorem 5.5.1 that

$$\text{span}\{1,\ 2\sin^2 x,\ -5\cos^2 x\} = \text{span}\{2\sin^2 x,\ -5\cos^2 x\}.$$

For more complicated situations, we must resort to Definitions 5.5.1 and 5.5.2, although conceptually it is helpful to keep in mind that the essence of the problem that we are solving is to determine whether a vector in a given set can be expressed as a linear combination of the remaining vectors in the set. We now consider several examples to illustrate the use of Definitions 5.5.1 and 5.5.2.

Example 5.5.1:   If $v_1 = (1, 2, -1)$, $v_2 = (2, -1, 1)$, and $v_3 = (8, 1, 1)$, show that $\{v_1, v_2, v_3\}$ is LD in $\mathbf{R}^3$, and determine the linear dependency relationship.

Solution   We must first establish that there are nonzero values of the scalars c_1, c_2, c_3 such that

$$c_1 v_1 + c_2 v_2 + c_3 v_3 = 0. \tag{5.5.1}$$

Substituting for the given vectors yields

$$c_1(1, 2, -1) + c_2(2, -1, 1) + c_3(8, 1, 1) = (0, 0, 0).$$

That is,

$$(c_1 + 2c_2 + 8c_3,\ 2c_1 - c_2 + c_3,\ -c_1 + c_2 + c_3) = (0, 0, 0).$$

Equating corresponding components on either side of this equation yields

$$c_1 + 2c_2 + 8c_3 = 0,$$
$$2c_1 - c_2 + c_3 = 0,$$
$$-c_1 + c_2 + c_3 = 0.$$

The RREF of the augmented matrix of this system is

$$\begin{bmatrix} 1 & 0 & 2 & 0 \\ 0 & 1 & 3 & 0 \\ 0 & 0 & 0 & 0 \end{bmatrix}.$$

Consequently, the system has an infinite number of solutions, and so, the vectors are LD. In order to express v_3 as a linear combination of v_1 and v_2, we need to solve the system to find $c_1, c_2,$ and c_3. Setting $c_3 = r$, it follows that $c_2 = -3r$, and $c_1 = -2r$. Substituting these values for c_1, c_2, c_3 into equation (5.5.1) we obtain the linear dependency relationship

$$-2v_1 - 3v_2 + v_3 = 0,$$

or equivalently,

$$v_1 = -\frac{3}{2} v_2 + \frac{1}{2} v_3,$$

which is easily verified using the given expressions for $v_1, v_2,$ and v_3. It follows from Theorem 5.5.1 that the vectors v_2 and v_3 span the same subspace of $\mathbf{R}^3$ as that spanned by $v_1, v_2,$ and v_3. What can we conclude geometrically about $v_1, v_2,$ and v_3?

Example 5.5.2   Determine whether the following matrices are LD or LI in $M_2(\mathbf{R})$

$$A_1 = \begin{bmatrix} 1 & -1 \\ 2 & 0 \end{bmatrix},\ A_2 = \begin{bmatrix} 2 & 1 \\ 0 & 3 \end{bmatrix},\ A_3 = \begin{bmatrix} 1 & -1 \\ 2 & 1 \end{bmatrix}.$$

Solution   The condition for determining LD or LI, namely,

$$c_1 A_1 + c_2 A_2 + c_3 A_3 = 0_2,$$

is equivalent, in this case, to

$$c_1 \begin{bmatrix} 1 & -1 \\ 2 & 0 \end{bmatrix} + c_2 \begin{bmatrix} 2 & 1 \\ 0 & 3 \end{bmatrix} + c_3 \begin{bmatrix} 1 & -1 \\ 2 & 1 \end{bmatrix} = \begin{bmatrix} 0 & 0 \\ 0 & 0 \end{bmatrix},$$

which is satisfied if and only if

$$\begin{aligned} c_1 + 2c_2 + c_3 &= 0, \\ -c_1 + c_2 - c_3 &= 0, \\ 2c_1 \qquad\quad + 2c_3 &= 0, \\ 3c_2 + c_3 &= 0. \end{aligned}$$

The RREF of the augmented matrix of this homogeneous system is

$$\begin{bmatrix} 1 & 0 & 0 & 0 \\ 0 & 1 & 0 & 0 \\ 0 & 0 & 1 & 0 \\ 0 & 0 & 0 & 0 \end{bmatrix},$$

which implies that the system has only the trivial solution $c_1 = c_2 = c_3 = 0$. It follows from Definition 5.5.2 that $\{A_1, A_2, A_3\}$ is LI.   ❏

The following general results are a direct consequence of the definition of linear dependency and are left for the exercises:

Let V be a vector space.

1. Any set of *two* vectors in V is LD if and only if the vectors are proportional.

2. Any set of vectors in V containing the zero vector is LD.

REMARK   We emphasize that the result (1) holds only for the case of two vectors. Do not try applying it to sets containing more than two vectors.

Example 5.5.3   If $v_1 = (1, 2, -9)$ and $v_2 = (-2, -4, 18)$ then $\{v_1, v_2\}$ is LD in $\mathbf{R}^3$, since $v_2 = -2v_1$. Geometrically, v_1 and v_2 lie on the same line.

Example 5.5.4   If $A_1 = \begin{bmatrix} 2 & 1 \\ 3 & 4 \end{bmatrix}$, $A_2 = \begin{bmatrix} 0 & 0 \\ 0 & 0 \end{bmatrix}$, and $A_3 = \begin{bmatrix} 2 & 5 \\ -3 & 2 \end{bmatrix}$, then $\{A_1, A_2, A_3\}$ is LD in $M_2(\mathbf{R})$, since it contains the zero vector from $M_2(\mathbf{R})$.   ❏

As a corollary to Theorem 5.5.1 we establish the following result.

Corollary 5.5.1:   Any nontrivial set of LD vectors in a vector space V contains a LI subset that has the same linear span as the given set of vectors.

PROOF   Since the given set is LD, at least one of the vectors in the set is a linear combination of the remaining vectors, and so, Theorem 5.5.1 implies that we can delete that vector from the set, and the resulting set of vectors will span the same subspace of V as the given set. If the resulting set is LI, then we are done. If not, then we can repeat the procedure to eliminate a further vector in the set. Continuing in this manner, we will obtain a LI set that spans the same subspace of V as the subspace spanned by the original set of vectors.   ∎

Note that the LI set that is obtained using the procedure given in the previous theorem is not unique, and therefore one question that arises is whether the number of vectors in any resulting LI set is independent of the manner in which the procedure is applied. We will give an affirmative answer to this question in Section 5.6.

Example 5.5.5   Let $v_1 = (1, 2, 3)$, $v_2 = (-1, 1, 4)$, $v_3 = (3, 3, 2)$, and $v_4 = (-2, -4, -6)$. Determine a LI set of vectors that spans the same subspace of $\mathbf{R}^3$ as span$\{v_1, v_2, v_3, v_4\}$.

Solution   We see by inspection that $v_4 = -2v_1$. Consequently, $\{v_1, v_2, v_3, v_4\}$ is LD. It follows from Theorem 5.5.1 that $\{v_1, v_2, v_3\}$ spans the same subspace of $\mathbf{R}^3$ as the original set of vectors. We must determine whether the new set is LD or LI. Setting

$$c_1 v_1 + c_2 v_2 + c_3 v_3 = 0$$

requires that

$$c_1(1, 2, 3) + c_2(-1, 1, 4) + c_3(3, 3, 2) = 0,$$

leading to the linear system

$$c_1 - c_2 + 3c_3 = 0,$$
$$2c_1 + c_2 + 3c_3 = 0,$$
$$3c_1 + 4c_2 + 2c_3 = 0.$$

The RREF of the augmented matrix of this system is

$$\begin{bmatrix} 1 & 0 & 2 & 0 \\ 0 & 1 & -1 & 0 \\ 0 & 0 & 0 & 0 \end{bmatrix}.$$

The system therefore has an infinite number of solutions, and so, $\{v_1, v_2, v_3\}$ is LD. Solving the system yields: $c_1 = -2r$, $c_2 = r$, $c_3 = r$. Consequently,

$$-2v_1 + v_2 + v_3 = 0,$$

or equivalently,

$$v_3 = 2v_1 - v_2.$$

Since v_3 is a linear combination of v_1 and v_2, according to Theorem 5.5.1, we can delete v_3 from the given set of vectors. Hence, $\{(1, 2, 3), (-1, 1, 4)\}$ spans the same subspace of $\mathbf{R}^3$ as the given set of vectors. Since v_1 and v_2 are not proportional, they are LI, and so, we have determined an appropriate set of vectors. Geometrically, the subspace of $\mathbf{R}^3$ spanned by v_1 and v_2 is a plane, and the vectors v_3 and v_4 lie in this plane.   ❑

We now investigate the question of LI or LD in the vector spaces $\mathbf{R}^n$ and $C^k(I)$. These two vector spaces will play a major role when we apply the vector space results to linear systems of DE later in the text.

LINEAR DEPENDENCE AND LINEAR INDEPENDENCE IN $\mathbf{R}^n$

Let $\{\mathbf{v}_1, \mathbf{v}_2, \ldots, \mathbf{v}_n\}$ be a set of vectors in $\mathbf{R}^m$, and let A denote the matrix that has $\mathbf{v}_1, \mathbf{v}_2, \ldots, \mathbf{v}_n$ as *column* vectors. Thus,

$$A = [\mathbf{v}_1, \mathbf{v}_2, \ldots, \mathbf{v}_n]. \tag{5.5.2}$$

Since each of the given vectors is in $\mathbf{R}^m$, it follows that A has m rows and is therefore an $m \times n$ matrix. The condition for determining linear dependence or linear independence of $\{\mathbf{v}_1, \mathbf{v}_2, \ldots, \mathbf{v}_n\}$ is

$$c_1\mathbf{v}_1 + c_2\mathbf{v}_2 + \cdots + c_n\mathbf{v}_n = \mathbf{0}.$$

However, this is just the column vector form of the homogeneous system of linear equations (see Theorem 3.2.1)

$$Ac = \mathbf{0}, \tag{5.5.3}$$

where A is given in equation (5.5.2) and $c = [c_1\ c_2\ \ldots\ c_n]^T$. Consequently, we can state the following theorem and corollary:

Theorem 5.5.4: Let $\mathbf{v}_1, \mathbf{v}_2, \ldots, \mathbf{v}_n$ be vectors in $\mathbf{R}^m$, and let $A = [\mathbf{v}_1, \mathbf{v}_2, \ldots, \mathbf{v}_n]$. Then $\{\mathbf{v}_1, \mathbf{v}_2, \ldots, \mathbf{v}_n\}$ is LD if and only if the linear system $Ac = \mathbf{0}$ has nontrivial solutions.

Corollary 5.5.2: Let $\mathbf{v}_1, \mathbf{v}_2, \ldots, \mathbf{v}_n$ be vectors in $\mathbf{R}^m$, and let $A = [\mathbf{v}_1, \mathbf{v}_2, \ldots, \mathbf{v}_n]$. Then
1. If $n > m$, $\{\mathbf{v}_1, \mathbf{v}_2, \ldots, \mathbf{v}_n\}$ is LD.
2. If $n = m$, $\{\mathbf{v}_1, \mathbf{v}_2, \ldots, \mathbf{v}_n\}$ is LD if and only if $\det(A) = 0$.

PROOF
1. If $n > m$ the system (5.5.3) has an infinite number of solutions (Corollary 3.5.2), and hence, the vectors are LD.
2. If $n = m$, the system (5.5.3) is $n \times n$, and hence, from Corollary 4.2.2, it has an infinite number of solutions if and only if $\det(A) = 0$. ∎

Example 5.5.6 Determine whether the given vectors are LD or LI in $\mathbf{R}^4$.
1. $\mathbf{v}_1 = (1, 3, -1, 0)$, $\mathbf{v}_2 = (2, 9, -1, 3)$, $\mathbf{v}_3 = (4, 5, 6, 11)$, $\mathbf{v}_4 = (1, -1, 2, 5)$, $\mathbf{v}_5 = (3, -2, 6, 7)$.
2. $\mathbf{v}_1 = (1, 4, 1, 7)$, $\mathbf{v}_2 = (3, -5, 2, 3)$, $\mathbf{v}_3 = (2, -1, 6, 9)$, $\mathbf{v}_4 = (-2, 3, 1, 6)$.

Solution
1. Since we have 5 vectors in $\mathbf{R}^4$, Corollary 5.5.2 implies that $\{\mathbf{v}_1, \mathbf{v}_2, \mathbf{v}_3, \mathbf{v}_4, \mathbf{v}_5\}$ is necessarily LD.
2. In this case, we have 4 vectors in $\mathbf{R}^4$, and, therefore, we can use the determinant. It is straightforward to show that

$$\det(A) = \det[\mathbf{v}_1, \mathbf{v}_2, \mathbf{v}_3, \mathbf{v}_4] = \begin{vmatrix} 1 & 3 & 2 & -2 \\ 4 & -5 & -1 & 3 \\ 1 & 2 & 6 & 1 \\ 7 & 3 & 9 & 6 \end{vmatrix} = -462.$$

Since this determinant is nonzero, it follows from Corollary 5.5.2 that the given vectors are LI.

LINEAR INDEPENDENCE OF FUNCTIONS

When considering solutions to second-order linear DE in Chapter 2, we used the Wronskian to derive an analytic criterion for determining the linear dependence or independence of a pair of solutions defined on an interval I. We now consider the general problem of whether a set of functions is LI or LD. Once more, we will see that the Wronskian is a useful analytical tool that will be indispensable when we return to study nth order linear DE in Chapter 7. We begin by specializing the general Definition 5.5.2 to the case of a set of functions defined on an interval I.

Definition 5.5.3:   The set of functions $\{f_1, f_2, \ldots, f_n\}$ is **LI on an interval I** if and only if the only values of the scalars $c_1, c_2, \ldots, c_n$ such that

$$c_1 f_1(x) + c_2 f_2(x) + \cdots + c_n f_n(x) = 0, \ \textit{for all } x \textit{ in } I, \tag{5.5.4}$$

are $c_1 = c_2 = \cdots = c_n = 0$.

The main point to notice is that the condition (5.5.4) must hold for all x in I. We next generalize the definition of the Wronskian when we have a set consisting of more than two functions.

Definition 5.5.4:   Let $f_1, f_2, \ldots, f_n$ be functions in $C^{n-1}(I)$. The **Wronskian** of these functions is the order n determinant defined by

$$W[f_1, f_2, \ldots, f_n](x) = \begin{vmatrix} f_1(x) & f_2(x) & \cdots & f_n(x) \\ f_1'(x) & f_2'(x) & \cdots & f_n'(x) \\ \vdots & \vdots & & \vdots \\ f_1^{(n-1)}(x) & f_2^{(n-1)}(x) & \cdots & f_n^{(n-1)}(x) \end{vmatrix}.$$

REMARK: Notice that the Wronskian is a function defined on I. Also note that the value of the Wronskian depends on the order of the functions. For example, using properties of determinants,

$$W[f_2, f_1, \ldots, f_n](x) = -W[f_1, f_2, \ldots, f_n](x).$$

Example 5.5.7   If $f_1(x) = x, f_2(x) = x^2, f_3(x) = x^3$ on $(-\infty, \infty)$, then

$$W[f_1, f_2, f_3](x) = \begin{vmatrix} x & x^2 & x^3 \\ 1 & 2x & 3x^2 \\ 0 & 2 & 6x \end{vmatrix} = x(12x^2 - 6x^2) - (6x^3 - 2x^3) = 2x^3.$$

◻

We can now state and prove the main result about the Wronskian.

Theorem 5.5.5:   Let $f_1, f_2, \ldots, f_n$ be functions in $C^{n-1}(I)$. If $W[f_1, f_2, \ldots, f_n](x)$ is nonzero at any point in I, then $\{f_1, f_2, \ldots, f_n\}$ is LI on I.

PROOF   If

$$c_1 f_1(x) + c_2 f_2(x) + \cdots + c_n f_n(x) = 0,$$

for all x in I, then, differentiating $n - 1$ times yields the linear system

$$\begin{array}{llll} c_1 f_1(x) & + c_2 f_2(x) & + \cdots + c_n f_n(x) & = 0, \\ c_1 f_1'(x) & + c_2 f_2'(x) & + \cdots + c_n f_n'(x) & = 0, \\ & & \vdots & \\ c_1 f_1^{(n-1)}(x) + c_2 f_2^{(n-1)}(x) + \cdots + c_n f_n^{(n-1)}(x) & = 0, \end{array}$$

where the unknowns in the system are $c_1, c_2, \ldots, c_n$. The determinant of the matrix of coefficients of this system is just $W[f_1, f_2, \ldots, f_n](x)$. Consequently, if $W[f_1, f_2, \ldots, f_n](x_0) \neq 0$ for some x_0 in I, then the determinant of the matrix of coefficients of the system is nonzero at that point, and, therefore, the only solution to the system is the trivial solution $c_1 = c_2 = \cdots = c_n = 0$. That is, the given set of functions is LI on I.   ■

REMARKS

1.   Notice that it is only necessary for $W[f_1, f_2, \ldots, f_n](x)$ to be nonzero at one point in I for $\{f_1, f_2, \ldots, f_n\}$ to be LI on I.

2.   The logical equivalent of the preceding theorem is: If $\{f_1, f_2, \ldots, f_n\}$ is LD on I, then $W[f_1, f_2, \ldots, f_n](x) = 0$ at every point of I. This does *not* say that if $W[f_1, f_2, \ldots, f_n](x) = 0$ at every point of I, then $\{f_1, f_2, \ldots, f_n\}$ is LD on I.

If $W[f_1, f_2, \ldots, f_n](x) = 0$ at every point of I, then Theorem 5.5.5 gives no information as to the linear dependence or linear independence of $\{f_1, f_2, \ldots, f_n\}$ on I.

Example 5.5.8   Determine whether the following functions are LD or LI on $I = (-\infty, \infty)$:

(a) $f_1(x) = e^x, f_2(x) = x^2 e^x$.

(b) $f_1(x) = x, f_2(x) = x + x^2, f_3(x) = 2x - x^2$.

(c) $f_1(x) = x^2, \quad -\infty < x < \infty, \quad f_2(x) = \begin{cases} 2x^2, & \text{if } x \geq 0, \\ -x^2, & \text{if } x < 0. \end{cases}$

Solution

(a)
$$W[f_1, f_2](x) = \begin{vmatrix} e^x & x^2 e^x \\ e^x & e^x(x^2 + 2x) \end{vmatrix} = e^{2x} \begin{vmatrix} 1 & x^2 \\ 1 & x^2 + 2x \end{vmatrix} = 2xe^{2x}.$$

Since $W[f_1, f_2](x) \neq 0$ (except at $x = 0$), the functions are LI on $(-\infty, \infty)$.

(b)
$$W[f_1, f_2, f_3](x) = \begin{vmatrix} x & x + x^2 & 2x - x^2 \\ 1 & 1 + 2x & 2 - 2x \\ 0 & 2 & -2 \end{vmatrix}$$

$$= x[(-2)(1 + 2x) - 2(2 - 2x)] - [(-2)(x + x^2) - 2(2x - x^2)]$$

$$= 0.$$

Thus, no conclusion can be drawn from Theorem 5.5.5. However, a closer inspection of the functions reveals, for example, that

$$f_2 = 3f_1 - f_3.$$

Consequently, the functions are LD on $(-\infty, \infty)$.

(c) If $x \geq 0$, then

$$W[f_1, f_2](x) = \begin{vmatrix} x^2 & 2x^2 \\ 2x & 4x \end{vmatrix} = 0,$$

whereas if $x < 0$, then

$$W[f_1, f_2](x) = \begin{vmatrix} x^2 & -x^2 \\ 2x & -2x \end{vmatrix} = 0.$$

Thus $W[f_1, f_2](x) = 0$ for all x in $(-\infty, \infty)$, and so, no conclusion can be drawn from Theorem 5.5.5. Again we take a closer at the given functions. In this case, we see that on the interval $(-\infty, 0)$ the functions are LD, since

$$f_1 + f_2 = 0.$$

They are also LD on $[0, \infty)$, since on this interval, we have

$$2f_1 - f_2 = 0.$$

The key point to realize is that there is no set of *nonzero* constants c_1, c_2 for which

$$c_1 f_1 + c_2 f_2 = 0$$

holds for all x in $(-\infty, \infty)$. Consequently, the given functions are *LI* on $(-\infty, \infty)$. Once more this emphasizes the importance of the role played by the interval when discussing linear dependence and linear independence of functions. Figure 5.5.3 contains a sketch of the given functions.

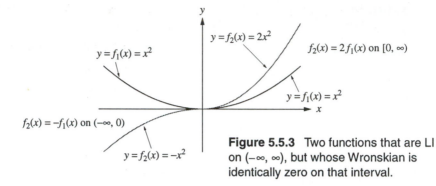

Figure 5.5.3 Two functions that are LI on $(-\infty, \infty)$, but whose Wronskian is identically zero on that interval.

It might appear at this stage that the usefulness of the Wronskian is questionable, since if $W[f_1, f_2, \ldots, f_n]$ vanishes on an interval I, then no conclusion can be drawn as to the linear dependence or linear independence of the functions $f_1, f_2, \ldots, f_n$ on I. However, the real power of the Wronskian is in its application to solutions of linear DE of the form

$$y^{(n)} + a_1(x)y^{(n-1)} + \cdots + a_{n-1}(x)y' + a_n(x)y = 0. \qquad (5.5.5)$$

For example, in Chapter 2, we showed that two solutions to $y'' + a_1 y' + a_2 y = 0$ are LD *if and only if* their Wronskian is zero at all points in the interval of interest. In Chapter 7, we will generalize this result to establish that if we have n functions that are *solutions of an equation of the form (5.5.5)* on an interval I, then if the Wronskian of these functions is identically zero on I, the functions are LD on I. Thus, the Wronskian does indeed completely characterize the linear dependence or linear independence of solutions of such equations. This is a fundamental result in the theory of linear DE.

EXERCISES 5.5

For problems 1–9, determine whether the given set of vectors is LI or LD in $\mathbf{R}^n$. In the case of linear dependence, find a dependency relationship.

1. $\{(1, -1), (1, 1)\}$.

2. $\{(2, -1), (3, 2), (0, 1)\}$.

3. $\{(1, -1, 0), (0, 1, -1), (1, 1, 1)\}$.

4. $\{(1, 2, 3), (1, -1, 2), (1, -4, 1)\}$.

5. $\{(-2, 4, -6), (3, -6, 9)\}$.

6. $\{(1, -1, 2), (2, 1, 0)\}$.

7. $\{(-1, 1, 2), (0, 2, -1), (3, 1, 2), (-1, -1, 1)\}$.

8. $\{(1, -1, 2, 3), (2, -1, 1, -1), (-1, 1, 1, 1)\}$.

9. $\{(2, -1, 0, 1), (1, 0, -1, 2), (0, 3, 1, 2), (-1, 1, 2, 1)\}$.

10. Determine whether $\{(1, 2, 3), (4, 5, 6), (7, 8, 9)\}$ is LI in $\mathbf{R}^3$. Describe span$\{(1, 2, 3), (4, 5, 6), (7, 8, 9)\}$ geometrically.

11. Consider the vectors $\mathbf{v}_1 = (2, -1, 5)$, $\mathbf{v}_2 = (1, 3, -4)$, and $\mathbf{v}_3 = (-3, -9, 12)$ in $\mathbf{R}^3$.

(a) Establish that $\{\mathbf{v}_1, \mathbf{v}_2, \mathbf{v}_3\}$ is LD.

(b) Is $\mathbf{v}_1 \in \text{span}\{\mathbf{v}_2, \mathbf{v}_3\}$? Draw a picture illustrating your answer.

12. Determine all values of the constant k for which the vectors $(1, 1, k)$, $(0, 2, k)$, and $(1, k, 6)$ are LD in $\mathbf{R}^3$.

For problems 13 and 14, determine all values of the constant k for which the given vectors are LI in $\mathbf{R}^4$.

13. $(1, 0, 1, k)$, $(-1, 0, k, 1)$, $(2, 0, 1, 3)$.

14. $(1, 1, 0, -1)$, $(1, k, 1, 1)$, $(2, 1, k, 1)$, $(-1, 1, 1, k)$.

For problems 15–17, determine whether the given vectors are LI in $M_2(\mathbf{R})$.

15. $A_1 = \begin{bmatrix} 1 & 1 \\ 0 & 1 \end{bmatrix}, A_2 = \begin{bmatrix} 2 & -1 \\ 0 & 1 \end{bmatrix},$

$A_3 = \begin{bmatrix} 3 & 6 \\ 0 & 4 \end{bmatrix}.$

16. $A_1 = \begin{bmatrix} 2 & -1 \\ 3 & 4 \end{bmatrix}, A_2 = \begin{bmatrix} -1 & 2 \\ 1 & 3 \end{bmatrix}.$

17. $A_1 = \begin{bmatrix} 1 & 0 \\ 1 & 2 \end{bmatrix}, A_2 = \begin{bmatrix} -1 & 1 \\ 2 & 1 \end{bmatrix},$

$A_3 = \begin{bmatrix} 2 & 1 \\ 5 & 7 \end{bmatrix}.$

18. Determine whether the following vectors are LI in P_2:

(a) $p_1(x) = 1 - x, p_2(x) = 1 + x.$

(b) $p_1(x) = 2 + 3x, p_2(x) = 4 + 6x.$

19. Show that the vectors $p_1(x) = a + bx, p_2(x) = c + dx$ are LI in P_2 if and only if the constants a, b, c, and d satisfy $ad - bc \neq 0$.

20. If $f_1(x) = \cos 2x, f_2(x) = \sin^2 x$, and $f_3(x) = \cos^2 x$, determine whether $\{f_1, f_2, f_3\}$ is LD or LI in $C^\infty(-\infty, \infty)$.

For problems 21–27, determine a LI set of vectors that spans the same subspace of V as that spanned by the original set of vectors.

21. $V = \mathbf{R}^3$, $\{(1, 2, 3), (-3, 4, 5), (1, -4/3, -5/3)\}$.

22. $V = \mathbf{R}^3$, $\{(1, 2, -1), (3, 1, 5), (0, 0, 0), (-1, 2, 3)\}$.

23. $V = \mathbf{R}^3$, $\{(1, 1, 1), (1, -1, 1), (1, -3, 1), (3, 1, 2)\}$.

24. $V = \mathbf{R}^4$, $\{(1, 1, -1, 1), (2, -1, 3, 1), (1, 1, 2, 1), (2, -1, 2, 1)\}$.

25. $V = M_2(\mathbf{R})$,

$\left\{ \begin{bmatrix} 1 & 2 \\ 3 & 4 \end{bmatrix}, \begin{bmatrix} -1 & 2 \\ 5 & 7 \end{bmatrix}, \begin{bmatrix} 3 & 2 \\ 1 & 1 \end{bmatrix} \right\}.$

26. $V = P_2$, $\{2 - 5x, 3 + 7x, 4 - x\}$.

27. $V = P_3$, $\{2 + x^2, 4 - 2x + 3x^2, 1 + x\}$.

For problems 28–32, use the Wronskian to show that the given functions are LI on the given interval I.

28. $f_1(x) = 1, f_2(x) = x, f_3(x) = x^2, I = (-\infty, \infty)$.

29. $f_1(x) = \sin x, f_2(x) = \cos x, f_3(x) = \tan x,$ $I = (-\pi/2, \pi/2)$.

30. $f_1(x) = 1, f_2(x) = 3x, f_3(x) = x^2 - 1,$ $I = (-\infty, \infty)$.

31. $f_1(x) = e^{2x}, f_2(x) = e^{3x}, f_3(x) = e^{-x}, I = (-\infty, \infty)$.

32. $f_1(x) = \begin{cases} x^2, \text{ if } x \geq 0, \\ 3x^3, \text{ if } x < 0, \end{cases} f_2(x) = 7x^2,$

$I = (-\infty, \infty)$.

For problems 33–35, show that the Wronskian of the given functions is identically zero on $(-\infty, \infty)$. Determine whether the functions are LI or LD on that interval.

33. $f_1(x) = 1, f_2(x) = x, f_3(x) = 2x - 1$.

34. $f_1(x) = e^x, f_2(x) = e^{-x}, f_3(x) = \cosh x$.

35. $f_1(x) = \begin{cases} 5x^3, \text{ if } x \geq 0, \\ -3x^3, \text{ if } x < 0, \end{cases} f_2(x) = 2x^3$.

36. Consider the functions

$$f_1(x) = x, f_2(x) = \begin{cases} x, \text{ if } x \geq 0, \\ -x, \text{ if } x < 0. \end{cases}$$

(a) Show that f_2 is not in $C^1(-\infty, \infty)$.

(b) Show that $\{f_1, f_2\}$ is LD on $(-\infty, 0)$ and $[0, \infty)$ and LI on $(-\infty, \infty)$. Justify your results by making a sketch showing both of the functions.

37. Determine whether the functions

$$f_1(x) = x, f_2(x) = \begin{cases} x, \text{ if } x \neq 0, \\ 1, \text{ if } x = 0, \end{cases}$$

are LD or LI on $I = (-\infty, \infty)$.

38. Show that the functions

$$f_1(x) = \begin{cases} x - 1, & \text{if } x \geq 1, \\ 2(x - 1), \text{ if } x < 1, \end{cases} f_2(x) = 2x,$$

$f_3(x) = 3$ are LI on $(-\infty, \infty)$. Determine all intervals on which they are LD.

39. (a) Show that the functions

$$f_0(x) = 1, \ f_1(x) = x, \ f_2(x) = x^2, \ f_3(x) = x^3$$

are LI on any interval.

(b) If $f_k(x) = x^k, k = 0, 1, 2, \ldots, n$, show that $\{f_0, f_1, \ldots, f_n\}$ is LI on any interval for all fixed n.

40. **(a)** Show that the functions

$$f_1(x) = e^{r_1 x}, \quad f_2(x) = e^{r_2 x}, \quad f_3(x) = e^{r_3 x}$$

have Wronskian

$$W[f_1, f_2, f_3](x) = e^{(r_1 + r_1 + r_3)x} \begin{vmatrix} 1 & 1 & 1 \\ r_1 & r_2 & r_3 \\ r_1^2 & r_2^2 & r_3^2 \end{vmatrix}$$

$$= e^{(r_1 + r_1 + r_3)x} (r_3 - r_1)(r_3 - r_2)(r_2 - r_1),$$

and hence determine the conditions on r_1, r_2, r_3, such that $\{f_1, f_2, f_3\}$ is LI on any interval.

(b) More generally, show that the functions

$$f_1(x) = e^{r_1 x}, \quad f_2(x) = e^{r_2 x}, \quad \dots, \quad f_n(x) = e^{r_n x}$$

are LI on any interval if and only if all the r_i are distinct. (Hint: Show that the Wronskian of the given functions is a multiple of the $n \times n$ Vandermonde determinant, and then use Exercise 21 in Section 4.3.)

41. Let v_1 and v_2 be LI vectors in a vector space V, and let $x = \alpha v_1 + v_2$, $y = v_1 + \alpha v_2$, where α is a constant. Use Definition 5.5.2 to determine all values of α for which $\{x, y\}$ is LI.

42. If v_1 and v_2 are LI vectors in a vector space V and u_1, u_2, u_3 are each linear combinations of them, prove that $\{u_1, u_2, u_3\}$ is LD.

43. Let $v_1, v_2, \dots, v_m$ be LI vectors in a vector space V, and suppose that the vectors $u_1, u_2, \dots, u_n$ are each linear combinations of them. It follows that we can write

$$u_k = \sum_{i=1}^{m} a_{ik} v_i, \quad k = 1, 2, \dots, n,$$

for appropriate constants a_{ik}.

(a) If $n > m$ prove that $\{u_1, u_2, \dots, u_n\}$ is necessarily LD in V.

(b) If $n = m$, prove that the $\{u_1, u_2, \dots, u_n\}$ is LI in V if and only if $\det [a_{ij}] \neq 0$.

(c) If $n < m$, prove that $\{u_1, u_2, \dots, u_n\}$ is LI in V if and only if $\text{rank}(A) = n$, where $A = [a_{ij}]$.

(d) Which theorem from this section do these results generalize ?

5.6   BASES AND DIMENSION

The results of the previous section show that if a minimal spanning set exists in a (nontrivial) vector space V, it cannot be LD. We therefore focus our attention on spanning sets that are LI. Any such spanning set is called a *basis* for V. One of the results of this section establishes that any basis for V is indeed a minimal spanning set. The basis is the key concept in the vector space.

Definition   5.6.1:   A set of vectors $\{v_1, v_2, \dots, v_k\}$ in a vector space V is called a **basis**[1] for V if

1. The vectors are LI.

2. The vectors span V.

A vector space V is called **finite-dimensional** if it has a basis consisting of a finite number of vectors.

Notice that if we have a spanning set for a vector space then we can always, in principle, determine a basis for V using the technique of Corollary 5.5.1. Furthermore, the computational aspects of determining a basis have been covered in the previous two sections, since all we are really doing is combining the two concepts of linear

---

[1] The plural of basis is bases.

independence and linear span. Consequently, this section is somewhat more theoretically oriented than the preceding ones. The reader is encouraged to not gloss over the theoretical aspects as these really are fundamental results.

Example 5.6.1   Determine a basis for the solution space to the DE

$$y'' + \omega^2 y = 0$$

on any interval I.

Solution   Our results from Chapter 2 tell us that all solutions to the given DE are of the form

$$y(x) = c_1 y_1(x) + c_2 y_2(x),$$

where y_1 and y_2 are any two LI solutions on I. Consequently, $\{\cos \omega x, \sin \omega x\}$ is a LI spanning set for the solution space of the DE and therefore a basis. ❏

Our results on spanning sets and linear independence in $\mathbf{R}^2$ and $\mathbf{R}^3$ immediately give the following characterizations of bases in these vector spaces.

Theorem 5.6.1:

(a)   Every basis in $\mathbf{R}^2$ contains precisely two vectors. Furthermore, $\{\mathbf{v}_1, \mathbf{v}_2\}$ is a basis for $\mathbf{R}^2$ if and only if $\mathbf{v}_1$ and $\mathbf{v}_2$ are noncolinear.

(b)   Every basis in $\mathbf{R}^3$ contains precisely three vectors. Furthermore, $\{\mathbf{v}_1, \mathbf{v}_2, \mathbf{v}_3\}$ is a basis for $\mathbf{R}^3$ if and only if $\mathbf{v}_1, \mathbf{v}_2, \mathbf{v}_3$ are noncoplanar.

Of special importance is the **standard basis** in $\mathbf{R}^3$, denoted $\{\mathbf{e}_1, \mathbf{e}_2, \mathbf{e}_3\}$, which is defined by

$$\mathbf{e}_1 = (1, 0, 0), \quad \mathbf{e}_2 = (0, 1, 0), \quad \mathbf{e}_3 = (0, 0, 1).$$

The corresponding geometric vectors are the familiar $\mathbf{i}, \mathbf{j}$, and $\mathbf{k}$. More generally, consider the set of vectors $\{\mathbf{e}_1, \mathbf{e}_2, ..., \mathbf{e}_n\}$ in $\mathbf{R}^n$ defined by

$$\mathbf{e}_1 = (1, 0, ..., 0), \quad \mathbf{e}_2 = (0, 1, ..., 0), \quad ..., \quad \mathbf{e}_n = (0, 0, ..., 1).$$

These vectors are LI, since

$$\det([\mathbf{e}_1, \mathbf{e}_2, ..., \mathbf{e}_n]) = \det(I_n) = 1 \neq 0.$$

Furthermore, the vectors span $\mathbf{R}^n$, since an arbitrary vector $\mathbf{x} = (x_1, x_2, ..., x_n)$ in $\mathbf{R}^n$ can be written as

$$\mathbf{x} = x_1(1, 0, ..., 0) + x_2(0, 1, ..., 0) + \cdots + x_n(0, 0, ..., 1)$$

$$= x_1 \mathbf{e}_1 + x_2 \mathbf{e}_2 + \cdots + x_n \mathbf{e}_n.$$

Consequently, $\{\mathbf{e}_1, \mathbf{e}_2, ..., \mathbf{e}_n\}$ is a basis for $\mathbf{R}^n$. We refer to this basis as the **standard basis** for $\mathbf{R}^n$.

The general vector in $\mathbf{R}^n$ has n components, and the standard basis vectors arise as the n vectors that are obtained by sequentially setting one component to the value 1 and the other components to zero. In general, this is how we obtain standard bases in vector spaces whose vectors are determined by the specification of n independent constants. We illustrate with some examples.

Example 5.6.2   Determine the standard basis for $M_2(\mathbf{R})$.

Solution   The general matrix in $M_2(\mathbf{R})$ is $\begin{bmatrix} a & b \\ c & d \end{bmatrix}$. Consequently, there are four independent parameters that give rise to the different vectors in $M_2(\mathbf{R})$. Sequentially setting one of these parameters to the value 1 and the others to zero generates the following four matrices:

$$A_1 = \begin{bmatrix} 1 & 0 \\ 0 & 0 \end{bmatrix}, \quad A_2 = \begin{bmatrix} 0 & 1 \\ 0 & 0 \end{bmatrix}, \quad A_3 = \begin{bmatrix} 0 & 0 \\ 1 & 0 \end{bmatrix}, \quad A_4 = \begin{bmatrix} 0 & 0 \\ 0 & 1 \end{bmatrix}.$$

We have shown in Example 5.4.3 that $\{A_1, A_2, A_3, A_4\}$ is a spanning set for $M_2(\mathbf{R})$. Furthermore,

$$c_1 A_1 + c_2 A_2 + c_3 A_3 + c_4 A_4 = 0_2$$

holds if and only if

$$c_1 \begin{bmatrix} 1 & 0 \\ 0 & 0 \end{bmatrix} + c_2 \begin{bmatrix} 0 & 1 \\ 0 & 0 \end{bmatrix} + c_3 \begin{bmatrix} 0 & 0 \\ 1 & 0 \end{bmatrix} + c_4 \begin{bmatrix} 0 & 0 \\ 0 & 1 \end{bmatrix} = \begin{bmatrix} 0 & 0 \\ 0 & 0 \end{bmatrix},$$

that is, if and only if $c_1 = c_2 = c_3 = c_4 = 0$. Consequently, $\{A_1, A_2, A_3, A_4\}$ is a LI spanning set for $M_2(\mathbf{R})$, and hence it is a basis. This is the standard basis for $M_2(\mathbf{R})$.

Example 5.6.3   Determine the standard basis for P_3.

Solution   We have

$$P_3 = \{a_0 + a_1 x + a_2 x^2 : a_0, a_1, a_2 \in \mathbf{R}\},$$

so that vectors in P_3 are determined by specifying values for the three parameters a_0, a_1, and a_2. Sequentially setting one of these parameters to the value 1 and the other two to zero, yields the following polynomials in P_3:

$$p_0(x) = 1, \quad p_1(x) = x, \quad p_2(x) = x^2.$$

We have shown in Example 5.4.4 that $\{p_0, p_1, p_2\}$ is a spanning set for P_3. Furthermore,

$$W[p_0, p_1, p_2](x) = \begin{vmatrix} 1 & x & x^2 \\ 0 & 1 & 2x \\ 0 & 0 & 2 \end{vmatrix} = 2 \neq 0,$$

which implies that $\{p_0, p_1, p_2\}$ is LI on any interval. Consequently, $\{p_0, p_1, p_2\}$ is a basis for P_3. This is the standard basis for P_3.

DIMENSION OF A FINITE-DIMENSIONAL VECTOR SPACE

We now develop the general theory for bases in a finite-dimensional vector space. If we have a basis for a vector space V, then, since the vectors in the basis span V, any vector in V can be expressed as a linear combination of the basis vectors. The next theorem establishes that there is only one way in which we can do this.

Theorem 5.6.2:   If V is a finite-dimensional vector space with basis $\{v_1, v_2, \ldots, v_n\}$, then any vector $v \in V$ can be written *uniquely* as a linear combination of $v_1, v_2, \ldots, v_n$.

PROOF   Since $v_1, v_2, \ldots, v_n$ span V, any vector $v \in V$ can be expressed as

$$v = a_1 v_1 + a_2 v_2 + \cdots + a_n v_n, \qquad (5.6.1)$$

for some scalars $a_1, a_2, \ldots, a_n$. Suppose also that

$$v = b_1 v_1 + b_2 v_2 + \cdots + b_n v_n \qquad (5.6.2)$$

or some scalars $b_1, b_2, \ldots, b_n$. We must show that $a_i = b_i$ for each i. Subtracting equation (5.6.2) from equation (5.6.1) yields

$$(a_1 - b_1)v_1 + (a_2 - b_2)v_2 + \cdots + (a_n - b_n)v_n = 0. \qquad (5.6.3)$$

But $\{v_1, v_2, \ldots, v_n\}$ is LI, and so, equation (5.6.3) implies that

$$a_1 - b_1 = 0, \quad a_2 - b_2 = 0, \quad \ldots, \quad a_n - b_n = 0.$$

That is, $a_i = b_i, \quad i = 1, 2, \ldots, n$.   ■

If $\{v_1, v_2, \ldots, v_n\}$ is a basis for V and v is any vector in V, then the scalars in the unique n-tuple $(c_1, c_2, \ldots, c_n)$ such that

$$v = c_1 v_1 + c_2 v_2 + \cdots + c_n v_n$$

are called the **components of v relative to the basis** $\{v_1, v_2, \ldots, v_n\}$.

Example 5.6.4   Determine the components of the vector $v = (1, 7)$ relative to the basis $\{(1, 2), (3, 1)\}$.

Solution   If we let $v_1 = (1, 2)$, and $v_2 = (3, 1)$, then, since these vectors are not colinear, $\{v_1, v_2\}$ is a basis for $\mathbf{R}^2$. We must determine constants c_1, c_2 such that

$$c_1 v_1 + c_2 v_2 = v,$$

that is, such that

$$c_1(1, 2) + c_2(3, 1) = (1, 7).$$

This requires that

$$c_1 + 3c_2 = 1 \quad \text{and} \quad 2c_1 + c_2 = 7.$$

The solution to this system is $(4, -1)$, which give the components of v relative to the basis $\{v_1, v_2\}$. (See Figure 5.6.1.) Thus,

$$v = 4v_1 - v_2.$$

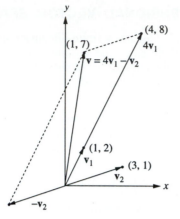

Figure 5.6.1 The components of the vector $\mathbf{v} = (1, 7)$ relative to the basis $\{(1, 2), (3, 1)\}$.

Example 5.6.5   In P_3, determine the components of $p(x) = 5 + 7x - 3x^2$ relative to the following:

(a) The standard basis $\{1, x, x^2\}$.   (b) The basis $\{1 + x, 2 + 3x, 5 + x + x^2\}$.

Solution

(a)   The given polynomial is already written as a linear combination of the standard basis vectors. Consequently, the components of $p(x) = 5 + 7x - 3x^2$ relative to the standard basis are 5, 7, –3.

(b)   The components of $p(x) = 5 + 7x - 3x^2$ relative to the basis

$$\{1 + x, 2 + 3x, 5 + x + x^2\}$$

are c_1, c_2, and c_3, where

$$c_1(1 + x) + c_2(2 + 3x) + c_3(5 + x + x^2) = 5 + 7x - 3x^2.$$

That is,

$$(c_1 + 2c_2 + 5c_3) + (c_1 + 3c_2 + c_3)x + c_3x^2 = 5 + 7x - 3x^2.$$

Hence, c_1, c_2, and c_3 satisfy

$$\begin{aligned} c_1 + 2c_2 + 5c_3 &= 5, \\ c_1 + 3c_2 + c_3 &= 7, \\ c_3 &= -3. \end{aligned}$$

The augmented matrix of this system has RREF

$$\begin{bmatrix} 1 & 0 & 0 & 40 \\ 0 & 1 & 0 & -10 \\ 0 & 0 & 1 & -3 \end{bmatrix},$$

so that the system has solution $(40, -10, -3)$, which gives the required components. Hence, we can write

$$5 + 7x - 3x^2 = 40(1 + x) - 10(2 + 3x) - 3(5 + x + x^2). \qquad \square$$

The following theorem and its corollaries are fundamental results in vector space theory.

Theorem 5.6.3:   If a finite-dimensional vector space has a basis consisting of m vectors, then any set of more than m vectors is LD.

PROOF   Let $\{\mathbf{v}_1, \mathbf{v}_2, \ldots, \mathbf{v}_m\}$ be a basis for V, and consider an arbitrary set of vectors in V, say, $\{\mathbf{u}_1, \mathbf{u}_2, \ldots, \mathbf{u}_n\}$, with $n > m$. We wish to prove that $\{\mathbf{u}_1, \mathbf{u}_2, \ldots, \mathbf{u}_n\}$ is necessarily LD. Since $\{\mathbf{v}_1, \mathbf{v}_2, \ldots, \mathbf{v}_m\}$ is a basis for V, it follows that each $\mathbf{u}_j$ can be written as a linear combination of $\mathbf{v}_1, \mathbf{v}_2, \ldots, \mathbf{v}_m$. Thus, there exist constants a_{ij} such that

$$\mathbf{u}_1 = a_{11}\mathbf{v}_1 + a_{21}\mathbf{v}_2 + \cdots + a_{m1}\mathbf{v}_m,$$
$$\mathbf{u}_2 = a_{12}\mathbf{v}_1 + a_{22}\mathbf{v}_2 + \cdots + a_{m2}\mathbf{v}_m,$$
$$\vdots$$
$$\mathbf{u}_n = a_{1n}\mathbf{v}_1 + a_{2n}\mathbf{v}_2 + \cdots + a_{mn}\mathbf{v}_m.$$

To prove that $\{\mathbf{u}_1, \mathbf{u}_2, \ldots, \mathbf{u}_n\}$ is LD, we must show that there exist scalars $c_1, c_2, \ldots, c_n$, not all zero, such that

$$c_1\mathbf{u}_1 + c_2\mathbf{u}_2 + \cdots + c_n\mathbf{u}_n = \mathbf{0}. \tag{5.6.4}$$

Inserting the expressions for $\mathbf{u}_1, \mathbf{u}_2, \ldots, \mathbf{u}_n$ into equation (5.6.4) yields

$$c_1(a_{11}\mathbf{v}_1 + a_{21}\mathbf{v}_2 + \cdots + a_{m1}\mathbf{v}_m) + c_2(a_{12}\mathbf{v}_1 + a_{22}\mathbf{v}_2 + \cdots + a_{m2}\mathbf{v}_m) + \cdots$$
$$+ c_n(a_{1n}\mathbf{v}_1 + a_{2n}\mathbf{v}_2 + \cdots + a_{mn}\mathbf{v}_m) = \mathbf{0}.$$

Rearranging terms, we have

$$(a_{11}c_1 + a_{12}c_2 + \cdots + a_{1n}c_n)\mathbf{v}_1 + (a_{21}c_1 + a_{22}c_2 + \cdots + a_{2n}c_n)\mathbf{v}_2 +$$
$$\cdots + (a_{m1}c_1 + a_{m2}c_2 + \cdots + a_{mn}c_n)\mathbf{v}_m = \mathbf{0}.$$

Since $\{\mathbf{v}_1, \mathbf{v}_2, \ldots, \mathbf{v}_m\}$ is LI, we can conclude that

$$a_{11}c_1 + a_{12}c_2 + \cdots + a_{1n}c_n = 0,$$
$$a_{21}c_1 + a_{22}c_2 + \cdots + a_{2n}c_n = 0,$$
$$\vdots$$
$$a_{m1}c_1 + a_{m2}c_2 + \cdots + a_{mn}c_n = 0.$$

This is an $m \times n$ homogeneous system of linear equations with $m < n$, and hence, from Corollary 3.5.2, it has nontrivial solutions for $c_1, c_2, \ldots, c_n$. It therefore follows from equation (5.6.4) that $\{\mathbf{u}_1, \mathbf{u}_2, \ldots, \mathbf{u}_n\}$ is LD. ∎

Corollary 5.6.1:   All bases in a finite-dimensional vector space V contain the same number of vectors.

PROOF   Suppose $\{\mathbf{v}_1, \mathbf{v}_2, \ldots, \mathbf{v}_n\}$ and $\{\mathbf{u}_1, \mathbf{u}_2, \ldots, \mathbf{u}_m\}$ are two bases for V. From Theorem 5.6.3 we know that we cannot have $m > n$ (otherwise $\{\mathbf{u}_1, \mathbf{u}_2, \ldots, \mathbf{u}_m\}$ would be a LD set and hence could not be a basis for V) or $n > m$ (otherwise $\{\mathbf{v}_1, \mathbf{v}_2, \ldots, \mathbf{v}_n\}$

would be a LD set and hence could not be a basis for V). Thus, it follows that we must have $m = n$.   ∎

We can now prove that any basis provides a minimal spanning set for V.

Corollary 5.6.2:   If a finite-dimensional vector space V has a basis consisting of n vectors, then any spanning set must contain at least n vectors.

PROOF   If the spanning set contained fewer than n vectors, then there would be a subset of less than n LI vectors that spanned V, that is, a basis consisting of less than n vectors. But this would contradict the previous corollary.   ∎

The number of vectors in a basis for a finite-dimensional vector space is clearly a fundamental property of the vector space. We call this number the *dimension* of the vector space.

Definition 5.6.2: The **dimension** of a finite-dimensional vector space V, written dim $[V]$, is the number of vectors in any basis for V. If V consists only of the zero vector, then we define its dimension to be zero.

REMARK   We say that the dimension of the world we live in is three for the very reason that the maximum number of independent directions that we can perceive is three. If a vector space has a basis containing n vectors, then, from Theorem 5.6.3, the maximum number of vectors in any LI set is n. Thus, we see that the terminology *dimension* used in an arbitrary vector space is a generalization of a familiar idea.

Example 5.6.6   It follows from Theorem 5.6.1 and Examples 5.6.1 and 5.6.2 that dim $[\mathbf{R}^3] = 3$, dim $[M_2(\mathbf{R})] = 4$, and dim$[P_3] = 3$, respectively.   □

More generally, the following dimensions should be remembered

$$\dim[\mathbf{R}^n] = n, \quad \dim[\mathrm{M}_n(\mathbf{R})] = n^2, \quad \text{and} \quad \dim[P_n] = n.$$

We have already established the first of these, since the standard (and hence any) basis in $\mathbf{R}^n$ contains n vectors. We leave it as an exercise to verify that the standard basis for $\mathrm{M}_n(\mathbf{R})$ is given by the n^2 LI matrices $\{ I_{ij} \}$, where I_{ij} denotes the $n \times n$ matrix that has a one in the ijth position and zeros elsewhere [see Example 5.6.2 for the case of $M_2(\mathbf{R})$] and that the standard basis for P_n is given by the n LI polynomials $\{1, x, x^2, \ldots, x^{n-1}\}$.

Next, consider the second-order linear homogeneous DE

$$y'' + a_1(x)y' + a_2(x)y = 0$$

on an interval I. According to Theorem 2.1.4, provided a_1 and a_2 are continuous on I, all solutions to the DE are of the form

$$y(x) = c_1 y_1(x) + c_2 y_2(x),$$

where $\{y_1, y_2\}$ is any LI set of solutions to the DE. Using the terminology introduced in this section, we can therefore conclude the following:

> The set of all solutions to $y'' + a_1(x)y' + a_2(x)y = 0$ on an interval I is a vector space of dimension two.

In Chapter 7, we will see how this result generalizes to the case of nth-order linear DE.

If a vector space has dimension n, then, from Theorem 5.6.3, the maximum number of vectors in any LI set is n, whereas, from Corollary 5.6.2, the minimum number of vectors that can span V is also n. We might therefore suspect that *any* set of n LI vectors is a basis for V. This is certainly the case for the solution space of linear homogeneous second-order DE, as well as for bases in $\mathbf{R}^2$ and $\mathbf{R}^3$ (see Theorem 5.6.1). The next theorem establishes the result in general.

Theorem 5.6.4:   If $\dim[V] = n$, then *any* set of n LI vectors in V is a basis for V.

PROOF   Let $\mathbf{v}_1, \mathbf{v}_2, \ldots, \mathbf{v}_n$ be n LI vectors in V. We must show that they span V. If $\mathbf{v}$ is an arbitrary vector in V, then, from Theorem 5.6.3, the set of vectors $\{\mathbf{v}, \mathbf{v}_1, \mathbf{v}_2, \ldots, \mathbf{v}_n\}$ is LD, and so, there exist scalars $c_0, c_1, c_2, \ldots, c_n$, not all zero, such that

$$c_0\mathbf{v} + c_1\mathbf{v}_1 + \cdots + c_n\mathbf{v}_n = \mathbf{0}. \tag{5.6.5}$$

Since $\{\mathbf{v}_1, \mathbf{v}_2, \ldots, \mathbf{v}_n\}$ is LI, it follows that necessarily $c_0 \neq 0$, and so, from equation (5.6.5),

$$\mathbf{v} = -\frac{1}{c_0}(c_1\mathbf{v}_1 + c_2\mathbf{v}_2 + \cdots + c_n\mathbf{v}_n).$$

That is, any vector can be written as a linear combination of $\mathbf{v}_1, \mathbf{v}_2, \ldots, \mathbf{v}_n$, and hence $\{\mathbf{v}_1, \mathbf{v}_2, \ldots, \mathbf{v}_n\}$ is a basis for V.   ■

Theorem 5.6.4 is one of the most important results of the section. In Chapter 7, we will explicitly construct a basis for the solution space to the DE

$$y^{(n)} + a_1(x)y^{(n-1)} + \cdots + a_{n-1}(x)y' + a_n(x)y = 0$$

consisting of n vectors. It will then follow immediately from Theorem 5.6.4 that every solution to this DE is of the form

$$y(x) = c_1y_1(x) + c_2y_2(x) + \cdots + c_ny_n(x),$$

where $\{y_1, y_2, \ldots, y_n\}$ is *any* LI set of n solutions to the DE. Therefore, determining all solutions to the DE will be reduced to determining any LI set of n solutions. A similar application of the theorem will be used to develop the theory for systems of DE in Chapter 8.

Example 5.6.7   Verify that $\{1 + x, 2 - 2x + x^2, 1 + x^2\}$ is a basis for P_3.

Solution   Since $\dim[P_3] = 3$, Theorem 5.6.4 implies that we need only verify that the given set of vectors is LI. The polynomials

$$p_1(x) = 1 + x, \quad p_2(x) = 2 - 2x + x^2, \quad p_3(x) = 1 + x^2$$

have Wronskian

$$W[p_1, p_2, p_3](x) = \begin{vmatrix} 1+x & 2-2x+x^2 & 1+x^2 \\ 1 & -2+2x & 2x \\ 0 & 2 & 2 \end{vmatrix} = -6 \neq 0.$$

Since the Wronskian is nonzero, the given set of vectors is LI on any interval. Consequently, $\{1 + x, 2 - 2x + x^2, 1 + x^2\}$ is indeed a basis for P_3. ☐

We establish a corollary to Theorem 5.6.4.

Corollary 5.6.3:   Let S be a subspace of a finite dimensional vector space V. If $\dim[V] = n$, then

$$\dim[S] \leq n.$$

Furthermore, if $\dim[S] = n$, then $S = V$.

PROOF   Suppose that $\dim[S] > n$. Then any basis for S would contain more than n LI vectors, and therefore, we would have a LI set of more than n vectors in V. This would contradict Theorem 5.6.3, and so, $\dim[S] \leq n$. Now consider the case when $\dim[S] = \dim[V] = n$. In this case, any basis for S consists of n LI vectors in S, and hence n LI vectors in V. Consequently, the span of these vectors would generate all vectors in V, and, therefore, $S = V$. ∎

Example 5.6.8   Give a geometric description of the subspaces of $\mathbf{R}^3$ of dimensions 0, 1, 2, 3.

Solution

Zero-dimensional subspace: This corresponds to the subspace $\{(0, 0, 0)\}$, and therefore, it is represented geometrically by the origin of a Cartesian coordinate system.

One-dimensional subspaces: These are subspaces generated by a single nonzero vector. Consequently, they correspond geometrically to lines through the origin.

Two-dimensional subspaces: These are the subspaces generated by any two noncolinear vectors and correspond geometrically to planes through the origin.

Three-dimensional subspace: Since $\dim[\mathbf{R}^3] = 3$, it follows from Corollary 5.6.3 that the only three-dimensional subspace of $\mathbf{R}^3$ is $\mathbf{R}^3$ itself.

Example 5.6.9   Determine a basis for the subspace of $\mathbf{R}^3$ consisting of all solutions to the equation $x_1 + 2x_2 - x_3 = 0$.

Solution   We can solve this problem geometrically. The given equation is that of a plane through the origin and therefore is a two-dimensional subspace of $\mathbf{R}^3$. In order to determine a basis for this subspace, we need only choose two noncolinear vectors that lie in the plane. A simple choice of vectors is $\mathbf{v}_1 = (1, 0, 1)$, $\mathbf{v}_2 = (2, -1, 0)$. Consequently, a basis for the subspace is $\{(1, 0, 1), (2, -1, 0)\}$. ☐

Corollary 5.6.3 has shown that if S is a subspace of a finite-dimensional vector space V with $\dim[S] = \dim[V]$, then $S = V$. Our next result establishes that, in general, a basis for a subspace of a finite-dimensional vector space V can be extended to a basis for V. This result will be required in the next section and also in Chapter 6.

Theorem 5.6.5: Let S be a nontrivial subspace of a finite-dimensional vector space V. Any basis for S is part of a basis for V.

PROOF Suppose $\dim[V] = n$ and $\dim[S] = k$. If $k = n$, then $S = V$, (Corollary 5.6.3) so that any basis for S is a basis for V. Suppose that $k < n$, and let $\{v_1, v_2, \ldots, v_k\}$ be a basis for S. Then there exists at least one vector, say, v_{k+1}, in V that is not in span$\{v_1, v_2, \ldots, v_k\}$. Thus, $\{v_1, v_2, \ldots, v_k, v_{k+1}\}$ is LI. If $k + 1 = n$, then we have a basis for V, and we are done. Otherwise, we can repeat the procedure to obtain the LI set $\{v_1, v_2, \ldots, v_k, v_{k+1}, v_{k+2}\}$. This process will terminate when we have a LI set containing n vectors. ∎

REMARK The process used in proving the previous theorem is referred to as **extending a basis**.

Example 5.6.10 Let S denote the subspace of $M_2(\mathbf{R})$ consisting of all symmetric 2×2 matrices. Determine a basis for S, and find $\dim[S]$. Extend your basis for S to obtain a basis for $M_2(\mathbf{R})$.

Solution We first express S in set notation as

$$S = \{A \in M_2(\mathbf{R}) : A^T = A\}.$$

In order to determine a basis for S we need to obtain the element form of the matrices in S. We can write

$$S = \left\{ \begin{bmatrix} a & b \\ b & c \end{bmatrix} : a, b, c \in \mathbf{R} \right\}.$$

Since

$$\begin{bmatrix} a & b \\ b & c \end{bmatrix} = a \begin{bmatrix} 1 & 0 \\ 0 & 0 \end{bmatrix} + b \begin{bmatrix} 0 & 1 \\ 1 & 0 \end{bmatrix} + c \begin{bmatrix} 0 & 0 \\ 0 & 1 \end{bmatrix},$$

it follows that

$$S = \text{span} \left\{ \begin{bmatrix} 1 & 0 \\ 0 & 0 \end{bmatrix}, \begin{bmatrix} 0 & 1 \\ 1 & 0 \end{bmatrix}, \begin{bmatrix} 0 & 0 \\ 0 & 1 \end{bmatrix} \right\}.$$

Furthermore, it is easily shown that the matrices in this spanning set are LI. Consequently, a basis for S is $\left\{ \begin{bmatrix} 1 & 0 \\ 0 & 0 \end{bmatrix}, \begin{bmatrix} 0 & 1 \\ 1 & 0 \end{bmatrix}, \begin{bmatrix} 0 & 0 \\ 0 & 1 \end{bmatrix} \right\}$, so that $\dim[S] = 3$.

Since $\dim[M_2(\mathbf{R})] = 4$, in order to extend the basis for S to a basis for $M_2(\mathbf{R})$, we need to add in one further matrix from $M_2(\mathbf{R})$ such that the resulting set is LI. Clearly we must choose a nonsymmetric matrix. A simple choice is $\begin{bmatrix} 0 & 1 \\ 0 & 0 \end{bmatrix}$. Adding this vector to the basis for S yields the LI set

$$\left\{ \begin{bmatrix} 1 & 0 \\ 0 & 0 \end{bmatrix}, \begin{bmatrix} 0 & 1 \\ 1 & 0 \end{bmatrix}, \begin{bmatrix} 0 & 0 \\ 0 & 1 \end{bmatrix}, \begin{bmatrix} 0 & 1 \\ 0 & 0 \end{bmatrix} \right\}. \tag{5.6.6}$$

Since $\dim[M_2(\mathbf{R})] = 4$, (5.6.6) gives a basis for $M_2(\mathbf{R})$. ☐

It is important to realize that not all vector spaces are finite-dimensional. A vector space in which we can find an arbitrarily large number of LI vectors is said to be **infinite-dimensional**. Indeed, the vector space of primary importance in DE theory, $C^n(I)$ is an infinite-dimensional vector space, as we now show.

Example 5.6.11   Show that the vector space $C^n(I)$ is an infinite-dimensional vector space.

Solution   Consider the functions $1, x, x^2, \ldots, x^k$. For each fixed k, the Wronskian of these functions is nonzero, and hence, the functions are LI on I. Since we can choose k arbitrarily it follows that there are an arbitrarily large number of LI vectors in $C^n(I)$, and hence, $C^n(I)$ is infinite-dimensional.   ❏

In the previous example, we showed that $C^n(I)$ is an infinite-dimensional vector space. Consequently, the use of our finite-dimensional vector space theory in the analysis of DE appears questionable. However, the key theoretical result that we will establish in Chapter 7 is that the *solution set* of certain linear DE is a *finite-dimensional subspace* of $C^n(I)$, and therefore, our basis results will be applicable in this solution set. The special case when $n = 2$ was considered earlier in the section.

EXERCISES 5.6

For problems 1–5, determine whether the given set of vectors is a basis for $\mathbf{R}^n$.

1. $\{(1, 1), (-1, 1)\}$.

2. $\{(1, 2, 1), (3, -1, 2), (1, 1, -1)\}$.

3. $\{(1, -1, 1), (2, 5, -2), (3, 11, -5)\}$.

4. $\{(1, 1, -1, 2), (1, 0, 1, -1), (2, -1, 1, -1)\}$.

5. $\{(1, 1, 0, 2), (2, 1, 3, -1), (-1, 1, 1, -2), (2, -1, 1, 2)\}$.

6. Determine all values of the constant k for which the set of vectors $\{(0, -1, 0, k), (1, 0, 1, 0), (0, 1, 1, 0), (k, 0, 2, 1)\}$ is a basis for $\mathbf{R}^4$.

7. Determine a basis for P_4, and hence, prove that $\dim[P_4] = 4$.

For problems 8–10, find the dimension of nullspace(A).

8. $A = \begin{bmatrix} 1 & 3 \\ -2 & -6 \end{bmatrix}$.

9. $A = \begin{bmatrix} 1 & -1 & 4 \\ 2 & 3 & -2 \\ 1 & 2 & -2 \end{bmatrix}$.

10. $A = \begin{bmatrix} 1 & -1 & 2 & 3 \\ 2 & -1 & 3 & 4 \\ 1 & 0 & 1 & 1 \\ 3 & -1 & 4 & 5 \end{bmatrix}$.

11. Let S be the subspace of $\mathbf{R}^3$ that consists of all solutions to the equation $x_1 - 3x_2 + x_3 = 0$. Determine a basis for S, and hence, find $\dim[S]$.

12. Let S be the subspace of $\mathbf{R}^3$ consisting of all vectors of the form $(r, r - 2s, 3s - 5r)$, where r and s are real numbers. Determine a basis for S, and hence, find $\dim[S]$.

13. Let S be the subspace of $M_2(\mathbf{R})$ consisting of all upper triangular matrices. Determine a basis for S, and hence, find $\dim[S]$.

14. Let S be the subspace of $M_2(\mathbf{R})$ consisting of all 2×2 matrices with zero trace. Determine a basis for S, and hence, find $\dim[S]$.

15. Let S be the subspace of $\mathbf{R}^3$ spanned by the vectors $\mathbf{v}_1 = (1, 0, 1)$, $\mathbf{v}_2 = (0, 1, 1)$, and $\mathbf{v}_3 = (2, 0, 2)$. Determine a basis for S, and hence, find $\dim[S]$.

16. Let S be the vector space consisting of the set of all linear combinations of the functions

$f_1(x) = e^x, f_2(x) = e^{-x}, f_3(x) = \sinh(x)$. Determine a basis for S, and hence, find dim[S].

17. Determine a basis for the subspace of $M_2(\mathbf{R})$ spanned by

$$\begin{bmatrix} 1 & 3 \\ -1 & 2 \end{bmatrix}, \begin{bmatrix} 0 & 0 \\ 0 & 0 \end{bmatrix}, \begin{bmatrix} -1 & 4 \\ 1 & 1 \end{bmatrix}, \begin{bmatrix} 5 & -6 \\ -5 & 1 \end{bmatrix}.$$

18. If $v_1 = (1, 1)$ and $v_2 = (-1, 1)$, show that $\{v_1, v_2\}$ is a basis for $\mathbf{R}^2$. Determine the components of $v = (5, -1)$ relative to the basis $\{v_1, v_2\}$.

19. Let $v_1 = (2, 1)$ and $v_2 = (3, 1)$. Show that $\{v_1, v_2\}$ is a basis for $\mathbf{R}^2$ and determine the components of each of the standard basis vectors $e_1 = (1, 0)$ and $e_2 = (0, 1)$ relative to this basis.

20. If $v_1 = (0, 6, 3)$, $v_2 = (3, 0, 3)$, and $v_3 = (6, -3, 0)$, show that $\{v_1, v_2, v_3\}$ is a basis for $\mathbf{R}^3$, and determine the components of an arbitrary vector $x = (x_1, x_2, x_3)$ relative to this basis.

21. Determine all values of the constant α for which $\{1 + \alpha x^2, 1 + x + x^2, 2 + x\}$ is a basis for P_3.

22. Let $p_1(x) = 1 + x, p_2(x) = x(x - 1), p_3(x) = 1 + 2x^2$. Verify that $\{p_1, p_2, p_3\}$ is a basis for P_3 and determine the components of an arbitrary polynomial $p(x) = a_0 + a_1 x + a_2 x^2$ relative to this basis.

23. The Legendre polynomial of degree n, $p_n(x)$, is defined to be the polynomial solution of the DE

$$(1 - x^2)y'' - 2xy' + n(n + 1)y = 0,$$

which has been normalized so that $p_n(1) = 1$. The first three Legendre polynomials are $p_0(x) = 1$, $p_1(x) = x$, and $p_2(x) = \frac{1}{2}(3x^2 - 1)$. Show that $\{p_0, p_1, p_2\}$ is a basis for P_3, and determine the components of $p(x) = x^2 - x + 2$ relative to this basis.

24. Verify that

$$\left\{ \begin{bmatrix} -1 & 1 \\ 0 & 1 \end{bmatrix}, \begin{bmatrix} 1 & 3 \\ -1 & 0 \end{bmatrix}, \begin{bmatrix} 1 & 0 \\ 1 & 2 \end{bmatrix}, \begin{bmatrix} 0 & -1 \\ 2 & 3 \end{bmatrix} \right\}$$

is a basis for $M_2(\mathbf{R})$.

25. Let S denote the solution set of the homogeneous linear system $Ax = 0$, with

coefficient matrix $A = \begin{bmatrix} 1 & 1 & -1 & 1 \\ 2 & -3 & 5 & -6 \\ 5 & 0 & 2 & -3 \end{bmatrix}$.

(a) Verify that $x_1 = (-2, 7, 5, 0)$ and $x_2 = (3, -8, 0, 5)$ are in S, and show that $\{x_1, x_2\}$ is LI.

(b) Given that dim[S] = 2, write an expression for an arbitrary vector in S.

For problems 26 and 27, $Sym_n(\mathbf{R})$ and $Skew_n(\mathbf{R})$ denote the vector spaces consisting of all real $n \times n$ matrices that are symmetric and skew-symmetric respectively.

26. Find a basis for $Sym_2(\mathbf{R})$ and $Skew_2(\mathbf{R})$, and show that

$$\dim[Sym_2(\mathbf{R})] + \dim[Skew_2(\mathbf{R})] = \dim[M_2(\mathbf{R})].$$

27. Determine the dimensions of $Sym_n(\mathbf{R})$ and $Skew_n(\mathbf{R})$, and show that

$$\dim[Sym_n(\mathbf{R})] + \dim[Skew_n(\mathbf{R})] = \dim[M_n(\mathbf{R})].$$

28. Let S denote the subspace of $\mathbf{R}^3$ consisting of all points lying on the plane with Cartesian equation

$$x_1 + 4x_2 - 3x_3 = 0.$$

(a) Determine a basis for S.

(b) Extend your basis for S to obtain a basis for $\mathbf{R}^3$.

29. Determine a basis for the subspace of $M_2(\mathbf{R})$ consisting of all matrices of the form $\begin{bmatrix} a & b \\ b & a \end{bmatrix}$, and extend your basis to obtain a basis for $M_2(\mathbf{R})$.

30. Consider the subspace, S, of P_3 consisting of all polynomials of the form $(2a_1 + a_2)x^2 + (a_1 + a_2)x + (3a_1 - a_2)$. Determine a basis for S and extend this basis to obtain a basis for P_3.

For problems 31–33, determine a basis for the solution space of the given DE.

31. $y'' + 2y' - 3y = 0$.

32. $y'' + 6y' + 9y = 0$.

33. $y'' - 6y' + 25y = 0$.

34. Let S denote the subspace of the solution space to the DE $y'' + 16y = 0$, with basis $\{\sin 4x + 5 \cos 4x\}$. Write the general vector in S and extend the basis for S to a basis for the full solution space of the DE.

5.7   ROW SPACE AND COLUMN SPACE

In this section, we consider two vector spaces that can be associated with any $m \times n$ matrix. For simplicity, we will assume that the matrices have real elements, although the results that we establish can easily be extended to matrices with complex entries.

ROW SPACE

Let $A = [a_{ij}]$ be an $m \times n$ real matrix. The row vectors of this matrix are row n-vectors, and therefore they can be associated with vectors in $\mathbf{R}^n$. The subspace of $\mathbf{R}^n$ spanned by these vectors is called the **row space** of A and denoted rowspace(A). For example, if $A = \begin{bmatrix} 2 & -1 & 3 \\ 5 & 9 & -7 \end{bmatrix}$, then

$$\text{rowspace}(A) = \text{span}\{(2, -1, 3), (5, 9, -7)\}.$$

We wish to determine an efficient method for obtaining a basis for the row space of an $m \times n$ matrix A. If we perform ERO on A, then we are merely taking linear combinations of vectors in rowspace(A) and would therefore probably suspect that the row space of the resulting matrix coincides with the row space of A. This is the content of the following theorem.

Theorem 5.7.1:   If A and B are row-equivalent matrices, then

$$\text{rowspace}(B) = \text{rowspace}(A).$$

PROOF   We establish that the matrix resulting from performing either of the three ERO on a matrix A has the same row space as the row space of A. If we interchange rows in A, then clearly we have not altered the row space, since we still have the same set of vectors, only their ordering has been changed. Now let $\mathbf{a}_1, \mathbf{a}_2, \ldots, \mathbf{a}_m$ denote the row vectors of A. We combine the remaining two types of ERO by considering the result of replacing $\mathbf{a}_i$ by the vector $r\mathbf{a}_i + s\mathbf{a}_j$, where $r \, (\neq 0)$ and s are real numbers. If $s = 0$, then this corresponds to scaling $\mathbf{a}_i$, whereas if $r = 1$ and $s \neq 0$, we are adding a multiple of row j to row i. If B denotes the resulting matrix, then

$$\begin{aligned} \text{rowspace}(B) &= \{c_1 \mathbf{a}_1 + c_2 \mathbf{a}_2 + \cdots + c_i (r\mathbf{a}_i + s\mathbf{a}_j) + \cdots + c_m \mathbf{a}_m\} \\ &= \{c_1 \mathbf{a}_1 + c_2 \mathbf{a}_2 + \cdots + (rc_i)\mathbf{a}_i + \cdots + (c_j + sc_i)\mathbf{a}_j + \cdots + c_m \mathbf{a}_m\} \\ &= \{c_1 \mathbf{a}_1 + c_2 \mathbf{a}_2 + \cdots + d_i \mathbf{a}_i + \cdots + d_j \mathbf{a}_j + \cdots + c_m \mathbf{a}_m\}, \end{aligned}$$

where $d_i = rc_i$, $d_j = c_j + sc_i$ are arbitrary constants. Hence,

$$\text{rowspace}(B) = \text{span}\{\mathbf{a}_1, \mathbf{a}_2, \ldots, \mathbf{a}_m\} = \text{rowspace}(A). \qquad \blacksquare$$

The previous theorem is the key to determining a basis for rowspace(A). The idea we use is to reduce A to REF. If $\mathbf{d}_1, \mathbf{d}_2, \ldots, \mathbf{d}_k$ denote the nonzero row vectors in this REF, then from the previous theorem

$$\text{rowspace}(A) = \text{span}\{\mathbf{d}_1, \mathbf{d}_2, \ldots, \mathbf{d}_k\}.$$

We now establish that $\{\mathbf{d}_1, \mathbf{d}_2, \ldots, \mathbf{d}_k\}$ is LI. Consider

$$c_1 \mathbf{d}_1 + c_2 \mathbf{d}_2 + \cdots + c_k \mathbf{d}_k = \mathbf{0}. \qquad (5.7.1)$$

Due to the positioning of the leading ones in a row-echelon matrix, each of the row vectors $\mathbf{d}_1, \mathbf{d}_2, \ldots, \mathbf{d}_{k-1}$ will have a leading one in a position where each succeeding row vector in the REF has a zero. Hence equation (5.7.1) is satisfied only if

$$c_1 = c_2 = \ldots = c_{k-1} = 0,$$

and therefore, it reduces to

$$c_k \mathbf{d}_k = \mathbf{0}.$$

However, $\mathbf{d}_k$ is a nonzero vector, and so we must have $c_k = 0$. Consequently, all of the constants in equation (5.7.1) must be zero and therefore, $\{\mathbf{d}_1, \mathbf{d}_2, \ldots, \mathbf{d}_k\}$ not only spans rowspace(A), but is also LI. Hence, $\{\mathbf{d}_1, \mathbf{d}_2, \ldots, \mathbf{d}_k\}$ is a basis for rowspace(A). We have therefore established the next theorem.

Theorem 5.7.2: The set of nonzero row vectors in any REF of an $m \times n$ matrix A is a basis for rowspace(A).

As a consequence of the preceding theorem, we can conclude that *all* REF of A have the same number of nonzero rows. For if this were not the case, then we could find two bases for rowspace(A) containing a different number of vectors, which would contradict Corollary 5.6.1. We can therefore consider Theorem 3.4.3 a direct consequence of Theorem 5.7.2.

Example 5.7.1 Determine a basis for the row space of

$$A = \begin{bmatrix} 1 & -1 & 1 & 3 & 2 \\ 2 & -1 & 1 & 5 & 1 \\ 3 & -1 & 1 & 7 & 0 \\ 0 & 1 & -1 & -1 & -3 \end{bmatrix}.$$

Solution We first reduce A to REF.

$$\begin{bmatrix} 1 & -1 & 1 & 3 & 2 \\ 2 & -1 & 1 & 5 & 1 \\ 3 & -1 & 1 & 7 & 0 \\ 0 & 1 & -1 & -1 & -3 \end{bmatrix} \sim \begin{bmatrix} 1 & -1 & 1 & 3 & 2 \\ 0 & 1 & -1 & -1 & -3 \\ 0 & 2 & -2 & -2 & -6 \\ 0 & 1 & -1 & -1 & -3 \end{bmatrix} \sim \begin{bmatrix} 1 & -1 & 1 & 3 & 2 \\ 0 & 1 & -1 & -1 & -3 \\ 0 & 0 & 0 & 0 & 0 \\ 0 & 0 & 0 & 0 & 0 \end{bmatrix}.$$

Consequently, a basis for rowspace(A) is $\{(1, -1, 1, 3, 2), (0, 1, -1, -1, -3)\}$, and therefore, rowspace(A) is a two-dimensional subspace of $\mathbf{R}^5$. ☐

Theorem 5.7.2 also gives an efficient method for determining a basis for the subspace of $\mathbf{R}^n$ spanned by a given set of vectors. If we let A be the matrix whose row vectors are the given vectors from $\mathbf{R}^n$, then rowspace(A) coincides with the subspace of $\mathbf{R}^n$ spanned by those vectors. Consequently the nonzero row vectors in any REF of A will be a basis for the subspace spanned by the given set of vectors.

Example 5.7.2: Determine a basis for the subspace of $\mathbf{R}^4$ spanned by $\{(1, 2, 3, 4), (4, 5, 6, 7), (7, 8, 9, 10)\}$.

Solution   We first let A denote the matrix that has the given vectors as row vectors. Thus,

$$A = \begin{bmatrix} 1 & 2 & 3 & 4 \\ 4 & 5 & 6 & 7 \\ 7 & 8 & 9 & 10 \end{bmatrix}.$$

We now reduce A to REF:

$$\begin{bmatrix} 1 & 2 & 3 & 4 \\ 0 & -3 & -6 & -9 \\ 0 & -6 & -12 & -18 \end{bmatrix} \sim \begin{bmatrix} 1 & 2 & 3 & 4 \\ 0 & 1 & 2 & 3 \\ 0 & -6 & -12 & -18 \end{bmatrix} \sim \begin{bmatrix} 1 & 2 & 3 & 4 \\ 0 & 1 & 2 & 3 \\ 0 & 0 & 0 & 0 \end{bmatrix}.$$

Consequently, a basis for the subspace of $\mathbf{R}^4$ spanned by the given vectors is $\{(1, 2, 3, 4), (0, 1, 2, 3)\}$. We see that the given vectors span a two-dimensional subspace of $\mathbf{R}^4$.

COLUMN SPACE

If A is an $m \times n$ matrix, the column vectors of A are column m-vectors and therefore can be associated with vectors in $\mathbf{R}^m$. The subspace of $\mathbf{R}^m$ spanned by these vectors is called the **column space** of A and denoted colspace(A). For the matrix $A = \begin{bmatrix} 2 & -1 & 3 \\ 5 & 9 & -7 \end{bmatrix}$,

$$\text{colspace}(A) = \text{span}\{(2, 5), (-1, 9), (3, -7)\}.$$

We now consider the problem of determining a basis for the column space of an $m \times n$ matrix A. Since the column vectors of A coincide with the row vectors of A^T, it follows that

$$\text{colspace}(A) = \text{rowspace}(A^T).$$

Hence one way to obtain a basis for colspace(A) would be to reduce A^T to REF and then the nonzero row vectors in the resulting matrix would give a basis for colspace(A). However, there is a better method for determining a basis for colspace(A) directly from any REF of A. The derivation of this technique is somewhat involved and will require full attention. We begin by determining the column space of an $m \times n$ *reduced* row-echelon matrix. In order to introduce the basic ideas, consider the particular reduced row-echelon matrix

$$E = \begin{bmatrix} 1 & 2 & 0 & 3 & 0 \\ 0 & 0 & 1 & 5 & 0 \\ 0 & 0 & 0 & 0 & 1 \\ 0 & 0 & 0 & 0 & 0 \end{bmatrix}.$$

In this case, we see that the first, third, and fifth column vectors, which are the column vectors containing the leading ones, coincide with the first three standard basis vectors in $\mathbf{R}^4$ (written as column vectors):

$$\mathbf{e}_1 = \begin{bmatrix} 1 \\ 0 \\ 0 \\ 0 \end{bmatrix}, \quad \mathbf{e}_2 = \begin{bmatrix} 0 \\ 1 \\ 0 \\ 0 \end{bmatrix}, \quad \mathbf{e}_3 = \begin{bmatrix} 0 \\ 0 \\ 1 \\ 0 \end{bmatrix}.$$

Consequently, these column vectors are LI. Furthermore, the remaining column vectors in E (those that do not contain leading ones) are both linear combinations of $\mathbf{e}_1$ and $\mathbf{e}_2$. Therefore, $\{\mathbf{e}_1, \mathbf{e}_2, \mathbf{e}_3\}$ is a LI set of vectors that spans colspace(E), and so a basis for colspace(E) is

$$\{(1, 0, 0, 0), (0, 1, 0, 0), (0, 0, 1, 0)\}.$$

Clearly, the same arguments apply to any reduced row echelon matrix E. Thus, if E contains k (necessarily $\leq n$) leading ones a basis for colspace(E) is $\{\mathbf{e}_1, \mathbf{e}_2, ..., \mathbf{e}_k\}$.

Now consider an arbitrary $m \times n$ matrix A, and let E denote the RREF of A. Since E can be obtained from A by performing ERO, it follows from the results of Chapter 3 that the two homogeneous systems of equations

$$A\mathbf{c} = \mathbf{0}, \qquad E\mathbf{c} = \mathbf{0} \tag{5.7.2}$$

have the same solution sets. If we write A and E in column vector form as $A = [\mathbf{a}_1, \mathbf{a}_2, ..., \mathbf{a}_n]$ and $E = [\mathbf{d}_1, \mathbf{d}_2, ..., \mathbf{d}_n]$, respectively, then the two systems in (5.7.2) can be written as

$$c_1\mathbf{a}_1 + c_2\mathbf{a}_2 + \cdots + c_n\mathbf{a}_n = \mathbf{0},$$

$$c_1\mathbf{d}_1 + c_2\mathbf{d}_2 + \cdots + c_n\mathbf{d}_n = \mathbf{0},$$

respectively. Thus, the fact that these two systems have the same solution set means that a linear dependence relationship will hold between the column vectors of E if and only if precisely the same linear dependence relation holds between the *corresponding* column vectors of A. In particular, since, as previously shown, the column vectors in E that contain leading ones give a basis for colspace(E), they give a maximal LI set in colspace(E). Therefore, the corresponding column vectors in A will also be a maximal LI set in colspace(A). Consequently, this set of vectors from A will be a basis for colspace(A). We have therefore shown that the set of column vectors of A corresponding to those column vectors containing leading ones in the RREF of A is a basis for colspace(A). But do we have to reduce A to RREF? The answer is no. We only need reduce A to REF. The reason is that reducing a matrix from REF to RREF does not alter the position or number of leading ones in the matrix, and therefore, the column vectors containing leading ones in any REF of A will correspond to the column vectors containing leading ones in the RREF of A. Consequently, we have established the following result.

Theorem 5.7.3:   Let A be an $m \times n$ matrix. The set of column vectors of A corresponding to those column vectors containing leading ones in any REF of A is a basis for colspace(A).

Example 5.7.3   Determine a basis for colspace(A) if

$$A = \begin{bmatrix} 1 & 2 & -1 & -2 & -1 \\ 2 & 4 & -2 & -3 & -1 \\ 5 & 10 & -5 & -3 & -1 \\ -3 & -6 & 3 & 2 & 1 \end{bmatrix}.$$

Solution   We first reduce A to REF

$$\begin{bmatrix} 1 & 2 & -1 & -2 & -1 \\ 2 & 4 & -2 & -3 & -1 \\ 5 & 10 & -5 & -3 & -1 \\ -3 & -6 & 3 & 2 & 1 \end{bmatrix} \sim \begin{bmatrix} 1 & 2 & -1 & -2 & -1 \\ 0 & 0 & 0 & 1 & 1 \\ 0 & 0 & 0 & 7 & 4 \\ 0 & 0 & 0 & -4 & -2 \end{bmatrix}$$

$$\sim \begin{bmatrix} 1 & 2 & -1 & -2 & -1 \\ 0 & 0 & 0 & 1 & 1 \\ 0 & 0 & 0 & 0 & -3 \\ 0 & 0 & 0 & 0 & 2 \end{bmatrix} \sim \begin{bmatrix} 1 & 2 & -1 & -2 & -1 \\ 0 & 0 & 0 & 1 & 1 \\ 0 & 0 & 0 & 0 & 1 \\ 0 & 0 & 0 & 0 & 0 \end{bmatrix}.$$

Since the first, fourth, and fifth column vectors in this REF of A contain the leading ones, it follows from Theorem 5.7.3 that the set of *corresponding* column vectors in A is a basis for colspace(A). Consequently, a basis for colspace(A) is

$$\{(1, 2, 5, -3), (-2, -3, -3, 2), (-1, -1, -1, 1)\}.$$

Hence, colspace(A) is a three-dimensional subspace of $\mathbf{R}^4$. Notice from the REF of A that a basis for rowspace(A) is $\{(1, 2, -1, -2, -1), (0, 0, 0, 1, 1), (0, 0, 0, 0, 1)\}$, so that rowspace($A$) is a three-dimensional subspace of $\mathbf{R}^5$.   ❏

We now summarize the discussion of row space and column space:

> Let A be an $m \times n$ matrix. In order to determine a basis for rowspace(A) and a basis for colspace(A), we reduce A to REF. The *row* vectors containing the leading ones in the REF give a basis for rowspace(A) (a subspace of $\mathbf{R}^n$), whereas the column vectors of A corresponding to the *column* vectors containing the leading ones in the REF give a basis for colspace(A) (a subspace of $\mathbf{R}^m$).

Previously, we defined the rank of an $m \times n$ matrix to be the number of nonzero rows in any REF of A. Often, this number is called the **row rank** of A and is denoted rowrank(A) to distinguish it from the **column rank** of A, denoted colrank(A), which is defined to be the number of nonzero rows in any REF of A^T. The reason that we have not distinguished these two ranks is that a fundamental result in linear algebra states that they are equal. We now establish this result by considering the relationship between rowspace(A) and colspace(A). The first point to emphasize is that in general these are subspaces of *different* vector spaces. The former is a subspace of $\mathbf{R}^n$ and the latter is a subspace of $\mathbf{R}^m$. However, since the number of vectors in a basis for rowspace(A) or in a basis for colspace(A) is equal to the number of leading ones in any REF of A, it follows that

$$\dim[\text{rowspace}(A)] = \dim[\text{colspace}(A)].$$

Now for the key point. As mentioned earlier, if we reduce A^T to REF, then the nonzero row vectors of the resulting matrix determine a basis for colspace(A), and the dimension of colspace(A) is equal to the number vectors in this basis. But, the number of nonzero rows vectors in the REF of A^T is just colrank(A). Consequently, we have

$$\text{colrank}(A) = \dim[\text{colspace}(A)] = \dim[\text{rowspace}(A)] = \text{rowrank}(A).$$

We therefore have established that

> $$\text{colrank}(A) = \text{rowrank}(A)$$

which holds for any $m \times n$ matrix. In keeping with our previous terminology, we will refer to the common value of these ranks simply as the rank of A.

EXERCISES 5.7

For problems $1-5$, determine a basis for rowspace(A) and a basis for colspace(A).

1. $A = \begin{bmatrix} 1 & -2 \\ -3 & 6 \end{bmatrix}.$

2. $A = \begin{bmatrix} 1 & 1 & -3 & 2 \\ 3 & 4 & -11 & 7 \end{bmatrix}.$

3. $A = \begin{bmatrix} 1 & 2 & 3 \\ 5 & 6 & 7 \\ 9 & 10 & 11 \end{bmatrix}.$

4. $A = \begin{bmatrix} 0 & 3 & 1 \\ 0 & -6 & -2 \\ 0 & 12 & 4 \end{bmatrix}.$

5. $A = \begin{bmatrix} 1 & 2 & -1 & 3 \\ 3 & 6 & -3 & 5 \\ 1 & 2 & -1 & -1 \\ 5 & 10 & -5 & 7 \end{bmatrix}.$

For problems $6-9$, determine a basis for the subspace of $\mathbf{R}^n$ spanned by the given set of vectors.

6. $\{(1, -1, 2), (5, -4, 1), (7, -5, -4)\}.$

7. $\{(1, 3, 3), (1, 5, -1), (2, 7, 4), (1, 4, 1)\}.$

8. $\{(1, 1, -1, 2), (2, 1, 3, -4), (1, 2, -6, 10)\}.$

9. $\{(1, 4, 1, 3), (2, 8, 3, 5), (1, 4, 0, 4), (2, 8, 2, 6)\}.$

10. Let $A = \begin{bmatrix} -3 & 9 \\ 1 & -3 \end{bmatrix}.$ Determine a basis for rowspace(A) and colspace(A). Make a sketch to show each subspace in the xy-plane.

11. Let $A = \begin{bmatrix} 1 & 2 & 4 \\ 5 & 11 & 21 \\ 3 & 7 & 13 \end{bmatrix}.$ Determine a basis for rowspace(A) and colspace(A). Show that rowspace(A) corresponds to the plane with Cartesian equation $2x + y - z = 0$, whereas colspace(A) corresponds to the plane with Cartesian equation $2x - y + z = 0$.

5.8*   THE RANK-NULLITY THEOREM

In Section 5.3, we defined the nullspace of a real $m \times n$ matrix A to be the set of all real solutions to the associated homogeneous linear system $A\mathbf{x} = \mathbf{0}$. Thus,

$$\text{nullspace}(A) = \{\mathbf{x} \in \mathbf{R}^n : A\mathbf{x} = \mathbf{0}\}.$$

The dimension of nullspace(A) is referred to as the **nullity** of A and is denoted nullity(A). In order to find nullity(A), we need to determine a basis for nullspace(A). Our results on systems of linear equations tell us that if rank(A) $= r$, then there are $n - r$ free variables in the solution of the system $A\mathbf{x} = \mathbf{0}$. We might therefore suspect that nullity(A) $= n - r$. Our next theorem, which is often referred to as the Rank-Nullity Theorem, establishes that this is indeed the case.

---

* The material in this section is not used later in the text.

Theorem 5.8.1 (Rank-Nullity Theorem):   For any $m \times n$ matrix A,

$$\text{rank}(A) + \text{nullity}(A) = n. \tag{5.8.1}$$

PROOF   We need to distinguish three cases depending on the value of rank(A).

Case 1: rank(A) = n. $A\mathbf{x} = \mathbf{0}$ has only the trivial solution, and so, nullity(A) = 0. Consequently, equation (5.8.1) holds.

Case 2: rank(A) = 0. In this case, A is the zero matrix and so every vector in $\mathbf{R}^n$ solves $A\mathbf{x} = \mathbf{0}$. Consequently, nullity(A) = n and so, once more, equation (5.8.1) holds.

Case 3: rank(A) = $r < n$. In this case, there are $n - r$ free variables in the solution to $A\mathbf{x} = \mathbf{0}$. Let $t_1, t_2, \ldots, t_{n-r}$ denote these free variables (chosen as those variables not attached to a leading one in any REF of A), and let $\mathbf{x}_1, \mathbf{x}_2, \ldots, \mathbf{x}_{n-r}$ denote the solutions obtained by sequentially setting each free variable to 1 and the remaining free variables to zero. Then every solution to $A\mathbf{x} = \mathbf{0}$ is of the form

$$\mathbf{x} = t_1\mathbf{x}_1 + t_2\mathbf{x}_2 + \cdots + t_{n-r}\mathbf{x}_{n-r}.$$

Furthermore, due to the choice of free variables, each vector in the set $\{\mathbf{x}_1, \mathbf{x}_2, \ldots, \mathbf{x}_{n-r}\}$ has a 1 in a position where all other vectors in the set have a zero. Consequently, $\mathbf{x} = \mathbf{0}$ if and only if $t_1 = t_2 = \cdots = t_{n-r} = 0$, so that $\{\mathbf{x}_1, \mathbf{x}_2, \ldots, \mathbf{x}_{n-r}\}$ is LI. Since $\{\mathbf{x}_1, \mathbf{x}_2, \ldots, \mathbf{x}_{n-r}\}$ spans nullspace(A) and is LI, it is a basis for nullspace(A). Therefore, nullity(A) = $n - r$. Rearranging this expression yields equation (5.8.1). ∎

Example 5.8.1   If $A = \begin{bmatrix} 1 & 1 & 2 & 3 \\ 3 & 4 & -1 & 2 \\ -1 & -2 & 5 & 4 \end{bmatrix}$, determine a basis for nullspace(A) and verify Theorem 5.8.1.

Solution   We must find all solutions to $A\mathbf{x} = \mathbf{0}$. Reducing the augmented matrix of this system yields

$$\begin{bmatrix} 1 & 1 & 2 & 3 & 0 \\ 3 & 4 & -1 & 2 & 0 \\ -1 & -2 & 5 & 4 & 0 \end{bmatrix} \sim \begin{bmatrix} 1 & 1 & 2 & 3 & 0 \\ 0 & 1 & -7 & -7 & 0 \\ 0 & -1 & 7 & 7 & 0 \end{bmatrix} \sim \begin{bmatrix} 1 & 1 & 2 & 3 & 0 \\ 0 & 1 & -7 & -7 & 0 \\ 0 & 0 & 0 & 0 & 0 \end{bmatrix}.$$

Consequently, there are two free variables

$$x_3 = t_1, \quad x_4 = t_2,$$

so that

$$x_2 = 7t_1 + 7t_2, \quad x_1 = -9t_1 - 10t_2.$$

Hence,

$$\text{nullspace}(A) = \{(-9t_1 - 10t_2, 7t_1 + 7t_2, t_1, t_2) : t_1, t_2 \in \mathbf{R}\}$$

$$= \{t_1(-9, 7, 1, 0) + t_2(-10, 7, 0, 1) : t_1, t_2 \in \mathbf{R}\}.$$

$$= \text{span}\{(-9, 7, 1, 0), (-10, 7, 0, 1)\}.$$

Since the two vectors in this spanning set are not proportional they are LI. Consequently, a basis for nullspace(A) is $\{(-9, 7, 1, 0), (-10, 7, 0, 1)\}$, so that nullity($A$) = 2. In this problem, A is a 3×4 matrix, and so, in the rank-nullity theorem, $n = 4$.

Further, from the foregoing REF of the augmented matrix of the system $A\mathbf{x} = \mathbf{0}$, we see that $\text{rank}(A) = 2$. Hence,

$$\text{rank}(A) + \text{nullity}(A) = 2 + 2 = 4 = n,$$

and the Rank-Nullity Theorem is verified.

SYSTEMS OF LINEAR ALGEBRAIC EQUATIONS

The major theoretical results of Chapter 2 were as follows:

1. All solutions to the homogeneous linear DE

$$y'' + a_1 y' + a_2 y = 0$$

are of the form

$$y(x) = c_1 y_1(x) + c_2 y_2(x),$$

where $\{y_1, y_2\}$ is any LI set of solutions.

2. All solutions to the nonhomogeneous DE

$$y'' + a_1 y' + a_2 y = F(x)$$

are of the form

$$y(x) = c_1 y_1(x) + c_2 y_2(x) + y_p(x),$$

where $\{y_1, y_2\}$ are as defined in (1) and $y = y_p(x)$ is a particular solution to the nonhomogeneous DE.

We now illustrate that the same linear structure holds for the solution set to $A\mathbf{x} = \mathbf{b}$. First we consider the homogeneous case.

Corollary 5.8.1:   Let A be an $m \times n$ matrix, and consider the corresponding homogeneous linear system $A\mathbf{x} = \mathbf{0}$.

1. If $\text{rank}(A) = n$, then $A\mathbf{x} = \mathbf{0}$ has only the trivial solution, and so, $\text{nullspace}(A) = \{\mathbf{0}\}$.

2. If $\text{rank}(A) = r < n$, then $A\mathbf{x} = \mathbf{0}$ has an infinite number of solutions, all of which can be obtained from

$$\mathbf{x} = c_1 \mathbf{x}_1 + c_2 \mathbf{x}_2 + \cdots + c_{n-r} \mathbf{x}_{n-r}, \tag{5.8.2}$$

where $\{\mathbf{x}_1, \mathbf{x}_2, \ldots, \mathbf{x}_{n-r}\}$ is any LI set of $n - r$ solutions to $A\mathbf{x} = \mathbf{0}$.

PROOF
(1) A restatement of our previous results.

(2) If $\text{rank}(A) = r < n$, then, by the rank-nullity theorem, $\text{nullity}(A) = n - r$. Thus, from Theorem 5.6.4, if $\{\mathbf{x}_1, \mathbf{x}_2, \ldots, \mathbf{x}_{n-r}\}$ is any set of $n - r$ LI solutions to $A\mathbf{x} = \mathbf{0}$ it is a basis for $\text{nullspace}(A)$, and so all vectors in $\text{nullspace}(A)$ can be written as

$$\mathbf{x} = c_1 \mathbf{x}_1 + c_2 \mathbf{x}_2 + \cdots + c_{n-r} \mathbf{x}_{n-r},$$

for appropriate values of the constants $c_1, c_2, \ldots, c_{n-r}$. (In the proof of Theorem 5.8.1 we made a specific choice of basis.)   ∎

REMARK   The expression (5.8.2) is referred to as the **general solution** to the system $A\mathbf{x} = \mathbf{0}$.

We now turn our attention to nonhomogeneous linear systems. We begin by formulating Theorem 3.5.1 in terms of colspace(A).

Theorem 5.8.2:   Let A be an $m \times n$ matrix and consider the linear system $A\mathbf{x} = \mathbf{b}$.

1. If $\mathbf{b}$ is not in colspace(A), then the system is inconsistent.

2. If $\mathbf{b} \in$ colspace(A) then the system is consistent and has the following:
   (a)   a unique solution if and only if dim[colspace(A)] $= n$.
   (b)   an infinite number of solutions if and only if dim[colspace(A)] $< n$.

PROOF
(1)   If we write A in terms of its column vectors as $A = [\mathbf{a}_1, \mathbf{a}_2, \ldots, \mathbf{a}_n]$, then the linear system $A\mathbf{x} = \mathbf{b}$ can be written as

$$x_1\mathbf{a}_1 + x_2\mathbf{a}_2 + \cdots + x_n\mathbf{a}_n = \mathbf{b}.$$

Consequently, the linear system is consistent if and only if the vector $\mathbf{b}$ is a linear combination of the column vectors of A. That is, $\mathbf{b}$ must lie in colspace(A).

(2) Parts 2a and 2b follow directly from Theorem 3.5.1, since rank(A) = dim[colspace(A)]. ∎

As we have stated earlier, the set of all solutions to a nonhomogeneous linear system is not a vector space. However, the linear structure of nullspace(A) can be used to determine the general form of the solution of a nonhomogeneous system.

Theorem 5.8.3:   Let A be an $m \times n$ matrix. If rank(A) $= r < n$ and $\mathbf{b} \in$ colspace(A), then all solutions to $A\mathbf{x} = \mathbf{b}$ are of the form

$$\mathbf{x} = c_1\mathbf{x}_1 + c_2\mathbf{x}_2 + \cdots + c_{n-r}\mathbf{x}_{n-r} + \mathbf{x}_p, \qquad (5.8.3)$$

where $\mathbf{x}_p$ is any particular solution to $A\mathbf{x} = \mathbf{b}$, and $\{\mathbf{x}_1, \mathbf{x}_2, \ldots, \mathbf{x}_{n-r}\}$ is a basis for nullspace(A).

PROOF   Since $\mathbf{x} = \mathbf{x}_p$ is a solution to $A\mathbf{x} = \mathbf{b}$, we have

$$A\mathbf{x}_p = \mathbf{b}. \qquad (5.8.4)$$

Let $\mathbf{x} = \mathbf{u}$ be any solution to $A\mathbf{x} = \mathbf{b}$. Then we also have

$$A\mathbf{u} = \mathbf{b}. \qquad (5.8.5)$$

Subtracting (5.8.4) from (5.8.5) yields

$$A\mathbf{u} - A\mathbf{x}_p = \mathbf{0},$$

or equivalently,

$$A(\mathbf{u} - \mathbf{x}_p) = \mathbf{0}.$$

Consequently, the vector $\mathbf{u} - \mathbf{x}_p$ is in nullspace(A), and, therefore, there exist scalars c_1, $c_2, \ldots, c_k$ such that

$$\mathbf{u} - \mathbf{x}_p = c_1\mathbf{x}_1 + c_2\mathbf{x}_2 + \cdots + c_{n-r}\mathbf{x}_{n-r},$$

where $\{\mathbf{x}_1, \mathbf{x}_2, \ldots, \mathbf{x}_{n-r}\}$ is a basis for nullspace(A). Hence,

$$\mathbf{u} = c_1 \mathbf{x}_1 + c_2 \mathbf{x}_2 + \cdots + c_{n-r} \mathbf{x}_{n-r} + \mathbf{x}_p,$$

as required.   ∎

REMARK   The expression given in equation (5.8.3) is called the **general solution** to $A\mathbf{x} = \mathbf{b}$. It has the structure

$$\mathbf{x} = \mathbf{x}_c + \mathbf{x}_p,$$

where

$$\mathbf{x}_c = c_1 \mathbf{x}_1 + c_2 \mathbf{x}_2 + \cdots + c_{n-r} \mathbf{x}_{n-r}$$

is the general solution of the associated homogenous system and $\mathbf{x}_p$ is one particular solution of the nonhomogeneous system. We see that the solution to the linear algebraic system of equations has exactly the same structure as that of a second-order linear DE. In the later chapters, we will see that this structure is reflected in the solution of all linear DE and also linear systems of DE. It is a result of the linearity inherent in the problem, rather than the specific problem that we are studying. The unifying concept, in addition to the vector space, is the idea of a linear transformation to be studied in the next chapter.

Example 5.8.2   Let $A = \begin{bmatrix} 1 & 1 & 2 & 3 \\ 3 & 4 & -1 & 2 \\ -1 & -2 & 5 & 4 \end{bmatrix}$ and $\mathbf{b} = \begin{bmatrix} 3 \\ 10 \\ -4 \end{bmatrix}$. Verify that $\mathbf{x}_p = (1, 1, -1, 1)$ is a particular solution to $A\mathbf{x} = \mathbf{b}$, and use Theorem 5.8.3 to determine the general solution to the system.

Solution   For the given $\mathbf{x}_p$, we have

$$A\mathbf{x}_p = \begin{bmatrix} 1 & 1 & 2 & 3 \\ 3 & 4 & -1 & 2 \\ -1 & -2 & 5 & 4 \end{bmatrix} \begin{bmatrix} 1 \\ 1 \\ -1 \\ 1 \end{bmatrix} = \begin{bmatrix} 3 \\ 10 \\ -4 \end{bmatrix} = \mathbf{b}.$$

Consequently, $\mathbf{x}_p = (1, 1, -1, 1)$ is a particular solution to $A\mathbf{x} = \mathbf{b}$. Further, from Example 5.8.1, a basis for nullspace(A) is $\{\mathbf{x}_1, \mathbf{x}_2\}$, where $\mathbf{x}_1 = (-9, 7, 1, 0)$, and $\mathbf{x}_2 = (-10, 7, 0, 1)$. Thus, the general solution to $A\mathbf{x} = \mathbf{0}$ is

$$\mathbf{x}_c = c_1 \mathbf{x}_1 + c_2 \mathbf{x}_2,$$

and, therefore, from Theorem 5.8.3, the general solution to $A\mathbf{x} = \mathbf{b}$ is

$$\mathbf{x} = c_1 \mathbf{x}_1 + c_2 \mathbf{x}_2 + \mathbf{x}_p = c_1(-9, 7, 1, 0) + c_2(-10, 7, 0, 1) + (1, 1, -1, 1),$$

which can be written as

$$\mathbf{x} = (-9c_1 - 10c_2 + 1, \; 7c_1 + 7c_2 + 1, \; c_1 - 1, \; c_2 + 1).$$

EXERCISES 5.8

For problems 1–3, determine nullspace(A) and verify the Rank-Nullity Theorem

1. $A = \begin{bmatrix} 2 & -1 \\ -4 & 2 \end{bmatrix}$.

2. $A = \begin{bmatrix} 1 & 1 & -1 \\ 3 & 4 & 4 \\ 1 & 1 & 0 \end{bmatrix}$.

3. $A = \begin{bmatrix} 1 & 4 & -1 & 3 \\ 2 & 9 & -1 & 7 \\ 2 & 8 & -2 & 8 \end{bmatrix}$.

4. $A = \begin{bmatrix} 1 & 3 & -1 \\ 2 & 7 & 9 \\ 1 & 5 & 21 \end{bmatrix}$, $b = \begin{bmatrix} 4 \\ 11 \\ 10 \end{bmatrix}$.

5. $A = \begin{bmatrix} 2 & -1 & 1 & 4 \\ 1 & -1 & 2 & 3 \\ 1 & -2 & 5 & 5 \end{bmatrix}$, $b = \begin{bmatrix} 5 \\ 6 \\ 13 \end{bmatrix}$.

For problems 4 and 5, determine the solution set to $A\mathbf{x} = \mathbf{b}$, and show that all solutions are of the form (5.8.3).

5.9   INNER PRODUCT SPACES

We now extend the familiar idea of the dot product for geometric vectors to an arbitrary vector space V. This enables us to associate a magnitude with each vector in V and also to define the angle between any two vectors in V. The major reason that we want to do this is that, as we will see in the next section, it enables us to construct orthogonal bases in a vector space and use of such a basis often simplifies the representation of vectors. We begin with a brief review of the dot product.

Let $\mathbf{x} = (x_1, x_2, x_3)$ and $\mathbf{y} = (y_1, y_2, y_3)$ be two arbitrary vectors in $\mathbf{R}^3$, and consider the corresponding geometric vectors

$$\mathbf{x} = x_1\mathbf{i} + x_2\mathbf{j} + x_3\mathbf{k}, \quad \mathbf{y} = y_1\mathbf{i} + y_2\mathbf{j} + y_3\mathbf{k}.$$

The dot product of $\mathbf{x}$ and $\mathbf{y}$ can be defined in terms of the components of these vectors as

$$\mathbf{x} \cdot \mathbf{y} = x_1y_1 + x_2y_2 + x_3y_3. \tag{5.9.1}$$

An equivalent geometric definition of the dot product is

$$\mathbf{x} \cdot \mathbf{y} = \| \mathbf{x} \| \, \| \mathbf{y} \| \cos \theta \tag{5.9.2}$$

where $\| \mathbf{x} \|$, $\| \mathbf{y} \|$ denote the lengths of $\mathbf{x}$ and $\mathbf{y}$ respectively and $0 \leq \theta \leq \pi$ is the angle between them. (See Figure 5.9.1.)

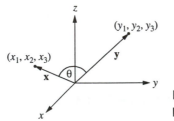

Figure 5.9.1 Defining the dot product in $\mathbf{R}^3$.

Taking $\mathbf{y} = \mathbf{x}$ in equations (5.9.1) and (5.9.2) yields

$$\| \mathbf{x} \|^2 = \mathbf{x} \cdot \mathbf{x} = x_1{}^2 + x_2{}^2 + x_3{}^2,$$

so that the length of a geometric vector is given in terms of the dot product by

$$\| \mathbf{x} \| = \sqrt{\mathbf{x} \cdot \mathbf{x}} = \sqrt{x_1{}^2 + x_2{}^2 + x_3{}^2}.$$

Furthermore, from equation (5.9.2), the angle between any two nonzero vectors $\mathbf{x}$ and $\mathbf{y}$ is

$$\cos \theta = \frac{\mathbf{x} \cdot \mathbf{y}}{\| \mathbf{x} \| \| \mathbf{y} \|}, \tag{5.9.3}$$

which implies that $\mathbf{x}$ and $\mathbf{y}$ are orthogonal (perpendicular) if and only if

$$\mathbf{x} \cdot \mathbf{y} = 0.$$

In a general vector space, we do not have a geometrical picture to guide us in defining the dot product, and hence, our definitions must be purely algebraic. We begin by considering the vector space $\mathbf{R}^n$, since there is a natural way to extend equation (5.9.1) in this case. Before proceeding, we note that from now on we will use the standard terms *inner product* and *norm* in place of dot product and length, respectively.

Definition   5.9.1:   Let $\mathbf{x} = (x_1, x_2, \ldots, x_n)$ and $\mathbf{y} = (y_1, y_2, \ldots, y_n)$ be vectors in $\mathbf{R}^n$. We define the standard **inner product** in $\mathbf{R}^n$, denoted $<\mathbf{x}, \mathbf{y}>$, by

$$<\mathbf{x}, \mathbf{y}> = x_1 y_1 + x_2 y_2 + x_3 y_3 + \cdots + x_n y_n.$$

The **norm** of $\mathbf{x}$ is

$$\| \mathbf{x} \| = \sqrt{<\mathbf{x}, \mathbf{x}>} = \sqrt{x_1^2 + x_2^2 + \cdots + x_n^2}.$$

Example 5.9.1   If $\mathbf{x} = (1, -1, 0, 2, 4)$ and $\mathbf{y} = (2, 1, 1, 3, 0)$ in $\mathbf{R}^5$, then

$$<\mathbf{x}, \mathbf{y}> = 2 - 1 + 0 + 6 + 0 = 7,$$

$$\| \mathbf{x} \| = \sqrt{1 + 1 + 0 + 4 + 16} = \sqrt{22},$$

$$\| \mathbf{y} \| = \sqrt{4 + 1 + 1 + 9 + 0} = \sqrt{15}.$$

BASIC PROPERTIES OF THE STANDARD INNER PRODUCT IN $\mathbf{R}^n$

In the case of $\mathbf{R}^n$, the definition of the standard inner product was a natural extension of the familiar dot product in $\mathbf{R}^3$. To generalize this definition to an arbitrary vector space we isolate the most important properties of the standard inner product in $\mathbf{R}^n$. We view this inner product as a mapping that associates with any two vectors $\mathbf{x} = (x_1, x_2, \ldots, x_n)$ and $\mathbf{y} = (y_1, y_2, \ldots, y_n)$ in $\mathbf{R}^n$, the real number

$$<\mathbf{x}, \mathbf{y}> = x_1 y_1 + x_2 y_2 + x_3 y_3 + \cdots + x_n y_n.$$

This mapping has the following properties.

For all $\mathbf{x}$, $\mathbf{y}$ and $\mathbf{z}$ in $\mathbf{R}^n$ and all real numbers k,

1. $<\mathbf{x}, \mathbf{x}> \geq 0$. Furthermore, $<\mathbf{x}, \mathbf{x}> = 0$ if and only if $\mathbf{x} = \mathbf{0}$.
2. $<\mathbf{y}, \mathbf{x}> = <\mathbf{x}, \mathbf{y}>$.
3. $<k\mathbf{x}, \mathbf{y}> = k<\mathbf{x}, \mathbf{y}>$.
4. $<\mathbf{x} + \mathbf{y}, \mathbf{z}> = <\mathbf{x}, \mathbf{z}> + <\mathbf{y}, \mathbf{z}>$.

These properties are easily established using Definition 5.9.1. For example, to prove (1), we proceed as follows. From Definition 5.9.1,

$$<\mathbf{x}, \mathbf{x}> = x_1^2 + x_2^2 + \cdots + x_n^2.$$

Since this is a sum of squares of real numbers it is necessarily nonnegative. Further, $<\mathbf{x}, \mathbf{x}> = 0$ if and only if $x_1 = x_2 = \cdots = x_n = 0$, that is, if and only if $\mathbf{x} = \mathbf{0}$.
Similarly,

$$<\mathbf{y}, \mathbf{x}> = y_1 x_1 + y_2 x_2 + \cdots + y_n x_n = x_1 y_1 + x_2 y_2 + \cdots + x_n y_n = <\mathbf{x}, \mathbf{y}>,$$

so that (2) is satisfied. We leave the verification of properties (3) and (4) as exercises.

DEFINITION OF A REAL INNER PRODUCT SPACE

We now use properties (1)–(4) as the basic defining properties of an inner product in a real vector space.

Definition 5.9.2: Let V be a real vector space. A mapping that associates with each pair of vectors $\mathbf{u}$ and $\mathbf{v}$ in V a real number, denoted $<\mathbf{u}, \mathbf{v}>$, is called an **inner product** in V provided it satisfies the following properties. For all $\mathbf{u}, \mathbf{v}$, and $\mathbf{w}$ in V, and all real numbers k,

1. $<\mathbf{u}, \mathbf{u}> \geq 0$. Furthermore, $<\mathbf{u}, \mathbf{u}> = 0$ if and only if $\mathbf{u} = \mathbf{0}$.

2. $<\mathbf{v}, \mathbf{u}> = <\mathbf{u}, \mathbf{v}>$.

3. $<k\mathbf{u}, \mathbf{v}> = k<\mathbf{u}, \mathbf{v}>$.

4. $<\mathbf{u} + \mathbf{v}, \mathbf{w}> = <\mathbf{u}, \mathbf{w}> + <\mathbf{v}, \mathbf{w}>$.

The **norm** of $\mathbf{u}$ is defined in terms of an inner product by

$$\| \mathbf{u} \| = \sqrt{<\mathbf{u}, \mathbf{u}>}.$$

A real vector space together with an inner product defined in it is called a **real inner product space**.

It follows from the preceding discussion that $\mathbf{R}^n$ together with the inner product defined in Definition 5.9.1 is an example of a real inner product space.

One of the fundamental inner products arises in the vector space $C^0[a, b]$ of all real-valued functions that are *continuous* on the interval $[a, b]$. In this vector space, we define the mapping $<f, g>$ by

$$<f, g> = \int_a^b f(x)g(x)dx, \tag{5.9.4}$$

for all f and g in $C^0[a, b]$. We establish that this mapping defines an inner product in $C^0[a, b]$ by verifying properties (1)–(4) of Definition 5.9.2. If f is in $C^0[a, b]$, then

$$<f, f> = \int_a^b [f(x)]^2 \, dx.$$

Since the integrand, $[f(x)]^2$, is a nonnegative continuous function, it follows that $<f, f>$ measures the area between the graph $y = [f(x)]^2$ and the x-axis on the interval $[a, b]$. (See Figure 5.9.2.) Consequently, $<f, f> \geq 0$. Furthermore, $<f, f> = 0$ if and only if there is zero area between the graph $y = [f(x)]^2$ and the x-axis, that is, if and only if

$$[f(x)]^2 = 0 \text{ for all } x \text{ in } [a, b].$$

Hence, $<f, f> = 0$ if and only if $f(x) = 0$, for all x in $[a, b]$, so f must be the zero function. (See Figure 5.9.3.) Consequently, property (1) of Definition 5.9.2 is satisfied. Now let f, g, and h be in $C^0[a, b]$. Then,

$$<g, f> = \int_a^b g(x)f(x)\, dx = \int_a^b f(x)g(x)\, dx = <f, g>.$$

Hence, property (2) of Definition 5.9.2 is satisfied. Further, if k is an arbitrary real number, then

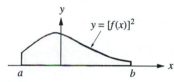

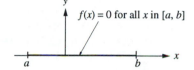

Figure 5.9.2 $<f, f>$ gives the area between the graph $y = f^2(x)$ and the x-axis, lying over the interval $[a, b]$.

Figure 5.9.3 $<f, f> = 0$ if and only if f is the zero function.

$$<kf, g> = \int_a^b (kf)(x)g(x)\, dx = \int_a^b kf(x)g(x)\, dx = k \int_a^b f(x)g(x)\, dx = k<f, g>,$$

so that property (3) of Definition 5.9.2 is also satisfied. Finally,

$$<f + g, h> = \int_a^b (f + g)(x)h(x)\, dx = \int_a^b [f(x) + g(x)]h(x)\, dx$$

$$= \int_a^b f(x)h(x)\, dx + \int_a^b g(x)h(x)\, dx = <f, h> + <g, h>,$$

so that property (4) of Definition 5.9.2 is satisfied. We can now conclude that equation (5.9.4) does define an inner product in the vector space $C^0[a, b]$.

Example 5.9.2   Use equation (5.9.4) to determine the inner product of the following functions in $C^0[0, 1]$:

$$f(x) = 8x, \quad g(x) = x^2 - 1.$$

Also find $\| f \|$ and $\| g \|$.

Solution   From equation (5.9.4),

$$<f, g> = \int_0^1 8x(x^2 - 1)\, dx = [2x^4 - 4x^2]_0^1 = -2.$$

$$\| f \| = \left[\int_0^1 64x^2\, dx \right]^{1/2} = \frac{8}{\sqrt{3}}.$$

$$\| g \| = \left[\int_0^1 (x^2 - 1)^2 \, dx \right]^{1/2} = \left[\int_0^1 (x^4 - 2x^2 + 1) \, dx \right]^{1/2} = \sqrt{\frac{8}{15}} .$$

Property (1) in the definition of an inner product indicates that the norm of a vector is a generalization of the length of a geometric vector. We now show how an inner product enables us to define the angle between two vectors. The key result is the *Cauchy–Schwartz inequality* established in the next theorem.

Theorem 5.9.1 (The Cauchy–Schwartz inequality): Let **u** and **v** be arbitrary vectors in a real inner product space V. Then

$$|<\mathbf{u}, \mathbf{v}>| \leq \| \mathbf{u} \| \| \mathbf{v} \|. \tag{5.9.5}$$

PROOF   If $<\mathbf{u}, \mathbf{v}> = 0$, then certainly equation (5.9.5) is satisfied. Now suppose $<\mathbf{u}, \mathbf{v}> \neq 0$, and let k be an arbitrary real number. For the vector $\mathbf{u} + k\mathbf{v}$, we have

$$0 \leq \| \mathbf{u} + k\mathbf{v} \|^2 = <\mathbf{u} + k\mathbf{v}, \mathbf{u} + k\mathbf{v}>. \tag{5.9.6}$$

But, using the properties of a real inner product,

$$<\mathbf{u} + k\mathbf{v}, \mathbf{u} + k\mathbf{v}> = <\mathbf{u}, \mathbf{u} + k\mathbf{v}> + <k\mathbf{v}, \mathbf{u} + k\mathbf{v}>$$

$$= <\mathbf{u} + k\mathbf{v}, \mathbf{u}> + <\mathbf{u} + k\mathbf{v}, k\mathbf{v}>$$

$$= <\mathbf{u}, \mathbf{u}> + <k\mathbf{v}, \mathbf{u}> + <\mathbf{u}, k\mathbf{v}> + <k\mathbf{v}, k\mathbf{v}>$$

$$= \| \mathbf{u} \|^2 + k<\mathbf{v}, \mathbf{u}> + <k\mathbf{v}, \mathbf{u}> + k<\mathbf{v}, k\mathbf{v}>$$

$$= \| \mathbf{u} \|^2 + 2k<\mathbf{v}, \mathbf{u}> + k<k\mathbf{v}, \mathbf{v}>$$

$$= \| \mathbf{u} \|^2 + 2k<\mathbf{v}, \mathbf{u}> + k^2\| \mathbf{v} \|^2.$$

Consequently, (5.9.6) implies that

$$\| \mathbf{v} \|^2 k^2 + 2<\mathbf{u}, \mathbf{v}>k + \| \mathbf{u} \|^2 \geq 0. \tag{5.9.7}$$

The left-hand side of this inequality defines the quadratic expression

$$P(k) = \| \mathbf{v} \|^2 k^2 + 2<\mathbf{u}, \mathbf{v}>k + \| \mathbf{u} \|^2.$$

The discriminant of this quadratic is

$$\Delta = 4(<\mathbf{u}, \mathbf{v}>)^2 - 4\| \mathbf{u} \|^2 \| \mathbf{v} \|^2.$$

If $\Delta > 0$, then $P(k)$ has two real and distinct zeros. This would imply that the graph of P crossed the k–axis, and, therefore, P would assume negative values. This would contradict (5.9.7). Consequently, we must have $\Delta \leq 0$. That is,

$$4(<\mathbf{u}, \mathbf{v}>)^2 - 4\| \mathbf{u} \|^2 \| \mathbf{v} \|^2 \leq 0,$$

or equivalently,

$$(<\mathbf{u}, \mathbf{v}>)^2 \leq \| \mathbf{u} \|^2 \| \mathbf{v} \|^2.$$

Hence,

$$|<\mathbf{u}, \mathbf{v}>| \leq \| \mathbf{u} \| \| \mathbf{v} \|. \qquad\blacksquare$$

If **u** and **v** are arbitrary vectors in a real inner product space V, then $<\mathbf{u}, \mathbf{v}>$ is a real number, and therefore, (5.9.5) can be written in the equivalent form

$$-\| \mathbf{u} \| \| \mathbf{v} \| \leq <\mathbf{u}, \mathbf{v}> \leq \| \mathbf{u} \| \| \mathbf{v} \|.$$

Consequently, provided that **u** and **v** are nonzero vectors, we have

$$-1 \leq \frac{<\mathbf{u}, \mathbf{v}>}{\|\mathbf{u}\| \|\mathbf{v}\|} \leq 1.$$

Thus, each pair of nonzero vectors in a *real* inner product space V determines a unique angle θ by

$$\cos \theta = \frac{<\mathbf{u}, \mathbf{v}>}{\|\mathbf{u}\| \|\mathbf{v}\|}, \quad 0 \leq \theta \leq \pi. \tag{5.9.8}$$

We call θ the angle between **u** and **v**. In the case when **u** and **v** are geometric vectors, the formula (5.9.8) coincides with equation (5.9.3).

Example 5.9.3   Determine the angle between the vectors $\mathbf{u} = (1, -1, 2, 3)$, $\mathbf{v} = (-2, 1, 2, -2)$ in $\mathbf{R}^4$.

Solution   Using the standard inner product in $\mathbf{R}^4$ yields

$$<\mathbf{u}, \mathbf{v}> = -5, \quad \|\mathbf{u}\| = \sqrt{15}, \quad \|\mathbf{v}\| = \sqrt{13},$$

so that the angle between **u** and **v** is given by

$$\cos \theta = -\frac{5}{\sqrt{15}\sqrt{13}} = -\frac{\sqrt{195}}{39}, \quad 0 \leq \theta \leq \pi.$$

Hence,

$$\theta = \arccos\left(-\frac{\sqrt{195}}{39}\right) \approx 1.937 \text{ radians} \approx 110° \ 58'.$$

Example 5.9.4   Use the inner product (5.9.4) to determine the angle between the functions $f_1(x) = \sin 2x$, and $f_2(x) = \cos 2x$ on the interval $[-\pi, \pi]$.

Solution Using the inner product (5.9.4), we have

$$<f_1, f_2> = \int_{-\pi}^{\pi} \sin 2x \cos 2x \, dx = \frac{1}{2} \int_{-\pi}^{\pi} \sin 4x \, dx = \frac{1}{8}[-\cos 4x]_{-\pi}^{\pi} = 0.$$

Consequently, the angle between the two functions satisfies

$$\cos \theta = 0, \quad 0 \leq \theta \leq \pi,$$

which implies that $\theta = \pi/2$. We say that the functions are orthogonal on the interval $[-\pi, \pi]$, relative to the inner product (5.9.4). In the next section, we will have much more to say about orthogonality of vectors.

COMPLEX INNER PRODUCTS[1]

The preceding discussion has been concerned with real vector spaces. In order to generalize the definition of an inner product to a complex vector space, we first consider the case of $\mathbf{C}^n$. By analogy with Definition 5.9.1, it might be thought that an

---

[1]In the remainder of the text the only complex inner product that we will require is the standard inner product in $\mathbf{C}^n$, and this is only needed in Section 6.8.

appropriate inner product for $\mathbf{C}^n$ would be obtained by summing the products of corresponding components of vectors in $\mathbf{C}^n$, that is, by applying the standard inner product from $\mathbf{R}^n$ to vectors in $\mathbf{C}^n$. However, one of the reasons for introducing an inner product is so that we can obtain a concept of "length" of a vector. In order for a quantity to be considered a reasonable measure of length, we would want it to be a nonnegative real number that vanishes if and only if the vector itself is the zero vector [property (1) of a real inner product]. But, if we apply the $\mathbf{R}^n$ inner product directly in $\mathbf{C}^n$ then, since the components of vectors in $\mathbf{C}^n$ are complex numbers, it follows that the resulting norm of a vector in $\mathbf{C}^n$ would be a complex number also. Furthermore, applying the $\mathbf{R}^2$ inner product to, for example, the vector $\mathbf{u} = (1 - i, 1 + i)$, we obtain

$$\| \mathbf{u} \|^2 = (1 - i)^2 + (1 + i)^2 = 0,$$

that is, a nonzero vector having zero "length". To rectify this situation, we extend the definition of an inner product in $\mathbf{C}^n$ as follows.

Definition 5.9.3:   If $\mathbf{u} = (u_1, u_2, \ldots, u_n)$ and $\mathbf{v} = (v_1, v_2, \ldots, v_n)$ are vectors in $\mathbf{C}^n$ we define the **standard inner product** in $\mathbf{C}^n$ by[1]

$$<\mathbf{u}, \mathbf{v}> = u_1 \bar{v}_1 + u_2 \bar{v}_2 + \cdots + u_n \bar{v}_n.$$

The **norm** of $\mathbf{u}$ is defined to be the *real number*

$$\| \mathbf{u} \| = \sqrt{<\mathbf{u}, \mathbf{u}>} = \sqrt{|u_1|^2 + |u_2|^2 + \cdots + |u_n|^2}\ .$$

The preceding inner product is a mapping that associates with the two vectors $\mathbf{u} = (u_1, u_2, \ldots, u_n)$ and $\mathbf{v} = (v_1, v_2, \ldots, v_n)$ in $\mathbf{C}^n$ the *scalar*

$$<\mathbf{u}, \mathbf{v}> = u_1 \bar{v}_1 + u_2 \bar{v}_2 + \cdots + u_n \bar{v}_n.$$

In general, $<\mathbf{u}, \mathbf{v}>$ will have a nonzero imaginary part. The key point to notice is that the norm of $\mathbf{u}$ is always a *real* number, even though the separate components of $\mathbf{u}$ are complex numbers.

Example 5.9.5   If $\mathbf{u} = (1 + 2i, 2 - 3i)$ and $\mathbf{v} = (2 - i, 3 + 4i)$, find $<\mathbf{u}, \mathbf{v}>$ and $\| \mathbf{u} \|$.

Solution   Using Definition 5.9.3,

$$<\mathbf{u}, \mathbf{v}> = (1 + 2i)(2 + i) + (2 - 3i)(3 - 4i) = 5i - 6 - 17i = -6 - 12i,$$

$$\| \mathbf{u} \| = \sqrt{<\mathbf{u}, \mathbf{u}>} = \sqrt{(1 + 2i)(1 - 2i) + (2 - 3i)(2 + 3i)} = \sqrt{18} = 3\sqrt{2}. \qquad \square$$

The standard inner product in $\mathbf{C}^n$ satisfies properties (1), (3), and (4), but not property (2), of Definition 5.9.2. We now derive the appropriate generalization of property (2) when using the standard inner product in $\mathbf{C}^n$. Let $\mathbf{u} = (u_1, u_2, \ldots, u_n)$ and $\mathbf{v} = (v_1, v_2, \ldots, v_n)$ be vectors in $\mathbf{C}^n$. Then, from Definition 5.9.3,

$$<\mathbf{v}, \mathbf{u}> = v_1 \bar{u}_1 + v_2 \bar{u}_2 + \cdots + v_n \bar{u}_n = \overline{u_1 \bar{v}_1 + u_2 \bar{v}_2 + \cdots + u_n \bar{v}_n} = \overline{<\mathbf{u}, \mathbf{v}>}$$

Thus,

---

[1]Recall that if $u = a + ib$, then $\bar{u} = a - ib$ and $|u|^2 = u\bar{u} = (a + ib)(a - ib) = a^2 + b^2$.

$$<\mathbf{v}, \mathbf{u}> = \overline{<\mathbf{u}, \mathbf{v}>} .$$

We now use the properties satisfied by the standard inner product in $\mathbf{C}^n$ to define an inner product in an arbitrary (that is, real or complex) vector space.

Definition 5.9.4: Let V be a vector space. A mapping that associates with each pair of vectors $\mathbf{u}, \mathbf{v}$ in V a scalar, denoted $<\mathbf{u}, \mathbf{v}>$, is called an **inner product in V** provided it satisfies the following properties. For all $\mathbf{u}, \mathbf{v}$ and $\mathbf{w}$ in V and all scalars k,

1. $<\mathbf{u}, \mathbf{u}> \geq 0$. Furthermore, $<\mathbf{u}, \mathbf{u}> = 0$ if and only if $\mathbf{u} = \mathbf{0}$.

2. $<\mathbf{v}, \mathbf{u}> = \overline{<\mathbf{u}, \mathbf{v}>}$.

3. $<k\mathbf{u}, \mathbf{v}> = k<\mathbf{u}, \mathbf{v}>$.

4. $<\mathbf{u} + \mathbf{v}, \mathbf{w}> = <\mathbf{u}, \mathbf{w}> + <\mathbf{v}, \mathbf{w}>$.

The norm of $\mathbf{u}$ is defined in terms of the inner product by

$$\| \mathbf{u} \| = \sqrt{<\mathbf{u}, \mathbf{u}>}.$$

A vector space together with an inner product defined in it is called an **inner product space**.

REMARK   Notice that the properties in the preceding definition reduce to those in Definition 5.9.2 in the case that V is a *real* vector space, since in such a case, the complex conjugates are redundant.

Example 5.9.6   Use properties (2) and (3) of Definition 5.9.4 to prove that in an inner product space $<\mathbf{u}, k\mathbf{v}> = \bar{k}<\mathbf{u}, \mathbf{v}>$ for all vectors $\mathbf{u}, \mathbf{v}$, and all scalars k.

Solution   From property (2), we have

$$<\mathbf{u}, k\mathbf{v}> = \overline{<k\mathbf{v}, \mathbf{u}>} ,$$

so that, using property (3),

$$<\mathbf{u}, k\mathbf{v}> = \bar{k} \ \overline{<\mathbf{v}, \mathbf{u}>} = \bar{k} <\mathbf{u}, \mathbf{v}>,$$

where we have once more used property (2) in the final step. Notice, that in the particular case of a real vector space the foregoing result reduces to

$$<\mathbf{u}, k\mathbf{v}> = k<\mathbf{u}, \mathbf{v}>,$$

since in such a case the scalars are real numbers.   ◻

The Cauchy–Schwartz inequality

$$|<\mathbf{u}, \mathbf{v}>| \leq \| \mathbf{u} \| \| \mathbf{v} \|$$

also holds in a complex inner product space, although the proof is somewhat more complicated than that given in Theorem 5.9.1. In this case the quantity on the left-hand side of the inequality is the modulus of the complex number $<\mathbf{u}, \mathbf{v}>$. We therefore have

$$0 \leq |<\mathbf{u}, \mathbf{v}>| \leq \| \mathbf{u} \| \| \mathbf{v} \|,$$

which, for nonzero vectors, can be written as

$$0 \leq \frac{|<\mathbf{u}, \mathbf{v}>|}{\| \mathbf{u} \| \| \mathbf{v} \|} \leq 1.$$

Hence we can determine a unique angle θ by

$$\cos \theta = \frac{|<\mathbf{u}, \mathbf{v}>|}{\| \mathbf{u} \| \| \mathbf{v} \|}, \quad 0 \leq \theta \leq \pi/2.$$

Once more, we refer to the angle θ defined in this manner as the angle between **u** and **v**.

Example 5.9.7   Determine the angle between the vectors $\mathbf{u} = (2 + i, 3 - 2i)$ and $\mathbf{v} = (5 + i, 1 - 4i)$ in $\mathbf{C}^2$.

Solution   Using the standard inner product in $\mathbf{C}^2$ we have

$$<\mathbf{u}, \mathbf{v}> = (2 + i)(5 - i) + (3 - 2i)(1 + 4i) = 22 + 13i,$$

which implies that $|<\mathbf{u}, \mathbf{v}>| = \sqrt{653}$. Further,

$$\| \mathbf{u} \| = \sqrt{18}, \quad \| \mathbf{v} \| = \sqrt{43}.$$

Therefore, the angle between **u** and **v** is θ, where

$$\cos \theta = (653/774)^{1/2}, \quad 0 \leq \theta \leq \pi/2.$$

Consequently,

$$\theta = \text{Arccos}(653/774)^{1/2} \approx 0.4065 \text{ radians} \approx 23° \, 17'.$$

EXERCISES 5.9

1. Use the standard inner product in $\mathbf{R}^4$ to determine the angle between the vectors $\mathbf{x} = (1, 3, -1, 4)$ and $\mathbf{y} = (-1, 1, -2, 1)$.

2. If $f(x) = \sin x$ and $g(x) = x$ on $[0, \pi]$, use the function inner product defined in the text to determine the angle between f and g.

3. If $\mathbf{u} = (2 + i, 3 - 2i, 4 + i)$ and $\mathbf{v} = (-1 + i, 1 - 3i, 3 - i)$, use the standard inner product in $\mathbf{C}^3$ to determine $<\mathbf{u}, \mathbf{v}>$, $\| \mathbf{u} \|$ and $\| \mathbf{v} \|$. Also find the angle between **u** and **v**.

4. Let $A = \begin{bmatrix} a_{11} & a_{12} \\ a_{21} & a_{22} \end{bmatrix}$ and $B = \begin{bmatrix} b_{11} & b_{12} \\ b_{21} & b_{22} \end{bmatrix}$ be arbitrary vectors in $M_2(\mathbf{R})$. Prove that the mapping

$$<A, B> = a_{11}b_{11} + a_{12}b_{12} + a_{21}b_{21} + a_{22}b_{22} \quad (4.1)$$

defines an inner product in $M_2(\mathbf{R})$.

For problems 5 and 6, use the inner product (4.1) to determine $<A, B>$, $\| A \|$, $\| B \|$. Also determine the angle between the given matrices.

5. $A = \begin{bmatrix} 2 & -1 \\ 3 & 5 \end{bmatrix}$, $B = \begin{bmatrix} 3 & 1 \\ -1 & 2 \end{bmatrix}$.

6. $A = \begin{bmatrix} 3 & 2 \\ -2 & 4 \end{bmatrix}$, $B = \begin{bmatrix} 1 & 1 \\ -2 & 1 \end{bmatrix}$.

7. Let $p_1(x) = a + bx$ and $p_2(x) = c + dx$ be arbitrary vectors in P_2. Determine a mapping $<p_1, p_2>$ that defines an inner product on P_2.

Consider the vector space $\mathbf{R}^2$. Define the mapping $< , >$ by

$$<\mathbf{x}, \mathbf{y}> = 2x_1y_1 + x_1y_2 + x_2y_1 + 2x_2y_2 \quad (5.9.9)$$

for all vectors $\mathbf{x} = (x_1, x_2)$ and $\mathbf{y} = (y_1, y_2)$ in $\mathbf{R}^2$. This mapping is required for problems 8–11.

8. Verify that equation (5.9.9) defines an inner product on $\mathbf{R}^2$.

For problems 9–11, determine the inner product of the given vectors using (a) the inner product (5.9.9), (b) the standard inner product in $\mathbf{R}^2$.

9. $\mathbf{x} = (1, 0)$, $\mathbf{y} = (-1, 2)$.

10. $\mathbf{x} = (2, -1)$, $\mathbf{y} = (3, 6)$.

11. $\mathbf{x} = (1, -2)$, $\mathbf{y} = (2, 1)$.

12. Consider the vector space $\mathbf{R}^2$. Define the mapping $< , >$ by

$$<\mathbf{x}, \mathbf{y}> = x_1 y_1 - x_2 y_2, \qquad (12.1)$$

for all vectors $\mathbf{x} = (x_1, x_2)$, $\mathbf{y} = (y_1, y_2)$. Verify that all of the properties in Definition 5.9.2 except (1) are satisfied by (12.1).

The mapping (12.1) is called a **pseudo-inner product** in $\mathbf{R}^2$ and, when generalized to $\mathbf{R}^4$, is of fundamental importance in Einstein's special relativity theory.

13. Using equation (12.1), determine all nonzero vectors satisfying $<\mathbf{x}, \mathbf{x}> = 0$. Such vectors are called *null* vectors.

14. Using equation (12.1), determine all vectors satisfying $<\mathbf{x}, \mathbf{x}> < 0$. Such vectors are called *timelike* vectors.

15. Using equation (12.1), determine all vectors satisfying $<\mathbf{x}, \mathbf{x}> > 0$. Such vectors are called *spacelike* vectors.

16. Make a sketch of $\mathbf{R}^2$ and indicate the position of the null, timelike and spacelike vectors.

17. Let V be a real inner product space.

(a) Prove that for all $\mathbf{x}$, $\mathbf{y}$ in V,

$$\| \mathbf{x} + \mathbf{y} \|^2 = \| \mathbf{x} \|^2 + 2<\mathbf{x}, \mathbf{y}> + \| \mathbf{y} \|^2.$$

(Hint: $\| \mathbf{x} + \mathbf{y} \|^2 = <\mathbf{x} + \mathbf{y}, \mathbf{x} + \mathbf{y}>$.)

(b) Prove that for all $\mathbf{x}$, $\mathbf{y}$ in V,

(i) $\| \mathbf{x} + \mathbf{y} \|^2 - \| \mathbf{x} - \mathbf{y} \|^2 = 4<\mathbf{x}, \mathbf{y}>$.

(ii) $\| \mathbf{x} + \mathbf{y} \|^2 + \| \mathbf{x} - \mathbf{y} \|^2 = 2(\| \mathbf{x} \|^2 + \| \mathbf{y} \|^2)$.

17. Let V be a complex inner product space. Prove that for all $\mathbf{u}$, $\mathbf{v}$ in V

$$\| \mathbf{u} + \mathbf{v} \|^2 = \| \mathbf{u} \|^2 + 2\,\mathrm{Re}(<\mathbf{u}, \mathbf{v}>) + \| \mathbf{v} \|^2,$$

where Re denotes the real part of a complex number.

5.10   ORTHOGONAL SETS OF VECTORS AND THE GRAM–SCHMIDT PROCEDURE

The discussion in the previous section has shown how an inner product can be used to define the angle between two nonzero vectors. In particular, if the inner product of two nonzero vectors is zero then the angle between those two vectors is $\pi/2$ radians, and therefore it is natural to call such vectors orthogonal (perpendicular). The following definition extends the idea of orthogonality into an arbitrary inner product space.

Definition 5.10.1:   Let V be an inner product space.

1. Two vectors $\mathbf{u}$ and $\mathbf{v}$ in V are said to be **orthogonal** if and only if $<\mathbf{u}, \mathbf{v}> = 0$.

2. A set of nonzero vectors $\{v_1, v_2, \ldots, v_k\}$ in V is called an **orthogonal set** of vectors if

$$<v_i, v_j> = 0, \qquad \text{whenever } i \neq j.$$

(That is, each vector is orthogonal to every other vector in the set.)

3. A vector $\mathbf{v}$ in V is called a **unit vector** if $\| \mathbf{v} \| = 1$.

4. An *orthogonal* set of *unit* vectors is called an **orthonormal set** of vectors. Thus, $\{v_1, v_2, \ldots, v_k\}$ in V is an orthonormal set if and only if

(a) $<v_i, v_j> = 0$, whenever $i \neq j$.

(b) $<v_i, v_i> = 1$, $i = 1, 2, \ldots, k$.

These two conditions can be written compactly in terms of the Kronecker delta symbol as

$$\langle \mathbf{v}_i, \mathbf{v}_j \rangle = \delta_{ij}, \quad i, j = 1, 2, \ldots, k.$$

REMARKS

1. Note that the inner products occurring in (1)–(4) of Definition 5.10.1 will depend upon in which inner product space we are working.

2. If $\mathbf{v}$ is any nonzero vector then $\dfrac{\mathbf{v}}{\|\ \mathbf{v}\ \|}$ is a unit vector, since the properties of an inner product imply that

$$\left\langle \frac{\mathbf{v}}{\|\ \mathbf{v}\ \|}, \frac{\mathbf{v}}{\|\ \mathbf{v}\ \|} \right\rangle = \frac{1}{\|\ \mathbf{v}\ \|^2} \langle \mathbf{v}, \mathbf{v} \rangle = \frac{1}{\|\ \mathbf{v}\ \|^2} \|\ \mathbf{v}\ \|^2 = 1.$$

Example 5.10.1   Verify that $\{(-2, 1, 3, 0), (0, -3, 1, -6), (-2, -4, 0, 2)\}$ is an orthogonal set of vectors in $\mathbf{R}^4$, and use it to construct an orthonormal set of vectors in $\mathbf{R}^4$.

Solution   Let $\mathbf{v}_1 = (-2, 1, 3, 0)$, $\mathbf{v}_2 = (0, -3, 1, -6)$, $\mathbf{v}_3 = (-2, -4, 0, 2)$. Then

$$\langle \mathbf{v}_1, \mathbf{v}_2 \rangle = 0, \quad \langle \mathbf{v}_1, \mathbf{v}_3 \rangle = 0, \quad \langle \mathbf{v}_2, \mathbf{v}_3 \rangle = 0,$$

so that the given set of vectors is an orthogonal set. Dividing each vector in the set by its norm yields the following orthonormal set

$$\left\{ \frac{1}{\sqrt{14}}\mathbf{v}_1, \frac{1}{\sqrt{46}}\mathbf{v}_2, \frac{1}{2\sqrt{6}}\mathbf{v}_3 \right\}.$$

Example 5.10.2   Verify that the functions $f_1(x) = 1$, $f_2(x) = \sin x$, $f_3(x) = \cos x$ are orthogonal in $C^0[-\pi, \pi]$, and use them to construct an orthonormal set of functions in $C^0[-\pi, \pi]$.

Solution   In this case, we have:

$$\langle f_1, f_2 \rangle = \int_{-\pi}^{\pi} \sin x \, dx = 0, \quad \langle f_1, f_3 \rangle = \int_{-\pi}^{\pi} \cos x \, dx = 0,$$

$$\langle f_2, f_3 \rangle = \int_{-\pi}^{\pi} \sin x \cos x \, dx = \left[\frac{1}{2} \sin^2 x \right]_{-\pi}^{\pi} = 0,$$

so that the functions are indeed orthogonal on $[-\pi, \pi]$. Taking the norm of each function, we obtain

$$\|\ f_1\ \| = \left[\int_{-\pi}^{\pi} 1 \, dx \right]^{1/2} = \sqrt{2\pi},$$

$$\| f_2 \| \;=\; \left[\int_{-\pi}^{\pi} \sin^2 x \, dx \right]^{1/2} = \left[\int_{-\pi}^{\pi} \frac{1}{2}(1 - \cos 2x) \, dx \right]^{1/2} = \sqrt{\pi} \; ,$$

$$\| f_3 \| \;=\; \left[\int_{-\pi}^{\pi} \cos^2 x \, dx \right]^{1/2} = \left[\int_{-\pi}^{\pi} \frac{1}{2}(1 + \cos 2x) \, dx \right]^{1/2} = \sqrt{\pi} .$$

Thus an orthonormal set of functions on $[-\pi, \pi]$ is

$$\left\{ \frac{1}{\sqrt{2\pi}}, \; \frac{1}{\sqrt{\pi}} \sin x, \; \frac{1}{\sqrt{\pi}} \cos x \right\} .$$

ORTHOGONAL BASES AND THE GRAM–SCHMIDT PROCEDURE

In the analysis of geometric vectors in elementary calculus courses, it is usual to use the standard basis $\{\mathbf{i}, \mathbf{j}, \mathbf{k}\}$. Notice that this set of vectors is in fact an orthonormal set. The introduction of an inner product in a vector space opens up the possibility of using similar bases in a general finite-dimensional vector space. The next definition introduces the appropriate terminology.

Definition 5.10.2:   A basis $\{\mathbf{v}_1, \mathbf{v}_2, \ldots, \mathbf{v}_m\}$ for a (finite-dimensional) inner product space is called an **orthogonal** basis if

$$\langle \mathbf{v}_i, \mathbf{v}_j \rangle = 0 \quad \text{whenever } i \neq j,$$

and is called an **orthonormal** basis if

$$\langle \mathbf{v}_i, \mathbf{v}_j \rangle = \delta_{ij}, \quad i, j = 1, 2, \ldots, m.$$

We now show that if we have a basis for a finite-dimensional inner product space, then it is always possible to construct an orthogonal basis for this vector space. Once more the motivation behind the result that we wish to derive comes from studying geometric vectors. If $\mathbf{v}_1$ and $\mathbf{v}_2$ are any two LI (noncolinear) geometric vectors, then the **orthogonal projection** of $\mathbf{v}_2$ on $\mathbf{v}_1$ is the vector $\mathbf{P}(\mathbf{v}_2, \mathbf{v}_1)$ shown in Figure 5.10.2. We see from the figure that an orthogonal basis for the subspace (plane) of three-space spanned by $\mathbf{v}_1$ and $\mathbf{v}_2$ is $\{\mathbf{u}_1, \mathbf{u}_2\}$, where $\mathbf{u}_1 = \mathbf{v}_1$ and

$$\mathbf{u}_2 = \mathbf{v}_2 - \mathbf{P}(\mathbf{v}_2, \mathbf{v}_1). \tag{5.10.1}$$

In order to understand the generalization of this result to the case of an arbitrary inner product space, it is useful to derive an expression for $\mathbf{P}(\mathbf{v}_2, \mathbf{v}_1)$ in terms of the dot product. We see from Figure 5.10.1 that the norm (length) of $\mathbf{P}(\mathbf{v}_2, \mathbf{v}_1)$ is

$$\| \mathbf{P}(\mathbf{v}_2, \mathbf{v}_1) \| = \| \mathbf{v}_2 \| \cos \theta,$$

so that

$$\mathbf{P}(\mathbf{v}_2, \mathbf{v}_1) = \| \mathbf{v}_2 \| \cos \theta \, \frac{\mathbf{v}_1}{\| \mathbf{v}_1 \|},$$

which we can write as

$$\mathbf{P}(\mathbf{v}_2, \mathbf{v}_1) = \frac{\| \mathbf{v}_2 \| \| \mathbf{v}_1 \|}{\| \mathbf{v}_1 \|^2} \cos \theta \, \mathbf{v}_1. \tag{5.10.2}$$

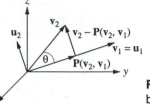

Figure 5.10.1 Obtaining an orthogonal basis for a subspace of $\mathbf{R}^3$.

Recalling that the dot product of the vectors $\mathbf{v}_1$, $\mathbf{v}_2$ is defined by

$$\mathbf{v}_2 \cdot \mathbf{v}_1 = \| \mathbf{v}_2 \| \, \| \mathbf{v}_1 \| \cos \theta,$$

it follows from equation (5.10.2) that

$$\mathbf{P}(\mathbf{v}_2, \mathbf{v}_1) = \frac{(\mathbf{v}_2 \cdot \mathbf{v}_1)}{\| \mathbf{v}_1 \|^2} \, \mathbf{v}_1,$$

or equivalently, using the notation for the inner product introduced in the previous section,

$$\mathbf{P}(\mathbf{v}_2, \mathbf{v}_1) = \frac{<\mathbf{v}_2, \mathbf{v}_1>}{\| \mathbf{v}_1 \|^2} \, \mathbf{v}_1. \tag{5.10.3}$$

Now let $\mathbf{v}_1$ and $\mathbf{v}_2$ be LI vectors in an arbitrary inner product space V. We show next that the foregoing formula can also be applied in V to obtain an orthogonal basis $\{\mathbf{u}_1, \mathbf{u}_2\}$ for the subspace of V spanned by $\mathbf{v}_1$ and $\mathbf{v}_2$. In order for $\mathbf{u}_1$ and $\mathbf{u}_2$ to remain in the subspace spanned by $\mathbf{v}_1$ and $\mathbf{v}_2$, they must be linear combinations of $\mathbf{v}_1$ and $\mathbf{v}_2$. We take $\mathbf{u}_1 = \mathbf{v}_1$, and let $\mathbf{u}_2 = \mathbf{v}_2 + \lambda \mathbf{u}_1$, where λ is a scalar. We wish to choose λ such that $\mathbf{u}_1$ and $\mathbf{u}_2$ are orthogonal, that is, such that $<\mathbf{u}_2, \mathbf{u}_1> = 0$. But this latter condition is satisfied if and only if

$$<\mathbf{v}_2 + \lambda \mathbf{u}_1, \mathbf{u}_1> = 0.$$

Using the properties of an inner product, the preceding condition can be written as

$$<\mathbf{v}_2, \mathbf{u}_1> + \lambda <\mathbf{u}_1, \mathbf{u}_1> = 0.$$

That is,

$$<\mathbf{v}_2, \mathbf{u}_1> + \lambda \| \mathbf{u}_1 \|^2 = 0.$$

Solving for λ yields

$$\lambda = -\frac{<\mathbf{v}_2, \mathbf{u}_1>}{\| \mathbf{u}_1 \|^2}.$$

Thus, if we choose $\mathbf{u}_1 = \mathbf{v}_1$ and

$$\mathbf{u}_2 = \mathbf{v}_2 - \frac{<\mathbf{v}_2, \mathbf{u}_1>}{\| \mathbf{u}_1 \|^2} \, \mathbf{u}_1, \tag{5.10.4}$$

then $\{\mathbf{u}_1, \mathbf{u}_2\}$ is an orthogonal basis for the subspace of V spanned by $\mathbf{v}_1$ and $\mathbf{v}_2$. We define the orthogonal projection of $\mathbf{v}_2$ on $\mathbf{v}_1$ in an arbitrary inner product space by the formula (5.10.3). Then the expression for $\mathbf{u}_2$ given in equation (5.10.4) is obtained by subtracting $\mathbf{P}(\mathbf{v}_2, \mathbf{u}_1)$ from $\mathbf{v}_2$. This is exactly the same formula as we derived earlier for $\mathbf{R}^3$. (See (5.10.1).)

More generally, suppose we are given a LI set of vectors $\{\mathbf{v}_1, \mathbf{v}_2, \ldots, \mathbf{v}_m\}$ in an inner product space V. In a similar manner, we can construct an orthogonal basis for the subspace of V spanned by these vectors. We let $\mathbf{u}_1 = \mathbf{v}_1$ and then successively define the remaining orthogonal vectors $\mathbf{u}_2$, $\mathbf{u}_3$ ..., $\mathbf{u}_m$ by subtracting off appropriate projections on the previous vectors. The resulting procedure is called the **Gram–Schmidt orthogonalization procedure**. The formal statement of the result is as follows.

   Theorem 5.10.1:   Let $\{\mathbf{v}_1, \mathbf{v}_2, ..., \mathbf{v}_m\}$ be a LI set of vectors in an inner product space V. Then an *orthogonal basis* for the subspace of V spanned by these vectors is $\{\mathbf{u}_1, \mathbf{u}_2, ..., \mathbf{u}_m\}$, where

$$\mathbf{u}_1 = \mathbf{v}_1,$$

$$\mathbf{u}_2 = \mathbf{v}_2 - \frac{<\mathbf{v}_2, \mathbf{u}_1>}{\|\mathbf{u}_1\|^2}\mathbf{u}_1 \qquad \leftarrow \mathbf{v}_2 - \mathbf{P}(\mathbf{v}_2, \mathbf{u}_1)$$

$$\mathbf{u}_3 = \mathbf{v}_3 - \frac{<\mathbf{v}_3, \mathbf{u}_1>}{\|\mathbf{u}_1\|^2}\mathbf{u}_1 - \frac{<\mathbf{v}_3, \mathbf{u}_2>}{\|\mathbf{u}_2\|^2}\mathbf{u}_2 \qquad \leftarrow \mathbf{v}_3 - \mathbf{P}(\mathbf{v}_3, \mathbf{u}_1) - \mathbf{P}(\mathbf{v}_3, \mathbf{u}_2)$$

$$\vdots$$

$$\mathbf{u}_i = \mathbf{v}_i - \sum_{k=1}^{i-1} \frac{<\mathbf{v}_i, \mathbf{u}_k>}{\|\mathbf{u}_k\|^2}\mathbf{u}_k \qquad \leftarrow \text{general term } \mathbf{v}_i - \sum_{k=1}^{i-1} \mathbf{P}(\mathbf{v}_i, \mathbf{u}_k)$$

$$\vdots$$

$$\mathbf{u}_m = \mathbf{v}_m - \frac{<\mathbf{v}_m, \mathbf{u}_1>}{\|\mathbf{u}_1\|^2}\mathbf{u}_1 - \frac{<\mathbf{v}_m, \mathbf{u}_2>}{\|\mathbf{u}_2\|^2}\mathbf{u}_2 - \cdots - \frac{<\mathbf{v}_m, \mathbf{u}_{m-1}>}{\|\mathbf{u}_{m-1}\|^2}\mathbf{u}_{m-1}.$$

PROOF (Outline)   The proof is constructional. We begin by choosing

$$\mathbf{u}_1 = \mathbf{v}_1 \quad \text{and} \quad \mathbf{u}_2 = \mathbf{v}_2 - \frac{<\mathbf{v}_2, \mathbf{u}_1>}{\|\mathbf{u}_1\|^2}\mathbf{u}_1.$$

We have already shown that these two vectors are orthogonal. Now use mathematical induction. Assume we have k orthogonal vectors $\mathbf{u}_1, \mathbf{u}_2, ..., \mathbf{u}_k$, and let

$$\mathbf{u}_{k+1} = \mathbf{v}_{k+1} + \sum_{i=1}^{k} \lambda_i \mathbf{u}_i.$$

The orthogonality conditions $<\mathbf{u}_{k+1}, \mathbf{u}_i> = 0$ for $1 \leq i \leq k$ imply that

$$\lambda_i = -\frac{<\mathbf{v}_{k+1}, \mathbf{u}_i>}{\|\mathbf{u}_i\|^2}, \quad i = 1, 2, ..., k,$$

and so lead to the formula given.   ■

REMARKS

   1. If $\{\mathbf{u}_1, \mathbf{u}_2, ..., \mathbf{u}_m\}$ is an orthogonal basis for a subspace of V then

$$\left\{ \frac{1}{\|\mathbf{u}_1\|}\mathbf{u}_1, \frac{1}{\|\mathbf{u}_2\|}\mathbf{u}_2, \cdots, \frac{1}{\|\mathbf{u}_m\|}\mathbf{u}_m \right\}$$

is an orthonormal basis for the subspace.

   2. In applying the Gram–Schmidt procedure, you must remember to use the appropriate inner product in the space that you are dealing with.

Example 5.10.3   Obtain an orthogonal basis for the subspace of $\mathbf{R}^4$ spanned by

$$\mathbf{v}_1 = (1, 0, 1, 0), \ \mathbf{v}_2 = (1, 1, 1, 1), \ \mathbf{v}_3 = (-1, 2, 0, 1).$$

Solution   Using the Gram–Schmidt procedure, we choose

$$\mathbf{u}_1 = \mathbf{v}_1 = (1, 0, 1, 0).$$

Then,

$$\langle \mathbf{v}_2, \mathbf{u}_1 \rangle = 1 + 0 + 1 + 0 = 2, \ \| \mathbf{u}_1 \|^2 = 2,$$

so that

$$\mathbf{u}_2 = \mathbf{v}_2 - \frac{\langle \mathbf{v}_2, \mathbf{u}_1 \rangle}{\| \mathbf{u}_1 \|^2} \mathbf{u}_1 = (1, 1, 1, 1) - \frac{2}{2}(1, 0, 1, 0) = (0, 1, 0, 1).$$

Continuing, we have

$$\langle \mathbf{v}_3, \mathbf{u}_1 \rangle = -1 + 0 + 0 + 0 = -1,$$
$$\langle \mathbf{v}_3, \mathbf{u}_2 \rangle = 0 + 2 + 0 + 1 = 3,$$

and

$$\| \mathbf{u}_2 \|^2 = \langle \mathbf{u}_2, \mathbf{u}_2 \rangle = 2,$$

so that

$$\mathbf{u}_3 = \mathbf{v}_3 - \frac{\langle \mathbf{v}_3, \mathbf{u}_1 \rangle}{\| \mathbf{u}_1 \|^2} \mathbf{u}_1 - \frac{\langle \mathbf{v}_3, \mathbf{u}_2 \rangle}{\| \mathbf{u}_2 \|^2} \mathbf{u}_2$$

$$= (-1, 2, 0, 1) + \frac{1}{2}(1, 0, 1, 0) - \frac{3}{2}(0, 1, 0, 1)$$

$$= \frac{1}{2}(-1, 1, 1, -1).$$

Thus, an orthogonal basis for the subspace spanned by $\mathbf{v}_1, \mathbf{v}_2, \mathbf{v}_3$ is

$$\left\{ (1, 0, 1, 0), \ (0, 1, 0, 1), \ \frac{1}{2}(-1, 1, 1, -1) \right\},$$

and an orthonormal basis is

$$\left\{ \frac{\sqrt{2}}{2}(1, 0, 1, 0), \ \frac{\sqrt{2}}{2}(0, 1, 0, 1), \ \frac{1}{2}(-1, 1, 1, -1) \right\}.$$

Example 5.10.4   Determine an orthogonal basis for the subspace of $C^0[-1, 1]$ spanned by the functions $f_1(x) = x, f_2(x) = x^3, f_3(x) = x^5$.

Solution   In this case, we let $\{g_1, g_2, g_3\}$ denote the orthogonal basis, and we apply the Gram–Schmidt procedure. Thus, $g_1(x) = x$, and

$$g_2(x) = f_2(x) - \frac{\langle f_2, g_1 \rangle}{\| g_1 \|^2} g_1(x). \tag{5.10.5}$$

Using the inner product for $C^0[-1, 1]$ defined in the previous section, we have

$$\langle f_2, g_1 \rangle = \int_{-1}^{1} f_2(x) \, g_1(x) \, dx = \int_{-1}^{1} x^4 \, dx = \frac{2}{5}.$$

$$\| g_1 \|^2 = \langle g_1, g_1 \rangle = \int_{-1}^{1} x^2 \, dx = \frac{2}{3}.$$

Substituting into equation (5.10.5) yields

$$g_2(x) = x^3 - \frac{3}{5}x.$$

That is,

$$g_2(x) = \frac{1}{5}x(5x^2 - 3).$$

We now compute $g_3(x)$. According to the Gram–Schmidt procedure,

$$g_3(x) = f_3(x) - \frac{<f_3, g_1>}{\| g_1 \|^2} g_1(x) - \frac{<f_3, g_2>}{\| g_2 \|^2} g_2(x). \tag{5.10.6}$$

We first evaluate the required inner products:

$$<f_3, g_1> = \int_{-1}^{1} f_3(x) g_1(x) \, dx = \int_{-1}^{1} x^6 \, dx = \frac{2}{7},$$

$$<f_3, g_2> = \int_{-1}^{1} f_3(x) g_2(x) \, dx = \frac{1}{5} \int_{-1}^{1} x^6(5x^2 - 3) \, dx = \frac{1}{5}(\frac{10}{9} - \frac{6}{7}) = \frac{16}{315},$$

$$\| g_2 \|^2 = \int_{-1}^{1} [g_2(x)]^2 \, dx = \frac{1}{25} \int_{-1}^{1} x^2(5x^2 - 3)^2 \, dx$$

$$= \frac{1}{25} \int_{-1}^{1} (25x^6 - 30x^4 + 9x^2) \, dx = \frac{8}{175}.$$

Substituting into equation (5.10.6) yields

$$g_3(x) = x^5 - \frac{3}{7}x - \frac{2}{9}x(5x^2 - 3) = \frac{1}{63}(63x^5 - 70x^3 + 15x).$$

Thus an orthogonal basis for the subspace of $C^0[-1, 1]$ spanned by f_1, f_2, f_3 is

$$\left\{ x, \frac{1}{5}x(5x^2 - 3), \frac{1}{63}x(63x^4 - 70x^2 + 15) \right\}.$$

EXERCISES 5.10

For problems 1–4, determine whether the given set of vectors is an orthogonal set in $\mathbf{R}^n$. For those that are, determine a corresponding orthonormal set of vectors.

1. $\{(2, -1, 1), (1, 1, -1), (0, 1, 1)\}$.

2. $\{1, 3, -1, 1), (-1, 1, 1, -1), (1, 0, 2, 1)\}$.

3. $\{(1, 2, -1, 0), (1, 0, 1, 2), (-1, 1, 1, 0), (1, -1, -1, 0)\}$.

4. $\{(1, 2, -1, 0, 3), (1, 1, 0, 2, -1), (4, 2, -4, -5, -4)\}$.

5. Determine all nonzero vectors $\mathbf{x}_3$ such that $\{\mathbf{x}_1, \mathbf{x}_2, \mathbf{x}_3\}$ is an orthogonal set, where $\mathbf{x}_1 = (1, 2, 3)$, $\mathbf{x}_2 = (1, 1, -1)$. Hence obtain an orthonormal set of vectors in $\mathbf{R}^3$.

For problems 6 and 7, show that the given set of vectors is an orthogonal set in $\mathbf{C}^n$, and hence, determine a corresponding orthonormal set of vectors in each case.

6. $\{(1 - i, 3 + 2i), (2 + 3i, 1 - i)\}$.

7. $\{(1 - i, 1 + i, i), (0, i, 1 - i), (-3 + 3i, 2 + 2i, 2i)\}$.

8 . Consider the vectors $\mathbf{x}_1 = (1 - i, 1 + 2i)$, $\mathbf{x}_2 = (2 + i, z)$ in $\mathbf{C}^2$. Determine the complex number z such that $\{\mathbf{x}_1, \mathbf{x}_2\}$ is an orthogonal set of vectors and hence obtain an orthonormal set of vectors in $\mathbf{C}^2$.

For problems 9 and 10, show that the given functions in $C^0[-1, 1]$ are orthogonal, and use them to construct an orthonormal set of functions in $C^0[-1, 1]$.

9 . $f_1(x) = 1, f_2(x) = \sin \pi x, f_3(x) = \cos \pi x$.

10. $f_1(x) = 1, f_2(x) = x, f_3(x) = \frac{1}{2} (3x^2 - 1)$.

These are the Legendre polynomials that arise as solutions of the Legendre differential equation

$$(1 - x^2)y'' - 2xy' + n(n + 1)y = 0,$$

when $n = 0, 1, 2$, respectively.

For problems 11 and 12, show that the given functions are orthonormal on $[-1, 1]$.

11. $f_1(x) = \sin \pi x, f_2(x) = \sin 2\pi x$,

$f_3(x) = \sin 3\pi x$.

(Hint: $\sin a \sin b = \frac{1}{2} [\cos(a + b) - \cos(a - b)]$.)

12. $f_1(x) = \cos \pi x, f_2(x) = \cos 2\pi x$,

$f_3(x) = \cos 3\pi x$.

13. Let $A_1 = \begin{bmatrix} 1 & 1 \\ -1 & 2 \end{bmatrix}$, $A_2 = \begin{bmatrix} -1 & 1 \\ 2 & 1 \end{bmatrix}$, and

$A_3 = \begin{bmatrix} -1 & -3 \\ 0 & 2 \end{bmatrix}$. Use the inner product

$$<A, B> = a_{11}b_{11} + a_{12}b_{12} + a_{21}b_{21} + a_{22}b_{22}$$

to determine all matrices $A_4 = \begin{bmatrix} a & b \\ c & d \end{bmatrix}$ such that

$\{A_1, A_2, A_3, A_4\}$ is an orthogonal set of matrices in $M_2(\mathbf{R})$.

For problems 14–19, use the Gram–Schmidt process to determine an orthonormal basis for the subspace of $\mathbf{R}^n$ spanned by the given set of vectors.

14. $\{(1, -1, -1), (2, 1, -1)\}$.

15. $\{(2, 1, -2), (1, 3, -1)\}$.

16. $\{(-1, 1, 1, 1), (1, 2, 1, 2)\}$.

17. $\{(1, 0, -1, 0), (1, 1, -1, 0), (-1, 1, 0, 1)\}$.

18. $\{(1, 2, 0, 1), (2, 1, 1, 0), (1, 0, 2, 1)\}$.

19. $\{(1, 1, -1, 0), (-1, 0, 1, 1), (2, -1, 2, 1)\}$.

20. If $A = \begin{bmatrix} 3 & 1 & 4 \\ 1 & -2 & 1 \\ 1 & 5 & 2 \end{bmatrix}$, determine an orthogonal basis for rowspace(A).

For problems 21 and 22, determine an orthonormal basis for the subspace of $\mathbf{C}^3$ spanned by the given set of vectors. Make sure that you use the appropriate inner product in $\mathbf{C}^3$.

21. $\{(1 - i, 0, i), (1, 1 + i, 0)\}$.

22. $\{(1 + i, i, 2 - i), (1 + 2i, 1 - i, i)\}$.

For problems 23–25, determine an orthogonal basis for the subspace of $C^0[a, b]$ spanned by the given vectors, for the given interval $[a, b]$.

23. $f_1(x) = 1, f_2(x) = x, f_3(x) = x^2, a = 0, b = 1$.

24. $f_1(x) = 1, f_2(x) = x^2, f_3(x) = x^4, a = -1, b = 1$.

25. $f_1(x) = 1, f_2(x) = \sin x, f_3(x) = \cos x$, $a = -\pi/2, b = \pi/2$.

On $M_2(\mathbf{R})$ define the inner product $<A, B>$ by

$$<A, B> = 5a_{11}b_{11} + 2a_{12}b_{12} + 3a_{21}b_{21} + 5a_{22}b_{22}$$

for all matrices $A = [a_{ij}]$ and $B = [b_{ij}]$. For problems 26 and 27, use this inner product in the Gram–Schmidt procedure to determine an orthogonal basis for the subspace of $M_2(\mathbf{R})$ spanned by the given matrices.

26. $A_1 = \begin{bmatrix} 1 & -1 \\ 2 & 1 \end{bmatrix}$, $A_2 = \begin{bmatrix} 2 & -3 \\ 4 & 1 \end{bmatrix}$.

27. $A_1 = \begin{bmatrix} 0 & 1 \\ 1 & 0 \end{bmatrix}$, $A_2 = \begin{bmatrix} 0 & 1 \\ 1 & 1 \end{bmatrix}$,

$A_3 = \begin{bmatrix} 1 & 1 \\ 1 & 0 \end{bmatrix}$. Also identify the subspace of $M_2(\mathbf{R})$ spanned by A_1, A_2, A_3.

On P_n define the inner product $<p_1, p_2>$ by

$$<p_1, p_2> = a_0 b_0 + a_1 b_1 + \cdots + a_{n-1} b_{n-1},$$

for all polynomials

$$p_1(x) = a_0 + a_1 x + \cdots + a_{n-1} x^{n-1},$$
$$p_2(x) = b_0 + b_1 x + \cdots + b_{n-1} x^{n-1}.$$

For problems 28 and 29, use this inner product to determine an orthogonal basis for the subspace of P_n spanned by the given polynomials.

28. $p_1(x) = 1 - 2x + 2x^2, p_2(x) = 2 - x - x^2$.

29. $p_1(x) = 1 + x^2, p_2(x) = 2 - x + x^3$,

$p_3(x) = 2x^2 - x$.

30. Let $\{\mathbf{u}_1, \mathbf{u}_2, \mathbf{v}\}$ be LI vectors in an inner product space V, and suppose that $\mathbf{u}_1$ and $\mathbf{u}_2$ are orthogonal. Define the vector $\mathbf{u}_3$ in V by

$$\mathbf{u}_3 = \mathbf{v} + \lambda \mathbf{u}_1 + \mu \mathbf{u}_2$$

where λ and μ are scalars. Derive the values of λ and μ such that $\{\mathbf{u}_1, \mathbf{u}_2, \mathbf{u}_3\}$ is an orthogonal basis for the subspace of V spanned by $\{\mathbf{u}_1, \mathbf{u}_2, \mathbf{v}\}$.

31. Let $\{\mathbf{v}_1, \mathbf{v}_2, \dots, \mathbf{v}_k\}$ be an *orthogonal* set of *nonzero* vectors in an inner product space. Show that this set of vectors is necessarily LI. (Hint: Write out the condition for determining LI or LD, and take the inner product with $\mathbf{v}_j$.)

32. Let $\{\mathbf{v}_1, \mathbf{v}_2, \dots, \mathbf{v}_k\}$ be an orthogonal basis in a subspace of an real inner product space V, and let

$$\mathbf{u} = c_1 \mathbf{v}_1 + c_2 \mathbf{v}_2 + \cdots + c_k \mathbf{v}_k.$$

Show that

$$c_i = \frac{\langle \mathbf{u}, \mathbf{u}_i \rangle}{\|\mathbf{u}_i\|^2}.$$

33. The subject of Fourier series is concerned with the representation of a 2π-periodic function f as the following *infinite* linear combination of the set of functions $\{1, \sin nx, \cos nx\}_{n=1}^{\infty}$:

$$f(x) = \frac{1}{2} a_0 + \sum_{n=1}^{\infty} (a_n \cos nx + b_n \sin nx). \quad (35.1)$$

In this problem, we investigate the possibility of performing such a representation.

(a) Use appropriate trigonometric identities, or some form of technology, to verify that the set of functions $\{1, \sin nx, \cos nx\}_{n=1}^{\infty}$ is orthogonal on the interval $[-\pi, \pi]$.

(b) By multiplying (35.1) by $\cos mx$ and integrating over the interval $[-\pi, \pi]$, show that

$$a_0 = \frac{1}{\pi} \int_{-\pi}^{\pi} f(x)\, dx, \qquad a_m = \frac{1}{\pi} \int_{-\pi}^{\pi} f(x) \cos mx\, dx$$

(You may assume that interchange of the infinite summation with the integral is permissible).

(c) Use a similar procedure to show that

$$b_m = \frac{1}{\pi} \int_{-\pi}^{\pi} f(x) \sin mx\, dx.$$

It can be shown that if f is in $C^1(-\pi, \pi)$, then equation (35.1) holds for each $x \in (-\pi, \pi)$. The series appearing on the right-hand side of (35.1) is called the **Fourier series of** f, and the constants in the summation are called the **Fourier coefficients for** f.

(d) Show that the Fourier coefficients for the function $f(x) = x$, $-\pi < x \le \pi$, $f(x + 2\pi) = f(x)$, are

$$a_n = 0, n = 0, 1, \dots,$$

$$b_n = -\frac{2}{n} \cos n\pi, n = 1, 2, \dots$$

and thereby determine the Fourier series of f.

◆ (e) Using some form of technology, sketch the approximations to $f(x) = x$ on the interval $(-\pi, \pi)$ obtained by considering the first 3 terms, first 5 terms, and first 10 terms in the Fourier series for f. What do you conclude?

5.11 SUMMARY OF RESULTS

In this chapter we have derived some basic results in linear algebra regarding vector spaces. These results form the framework for much of linear mathematics. Following are listed some of the highlights of the chapter.

THE DEFINITION OF A VECTOR SPACE

A vector space consists of the following four different components:

1. A set of vectors V.

2. A set of scalars F (either the set of all real numbers $\mathbf{R}$, or the set of all complex numbers $\mathbf{C}$).

3. A rule, +, for adding vectors in V.

4. A rule, ·, for multiplying vectors in V by scalars in F.

Then $(V, F, +, \cdot)$ is a vector space if and only if axioms A1 – A10 of Definition 5.2.1 are satisfied. If F is the set of all real numbers then $(V, \mathbf{R}, +, \cdot)$ is called a *real* vector space, whereas if F is the set of all complex numbers then $(V, \mathbf{C}, +, \cdot)$ is called a *complex* vector space. Since it is usually quite clear what the addition and scalar multiplication operations are, we usually specify a vector space by just giving the set of vectors V. The vector spaces that we have dealt with are the following:

$\mathbf{R}^n$, the (real) vector space of all ordered n-tuples of real numbers.

$\mathbf{C}^n$, the (complex) vector space of all ordered n-tuples of complex numbers.

$M_n(\mathbf{R})$, the (real) vector space of all $n \times n$ matrices with real elements.

$C^k(I)$, the vector space of all real-valued functions that are continuous and have (at least) k continuous derivatives on I.

P_n, the vector space of all polynomials of degree less than n with real coefficients.

SUBSPACES

Usually the vector space, V, that underlies a given problem is known; for example, it may be $\mathbf{R}^n$ or $C^n(I)$. However the solution of a given problem in general only involves a subset of vectors from this vector space. The question that then arises is whether this subset of vectors is itself a vector space under the same operations of addition and scalar multiplication as in V. In order to answer this question we have the basic Theorem 5.3.1, that tells us that *a nonempty subset of a vector space V is a subspace of V if and only if the subset is closed under addition and closed under scalar multiplication.*

SPANNING SETS

A set of vectors $\{\mathbf{v}_1, \mathbf{v}_2, \ldots, \mathbf{v}_k\}$ in a vector space V is said to *span* V if *every* vector in V can be written as a linear combination of $\mathbf{v}_1, \mathbf{v}_2, \ldots, \mathbf{v}_k$, that is, if for any $\mathbf{v} \in V$ there exist scalars $c_1, c_2, \ldots, c_k$ such that

$$\mathbf{v} = c_1\mathbf{v}_1 + c_2\mathbf{v}_2 + \cdots + c_k\mathbf{v}_k.$$

Given a set of vectors $\{\mathbf{v}_1, \mathbf{v}_2, \ldots, \mathbf{v}_k\}$ in a vector space V, we can form the set of *all* vectors that can be written as a linear combination of $\mathbf{v}_1, \mathbf{v}_2, \ldots, \mathbf{v}_k$. This set of vectors is a subspace of V called the *subspace spanned by* $\{\mathbf{v}_1, \mathbf{v}_2, \ldots, \mathbf{v}_k\}$, and denoted $\mathrm{span}\{\mathbf{v}_1, \mathbf{v}_2, \ldots, \mathbf{v}_k\}$. Thus

$$\mathrm{span}\{\mathbf{v}_1, \mathbf{v}_2, \ldots, \mathbf{v}_k\} = \{\mathbf{v} \in V : \mathbf{v} = c_1\mathbf{v}_1 + c_2\mathbf{v}_2 + \cdots + c_k\mathbf{v}_k\}.$$

LINEAR DEPENDENCE AND LINEAR INDEPENDENCE

Let $\{\mathbf{v}_1, \mathbf{v}_2, \ldots, \mathbf{v}_k\}$ be a set of vectors in a vector space V, and consider the vector equation

$$c_1\mathbf{v}_1 + c_2\mathbf{v}_2 + \cdots + c_k\mathbf{v}_k = \mathbf{0}. \qquad (5.11.1)$$

Clearly this equation will hold if $c_1 = c_2 = \cdots = c_k = 0$. The question of interest is whether there are nonzero values of the scalars $c_1, c_2, \ldots, c_k$ such that (5.11.1) holds. This leads to the following two ideas:

Linear dependence: There exist scalars $c_1, c_2, \ldots, c_k$, *not all zero*, such that (5.11.1) holds.

Linear Independence: The *only* values of the scalars $c_1, c_2, \ldots, c_k$ such that (5.11.1) holds are $c_1 = c_2 = \cdots = c_k = 0$.

To determine whether a set of vectors is LD or LI we usually have to use (5.11.1). However, if the vectors are from $\mathbf{R}^n$, then we can use Corollary 5.5.2, whereas for vectors in $C^{n-1}(I)$ the Wronskian can be useful.

BASES AND DIMENSION

A LI set of vectors that spans a vector space V is called a *basis* for V. If $\{\mathbf{v}_1, \mathbf{v}_2, \ldots, \mathbf{v}_k\}$ is a basis for V, then any vector in V can be written uniquely as

$$\mathbf{v} = c_1\mathbf{v}_1 + c_2\mathbf{v}_2 + \cdots + c_k\mathbf{v}_k,$$

for appropriate values of the scalars $c_1, c_2, \ldots, c_k$.

1.   All bases in a finite-dimensional vector space V contain the same number of vectors, and this number is called the *dimension* of V, denoted by $\dim[V]$.

2.   We can view the dimension of a finite-dimensional vector space V in two different ways. First, it gives the minimum number of vectors that span V. Alternatively, we can regard $\dim[V]$ as determining the maximum number of vectors that a LI set can contain.

3.   If $\dim[V] = n$, then *any* LI set of n vectors in V is a basis for V.

INNER PRODUCT SPACES

An inner product is a mapping that associates with any two vectors, $\mathbf{u}, \mathbf{v}$, in a vector space V a scalar that we denote by $<\mathbf{u}, \mathbf{v}>$. This mapping must satisfy the properties given in Definition 5.9.4. The main reason for introducing the idea of an inner product is that it enables us to extend the familiar idea of orthogonality of vectors in $\mathbf{R}^3$ to a general vector space. Thus $\mathbf{u}$ and $\mathbf{v}$ are said to be orthogonal in an inner product space if and only if

$$<\mathbf{u}, \mathbf{v}> = 0.$$

THE GRAM–SCHMIDT ORTHOGONALIZATION PROCEDURE

The Gram–Schmidt procedure is a process that takes a LI set of vectors $\{\mathbf{v}_1, \mathbf{v}_2, \ldots, \mathbf{v}_k\}$ in an inner product space V and returns an *orthogonal* basis $\{\mathbf{u}_1, \mathbf{u}_2, \ldots, \mathbf{u}_k\}$ for $\text{span}\{\mathbf{v}_1, \mathbf{v}_2, \ldots, \mathbf{v}_k\}$.

Linear Transformations and the Eigenvalue/Eigenvector Problem

In the preceding chapter, we began building a general framework for studying linear problems. We can view the vector space as giving us the set from which the unknowns in a problem come from, and, for certain homogeneous problems, we have seen that the solution set is a subspace of that vector space. Consequently, all solutions to such a problem can be generated once a basis for the solution set has been determined. In developing these ideas, we have been restricted in the problems that we can use to illustrate the concepts. Basically, we have the case of second-order linear homogeneous DE and linear homogeneous algebraic systems. Our overall aim, however, has been to build a framework that can be applied to obtain results about other types of linear problems. What is missing is a general method of discussing linear equations. In order to motivate the direction we now take, consider the DE

$$y'' + y = 0 \qquad\qquad (6.0.1)$$

and the associated "mapping of functions" T defined by

$$T(y) = y'' + y.$$

Given a function y, T maps y to the function $y'' + y$. For example,

$$T(x^2) = (x^2)'' + (x^2) = 2 + x^2,$$

$$T(\ln x) = (\ln x)'' + (\ln x) = -\frac{1}{x^2} + \ln x.$$

In terms of the mapping T, the solution set S, to the DE (6.0.1) consists of all those functions y that are mapped to the zero function

$$S = \{y : T(y) = 0\}.$$

Now let A be an $m \times n$ matrix, and define a corresponding mapping T by

$$T(\mathbf{x}) = A\mathbf{x},$$

for all $\mathbf{x}$ in $\mathbf{R}^n$. In terms of this mapping the solution set to the homogeneous linear system $A\mathbf{x} = 0$ consists of all those vectors $\mathbf{x}$ with the property that $T(\mathbf{x}) = 0$. The point that we are making is that the two homogeneous linear problems we have studied to this point in the text are both special cases of the general problem of finding all vectors $\mathbf{v}$ with the property that $T(\mathbf{v}) = 0$, where T is a mapping from a vector space V into a vector space W. But where is the linearity? The answer is in the mapping T. Both of the specific mappings that we have considered satisfy the linearity properties

$$T(\mathbf{u} + \mathbf{v}) = T(\mathbf{u}) + T(\mathbf{v}) \text{ for all } \mathbf{u}, \mathbf{v} \in V,$$

$$T(c\mathbf{v}) = cT(\mathbf{v}) \text{ for all } \mathbf{v} \in V \text{ and all scalars } c.$$

Any mapping that satisfies these properties is called a linear function, or a linear transformation. We will see in this chapter that the general linear framework that we have been aiming for is indeed completed once an appropriate linear transformation is defined on the vector space of unknowns in our problem. We will show that the set of all solutions to the corresponding homogeneous linear problem $T(\mathbf{v}) = 0$ is a subspace of the vector space from which we are mapping. Consequently, once we have determined the dimension of that solution space, we will know how many LI solutions to $T(\mathbf{v}) = 0$ are required to determine all of its solutions.

6.1   DEFINITION OF A LINEAR TRANSFORMATION

We begin with a precise definition of a mapping between two vector spaces.

> *Definition 6.1.1:*   Let V and W be vector spaces. A **mapping** T from V into W is a rule that associates with each vector $\mathbf{v}$ in V precisely one vector $\mathbf{w} = T(\mathbf{v})$ in W. We denote such a mapping by $T : V \rightarrow W$.

Example 6.1.1   The following are examples of mappings between vector spaces:

$T : M_n(\mathbf{R}) \rightarrow M_n(\mathbf{R})$ defined by $T(A) = A^T$.

$T : M_n(\mathbf{R}) \rightarrow \mathbf{R}$ defined by $T(A) = \det(A)$.

$T : C^0[a, b] \rightarrow \mathbf{R}$ defined by $T(f) = \displaystyle\int_a^b f(x)dx.$

$T : P_2 \rightarrow P_3$ defined by $T(a_0 + a_1x) = 2a_0 + a_1 + (a_0 + 3a_1)x + 4a_1x^2$ ❑

The basic operations of addition and scalar multiplication in a vector space V enable us to form only linear combinations of vectors in V. In keeping with the aim of studying linear mathematics, it is natural to restrict attention to mappings that preserve such linear combinations of vectors in the sense that

$$T(c_1 \mathbf{v}_1 + c_2 \mathbf{v}_2) = c_1 T(\mathbf{v}_1) + c_2 T(\mathbf{v}_2),$$

for all vectors $\mathbf{v}_1$, $\mathbf{v}_2$ in V and all scalars c_1, c_2. The most general type of mapping that does this is called a linear transformation.

Definition 6.1.2:   Let V and W be vector spaces.[1] A mapping $T : V \to W$ is called a **linear transformation** from V into W if it satisfies the following properties:

(1) $T(\mathbf{u} + \mathbf{v}) = T(\mathbf{u}) + T(\mathbf{v})$ for *all* $\mathbf{u}$, $\mathbf{v} \in V$.

(2) $T(c\mathbf{v}) = cT(\mathbf{v})$ for *all* $\mathbf{v} \in V$ and *all* scalars c.

We refer to these properties as the **linearity properties.**

A mapping $T : V \to W$ that does not satisfy Definition 6.1.2 is called a **nonlinear transformation.**

Theorem 6.1.1:   A mapping $T : V \to W$ satisfies

$$T(c_1 \mathbf{v}_1 + c_2 \mathbf{v}_2) = c_1 T(\mathbf{v}_1) + c_2 T(\mathbf{v}_2), \tag{6.1.1}$$

for all $\mathbf{v}_1$, $\mathbf{v}_2$ in V and all scalars c_1, c_2, if and only if T is a linear transformation.

PROOF   Suppose that T satisfies equation (6.1.1). Then property (1) of Definition 6.1.2 arises as the special case $c_1 = c_2 = 1$, $\mathbf{v}_1 = \mathbf{u}$, $\mathbf{v}_2 = \mathbf{v}$. Further, property (2) of Definition 6.1.2 is the special case $c_1 = c$, $\mathbf{v}_1 = \mathbf{v}$, $c_2 = 0$. Consequently, T is a linear transformation.
      Conversely, if T is a linear transformation, then, using properties (1) and (2) from Definition 6.1.2 yields

$$T(c_1 \mathbf{v}_1 + c_2 \mathbf{v}_2) = T(c_1 \mathbf{v}_1) + T(c_2 \mathbf{v}_2) = c_1 T(\mathbf{v}_1) + c_2 T(\mathbf{v}_2),$$

so that equation (6.1.1) is satisfied.   ■

Repeated application of the linearity properties can now be used to establish that if $T : V \to W$ is a linear transformation, then for all $\mathbf{v}_1$, $\mathbf{v}_2$, ..., $\mathbf{v}_k$ in V and all scalars c_1, c_2, ..., c_k,

$$T(c_1 \mathbf{v}_1 + c_2 \mathbf{v}_2 + \cdots + c_k \mathbf{v}_k) = c_1 T(\mathbf{v}_1) + c_2 T(\mathbf{v}_2) + \cdots + c_k T(\mathbf{v}_k). \tag{6.1.2}$$

In particular, if $\{\mathbf{v}_1, \mathbf{v}_2, ..., \mathbf{v}_k\}$ is a basis for V, then any vector in V can be written as

$$\mathbf{v} = c_1 \mathbf{v}_1 + c_2 \mathbf{v}_2 + \cdots + c_k \mathbf{v}_k,$$

for appropriate scalars c_1, c_2, ..., c_k, so that from equation (6.1.2),

---

[1] V and W must either be both real vector spaces or both complex vector spaces in order that we use the same scalars in both spaces.

$$T(\mathbf{v}) = T(c_1\mathbf{v}_1 + c_2\mathbf{v}_2 + \cdots + c_k\mathbf{v}_k) = c_1T(\mathbf{v}_1) + c_2T(\mathbf{v}_2) + \cdots + c_kT(\mathbf{v}_k).$$

Consequently, if we know $T(\mathbf{v}_1)$, $T(\mathbf{v}_2)$, ..., $T(\mathbf{v}_k)$, then we know how every vector in V transforms. This once more emphasizes the importance of the basis in studying vector spaces.

Example 6.1.2   Define $T : C^1(I) \to C^0(I)$ by $T(f) = f'$. Verify that T is a linear transformation.

Solution   If f and g are in $C^1(I)$ and c is a real number, then

$$T(f + g) = (f + g)' = f' + g' = T(f) + T(g)$$

and

$$T(cf) = (cf)' = cf' = cT(f).$$

Thus, T satisfies both properties of Definition 6.1.2 and is a linear transformation.

Example 6.1.3:   Define $T : C^2(I) \to C^0(I)$ by $T(y) = y'' + y$. Verify that T is a linear transformation.

Solution   If y_1 and y_2 are in $C^2(I)$, then

$$
\begin{aligned}
T(y_1 + y_2) &= (y_1 + y_2)'' + (y_1 + y_2) \\
&= y_1'' + y_2'' + y_1 + y_2 \\
&= y_1'' + y_1 + y_2'' + y_2 \\
&= T(y_1) + T(y_2).
\end{aligned}
$$

Furthermore, if c is an arbitrary real number, then

$$
\begin{aligned}
T(cy_1) &= (cy_1)'' + (cy_1) = cy_1'' + cy_1 \\
&= c(y_1'' + y_1) = cT(y_1).
\end{aligned}
$$

Consequently, T is a linear transformation.

Example 6.1.4   Define $S : M_n(\mathbf{R}) \to M_n(\mathbf{R})$ by $S(A) = A - A^T$. Verify that S is a linear transformation.

Solution   Let A and B be any two matrices in $M_n(\mathbf{R})$. Then, using the properties of the transpose,

$$S(A + B) = (A + B) - (A + B)^T = (A - A^T) + (B - B^T) = S(A) + S(B)$$

and

$$S(cA) = (cA) - (cA)^T = c(A - A^T) = cS(A),$$

for any real number c. Since both conditions of Definition 6.1.2 are satisfied, it follows that S is a linear transformation.

Example 6.1.5   Let $T : P_3 \to P_3$ be the *linear transformation* satisfying

$$T(1) = 2 - 3x, \quad T(x) = 2x + 5x^2, \quad T(x^2) = 3 - x + x^2.$$

Determine $T(a_0 + a_1x + a_2x^2)$ for all vectors in P_3.

Solution   Since we have been given the transform of the (standard) basis $\{1, x, x^2\}$ for P_3, we can determine the transform of all vectors in P_3. Using the linearity properties in Definition 6.1.2 it follows that

$$T(a_0 + a_1 x + a_2 x^2) = T(a_0) + T(a_1 x) + T(a_2 x^2)$$

$$= a_0 T(1) + a_1 T(x) + a_2 T(x^2)$$

$$= a_0(2 - 3x) + a_1(2x + 5x^2) + a_2(3 - x + x^2)$$

$$= 2a_0 + 3a_2 + (-3a_0 + 2a_1 - a_2)x + (5a_1 + a_2)x^2. \qquad \square$$

The next theorem lists two basic properties of linear transformations. In this theorem, we distinguish the zero vector in V, denoted $\mathbf{0}_V$, from the zero vector in W, denoted $\mathbf{0}_W$.

Theorem 6.1.2:   Let $T : V \rightarrow W$ be a linear transformation. Then

1. $T(\mathbf{0}_V) = \mathbf{0}_W$,

2. $T(-\mathbf{v}) = -T(\mathbf{v})$ for all $\mathbf{v} \in V$.

PROOF

(1) If $\mathbf{v}$ is any vector in V, then $0\mathbf{v} = \mathbf{0}_V$. Consequently,

$$T(\mathbf{0}_V) = T(0\mathbf{v}) = 0T(\mathbf{v}) = \mathbf{0}_W.$$

(2) Since V is a vector space, we know that $-\mathbf{v} = -1\mathbf{v}$. Consequently,

$$T(-\mathbf{v}) = T(-1\mathbf{v}) = -1T(\mathbf{v}) = -T(\mathbf{v}). \qquad \blacksquare$$

LINEAR TRANSFORMATIONS FROM R^n TO R^m

Linear transformations between the vector spaces $\mathbf{R}^n$ and $\mathbf{R}^m$ play a very fundamental role in linear algebra and its applications. We now investigate some of their properties.

Example 6.1.6   Define $T : \mathbf{R}^3 \rightarrow \mathbf{R}^2$ as follows: If $\mathbf{x} = (x_1, x_2, x_3)$, then

$$T(\mathbf{x}) = (2x_1 + x_2, x_1 + x_2 - x_3).$$

Verify that T is a linear transformation from $\mathbf{R}^3$ to $\mathbf{R}^2$.

Solution   Let $\mathbf{x} = (x_1, x_2, x_3)$ and $\mathbf{y} = (y_1, y_2, y_3)$ be arbitrary vectors in $\mathbf{R}^3$. Then, using vector addition in $\mathbf{R}^3$, we have

$$\mathbf{x} + \mathbf{y} = (x_1 + y_1, x_2 + y_2, x_3 + y_3).$$

Consequently,

$$T(\mathbf{x} + \mathbf{y}) = T((x_1 + y_1, x_2 + y_2, x_3 + y_3))$$

$$= (2(x_1 + y_1) + (x_2 + y_2), (x_1 + y_1) + (x_2 + y_2) - (x_3 + y_3))$$

$$= ((2x_1 + x_2) + (2y_1 + y_2), (x_1 + x_2 - x_3) + (y_1 + y_2 - y_3))$$

$$= (2x_1 + x_2, x_1 + x_2 - x_3) + (2y_1 + y_2, y_1 + y_2 - y_3)$$

$$= T(\mathbf{x}) + T(\mathbf{y}).$$

Further, if c is any real number, then

$$c\mathbf{x} = (cx_1, cx_2, cx_3),$$

so that

$$T(c\mathbf{x}) = T((cx_1, cx_2, cx_3)) = (2cx_1 + cx_2, cx_1 + cx_2 - cx_3)$$
$$= c(2x_1 + x_2, x_1 + x_2 - x_3)$$
$$= cT(\mathbf{x}).$$

Hence Properties 1 and 2 of Definition 6.1.2 are satisfied and so T is a linear transformation from $\mathbf{R}^3$ to $\mathbf{R}^2$.   ❑

Our next theorem introduces one way in which linear transformations from $\mathbf{R}^n$ to $\mathbf{R}^m$ arise.

Theorem 6.1.3:   Let A be an $m \times n$ real matrix, and define $T : \mathbf{R}^n \to \mathbf{R}^m$ by $T(\mathbf{x}) = A\mathbf{x}$. Then T is a linear transformation.

PROOF   We need only verify the two linearity properties in Definition 6.1.2. Let $\mathbf{x}$ and $\mathbf{y}$ be arbitrary vectors in $\mathbf{R}^n$, and let c be an arbitrary real number. Then

$$T(\mathbf{x} + \mathbf{y}) = A(\mathbf{x} + \mathbf{y}) = A\mathbf{x} + A\mathbf{y} = T(\mathbf{x}) + T(\mathbf{y}),$$
$$T(c\mathbf{x}) = A(c\mathbf{x}) = cA\mathbf{x} = cT(\mathbf{x}). \qquad \blacksquare$$

A linear transformation $T : \mathbf{R}^n \to \mathbf{R}^m$ defined by $T(\mathbf{x}) = A\mathbf{x}$ where A is an $m \times n$ matrix is called a **matrix transformation**.

Example 6.1.7 Determine the matrix transformation $T : \mathbf{R}^3 \to \mathbf{R}^2$ if

$$A = \begin{bmatrix} 2 & 1 & 0 \\ 1 & 1 & -1 \end{bmatrix}.$$

Solution

$$T(\mathbf{x}) = A\mathbf{x} = \begin{bmatrix} 2 & 1 & 0 \\ 1 & 1 & -1 \end{bmatrix} \begin{bmatrix} x_1 \\ x_2 \\ x_2 \end{bmatrix} = \begin{bmatrix} 2x_1 + x_2 \\ x_1 + x_2 - x_3 \end{bmatrix},$$

which we write as

$$T(x_1, x_2, x_3) = (2x_1 + x_2, x_1 + x_2 - x_3). \qquad ❑$$

If we compare the previous two examples, we see that they contain the same linear transformation, but defined in two different ways. Certainly it was easier to establish Theorem 6.1.3 than it was to verify the linearity in Example 6.1.6. This leads to the question as to whether we can always describe a linear transformation from $\mathbf{R}^n$ to $\mathbf{R}^m$ as a matrix transformation. The following theorem answers this in the affirmative:

Theorem 6.1.4:   Let $T : \mathbf{R}^n \to \mathbf{R}^m$ be a linear transformation. Then T is described by the matrix transformation

$$T(\mathbf{x}) = A\mathbf{x},$$

where

$$A = [T(\mathbf{e}_1), T(\mathbf{e}_2), \ldots, T(\mathbf{e}_n)]$$

and $\mathbf{e}_1, \mathbf{e}_2, \ldots, \mathbf{e}_n$ denote the standard basis vectors in $\mathbf{R}^n$.

PROOF   Any vector $\mathbf{x} = (x_1, x_2, \ldots, x_n)$ in $\mathbf{R}^n$ can be expressed in terms of the standard basis as

$$\mathbf{x} = x_1\mathbf{e}_1 + x_2\mathbf{e}_2 + \cdots + x_n\mathbf{e}_n.$$

Thus, since $T: \mathbf{R}^n \to \mathbf{R}^m$ is a linear transformation, properties (1) and (2) of Definition 6.1.2 imply that

$$T(\mathbf{x}) = T(x_1\mathbf{e}_1 + x_2\mathbf{e}_2 + \cdots + x_n\mathbf{e}_n)$$

$$= x_1 T(\mathbf{e}_1) + x_2 T(\mathbf{e}_2) + \cdots + x_n T(\mathbf{e}_n).$$

Consequently, $T(\mathbf{x})$ is a linear combination of the transformed vectors $T(\mathbf{e}_1)$, $T(\mathbf{e}_2)$, ..., $T(\mathbf{e}_n)$, each of which is a vector in $\mathbf{R}^m$. Therefore, the preceding expression for $T(\mathbf{x})$ can be written as the matrix product

$$T(\mathbf{x}) = [T(\mathbf{e}_1), T(\mathbf{e}_2), \cdots, T(\mathbf{e}_n)] \begin{bmatrix} x_1 \\ x_2 \\ \vdots \\ x_n \end{bmatrix} = A\mathbf{x},$$

where $A = [T(\mathbf{e}_1), T(\mathbf{e}_2), \ldots, T(\mathbf{e}_n)]$. ∎

Definition 6.1.3:   If $T: \mathbf{R}^n \to \mathbf{R}^m$ is a linear transformation, then the $m \times n$ matrix

$$A = [T(\mathbf{e}_1), T(\mathbf{e}_2), \cdots, T(\mathbf{e}_n)]$$

is called the **matrix of T**.

Example 6.1.8   Determine the matrix of the linear transformation $T: \mathbf{R}^4 \to \mathbf{R}^3$ defined by

$$T(x_1, x_2, x_3, x_4) = (2x_1 + 3x_2 + x_4, \ 5x_1 + 9x_3 - x_4, \ 4x_1 + 2x_2 - x_3 + 7x_4). \qquad (6.1.3)$$

Solution   The standard basis vectors in $\mathbf{R}^4$ are

$$\mathbf{e}_1 = (1, 0, 0, 0), \ \mathbf{e}_2 = (0, 1, 0, 0), \ \mathbf{e}_3 = (0, 0, 1, 0), \ \mathbf{e}_4 = (0, 0, 0, 1).$$

Consequently, from (6.1.3),

$$T(\mathbf{e}_1) = (2, 5, 4), \ T(\mathbf{e}_2) = (3, 0, 2), \ T(\mathbf{e}_3) = (0, 9, -1), \ T(\mathbf{e}_4) = (1, -1, 7),$$

so that the matrix of the transformation is

$$A = [T(\mathbf{e}_1), T(\mathbf{e}_2), T(\mathbf{e}_3), T(\mathbf{e}_4)] = \begin{bmatrix} 2 & 3 & 0 & 1 \\ 5 & 0 & 9 & -1 \\ 4 & 2 & -1 & 7 \end{bmatrix}.$$

Example 6.1.9   Determine the linear transformation $T: \mathbf{R}^2 \to \mathbf{R}^3$ satisfying

$$T(1, 0) = (2, 3, -1), \quad T(0, 1) = (5, -4, 7).$$

Solution   The matrix of the transformation is

$$A = [T(\mathbf{e}_1), T(\mathbf{e}_2)] = \begin{bmatrix} 2 & 5 \\ 3 & -4 \\ -1 & 7 \end{bmatrix},$$

so that

$$T(\mathbf{x}) = \begin{bmatrix} 2 & 5 \\ 3 & -4 \\ -1 & 7 \end{bmatrix}\begin{bmatrix} x_1 \\ x_2 \end{bmatrix} = \begin{bmatrix} 2x_1 + 5x_2 \\ 3x_1 - 4x_2 \\ -x_1 + 7x_2 \end{bmatrix},$$

which we write as

$$T(x_1, x_2, x_3) = (2x_1 + 5x_2, 3x_1 - 4x_2, -x_1 + 7x_2). \qquad \square$$

Finally in this section, consider the mapping $T : \mathbf{R}^2 \to \mathbf{R}^2$, which rotates each point in the plane through an angle θ in the counter-clockwise direction, where $0 \le \theta < 2\pi$. As Figure 6.1.1 illustrates, T is a *linear* transformation. In order to determine the

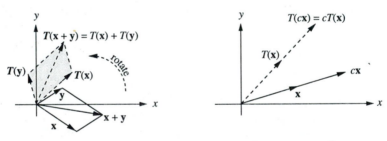

Figure 6.1.1 Rotation in the plane satisfies the basic linearity properties and therefore is a linear transformation.

matrix of this transformation, all we need do is obtain $T(\mathbf{e}_1)$, $T(\mathbf{e}_2)$. From Figure 6.1.2, we see that

$$T(\mathbf{e}_1) = (\cos \theta, \sin \theta), \quad T(\mathbf{e}_2) = (-\sin \theta, \cos \theta).$$

Consequently, the matrix of the transformation is

$$R(\theta) = \begin{bmatrix} \cos \theta & -\sin \theta \\ \sin \theta & \cos \theta \end{bmatrix}.$$

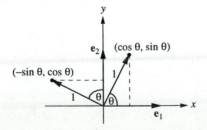

Figure 6.1.2 Determining the transformation matrix corresponding to a rotation in the *xy*-plane.

EXERCISES 6.1

For problems 1–8, show that the given mapping is a linear transformation

1. $T : \mathbf{R}^2 \to \mathbf{R}^2$ defined by
$$T(x_1, x_2) = (x_1 + 2x_2, 2x_1 - x_2).$$

2. $T : \mathbf{R}^2 \to \mathbf{R}^2$ defined by
$$T(x_1, x_2, x_3) = (x_1 + 3x_2 + x_3, x_1 - x_2).$$

3. $T : C^2(I) \to C^0(I)$ defined by
$$T(y) = y'' - 16y.$$

4. $T : C^2(I) \to C^0(I)$ defined by
$$T(y) = y'' + a_1 y' + a_2 y,$$
where a_1 and a_2 are functions defined on I.

5. $T : C^0[a, b] \to \mathbf{R}$ defined by
$$T(f) = \int_a^b f(x)\, dx.$$

6. $T : M_n(\mathbf{R}) \to M_n(\mathbf{R})$ defined by
$$T(A) = AB - BA,$$
where B is a fixed $n \times n$ matrix.

7. $S : M_n(\mathbf{R}) \to M_n(\mathbf{R})$ defined by
$$S(A) = A + A^T.$$

8. $T : M_n(\mathbf{R}) \to \mathbf{R}$ defined by $T(A) = \text{tr}(A)$, where $\text{tr}(A)$ denotes the trace of A.

For problems 9 and 10, show that the given mapping is a nonlinear transformation.

9. $T : \mathbf{R}^2 \to \mathbf{R}^2$ defined by
$$T(x_1, x_2) = (x_1 + x_2, 2).$$

10. $T : M_2(\mathbf{R}) \to \mathbf{R}$ by $T(A) = \det(A)$.

For problems 11–14, determine the matrix of the given transformation $T : \mathbf{R}^n \to \mathbf{R}^m$.

11. $T(x_1, x_2) = (3x_1 - 2x_2, x_1 + 5x_2)$.

12. $T(x_1, x_2) = (x_1 + 3x_2, 2x_1 - 7x_2, x_1)$.

13. $T(x_1, x_2, x_3) = (x_1 - x_2 + x_3, x_3 - x_1)$.

14. $T(x_1, x_2, x_3) = x_1 + 5x_2 - 3x_3$.

For problems 15–17, determine the linear transformation $T : \mathbf{R}^n \to \mathbf{R}^m$ that has the given matrix.

15. $A = \begin{bmatrix} 1 & 3 \\ -4 & 7 \end{bmatrix}$.

16. $A = \begin{bmatrix} 2 & -1 & 5 \\ 3 & 1 & -2 \end{bmatrix}$.

17. $A = \begin{bmatrix} 2 & 2 & -3 \\ 4 & -1 & 2 \\ 5 & 7 & -8 \end{bmatrix}$.

18. Let V be a real inner product space, and let $\mathbf{u}$ be a fixed (nonzero) vector in V. Define $T : V \to \mathbf{R}$ by
$$T(\mathbf{v}) = <\mathbf{u}, \mathbf{v}>$$
Use properties of the inner product to show that T is a linear transformation.

19. (a) Let $\mathbf{v}_1 = (1, 1)$ and $\mathbf{v}_2 = (1, -1)$. Show that $\{\mathbf{v}_1, \mathbf{v}_2\}$ is a basis for $\mathbf{R}^2$.

(b) Let $T : \mathbf{R}^2 \to \mathbf{R}^2$ be the *linear* transformation satisfying
$$T(\mathbf{v}_1) = (2, 3), \quad T(\mathbf{v}_2) = (-1, 1),$$
where $\mathbf{v}_1$ and $\mathbf{v}_2$ are the basis vectors given in (a). Find $T(x_1, x_2)$ for an arbitrary vector (x_1, x_2) in $\mathbf{R}^2$. What is $T(4, -2)$?

20. Let $T : P_3 \to P_3$ be the *linear* transformation satisfying
$$T(1) = x + 1, \quad T(x) = x^2 - 1, \quad T(x^2) = 3x + 2.$$
Determine $T(ax^2 + bx + c)$, where a, b, and c are arbitrary real numbers.

21. Let $T : V \to V$ be a linear transformation, and suppose that
$$T(2\mathbf{v}_1 + 3\mathbf{v}_2) = \mathbf{v}_1 + \mathbf{v}_2, \quad T(\mathbf{v}_1 + \mathbf{v}_2) = 3\mathbf{v}_1 - \mathbf{v}_2.$$
Find $T(\mathbf{v}_1)$ and $T(\mathbf{v}_2)$.

22. Let $T : P_3 \to P_3$ be the *linear* transformation satisfying:
$$T(x^2 - 1) = x^2 + x - 3, \quad T(2x) = 4x,$$
$$T(3x + 2) = 2(x + 3).$$
Find $T(1)$, $T(x)$, and $T(x^2)$, and hence, show that
$$T(ax^2 + bx + c) = ax^2 - (a - 2b + 2c)x + 3c,$$
where a, b, and c are arbitrary real numbers.

23. Let $\{v_1, v_2\}$ be a basis for the vector space V. If $T : V \to V$ is the linear transformation satisfying

$$T(v_1) = 3v_1 - v_2, \quad T(v_2) = v_1 + 2v_2,$$

find $T(v)$ for an arbitrary vector in V.

Let $T_1 : V \to W$ and $T_2 : V \to W$ be linear transformations, and let c be a scalar. We define the **sum** $T_1 + T_2$ and the **scalar product** cT_1 by

$$(T_1 + T_2)(v) = T_1(v) + T_2(v), \quad \text{for all } v \in V,$$

$$(cT_1)(v) = cT_1(v) \qquad \text{for all } v \in V.$$

The remaining problems in this section consider the properties of these mappings.

24. Verify that $T_1 + T_2$ and cT_1 are *linear transformations*.

25. Let $T_1 : \mathbf{R} \to \mathbf{R}$ and $T_2 : \mathbf{R} \to \mathbf{R}$ be the linear transformations with matrices

$$A = \begin{bmatrix} 3 & 1 \\ -1 & 2 \end{bmatrix}, B = \begin{bmatrix} 2 & 5 \\ 3 & -4 \end{bmatrix}.$$

Find $T_1 + T_2$ and cT_1.

26. Let $T_1 : \mathbf{R}^n \to \mathbf{R}^m$ and $T_2 : \mathbf{R}^n \to \mathbf{R}^m$ be the linear transformations with matrices A and B respectively. Show that $T_1 + T_2$ and cT_1 are the linear transformations with matrices $A + B$ and cA respectively.

27. Let V and W be vector spaces, and let $L(U, V)$ denote the set of all linear transformations from V into W. Verify that $L(U, V)$ together with the operations of addition and scalar multiplication just defined for linear transformations is a vector space.

6.2*   TRANSFORMATIONS OF $\mathbf{R}^2$

To gain some geometric insight into linear transformations, we consider the particular case of linear transformations from $\mathbf{R}^2$ to $\mathbf{R}^2$. Any such transformation is called a **transformation of $\mathbf{R}^2$**. Geometrically, the action of a transformation of $\mathbf{R}^2$ can be represented by its effect on an arbitrary point in the Cartesian plane. We first establish that transformations of $\mathbf{R}^2$ map lines into lines. Recall that the parametric equations of the line passing through the point (x_1, y_1) in the direction of the vector with components (a, b) are

$$x = x_1 + at, \quad y = y_1 + bt,$$

which can be written as

$$\mathbf{x} = \mathbf{x}_1 + t\mathbf{v}, \tag{6.2.1}$$

where $\mathbf{x}_1 = (x_1, y_1)$, $\mathbf{v} = (a, b)$. The transformation $T(\mathbf{x}) = A\mathbf{x}$ therefore transforms the points along this line into

$$T(\mathbf{x}) = A(\mathbf{x}_1 + t\mathbf{v}) = A\mathbf{x}_1 + tA\mathbf{v}. \tag{6.2.2}$$

Consequently, the transformed points lie along the line with parametric equations

$$\mathbf{x} = \mathbf{y}_1 + t\mathbf{w},$$

---

* This section can be omitted without loss of continuity.

where $\mathbf{y}_1 = A\mathbf{x}_1$ and $\mathbf{w} = A\mathbf{v}$. This is illustrated in Figure 6.2.1. It follows that if we

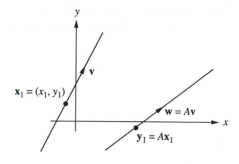

Figure 6.2.1 A transformation of $\mathbf{R}^2$ maps the line with vector parametric equation $\mathbf{x} = \mathbf{x}_1 + t\mathbf{v}$ into the line $\mathbf{x} = A\mathbf{x}_1 + tA\mathbf{v}$.

know how two points in $\mathbf{R}^2$ transform, then we can determine the transformation of all points along the line joining those two points. Further, the linearity properties (1) and (2) of Definition 6.1.2 are the statement that the parallelogram law for vector addition is preserved under the transformation. This is illustrated in Figure 6.2.2.

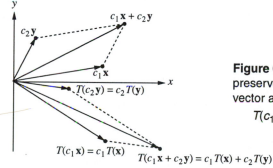

Figure 6.2.2 A transformation of $\mathbf{R}^2$ preserves the parallelogram law of vector addition in the sense that

$$T(c_1\mathbf{x} + c_2\mathbf{y}) = c_1 T(\mathbf{x}) + c_2 T(\mathbf{y}).$$

From equations (6.2.1) and (6.2.2), we see that *any* line with direction vector $\mathbf{v}$ is mapped into a line with direction vector $A\mathbf{v}$, so that parallel lines are mapped to parallel lines by a transformation of $\mathbf{R}^2$. In describing specific transformations of $\mathbf{R}^2$, it is often useful to determine the effect of such a transformation on the points lying inside a rectangle. Due to the fact that a transformation of $\mathbf{R}^2$ maps parallel lines into parallel lines, it follows that the transform of a rectangle will be the parallelogram whose vertices are the transforms of the vertices of the rectangle. This is illustrated in Figure 6.2.3.

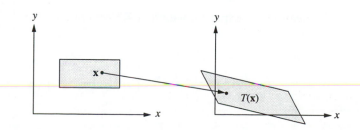

Figure 6.2.3 Transformations of $\mathbf{R}^2$ map rectangles into parallelograms.

SIMPLE TRANSFORMATIONS

We next introduce some simple transformations of $\mathbf{R}^2$ and their geometrical interpretation. We then show how more complicated transformations can be considered as a combination of these simple ones.

I: Reflections   Consider the transformation of $\mathbf{R}^2$ with matrix

$$R_x = \begin{bmatrix} 1 & 0 \\ 0 & -1 \end{bmatrix}.$$

If $\mathbf{x} = (x, y)$ is an arbitrary point in $\mathbf{R}^2$, then

$$T(\mathbf{x}) = R_x\mathbf{x} = \begin{bmatrix} 1 & 0 \\ 0 & -1 \end{bmatrix}\begin{bmatrix} x \\ y \end{bmatrix} = \begin{bmatrix} x \\ -y \end{bmatrix}.$$

Thus,

$$T(x, y) = (x, -y).$$

We see that each point in the Cartesian plane is mapped to its mirror image in the x-axis. Hence, R_x describes a reflection in the x-axis. This can be illustrated by considering the rectangle in the xy-plane through the points $(0, 0)$, $(0, a)$, and $(b, 0)$. In order to determine the transform of this rectangle, all we need do is determine the transform of the vertices. (See Figure 6.2.4.)

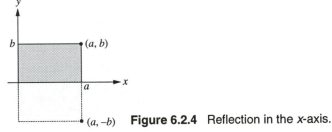

Figure 6.2.4   Reflection in the x-axis.

We leave it as an exercise to show that the transformation of $\mathbf{R}^2$ with matrix

$$R_y = \begin{bmatrix} -1 & 0 \\ 0 & 1 \end{bmatrix}$$

describes a reflection in the y-axis.

Now consider the transformation of $\mathbf{R}^2$ with matrix $R_{xy} = \begin{bmatrix} 0 & 1 \\ 1 & 0 \end{bmatrix}$. If $\mathbf{x} = (x, y)$,

then

$$T(\mathbf{x}) = R_{xy}\mathbf{x} = \begin{bmatrix} 0 & 1 \\ 1 & 0 \end{bmatrix}\begin{bmatrix} x \\ y \end{bmatrix} = \begin{bmatrix} y \\ x \end{bmatrix}.$$

Thus,

$$T(x, y) = (y, x).$$

We see that the x- and y- coordinates have been interchanged. Geometrically, all points along the line $y = x$ remain fixed, and all other points are transformed into their mirror image in the line $y = x$. Consequently, R_{xy} describes a reflection in the line $y = x$. This is illustrated in Figure 6.2.5.

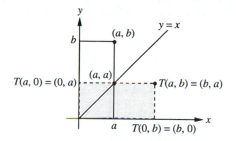

Figure 6.2.5 Reflection in the line $y = x$. The rectangle with vertices $(0, 0)$, $(a, 0)$, $(0, b)$, (a, b) is transformed into the rectangle with vertices $(0, 0)$, $(0, a)$, $(b, 0)$, (b, a).

II   Stretches: The next transformations we analyze are those with matrices

$$(a) \quad LS_x = \begin{bmatrix} k & 0 \\ 0 & 1 \end{bmatrix}, k > 0. \quad (b) \quad LS_y = \begin{bmatrix} 1 & 0 \\ 0 & k \end{bmatrix}, k > 0.$$

Consider first (a). If $\mathbf{x} = (x, y)$, then

$$T(\mathbf{x}) = LS_x \mathbf{x} = \begin{bmatrix} k & 0 \\ 0 & 1 \end{bmatrix} \begin{bmatrix} x \\ y \end{bmatrix} = \begin{bmatrix} kx \\ y \end{bmatrix}.$$

Thus,

$$T(x, y) = (kx, y).$$

We see that the x-coordinate of each point in the plane is scaled by a factor k, whereas the y-coordinate is unaltered. Thus, points along the y-axis ($x = 0$) remain fixed, whereas points with positive (negative) x-coordinate are moved horizontally to the right (left). This transformation is called a **linear stretch in the x-direction** (hence, the notation LS_x). The effect of this transformation is illustrated in Figure 6.2.6. If $k > 1$, then the transformation is an **expansion**, whereas if $0 < k < 1$, we have a **compression**. If $k = 1$, then $LS_x = I_2$, and all points remain fixed, this is the **identity transformation**.

In a similar manner, it is easily verified that the transformation with matrix LS_y corresponds to a **linear stretch in the y-direction**.

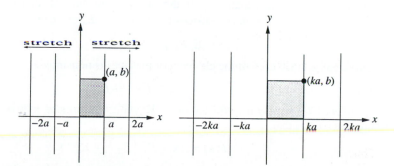

Figure 6.2.6 The effect of a linear stretch in the x-direction. Points along the y-axis remain fixed. All other points are moved parallel to the x-axis.

III   Shears: Now consider transformations of $\mathbf{R}^2$ with matrices

$$\text{(a) } S_x = \begin{bmatrix} 1 & k \\ 0 & 1 \end{bmatrix}, \quad \text{(b) } S_y = \begin{bmatrix} 1 & 0 \\ k & 1 \end{bmatrix}.$$

For the matrix in (a), if $\mathbf{x} = (x, y)$, then

$$T(\mathbf{x}) = S_x\mathbf{x} = \begin{bmatrix} 1 & k \\ 0 & 1 \end{bmatrix}\begin{bmatrix} x \\ y \end{bmatrix} = \begin{bmatrix} x + ky \\ y \end{bmatrix},$$

so that

$$T(x, y) = (x + ky, y).$$

In this case, each point in the plane is moved parallel to the x-axis a distance proportional to its y-coordinate (points along the x-axis remain fixed). This is referred to as a **shear parallel to the x-axis** and is illustrated in Figure 6.2.7 for the case $k > 0$. We leave it as an exercise to verify that the transformation of $\mathbf{R}^2$ with matrix S_y corresponds to a **shear parallel to the y-axis**.

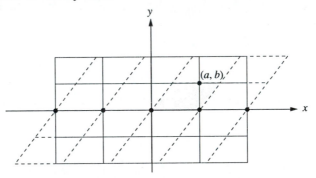

Figure 6.2.7 A shear parallel to the x-axis. Points on the x-axis remain fixed. Rectangles are transformed into parallelograms.

NONSINGULAR TRANSFORMATIONS OF $\mathbf{R}^2$

We now show that any transformation of $\mathbf{R}^2$ with a *nonsingular* matrix can be obtained by combining the basic transformations I–III just described. To do so, we recall that any 2×2 matrix A can be reduced to RREF through multiplication by an appropriate sequence of elementary matrices. In the case of a nonsingular matrix, the RREF is I_2, so that, denoting the elementary matrices by $E_1, E_2, \ldots, E_n$, we can write

$$E_n E_{n-1} \cdots E_2 E_1 A = I_2.$$

Equivalently, since each of the elementary matrices is nonsingular,

$$A = E_1^{-1} E_2^{-1} \cdots E_n^{-1}. \tag{6.2.3}$$

Now let T be any transformation of $\mathbf{R}^2$ with *nonsingular* matrix A. It follows from equation (6.2.3) that

$$T(\mathbf{x}) = A\mathbf{x} = E_1^{-1} E_2^{-1} \cdots E_n^{-1}\mathbf{x}. \tag{6.2.4}$$

Since, as we have shown in Section 3.7, the inverse of an elementary matrix is also an elementary matrix, (6.2.4) implies that a general transformation of $\mathbf{R}^2$ with nonsingular matrix can be obtained by applying a sequence of transformations corresponding to appropriate elementary matrices. Now for the key point. A closer look at the elementary matrices shows that the following relationships hold between them and the matrices of the simple transformations introduced previously in this section.

(1) $P_{12} = \begin{bmatrix} 0 & 1 \\ 1 & 0 \end{bmatrix}$ corresponds to a reflection in $y = x$.

(2a) $M_1(k) = \begin{bmatrix} k & 0 \\ 0 & 1 \end{bmatrix}$, $k > 0$ corresponds to a linear stretch in the x-direction.

(2b) $M_1(k) = \begin{bmatrix} k & 0 \\ 0 & 1 \end{bmatrix}$, $k < 0$. In this case, we can write

$$\begin{bmatrix} k & 0 \\ 0 & 1 \end{bmatrix} = \begin{bmatrix} -k & 0 \\ 0 & 1 \end{bmatrix} \begin{bmatrix} -1 & 0 \\ 0 & 1 \end{bmatrix},$$

which corresponds to a reflection in the y-axis followed by a linear stretch (stretch factor $-k > 0$) in the x-direction.

(3a) $M_2(k) = \begin{bmatrix} 1 & 0 \\ 0 & k \end{bmatrix}$, $k > 0$ corresponds to a linear stretch in the y-direction.

(3b) $M_2(k) = \begin{bmatrix} 1 & 0 \\ 0 & k \end{bmatrix}$, $k < 0$. This corresponds to a reflection in the x-axis followed by a linear stretch in the y-direction.

(4) $A_{12}(k) = \begin{bmatrix} 1 & 0 \\ k & 1 \end{bmatrix}$ corresponds to a shear parallel to the y-axis.

(5) $A_{21}(k) = \begin{bmatrix} 1 & k \\ 0 & 1 \end{bmatrix}$ corresponds to a shear parallel to the x-axis.

We can therefore conclude that any transformation of $\mathbf{R}^2$ with nonsingular matrix can be obtained by applying an appropriate sequence of reflections, shears, and stretches.

Example 6.2.1 Let $T : \mathbf{R}^2 \to \mathbf{R}^2$ be the transformation of $\mathbf{R}^2$ with nonsingular matrix $A = \begin{bmatrix} 3 & 9 \\ 1 & 2 \end{bmatrix}$. Describe T as a combination of reflections, shears, and stretches.

Solution   Reducing A to RREF in the usual manner yields

$$\begin{bmatrix} 3 & 9 \\ 1 & 2 \end{bmatrix} \sim \begin{bmatrix} 1 & 2 \\ 3 & 9 \end{bmatrix} \sim \begin{bmatrix} 1 & 2 \\ 0 & 3 \end{bmatrix} \sim \begin{bmatrix} 1 & 2 \\ 0 & 1 \end{bmatrix} \sim \begin{bmatrix} 1 & 0 \\ 0 & 1 \end{bmatrix}.$$

The corresponding elementary matrices that accomplish this reduction are

$$P_{12} = \begin{bmatrix} 0 & 1 \\ 1 & 0 \end{bmatrix}, \quad A_{12}(-3) = \begin{bmatrix} 1 & 0 \\ -3 & 1 \end{bmatrix},$$

$$M_2(1/3) = \begin{bmatrix} 1 & 0 \\ 0 & 1/3 \end{bmatrix}, \quad A_{21}(-2) = \begin{bmatrix} 1 & -2 \\ 0 & 1 \end{bmatrix},$$

with inverses

$$P_{12}, \quad A_{12}(3), \quad M_2(3), \quad A_{21}(2),$$

respectively. Consequently,

$$A = \begin{bmatrix} 0 & 1 \\ 1 & 0 \end{bmatrix} \begin{bmatrix} 1 & 0 \\ 3 & 1 \end{bmatrix} \begin{bmatrix} 1 & 0 \\ 0 & 3 \end{bmatrix} \begin{bmatrix} 1 & 2 \\ 0 & 1 \end{bmatrix}.$$

We can therefore write

$$T(\mathbf{x}) = A\mathbf{x} = \underbrace{\begin{bmatrix} 0 & 1 \\ 1 & 0 \end{bmatrix}}_{\substack{\text{reflection in} \\ y = x}} \underbrace{\begin{bmatrix} 1 & 0 \\ 3 & 1 \end{bmatrix}}_{\substack{\text{shear parallel} \\ \text{to } y\text{-axis}}} \underbrace{\begin{bmatrix} 1 & 0 \\ 0 & 3 \end{bmatrix}}_{\substack{\text{stretch in} \\ y\text{-direction}}} \underbrace{\begin{bmatrix} 1 & 2 \\ 0 & 1 \end{bmatrix}}_{\substack{\text{shear parallel} \\ \text{to } x\text{-axis}}} \mathbf{x}.$$

We see that T consists of a shear parallel to the x-axis, followed by a stretch in the y-direction, followed by a shear parallel to the y-axis, followed by a reflection in $y = x$. □

In the previous section we derived the matrix of the transformation of $\mathbf{R}^2$ that corresponds to a rotation through an angle θ in the counter-clockwise direction, namely,

$$R(\theta) = \begin{bmatrix} \cos \theta & -\sin \theta \\ \sin \theta & \cos \theta \end{bmatrix}.$$

Since this is a nonsingular matrix, it follows from the analysis in this section that a rotation is an appropriate combination of reflections, stretches, and shears. Indeed we leave it as an exercise to verify that for $\theta \neq \pi/2, 3\pi/2$,

$$R(\theta) = \begin{bmatrix} \cos \theta & 0 \\ 0 & 1 \end{bmatrix} \begin{bmatrix} 1 & 0 \\ \sin \theta & 1 \end{bmatrix} \begin{bmatrix} 1 & 0 \\ 0 & \sec \theta \end{bmatrix} \begin{bmatrix} 1 & -\tan \theta \\ 0 & 1 \end{bmatrix}. \qquad (6.3.5)$$

EXERCISES 6.2

For problems 1–4, for the transformation of $\mathbf{R}^2$ with the given matrix, sketch the transform of the square with vertices (1, 1), (2, 1), (2, 2), and (1, 2).

1. $A = \begin{bmatrix} 0 & 1 \\ -1 & 0 \end{bmatrix}$.

2. $A = \begin{bmatrix} 1 & -1 \\ 1 & 2 \end{bmatrix}$.

3. $A = \begin{bmatrix} 1 & 1 \\ -1 & 1 \end{bmatrix}$.

4. $A = \begin{bmatrix} -2 & -2 \\ -2 & 0 \end{bmatrix}$.

For problems 5–12, describe the transformation of $\mathbf{R}^2$ with the given matrix as a product of reflections, strectches, and shears.

5. $A = \begin{bmatrix} 1 & 2 \\ 0 & 1 \end{bmatrix}$.

6. $A = \begin{bmatrix} 0 & 2 \\ 2 & 0 \end{bmatrix}$.

7. $A = \begin{bmatrix} 1 & 0 \\ 3 & 1 \end{bmatrix}$.

8. $A = \begin{bmatrix} -1 & 0 \\ 0 & -1 \end{bmatrix}$.

9. $A = \begin{bmatrix} 1 & -3 \\ -2 & 8 \end{bmatrix}$.

10. $A = \begin{bmatrix} 1 & 2 \\ 3 & 4 \end{bmatrix}$

11. $A = \begin{bmatrix} 1 & 0 \\ 0 & -2 \end{bmatrix}$.

12. $A = \begin{bmatrix} -1 & -1 \\ -1 & 0 \end{bmatrix}$.

13. Consider the transformation of $\mathbf{R}^2$ corresponding to a counterclockwise rotation through angle $\theta \in [0, 2\pi)$. For $\theta \neq \pi/2, 3\pi/2$, verify that the matrix of the transformation is given by (6.3.5), and describe it in terms of reflections, stretches, and shears.

14. Express the transformation of $\mathbf{R}^2$ corresponding to a counterclockwise rotation through an angle $\theta = \pi/2$ as product of reflections, stretches, and shears. Repeat for the case $\theta = 3\pi/2$.

6.3 THE KERNEL AND RANGE OF A LINEAR TRANSFORMATION

If $T : V \to W$ is any linear transformation, there is an associated homogeneous linear vector equation, namely,

$$T(\mathbf{v}) = \mathbf{0}.$$

The solution set to this vector equation is called the kernel of T.

Definition 6.3.1: Let $T : V \to W$ be a linear transformation. The set of *all* vectors $\mathbf{v} \in V$ such that $T(\mathbf{v}) = \mathbf{0}$ is called the **kernel of T** and is denoted $\text{Ker}(T)$. Thus,
$$\text{Ker}(T) = \{\mathbf{v} \in V : T(\mathbf{v}) = \mathbf{0}\}.$$

Example 6.3.1 Determine $\text{Ker}(T)$ for the linear transformation $T : C^2(I) \to C^0(I)$ defined by $T(y) = y'' + y$.

Solution

$$\text{Ker}(T) = \{y \in C^2(I) : T(y) = 0\} = \{y \in C^2(I) : y'' + y = 0 \text{ for all } x \in I\}.$$

Hence, in this case, $\text{Ker}(T)$ is the solution set to the DE

$$y'' + y = 0.$$

Since this DE has general solution $y(x) = c_1 \cos x + c_2 \sin x$, we have

$$\text{Ker}(T) = \{y \in C^2(I) : y(x) = c_1 \cos x + c_2 \sin x\}.$$

This is the subspace of $C^2(I)$ spanned by $\{\cos x, \sin x\}$. ❏

The set of all vectors in W that we map onto when T is applied to *all* vectors in V is called the range of T. We can think of the range of T as being the set of function values. A formal definition follows.

> **Definition 6.3.2:**   The **range** of the linear transformation $T : V \to W$ is the subset of W consisting of all transformed vectors from V. We denote the range of T by Rng(T). Thus,
>
> $$\text{Rng}(T) = \{T(\mathbf{v}) : \mathbf{v} \in V\}.$$

A schematic representation of Ker(T) and Rng(T) is given in Figure 6.3.1.

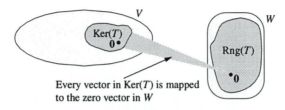

Figure 6.3.1 Schematic representation of the kernel and range of a linear transformation.

Now let us focus our attention on linear transformations of the form $T : \mathbf{R}^n \to \mathbf{R}^m$. In this particular case,

$$\text{Ker}(T) = \{\mathbf{x} \in \mathbf{R}^n : T(\mathbf{x}) = \mathbf{0}\}.$$

If we let A denote the matrix of T, then $T(\mathbf{x}) = A\mathbf{x}$, so that

$$\text{Ker}(T) = \{\mathbf{x} \in \mathbf{R}^n : A\mathbf{x} = \mathbf{0}\}.$$

Consequently,

> If $T : \mathbf{R}^n \to \mathbf{R}^m$ is the linear transformation with matrix A then Ker(T) is the solution set to the homogeneous linear system $A\mathbf{x} = \mathbf{0}$.

In Section 5.4, we defined the solution set to $A\mathbf{x} = \mathbf{0}$ to be nullspace(A). Therefore, we have

$$\text{Ker}(T) = \text{nullspace}(A)$$

from which it follows directly that Ker(T) is a *subspace* of $\mathbf{R}^n$. Furthermore, for a linear transformation $T : \mathbf{R}^n \to \mathbf{R}^m$,

$$\text{Rng}(T) = \{T(\mathbf{x}) : \mathbf{x} \in \mathbf{R}^n\}.$$

If $A = [\mathbf{a}_1, \mathbf{a}_2, \ldots, \mathbf{a}_n]$ denotes the matrix of T, then

$$
\begin{aligned}
\text{Rng}(T) \ &= \{A\mathbf{x} : \mathbf{x} \in \mathbf{R}^n\} \\
&= \{x_1\mathbf{a}_1 + x_2\mathbf{a}_2 + \cdots + x_n\mathbf{a}_n : x_1, x_2, \ldots, x_n \in \mathbf{R}\} \\
&= \text{colspace}(A).
\end{aligned}
$$

Consequently, in this case Rng(T) is a *subspace of* $\mathbf{R}^m$. We illustrate these results with an example.

Example 6.3.2   Let $T : \mathbf{R}^3 \to \mathbf{R}^2$ be the linear transformation with matrix $A = \begin{bmatrix} 1 & -2 & 5 \\ -2 & 4 & -10 \end{bmatrix}$. Determine Ker($T$) and Rng($T$).

Solution

$$\text{Ker}(T) = \{\mathbf{x} \in \mathbf{R}^3 : T(\mathbf{x}) = \mathbf{0}\} = \{\mathbf{x} \in \mathbf{R}^3 : A\mathbf{x} = \mathbf{0}\} = \text{nullspace}(A)$$

We therefore must determine the solution set to the system $A\mathbf{x} = \mathbf{0}$. The RREF of the augmented matrix of this system is

$$\begin{bmatrix} 1 & -2 & 5 & 0 \\ 0 & 0 & 0 & 0 \end{bmatrix},$$

so that there are two free variables. Setting $x_2 = r$, $x_3 = s$, it follows that $x_1 = 2r - 5s$, so that $\mathbf{x} = (2r - 5s, r, s)$. Hence,

$$\text{Ker}(T) = \{\mathbf{x} \in \mathbf{R}^3 : \mathbf{x} = (2r - 5s, r, s) : r, s \in \mathbf{R}\}$$
$$= \{\mathbf{x} \in \mathbf{R}^3 : \mathbf{x} = r(2, 1, 0) + s(-5, 0, 1), r, s \in \mathbf{R}\}.$$

We see that Ker(T) is the two-dimensional subspace of $\mathbf{R}^3$ spanned by the LI vectors $(2, 1, 0)$ and $(-5, 0, 1)$, and therefore, it consists of all points lying on the plane through the origin that contains these vectors. We leave it as an exercise to verify that the equation of this plane is $x_1 - 2x_2 + 5x_3 = 0$. The linear transformation T maps all points lying on this plane to the zero vector in $\mathbf{R}^2$. (See Figure 6.3.2.) For the given transformation, we have

$$\text{Rng}(T) = \text{colspace}(A).$$

From the foregoing RREF of A we see that colspace(A) is generated by the first column vector in A. Consequently,

$$\text{Rng}(T) = \{\mathbf{y} \in \mathbf{R}^2 : \mathbf{y} = r(1, -2), r \in \mathbf{R}\}.$$

Hence, the points in Rng(T) lie along the line through the origin in $\mathbf{R}^2$ whose direction is determined by $\mathbf{v} = (1, -2)$. The Cartesian equation of this line is $y_2 = -2y_1$. Consequently, T maps all points in $\mathbf{R}^3$ onto this line, and therefore Rng(T) is a 1-dimensional subspace of $\mathbf{R}^2$. This is illustrated in Figure 6.3.2.

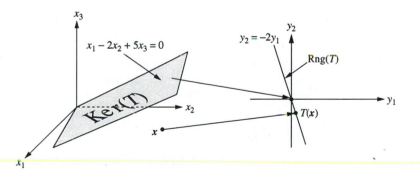

Figure 6.3.2 The kernel and range of the linear transformation in Example 6.3.2.

□

Now let us return to arbitrary linear transformations. The preceding discussion has shown that both the kernel and range of any linear transformation from $\mathbf{R}^n$ to $\mathbf{R}^m$ are subspaces of the appropriate vector space. Our next result, which is fundamental, establishes that this is true in general.

Theorem 6.3.1:   If $T : V \to W$ is a linear transformation, then

1. $\text{Ker}(T)$ is a subspace of V,
2. $\text{Rng}(T)$ is a subspace of W.

PROOF   In this proof, we once more denote the zero vector in W by $\mathbf{0}_W$. Both $\text{Ker}(T)$ and $\text{Rng}(T)$ are necessarily nonempty, since, as we verified in Section 6.1, any linear transformation maps the zero vector in V to the zero vector in W. We must now establish that $\text{Ker}(T)$ and $\text{Rng}(T)$ are both closed under addition and closed under scalar multiplication in the appropriate vector space.

(1) If $\mathbf{u}$ and $\mathbf{v}$ are in $\text{Ker}(T)$, then $T(\mathbf{u}) = \mathbf{0}_W$ and $T(\mathbf{v}) = \mathbf{0}_W$. Hence,

$$T(\mathbf{u} + \mathbf{v}) = T(\mathbf{u}) + T(\mathbf{v}) = \mathbf{0}_W + \mathbf{0}_W = \mathbf{0}_W,$$

so that $\text{Ker}(T)$ is closed under addition. Further, if c is any scalar,

$$T(c\mathbf{u}) = cT(\mathbf{u}) = c\mathbf{0}_W = \mathbf{0}_W$$

so that $\text{Ker}(T)$ is also closed under scalar multiplication. Thus, $\text{Ker}(T)$ is a subspace of V.

(2) If $\mathbf{w}_1$ and $\mathbf{w}_2$ are in $\text{Rng}(T)$, then $\mathbf{w}_1 = T(\mathbf{v}_1)$ and $\mathbf{w}_2 = T(\mathbf{v}_2)$ for some $\mathbf{v}_1$ and $\mathbf{v}_2$ in V. Thus,

$$\mathbf{w}_1 + \mathbf{w}_2 = T(\mathbf{v}_1) + T(\mathbf{v}_2) = T(\mathbf{v}_1 + \mathbf{v}_2).$$

That is, $\mathbf{w}_1 + \mathbf{w}_2$ is the transform of the vector $\mathbf{v}_1 + \mathbf{v}_2$, and, therefore, $\mathbf{w}_1 + \mathbf{w}_2$ is in $\text{Rng}(T)$. Further, if c is any scalar, then

$$c\mathbf{w}_1 = cT(\mathbf{v}_1) = T(c\mathbf{v}_1),$$

so that $c\mathbf{w}_1$ is the transform of $c\mathbf{v}_1$ and therefore $c\mathbf{w}_1$ is in $\text{Rng}(T)$. Consequently, $\text{Rng}(T)$ is a subspace of W. ∎

REMARK   We can interpret the first part of the preceding theorem as telling us that if T is a linear transformation, then the solution set to the corresponding linear homogenous problem

$$T(\mathbf{v}) = \mathbf{0}$$

is a vector space. Consequently, if we can determine the dimension of this vector space then we know how many LI solutions are required to build every solution to the problem. This is the formulation for linear problems that we have been looking for.

Example 6.3.3   Find $\text{Ker}(T)$, $\text{Rng}(T)$, and their dimensions for the linear trans–formation $T : M_2(\mathbf{R}) \to M_2(\mathbf{R})$ defined by

$$T(A) = A - A^T.$$

Solution   In this case

$$\text{Ker}(T) = \{A \in M_2(\mathbf{R}) : T(A) = 0_2\} = \{A \in M_2(\mathbf{R}) : A - A^T = 0_2\}.$$

Thus, $\text{Ker}(T)$ is the solution set of the matrix equation

$$A - A^T = 0_2,$$

so that the matrices in Ker(T) satisfy

$$A^T = A.$$

Hence, Ker(T) is the subspace of $M_2(\mathbf{R})$ consisting of all symmetric 2×2 matrices. We have shown previously that a basis for this subspace is

$$\left\{ \begin{bmatrix} 1 & 0 \\ 0 & 0 \end{bmatrix}, \begin{bmatrix} 0 & 1 \\ 1 & 0 \end{bmatrix}, \begin{bmatrix} 0 & 0 \\ 0 & 1 \end{bmatrix} \right\},$$

so that dim[Ker(T)] = 3. We now determine the range of T.

$$\mathrm{Rng}(T) \ = \ \{T(A) : A \in M_2(\mathbf{R})\} = \{A - A^T : A \in M_2(\mathbf{R})\}$$

$$= \left\{ \begin{bmatrix} a & b \\ c & d \end{bmatrix} - \begin{bmatrix} a & c \\ b & d \end{bmatrix} : a, b, c, d \in \mathbf{R} \right\}$$

$$= \left\{ \begin{bmatrix} 0 & b - c \\ -(b-c) & 0 \end{bmatrix} : b, c \in \mathbf{R} \right\}.$$

Thus, setting $e = b - c$,

$$\mathrm{Rng}(T) = \left\{ \begin{bmatrix} 0 & e \\ -e & 0 \end{bmatrix} : e \in \mathbf{R} \right\} = \mathrm{span} \left\{ \begin{bmatrix} 0 & 1 \\ -1 & 0 \end{bmatrix} \right\}.$$

Consequently, Rng(T) consists of all skew-symmetric 2×2 matrices with real elements. Since Rng(T) is generated by the single nonzero matrix $\begin{bmatrix} 0 & 1 \\ -1 & 0 \end{bmatrix}$, it follows that a basis for Rng(T) is $\left\{ \begin{bmatrix} 0 & 1 \\ -1 & 0 \end{bmatrix} \right\}$, so that dim[Rng($T$)] = 1.

Example 6.3.4   Let $T : P_2 \to P_3$ be the linear transformation defined by

$$T(ax + b) = (a + b)x^2 + (a - 5b)x + (2b - 3a).$$

Find Ker(T), Rng(T), and their dimensions.

Solution   From Definition 6.3.1,

$$\mathrm{Ker}(T) \ = \ \{ p \in P_2 : T(p) = 0\}$$
$$= \{ax + b : (a + b)x^2 + (a - 5b)x + (2b - 3a) = 0 \text{ for all } x\}$$
$$= \{ax + b : a + b = 0, a - 5b = 0, 2b - 3a = 0\}.$$

But the only values of a and b that satisfy the conditions

$$a + b = 0, \quad a - 5b = 0, \quad 2b - 3a = 0$$

are

$$a = b = 0.$$

Consequently, Ker(T) contains only the zero polynomial. Hence, we write

$$\mathrm{Ker}(T) = \{\mathbf{0}\}.$$

It follows from Definition 5.6.2 that dim[Ker(T)] = 0. Furthermore,

$$\mathrm{Rng}(T) = \{T(ax + b) : a, b \in \mathbf{R}\} = \{(a + b)x^2 + (a - 5b)x + (2b - 3a) : a, b \in \mathbf{R}\}.$$

To determine a basis for $\text{Rng}(T)$, we write this as

$$\text{Rng}(T) = \{a(x^2 + x - 3) + b(x^2 - 5x + 2) : a, b \in \mathbf{R}\}$$
$$= \text{span}\{x^2 + x - 3, x^2 - 5x + 2\}.$$

Thus, $\text{Rng}(T)$ is spanned by the polynomials $p_1(x) = x^2 + x - 3$ and $p_2(x) = x^2 - 5x + 2$. Since p_1 and p_2 are not proportional to one another, they are LI on any interval. Consequently, a basis for $\text{Rng}(T)$ is $\{x^2 + x - 3, x^2 - 5x + 2\}$, so that $\dim[\text{Rng}(T)] = 2$.

$\square$

THE GENERAL RANK–NULLITY THEOREM

Finally in this section, we consider a fundamental theorem for linear transformations $T : V \to W$ that gives a relationship between the dimensions of $\text{Ker}(T)$, $\text{Rng}(T)$, and V. This is a generalization of the Rank–Nullity Theorem considered in Chapter 5 to which it reduces when T is a linear transformation from $\mathbf{R}^n$ to $\mathbf{R}^m$ with matrix A. Suppose that $\dim[V] = n$ and that $\dim[\text{Ker}(T)] = k$. Then k-dimensions worth of the vectors in V are all mapped onto the zero vector in W. Consequently, we only have $n - k$ dimensions worth of vectors to map onto the remaining vectors in W. This *suggests* that

$$\dim[\text{Rng}(T)] = \dim(V) - \dim[\text{Ker}(T)].$$

This is indeed correct, although a rigorous proof is somewhat involved. We state the result as a theorem.

Theorem 6.3.2:   If $T : V \to W$ is a linear transformation and V is finite-dimensional, then

$$\dim[\text{Ker}(T)] + \dim[\text{Rng}(T)] = \dim[V].$$

PROOF   Suppose that $\dim[V] = n$. There are three different cases to consider.

Case 1: If $\text{Ker}(T) = V$, then $\text{Rng}(T) = \{\mathbf{0}\}$, and so

$$\dim[\text{Ker}(T)] + \dim[\text{Rng}(T)] = n + 0 = n,$$

as required.

Case 2: Suppose $\dim[\text{Ker}(T)] = k$, where $0 < k < n$. Let $\{\mathbf{v}_1, \mathbf{v}_2, ..., \mathbf{v}_k\}$ be a basis for $\text{Ker}(T)$. Then, using Theorem 5.6.5, we can extend this basis to a basis for V, which we denote by $\{\mathbf{v}_1, \mathbf{v}_2, ..., \mathbf{v}_k, \mathbf{v}_{k+1}, ..., \mathbf{v}_n\}$. We prove that $\{T(\mathbf{v}_{k+1}), T(\mathbf{v}_{k+2}), ..., T(\mathbf{v}_n)\}$ is a basis for $\text{Rng}(T)$. Let $\mathbf{w}$ be any vector in $\text{Rng}(T)$. Then $\mathbf{w} = T(\mathbf{v})$, for some $\mathbf{v} \in V$. Consequently, there exist scalars $c_1, c_2, .., c_n$ such that

$$\mathbf{w} = T(c_1\mathbf{v}_1 + c_2\mathbf{v}_2 + \cdots + c_n\mathbf{v}_n) = c_1 T(\mathbf{v}_1) + c_2 T(\mathbf{v}_2) + \cdots + c_n T(\mathbf{v}_n).$$

Since $\mathbf{v}_1, \mathbf{v}_2, ..., \mathbf{v}_k$ are in $\text{Ker}(T)$, this reduces to

$$\mathbf{w} = c_{k+1} T(\mathbf{v}_{k+1}) + c_{k+2} T(\mathbf{v}_{k+2}) + \cdots + c_n T(\mathbf{v}_n).$$

Thus,

$$\text{Rng}(T) = \text{span}\{T(\mathbf{v}_{k+1}), T(\mathbf{v}_{k+2}), ..., T(\mathbf{v}_n)\}.$$

We must also show that $\{T(\mathbf{v}_{k+1}), T(\mathbf{v}_{k+2}), ..., T(\mathbf{v}_n)\}$ is LI. Suppose that

$$d_{k+1} T(\mathbf{v}_{k+1}) + d_{k+2} T(\mathbf{v}_{k+2}) + \cdots + d_n T(\mathbf{v}_n) = \mathbf{0}, \qquad (6.3.1)$$

where $d_{k+1}, d_{k+2}, ..., d_n$ are scalars. Then, using the linearity of T,

$$T(d_{k+1} \mathbf{v}_{k+1} + d_{k+2} \mathbf{v}_{k+2} + \cdots + d_n \mathbf{v}_n) = \mathbf{0},$$

which implies that the vector $d_{k+1}\mathbf{v}_{k+1} + d_{k+2}\mathbf{v}_{k+2} + \cdots + d_n\mathbf{v}_n$ is in $\text{Ker}(T)$. Consequently, there exist scalars $d_1, d_2, \ldots, d_k$ such that

$$d_{k+1}\mathbf{v}_{k+1} + d_{k+2}\mathbf{v}_{k+2} + \cdots + d_n\mathbf{v}_n = d_1\mathbf{v}_1 + d_2\mathbf{v}_2 + \ldots + d_k\mathbf{v}_k,$$

that is, such that

$$d_1\mathbf{v}_1 + d_2\mathbf{v}_2 + \cdots + d_k\mathbf{v}_k - (d_{k+1}\mathbf{v}_{k+1} + d_{k+2}\mathbf{v}_{k+2} + \cdots + d_n\mathbf{v}_n) = \mathbf{0}.$$

Since the set of vectors $\{\mathbf{v}_1, \mathbf{v}_2, \ldots, \mathbf{v}_k, \mathbf{v}_{k+1}, \ldots, \mathbf{v}_n\}$ is LI, we must have

$$d_1 = d_2 = \cdots = d_k = d_{k+1} = \cdots = d_n = 0.$$

Thus, from equation (6.3.1), $\{T(\mathbf{v}_{k+1}), T(\mathbf{v}_{k+2}), \ldots, T(\mathbf{v}_n)\}$ is LI, and therefore $\{T(\mathbf{v}_{k+1}), T(\mathbf{v}_{k+2}), \ldots, T(\mathbf{v}_n)\}$ is a basis for $\text{Rng}(T)$. Since there are $n - k$ vectors in this basis, it follows that $\dim[\text{Rng}(T)] = n - k$. Consequently,

$$\dim[\text{Ker}(T)] + \dim[\text{Rng}(T)] = k + (n - k) = n$$

as required.

Case 3:   Finally, if $\text{Ker}(T) = \{\mathbf{0}\}$, let $\{\mathbf{v}_1, \mathbf{v}_2, \ldots, \mathbf{v}_n\}$ be any basis for V. By a similar argument to that used above, it can be shown that $\{T(\mathbf{v}_1), T(\mathbf{v}_2), \ldots, T(\mathbf{v}_n)\}$ is a basis for $\text{Rng}(T)$, and so once more we have

$$\dim[\text{Ker}(T)] + \dim[\text{Rng}(T)] = n. \qquad \blacksquare$$

The preceding theorem is useful for checking that we have the correct dimensions when determining the kernel and range of a linear transformation. Furthermore, it can also be used to determine, say, the dimension of $\text{Rng}(T)$ once we know $\dim[\text{Ker}(T)]$. For example, consider the linear transformation discussed in Example 6.3.3. Theorem 6.3.2 tells us that

$$\dim[\text{Ker}(T)] + \dim[\text{Rng}(T)] = \dim[M_2(\mathbf{R})]$$

so that once we had determined that $\dim[\text{Ker}(T)] = 3$, it immediately follows that

$$3 + \dim[\text{Rng}(T)] = 4$$

so that

$$\dim[\text{Rng}(T)] = 1.$$

EXERCISES 6.3

1. Consider $T: \mathbf{R}^3 \rightarrow \mathbf{R}^2$ defined by $T(\mathbf{x}) = A\mathbf{x}$, where $A = \begin{bmatrix} 1 & -1 & 2 \\ 1 & -2 & -3 \end{bmatrix}$. Find $T(\mathbf{x})$ and thereby determine whether $\mathbf{x}$ is in $\text{Ker}(T)$.

(a) $\mathbf{x} = (7, 5, -1)$.

(b) $\mathbf{x} = (-21, -15, 2)$.

(c) $\mathbf{x} = (35, 25, -5)$.

For problems 2–6, find $\text{Ker}(T)$ and $\text{Rng}(T)$, and give a geometrical description of each. Also find $\dim[\text{Ker}(T)]$ and $\dim[\text{Rng}(T)]$, and verify Theorem 6.3.2.

2. $T: \mathbf{R}^2 \rightarrow \mathbf{R}^2$ defined by $T(\mathbf{x}) = A\mathbf{x}$, where

$$A = \begin{bmatrix} 3 & 6 \\ 1 & 2 \end{bmatrix}.$$

3. $T : \mathbf{R}^3 \to \mathbf{R}^3$ defined by $T(\mathbf{x}) = A\mathbf{x}$, where

$$A = \begin{bmatrix} 1 & -1 & 0 \\ 0 & 1 & 2 \\ 2 & -1 & 1 \end{bmatrix}.$$

4. $T : \mathbf{R}^3 \to \mathbf{R}^3$ defined by $T(\mathbf{x}) = A\mathbf{x}$, where

$$A = \begin{bmatrix} 1 & -2 & 1 \\ 2 & -3 & -1 \\ 5 & -8 & -1 \end{bmatrix}.$$

5. $T : \mathbf{R}^3 \to \mathbf{R}^2$ defined by $T(\mathbf{x}) = A\mathbf{x}$, where

$$A = \begin{bmatrix} 1 & -1 & 2 \\ -3 & 3 & -6 \end{bmatrix}.$$

6. $T : \mathbf{R}^3 \to \mathbf{R}^2$ defined by $T(\mathbf{x}) = A\mathbf{x}$, where

$$A = \begin{bmatrix} 1 & 3 & 2 \\ 2 & 6 & 5 \end{bmatrix}.$$

For problems 7–10, determine a basis for the kernel of the given linear transformation.

7. $T(y) = y'' - y$.

8. $T(y) = y'' + y' - 12y$.

9. $T(y) = y'' + 9y$.

10. $T(y) = y'' - 4y' + 29y$.

11. Consider the linear transformation $T : \mathbf{R}^3 \to \mathbf{R}$ defined by

$$T(\mathbf{v}) = \langle \mathbf{u}, \mathbf{v} \rangle,$$

where $\mathbf{u}$ is a fixed vector in $\mathbf{R}^3$.

(a) Find Ker(T) and dim[Ker(T)], and interpret geometrically.

(b) Find Rng(T) and dim[Rng(T)].

12. Consider the linear transformation $S : M_n(\mathbf{R}) \to M_n(\mathbf{R})$ defined by $S(A) = A - A^T$, where A is an $n \times n$ matrix.

(a) Find Ker(S) and describe it.

(b) In the particular case when A is a 2×2 matrix, determine a basis for Ker(S), and hence, find its dimension.

13. Consider the linear transformation $T : M_n(\mathbf{R}) \to M_n(\mathbf{R})$ defined by

$$T(A) = AB - BA,$$

where B is a fixed $n \times n$ matrix. Find Ker(T) and describe it.

14. Consider the *linear* transformation $T : P_3 \to P_3$ defined by:

$$T(ax^2 + bx + c) = ax^2 + (a + 2b + c)x + (3a - 2b - c),$$

where a, b, and c are arbitrary constants.

(a) Show that Ker(T) consists of all polynomials of the form $b(x - 2)$, and hence, find its dimension.

(b) Find Rng(T) and its dimension.

15. Consider the *linear* transformation $T : P_3 \to P_2$ defined by

$$T(ax^2 + bx + c) = (a + b) + (b - c)x$$

where a, b, and c are arbitrary real numbers. Determine Ker(T), Rng(T), and their dimensions.

16. Consider the *linear* transformation $T : P_2 \to P_3$ defined by

$$T(ax + b) = (b - a) + (2b - 3a)x + bx^2.$$

Determine Ker(T), Rng(T), and their dimensions.

17. Let $\{\mathbf{v}_1, \mathbf{v}_2, \mathbf{v}_3\}$ and $\{\mathbf{w}_1, \mathbf{w}_2\}$ be bases for the (real) vector spaces V and W, respectively, and let $T : V \to W$ be the linear transformation satisfying

$$T(\mathbf{v}_1) = 2\mathbf{w}_1 - \mathbf{w}_2, \quad T(\mathbf{v}_2) = \mathbf{w}_1 - \mathbf{w}_2,$$
$$T(\mathbf{v}_3) = \mathbf{w}_1 + 2\mathbf{w}_2.$$

Find Ker(T), Rng(T), and their dimensions.

18. Let $T : V \to W$ be a linear transformation, and suppose that dim$[V] = n$. If Ker(T) = $\{\mathbf{0}\}$ and $\{\mathbf{v}_1, \mathbf{v}_2, \ldots, \mathbf{v}_n\}$ is any basis for V, prove that $\{T(\mathbf{v}_1), T(\mathbf{v}_2), \ldots, T(\mathbf{v}_n)\}$ is a basis for Rng(T). (This fills in the missing details in the proof of Theorem 6.3.2.)

6.4   FURTHER PROPERTIES OF LINEAR TRANSFORMATIONS

One of the aims of this section is to establish that all real vector spaces of (finite) dimension n are essentially the same as $\mathbf{R}^n$. In order to do so, we need to consider the composition of linear transformations.

Definition   6.4.1:   Let $T_1 : U \to V$ and $T_2 : V \to W$ be two linear transformations.[1] We define the **composition, or product,** $T_2 T_1 : U \to W$ by

$$(T_2 T_1)(\mathbf{u}) = T_2(T_1(\mathbf{u})) \text{ for all } \mathbf{u} \in U.$$

The composition is illustrated in Figure 6.4.1.

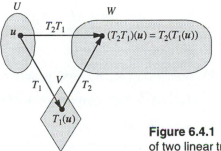

Figure 6.4.1 The composition of two linear transformations.

Our first result establishes that $T_2 T_1$ is a *linear* transformation.

Theorem 6.4.1:   Let $T_1 : U \to V$ and $T_2 : V \to W$ be linear transformations. Then $T_2 T_1 : U \to W$ is a linear transformation.

PROOF   Let $\mathbf{v}_1, \mathbf{v}_2$ be arbitrary vectors in U, and let c be a scalar. We must prove that

$$(T_2 T_1)(\mathbf{v}_1 + \mathbf{v}_2) = (T_2 T_1)(\mathbf{v}_1) + (T_2 T_1)(\mathbf{v}_2) \tag{6.4.1}$$

and that

$$(T_2 T_1)(c\mathbf{v}_1) = c(T_2 T_1)(\mathbf{v}_1). \tag{6.4.2}$$

Consider first equation (6.4.1). We have:

$$(T_2 T_1)(\mathbf{v}_1 + \mathbf{v}_2) \quad = \quad T_2(T_1(\mathbf{v}_1 + \mathbf{v}_2)) \quad = \quad T_2(T_1(\mathbf{v}_1) + T_1(\mathbf{v}_2))$$

$$\uparrow \qquad\qquad\qquad\qquad\qquad\qquad \uparrow$$

$$\text{Definition of } T_2 T_1 \qquad\qquad\qquad\qquad \text{Using the linearity of } T_1$$

$$= \quad T_2(T_1(\mathbf{v}_1)) + T_2(T_1(\mathbf{v}_2))$$

$$\uparrow$$

$$\text{Using the linearity of } T_2$$

$$= \quad (T_2 T_1)(\mathbf{v}_1) + (T_2 T_1)(\mathbf{v}_2),$$

$$\uparrow$$

$$\text{Definition of } T_2 T_1$$

so that equation (6.4.1) is satisfied. We leave the proof of (6.4.2) as an exercise.   ∎

Example 6.4.1   Let $T_1 : \mathbf{R}^n \to \mathbf{R}^m$ and $T_2 : \mathbf{R}^m \to \mathbf{R}^p$ be linear transformations with matrices A and B respectively. Determine the linear transformation $T_2 T_1 : \mathbf{R}^n \to \mathbf{R}^p$.

---

[1] U, V, and W must be either all real vector spaces or all complex vector spaces.

Solution   From Definition 6.4.2,

$$(T_2 T_1)(\mathbf{x}) = T_2(T_1(\mathbf{x})) = T_2(A\mathbf{x}) = B(A\mathbf{x}) = (BA)\mathbf{x}.$$

Consequently, $T_2 T_1$ is the linear transformation with matrix BA. Note that A is an $m \times n$ matrix and B is a $p \times m$, matrix so that the matrix product BA is defined.

Example 6.4.2   Let $T_1 : M_n(\mathbf{R}) \to M_n(\mathbf{R})$ and $T_2 : M_n(\mathbf{R}) \to \mathbf{R}$ be the linear transformations defined by

$$T_1(A) = A + A^T, \qquad T_2(A) = \mathrm{tr}(A).$$

Determine $T_2 T_1$.

Solution   In this case, $T_2 T_1 : M_n(\mathbf{R}) \to \mathbf{R}$ is defined by

$$(T_2 T_1)(A) = T_2(T_1(A)) = T_2(A + A^T) = \mathrm{tr}(A + A^T).$$

This can be written in the equivalent form

$$(T_2 T_1)(A) = 2\,\mathrm{tr}(A).$$

As a specific example, consider the matrix $A = \begin{bmatrix} 2 & -1 \\ -3 & 6 \end{bmatrix}$. We have

$$T_1(A) = A + A^T = \begin{bmatrix} 4 & -4 \\ -4 & 12 \end{bmatrix}$$

so that

$$T_2(T_1(A)) = \mathrm{tr}\left(\begin{bmatrix} 4 & -4 \\ -4 & 12 \end{bmatrix} \right) = 16.$$

◻

We now extend the definitions of *one-to-one* and *onto*, which should be familiar in the case of a function of a single variable $f : \mathbf{R} \to \mathbf{R}$, to the case of arbitrary linear transformations.

Definition 6.4.2:   A linear transformation $T : V \to W$ is said to be

1. **one-to-one** if distinct elements in V are mapped to distinct elements in W. That is, if $\mathbf{v}_1 \neq \mathbf{v}_2$, then $T(\mathbf{v}_1) \neq T(\mathbf{v}_2)$, for all $\mathbf{v}_1$ and $\mathbf{v}_2 \in V$.

2. **onto** if the range of T is the whole of W, that is, if *every* $\mathbf{w} \in W$ is the image of at least one vector $\mathbf{v} \in V$.

The following theorem can be helpful in determining whether a given linear transformation is one-to-one.

Theorem 6.4.2:   Let $T : V \to W$ be a linear transformation. Then T is one-to-one if and only if $\mathrm{Ker}(T)$ consists only of the zero vector.

PROOF   Since T is a linear transformation, we have $T(\mathbf{0}) = \mathbf{0}$. Thus, if T is one-to-one, there can be no other $\mathbf{v}$ in V satisfying $T(\mathbf{v}) = \mathbf{0}$, and so, $\mathrm{Ker}(T) = \{\mathbf{0}\}$.

Conversely, suppose $\text{Ker}(T) = \{0\}$. If $\mathbf{v}_1 \neq \mathbf{v}_2$, then $\mathbf{v}_1 - \mathbf{v}_2 \neq 0$, and, since $\text{Ker}(T) = \{0\}$, $T(\mathbf{v}_1 - \mathbf{v}_2) \neq 0$. Hence, $T(\mathbf{v}_1) - T(\mathbf{v}_2) \neq 0$ or, equivalently, $T(\mathbf{v}_1) \neq T(\mathbf{v}_2)$. Thus, if $\text{Ker}(T) = \{0\}$, then T is one-to-one.   ■

We now have the following characterization of one-to-one and onto in terms of the kernel and range of T:

The linear transformation $T : V \to W$ is

1. one-to-one if and only if $\text{Ker}(T) = \{0\}$.

2. onto if and only if $\text{Rng}(T) = W$.

Example 6.4.3   Consider the linear transformation $T : P_3 \to P_3$ defined by

$$T(ax^2 + bx + c) = ax^2 + (b - 2c)x + (a - b + 2c).$$

Determine whether T is one-to-one or onto.

Solution   To determine whether T is one-to-one, we find $\text{Ker}(T)$. For the given transformation we have

$$\text{Ker}(T) = \{p \in P_3 : T(p) = 0\}$$
$$= \{ax^2 + bx + c : T(ax^2 + bx + c) = 0 \text{ for all } x\}$$
$$= \{ax^2 + bx + c : ax^2 + (b - 2c)x + (a - b + 2c) = 0 \text{ for all } x\}.$$

But,

$$ax^2 + (b - 2c)x + (a - b + 2c) = 0,$$

for all real x, if and only if

$$a = 0, \quad b - 2c = 0, \quad a - b + 2c = 0.$$

These equations are satisfied if and only if

$$a = 0, \ b = 2c,$$

so that

$$\text{Ker}(T) = \{ax^2 + bx + c : a = 0, b = 2c\}$$
$$= \{c(2x + 1) : c \in \mathbf{R}\}.$$

Since the kernel of T does not consist of just the zero vector, Theorem 6.4.2 implies that T is *not* one-to-one. Rather than computing $\text{Rng}(T)$ to determine whether T is onto, we use Theorem 6.3.2. The vectors in $\text{Ker}(T)$ consist of all scalar multiples of the nonzero (and hence LI) polynomial $p_1(x) = 2x + 1$. Thus $\{2x + 1\}$ is a basis for $\text{Ker}(T)$, and hence, $\dim[\text{Ker}(T)] = 1$. Further, we have seen in Chapter 5 that $\dim[P_3] = 3$. Therefore, from Theorem 6.3.2, we have

$$1 + \dim[\text{Rng}(T)] = 3,$$

which implies that

$$\dim[\text{Rng}(T)] = 2.$$

Thus $\text{Rng}(T)$ is a two-dimensional subspace of the three-dimensional vector space P_3, and so, $\text{Rng}(T) \neq P_3$. Hence, T is *not* onto.   ❑

Corollary 6.4.1:   If $T : V \to W$ is one-to-one and onto, then

$$\dim[V] = \dim[W].$$

PROOF   In general, we have

$$\dim[\mathrm{Ker}(T)] + \dim[\mathrm{Rng}(T)] = \dim[V].$$

If T is one-to-one, then $\dim[\mathrm{Ker}(T)] = 0$, and if T is onto, then $\dim[\mathrm{Rng}(T)] = \dim[W]$. Substituting these results into the previous equation yields

$$\dim[W] = \dim[V]. \qquad \blacksquare$$

If $T : V \to W$ is both one-to-one and onto, then for each $\mathbf{w} \in W$, there is a *unique* $\mathbf{v} \in V$ such that $T(\mathbf{v}) = \mathbf{w}$. We can therefore define a transformation $T^{-1} : W \to V$ by

$$T^{-1}(\mathbf{w}) = \mathbf{v} \text{ if and only if } \mathbf{w} = T(\mathbf{v}).$$

This transformation satisfies the basic properties of an inverse, namely,

$$T^{-1}(T(\mathbf{v})) = \mathbf{v} \text{ for all } \mathbf{v} \in V$$

and

$$T(T^{-1}(\mathbf{w})) = \mathbf{w} \text{ for all } \mathbf{w} \in W.$$

We call T^{-1} the *inverse* transformation. Again we stress that T^{-1} exists if and only if T is both one-to-one *and* onto. We leave it as an exercise to verify that T^{-1} is a *linear* transformation.

Definition 6.4.3:   Let $T : V \to W$ be a linear transformation. If T is both one-to-one and onto, then the linear transformation $T^{-1} : W \to V$ defined by

$$T^{-1}(\mathbf{w}) = \mathbf{v} \text{ if and only if } \mathbf{w} = T(\mathbf{v})$$

is called the **inverse transformation** to T.

Since the transformation in the previous example was neither one-to-one or onto, it follows that T^{-1} does not exist. For linear transformations $T : \mathbf{R}^n \to \mathbf{R}^n$, the following theorem characterizes the existence of an inverse transformation.

Theorem 6.4.3:   Let $T : \mathbf{R}^n \to \mathbf{R}^n$ be the linear transformation with matrix A. Then T^{-1} exists if and only if $\det(A) \neq 0$. Furthermore, $T^{-1} : \mathbf{R}^n \to \mathbf{R}^n$ is the linear transformation with matrix A^{-1}.

PROOF

$$\mathrm{Ker}(T) = \{\mathbf{x} \in \mathbf{R}^n : A\mathbf{x} = \mathbf{0}\}.$$

Hence T is one-to-one if and only if the homogeneous linear system $A\mathbf{x} = \mathbf{0}$ has just the trivial solution. But this is true if and only if $\det(A) \neq 0$. Furthermore,

$$\mathrm{Rng}(T) = \{A\mathbf{x} : \mathbf{x} \in \mathbf{R}^n\} = \mathrm{colspace}(A).$$

Consequently, T is onto if and only if the column vectors of A are LI. But this is true if and only if $\det(A) \neq 0$. Hence, T is both one-to-one and onto if and only if $\det(A) \neq 0$.
    Finally, if $\det(A) \neq 0$, then A^{-1} exists, so that

$$T(\mathbf{x}) = \mathbf{y} \Leftrightarrow A\mathbf{x} = \mathbf{y} \Leftrightarrow \mathbf{x} = A^{-1}\mathbf{y}.$$

Consequently, the inverse transformation is

$$T^{-1}(\mathbf{y}) = A^{-1}\mathbf{y},$$

from which it follows that T^{-1} is itself a *linear* transformation with matrix A^{-1}. ∎

Example 6.4.4 If $T : \mathbf{R}^3 \to \mathbf{R}^3$ has matrix $A = \begin{bmatrix} 2 & 3 & 1 \\ -1 & 2 & 3 \\ 4 & 1 & 6 \end{bmatrix}$, show that T^{-1} exists and find it.

Solution It is easily shown that $\det(A) = 63 \neq 0$, so that A is nonsingular. Consequently, T^{-1} exists and is given by

$$T^{-1}(\mathbf{x}) = A^{-1}\mathbf{x}.$$

Using the Gauss–Jordan technique or the adjoint method, it is found that

$$A^{-1} = \begin{bmatrix} 1/7 & -17/63 & 1/9 \\ 2/7 & 8/63 & -1/9 \\ -1/7 & 10/63 & 1/9 \end{bmatrix}.$$

Hence,

$$T^{-1}(x_1, x_2, x_3) = \left(\frac{1}{7}x_1 - \frac{17}{63}x_2 + \frac{1}{9}x_3, \frac{2}{7}x_1 + \frac{8}{63}x_2 - \frac{1}{9}x_3, -\frac{1}{7}x_1 + \frac{10}{63}x_2 + \frac{1}{9}x_3 \right).$$

ISOMORPHISM

Now let V be a (real) vector space of dimension n, and let $\{\mathbf{v}_1, \mathbf{v}_2, \ldots, \mathbf{v}_n\}$ be a basis for V. Then every vector in V can be written uniquely as

$$\mathbf{v} = c_1\mathbf{v}_1 + c_2\mathbf{v}_2 + \cdots + c_n\mathbf{v}_n.$$

We define the linear transformation $T : \mathbf{R}^n \to V$ by

$$T(c_1, c_2, \ldots, c_n) = c_1\mathbf{v}_1 + c_2\mathbf{v}_2 + \cdots + c_n\mathbf{v}_n.$$

Since $\{\mathbf{v}_1, \mathbf{v}_2, \ldots, \mathbf{v}_n\}$ is LI, it follows that $\mathrm{Ker}(T) = \{\mathbf{0}\}$, so that T is one-to-one. Furthermore, $\mathrm{Rng}(T) = \mathrm{span}\{\mathbf{v}_1, \mathbf{v}_2, \ldots, \mathbf{v}_n\} = V$, so that T is also onto. This trans–formation has therefore matched up vectors in $\mathbf{R}^n$ with vectors in V in such a manner that linear combinations are preserved under the mapping. Such a transformation is called an **isomorphism** between $\mathbf{R}^n$ and V. We say that the two vector spaces are **isomorphic**. The importance of this concept stems from the fact that it implies that all n-dimensional (real) vector spaces are isomorphic to $\mathbf{R}^n$, and therefore, by studying properties in $\mathbf{R}^n$, we are really studying all (real) vector spaces of dimension n. This illustrates the importance of the vector space $\mathbf{R}^n$.

Example 6.4.5 Determine an isomorphism between $\mathbf{R}^3$ and the vector space P_3.

Solution An arbitrary vector in P_3 can be expressed relative to the standard basis as

$$a_0 + a_1x + a_2x^2.$$

Consequently, an isomorphism between $\mathbf{R}^3$ and P_3 is $T : \mathbf{R}^3 \to P_3$ defined by

$$T(a_0, a_1, a_2) = a_0 + a_1x + a_2x^2.$$

Example 6.4.6   Determine an isomorphism between $\mathbf{R}^4$ and $M_2(\mathbf{R})$.

Solution   An arbitrary vector $A = \begin{bmatrix} a & b \\ c & d \end{bmatrix}$ in $M_2(\mathbf{R})$ can be written relative to the standard basis as

$$A = a\begin{bmatrix} 1 & 0 \\ 0 & 0 \end{bmatrix} + b\begin{bmatrix} 0 & 1 \\ 0 & 0 \end{bmatrix} + c\begin{bmatrix} 0 & 0 \\ 1 & 0 \end{bmatrix} + d\begin{bmatrix} 0 & 0 \\ 0 & 1 \end{bmatrix}.$$

Hence, we can define an isomorphism $T : \mathbf{R}^4 \rightarrow M_2(\mathbf{R})$ by

$$T(a, b, c, d) = \begin{bmatrix} a & b \\ c & d \end{bmatrix}.$$

EXERCISES 6.4

1. Let $T_1 : \mathbf{R}^2 \rightarrow \mathbf{R}^2$ and $T_2 : \mathbf{R}^2 \rightarrow \mathbf{R}^2$ be the linear transformations with matrices

$$A = \begin{bmatrix} -1 & 2 \\ 3 & 1 \end{bmatrix}, B = \begin{bmatrix} 1 & 5 \\ -2 & 0 \end{bmatrix},$$

respectively. Find T_1T_2 and T_2T_1. Does $T_1T_2 = T_2T_1$?

2. Let $T_1 : \mathbf{R}^2 \rightarrow \mathbf{R}^2$ and $T_2 : \mathbf{R}^2 \rightarrow \mathbf{R}$ be the linear transformations with matrices

$$A = \begin{bmatrix} 1 & -1 \\ 3 & 2 \end{bmatrix}, \quad B = [-1 \ \ 1],$$

respectively. Find T_2T_1. Does T_1T_2 exist? Explain.

3. Let $T_1 : \mathbf{R}^2 \rightarrow \mathbf{R}^2$, and $T_2 : \mathbf{R}^2 \rightarrow \mathbf{R}^2$ be the linear transformations with matrices

$$A = \begin{bmatrix} 1 & -1 \\ 2 & -2 \end{bmatrix} \text{ and } B = \begin{bmatrix} 2 & 1 \\ 3 & -1 \end{bmatrix},$$

respectively. Find: $\text{Ker}(T_1)$, $\text{Ker}(T_2)$, $\text{Ker}(T_1T_2)$, and $\text{Ker}(T_2T_1)$.

4. Let $T_1 : M_n(\mathbf{R}) \rightarrow M_n(\mathbf{R})$ and $T_2 : M_n(\mathbf{R}) \rightarrow M_n(\mathbf{R})$ be the linear transformations defined by $T_1(A) = A - A^T$ and $T_2(A) = A + A^T$. Show that T_2T_1 is the zero transformation.

5. Define $T_1 : C^1[a, b] \rightarrow C^0[a, b]$ and $T_2 : C^0[a, b] \rightarrow C^1[a, b]$ by

$$T_1(f) = f', \quad [T_2(f)](x) = \int_a^x f(t) \, dt, \quad a \leq x \leq b.$$

(a) If $f(x) = \sin(x - a)$, find $[T_1(f)](x)$, $[T_2(f)](x)$, and show that, for the given function,

$$[T_1T_2](f) = [T_2T_1](f) = f.$$

(b) Show that for general functions f and g,

$$[T_1T_2](f) = f, \quad \{[T_2T_1](g)\}(x) = g(x) - g(a).$$

6. Let $\{\mathbf{v}_1, \mathbf{v}_2\}$ be a basis for the vector space V, and suppose that $T_1 : V \rightarrow V$ and $T_2 : V \rightarrow V$ are the linear transformations satisfying

$$T_1(\mathbf{v}_1) = \mathbf{v}_1 - \mathbf{v}_2, \qquad T_1(\mathbf{v}_2) = 2\mathbf{v}_1 + \mathbf{v}_2,$$
$$T_2(\mathbf{v}_1) = \mathbf{v}_1 + 2\mathbf{v}_2, \qquad T_2(\mathbf{v}_2) = 3\mathbf{v}_1 - \mathbf{v}_2.$$

Determine $(T_2T_1)(\mathbf{v})$ for an arbitrary vector in V.

7. Let $T_1 : U \rightarrow V$ and $T_2 : V \rightarrow W$ be linear transformations. Complete the proof of Theorem 6.4.1 by showing that $(T_2T_1)(c\mathbf{v}_1) = c(T_2T_1)(\mathbf{v}_1)$.

For problems 8–10, find $\text{Ker}(T)$ and $\text{Rng}(T)$, and hence, determine whether the given transformation is one-to-one or onto. If T^{-1} exists, find it.

8. $T(\mathbf{x}) = A\mathbf{x}$ where $A = \begin{bmatrix} 4 & 2 \\ 1 & 3 \end{bmatrix}$.

9. $T(\mathbf{x}) = A\mathbf{x}$ where $A = \begin{bmatrix} 1 & 2 \\ -2 & -4 \end{bmatrix}$.

10. $T(\mathbf{x}) = A\mathbf{x}$ where $A = \begin{bmatrix} 1 & 2 & -1 \\ 2 & 5 & 1 \end{bmatrix}$.

11. Let V be a vector space and define $T : V \rightarrow V$ by $T(\mathbf{x}) = \lambda \mathbf{x}$, where λ is a (nonzero) scalar.

Show that T is a linear transformation that is one-to-one and onto, and find T^{-1}.

12. Define $T : P_2 \to P_2$ by:

$$T(ax + b) = (2b - a)x + (b + a).$$

Show that T is both one-to-one and onto, and find T^{-1}.

13. Define $T : P_3 \to P_2$ by

$$T(ax^2 + bx + c) = (a - b)x + c.$$

Determine whether T is one-to-one or onto. Does T^{-1} exist?

14. Let $\{\mathbf{v}_1, \mathbf{v}_2\}$ be a basis for the vector space V, and suppose that $T : V \to V$ is a linear transformation. If $T(\mathbf{v}_1) = \mathbf{v}_1 + 2\mathbf{v}_2$ and $T(\mathbf{v}_2) = 2\mathbf{v}_1 - 3\mathbf{v}_2$, show that T is one-to-one and onto and find T^{-1}.

15. Let $\mathbf{v}_1$, and $\mathbf{v}_2$ be a basis for the vector space V, and suppose that $T_1 : V \to V$ and $T_2 : V \to V$ are the linear transformations satisfying

$$T_1(\mathbf{v}_1) = \mathbf{v}_1 + \mathbf{v}_2, \quad T_1(\mathbf{v}_2) = \mathbf{v}_1 - \mathbf{v}_2,$$

$$T_2(\mathbf{v}_1) = \frac{1}{2}(\mathbf{v}_1 + \mathbf{v}_2), \quad T_2(\mathbf{v}_2) = \frac{1}{2}(\mathbf{v}_1 - \mathbf{v}_2).$$

Find $(T_1 T_2)(\mathbf{v})$ and $(T_2 T_1)(\mathbf{v})$ for an arbitrary vector in V and show that $T_2 = T_1^{-1}$.

16. Determine an isomorphism between $\mathbf{R}^2$ and the vector space P_2.

17. Determine an isomorphism between $\mathbf{R}^3$ and the subspace of $M_2(\mathbf{R})$ consisting of all upper triangular matrices.

18. Determine an isomorphism between $\mathbf{R}$ and the subspace of $M_2(\mathbf{R})$ consisting of all skew-symmetric matrices.

19. Determine an isomorphism between $\mathbf{R}^3$ and the subspace of $M_2(\mathbf{R})$ consisting of all symmetric matrices.

20. If $T : V \to W$ is an invertible linear transformation (that is, T^{-1} exists), show that $T^{-1} : W \to V$ is also a linear transformation.

21. Let $T : V \to W$ be a linear transformation. Prove that if $\dim[W] < \dim[V]$, then T cannot be one-to-one.

22. Let $T : V \to W$ be a linear transformation. Prove that if $\dim[W] > \dim[V]$, then T cannot be onto.

23. Prove that if $T : V \to V$ is a one-to-one linear transformation, then T^{-1} exists.

24 Let $T : V \to W$ be a linear transformation, and suppose that $\dim[W] = n$. Prove that if $\dim[\text{Rng}(T)] = n$, then T is onto.

25. Let $T_1 : V \to V$, and $T_2 : V \to V$ be linear transformations and suppose that T_2 is one-to-one. If

$$(T_1 T_2)(\mathbf{v}) = \mathbf{v} \text{ for all } \mathbf{v} \text{ in } V,$$

prove that

$$(T_2 T_1)(\mathbf{v}) = \mathbf{v} \text{ for all } \mathbf{v} \text{ in } V.$$

6.5 THE ALGEBRAIC EIGENVALUE/EIGENVECTOR PROBLEM

In order to motivate the problem to be studied in the next several sections, we recall from Chapter 1 that the DE

$$\frac{dx}{dt} = ax, \tag{6.5.1}$$

where a is a constant has general solution

$$x(t) = ce^{at}. \tag{6.5.2}$$

Now consider the pair of linear differential equations

$$\frac{dx_1}{dt} = a_{11}x_1 + a_{12}x_2,$$

$$\frac{dx_2}{dt} = a_{21}x_1 + a_{22}x_2,$$

where x_1 and x_2 are functions of the independent variable t, and the a_{ij} are constants. As we have seen in Section 3.3, this system can be written more succinctly as the vector equation

$$\frac{d\mathbf{x}}{dt} = A\mathbf{x}(t), \tag{6.5.3}$$

where

$$\mathbf{x}(t) = \begin{bmatrix} x_1(t) \\ x_2(t) \end{bmatrix}, \qquad A = [a_{ij}], \qquad \frac{d\mathbf{x}}{dt} = \begin{bmatrix} \dfrac{dx_1}{dt} \\ \dfrac{dx_2}{dt} \end{bmatrix}.$$

By a solution to this system we mean a vector

$$\mathbf{x}(t) = \begin{bmatrix} f_1(t) \\ f_2(t) \end{bmatrix},$$

that satisfies equation (6.5.3) for all t. Based on the solution (6.5.2) to the DE (6.5.1), we might suspect that the system (6.5.3) may have solutions of the form

$$\mathbf{x}(t) = \begin{bmatrix} e^{\lambda t}v_1 \\ e^{\lambda t}v_2 \end{bmatrix} = e^{\lambda t}\mathbf{v},$$

where λ, v_1, and v_2 are constants and $\mathbf{v} = \begin{bmatrix} v_1 \\ v_2 \end{bmatrix}$. Indeed, substituting this expression for $\mathbf{x}(t)$ into (6.5.3) yields

$$\lambda e^{\lambda t}\mathbf{v} = A(e^{\lambda t}\mathbf{v})$$

or equivalently,

$$\boxed{A\mathbf{v} = \lambda\mathbf{v}.} \tag{6.5.4}$$

We have therefore shown that

$$\mathbf{x}(t) = e^{\lambda t}\mathbf{v}$$

is a solution to the system of differential equations (6.5.3), provided that λ and $\mathbf{v}$ satisfy equation (6.5.4). In Chapter 8, we will pursue this technique for determining solutions to general linear systems of differential equations. For the remainder of the present chapter, however, we focus our attention on the mathematical problem of finding all scalars λ and all nonzero vectors $\mathbf{v}$ satisfying equation (6.5.4) for a given $n \times n$ matrix A. Although we used systems of DE to motivate the study of this problem, it is important to realize that it arises also in many areas of applications.

We begin by introducing the appropriate terminology.

> **Definition 6.5.1:**   Let A be an $n \times n$ matrix. Any values of λ for which
>
> $$A\mathbf{v} = \lambda\mathbf{v}$$
>
> has *nontrivial* solutions are called **eigenvalues** of A. The corresponding *nonzero* vectors $\mathbf{v}$ are called **eigenvectors** of A.

REMARK Eigenvalues and eigenvectors are also often referred to as **characteristic values** and **characteristic vectors**, respectively.

In order to formulate the eigenvalue/eigenvector problem within the vector space framework, we will interpret A as the matrix of a linear transformation $T : \mathbf{C}^n \to \mathbf{C}^n$ in the usual manner, that is $T(\mathbf{v}) = A\mathbf{v}$. In many of our problems, A and λ will both be real, which will enable us to restrict attention to $\mathbf{R}^n$, although we will often require the complex vector space $\mathbf{C}^n$. Indeed, we will see in the later chapters that complex eigenvalues and eigenvectors are required to describe linear physical systems that exhibit oscillatory behavior (in a similar manner that oscillatory behavior in a spring-mass system arises only if the auxiliary polynomial has complex conjugate roots).

It is always helpful to have a geometric interpretation of the problem under consideration. According to equation (6.5.4), the eigenvectors of A are those nonzero vectors that are mapped into themselves up to a constant scalar multiple by the linear transformation $T(\mathbf{v}) = A\mathbf{v}$. This is illustrated for the case of $\mathbf{R}^2$ in Figure 6.5.1.

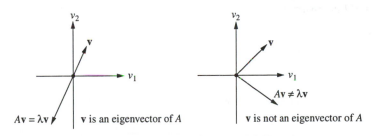

Figure 6.5.1 A geometrical description of the eigenvalue/eigenvector problem in $\mathbf{R}^2$.

Note that if $A\mathbf{v} = \lambda\mathbf{v}$ and c is an arbitrary scalar, then

$$A(c\mathbf{v}) = cA\mathbf{v} = c(\lambda\mathbf{v}) = \lambda(c\mathbf{v}).$$

Consequently, if $\mathbf{v}$ is an eigenvector of A, then so is $\mathbf{w} = c\mathbf{v}$, for any nonzero scalar c.

Example 6.5.1:   Verify that $\mathbf{v} = (1, 3)$ is an eigenvector of the matrix $A = \begin{bmatrix} 1 & 1 \\ -3 & 5 \end{bmatrix}$ corresponding to the eigenvalue $\lambda = 4$.

Solution   For the given vector, we have the following:[1]

$$A\mathbf{v} = \begin{bmatrix} 1 & 1 \\ -3 & 5 \end{bmatrix}\begin{bmatrix} 1 \\ 3 \end{bmatrix} = \begin{bmatrix} 4 \\ 12 \end{bmatrix} = 4\begin{bmatrix} 1 \\ 3 \end{bmatrix} = 4\mathbf{v}.$$

---

[1]Notice that once more we will switch between vectors in R^n and column n–vectors.

Consequently $\mathbf{v}$ is an eigenvector of A corresponding to the eigenvalue $\lambda = 4$.   ❑

SOLUTION OF THE PROBLEM

The solution of the eigenvalue/eigenvector problem hinges on the observation that (6.5.4) can be written in the equivalent form

$$(A - \lambda I)\mathbf{v} = \mathbf{0}, \tag{6.5.5}$$

where I denotes the identity matrix. Consequently, the eigenvalues of A are those values of λ for which the $n \times n$ linear system (6.5.5) has nontrivial solutions and the eigenvectors are the corresponding solutions. But, according to Corollary 4.2.2, the system (6.5.5) has nontrivial solutions if and only if

$$\det(A - \lambda I) = 0.$$

To solve the eigenvalue/eigenvector problem, we therefore proceed as follows:

---

1. Find all scalars λ such that $\det(A - \lambda I) = 0$. These are the eigenvalues of A.

2. If $\lambda_1, \lambda_2, \ldots, \lambda_k$ are the *distinct* eigenvalues obtained in (1) then solve the k systems of linear equations

$$(A - \lambda_i I)\mathbf{v}_i = 0, \quad i = 1, 2, 3, \ldots, k$$

to find all eigenvectors $\mathbf{v}_i$ corresponding to each eigenvalue.

For a given $n \times n$ matrix A, the degree n polynomial $p(\lambda)$ defined by

$$p(\lambda) = \det(A - \lambda I)$$

is called the **characteristic polynomial of** A, and the equation

$$p(\lambda) = 0$$

is called the **characteristic equation of** A. It follows from (1) that the eigenvalues of A are the roots of the characteristic equation.

SOME EXAMPLES

We now consider three examples to illustrate some of the possibilities that can arise in the eigenvalue/eigenvector problem and also to motivate some of the theoretical results that will be established in the next section.

Example 6.5.2   Find all eigenvalues and eigenvectors of $A = \begin{bmatrix} 5 & -4 \\ 8 & -7 \end{bmatrix}$.

Solution   The linear system for determining the eigenvalues and eigenvectors is $(A - \lambda I)\mathbf{v} = \mathbf{0}$, that is,

$$\begin{bmatrix} 5 - \lambda & -4 \\ 8 & -7 - \lambda \end{bmatrix} \begin{bmatrix} v_1 \\ v_2 \end{bmatrix} = \begin{bmatrix} 0 \\ 0 \end{bmatrix}. \tag{6.5.6}$$

This system has nontrivial solutions if and only if λ satisfies the characteristic equation

$$\begin{vmatrix} 5-\lambda & -4 \\ 8 & -7-\lambda \end{vmatrix} = 0.$$

Expanding the determinant yields

$$(\lambda - 5)(\lambda + 7) + 32 = 0.$$

That is,

$$\lambda^2 + 2\lambda - 3 = 0,$$

which has factorization

$$(\lambda + 3)(\lambda - 1) = 0.$$

Consequently, the eigenvalues of A are

$$\lambda_1 = -3, \quad \lambda_2 = 1.$$

Eigenvectors The corresponding eigenvectors are obtained by successively substituting the foregoing eigenvalues into (6.5.6) and solving the resulting system.

$\lambda_1 = -3$: The augmented matrix of the linear system $(A - \lambda_1 I)\mathbf{v} = \mathbf{0}$ is

$$\begin{bmatrix} 8 & -4 & 0 \\ 8 & -4 & 0 \end{bmatrix}$$

with RREF

$$\begin{bmatrix} 1 & -1/2 & 0 \\ 0 & 0 & 0 \end{bmatrix}.$$

The solution to the system can therefore be written in the form $\mathbf{v} = (r, 2r)$, where r is a free variable. It follows that the eigenvectors corresponding to $\lambda_1 = -3$ are those vectors in $\mathbf{R}^2$ of the form

$$\mathbf{v} = r(1, 2),$$

where r is any *nonzero* real number. Notice that there is only one *LI* eigenvector corresponding to the eigenvalue $\lambda_1 = -3$, which we may choose as $\mathbf{v}_1 = (1, 2)$. All other eigenvectors (corresponding to $\lambda_1 = -3$) are scalar multiples of $\mathbf{v}_1$.

$\lambda_2 = 1$: The augmented matrix of the system $(A - \lambda_2 I)\mathbf{v} = \mathbf{0}$ is

$$\begin{bmatrix} 4 & -4 & 0 \\ 8 & -8 & 0 \end{bmatrix}$$

with RREF

$$\begin{bmatrix} 1 & -1 & 0 \\ 0 & 0 & 0 \end{bmatrix}.$$

Consequently, the system has solutions $\mathbf{v} = (s, s)$, where s is a free variable. It follows that the eigenvectors corresponding to $\lambda = 1$ are those vectors in $\mathbf{R}^2$ of the form

$$\mathbf{v} = s(1, 1)$$

where s is any (nonzero) real number. Once more there is only one *LI* eigenvector corresponding to the eigenvalue $\lambda_2 = 1$, which we may choose as $\mathbf{v}_2 = (1, 1)$. All other eigenvectors (corresponding to $\lambda_2 = 1$) are scalar multiples of $\mathbf{v}_2$.

Notice that the eigenvectors $\mathbf{v}_1 = (1, 2)$ and $\mathbf{v}_2 = (1, 1)$ (which correspond to different eigenvalues) are nonproportional and therefore are LI in $\mathbf{R}^2$. The matrix A has therefore picked out a basis for $\mathbf{R}^2$. This is illustrated in Figure 6.5.2.

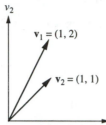

Figure 6.5.2 Two LI eigenvectors for the matrix in Example 6.5.2.

Example 6.5.3   Find all eigenvalues and eigenvectors of $A = \begin{bmatrix} 5 & 12 & -6 \\ -3 & -10 & 6 \\ -3 & -12 & 8 \end{bmatrix}$.

Solution   The system $(A - \lambda I)\mathbf{v} = \mathbf{0}$ has nontrivial solutions if and only if $\det(A - \lambda I) = 0$. For the given matrix,

$$\det(A - \lambda I) = \begin{vmatrix} 5 - \lambda & 12 & -6 \\ -3 & -10 - \lambda & 6 \\ -3 & -12 & 8 - \lambda \end{vmatrix}.$$

Using the cofactor expansion theorem along row 1 yields

$$\det(A - \lambda I) = (5 - \lambda)[(\lambda - 8)(\lambda + 10) + 72] - 12[3(\lambda - 8) + 18]$$
$$- 6[36 - 3(10 + \lambda)]$$
$$= (5 - \lambda)(\lambda^2 + 2\lambda - 8) + 18(2 - \lambda)$$
$$= (5 - \lambda)((\lambda - 2)(\lambda + 4) + 18(2 - \lambda)$$
$$= (2 - \lambda)[(\lambda - 5)(\lambda + 4) + 18] = (2 - \lambda)(\lambda^2 - \lambda - 2)$$
$$= (2 - \lambda)(\lambda - 2)(\lambda + 1) = -(\lambda - 2)^2(\lambda + 1).$$

Consequently, A has eigenvalues

$$\lambda_1 = 2 \text{ (multiplicity 2)}, \quad \lambda_2 = -1 \text{ (multiplicity 1)}.$$

Eigenvectors   To determine the corresponding eigenvectors of A, we must solve each of the homogeneous linear systems

$$(A - \lambda_1 I)\mathbf{v} = \mathbf{0}, \quad (A - \lambda_2 I)\mathbf{v} = \mathbf{0}.$$

$\lambda_1 = 2$:   The augmented matrix of the homogeneous linear system $(A - \lambda_1 I)\mathbf{v} = \mathbf{0}$ is

$$\begin{bmatrix} 3 & 12 & -6 & 0 \\ -3 & -12 & 6 & 0 \\ -3 & 12 & 6 & 0 \end{bmatrix}$$

with RREF

$$\begin{bmatrix} 1 & 4 & -2 & 0 \\ 0 & 0 & 0 & 0 \\ 0 & 0 & 0 & 0 \end{bmatrix}.$$

Hence, the eigenvectors of A corresponding to the eigenvalue $\lambda_1 = 2$ are

$$\mathbf{v} = (-4r + 2s, r, s),$$

where r and s are free variables, which cannot be zero simultaneously. (Why not?) Writing $\mathbf{v}$ in the equivalent form

$$\mathbf{v} = r(-4, 1, 0) + s(2, 0, 1),$$

we see that there are two LI eigenvectors corresponding to $\lambda_1 = 2$, which we may choose as $\mathbf{v}_1 = (-4, 1, 0)$ and $\mathbf{v}_2 = (2, 0, 1)$. All other eigenvectors corresponding to $\lambda_1 = 2$ are obtained by taking nontrivial linear combinations of $\mathbf{v}_1, \mathbf{v}_2$. Therefore, geometrically, these eigenvectors all lie in the plane through the origin of a Cartesian coordinate system that contains $\mathbf{v}_1$ and $\mathbf{v}_2$.

$\lambda_2 = -1$:   The augmented matrix of the linear system $(A - \lambda_2 I)\mathbf{v} = \mathbf{0}$ is

$$\begin{bmatrix} 6 & 12 & -6 & 0 \\ -3 & -9 & 6 & 0 \\ -3 & -12 & 9 & 0 \end{bmatrix}.$$

with RREF

$$\begin{bmatrix} 1 & 0 & 1 & 0 \\ 0 & 1 & -1 & 0 \\ 0 & 0 & 0 & 0 \end{bmatrix}.$$

Consequently, the eigenvectors of A corresponding to the eigenvalue $\lambda_2 = -1$ are those nonzero vectors in $\mathbf{R}^3$ of the form

$$\mathbf{v} = t(-1, 1, 1),$$

where t is a free variable ($t \neq 0$). We see that there is only one LI eigenvector corresponding to $\lambda_2 = -1$, which we may take to be $\mathbf{v}_3 = (-1, 1, 1)$. All other eigenvectors of A corresponding to $\lambda_2 = -1$ are scalar multiples of $\mathbf{v}_3$. Geometrically, these eigenvectors lie on the line with direction vector $\mathbf{v}_3$ which passes through the origin of a Cartesian coordinate system.

We note that the eigenvectors of A have given rise to the set of vectors $\{(-4, 1, 0), (2, 0, 1), (-1, 1, 1)\}$. Furthermore, since

$$\det([\mathbf{v}_1, \mathbf{v}_2, \mathbf{v}_3]) = \begin{vmatrix} -4 & 2 & -1 \\ 1 & 0 & 1 \\ 0 & 1 & 1 \end{vmatrix} = 1 \neq 0$$

the set of eigenvectors $\{(-4, 1, 0), (2, 0, 1), (-1, 1, 1)\}$ is LI and therefore is a basis for $\mathbf{R}^3$.   □

The matrices and eigenvalues in the previous examples have all been *real*, and this enabled us to regard the eigenvectors as being vectors in $\mathbf{R}^n$. We now consider the case when some or all of the eigenvalues and eigenvectors are complex. The steps in determining the eigenvalues and eigenvectors do not change, although they can be more complicated algebraically, since the equations determining the eigenvectors will have complex coefficients. For the majority of matrices that we consider, the matrix A will have only *real* elements. In these cases, the following theorem can save some work in determining any complex eigenvectors.

Theorem 6.5.1:   Let A be an $n \times n$ matrix with *real* elements. If λ is a complex eigenvalue of A with corresponding eigenvector $\mathbf{v}$, then $\bar{\lambda}$ is an eigenvalue of A with corresponding eigenvector $\bar{\mathbf{v}}$.

PROOF   If $A\mathbf{v} = \lambda\mathbf{v}$,[1] then $\overline{A\mathbf{v}} = \overline{\lambda\mathbf{v}}$, which implies that $A\bar{\mathbf{v}} = \bar{\lambda}\bar{\mathbf{v}}$, since A has real entries.   ∎

REMARK   According to the previous theorem, if we find the eigenvectors of a real matrix A corresponding to a complex eigenvalue λ, then we can obtain the eigenvectors corresponding to the eigenvalue $\bar{\lambda}$ *without having to solve a linear system*.

Example 6.5.4   Find all eigenvalues and eigenvectors of $A = \begin{bmatrix} -2 & -6 \\ 3 & 4 \end{bmatrix}$.

Solution A has characteristic polynomial

$$p(\lambda) = \begin{vmatrix} -2-\lambda & -6 \\ 3 & 4-\lambda \end{vmatrix} = \lambda^2 - 2\lambda + 10.$$

Consequently the eigenvalues of A are

$$\lambda_1 = 1 + 3i, \ \lambda_2 = \bar{\lambda}_1 = 1 - 3i.$$

Since these are complex eigenvalues, we take the underlying vector space as being $\mathbf{C}^2$, and hence, any scalars that arise in the solution of the problem will be complex.

Eigenvectors

$\lambda_1 = 1 + 3i$:   Reducing the augmented matrix of the system $(A - \lambda_1 I)\mathbf{v} = \mathbf{0}$ yields

$$\begin{bmatrix} -3 - 3i & -6 & 0 \\ 3 & 3 - 3i & 0 \end{bmatrix} \sim \begin{bmatrix} 18 & 18 - 18i & 0 \\ 3 & 3 - 3i & 0 \end{bmatrix} \sim \begin{bmatrix} 1 & 1 - i & 0 \\ 0 & 0 & 0 \end{bmatrix},$$

so that the eigenvectors of A corresponding to $\lambda_1 = 1 + 3i$ are those vectors in $\mathbf{C}^2$ of the form

$$\mathbf{v} = r(-(1 - i), 1),$$

where r is an arbitrary (nonzero) *complex* number.

$\lambda_2 = 1 - 3i$:   From Theorem 6.5.1, the eigenvectors in this case are those vectors in $\mathbf{C}^2$ of the form $\mathbf{v} = s(-(1 + i), 1)$, where s is an arbitrary (nonzero) complex number.

Notice that the eigenvectors corresponding to different eigenvalues are LI vectors in $\mathbf{C}^2$. For example, a LI set of eigenvectors is $\{(-(1 - i), 1), (-(1 + i), 1)\}$. Once more the eigenvectors have determined a basis in the underlying vector space (in this case $\mathbf{C}^2$).   ❑

---

[1] If $A = [a_{ij}]$ then, by definition, $\bar{A} = [\bar{a}_{ij}]$.

When first encountering the eigenvalue/eigenvector problem, students often focus so much attention on the computational aspects of the problem that they lose sight of the original equation that defines the eigenvalues and eigenvectors of A. The following examples illustrate the importance of the defining equation when establishing theoretical results.

Example 6.5.5 Let λ be an eigenvalue of the matrix A with corresponding eigenvector $\mathbf{v}$. Prove that λ^2 is an eigenvalue of A^2 with corresponding eigenvector $\mathbf{v}$.

Solution We are given that

$$A\mathbf{v} = \lambda\mathbf{v}, \qquad (6.5.7)$$

and we must establish that

$$A^2\mathbf{v} = \lambda^2\mathbf{v}.$$

Equation (6.5.7) provides a starting point. If we premultiply both sides of this equation by A, then

$$A^2\mathbf{v} = A(\lambda\mathbf{v}) = \lambda(A\mathbf{v}) = \lambda(\lambda\mathbf{v}) = \lambda^2\mathbf{v},$$

and the result is established.

Example 6.5.6 Let λ and $\mathbf{v}$ be an eigenvalue/eigenvector pair for the $n \times n$ matrix A. If k is an arbitrary real number, prove that $\mathbf{v}$ is also an eigenvector of the matrix $A - kI$ corresponding to the eigenvalue $\lambda - k$.

Solution Once more, the only information that we are given is that

$$A\mathbf{v} = \lambda\mathbf{v}.$$

If we let $B = A - kI$, then we must establish that

$$B\mathbf{v} = (\lambda - k)\mathbf{v}.$$

But,

$$B\mathbf{v} = (A - kI)\mathbf{v} = A\mathbf{v} - k\mathbf{v} = \lambda\mathbf{v} - k\mathbf{v} = (\lambda - k)\mathbf{v},$$

as required.

EXERCISES 6.5

For problems 1–3, use equation (6.5.4) to verify that λ and $\mathbf{v}$ are an eigenvalue/eigenvector pair for the given matrix A.

$$A = \begin{bmatrix} 1 & -2 & -6 \\ -2 & 2 & -5 \\ 2 & 1 & 8 \end{bmatrix}.$$

1. $\lambda = 4$, $\mathbf{v} = (1, 1)$, $A = \begin{bmatrix} 1 & 3 \\ 2 & 2 \end{bmatrix}$.

2. $\lambda = 3$, $\mathbf{v} = (2, 1, -1)$,

3. $\lambda = -2$, $\mathbf{v} = \alpha(1, 0, -3) + \beta(4, -3, 0)$,

$$A = \begin{bmatrix} 1 & 4 & 1 \\ 3 & 2 & 1 \\ 3 & 4 & -1 \end{bmatrix}, \text{ where } \alpha \text{ and } \beta \text{ are constants.}$$

4. Given that $\mathbf{v}_1 = (1, -2)$ and $\mathbf{v}_2 = (1, 1)$ are eigenvectors of $A = \begin{bmatrix} 4 & 1 \\ 2 & 3 \end{bmatrix}$, determine the eigenvalues of A.

5. The effect of the linear transformation $T : \mathbf{R}^2 \rightarrow \mathbf{R}^2$ with matrix $A = \begin{bmatrix} 1 & 0 \\ 0 & -1 \end{bmatrix}$ is to reflect each vector in the x-axis. By arguing geometrically, determine all eigenvalues and eigenvectors of A.

6. The linear transformation $T : \mathbf{R}^2 \rightarrow \mathbf{R}^2$ with matrix $A = \begin{bmatrix} \cos \theta & -\sin \theta \\ \sin \theta & \cos \theta \end{bmatrix}$ rotates vectors in the xy-plane counterclockwise through an angle θ, where $0 \leq \theta < 2\pi$. By arguing geometrically, determine all values of θ for which A has *real* eigenvalues. Find the real eigenvalues and the corresponding eigenvectors.

For problems 7–25, determine all eigenvalues and corresponding eigenvectors of the given matrix.

7. $\begin{bmatrix} 3 & -1 \\ -5 & -1 \end{bmatrix}$.

8. $\begin{bmatrix} 1 & 6 \\ 2 & -3 \end{bmatrix}$.

9. $\begin{bmatrix} 7 & 4 \\ -1 & 3 \end{bmatrix}$.

10. $\begin{bmatrix} 2 & 0 \\ 0 & 2 \end{bmatrix}$.

11. $\begin{bmatrix} 3 & -2 \\ 4 & -1 \end{bmatrix}$.

12. $\begin{bmatrix} 2 & 3 \\ -3 & 2 \end{bmatrix}$.

13. $\begin{bmatrix} 10 & -12 & 8 \\ 0 & 2 & 0 \\ -8 & 12 & -6 \end{bmatrix}$.

14. $\begin{bmatrix} 3 & 0 & 0 \\ 0 & 2 & -1 \\ 1 & -1 & 2 \end{bmatrix}$.

15. $\begin{bmatrix} 1 & 0 & 0 \\ 0 & 3 & 2 \\ 2 & -2 & -1 \end{bmatrix}$.

16. $\begin{bmatrix} 6 & 3 & -4 \\ -5 & -2 & 2 \\ 0 & 0 & -1 \end{bmatrix}$.

17. $\begin{bmatrix} 7 & -8 & 6 \\ 8 & -9 & 6 \\ 0 & 0 & -1 \end{bmatrix}$.

18. $\begin{bmatrix} 0 & 1 & -1 \\ 0 & 2 & 0 \\ 2 & -1 & 3 \end{bmatrix}$.

19. $\begin{bmatrix} 1 & 0 & 0 \\ 0 & 0 & 1 \\ 0 & -1 & 0 \end{bmatrix}$.

20. $\begin{bmatrix} -2 & 1 & 0 \\ 1 & -1 & -1 \\ 1 & 3 & -3 \end{bmatrix}$.

21. $\begin{bmatrix} 2 & -1 & 3 \\ 3 & 1 & 0 \\ 2 & -1 & 3 \end{bmatrix}$.

22. $\begin{bmatrix} 5 & 0 & 0 \\ 0 & 5 & 0 \\ 0 & 0 & 5 \end{bmatrix}$.

23. $\begin{bmatrix} 0 & 2 & 2 \\ 2 & 0 & 2 \\ 2 & 2 & 0 \end{bmatrix}$.

24. $\begin{bmatrix} 1 & 2 & 3 & 4 \\ 4 & 3 & 2 & 1 \\ 4 & 5 & 6 & 7 \\ 7 & 6 & 5 & 4 \end{bmatrix}$.

25. $\begin{bmatrix} 0 & 1 & 0 & 0 \\ -1 & 0 & 0 & 0 \\ 0 & 0 & 0 & -1 \\ 0 & 0 & 1 & 0 \end{bmatrix}$.

26. Find all eigenvalues and corresponding eigenvectors of

$$A = \begin{bmatrix} 1+i & 0 & 0 \\ 2-2i & 1-i & 0 \\ 2i & 0 & 1 \end{bmatrix}.$$

Note that the eigenvectors do not occur in complex conjugate pairs. Does this contradict Theorem 6.5.1? Explain.

27. Consider the matrix $A = \begin{bmatrix} 1 & -1 \\ 2 & 4 \end{bmatrix}$.

(a) Show that the characteristic polynomial of A is $p(\lambda) = \lambda^2 - 5\lambda + 6$.

(b) Show that A satisfies its characteristic equation. That is, $A^2 - 5A + 6I_2 = 0_2$. (This result is known as the Cayley–Hamilton Theorem and is true for a general $n \times n$ matrix.)

(c) Use the result from (b) to find A^{-1} [Hint: Multiply the equation in (b) by A^{-1}.]

28. Let $A = \begin{bmatrix} 1 & 2 \\ 2 & -2 \end{bmatrix}$.

(a) Determine all eigenvalues of A.

(b) Reduce A to row-echelon form, and determine the eigenvalues of the resulting matrix. Are these the same as the eigenvalues of A ?

29. If $v_1 = (1, -1)$ and $v_2 = (2, 1)$ are eigenvectors of the matrix A corresponding to the eigenvalues $\lambda_1 = 2$, $\lambda_2 = -3$, respectively, find $A(3v_1 - v_2)$.

30. Let $v_1 = (1, -1, 1)$, $v_2 = (2, 1, 3)$, and $v_3 = (-1, -1, 2)$ be eigenvectors of the matrix A which correspond to the eigenvalues $\lambda_1 = 2$, $\lambda_2 = -2$, and $\lambda_3 = 3$, respectively, and let $v = (5, 0, 3)$.
(a) Express v as a linear combination of v_1, v_2, v_3.

(b) Find Av.

31. If v_1, v_2, v_3 are eigenvectors of A corresponding to the eigenvalue λ, and c_1, c_2, c_3 are scalars (not all zero), show that $c_1v_1 + c_2v_2 + c_3v_3$ is also an eigenvector of A corresponding to the eigenvalue λ.

32. Prove that the eigenvalues of an upper (or lower) triangular matrix are just the diagonal elements of the matrix.

33. Let A be an $n \times n$ nonsingular matrix. Prove that if λ is an eigenvalue of A, then $1/\lambda$ is an eigenvalue of A^{-1}.

34. Let A be an $n \times n$ matrix. Prove that $\lambda = 0$ is an eigenvalue of A if and only if $\det(A) = 0$.

35. Let A be an $n \times n$ matrix. Prove that A and A^T have the same eigenvalues. [Hint: Show that $\det(A^T - \lambda I) = \det(A - \lambda I)$.]

36. Let A be an $n \times n$ real matrix with complex eigenvalue $\lambda = a + ib$, where $b \neq 0$, and let $v = r + is$ be a corresponding eigenvector of A.
(a) Prove that r and s are *nonzero* vectors in $\mathbf{R}^n$.
(b) Prove that $\{r, s\}$ is LI in $\mathbf{R}^n$.

For problems 37–42, use some form of technology to determine the eigenvalues and eigenvectors of A in the following manner.

(1) Form the matrix $A - \lambda I$.

(2) Solve the characteristic equation $\det(A - \lambda I) = 0$ to determine the eigenvalues of A.

(3) Solve each of the systems $(A - \lambda_i I)v = 0$ to determine the eigenvectors of A.

◆ **37.** $A = \begin{bmatrix} 3 & 1 \\ 2 & 4 \end{bmatrix}$.

◆ **38.** $A = \begin{bmatrix} 5 & 34 & -41 \\ 4 & 17 & -23 \\ 5 & 24 & -31 \end{bmatrix}$.

◆ **39.** $A = \begin{bmatrix} 4 & 1 & 1 \\ 1 & 4 & 1 \\ 1 & 1 & 4 \end{bmatrix}$.

◆ **40.** $A = \begin{bmatrix} 1 & 1 & 1 \\ 3 & -1 & 2 \\ 3 & 1 & 4 \end{bmatrix}$.

◆ **41.** $A = \begin{bmatrix} 0 & 1 & -2 \\ -1 & 0 & 2 \\ 2 & -2 & 0 \end{bmatrix}$.

◆ **42.** $A = \begin{bmatrix} 0 & 1 & 1 & 1 & 1 \\ 1 & 0 & 1 & 1 & 1 \\ 1 & 1 & 0 & 1 & 1 \\ 1 & 1 & 1 & 0 & 1 \\ 1 & 1 & 1 & 1 & 0 \end{bmatrix}$.

For problems 43–48, use some form of technology to directly determine the eigenvalues and eigenvectors of the given matrix.

◆ **43.** The matrix in problem 37.

◆ **44.** The matrix in problem 38.

◆ **45.** The matrix in problem 39.

◆ **46.** The matrix in problem 40.

◆ **47.** The matrix in problem 41.

◆ **48.** The matrix in problem 42.

6.6   GENERAL RESULTS FOR EIGENVALUES AND EIGENVECTORS

In this section, we look more closely at the relationship between the eigenvalues and eigenvectors of an $n \times n$ matrix. Our aim is to formalize several of the ideas introduced via the examples of the previous section.

For a given $n \times n$ matrix $A = [a_{ij}]$, the characteristic polynomial $p(\lambda)$ assumes the form

$$p(\lambda) = \det(A - \lambda I) = \begin{vmatrix} a_{11} - \lambda & a_{12} & \cdots & a_{1n} \\ a_{21} & a_{22} - \lambda & \cdots & a_{2n} \\ \vdots & \vdots & & \vdots \\ a_{n1} & a_{n2} & \cdots & a_{nn} - \lambda \end{vmatrix}$$

Expanding this determinant yields a polynomial of degree n in λ with leading coefficient $(-1)^n$. It follows that $p(\lambda)$ can be written in the form

$$p(\lambda) = (-1)^n \lambda^n + b_1 \lambda^{n-1} + b_2 \lambda^{n-2} + \cdots + b_n,$$

where $b_1, b_2, \ldots, b_n$ are scalars. Since we consider the underlying vector space to be $\mathbf{C}^n$, the fundamental theorem of algebra guarantees that $p(\lambda)$ will have precisely n zeros (not necessarily distinct), and hence, A will have n eigenvalues. If we let $\lambda_1, \lambda_2, \ldots, \lambda_k$ denote the *distinct* eigenvalues of A, then $p(\lambda)$ can be factored as

$$p(\lambda) = (-1)^n (\lambda - \lambda_1)^{m_1} (\lambda - \lambda_2)^{m_2} (\lambda - \lambda_3)^{m_3} \cdots (\lambda - \lambda_k)^{m_k},$$

where, since $p(\lambda)$ has degree n,

$$m_1 + m_2 + m_3 + \cdots + m_k = n.$$

Thus, associated with each eigenvalue λ_i is a number m_i, called the **multiplicity** of λ_i.

We now focus our attention on the eigenvectors of A.

Definition 6.6.1:   Let A be an $n \times n$ matrix. For a given eigenvalue λ_i, let E_i denote the set of *all* vectors satisfying $A\mathbf{v} = \lambda_i \mathbf{v}$. Then E_i is called the **eigenspace** of A corresponding to the eigenvalue λ_i. Thus, E_i is the solution set to the linear system $(A - \lambda_i I)\mathbf{v} = \mathbf{0}$.

REMARKS

1. Equivalently, we can say that the eigenspace E_i is the kernel of the linear transformation $T_i : \mathbf{C}^n \to \mathbf{C}^n$ defined by $T_i(\mathbf{v}) = (A - \lambda_i I)\mathbf{v}$.

2. It is important to notice that there is one eigenspace associated with each eigenvalue of A.

3. Also note that the only difference between the eigenspace corresponding to a specific eigenvalue, and the set of all eigenvectors corresponding to that eigenvalue is that the eigenspace includes the zero vector.

Example 6.6.1   Determine all eigenspaces for the matrix $A = \begin{bmatrix} 5 & -4 \\ 8 & -7 \end{bmatrix}$.

Solution   We have already computed the eigenvalues and eigenvectors of A in Example 6.5.2. The eigenvalues of A are $\lambda_1 = -3$ and $\lambda_2 = 1$. The eigenvectors corresponding to $\lambda_1 = -3$ are all nonzero vectors of the form $\mathbf{v} = r(1, 2)$. Thus, the eigenspace corresponding to $\lambda_1 = -3$ is

$$E_1 = \{ \mathbf{v} \in \mathbf{R}^2 : \mathbf{v} = r(1, 2), \, r \in \mathbf{R} \}.$$

The eigenvectors corresponding to the eigenvalue $\lambda_2 = 1$ are of the form $\mathbf{v} = s(1, 1)$, where $s \neq 0$, so that the eigenspace corresponding to $\lambda_2 = 1$ is

$$E_2 = \{ \mathbf{v} \in \mathbf{R}^2 : \mathbf{v} = s(1, 1), \, s \in \mathbf{R} \}. \qquad \square$$

We have one main result for eigenspaces.

Theorem 6.6.1:   Let λ_i be an eigenvalue of A of multiplicity m_i and let E_i denote the corresponding eigenspace. Then

1. For each i, E_i is a vector space.

2. If n_i denotes the dimension of E_i, then $1 \le n_i \le m_i$ for each i. In words, the dimension of the eigenspace corresponding to λ_i is less than or equal to the multiplicity of λ_i.

PROOF
1. From Definition 6.6.1, E_i is the null space of the matrix $A - \lambda_i I$ and hence is a vector space.
2. The proof of this result requires some more advanced ideas about linear transformations than we have developed and is therefore omitted. (See, for example, Shilov, G.E., *Linear Algebra*, Dover Publications, 1977.) ∎

REMARK   The numbers m_i and n_i are called the **algebraic** multiplicity and **geometric** multiplicity of the eigenvalue λ_i respectively.

Example 6.6.2   Determine all eigenspaces and their dimensions for the matrix

$$A = \begin{bmatrix} 3 & -1 & 0 \\ 0 & 2 & 0 \\ -1 & 1 & 2 \end{bmatrix}.$$

Solution   A straightforward calculation yields the characteristic polynomial

$$p(\lambda) = -(\lambda - 2)^2(\lambda - 3),$$

so that the eigenvalues of A are $\lambda_1 = 2$ (with multiplicity 2) and $\lambda_2 = 3$. The eigenvectors corresponding to $\lambda_1 = 2$ are determined by solving

$$v_1 - v_2 = 0$$

and hence are of the form

$$\mathbf{v} = (r, r, s) = r(1, 1, 0) + s(0, 0, 1),$$

where r and s cannot be zero simultaneously (since eigenvectors are nonzero vectors). Thus, the eigenspace corresponding to $\lambda_1 = 2$ is

$$E_1 = \{\mathbf{v} \in \mathbf{R}^3 : \mathbf{v} = r(1, 1, 0) + s(0, 0, 1), \, r, \, s \in \mathbf{R}\}.$$

We see that the LI set $\{(1, 1, 0), (0, 0, 1)\}$ is a basis for the eigenspace E_1, and hence, the dimension of this eigenspace is 2 (that is, $n_1 = 2$).

It is easily shown that the eigenvectors corresponding to the eigenvalue $\lambda_2 = 3$ are of the form

$$\mathbf{v} = t(1, 0, -1),$$

where t is a nonzero real number. Thus, the eigenspace corresponding to $\lambda_2 = 3$ is

$$E_2 = \{\mathbf{v} \in \mathbf{R}^3 : \mathbf{v} = t(1, 0, -1), \, t \in \mathbf{R}\}.$$

Consequently, $\{(1, 0, -1)\}$ is a basis for this eigenspace and hence the dimension of the eigenspace is one (that is, $n_2 = 1$).

The eigenspaces E_1 and E_2 are sketched in Figure 6.6.1.

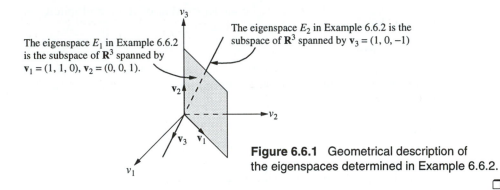

The eigenspace E_1 in Example 6.6.2 is the subspace of $\mathbf{R}^3$ spanned by $\mathbf{v}_1 = (1, 1, 0)$, $\mathbf{v}_2 = (0, 0, 1)$.

The eigenspace E_2 in Example 6.6.2 is the subspace of $\mathbf{R}^3$ spanned by $\mathbf{v}_3 = (1, 0, -1)$

Figure 6.6.1 Geometrical description of the eigenspaces determined in Example 6.6.2.

We now consider the relationship between eigenvectors corresponding to *distinct* eigenvalues. There is one key theorem which has already been illustrated in the previous section.

Theorem 6.6.2: Eigenvectors corresponding to *distinct* eigenvalues are LI.

PROOF   We use induction to prove the result. Let $\lambda_1, \lambda_2, \ldots, \lambda_m$ be distinct eigenvalues of A with corresponding eigenvectors $\mathbf{v}_1, \mathbf{v}_2, \ldots, \mathbf{v}_m$. It is certainly true that $\{\mathbf{v}_1\}$ is LI. Now suppose that $\{\mathbf{v}_1, \mathbf{v}_2, \ldots, \mathbf{v}_k\}$ is LI for some $k < m$, and consider the set $\{\mathbf{v}_1, \mathbf{v}_2, \ldots, \mathbf{v}_k, \mathbf{v}_{k+1}\}$. We wish to show that this set of vectors is LI. Consider

$$c_1\mathbf{v}_1 + c_2\mathbf{v}_2 + \cdots + c_k\mathbf{v}_k + c_{k+1}\mathbf{v}_{k+1} = \mathbf{0} . \tag{6.6.1}$$

Premultiplying both sides of this equation by A yields, using $A\mathbf{v}_i = \lambda_i\mathbf{v}_i$,

$$c_1\lambda_1\mathbf{v}_1 + c_2\lambda_2\mathbf{v}_2 + \cdots + c_k\lambda_k\mathbf{v}_k + c_{k+1}\lambda_{k+1}\mathbf{v}_{k+1} = \mathbf{0}. \tag{6.6.2}$$

But, from equation (6.6.1),

$$c_{k+1}\mathbf{v}_{k+1} = -(c_1\mathbf{v}_1 + c_2\mathbf{v}_2 + \cdots + c_k\mathbf{v}_k),$$

so that equation (6.6.2) can be written as

$$c_1\lambda_1\mathbf{v}_1 + c_2\lambda_2\mathbf{v}_2 + \cdots + c_k\lambda_k\mathbf{v}_k - \lambda_{k+1}(c_1\mathbf{v}_1 + c_2\mathbf{v}_2 + \cdots + c_k\mathbf{v}_k) = \mathbf{0}.$$

That is,

$$c_1(\lambda_1 - \lambda_{k+1})\mathbf{v}_1 + c_2(\lambda_2 - \lambda_{k+1})\mathbf{v}_2 + \cdots + c_k(\lambda_k - \lambda_{k+1})\mathbf{v}_k = \mathbf{0}.$$

Since $\mathbf{v}_1, \mathbf{v}_2, ..., \mathbf{v}_k$ are LI, this implies that

$$c_1(\lambda_1 - \lambda_{k+1}) = 0, \;\; c_2(\lambda_2 - \lambda_{k+1}) = 0, \;\; ..., \;\; c_k(\lambda_k - \lambda_{k+1}) = 0,$$

and hence, since the λ_i are distinct,

$$c_1 = c_2 = \cdots = c_k = 0.$$

But now, since $\mathbf{v}_{k+1} \neq \mathbf{0}$, it follows from equation (6.6.1) that $c_{k+1} = 0$ also, and so, $\{\mathbf{v}_1, \mathbf{v}_2, ..., \mathbf{v}_k, \mathbf{v}_{k+1}\}$ is LI. We have therefore shown that the desired result is true for $\{\mathbf{v}_1, \mathbf{v}_2, ..., \mathbf{v}_k, \mathbf{v}_{k+1}\}$ whenever it is true for $\{\mathbf{v}_1, \mathbf{v}_2, ..., \mathbf{v}_k\}$, and, since the result is true for a single eigenvector, it is true for $\{\mathbf{v}_1, \mathbf{v}_2, ..., \mathbf{v}_k\}$, $1 \leq k \leq m$. ■

Corollary 6.6.1:   Let $E_1, E_2, ..., E_k$ denote the eigenspaces of the $m \times n$ matrix A. In each eigenspace, choose a set of LI eigenvectors, and let $\{\mathbf{v}_1, \mathbf{v}_2, ..., \mathbf{v}_r\}$ denote the union of the LI sets. Then $\{\mathbf{v}_1, \mathbf{v}_2, ..., \mathbf{v}_r\}$ is LI.

PROOF   We argue by contradiction. Suppose that $\{\mathbf{v}_1, \mathbf{v}_2, ..., \mathbf{v}_r\}$ is LD. Then there exist scalars $c_1, c_2, ..., c_r$, not all zero, such that

$$c_1\mathbf{v}_1 + c_2\mathbf{v}_2 + \cdots + c_r\mathbf{v}_r = \mathbf{0}, \tag{6.6.3}$$

which can be written as

$$\mathbf{w}_1 + \mathbf{w}_2 + \cdots + \mathbf{w}_k = \mathbf{0},$$

where each $\mathbf{w}_i$ is a vector in E_i. Since not all the scalars are zero in equation (6.6.3), it follows that at least one of the $\mathbf{w}_i$ is nonzero. But this would imply that $\{\mathbf{w}_1, \mathbf{w}_2, ..., \mathbf{w}_k\}$ is LD, which contradicts Theorem 6.6.2. Consequently, all of the scalars in equation (6.6.3) must be zero, and so $\{\mathbf{v}_1, \mathbf{v}_2, ..., \mathbf{v}_r\}$ is indeed LI. ■

Since the dimension of $\mathbf{R}^n$ (or $\mathbf{C}^n$) is n, the maximum number of LI eigenvectors that A can have is n. In such a case, we say that A is nondefective. The following definition introduces the appropriate terminology.

Definition 6.6.2:   An $n \times n$ matrix A that has n LI eigenvectors is called **nondefective**. In such a case we say that A has a **complete set of eigenvectors**. If A has less than n LI eigenvectors it is called **defective**.

If A is nondefective, then any set of n LI eigenvectors of A is a basis for $\mathbf{R}^n$ (or $\mathbf{C}^n$). Such a basis is referred to as an **eigenbasis** of A.

Example 6.6.3   For the matrix in the previous example a complete set of eigenvectors is $\{(1, 1, 0), (0, 0, 1), (1, 0, -1)\}$. Consequently, the matrix is non-defective.

Example 6.6.4   Determine whether $A = \begin{bmatrix} 4 & -1 \\ 1 & 2 \end{bmatrix}$ is defective or nondefective.

Solution   The characteristic polynomial of A is

$$p(\lambda) = (4 - \lambda)(2 - \lambda) + 1 = \lambda^2 - 6\lambda + 9 = (\lambda - 3)^2.$$

Thus, $\lambda_1 = 3$ is an eigenvalue of multiplicity 2. The eigenvectors of A are determined by solving

$$v_1 - v_2 = 0,$$

so that the eigenspace corresponding to $\lambda_1 = 3$ is

$$E_1 = \{ \mathbf{v} \in \mathbf{R}^2 : \mathbf{v} = r(1, 1),\ r \in \mathbf{R} \}.$$

Thus, $\dim[E_1] = 1$ which implies that A is defective.   □

The next result is a direct consequence of the previous theorem.

Corollary 6.6.2:   If an $n \times n$ matrix A has n *distinct* eigenvalues, then it is nondefective.

PROOF   If A has n distinct eigenvalues, then, from the previous theorem, it has n LI eigenvectors.   ■

Note that if A does *not* have n distinct eigenvalues, it may *still* be nondefective. (In Example 6.5.3, there are only two distinct eigenvalues, but there are three LI eigenvectors, which gives a complete set.) The general result is as follows.

Theorem 6.6.3:   An $n \times n$ matrix A is nondefective if and only if the dimension of each eigenspace is the same as the algebraic multiplicity of the corresponding eigenvalue, that is, if and only if $n_i = m_i$ for each i.

PROOF   Suppose that A is nondefective. Then

$$n_1 + n_2 + \cdots + n_k = n.$$

If $n_i < m_i$ for some i, then

$$n_1 + n_2 + \cdots + n_k < m_1 + m_2 + \cdots + m_k = n,$$

which would imply that we had fewer than n eigenvectors, thereby contradicting the assumption that A is nondefective. Hence, we must have $n_i = m_i$ for each i.

Conversely, suppose that $n_i = m_i$ for each i. Then Corollary 6.6.1 can be applied to conclude that the union of the LI eigenvectors that span each eigenspace are LI, and therefore A has n LI eigenvectors.   ■

EXERCISES 6.6

For problems 1–15, determine the multiplicity of each eigenvalue and a basis for each eigenspace of the given matrix. Hence, determine the dimension of each eigenspace and state whether the matrix is defective or nondefective.

1. $\begin{bmatrix} 1 & 4 \\ 2 & 3 \end{bmatrix}$.

2. $\begin{bmatrix} 3 & 0 \\ 0 & 3 \end{bmatrix}$.

3. $\begin{bmatrix} 1 & 2 \\ -2 & 5 \end{bmatrix}$.

4. $\begin{bmatrix} 5 & 5 \\ -2 & -1 \end{bmatrix}$.

5. $\begin{bmatrix} 3 & -4 & -1 \\ 0 & -1 & -1 \\ 0 & -4 & 2 \end{bmatrix}$.

6. $\begin{bmatrix} 4 & 0 & 0 \\ 0 & 2 & -3 \\ 0 & -2 & 1 \end{bmatrix}$.

7. $\begin{bmatrix} 3 & 1 & 0 \\ -1 & 5 & 0 \\ 0 & 0 & 4 \end{bmatrix}$.

8. $\begin{bmatrix} 3 & 0 & 0 \\ 2 & 0 & -4 \\ 1 & 4 & 0 \end{bmatrix}$.

9. $\begin{bmatrix} 4 & 1 & 6 \\ -4 & 0 & -7 \\ 0 & 0 & -3 \end{bmatrix}$.

10. $\begin{bmatrix} 2 & 0 & 0 \\ 0 & 2 & 0 \\ 0 & 0 & 2 \end{bmatrix}$.

11. $\begin{bmatrix} 7 & -8 & 6 \\ 8 & -9 & 6 \\ 0 & 0 & -1 \end{bmatrix}$.

12. $\begin{bmatrix} 2 & 2 & -1 \\ 2 & 1 & -1 \\ 2 & 3 & -1 \end{bmatrix}$.

13. $\begin{bmatrix} 1 & -1 & 2 \\ 1 & -1 & 2 \\ 1 & -1 & 2 \end{bmatrix}$.

14. $\begin{bmatrix} 2 & 3 & 0 \\ -1 & 0 & 1 \\ -2 & -1 & 4 \end{bmatrix}$.

15. $\begin{bmatrix} 0 & -1 & -1 \\ -1 & 0 & -1 \\ -1 & -1 & 0 \end{bmatrix}$.

For problems 16–20, determine whether the given matrix is defective or nondefective.

16. $A = \begin{bmatrix} 2 & 3 \\ 2 & 1 \end{bmatrix}$, characteristic polynomial $p(\lambda) = (\lambda + 1)(\lambda - 4)$.

17. $A = \begin{bmatrix} 6 & 5 \\ -5 & -4 \end{bmatrix}$, characteristic polynomial $p(\lambda) = (\lambda - 1)^2$.

18. $A = \begin{bmatrix} 1 & -2 \\ 5 & 3 \end{bmatrix}$, characteristic polynomial $p(\lambda) = \lambda^2 - 4\lambda + 13$.

19. $A = \begin{bmatrix} 1 & -3 & 1 \\ -1 & -1 & 1 \\ -1 & -3 & 3 \end{bmatrix}$, characteristic polynomial $p(\lambda) = -(\lambda - 2)^2(\lambda + 1)$.

20. $A = \begin{bmatrix} -1 & 2 & 2 \\ -4 & 5 & 2 \\ -4 & 2 & 5 \end{bmatrix}$, characteristic polynomial $p(\lambda) = (3 - \lambda)^3$.

For problems 21–25, determine a basis for each eigenspace of A and sketch the eigenspaces.

21. $A = \begin{bmatrix} 2 & 1 \\ 3 & 4 \end{bmatrix}$.

22. $A = \begin{bmatrix} 2 & 3 \\ 0 & 2 \end{bmatrix}$.

23. $A = \begin{bmatrix} 5 & 0 \\ 0 & 5 \end{bmatrix}$.

24. $A = \begin{bmatrix} 3 & 1 & -1 \\ 1 & 3 & -1 \\ -1 & -1 & 3 \end{bmatrix}$. [You may assume that A has characteristic polynomial
$$p(\lambda) = (5 - \lambda)(\lambda - 2)^2.]$$

25. $\begin{bmatrix} -3 & 1 & 0 \\ -1 & -1 & 2 \\ 0 & 0 & -2 \end{bmatrix}$.

26. The matrix $A = \begin{bmatrix} 2 & -2 & 3 \\ 1 & -1 & 3 \\ 1 & -2 & 4 \end{bmatrix}$ has eigen-

values $\lambda_1 = 1$ and $\lambda_2 = 3$.

(a) Determine a basis for the eigenspace E_1 and then use the Gram–Schmidt procedure to obtain an orthogonal basis for E_1.

(b) Are the vectors in E_1 orthogonal to the vectors in E_2?

27. Repeat the previous question for

$A = \begin{bmatrix} 1 & -1 & 1 \\ -1 & 1 & 1 \\ 1 & 1 & 1 \end{bmatrix}$. ($A$ has eigenvalues $\lambda_1 = 2$,

$\lambda_2 = -1$.) What is special about the matrix in this problem? (See Section 6.8 for a generalization.)

28. The matrix $A = \begin{bmatrix} a & b & c \\ a & b & c \\ a & b & c \end{bmatrix}$ has eigenvalues

0, 0, and $a + b + c$. Determine all values of the constants $a, b,$ and c for which A is nondefective.

29. Consider the characteristic polynomial of an $n \times n$ matrix A, namely,

$p(\lambda) = \det(A - \lambda I)$

$$= \begin{vmatrix} a_{11} - \lambda & a_{12} & \cdots & a_{1n} \\ a_{21} & a_{22} - \lambda & \cdots & a_{2n} \\ \vdots & \vdots & & \vdots \\ a_{n1} & a_{n2} & \cdots & a_{nn} - \lambda \end{vmatrix} \quad (29.1)$$

which can be written in either of the following equivalent forms:

$p(\lambda) = (-1)^n \lambda^n + b_1 \lambda^{n-1} + \cdots + b_n, \quad (29.2)$

$p(\lambda) = (\lambda_1 - \lambda)(\lambda_2 - \lambda)(\lambda_3 - \lambda) \cdots (\lambda_n - \lambda), \quad (29.3)$

where $\lambda_1, \lambda_2, ..., \lambda_n$ (not necessarily distinct) are the eigenvalues of A.

(a) Use equations (29.1) and (29.2) to show that

$b_1 = (-1)^{n-1}(a_{11} + a_{22} + \cdots + a_{nn}), \quad b_n = \det(A)$.

Recall that the quantity $a_{11} + a_{22} + \cdots + a_{nn}$ is called the *trace* of the matrix A, denoted tr(A).

(b) Use equations (29.2) and (29.3) to show that

$b_1 = (-1)^{n-1}(\lambda_1 + \lambda_2 + \cdots + \lambda_n),$

$b_n = \lambda_1 \lambda_2 \cdots \lambda_n.$

(c) Use your results from (a) and (b) to show that:

det(A) = product of the eigenvalues of A

tr(A) = sum of the eigenvalues of A.

(d) Find the sum and the product of the eigenvalues of

$$A = \begin{bmatrix} 12 & 11 & 9 & -7 \\ 2 & 3 & -5 & 6 \\ 10 & 8 & 5 & 4 \\ 1 & 0 & 3 & 4 \end{bmatrix}.$$

30. Let E_i denote the eigenspace of A corresponding to the eigenvalue λ_i. Use Theorem 5.3.1 to prove that E_i is a subspace of $\mathbf{C}^n$.

31. Let v_1 and v_2 be eigenvectors of A corresponding to the distinct eigenvalues λ_1 and λ_2, respectively. Prove that v_1 and v_2 are LI. (Model your proof on the general case considered in Theorem 6.6.2.)

32. Let E_i denote the eigenspace of A corresponding to the eigenvalue λ_i. If $\{v_1\}$ is a basis for E_1 and $\{v_2, v_3\}$ is a basis for E_2, prove that $\{v_1, v_2, v_3\}$ is LI. (Model your proof on the general case considered in Theorem 6.6.2.)

For problems 33 – 37, use some form of technology to determine the eigenvalues and a basis for each eigenspace of the given matrix.

♦ 33. $A = \begin{bmatrix} 1 & -3 & 3 \\ -1 & -2 & 3 \\ -1 & -3 & 4 \end{bmatrix}$.

♦ 34. $A = \begin{bmatrix} 1 & 1 & 1 \\ 1 & 1 & 1 \\ 1 & 1 & 1 \end{bmatrix}$.

♦ 35. $A = \begin{bmatrix} 3 & \sqrt{2} & 3 \\ \sqrt{2} & 3 & \sqrt{2} \\ 3 & \sqrt{2} & 3 \end{bmatrix}$.

♦ 36. $A = \begin{bmatrix} \frac{3}{25} & \frac{\sqrt{2}}{-6} & \frac{3}{12} \\ 11 & 0 & 6 \\ -44 & 12 & -21 \end{bmatrix}$.

♦ 37. $A = \begin{bmatrix} 1 & 2 & 1 & 2 \\ 2 & 1 & 2 & 1 \\ 1 & 2 & 1 & 2 \\ 2 & 1 & 2 & 1 \end{bmatrix}$.

For problems 38 and 39, show that the given matrix is nondefective.

◆ **38.** $A = \begin{bmatrix} a & b & a \\ b & a & b \\ a & b & a \end{bmatrix}$.

◆ **39.** $A = \begin{bmatrix} a & a & b \\ a & 2a+b & a \\ b & a & a \end{bmatrix}$.

6.7   DIAGONALIZATION

As motivation for the problem to be introduced in this section, we once more consider the *system* of DE

$$\frac{dx_1}{dt} = a_{11}x_1 + a_{12}x_2, \tag{6.7.1}$$

$$\frac{dx_2}{dt} = a_{21}x_1 + a_{22}x_2. \tag{6.7.2}$$

written as the vector equation

$$\mathbf{x}' = A\mathbf{x}, \tag{6.7.3}$$

where

$$\mathbf{x} = \begin{bmatrix} x_1 \\ x_2 \end{bmatrix}, \quad \mathbf{x}' = \begin{bmatrix} x_1' \\ x_2' \end{bmatrix}, \quad A = [a_{ij}]$$

and a prime denotes differentiation with respect to t. In general, we cannot integrate the given system directly, because each equation involves both unknown functions. We say that the equations are *coupled*. Suppose, however, we make a *linear* change of variables defined by

$$\mathbf{x} = S\mathbf{y}, \tag{6.7.4}$$

where S is a nonsingular matrix of constants. Then,

$$\mathbf{x}' = S\mathbf{y}',$$

so that (6.7.3) is transformed to the equivalent system

$$S\mathbf{y}' = AS\mathbf{y}.$$

Multiplying by S^{-1} yields

$$\mathbf{y}' = B\mathbf{y}, \tag{6.7.5}$$

where $B = S^{-1}AS$. The question that now arises is whether it is possible to choose S such that the system (6.7.5) can be integrated. For if this is the case, then, upon performing the integration to find $\mathbf{y}$, the solution to (6.7.3) can be determined from (6.7.4). The results of this section will establish that, provided A is nondefective, this is indeed possible.

The aim of the section therefore is to investigate matrices that are related via $B = S^{-1}AS$. Of particular interest to us is the possibility of choosing S so that $S^{-1}AS$ has a simple structure. Of course, the question that needs answering is how simple a form should we aim for. First we introduce some terminology and a helpful result.

Definition 6.7.1:   Let A and B be $n \times n$ matrices. A is said to be **similar** to B if there exists a nonsingular matrix S such that $B = S^{-1}AS$.

Example 6.7.1   If $A = \begin{bmatrix} -4 & -10 \\ 3 & 7 \end{bmatrix}$ and $B = \begin{bmatrix} 1 & -1 \\ 0 & 2 \end{bmatrix}$, verify that $B = S^{-1}AS$,

where $S = \begin{bmatrix} 2 & -3 \\ -1 & 2 \end{bmatrix}$.

Solution   It is easily shown that $S^{-1} = \begin{bmatrix} 2 & 3 \\ 1 & 2 \end{bmatrix}$, so that

$$S^{-1}AS = \begin{bmatrix} 2 & 3 \\ 1 & 2 \end{bmatrix} \begin{bmatrix} -4 & -10 \\ 3 & 7 \end{bmatrix} \begin{bmatrix} 2 & -3 \\ -1 & 2 \end{bmatrix}$$

$$= \begin{bmatrix} 2 & 3 \\ 1 & 2 \end{bmatrix} \begin{bmatrix} 2 & -8 \\ -1 & 5 \end{bmatrix} = \begin{bmatrix} 1 & -1 \\ 0 & 2 \end{bmatrix}$$

that is,

$$S^{-1}AS = B. \qquad \qquad \square$$

Theorem 6.7.1:   Similar matrices have the same eigenvalues.

PROOF   If A is similar to B then $B = S^{-1}AS$, for some nonsingular matrix S. Thus,

$$\det(B - \lambda I) = \det(S^{-1}AS - \lambda I) = \det(S^{-1}AS - \lambda S^{-1}S)$$
$$= \det(S^{-1}(A - \lambda I)S) = \det(S^{-1})\det(S)\det(A - \lambda I)$$
$$= \det(A - \lambda I).$$

Consequently, A and B have the same characteristic polynomial and hence the same eigenvalues.   ∎

We now know from Theorem 6.7.1 that A and $S^{-1}AS$ have the same eigenvalues $\lambda_1, \lambda_2, \dots, \lambda_n$. Furthermore, the simplest matrix that has these eigenvalues is the diagonal matrix $D = \text{diag}(\lambda_1, \lambda_2, \dots, \lambda_n)$. Consequently, the simplest possible structure for $S^{-1}AS$ is

$$S^{-1}AS = \text{diag}(\lambda_1, \lambda_2, \dots, \lambda_n).$$

We have therefore been led to the question:

---

for a given $n \times n$ matrix A, when does there exist a matrix S such that
$$S^{-1}AS = \text{diag}(\lambda_1, \lambda_2, \dots, \lambda_n)?$$

The answer is provided in the next theorem.

Theorem 6.7.2:   An $n \times n$ matrix A is similar to a diagonal matrix if and only if A is nondefective. In such a case, if $\mathbf{v}_1, \mathbf{v}_2, \ldots, \mathbf{v}_n$ denote n LI eigenvectors of A and

$$S = [\mathbf{v}_1, \mathbf{v}_2, \ldots, \mathbf{v}_n],$$

then

$$S^{-1}AS = \text{diag}(\lambda_1, \lambda_2, \ldots, \lambda_n),$$

where $\lambda_1, \lambda_2, \ldots, \lambda_n$ are the eigenvalues of A (not necessarily distinct) corresponding to the eigenvectors $\mathbf{v}_1, \mathbf{v}_2, \ldots, \mathbf{v}_n$.

PROOF   If A is similar to a diagonal matrix, then there exists a nonsingular matrix $S = [\mathbf{v}_1, \mathbf{v}_2, \ldots, \mathbf{v}_n]$ such that

$$S^{-1}AS = D, \tag{6.7.6}$$

where $D = \text{diag}(\lambda_1, \lambda_2, \ldots, \lambda_n)$ and, from Theorem 6.7.1, $\lambda_1, \lambda_2, \ldots, \lambda_n$ are the eigenvalues of A. Premultiplying both sides of (6.7.6) by S yields

$$AS = SD,$$

or, equivalently,

$$[A\mathbf{v}_1, A\mathbf{v}_2, \ldots, A\mathbf{v}_n] = [\lambda_1\mathbf{v}_1, \lambda_2\mathbf{v}_2, \ldots, \lambda_n\mathbf{v}_n].$$

Equating corresponding column vectors we must have

$$A\mathbf{v}_1 = \lambda_1\mathbf{v}_1, \quad A\mathbf{v}_2 = \lambda_2\mathbf{v}_2, \quad \cdots, \quad A\mathbf{v}_n = \lambda_n\mathbf{v}_n.$$

Consequently, $\mathbf{v}_1, \mathbf{v}_2, \ldots, \mathbf{v}_n$ are eigenvectors of A corresponding to the eigenvalues $\lambda_1, \lambda_2, \ldots, \lambda_n$. Further, since $\det(S) \neq 0$, the eigenvectors are LI.

Conversely, suppose A is nondefective, and let $S = [\mathbf{v}_1, \mathbf{v}_2, \ldots, \mathbf{v}_n]$ where $\{\mathbf{v}_1, \mathbf{v}_2, \ldots, \mathbf{v}_n\}$ is any complete set of eigenvectors for A. Then

$$AS = A[\mathbf{v}_1, \mathbf{v}_2, \ldots, \mathbf{v}_n] = [A\mathbf{v}_1, A\mathbf{v}_2, \ldots, A\mathbf{v}_n]$$
$$= [\lambda_1\mathbf{v}_1, \lambda_2\mathbf{v}_2, \ldots, \lambda_n\mathbf{v}_n].$$

This can be written in the equivalent form

$$AS = SD, \tag{6.7.7}$$

where $D = \text{diag}(\lambda_1, \lambda_2, \ldots, \lambda_n)$. Since $\det(S) \neq 0$, S is nonsingular. Premultiplying both sides of (6.7.7) by S^{-1} yields

$$S^{-1}AS = D$$

so that S is indeed similar to a diagonal matrix.   ∎

Definition 6.7.2:   An $n \times n$ matrix that is similar to a diagonal matrix is said to be **diagonalizable**.

REMARK: It follows from Theorem 6.7.2 that the matrices in Examples 6.5.2, 6.5.3, 6.5.4, and 6.6.2 *are* diagonalizable, whereas the matrix in Example 6.6.3 is *not* diagonalizable. As an exercise, you should write down an appropriate matrix S in each of Examples 6.5.2, 6.5.3, 6.5.4, and 6.6.2 together with diagonal matrices to which the given matrices are similar.

Example 6.7.2   Verify that $A = \begin{bmatrix} 3 & -2 & -2 \\ -3 & -2 & -6 \\ 3 & 6 & 10 \end{bmatrix}$ is diagonalizable and find a matrix

S such that $S^{-1}AS = \text{diag}(\lambda_1, \lambda_2, \lambda_3)$.

Solution A has characteristic polynomial $p(\lambda) = -(\lambda - 4)^2(\lambda - 3)$, so that the eigenvalues of A are $\lambda = 4, 4, 3$. Corresponding LI eigenvectors are

$$\lambda = 4: \quad \mathbf{v}_1 = (-2, 0, 1), \ \mathbf{v}_2 = (-2, 1, 0),$$

$$\lambda = 3: \quad \mathbf{v}_3 = (1, 3, -3).$$

Consequently, A is nondefective and therefore is diagonalizable. If we set

$$S = \begin{bmatrix} -2 & -2 & 1 \\ 0 & 1 & 3 \\ 1 & 0 & -3 \end{bmatrix},$$

then, according to Theorem 6.7.2,

$$S^{-1}AS = \text{diag}(4, 4, 3).$$

It is important to note that the ordering of the eigenvalues in the diagonal matrix must be in correspondence with the ordering of the eigenvectors in the matrix S. For example, permuting columns 2 and 3 in S yields the matrix

$$\tilde{S} = \begin{bmatrix} -2 & 1 & -2 \\ 0 & 3 & 1 \\ 1 & -3 & 0 \end{bmatrix}.$$

Since the column vectors of $\tilde{S}$ are eigenvectors of A, Theorem 6.7.2 implies that

$$\tilde{S}^{-1}A\tilde{S} = \text{diag}(4, 3, 4). \qquad \qquad \square$$

Now we return to the system of DE example that motivated our discussion. We assume that A is nondefective and choose $S = [\mathbf{v}_1, \mathbf{v}_2]$ such that $S^{-1}AS = \text{diag}(\lambda_1, \lambda_2)$. Then the system of DE (6.7.5) reduces to

$$\mathbf{y}' = \text{diag}(\lambda_1, \lambda_2)\mathbf{y}.$$

That is,

$$\begin{bmatrix} y_1' \\ y_2' \end{bmatrix} = \begin{bmatrix} \lambda_1 & 0 \\ 0 & \lambda_2 \end{bmatrix}\begin{bmatrix} y_1 \\ y_2 \end{bmatrix},$$

so that

$$y_1' = \lambda_1 y_1, \quad y_2' = \lambda_2 y_2.$$

We see that the system of DE has decoupled, and both of the resulting DE are easily integrated to obtain

$$y_1(t) = c_1 e^{\lambda_1 t}, \ y_2(t) = c_2 e^{\lambda_2 t}.$$

From (6.7.4) we see that the solution in the original variables is

$$\mathbf{x}(t) = S\mathbf{y},$$

which can be written in the equivalent form

$$\mathbf{x}(t) = S\mathbf{y} = [\mathbf{v}_1, \mathbf{v}_2] \begin{bmatrix} c_1 e^{\lambda_1 t} \\ c_2 e^{\lambda_2 t} \end{bmatrix}.$$

That is

$$\mathbf{x}(t) = c_1 e^{\lambda_1 t} \mathbf{v}_1 + c_2 e^{\lambda_2 t} \mathbf{v}_2, \tag{6.7.8}$$

where λ_1, λ_2 are the eigenvalues of A *corresponding* to the eigenvectors $\mathbf{v}_1, \mathbf{v}_2$. The formula (6.7.8) looks suspiciously like a statement about the set of all solutions to the system of DE being generated by taking all linear combinations of a certain set (two in this case) of basic solutions. As mentioned previously, a full vector space formulation for systems of linear DE will be given in Chapter 8.

Example 6.7.3 Use the ideas introduced in this section to determine all solutions to

$$x_1' = 6x_1 - x_2, \quad x_2' = -5x_1 + 2x_2.$$

Solution The given system can be written as

$$\mathbf{x}' = A\mathbf{x},$$

where $A = \begin{bmatrix} 6 & -1 \\ -5 & 2 \end{bmatrix}$. The transformed system is

$$\mathbf{y}' = (S^{-1}AS)\mathbf{y}, \tag{6.7.9}$$

where

$$\mathbf{x} = S\mathbf{y}. \tag{6.7.10}$$

To determine S, we need the eigenvalues and eigenvectors of A. The characteristic polynomial of A is

$$p(\lambda) = \begin{vmatrix} 6 - \lambda & -1 \\ -5 & 2 - \lambda \end{vmatrix} = (\lambda - 1)(\lambda - 7).$$

Hence, A is nondefective. The eigenvectors are easily computed:

$$\lambda_1 = 1 \Rightarrow \mathbf{v} = r(1, 5); \quad \lambda_2 = 7 \Rightarrow \mathbf{v} = s(-1, 1).$$

We could now substitute into (6.7.8) to obtain the solution to the system, but it is more instructive to go through the steps that led to that equation. If we set

$$S = \begin{bmatrix} 1 & -1 \\ 5 & 1 \end{bmatrix},$$

then from Theorem 6.7.2

$$S^{-1}AS = \text{diag}(1, 7),$$

so that the system (6.7.9) is

$$\begin{bmatrix} y_1' \\ y_2' \end{bmatrix} = \begin{bmatrix} 1 & 0 \\ 0 & 7 \end{bmatrix} \begin{bmatrix} y_1 \\ y_2 \end{bmatrix}.$$

Hence,

$$y_1' = y_1, \quad y_2' = 7y_2.$$

Both of these equations can be integrated to obtain

$$y_1(t) = c_1 e^t, \quad y_2(t) = c_2 e^{7t}.$$

Using (6.7.10) to return to the original variables we have

$$\mathbf{x} = S\mathbf{y} = \begin{bmatrix} 1 & -1 \\ 5 & 1 \end{bmatrix} \begin{bmatrix} c_1 e^t \\ c_2 e^{7t} \end{bmatrix} = \begin{bmatrix} c_1 e^t - c_2 e^{7t} \\ 5c_1 e^t + c_2 e^{7t} \end{bmatrix}.$$

Consequently,

$$x_1(t) = c_1 e^t - c_2 e^{7t}, \quad x_2(t) = 5c_1 e^t + c_2 e^{7t}. \qquad \square$$

Finally in this section we note that if a matrix is defective then, from Theorem 6.7.2, there does *not* exist a matrix S such that $S^{-1}AS = \text{diag}(\lambda_1, \lambda_2, \ldots, \lambda_n)$. In more advanced treatments of the eigenvalue/eigenvector problem the possible "simple" forms of a defective matrix are discussed. Some special cases are considered in problems 31–33.

EXERCISES 6.7

For problems 1–12, determine whether the given matrix is diagonalizable. Where possible, find a matrix S such that $S^{-1}A\ S = \text{diag}(\lambda_1, \lambda_2, \ldots, \lambda_n)$.

1. $\begin{bmatrix} -1 & -2 \\ -2 & 2 \end{bmatrix}$.

2. $\begin{bmatrix} -7 & 4 \\ -4 & 1 \end{bmatrix}$.

3. $\begin{bmatrix} 1 & -8 \\ 2 & -7 \end{bmatrix}$.

4. $\begin{bmatrix} 0 & 4 \\ -4 & 0 \end{bmatrix}$.

5. $\begin{bmatrix} 1 & 0 & 0 \\ 0 & 3 & 7 \\ 1 & 1 & -3 \end{bmatrix}$.

6. $\begin{bmatrix} 1 & -2 & 0 \\ 2 & -3 & 0 \\ 2 & -2 & -1 \end{bmatrix}$.

7. $\begin{bmatrix} 0 & -2 & -2 \\ -2 & 0 & -2 \\ -2 & -2 & 0 \end{bmatrix}$.

8. $\begin{bmatrix} -2 & 1 & 4 \\ -2 & 1 & 4 \\ -2 & 1 & 4 \end{bmatrix}$.

9. $\begin{bmatrix} 2 & 0 & 0 \\ 0 & 1 & 0 \\ 2 & -1 & 1 \end{bmatrix}$.

10. $\begin{bmatrix} 4 & 0 & 0 \\ 3 & -1 & -1 \\ 0 & 2 & 1 \end{bmatrix}$.

11. $\begin{bmatrix} 0 & 2 & -1 \\ -2 & 0 & -2 \\ 1 & 2 & 0 \end{bmatrix}$.

12. $\begin{bmatrix} 1 & -2 & 0 \\ -2 & 1 & 0 \\ 0 & 0 & 3 \end{bmatrix}$.

◆ For problems 13 and 14, use some form of technology to determine a complete set of eigenvectors for the given matrix A. Construct a matrix, S, that diagonalizes A and explicitly verify that $S^{-1}AS = \text{diag}(\lambda_1, \lambda_2, \ldots, \lambda_n)$.

13. $A = \begin{bmatrix} 1 & -3 & 3 \\ -2 & -4 & 6 \\ -2 & -6 & 8 \end{bmatrix}$.

14. $A = \begin{bmatrix} 3 & -2 & 3 & -2 \\ -2 & 3 & -2 & 3 \\ 3 & -2 & 3 & -2 \\ -2 & 3 & -2 & 3 \end{bmatrix}$.

For problems 15–19, use the ideas introduced in this section to solve the given system of DE.

15. $x_1' = x_1 + 4x_2, \quad x_2' = 2x_1 + 3x_2.$

16. $x_1' = 6x_1 - 2x_2, \quad x_2' = -2x_1 + 6x_2.$

17. $x_1' = 9x_1 + 6x_2, \quad x_2' = -10x_1 - 7x_2.$

18. $x_1' = -12x_1 - 7x_2, \quad x_2' = 16x_1 + 10x_2.$

19. $x_1' = x_2, \quad x_2' = -x_1.$

For problems 20 and 21, first write the given system of DE in matrix form, and then use the ideas from this section to determine all solutions.

20. $x_1' = 3x_1 - 4x_2 - x_3, \quad x_2' = -x_2 - x_3$
$$x_3' = -4x_2 + 2x_3.$$

21. $x_1' = x_1 + x_2 - x_3, \quad x_2' = x_1 + x_2 + x_3$
$$x_3' = -x_1 + x_2 + x_3.$$

22. Let A be a nondefective matrix. Then
$$S^{-1}AS = D,$$
where D is a diagonal matrix. This can be written as
$$A = SDS^{-1}.$$
Use this result to show that
$$A^2 = SD^2S^{-1}$$
and that for k a positive integer,
$$A^k = SD^kS^{-1}.$$

23. If $D = \text{diag}(\lambda_1, \lambda_2, ..., \lambda_n)$, show that for each positive integer k,
$$D^k = \text{diag}(\lambda_1{}^k, \lambda_2{}^k, ..., \lambda_n{}^k).$$

24. Use the results of the preceding two problems to determine A^3, A^5, if $A = \begin{bmatrix} -7 & -4 \\ 18 & 11 \end{bmatrix}.$

25. Prove the following properties for similar matrices:

(a) If A is similar to B, then B is similar to A.

(b) A is similar to itself.

(c) If A is similar to B and B is similar to C, then A is similar to C.

26. If A is similar to B, prove that A^T is similar to B^T.

27. In Theorem 6.7.1 we proved that similar matrices have the same eigenvalues. This problem investigates the relationship between their eigenvectors. Let $\mathbf{v}$ be an eigenvector of A corresponding to the eigenvalue λ. Prove that if $B = S^{-1}AS$, then $S^{-1}\mathbf{v}$ is an eigenvector of B corresponding to the eigenvalue λ.

28. Let A be nondefective and let S be a matrix such that $S^{-1}AS = \text{diag}(\lambda_1, \lambda_2, ..., \lambda_n)$, where all λ_i are nonzero.

(a) Prove that A^{-1} exists.

(b) Prove that
$$S^{-1}A^{-1}S = \text{diag}\left(\frac{1}{\lambda_1}, \frac{1}{\lambda_2}, \cdots, \frac{1}{\lambda_n}\right).$$

29. Let A be nondefective and let S be a matrix such that $S^{-1}AS = \text{diag}(\lambda_1, \lambda_2, ..., \lambda_n)$.

(a) Prove that $Q^{-1}A^TQ = \text{diag}(\lambda_1, \lambda_2, ..., \lambda_n)$, where $Q = (S^T)^{-1}$. (This establishes that A^T is also nondefective.)

(b) If C denotes the matrix of cofactors of S, prove that the column vectors of C are LI eigenvectors of A^T. (Hint: Use the adjoint method to determine S^{-1}.)

30. If $A = \begin{bmatrix} -2 & 4 \\ 1 & 1 \end{bmatrix}$, determine S such that $S^{-1}AS = \text{diag}(-3, 2)$, and use the result from the previous problem to determine all eigenvectors of A^T.

Problems 31–33 deal with the generalization of the diagonalization problem to defective matrices.

31. Let A be a 2×2 *defective* matrix. It follows from Theorem 6.7.2 that A is not diagonalizable. However, it can be shown that A is similar to the (elementary) Jordan matrix $J_1(\lambda) = \begin{bmatrix} \lambda & 1 \\ 0 & \lambda \end{bmatrix}$.

Thus, there exists a matrix $S = [\mathbf{v}_1, \mathbf{v}_2]$, such that
$$S^{-1}AS = J_1(\lambda).$$
Prove that $\mathbf{v}_1$ and $\mathbf{v}_2$ must satisfy:
$$(A - \lambda I)\mathbf{v}_1 = \mathbf{0} \qquad (31.1)$$
$$(A - \lambda I)\mathbf{v}_2 = \mathbf{v}_1. \qquad (31.2)$$
Equation (31.1) is the statement that $\mathbf{v}_1$ must be an eigenvector of A corresponding to the eigenvalue λ. Any vectors that satisfy (31.2) are called *generalized eigenvectors* of A.

32. Show that $A = \begin{bmatrix} 2 & 1 \\ -1 & 4 \end{bmatrix}$ is defective and use the previous problem to determine a matrix S such that
$$S^{-1}AS = \begin{bmatrix} 3 & 1 \\ 0 & 3 \end{bmatrix}.$$

33. Let λ be an eigenvalue of the 3×3 matrix A of multiplicity 3, and suppose the corresponding eigenspace has dimension 1. It can be shown that, in this case, there exists a matrix $S = [\mathbf{v}_1, \mathbf{v}_2, \mathbf{v}_3]$ such that

$$S^{-1}AS = \begin{bmatrix} \lambda & 1 & 0 \\ 0 & \lambda & 1 \\ 0 & 0 & \lambda \end{bmatrix}.$$

Prove that $\mathbf{v}_1, \mathbf{v}_2, \mathbf{v}_3$ must satisfy

$$(A - \lambda I)\mathbf{v}_1 = \mathbf{0}$$
$$(A - \lambda I)\mathbf{v}_2 = \mathbf{v}_1$$
$$(A - \lambda I)\mathbf{v}_3 = \mathbf{v}_2.$$

34. In this problem, we establish that similar matrices describe the same linear transformation relative to different bases. Let $\{\mathbf{e}_1, \mathbf{e}_2, \dots, \mathbf{e}_n\}$ and $\{\mathbf{f}_1, \mathbf{f}_2, \dots, \mathbf{f}_n\}$ be bases for a vector space V and let $T : V \to V$ be a linear transformation. Define the $n \times n$ matrices $A = [a_{ik}]$ and $B = [b_{ik}]$ by

$$T(\mathbf{e}_k) = \sum_{i=1}^{n} a_{ik}\,\mathbf{e}_i, \quad k = 1, 2, \dots, n, \qquad (34.1)$$

$$T(\mathbf{f}_k) = \sum_{i=1}^{n} b_{ik}\,\mathbf{f}_i, \quad k = 1, 2, \dots, n. \qquad (34.2)$$

Expressing each of the basis vectors $\mathbf{f}_1, \mathbf{f}_2, \dots, \mathbf{f}_n$ in terms of the basis $\{\mathbf{e}_1, \mathbf{e}_2, \dots, \mathbf{e}_n\}$ yields

$$\mathbf{f}_i = \sum_{j=1}^{n} s_{ji}\,\mathbf{e}_j, \quad i = 1, 2, \dots, n, \qquad (34.3)$$

for appropriate scalars s_{ji}. Thus, the matrix $S = [s_{ji}]$ describes the relationship between the two bases.

(a) Prove that S is nonsingular. (Hint: Use the fact that $\mathbf{f}_1, \mathbf{f}_2, \dots, \mathbf{f}_n$ are LI.)

(b) Use (34.2) and (34.3) to show that

$$T(\mathbf{f}_k) = \sum_{j=1}^{n} \left(\sum_{i=1}^{n} s_{ji} b_{ik} \right)\mathbf{e}_j, \quad k = 1, 2, \dots, n,$$

or, equivalently,

$$T(\mathbf{f}_k) = \sum_{i=1}^{n} \left(\sum_{j=1}^{n} s_{ij} b_{jk} \right)\mathbf{e}_i, \quad k = 1, 2, \dots, n. \quad (34.4)$$

(c) Use (34.3) and (34.1) to show that

$$T(\mathbf{f}_k) = \sum_{i=1}^{n} \left(\sum_{j=1}^{n} a_{ij} s_{jk} \right)\mathbf{e}_i, \quad k = 1, 2, \dots, n. \quad (34.5)$$

(d) Use (34.4) and (34.5) together with the linear independence of $\mathbf{e}_1, \mathbf{e}_2, \dots, \mathbf{e}_n$ to show that

$$\sum_{j=1}^{n} s_{ij} b_{jk} = \sum_{j=1}^{n} a_{ij} s_{jk}, \quad 1 \le i \le n, \, 1 \le k \le n,$$

and hence that

$$SB = AS.$$

Finally, conclude that A and B are related by

$$B = S^{-1}AS.$$

6.8* ORTHOGONAL DIAGONALIZATION AND QUADRATIC FORMS

Symmetric matrices with real elements play an important role in many applications of linear algebra. In this section we study the special properties satisfied by the eigenvectors of such a matrix and show how this simplifies the diagonalization problem introduced in the previous section. We also give a brief introduction to one area of application of the theoretical results that are obtained. We begin with a definition

> **Definition 6.8.1:**   A real nonsingular matrix A is called **orthogonal** if
> $$A^{-1} = A^T.$$

---

* This section may be omitted without loss of continuity.

Example 6.8.1 Verify that the following matrix is an orthogonal matrix

$$A = \begin{bmatrix} 1/3 & -2/3 & 2/3 \\ 2/3 & -1/3 & -2/3 \\ 2/3 & 2/3 & 1/3 \end{bmatrix}.$$

Solution For the given matrix, we have

$$AA^T = \begin{bmatrix} 1/3 & -2/3 & 2/3 \\ 2/3 & -1/3 & -2/3 \\ 2/3 & 2/3 & 1/3 \end{bmatrix} \begin{bmatrix} 1/3 & 2/3 & 2/3 \\ -2/3 & -1/3 & 2/3 \\ 2/3 & -2/3 & 1/3 \end{bmatrix} = I_3.$$

Similarly, $A^T A = I_3$ so that $A^T = A^{-1}$. Consequently, A is an orthogonal matrix. ⃞

If we look more closely at the preceding example, we see that the column vectors of A and the row vectors of A are *orthonormal* vectors. The next theorem establishes that this is a basic characterizing property of all orthogonal matrices.

Theorem 6.8.1: A real $n \times n$ matrix A is an orthogonal matrix if and only if the row (or column) vectors are orthogonal unit vectors.

PROOF We leave this proof as an exercise. ∎

We can now state the main results of the section.

BASIC RESULTS FOR REAL SYMMETRIC MATRICES

Theorem 6.8.2: Let A be a real symmetric matrix. Then:

1. All eigenvalues of A are real.

2. Real eigenvectors of A that correspond to *distinct* eigenvalues are *orthogonal*.

3. A is nondefective.

4. A has a complete set of *orthonormal* eigenvectors.

5. A can be diagonalized with an *orthogonal* matrix.

PROOF The proofs of properties (1) and (2) are deferred to the end of the section, whereas the proof of (3) is somewhat lengthy and therefore omitted. (See, for example, Shilov, G.E., *Linear Algebra*, Dover Publications, 1977.) We establish properties (4) and (5). Eigenvectors corresponding to distinct eigenvalues are orthogonal from (2). Now suppose that the eigenvalue λ_i has multiplicity m_i. Then, since A is nondefective [from property (3)], it follows from Theorem 6.6.3 that we can find m_i LI eigenvectors corresponding to λ_i. These vectors span the eigenspace corresponding to λ_i, and hence, we can use the Gram–Schmidt process to find m_i *orthogonal* eigenvectors (corresponding to λ_i) that span this eigenspace. (See Figure 6.8.1 for an illustration of the 3×3 case.)

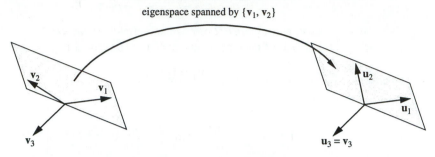

Figure 6.8.1 Construction of a complete set of orthogonal eigenvectors $\{\mathbf{u}_1, \mathbf{u}_2, \mathbf{u}_3\}$ for a real symmetric 3×3 matrix that has an eigenvalue of multiplicity 2.

Proceeding in this manner for each eigenvalue, we obtain a complete set of orthogonal eigenvectors, and hence, by normalization, we can find an orthonormal set. If we denote this orthonormal set of eigenvectors by $\{\mathbf{w}_1, \mathbf{w}_2, \ldots, \mathbf{w}_n\}$ and let $S = [\mathbf{w}_1, \mathbf{w}_2, \ldots, \mathbf{w}_n]$, then Theorem 6.8.1 implies that S is an orthogonal matrix. Consequently,

$$S^T A S = S^{-1} A S = \operatorname{diag}(\lambda_1, \lambda_2, \ldots, \lambda_n),$$

where we have used Theorem 6.7.2. ∎

We emphasize that the orthogonal diagonalization of a real symmetric matrix requires that we use a complete set of *orthonormal* eigenvectors in constructing S.

Example 6.8.2   Find a complete set of orthonormal eigenvectors for

$$A = \begin{bmatrix} 2 & 2 & 1 \\ 2 & 5 & 2 \\ 1 & 2 & 2 \end{bmatrix},$$

and determine an orthogonal matrix that diagonalizes A.

Solution   Since A is real and symmetric a complete orthonormal set of eigenvectors exists. We first find the eigenvalues of A. The characteristic polynomial of A is

$$\det(A - \lambda I) = \begin{vmatrix} 2 - \lambda & 2 & 1 \\ 2 & 5 - \lambda & 2 \\ 1 & 2 & 2 - \lambda \end{vmatrix} = -(\lambda - 1)^2(\lambda - 7).$$

Thus, A has eigenvalues

$$\lambda = 1, 1, 7.$$

Eigenvectors
$\lambda = 1$:   In this case, the system $(A - \lambda I)\mathbf{v} = \mathbf{0}$ reduces to the single equation

$$v_1 + 2v_2 + v_3 = 0$$

which has solution $(-2r - s, r, s)$, so that the corresponding eigenvectors are

$$\mathbf{v} = (-2r - s, r, s) = r(-2, 1, 0) + s(-1, 0, 1).$$

Two LI eigenvectors corresponding to the eigenvalue $\lambda = 1$ are

$$\mathbf{v}_1 = (-1, 0, 1), \qquad \mathbf{v}_2 = (-2, 1, 0).$$

The eigenvectors $\mathbf{v}_1$ and $\mathbf{v}_2$ are *not* orthogonal. However, we can use the Gram–Schmidt process to obtain a pair of orthogonal eigenvectors corresponding to $\lambda = 1$. We take

$$\mathbf{u}_1 = (-1, 0, 1).$$

Then $\mathbf{u}_2$ is given by

$$\mathbf{u}_2 = \mathbf{v}_2 - \frac{<\mathbf{v}_2, \mathbf{u}_1>}{\|\mathbf{u}_1\|^2}\mathbf{u}_1 = (-2, 1, 0) - \frac{2}{2}(-1, 0, 1).$$

That is,

$$\mathbf{u}_2 = (-1, 1, -1).$$

Thus, the vectors $\mathbf{u}_1$ and $\mathbf{u}_2$ are orthogonal eigenvectors corresponding to $\lambda = 1$. ($\{\mathbf{u}_1, \mathbf{u}_2\}$ is an orthogonal basis for the eigenspace corresponding to $\lambda = 1$.) Two orthonormal eigenvectors corresponding to $\lambda = 1$ are thus

$$\mathbf{w}_1 = \frac{\mathbf{u}_1}{\|\mathbf{u}_1\|} = \left(-\frac{1}{\sqrt{2}}, 0, \frac{1}{\sqrt{2}}\right), \tag{6.8.1}$$

$$\mathbf{w}_2 = \frac{\mathbf{u}_2}{\|\mathbf{u}_2\|} = \left(-\frac{1}{\sqrt{3}}, \frac{1}{\sqrt{3}}, -\frac{1}{\sqrt{3}}\right). \tag{6.8.2}$$

$\lambda = 7$:   In this case, the system $(A - \lambda I)\mathbf{v} = \mathbf{0}$ is

$$\begin{aligned}
-5v_1 + 2v_2 + v_3 &= 0, \\
2v_1 - 2v_2 + 2v_3 &= 0, \\
v_1 + 2v_2 - 5v_3 &= 0.
\end{aligned}$$

Solving using Gaussian elimination yields $(t, 2t, t)$, so that the eigenvectors corresponding to $\lambda = 7$ are

$$\mathbf{v} = t(1, 2, 1),$$

and a *unit* eigenvector corresponding to $\lambda = 7$ is

$$\mathbf{w}_3 = \frac{\mathbf{v}}{\|\mathbf{v}\|} = \left(\frac{1}{\sqrt{6}}, \frac{2}{\sqrt{6}}, \frac{1}{\sqrt{6}}\right). \tag{6.8.3}$$

Notice that $\mathbf{w}_3$ is orthogonal to both $\mathbf{w}_1$ and $\mathbf{w}_2$ as guaranteed by Theorem 6.8.2. It follows from (6.8.1), (6.8.2) and (6.8.3) that a complete set of orthonormal eigenvectors for the given matrix A is $\{\mathbf{w}_1, \mathbf{w}_2, \mathbf{w}_3\}$, that is,

$$\left\{\left(-\frac{1}{\sqrt{2}}, 0, \frac{1}{\sqrt{2}}\right), \left(-\frac{1}{\sqrt{3}}, \frac{1}{\sqrt{3}}, -\frac{1}{\sqrt{3}}\right), \left(\frac{1}{\sqrt{6}}, \frac{2}{\sqrt{6}}, \frac{1}{\sqrt{6}}\right)\right\}.$$

If we set

$$S = \begin{bmatrix} -\dfrac{1}{\sqrt{2}} & -\dfrac{1}{\sqrt{3}} & \dfrac{1}{\sqrt{6}} \\[2ex] 0 & \dfrac{1}{\sqrt{3}} & \dfrac{2}{\sqrt{6}} \\[2ex] \dfrac{1}{\sqrt{2}} & -\dfrac{1}{\sqrt{3}} & \dfrac{1}{\sqrt{6}} \end{bmatrix},$$

then S is an orthogonal matrix satisfying

$$S^T A S = S^{-1} A S = \text{diag } (1, 1, 7).$$

QUADRATIC FORMS

If A is a *symmetric* $n \times n$ real matrix and $\mathbf{x}$ is a vector in $\mathbf{R}^n$, then an expression of the form

$$\mathbf{x}^T A \mathbf{x}$$

is called a **quadratic form**. We can consider a quadratic form as defining a mapping from $\mathbf{R}^n$ to $\mathbf{R}$. For example, if $n = 2$, then

$$\mathbf{x}^T A \, \mathbf{x} = [x_1 \; x_2] \begin{bmatrix} a_{11} & a_{12} \\ a_{12} & a_{22} \end{bmatrix} \begin{bmatrix} x_1 \\ x_2 \end{bmatrix}$$

$$= [x_1 \; x_2] \begin{bmatrix} a_{11}x_1 + a_{12}x_2 \\ a_{12}x_1 + a_{22}x_2 \end{bmatrix}$$

$$= a_{11}x_1^2 + 2a_{12}x_1 x_2 + a_{22}x_2^2.$$

We see that this is indeed quadratic in x_1 and x_2. Quadratic forms arise in many applications. For example, in geometry, the conic sections have Cartesian equations that can be expressed as

$$\mathbf{x}^T A \mathbf{x} = c,$$

whereas quadric surfaces have equations of this same form where now A is a 3×3 matrix and $\mathbf{x}$ is a vector in $\mathbf{R}^3$. In mechanics, the kinetic energy K of a physical system with n degrees of freedom (and time independent constraints) can be written as

$$K = \mathbf{x}^T A \mathbf{x},$$

where A is an $n \times n$ matrix and $\mathbf{x}$ is the vector of (generalized) velocities. The question that we are going to address is whether it is possible to make a *linear* change of variables

$$\mathbf{x} = S\mathbf{y} \qquad (6.8.4)$$

that enables us to simplify a quadratic form by eliminating the cross terms. Equivalently, we want to reduce a quadratic form to a sum of squares. Making the change of variables (6.8.4) in the general quadratic form yields

$$\mathbf{x}^T A \mathbf{x} = (S\mathbf{y})^T A (S\mathbf{y}) = \mathbf{y}^T (S^T A S) \mathbf{y}. \qquad (6.8.5)$$

If we choose $S = [\mathbf{w}_1, \mathbf{w}_2, \ldots, \mathbf{w}_n]$, where $\{\mathbf{w}_1, \mathbf{w}_2, \ldots, \mathbf{w}_n\}$ is any complete set of orthonormal eigenvectors for A, then Theorem 6.8.2 implies that

$$S^T A S = \text{diag}(\lambda_1, \lambda_2, \ldots, \lambda_n).$$

where $\lambda_1, \lambda_2, \ldots, \lambda_n$ are the eigenvalues of A corresponding to $\mathbf{w}_1, \mathbf{w}_2, \ldots, \mathbf{w}_n$. With this choice of S the right-hand side of equation (6.8.5) reduces to

$$[y_1 \ y_2 \ \cdots \ y_n] \begin{bmatrix} \lambda_1 & & & \\ & \lambda_2 & & \\ & & \ddots & \\ & & & \lambda_n \end{bmatrix} \begin{bmatrix} y_1 \\ y_2 \\ \vdots \\ y_n \end{bmatrix} = \lambda_1 y_1^2 + \lambda_2 y_2^2 + \cdots + \lambda_n y_n^2,$$

and we have accomplished our goal of reducing the quadratic form to a sum of squares. This result is summarized in the following theorem.

Theorem 6.8.3 (Principal Axes Theorem):   Let A be an $n \times n$ real symmetric matrix. Then the quadratic form $\mathbf{x}^T A \mathbf{x}$ can be reduced to a sum of squares by the change of variables $\mathbf{x} = S\mathbf{y}$, where S is an orthogonal matrix whose column vectors $\{\mathbf{w}_1, \mathbf{w}_2, \ldots, \mathbf{w}_n\}$ are any complete orthonormal set of eigenvectors for A. The transformed quadratic form is

$$\lambda_1 y_1^2 + \lambda_2 y_2^2 + \cdots + \lambda_n y_n^2,$$

where $\lambda_1, \lambda_2, \ldots, \lambda_n$ are the eigenvalues of A corresponding to $\mathbf{w}_1, \mathbf{w}_2, \ldots, \mathbf{w}_n$. The column vectors of S are called **principal axes** for the quadratic form $\mathbf{x}^T A \mathbf{x}$.

Example 6.8.3   By transforming to principal axes, reduce the quadratic form $\mathbf{x}^T A \mathbf{x}$ for

$$A = \begin{bmatrix} 2 & 2 & 1 \\ 2 & 5 & 2 \\ 1 & 2 & 2 \end{bmatrix}$$

to a sum of squares.

Solution   In the previous example, we found an orthogonal matrix that diagonalizes A, namely,

$$S = \begin{bmatrix} -\dfrac{1}{\sqrt{2}} & -\dfrac{1}{\sqrt{3}} & \dfrac{1}{\sqrt{6}} \\[2ex] 0 & \dfrac{1}{\sqrt{3}} & \dfrac{2}{\sqrt{6}} \\[2ex] \dfrac{1}{\sqrt{2}} & -\dfrac{1}{\sqrt{3}} & \dfrac{1}{\sqrt{6}} \end{bmatrix}.$$

Furthermore, the corresponding eigenvalues of A are $(1, 1, 7)$. Consequently, the change of variables $\mathbf{x} = S\mathbf{y}$ reduces the quadratic form to

$$y_1^2 + y_2^2 + 7y_3^2.$$

This is quite a simplification from the quadratic form expressed in the original variables, namely,

$$\mathbf{x}^T A \mathbf{x} = \begin{bmatrix} x_1 & x_2 & x_3 \end{bmatrix} \begin{bmatrix} 2 & 2 & 1 \\ 2 & 5 & 2 \\ 1 & 2 & 2 \end{bmatrix} \begin{bmatrix} x_1 \\ x_2 \\ x_3 \end{bmatrix}$$

$$= 2x_1^2 + 5x_2^2 + 2x_3^2 + 4x_1x_2 + 2x_1x_3 + 4x_2x_3 .$$

Example 6.8.4   By transforming to principal axes, identify the conic section with Cartesian equation

$$5x_1^2 + 2x_1x_2 + 5x_2^2 = 1.$$

Solution We first write the quadratic form appearing on the left-hand side of the given equation in matrix form. If we set $A = \begin{bmatrix} 5 & 1 \\ 1 & 5 \end{bmatrix}$ then the given equation is

$$\mathbf{x}^T A \mathbf{x} = 1. \tag{6.8.6}$$

A has characteristic polynomial $p(\lambda) = (\lambda - 6)(\lambda - 4)$ so that the eigenvalues of A are $\lambda_1 = 6$ and $\lambda_2 = 4$ with corresponding orthonormal eigenvectors

$$\mathbf{w}_1 = \left(\frac{1}{\sqrt{2}}, \frac{1}{\sqrt{2}} \right), \ \mathbf{w}_2 = \left(-\frac{1}{\sqrt{2}}, \frac{1}{\sqrt{2}} \right).$$

Therefore, if we set $S = \begin{bmatrix} \dfrac{1}{\sqrt{2}} & -\dfrac{1}{\sqrt{2}} \\ \dfrac{1}{\sqrt{2}} & \dfrac{1}{\sqrt{2}} \end{bmatrix}$, then under the change of variables $\mathbf{x} = S\mathbf{y}$,

equation (6.8.6) reduces to

$$6y_1^2 + 4y_2^2 = 1, \tag{6.8.7}$$

which is the equation of an ellipse. Geometrically, vectors $\mathbf{w}_1$ and $\mathbf{w}_2$ are obtained by rotating the standard basis vectors $\mathbf{e}_1$ and $\mathbf{e}_2$, respectively, counterclockwise through an angle $\pi/4$ radians. Relative to the Cartesian coordinate system, (y_1, y_2), corresponding to the principal axes the ellipse has the simple equation (6.8.7). Figure 6.8.2 shows how the conic looks relative to the standard Cartesian axes and the rotated principal axes.

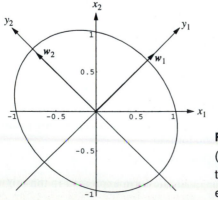

Figure 6.8.2 Relative to the Cartesian axes (x_1, x_2) cross terms come into the equation for the ellipse. Rotating to the principle axes (y_1, y_2) eliminates these cross terms.

PROOF OF RESULTS   We first establish a preliminary lemma.

Lemma 6.8.1:   Let $\mathbf{v}_1$ and $\mathbf{v}_2$ be vectors in $\mathbf{R}^n$ (or $\mathbf{C}^n$), and let A be an $n \times n$ real symmetric matrix. Then

$$<A\mathbf{v}_1, \mathbf{v}_2> = <\mathbf{v}_1, A\mathbf{v}_2>.$$

PROOF   The key to the proof is to note that the inner product of two vectors in $\mathbf{R}^n$ or $\mathbf{C}^n$ can be written in matrix form as

$$[<\mathbf{x}, \mathbf{y}>] = \mathbf{x}^T \bar{\mathbf{y}}.$$

Applying this to the vectors $A\mathbf{v}_1, \mathbf{v}_2$ yields

$$[<A\mathbf{v}_1, \mathbf{v}_2>] = (A\mathbf{v}_1)^T \bar{\mathbf{v}}_2 = \mathbf{v}_1^T A^T \bar{\mathbf{v}}_2 = \mathbf{v}_1^T \overline{(A\mathbf{v}_2)} = [<\mathbf{v}_1, A\mathbf{v}_2>],$$

from which the result follows directly.   ■

PROOF OF PARTS 1 AND 2 OF THEOREM 6.8.2

1. Suppose that $(\lambda_1, \mathbf{v}_1)$ are an eigenvalue/eigenvector pair for A, that is,

$$A\mathbf{v}_1 = \lambda_1\mathbf{v}_1. \tag{6.8.8}$$

We must prove that

$$\lambda_1 = \bar{\lambda}_1.$$

Taking the inner product of both sides of (6.8.8) with the vector $\mathbf{v}_1$ yields

$$<A\mathbf{v}_1, \mathbf{v}_1> = <\lambda_1\mathbf{v}_1, \mathbf{v}_1>.$$

That is, using the properties of the inner product,

$$<A\mathbf{v}_1, \mathbf{v}_1> = \lambda_1 \| \mathbf{v}_1 \|^2. \tag{6.8.9}$$

Taking the complex conjugate of (6.8.9) yields (remember that $\| \mathbf{v}_1 \|$ is a *real* number),

$$\overline{<A\mathbf{v}_1, \mathbf{v}_1>} = \bar{\lambda}_1 \| \mathbf{v}_1 \|^2.$$

That is, using the fact that $\overline{<\mathbf{u}, \mathbf{v}>} = <\mathbf{v}, \mathbf{u}>$

$$<\mathbf{v}_1, A\mathbf{v}_1> = \bar{\lambda}_1 \| \mathbf{v}_1 \|^2. \tag{6.8.10}$$

Subtracting equation (6.8.10) from equation (6.8.9) and using Lemma 6.8.1 with $\mathbf{v}_2 = \mathbf{v}_1$ yields

$$(\lambda_1 - \bar{\lambda}_1) \| \mathbf{v}_1 \|^2 = 0. \tag{6.8.11}$$

However, $\mathbf{v}_1 \neq \mathbf{0}$ since it is an eigenvector of A. Consequently, from equation (6.8.11), we must have

$$\lambda_1 = \bar{\lambda}_1.$$

2. It follows from part (1) that all eigenvalues of A are necessarily real. We can therefore take the underlying vector space as being $\mathbf{R}^n$, so that the corresponding eigenvectors of A will also be real. Let $(\lambda_1, \mathbf{v}_1)$ and $(\lambda_2, \mathbf{v}_2)$ be two such real eigenvalue/eigenvector pairs with $\lambda_1 \neq \lambda_2$. Then, in addition to (6.8.8) we also have

$$A\mathbf{v}_2 = \lambda_2 \mathbf{v}_2. \qquad (6.8.12)$$

We must prove that

$$<\mathbf{v}_1, \mathbf{v}_2> = 0.$$

Taking the inner product of (6.8.8) with $\mathbf{v}_2$ and the inner product of (6.8.12) with $\mathbf{v}_1$ and using $<\mathbf{v}_1, \lambda_2 \mathbf{v}_2> = \lambda_2 <\mathbf{v}_1, \mathbf{v}_2>$ yields, respectively,

$$<A\mathbf{v}_1, \mathbf{v}_2> = \lambda_1 <\mathbf{v}_1, \mathbf{v}_2>, \qquad <\mathbf{v}_1, A\mathbf{v}_2> = \lambda_2 <\mathbf{v}_1, \mathbf{v}_2>.$$

Subtracting the second equation from the first and using Lemma 6.8.1 we obtain

$$(\lambda_1 - \lambda_2) <\mathbf{v}_1, \mathbf{v}_2> = 0.$$

Since $\lambda_1 - \lambda_2 \neq 0$ by assumption, it follows that $<\mathbf{v}_1, \mathbf{v}_2> = 0$.

EXERCISES 6.8

For problems 1–12, determine an orthogonal matrix S such that $S^T A S = \operatorname{diag}(\lambda_1, \lambda_2, ..., \lambda_n)$, where A denotes the given matrix.

1. $\begin{bmatrix} 2 & 2 \\ 2 & -1 \end{bmatrix}$.

2. $\begin{bmatrix} 4 & 6 \\ 6 & 9 \end{bmatrix}$.

3. $\begin{bmatrix} 1 & 2 \\ 2 & 1 \end{bmatrix}$.

4. $\begin{bmatrix} 0 & 0 & 3 \\ 0 & -2 & 0 \\ 3 & 0 & 0 \end{bmatrix}$.

5. $\begin{bmatrix} 1 & 2 & 1 \\ 2 & 4 & 2 \\ 1 & 2 & 1 \end{bmatrix}$.

6. $\begin{bmatrix} 2 & 0 & 0 \\ 0 & 3 & 1 \\ 0 & 1 & 3 \end{bmatrix}$.

7. $\begin{bmatrix} 0 & 1 & 0 \\ 1 & 0 & 0 \\ 0 & 0 & 1 \end{bmatrix}$.

8. $\begin{bmatrix} 1 & 1 & -1 \\ 1 & 1 & 1 \\ -1 & 1 & 1 \end{bmatrix}$.

9. $\begin{bmatrix} 1 & 0 & -1 \\ 0 & 1 & 1 \\ -1 & 1 & 0 \end{bmatrix}$.

10. $\begin{bmatrix} 3 & 3 & 4 \\ 3 & 3 & 0 \\ 4 & 0 & 3 \end{bmatrix}$. You may assume that $p(\lambda) = (\lambda + 2)(\lambda - 3)(8 - \lambda)$.

11. $\begin{bmatrix} -3 & 2 & 2 \\ 2 & -3 & 2 \\ 2 & 2 & -3 \end{bmatrix}$. You may assume that $p(\lambda) = (1 - \lambda)(\lambda + 5)^2$.

12. $\begin{bmatrix} 0 & 1 & 1 \\ 1 & 0 & 1 \\ 1 & 1 & 0 \end{bmatrix}$. You may assume that $p(\lambda) = (\lambda + 1)^2(2 - \lambda)$.

For problems 13–16, determine a set of principal axes for the given quadratic form, and reduce the quadratic form to a sum of squares.

13. $\mathbf{x}^T A \mathbf{x}, A = \begin{bmatrix} 1 & 3 \\ 3 & 1 \end{bmatrix}$.

14. $\mathbf{x}^T A \mathbf{x}, A = \begin{bmatrix} 5 & 2 \\ 2 & 5 \end{bmatrix}$.

15. $\mathbf{x}^T A \mathbf{x}, A = \begin{bmatrix} 1 & 1 & -1 \\ 1 & 1 & 1 \\ -1 & 1 & 1 \end{bmatrix}$.

◆ **16.** $\mathbf{x}^T A \mathbf{x}$, $A = \begin{bmatrix} 3 & 1 & 3 & 1 \\ 1 & 3 & 1 & 3 \\ 3 & 1 & 3 & 1 \\ 1 & 3 & 1 & 3 \end{bmatrix}$.

17. Consider the general 2×2 real symmetric matrix $A = \begin{bmatrix} a & b \\ b & c \end{bmatrix}$. Prove that A has an eigenvalue of multiplicity two if and only if it is a scalar matrix (that is a matrix of the form dI_2, where d is a constant).

18. (a) Let A be an $n \times n$ real symmetric matrix. Prove that if λ is an eigenvalue of A of multiplicity n then A is a scalar matrix. (Hint: Prove that there exists an orthogonal matrix S such that $S^T A S = \lambda I_n$, and then solve for A.)

(b) State and prove the corresponding result for general $n \times n$ matrices.

19. The 2×2 real symmetric matrix A has two distinct eigenvalues λ_1 and λ_2. If $\mathbf{v}_1 = (1, 2)$ is an eigenvector of A corresponding to the eigenvalue λ_1, determine an eigenvector corresponding to λ_2.

20. The 2×2 real symmetric matrix A has two distinct eigenvalues λ_1 and λ_2.

(a) If $\mathbf{v}_1 = (a, b)$ is an eigenvector of A corresponding to the eigenvalue λ_1, determine an eigenvector corresponding to λ_2, and hence find an orthogonal matrix S such that $S^T A S = \text{diag}(\lambda_1, \lambda_2)$.

(b) Use your result from (a) to find A. (Your answer will involve λ_1, λ_2, a and b.)

21. The 3×3 real symmetric matrix A has eigenvalues λ_1 and λ_2 (multiplicity 2).

(a) If $\mathbf{v}_1 = (1, -1, 1)$ spans the eigenspace E_1, determine a basis for E_2 and hence find an orthogonal matrix S, such that $S^T A S = \text{diag}(\lambda_1, \lambda_2, \lambda_2)$.

(b) Use your result from (a) to find A.

22. Prove that a real $n \times n$ matrix A is orthogonal if and only if its row (or column) vectors are orthogonal unit vectors.

23. Prove that if A and B are $n \times n$ orthogonal matrices, then AB is also an orthogonal matrix.

Problems $24 - 26$ deal with the eigenvalue/eigenvector problem for $n \times n$ real *skew-symmetric* matrices.

24. Let A be an $n \times n$ real skew-symmetric matrix.

(a) Prove that for all $\mathbf{v}_1$ and $\mathbf{v}_2$ in $\mathbf{C}^n$,

$$<A\mathbf{v}_1, \mathbf{v}_2> = - <\mathbf{v}_1, A\mathbf{v}_2>,$$

where $< , >$ denotes the standard inner product in $\mathbf{C}^n$. (Hint: See Lemma 6.8.1.)

(b) Prove that all *nonzero* eigenvalues of A are pure imaginary ($\lambda = -\bar{\lambda}$). [Hint: Model your proof after that of (1) in Theorem 6.8.2.]

25. It follows from the previous problem that the only real eigenvalue that a real skew-symmetric matrix can possess is $\lambda = 0$. Use this to prove that if A is an $n \times n$ real skew-symmetric matrix, *with n odd*, then A necessarily has zero as one of its eigenvalues.

26. Determine all eigenvalues and corresponding eigenvectors of the matrix

$$A = \begin{bmatrix} 0 & 4 & -4 \\ -4 & 0 & -2 \\ 4 & 2 & 0 \end{bmatrix}.$$

6.9   SUMMARY OF RESULTS

LINEAR TRANSFORMATIONS

In this chapter we have considered mappings $T : V \to W$ between vector spaces V and W that satisfy the basic linearity properties

$$T(\mathbf{x} + \mathbf{y}) = T(\mathbf{x}) + T(\mathbf{y}), \text{ for all } \mathbf{x} \text{ and } \mathbf{y} \text{ in } V,$$
$$T(c\mathbf{x}) = cT(\mathbf{x}), \text{ for all } \mathbf{x} \text{ in } V \text{ and all scalars } c.$$

We now list some of the key definitions and theorems for linear transformations.

We have identified the following two important subsets of vectors associated with a linear transformation:

1. The *kernel* of T, denoted Ker(T). This is the set of all vectors in V that are mapped to the zero vector in W.

2. The *range* of T, denoted Rng(T). This is the set of vectors in W that we obtain when we allow T to act on every vector in V. Equivalently, Rng(T) is the set of all transformed vectors.

The key results about Ker(T) and Rng(T) are as follows.

Let $T : V \to W$ be a linear transformation. Then,

1. Ker(T) is a subspace of V.
2. Rng(T) is a subspace of W.
3. If V is finite-dimensional, dim[Ker(T)] + dim[Rng(T)] = dim[V].
4. T is one-to-one if and only if Ker(T) = {$\mathbf{0}$}.
5. T is onto if and only if Rng(T) = W.

Finally, if $T : V \to W$ is a linear transformation then the inverse transformation $T^{-1} : W \to V$, exists if and only if T is both one-to-one and onto.

THE ALGEBRAIC EIGENVALUE/EIGENVECTOR PROBLEM

We next summarize the main results obtained in this chapter regarding the algebraic eigenvalue/eigenvector problem.

1. For a given $n \times n$ matrix A, the eigenvalue/eigenvector problem consists of determining all scalars λ and all *nonzero* vectors $\mathbf{v}$ such that
$$A\mathbf{v} = \lambda\mathbf{v}.$$

2. The eigenvalues of A are the roots of the characteristic polynomial
$$p(\lambda) = \det(A - \lambda I) = 0, \tag{6.9.1}$$
and the eigenvectors of A are obtained by solving the linear systems
$$(A - \lambda I)\mathbf{v} = \mathbf{0}, \tag{6.9.2}$$
when λ assumes the values obtained in (6.9.1).

3. If A is a matrix with real elements, then complex eigenvalues and eigenvectors occur in conjugate pairs.

4. Associated with each eigenvalue λ there is a vector space called the eigenspace of λ. This is the set of all eigenvectors corresponding to λ together with the zero vector. Equivalently, it can be considered as the set of *all* solutions to the linear system (6.9.2).

5. If m denotes the multiplicity of the eigenvalue λ and n denotes the dimension of the corresponding eigenspace, then

$$1 \le n \le m.$$

6. Eigenvectors corresponding to distinct eigenvalues are LI.

7. An $n \times n$ matrix that has n LI eigenvectors is said to have a *complete set of eigenvectors*, and we call such a matrix *nondefective*.

8. Two $n \times n$ matrices A and B are said to be *similar* if there exists a matrix S such that

$$B = S^{-1}AS.$$

A matrix that is similar to a diagonal matrix is said to be diagonalizable. We have shown that A is diagonalizable if and only if it is nondefective.

9. If A is nondefective and $S = [\mathbf{v}_1, \mathbf{v}_2, \ldots, \mathbf{v}_n]$, where $\mathbf{v}_1, \mathbf{v}_2, \ldots, \mathbf{v}_n$ are LI eigenvectors of A, then

$$S^{-1}AS = \mathrm{diag}(\lambda_1, \lambda_2, \ldots, \lambda_n),$$

where $\lambda_1, \lambda_2, \ldots, \lambda_n$ are the eigenvalues of A corresponding to the eigenvectors $\mathbf{v}_1, \mathbf{v}_2, \ldots, \mathbf{v}_n$.

10. If A is a *real symmetric* matrix, then

(a)   All eigenvalues of A are real.

(b)   Eigenvectors corresponding to different eigenvalues are *orthogonal*.

(c) A is nondefective.

(d) A has a complete set of orthonormal eigenvectors, say $\mathbf{w}_1, \mathbf{w}_2, \ldots, \mathbf{w}_n$.

(e)   If we let $S = [\mathbf{w}_1, \mathbf{w}_2, \ldots, \mathbf{w}_n]$, where $\mathbf{w}_1, \mathbf{w}_2, \ldots, \mathbf{w}_n$ are the orthonormal eigenvectors determined in (d), then S is an orthogonal matrix ($S^{-1} = S^T$), and hence, from (9), for this matrix,

$$S^T AS = \mathrm{diag}(\lambda_1, \lambda_2, \ldots, \lambda_n).$$

7

Linear Differential Equations of Order *n*

In Chapter 2, we considered second-order linear DE, and were able to establish that the general solution to the homogeneous DE

$$y'' + a_1(x)y' + a_2(x)y = 0$$

is of the form

$$y(x) = c_1 y_1(x) + c_2 y_2(x),$$

where y_1 and y_2 are any two LI solutions to the DE. Consequently, the solution set to the preceding DE is a vector space of dimension two. The power of this result is that it tells us how many LI solutions we need to find all solutions of the DE. In the case of constant coefficient DE, we derived techniques for actually finding these solutions. Furthermore, we showed that all solutions to the nonhomogeneous DE

$$y'' + a_1(x)y' + a_2(x)y = F(x)$$

are of the form

$$y(x) = y_c(x) + y_p(x),$$

where $y_c(x)$ denotes the general solution to the associated homogeneous DE, and $y_p(x)$ is any particular solution to the nonhomogeneous DE. Consequently, having determined the complementary function, we then needed to find only one particular solution to the nonhomogeneous DE to determine all of its solutions. Techniques were derived for finding such a particular solution.

With the power of the vector space methods at our disposal, we are now able to complete our discussion of linear DE by extending the foregoing results to linear DE of arbitrary order *n*. We accomplish this goal in the following three steps.

(1) Reformulate the problem of solving an nth-order linear DE

$$a_0(x)y^{(n)} + a_1(x)y^{(n-1)} + \cdots + a_{n-1}(x)y' + a_n(x)y = F(x)$$

in the equivalent form

$$Ly = F,$$

where L is an appropriate *linear transformation*.

(2) Establish that the set of all solutions to the associated homogeneous DE

$$Ly = 0$$

is a vector space *of dimension n*, so that every solution to the homogenous DE can be expressed as

$$y(x) = c_1 y_1(x) + c_2 y_2(x) + \cdots + c_n y_n(x),$$

where $\{y_1, y_2, \ldots, y_n\}$ is any LI set of n solutions to $Ly = 0$.

(3) Establish that every solution to the nonhomogeneous problem $Ly = F$ is of the form

$$y(x) = c_1 y_1(x) + c_2 y_2(x) + \cdots + c_n y_n(x) + y_p(x),$$

where $y_p(x)$ is any particular solution to the nonhomogeneous equation.

The hard work put into understanding the material in Chapters 5 and 6 is really seen to pay off in Section 7.1, where the power of the vector space methods enables us to build this general theory very quickly.

Having obtained the general theory for linear DE of order n, the remainder of the chapter is concerned with obtaining the requisite number of solutions needed to build the general solution. We will see how the methods introduced in Chapter 2 for the case of a second-order DE can be extended to arbitrary order n.

7.1   GENERAL THEORY FOR LINEAR DIFFERENTIAL EQUATIONS

Recall from Chapter 6 that the mapping $D : C^1(I) \to C^0(I)$ defined by $D(f) = f'$ is a *linear* transformation. We call D the *derivative operator*. Higher order derivative operators can be defined by composition. Thus, $D^k : C^k(I) \to C^0(I)$ is defined by

$$D^k = D(D^{k-1}), \quad k = 2, 3, \ldots,$$

so that

$$D^k(f) = \frac{d^k f}{dx^k}.$$

By taking a linear combination of the basic derivative operators, we obtain the general **linear differential operator of order** n,

$$L = D^n + a_1 D^{n-1} + \cdots + a_{n-1} D + a_n \qquad (7.1.1)$$

defined by

$$Ly = y^{(n)} + a_1 y^{(n-1)} + \cdots + a_{n-1} y' + a_n y,$$

where the a_i are, in general, functions of x. We leave it as an exercise to verify that for all $y_1, y_2 \in C^n(I)$, and all scalars c,

$$L(y_1 + y_2) = Ly_1 + Ly_2$$

$$L(cy_1) = cLy_1.$$

Consequently, L is a linear transformation from $C^n(I)$ into $C^0(I)$.

Example 7.1.1   If $L = D^2 + 4xD - 3x$, then

$$Ly = y'' + 4xy' - 3xy,$$

so that, for example,

$$L(\sin x) = -\sin x + 4x\cos x - 3x\sin x,$$

whereas

$$L(x^2) = 2 + 8x^2 - 3x^3. \qquad \square$$

Now recall from Chapter 1 that a DE of the form

$$a_0(x)y^{(n)} + a_1(x)y^{(n-1)} + \cdots + a_{n-1}(x)y' + a_n(x)y = F(x), \qquad (7.1.2)$$

where a_0 ($\neq 0$), a_1, ..., a_n, and F are functions specified on an interval I, is called a linear DE of order n. If $F(x)$ is identically zero on I, then the DE is called **homogeneous**. Otherwise, it is called **nonhomogeneous**. We will assume that $a_0(x)$ is nonzero on I, in which case we can divide equation (7.1.2) by a_0 and redefine the remaining functions to obtain the following standard form

$$y^{(n)} + a_1(x)y^{(n-1)} + \cdots + a_{n-1}(x)y' + a_n(x)y = F(x). \qquad (7.1.3)$$

This can be written in the equivalent form

$$Ly = F(x),$$

where L is given in equation (7.1.1). The key result that we require in developing the theory for linear DE is the following existence and uniqueness theorem.

Theorem 7.1.1:   Let a_1, ..., a_n, F be functions that are continuous on an interval I. Then, for any x_0 in I, the IVP

$$Ly = F(x),$$

$$y(x_0) = y_0, \, y'(x_0) = y_1, \, ..., \, y^{(n-1)}(x_0) = y_{n-1}$$

has a unique solution on I.

PROOF   The proof of this theorem requires concepts from advanced calculus and is best left for a second course in DE. See, for example, Coddington, E.A. and Levinson, N., *Theory of Differential Equations*, McGraw–Hill, 1955.   ∎

The DE (7.1.3) is said to be *regular* on I if the functions $a_1, \ldots, a_n, F$ are continuous on I. In developing the theory for linear DE we will always assume that our DE are regular on the interval of interest so that the existence and uniqueness theorem can be applied on that interval.

HOMOGENOUS LINEAR DIFFERENTIAL EQUATIONS

We first consider the nth-order linear homogeneous DE

$$y^{(n)} + a_1(x)y^{(n-1)} + \cdots + a_{n-1}(x)y' + a_n(x)y = 0 \tag{7.1.4}$$

on an interval I. This DE can be written as the operator equation

$$Ly = 0,$$

where $L : C^n(I) \rightarrow C^0(I)$ is the nth-order linear differential operator

$$L = D^n + a_1 D^{n-1} + \cdots + a_{n-1}D + a_n.$$

If we let S denote the set of all solutions to the DE (7.1.4), then

$$S = \{y \in C^n(I) : Ly = 0\}.$$

That is,

$$S = \text{Ker}(L).$$

In Chapter 6, we proved that the kernel of any linear transformation $T : V \rightarrow W$ is a subspace of V. It follows directly from this result that S, the set of all solutions to equation (7.1.4), is a subspace of $C^n(I)$. We will refer to this subspace as the **solution space** of the DE. If we can determine the dimension of S, then we will know how many LI solutions are required to span the solution space. For example, in the second order case, the results from Chapter 2 tell us that all solutions are of the form $y(x) = c_1 y_1(x) + c_2 y_2(x)$, where y_1 and y_2 are any two LI solutions to the DE. Hence, as noted in Section 5.6, the dimension of the solution space is two in this case. The following theorem generalizes this result to arbitrary n.

   Theorem 7.1.2:   The set of all solutions to a regular nth-order homogeneous linear DE

$$y^{(n)} + a_1(x)y^{(n-1)} + \cdots + a_{n-1}(x)y' + a_n(x)y = 0$$

on an interval I is a vector space of dimension n.

PROOF   The given DE can be written in operator form as

$$Ly = 0.$$

We have already shown that the set of all solutions to this DE is a vector space. To prove that the dimension of the solution space is n, we must establish the existence of a basis consisting of n solutions. For simplicity, we provide the details only for the case $n = 3$.

   Let $y_1, y_2,$ and y_3 be the unique solutions to the three IVP

$$Ly = 0, \ y(x_0) = 1, \ y'(x_0) = 0, \ y''(x_0) = 0,$$
$$Ly = 0, \ y(x_0) = 0, \ y'(x_0) = 1, \ y''(x_0) = 0,$$
$$Ly = 0, \ y(x_0) = 0, \ y'(x_0) = 0, \ y''(x_0) = 1,$$

respectively, where

$$L = D^3 + a_1(x)D^2 + a_2(x)D + a_3(x).$$

The Wronskian of these solutions at $x_0 \in I$ is $W[y_1, y_2, y_3](x_0) = \det[I_3] = 1 \neq 0$ so that the solutions are LI on I. To be a basis for the solution space, $y_1, y_2,$ and y_3 must also span the solution space. Let $y = u(x)$ be any solution to the DE $Ly = 0$ on I, and suppose that

$$u(x_0) = c_1, \; u'(x_0) = c_2, \; u''(x_0) = c_3,$$

where c_1, c_2, c_3 are constants. Then $y = u(x)$ is the unique solution to the IVP

$$Ly = 0, \; y(x_0) = c_1, \; y'(x_0) = c_2, \; y''(x_0) = c_3. \tag{7.1.5}$$

However, if we define

$$w(x) = c_1 y_1(x) + c_2 y_2(x) + c_3 y_3(x),$$

then $w(x)$ also satisfies the IVP (7.1.5). Thus, by uniqueness, we must have

$$u(x) = w(x).$$

That is,

$$u(x) = c_1 y_1(x) + c_2 y_2(x) + c_3 y_3(x).$$

Therefore, we have shown that any solution to $Ly = 0$ can be written as a linear combination of the LI solutions $y_1, y_2,$ and y_3 and hence, these solutions do span the solution space. It follows that $\{y_1, y_2, y_3\}$ is a basis for the solution space and, since the basis consists of three vectors, the dimension of this solution space is three. The extension of the foregoing proof to arbitrary n is left as an exercise. ∎

It follows from the previous theorem and Theorem 5.6.4, that *any* set of n LI solutions, say, $\{y_1, y_2, \ldots, y_n\}$, to

$$y^{(n)} + a_1(x)y^{(n-1)} + \cdots + a_{n-1}(x)y' + a_n(x)y = 0 \tag{7.1.6}$$

is a basis for the solution space of this DE. Consequently, every solution to the DE can be written as

$$y(x) = c_1 y_1(x) + c_2 y_2(x) + \cdots + c_n y_n(x), \tag{7.1.7}$$

for appropriate constants $c_1, c_2, \ldots, c_n$. We refer to (7.1.7) as the **general solution** to the DE (7.1.6).

Example 7.1.2   Determine the general solution to $y''' + 2y'' - y' - 2y = 0$.

Solution   Since the DE has constant coefficients, our prior experience from Chapter 2 suggests that there may be solutions of the form $y(x) = e^{rx}$. Substituting this proposed expression into the DE yields

$$e^{rx}(r^3 + 2r^2 - r - 2) = 0,$$

so that we should choose r to satisfy

$$r^3 + 2r^2 - r - 2 = 0.$$

By inspection we see that $r = 1$ is a root of this equation, so that $r - 1$ is a factor of the polynomial on the left-hand side. Consequently, the equation can be written as

$$(r-1)(r^2 + 3r + 2) = 0,$$

or equivalently,

$$(r - 1)(r + 1)(r + 2) = 0.$$

Hence, three solutions to the given DE are

$$y_1(x) = e^x, \quad y_2(x) = e^{-x}, \quad y_3(x) = e^{-2x}.$$

Further,

$$W[y_1, y_2, y_3](x) = \begin{vmatrix} e^x & e^{-x} & e^{-2x} \\ e^x & -e^{-x} & -2e^{-2x} \\ e^x & e^{-x} & 4e^{-2x} \end{vmatrix} = e^{-2x} \begin{vmatrix} 1 & 1 & 1 \\ 1 & -1 & -2 \\ 1 & 1 & 4 \end{vmatrix} = -6e^{-2x} \neq 0,$$

so that the solutions are LI on any interval. It follows that the general solution to the given DE is

$$y(x) = c_1 e^x + c_2 e^{-x} + c_3 e^{-2x}. \qquad \square$$

In the previous example, we were able to determine that the solutions y_1, y_2, and y_3 were LI on any interval since their Wronskian was nonzero. What would have happened if their Wronskian had been identically zero? Based on Theorem 5.5.5, we would not have been able to draw any conclusion as to the linear dependence or linear independence of the solutions. We now show, however, that *when dealing with solutions of an nth-order homogeneous linear DE*, if the Wronskian of the solutions is zero at any point in I then the solutions are LD on I.

Theorem 7.1.3:   Let $y_1, y_2, \ldots, y_n$ be solutions to the regular nth-order DE $Ly = 0$ on an interval I, and let $W[y_1, y_2, \ldots, y_n](x)$ denote their Wronskian. If $W[y_1, y_2, \ldots, y_n](x_0) = 0$ at any point x_0 in I, then $\{y_1, y_2, \ldots, y_n\}$ is LD on I.

PROOF   We provide details for the case $n = 3$ and leave the extension to arbitrary n as an exercise. Once more, the proof depends on the existence–uniqueness theorem. Let x_0 be a point in I at which $W[y_1, y_2, y_3](x_0) = 0$, and consider the linear system

$$\begin{aligned} c_1 y_1(x_0) \quad + c_2 y_2(x_0) \quad + c_3 y_3(x_0) \quad &= 0, \\ c_1 y_1{'}(x_0) \quad + c_2 y_2{'}(x_0) + c_3 y_3{'}(x_0) \quad &= 0, \\ c_1 y_1{''}(x_0) + c_2 y_2{''}(x_0) + c_3 y_3{''}(x_0) &= 0, \end{aligned}$$

where the unknowns are c_1, c_2, and c_3. The determinant of the matrix of coefficients of this system is $W[y_1, y_2, y_3](x_0) = 0$, so that the system has nontrivial solutions. Let $(\alpha_1, \alpha_2, \alpha_3)$ be one such *nontrivial* solution, and define the function $u(x)$ by

$$u(x) = \alpha_1 y_1(x) + \alpha_2 y_2(x) + \alpha_3 y_3(x).$$

Then $y = u(x)$ satisfies the IVP

$$Ly = 0, \quad y(x_0) = 0, \, y{'}(x_0) = 0, \, y{''}(x_0) = 0$$

However, $y(x) = 0$ also satisfies the IVP preceding it, and hence, by uniqueness, we must have $u(x) = 0$, that is,

$$\alpha_1 y_1(x) + \alpha_2 y_2(x) + \alpha_3 y_3(x) = 0,$$

where at least one of α_1, α_2, α_3 is nonzero. Consequently, $\{y_1, y_2, y_3\}$ is LD on I.

■

To summarize, the vanishing or nonvanishing of the Wronskian on an interval I completely characterizes the linear dependence or independence on I of *solutions* to $Ly = 0$.

NONHOMOGENEOUS LINEAR DIFFERENTIAL EQUATIONS

We now consider the nonhomogeneous linear DE

$$y^{(n)} + a_1(x)y^{(n-1)} + \cdots + a_{n-1}(x)y' + a_n(x)y = F(x), \qquad (7.1.8)$$

where $F(x)$ is not identically zero on the interval of interest. If we set $F(x) = 0$ in (7.1.8), we obtain the **associated homogeneous equation**

$$y^{(n)} + a_1(x)y^{(n-1)} + \cdots + a_{n-1}(x)y' + a_n(x)y = 0. \qquad (7.1.9)$$

Equations (7.1.8) and (7.1.9) can be written in operator form as

$$Ly = F \quad \text{and} \quad Ly = 0,$$

respectively, where

$$L = D^n + a_1(x)D^{n-1} + \cdots + a_{n-1}(x)D + a_n(x).$$

The main theoretical result for nonhomogeneous linear DE is given in the following theorem:

Theorem 7.1.4:   Let $\{y_1, y_2, \ldots, y_n\}$ be a LI set of solutions to $Ly = 0$ on an interval I, and let $y = y_p$ be any *particular* solution to $Ly = F$ on I. Then every solution to $Ly = F$ on I is of the form

$$y = c_1 y_1 + c_2 y_2 + \cdots + c_n y_n + y_p,$$

for appropriate constants $c_1, c_2, \ldots, c_n$.

PROOF   Since $y = y_p$ satisfies equation (7.1.8) we have

$$Ly_p = F. \qquad (7.1.10)$$

Let $y = u$ be any solution to equation (7.1.8). Then we also have

$$Lu = F. \qquad (7.1.11)$$

Subtracting equation (7.1.10) from equation (7.1.11) and using the linearity of L, yields

$$L(u - y_p) = 0.$$

Thus, $y = u - y_p$ is a solution to the associated homogeneous equation $Ly = 0$ and therefore can be written as

$$u - y_p = c_1 y_1 + c_2 y_2 + \cdots + c_n y_n,$$

for appropriately chosen $c_1, c_2, \ldots, c_n$. Consequently,

$$u = c_1 y_1 + c_2 y_2 + \cdots + c_n y_n + y_p. \qquad \blacksquare$$

According to the previous theorem, the general solution to the nonhomogeneous DE $Ly = F$ is of the form

$$y(x) = y_c(x) + y_p(x),$$

where

$$y_c(x) = c_1 y_1(x) + c_2 y_2(x) + \cdots + c_n y_n(x)$$

is the general solution to the associated homogeneous equation $Ly = 0$, and y_p is a particular solution to $Ly = F$. We refer to y_c as the **complementary function** for $Ly = F$.

We need one final result regarding solutions to nonhomogeneous DE.

Theorem 7.1.5:   If $y = u_p$ and $y = v_p$ are particular solutions of $Ly = f(x)$ and $Ly = g(x)$, respectively, then $y = u_p + v_p$ is a solution to $Ly = f(x) + g(x)$.

PROOF $L(u_p + v_p) = L(u_p) + L(v_p) = f(x) + g(x)$.   ∎

We have now derived the fundamental theory for linear DE. For the remainder of the chapter, we focus our attention on extending the techniques used in the second-order case to linear DE of arbitrary order n.

EXERCISES 7.1

For problems 1–3, find Ly for the given differential operator if $y(x) = 2x - 3e^{2x}$.

1. $L = D - x$.

2. $L = D^2 - x^2 D + x$.

3. $L = D^3 - 2x D^2$.

For problems 4 and 5, verify that the given function is in the kernel of L.

4. $y(x) = x^{-2}$, $L = x^2 D^2 + 2xD - 2$.

5. $y(x) = \sin(x^2)$, $L = D^2 - x^{-1} D + 4x^2$.

6. Find Ker(L) if $L = D - 2x$.

For problems 7 and 8, find $L_1 L_2$ and $L_2 L_1$ for the given differential operators, and determine whether $L_1 L_2 = L_2 L_1$.

7. $L_1 = D + 1$, $L_2 = D - 2x^2$.

8. $L_1 = D + x$, $L_2 = D + (2x - 1)$.

9. If $L_1 = D + a_1(x)$, determine all differential operators of the form $L_2 = D + b_1(x)$ such that $L_1 L_2 = L_2 L_1$.

For problems 10 and 11, write the given non-homogeneous DE as an operator equation, and give the associated homogeneous DE.

10. $y''' + x^2 y'' - \sin x \, y' + e^x y = x^3$.

11. $y'' + 4xy' - 6x^2 y = x^2 \sin x$.

12. Use the existence and uniqueness theorem to prove that the only solution to the IVP

$$y'' + x^2 y + e^x y = 0, \ y(0) = 0, \ y'(0) = 0$$

is the trivial solution $y(x) = 0$.

13. Use the existence and uniqueness theorem to formulate and prove a general theorem regarding the solution to the IVP

$$Ly = 0,$$

$$y(x_0) = 0, \ y'(x_0) = 0, \ \ldots, \ y^{(n-1)}(x_0) = 0.$$

For problems 14–16, determine three LI solutions to the given DE of the form $y(x) = e^{rx}$, and thereby determine the general solution to the DE.

14. $y''' - 3y'' - y' + 3y = 0$.

15. $y''' + 3y'' - 4y' - 12y = 0$.

16. $y''' + 3y'' - 18y' - 40y = 0$.

17. Determine three LI solutions to the given DE of the form $y(x) = x^r$, and thereby determine the general solution to the DE

$$x^3 y''' + x^2 y'' - 2xy' + 2y = 0, \ x > 0.$$

18. Determine a particular solution to the given DE of the form $y(x) = A_0 e^{3x}$. Also find the general solution to the DE.

$$y''' + 2y'' - y' - 2y = 4e^{3x}.$$

19. Extend the proof of Theorem 7.1.2 to arbitrary n.

20. Extend the proof of Theorem 7.1.3 to arbitrary n.

21. Let $T: V \rightarrow W$ be a linear transformation, and suppose that $\{v_1, v_2, \ldots, v_n\}$ is a basis for $\text{Ker}(T)$. Prove that every solution to the operator equation

$$T(v) = w \qquad (21.1)$$

is of the form

$$v = c_1 v_1 + c_2 v_2 + \cdots + c_n v_n + v_p$$

where v_p is any particular solution to equation (21.1).

7.2   CONSTANT COEFFICIENT HOMOGENEOUS LINEAR DE

In Chapter 2, we saw that constant coefficient homogeneous linear DE have solutions of the form $y(x) = e^{rx}$. This observation enabled us to derive the two LI solutions needed to determine all solutions of the DE. In this section we show that a similar technique can be applied in the nth-order case.

Consider the DE

$$y^{(n)} + a_1 y^{(n-1)} + \cdots + a_{n-1} y' + a_n y = 0,$$

where $a_1, a_2, \ldots, a_n$ are *constants*. In operator form, we write this as

$$P(D)y = 0,$$

where

$$P(D) = D^n + a_1 D^{n-1} + \cdots + a_{n-1}D + a_n.$$

The operator $P(D)$ is called a **polynomial differential operator**. It is the special case of the general linear differential operator introduced in the previous section that arises when the coefficients are constant. Associated with any polynomial differential operator is the real polynomial

$$P(r) = r^n + a_1 r^{n-1} + \cdots + a_{n-1}r + a_n,$$

referred to as the **auxiliary polynomial**. The corresponding polynomial equation

$$P(r) = 0$$

is called the **auxiliary equation**.

In general, the composition of two linear transformations is *not* commutative, so that, in particular, if L_1 and L_2 are two linear differential operators, then, in general, $L_1 L_2 \neq L_2 L_1$. According to the next theorem, however, commutativity does hold for *polynomial differential operators*. This is a key result in determining all solutions to $P(D)y = 0$.

Theorem 7.2.1:   If $P(D)$ and $Q(D)$ are polynomial differential operators, then

$$P(D)Q(D) = Q(D)P(D).$$

PROOF   The proof consists of a straightforward verification and is left for the exercises.   ∎

Example 7.2.1   If $P(D) = D - 5$ and $Q(D) = D + 7$, verify that

$$P(D)Q(D) = Q(D)P(D).$$

Solution   For any twice differentiable function f, we have

$$P(D)Q(D)f = (D - 5)(D + 7)f = (D - 5)(f' + 7f)$$
$$= f'' + 2f' - 35f = (D^2 + 2D - 35)f.$$

Consequently,

$$P(D)Q(D) = D^2 + 2D - 35.$$

Similarly,

$$Q(D)P(D)f = (D + 7)(D - 5)f = (D + 7)(f' - 5f)$$
$$= f'' + 2f' - 35f = (D^2 + 2D - 35)f,$$

so that

$$Q(D)P(D) = D^2 + 2D - 35 = P(D)Q(D). \qquad \square$$

The importance of the preceding theorem is that it enables us to factor polynomial differential operators in the same way that we can factor real polynomials. More specifically, if $P(D)$ is a polynomial differential operator, then the auxiliary polynomial $P(r)$ can be factored as

$$P(r) = (r - r_1)^{m_1}(r - r_2)^{m_2} \cdots (r - r_k)^{m_k},$$

where m_i denotes the multiplicity of the root r_i, and

$$m_1 + m_2 + \cdots + m_k = n.$$

Consequently, $P(D)$ has the corresponding factorization

$$P(D) = (D - r_1)^{m_1}(D - r_2)^{m_2} \cdots (D - r_k)^{m_k},$$

and Theorem 7.2.1 tells us that the ordering of the terms in this factored form of $P(D)$ does not matter. It follows that the DE $P(D)y = 0$ can be written as

$$(D - r_1)^{m_1}(D - r_2)^{m_2} \cdots (D - r_k)^{m_k}y = 0. \tag{7.2.1}$$

The next step is to establish the following theorem.

Theorem 7.2.2:   If $P(D) = P_1(D) P_2(D) \cdots P_k(D)$ where each $P_i(D)$ is a polynomial differential operator, then any solution to $P_i(D)y = 0$ is also a solution to $P(D)y = 0$.

PROOF   Suppose $P_i(D)u = 0$. Then, since we can change the order of the factors in a polynomial differential operator (with constant coefficients), it follows that the expression for $P(D)$ can be rearranged as follows:

$$P(D) = P_1(D) \cdots P_{i-1}(D)P_{i+1}(D) \cdots P_k(D)P_i(D).$$

Hence,

$$P(D)u = P_1(D) \cdots P_{i-1}(D)P_{i+1}(D) \cdots P_k(D)P_i(D)u = 0. \qquad \blacksquare$$

Applying the preceding theorem to equation (7.2.1), we see that any solutions to

$$(D - r_i)^{m_i}y = 0 \tag{7.2.2}$$

will also be solutions to the full DE. Therefore we first focus our attention on DE of the form (7.2.2). A set of LI solutions to this DE is given in the next theorem.

Theorem 7.2.3:   The DE $(D - r)^m y = 0$ where m is a positive integer and r is a real or complex number has the following m solutions that are LI on any interval:

$$e^{rx}, \ x e^{rx}, \ x^2 e^{rx}, \ \dots, \ x^{m-1} e^{rx}.$$

PROOF   Consider the effect of applying the differential operator $(D - r)^m$ to the function $e^{rx} u(x)$, where u is an arbitrary (but sufficiently smooth) function. When $m = 1$, we obtain

$$(D - r)(e^{rx} u) = e^{rx} u' + r e^{rx} u - r e^{rx} u.$$

Thus,

$$(D - r)(e^{rx} u) = e^{rx} u'.$$

Repeating this procedure yields

$$(D - r)^2 (e^{rx} u) = (D - r)(e^{rx} u') = e^{rx} u'',$$

so that in general

$$(D - r)^m (e^{rx} u) = e^{rx} D^m (u). \tag{7.2.3}$$

Choosing $u(x) = x^k$ and using the fact that $D^m(x^k) = 0$, for $k = 0, 1, \dots, m - 1$, it follows from equation (7.2.3) that

$$(D - r)^m (e^{rx} x^k) = 0, \ \ k = 0, 1, \dots, m - 1,$$

and hence, $e^{rx}, \ x e^{rx}, \ x^2 e^{rx}, \ \dots, \ x^{m-1} e^{rx}$ are solutions to the DE $(D - r)^m y = 0$. We now prove that these solutions are LI on any interval. We must show that

$$c_1 e^{rx} + c_2 x e^{rx} + c_3 x^2 e^{rx} + \cdots + c_m x^{m-1} e^{rx} = 0,$$

for x in any interval if and only if $c_1 = c_2 = \cdots = c_m = 0$. Dividing by e^{rx}, we obtain the equivalent expression

$$c_1 + c_2 x + c_3 x^2 + \cdots + c_m x^{m-1} = 0.$$

Since the set of functions $\{1, x, x^2, \dots, x^{m-1}\}$ is LI on any interval (see problem 39 in Section 5.5) it follows that $c_1 = c_2 = \cdots = c_m = 0$. Hence, the given functions are indeed LI on any interval. ∎

We now apply the results of the previous two theorems to the DE

$$(D - r_1)^{m_1} (D - r_2)^{m_2} \cdots (D - r_k)^{m_k} y = 0. \tag{7.2.4}$$

The solutions that are obtained due to a term of the form $(D - r)^m$ depend on whether r is a real or complex number. We consider the two cases separately.

1.   Each term of the form $(D - r)^m$, where r is *real* contributes the LI solutions

$$e^{rx}, \ x e^{rx}, \ \dots, \ x^{m-1} e^{rx}.$$

2. $(D - r)^m$, $r = a + ib$, $b \neq 0$. In this case the term $(D - \bar{r})^m$ must also occur in equation (7.2.4). These complex conjugate terms contribute the complex-valued solutions

$$e^{(a \pm ib)x}, \ x \, e^{(a \pm ib)x}, \ x^2 e^{(a \pm ib)x}, \ \dots, \ x^{m-1} e^{(a \pm ib)x},$$

which we can write, using Euler's formula, as

$$e^{ax}(\cos bx \pm i \sin bx), \; x e^{ax}(\cos bx \pm i \sin bx), \; x^2 e^{ax}(\cos bx \pm i \sin bx),$$

$$\ldots, \; x^{m-1} e^{ax}(\cos bx \pm i \sin bx).$$

Taking appropriate linear combinations of these complex-valued solutions yields the following $2m$ *real-valued* solutions to equation (7.2.4):

$$e^{ax}\cos bx, \; x e^{ax}\cos bx, \; \ldots, \; x^{m-1} e^{ax}\cos bx, \; e^{ax}\sin bx, \; x e^{ax}\sin bx, \; \ldots, \; x^{m-1} e^{ax}\sin bx.$$

We leave the verification that these solutions are LI on any interval as an exercise.

By considering each term in equation (7.2.4) successively, we can therefore obtain n real-valued solutions to $P(D)y = 0$. The proof that the resulting set of solutions is LI on any interval is tedious and not particularly instructive. Consequently this proof is omitted (see, for example, Kaplan, W., *Differential Equations*, Addison–Wesley, 1958).

We now summarize our results.

Theorem 7.2.4:   Consider the DE

$$P(D)y = 0. \tag{7.2.5}$$

Let $r_1, r_2, \ldots, r_k$ be the distinct roots of the auxiliary equation, so that

$$P(r) = (r - r_1)^{m_1}(r - r_2)^{m_2} \cdots (r - r_k)^{m_k},$$

where m_i denotes the multiplicity of the root $r = r_i$.

1. If r_i is *real* then the functions $e^{r_i x}, x e^{r_i x}, \ldots, x^{m_i - 1} e^{r_i x}$ are LI solutions to equation (7.2.5) on any interval.

2. If r_j is *complex*, say, $r_j = a + ib$ (a, and b are real and $b \neq 0$), then the functions

$$e^{ax}\cos bx, \; x e^{ax}\cos bx, \; \ldots, \; x^{m_j - 1} e^{ax}\cos bx,$$

$$e^{ax}\sin bx, \; x e^{ax}\sin bx, \; \ldots, \; x^{m_j - 1} e^{ax}\sin bx$$

corresponding to the conjugate roots $r = a \pm ib$ are LI solutions to equation (7.2.5) on any interval.

3. The n real-valued solutions $y_1, y_2, \ldots, y_n$ to equation (7.2.5) that are obtained by considering the distinct roots $r_1, r_2, \ldots, r_k$ are LI on any interval. Consequently, the general solution to equation (7.2.5) is

$$y(x) = c_1 y_1(x) + c_2 y_2(x) + \cdots + c_n y_n(x).$$

Example 7.2.2   Find the general solution to $y''' + 2y'' + 3y' + 2y = 0$.

Solution   The auxiliary polynomial is $P(r) = r^3 + 2r^2 + 3r + 2$, which can be factored as $P(r) = (r + 1)(r^2 + r + 2)$. The roots of the auxiliary equation are therefore

$$r = -1 \; \text{and} \; r = \frac{-1 \pm i\sqrt{7}}{2}.$$

Hence, three LI solutions to the given DE are

$$y_1(x) = e^{-x}, \quad y_2(x) = e^{-x/2}\cos\left(\frac{\sqrt{7}x}{2}\right), \quad y_3(x) = e^{-x/2}\sin\left(\frac{\sqrt{7}x}{2}\right),$$

so that the general solution is

$$y(x) = c_1 e^{-x} + c_2 e^{-x/2}\cos\left(\frac{\sqrt{7}x}{2}\right) + c_3 e^{-x/2}\sin\left(\frac{\sqrt{7}x}{2}\right).$$

Example 7.2.3   Find the general solution to

$$(D - 3)(D^2 + 2D + 2)^2 y = 0. \tag{7.2.6}$$

Solution   The auxiliary polynomial is

$$P(r) = (r - 3)(r^2 + 2r + 2)^2,$$

so that the roots of the auxiliary equation are $r = 3$, and $r = -1 \pm i$ (multiplicity 2). The corresponding LI solutions to equation (7.2.6) are

$$y_1(x) = e^{3x}, \quad y_2(x) = e^{-x}\cos x, \quad y_3(x) = xe^{-x}\cos x, \quad y_4(x) = e^{-x}\sin x, \quad y_5(x) = xe^{-x}\sin x,$$

and hence, the general solution to equation (7.2.6) is

$$y(x) = c_1 e^{3x} + e^{-x}(c_2\cos x + c_3 x\cos x + c_4\sin x + c_5 x\sin x).$$

Example 7.2.4   Find the general solution to

$$D^3(D - 2)^2(D^2 + 1)^2 y = 0. \tag{7.2.7}$$

Solution   The auxiliary polynomial is $P(r) = r^3(r - 2)^2 (r^2 + 1)^2$, with zeros $r = 0$ (multiplicity 3), $r = 2$ (multiplicity 2), and $r = \pm i$ (multiplicity 2). We therefore obtain the following LI solutions to the given DE

$$y_1(x) = 1, \quad y_2(x) = x, \quad y_3(x) = x^2, \quad y_4(x) = e^{2x}, \quad y_5(x) = xe^{2x},$$

$$y_6(x) = \cos x, \quad y_7(x) = x\cos x, \quad y_8(x) = \sin x, \quad y_9(x) = x\sin x.$$

Hence, the general solution to equation (7.2.7) is

$$y(x) = c_1 + c_2 x + c_3 x^2 + c_4 e^{2x} + c_5 x e^{2x} + (c_6 + c_7 x)\cos x + (c_8 + c_9 x)\sin x.$$

EXERCISES 7.2

For problems 1–15, find the general solution to the given DE.

1. $y''' - y'' + y' - y = 0$.

2. $y''' - 2y'' - 4y' + 8y = 0$.

3. $(D - 2)(D^2 - 16)y = 0$.

4. $(D^2 + 2D + 10)^2 y = 0$.

5. $(D^2 + 4)^2(D + 1)y = 0$.

6. $(D^2 + 3)(D + 1)^2 y = 0$.

7. $D^2(D - 1)y = 0$.

8. $y^{(iv)} - 8y'' + 16y = 0$.

9. $y^{(iv)} - 16y = 0$.

10. $y''' + 8y'' + 22y' + 20y = 0$.

11. $y^{(iv)} - 16y'' + 40y' - 25y = 0$

12. $(D - 1)^3(D^2 + 9)y = 0$.

13. $(D^2 - 2D + 2)^2(D^2 - 1)y = 0$.

14. $(D + 3)(D - 1)(D + 5)^3 y = 0$.

15. $(D^2 + 9)^3 y = 0$.

For problems 16 and 17, solve the given IVP

16. $y''' - y'' + y' - y = 0$, $y(0) = 0$, $y'(0) = 1$, $y''(0) = 2$.

17. $y''' + 2y'' - 4y' - 8y = 0$, $y(0) = 0$, $y'(0) = 6$, $y''(0) = 8$.

18. Consider $P(D)y = 0$. What conditions on the roots of the auxiliary equation would guarantee that every solution to the DE satisfies

$$\lim_{x \to +\infty} y(x) = 0 \ ?$$

19. Prove that the set of functions

$\{e^{ax} \cos bx, \ xe^{ax} \cos bx, \ x^2 e^{ax} \cos bx, \ \dots, \ x^{m-1} e^{ax} \cos bx, \ e^{ax} \sin bx, \ xe^{ax} \sin bx, \ x^2 e^{ax} \sin bx, \ \dots, x^{m-1} e^{ax} \sin bx \ \}$

is LI on $(-\infty, \infty)$. [Hint: Show that the condition for determining linear dependence or independence can be written as

$$P(x) \cos bx + Q(x) \sin bx = 0,$$

where $P(x) = c_1 + c_2 x + \cdots + c_m x^{m-1}$ and $Q(x) = d_1 + d_2 x + \cdots + d_m x^{m-1}$. Then show that this implies $P(n\pi/b) = 0$, $Q((2n+1)\pi/b) = 0$, for all integers n, which means that P and Q must both be the zero polynomial.]

For problems 20–24, use some form of technology to factor the auxiliary polynomial of the given DE. Write the general solution to the DE,

◆ **20.** $y''' - 7 y'' - 193 y' - 665y = 0$.

◆ **21.** $y^{(iv)} + 4y''' - 3y'' - 64y' - 208y = 0$.

◆ **22.** $y^{(iv)} + 8y''' + 28y'' + 47y' + 36y = 0$.

◆ **23.** $y^{(v)} + 4y^{(iv)} + 50y''' + 200y''$
$$+ 625y' + 2500y = 0.$$

◆ **24.** $y^{(vii)} + 3y^{(vi)} + 3y^{(v)} + 9y^{(iv)}$
$$+ 3y''' + 9y'' + y' + 3y = 0.$$

◆ **25.** Use some form of technology to solve the IVP in problem 16. Also sketch the solution curve.

◆ **26.** Use some form of technology to solve the IVP in problem 17. Also sketch the solution curve.

7.3   THE METHOD OF UNDETERMINED COEFFICIENTS: ANNIHILATORS

In the next two sections, we turn our attention to nonhomogeneous linear DE, and illustrate how the methods for solving such DE introduced in Chapter 2 extend to the general nth-order case. In this section, we consider the method of undetermined coefficients. We therefore restrict our attention to determining a particular solution to any *constant coefficient* DE

$$P(D)y = F(x), \tag{7.3.1}$$

where $F(x)$ is restricted to be one of the following forms:

(1) $F(x) = \begin{cases} A e^{ax}, \\ A x^k, \\ A \cos bx + B \sin bx. \end{cases}$

(2)   Sums or products of functions given in (1).

In Chapter 2, we introduced the following "usual" trial solutions corresponding to each of the nonhomogeneous terms given in (1)

$$y_p(x) = \begin{cases} A_0 e^{ax}, \\ A_0 + A_1 x + \cdots + A_k x^k, \\ A_0 \cos bx + B_0 \sin bx. \end{cases}$$

Substitution of the appropriate trial solution into the given DE then enabled us to obtain a particular solution. If, however, the usual trial solution contained terms that solved the homogeneous DE, then we had to modify the trial solution by multiplying by x or x^2. Nonhomogeneous terms that were a sum or product of the forms given in (1) required trial solutions that were a corresponding sum or product of the basic trial solutions. In the nth-order case the usual trial solutions remain the same, since we are trying to build the same basic nonhomogeneous terms. However, for $n > 2$, due to the higher order of the DE, the modification procedure can involve multiplication by powers of x higher than the second. In such cases, the following rule tells us how to modify the usual trial solution corresponding to a given nonhomogeneous term $F(x)$.

Modification Rule: Multiply the usual trial solution by x^m, where m is the smallest positive integer such that the proposed trial solution has no terms that solve the homogeneous DE.

Justification of this rule, and indeed justification of the general method of undetermined coefficients itself, will be given after we have considered some examples.

Example 7.3.1   Determine a trial solution for $(D - 1)(D^2 + 1)^2 y = 4\cos x$.

Solution   The complementary function for this DE is

$$y_c(x) = c_1 e^x + c_2 \cos x + c_3 \sin x + x(c_4 \cos x + c_5 \sin x).$$

The usual trial solution corresponding to the nonhomogenous term $F(x) = 4\cos x$ is

$$y_p(x) = A_0 \cos x + B_0 \sin x.$$

However, we see that this y_p coincides with part of the complementary function and therefore solves the associated homogeneous equation. In this case, we need to modify the usual trial solution by multiplication by x^2. Hence, we obtain

$$y_p(x) = x^2(A_0 \cos x + B_0 \sin x).$$

The constants A_0 and B_0 can now be determined by substitution of the proposed trial solution into the given DE.

Example 7.3.2   Determine a trial solution for the DE
$$(D^2 + 2D + 5)(D - 1)^3 y = 5 e^x + 7 e^{-x} \sin 2x.$$

Solution   The complementary function is

$$y_c(x) = e^{-x}(c_1 \cos 2x + c_2 \sin 2x) + c_3 e^x + c_4 x e^x + c_5 x^2 e^x.$$

The usual trial solution corresponding to the nonhomogeneous term $F_1(x) = 5 e^x$ is $y_{p_1}(x) = A_0 e^x$. However, this coincides with part of the complementary function. According to the modification rule, we need to modify the trial solution by multiplication by x^3, thereby obtaining

$$y_{p_1}(x) = A_0 x^3 e^x.$$

The usual trial solution corresponding to the nonhomogeneous term $F_2(x) = 7e^{-x}\sin 2x$ is $y_{p_2}(x) = e^{-x}(A_1 \cos 2x + B_1 \sin 2x)$. Once more, this needs modifying due to coincidence with part of the complementary function. For this term, we only require multiplication by x. Hence,

$$y_{p_2}(x) = xe^{-x}(A_1 \cos 2x + B_1 \sin 2x).$$

Consequently, an appropriate trial solution for the given DE is

$$y_p(x) = y_{p_1}(x) + y_{p_2}(x) = A_0 x^3 e^x + xe^{-x}(A_1 \cos 2x + B_1 \sin 2x),$$

where the constants A_0, A_1, and B_1 could be obtained by substitution into the given DE.

ANNIHILATORS

We now give some justification for the trial solutions that we have been using in the method of undetermined coefficients. If we think about the types of nonhomogeneous terms for which the method works, we see that it is those functions that themselves solve a constant coefficient homogeneous linear DE. Consequently, the types of problems on which the method works are those that satisfy the following two conditions

$$P(D)y = F(x), \qquad\qquad (7.3.2)$$
$$Q(D)F = 0, \qquad\qquad (7.3.3)$$

where $Q(D)$ is a polynomial differential operator. Any polynomial differential operator $Q(D)$ that satisfies (7.3.3) is said to **annihilate** $F(x)$. The polynomial differential operator of lowest order that satisfies equation (7.3.3) is called the **annihilator** of F. An appropriate trial solution for any DE satisfying equations (7.3.2) and (7.3.3) can be derived as follows. Operating on (7.3.2) with $Q(D)$, and using (7.3.3), yields

$$Q(D)P(D)y = 0. \qquad\qquad (7.3.4)$$

This is a constant coefficient homogeneous linear DE and therefore can be solved using the technique of the previous section. The key point is the following. Any solution to equation (7.3.2) must also solve equation (7.3.4). Consequently, from the general solution to equation (7.3.4), we must be able to obtain a particular solution to equation (7.3.2) by choosing the arbitrary constants in this general solution appropriately. We note that the general solution to equation (7.3.4) will contain the complementary function for equation (7.3.2), since $P(D)$ is part of the overall differential operator in equation (7.3.4). Hence, in choosing the trial solution for equation (7.3.2), we should choose only that part of the general solution to equation (7.3.4) that does not include the complementary function. We illustrate these ideas with an example.

Example 7.3.3   Use the annihilator method to derive a trial solution for

$$(D - 3)^2 y = 7e^{3x}. \qquad\qquad (7.3.5)$$

Solution   The complementary function for the given DE is

$$y_c(x) = c_1 e^{3x} + c_2 x e^{3x}.$$

We next need to determine the annihilator for $F(x) = 7e^{3x}$. Since $(D - a)(e^{ax}) = 0$, it follows that we can choose $Q(D) = D - 3$. Operating on equation (7.3.5) with $Q(D)$ yields

$$(D - 3)^3 y = 0,$$

which has general solution

$$y(x) = c_1 e^{3x} + c_2 x e^{3x} + A_0 x^2 e^{3x} = y_c(x) + A_0 x^2 e^{3x}.$$

Consequently, an appropriate trial solution for equation (7.3.5) is

$$y_p(x) = A_0 x^2 e^{3x},$$

where A_0 could be determined by substitution into (7.3.5) in the usual manner. $\square$

We next derive appropriate annihilators to cover any case that might arise. Consider first $F(x) = x^k e^{ax}$, where a is a real number. Since the DE

$$(D - a)^{k+1} y = 0,$$

where a is a real number and k is a nonnegative integer, has the real-valued solutions

$$e^{ax}, x e^{ax}, \ldots, x^k e^{ax},$$

it follows that

---

1. $Q(D) = (D - a)^{k+1}$ annihilates each of the functions

$$e^{ax}, x\, e^{ax}, \ldots, x^k\, e^{ax},$$

and, therefore, it also annihilates

$$F(x) = (a_0 + a_1 x + \cdots + a_k x^k)\, e^{ax},$$

for all values of the constants $a_0, a_1, \ldots, a_k$.

REMARK   Note the special case of (1) that arises when $a = 0$, namely,

---

$Q(D) = D^{k+1}$ annihilates $F(x) = a_0 + a_1 x + \cdots + a_k x^k$.

Now consider the functions $e^{ax}\cos bx$, $e^{ax}\sin bx$, where a and b are real numbers. These functions arise as LI (real-valued) solutions to the DE

$$(D - \alpha)(D - \bar{\alpha})\, y = 0,$$

where $\alpha = a + ib$. Expanding the polynomial differential operator, we have

$$[D^2 - 2aD + a^2 + b^2] y = 0.$$

Consequently,

---

2. $Q(D) = D^2 - 2aD + a^2 + b^2$ annihilates both of the functions

$$e^{ax} \cos bx, \; e^{ax} \sin bx,$$

and, therefore, it annihilates

$$F(x) = e^{ax}(a_0 \cos bx + b_0 \sin bx),$$

for all values of the constants a_0, b_0. In particular,

$$Q(D) = D^2 + b^2$$

annihilates the functions $\cos bx$ and $\sin bx$.

Finally, the functions

$$e^{ax}\cos bx, \; xe^{ax}\cos bx, \; x^2 e^{ax}\cos bx, \; \ldots, \; x^k e^{ax}\cos bx,$$

$$e^{ax}\sin bx, \; xe^{ax}\sin bx, \; x^2 e^{ax}\sin bx, \; \ldots, \; x^k e^{ax}\sin bx,$$

arise as LI (real-valued) solutions to the DE

$$(D^2 - 2aD + a^2 + b^2)^{k+1}y = 0.$$

Equivalently, we can state that

---

3. $Q(D) = (D^2 - 2aD + a^2 + b^2)^{k+1}$ annihilates each of the functions

$$e^{ax}\cos bx, \; x\,e^{ax}\cos bx, \; x^2 e^{ax}\cos bx, \; \ldots, \; x^k e^{ax}\cos bx,$$

$$e^{ax}\sin bx, \; x\,e^{ax}\sin bx, \; x^2 e^{ax}\sin bx, \; \ldots, \; x^k e^{ax}\sin bx,$$

and hence, for all values of the constants $a_0, a_1, \ldots, a_k, b_0, b_1, \ldots, b_k$, it annihilates

$$F(x) = (a_0 + a_1 x + \cdots + a_k x^k)e^{ax}\cos bx$$
$$+ (b_0 + b_1 x + \cdots + b_k x^k)\,e^{ax}\sin bx.$$

The following examples give further illustrations of the annihilator technique:

Example 7.3.4   Use the annihilator method to determine a trial solution for

$$(D^2 + 1)y = 4\sin x.$$

Solution   The complementary function is

$$y_c(x) = c_1\cos x + c_2\sin x.$$

Furthermore, the annihilator for $F(x) = 4\sin x$ is $Q(D) = D^2 + 1$. Operating on the given DE with $Q(D)$ yields

$$(D^2 + 1)^2 y = 0,$$

which has general solution

$$y(x) = c_1\cos x + c_2\sin x + x(A_0\cos x + B_0\sin x).$$

Hence, a trial solution is

$$y_p(x) = x(A_0\cos x + B_0\sin x).$$

Example 7.3.5   Use the annihilator technique to determine a trial solution for

$$(D + 1)(D^2 + 9)y = 4xe^{-x} + 5e^{2x}\cos 3x.$$

Solution   The complementary function is

$$y_c(x) = c_1 e^{-x} + c_2\cos 3x + c_3\sin 3x.$$

An annihilator for $F_1(x) = 4xe^{-x}$ is

$$Q_1(D) = (D + 1)^2,$$

whereas an annihilator for $F_2(x) = 5e^{2x}\cos 3x$ is

$$Q_2(D) = D^2 - 4D + 13.$$

Hence, operating on the given DE with $Q(D) = (D^2 - 4D + 13)(D + 1)^2$ yields the homogeneous DE

$$(D^2 - 4D + 13)(D + 1)^3(D^2 + 9)y = 0,$$

which has general solution

$$y(x) = c_1 e^{-x} + c_2 \cos 3x + c_3 \sin 3x + A_0 x e^{-x} + A_1 x^2 e^{-x} + e^{2x}(B_0 \cos 3x + B_1 \sin 3x).$$

Consequently, a trial solution for the given DE is

$$y_p(x) = A_0 x e^{-x} + A_1 x^2 e^{-x} + e^{2x}(B_0 \cos 3x + B_1 \sin 3x). \qquad \square$$

More generally, we can derive appropriate trial solutions to cover any case that may arise. For example, consider

$$P(D)y = c x^k e^{ax}, \tag{7.3.6}$$

and let y_c denote the complementary function. The appropriate annihilator for equation (7.3.6) is $Q(D) = (D - a)^{k+1}$, and so a trial solution for equation (7.3.6) can be determined from the general solution to

$$Q(D)P(D)y = 0. \tag{7.3.7}$$

The following two cases arise:

Case 1: If $r = a$ is *not* a root of $P(r) = 0$, then the general solution to equation (7.3.7) will be of the form

$$y(x) = y_c(x) + e^{ax}(A_0 + A_1 x + \cdots + A_k x^k),$$

so that an appropriate trial solution is

$$y_p(x) = e^{ax}(A_0 + A_1 x + \cdots + A_k x^k).$$

This is the "usual" trial solution.

Case 2: If $r = a$ is a root of multiplicity m of $P(r) = 0$, then the complementary function $y_c(x)$ will contain the terms

$$e^{ax}(c_0 + c_1 x + \cdots + c_{m-1} x^{m-1}).$$

The operator $Q(D)P(D)$ will therefore contain the factor $(D - a)^{m+k+1}$, so that the terms in the general solution to equation (7.3.7) that do not arise in the complementary function are

$$y_p(x) = e^{ax} x^m (A_0 + A_1 x + \cdots + A_k x^k),$$

which is the "modified" trial solution.

The derivation of appropriate trial solutions for $P(D)y = F(x)$ in the case when

$$F(x) = c x^k e^{ax} \cos bx \qquad \text{or} \qquad F(x) = c x^k e^{ax} \sin bx$$

is left as an exercise. We summarize the results in a table.

$F(x)$	Usual trial solution	Modified trial solution
$cx^k e^{ax}$	If $P(a) \neq 0$: $y_p(x) = e^{ax}(A_0 + A_1 x + \cdots$ $+ A_k x^k)$.	If a is a root of $P(r) = 0$ of multiplicity m: $y_p(x) = x^m e^{ax}(A_0 + A_1 x + \cdots + A_k x^k)$.
$cx^k e^{ax} \cos bx$ or $cx^k e^{ax} \sin bx$	If $P(a + ib) \neq 0$: $y_p(x) = e^{ax}[A_0 \cos bx + B_0 \sin bx$ $+ x(A_1 \cos bx + B_1 \sin bx)$ $+ \cdots + x^k(A_k \cos bx + B_k \sin bx)]$.	If $a + ib$ is a root of $P(r)$ of multiplicity m: $y_p(x) = x^m e^{ax}[A_0 \cos bx + B_0 \sin bx$ $+ x(A_1 \cos bx + B_1 \sin bx)$ $+ \cdots + x^k(A_k \cos bx + B_k \sin bx)]$

If $F(x)$ is the sum of functions of the preceding form then the appropriate trial solution is the corresponding sum.

In the following table, we have specialized the foregoing trial solutions to the cases that arise most often in applications:

$F(x)$	Usual trial solution	Modified trial solution
ce^{ax}	If $P(a) \neq 0$: $y_p(x) = A_0 e^{ax}$.	If a is a root of $P(r) = 0$ of multiplicity m: $y_p(x) = A_0 x^m e^{ax}$.
$c \cos bx$ or $c \sin bx$	If $P(ib) \neq 0$: $y_p(x) = A_0 \cos bx + B_0 \sin bx$.	If ib is a root of $P(r) = 0$ of multiplicity m: $y_p(x) = x^m(A_0 \cos bx + B_0 \sin bx)$.
cx^k	If $P(0) \neq 0$: $y_p(x) = A_0 + A_1 x + \cdots + A_k x^k$.	If zero is a root of $P(r) = 0$ of multiplicity m: $y_p(x) = x^m(A_0 + A_1 x + \cdots + A_k x^k)$.

EXERCISES 7.3

For problems 1–5, determine the general solution to the given DE. Do not use annihilators in determining your trial solution.

1 . $y''' - 2y'' - y' + 2y = 4e^{3x}$.

2 . $y''' - 3y'' - 16y' + 48y = 6e^{3x}$.

3 . $y''' + 3y'' - 4y' - 12y = 4\cos x$.

4 . $D^2(D^2 + 1)y = 6 - 12x$.

5 . $(D + 1)^3 y = 5e^{-x}$.

For problems 6–15, determine the annihilator of the given function.

6 . $F(x) = 5e^{-3x}$.

7 . $F(x) = 2e^x - 3x$.

8 . $F(x) = \sin x + 3xe^{2x}$.

9 . $F(x) = x^3 e^{7x} + 5 \sin 4x$.

10. $F(x) = 4e^{-2x} \sin x$.

11. $F(x) = e^x \sin 2x + 3 \cos 2x$.

12. $F(x) = (1 - 3x)e^{4x} + 2x^2$.

13. $F(x) = e^{5x}(2 - x^2) \cos x$.

14. $F(x) = e^{-3x}(2 \sin x + 7 \cos x)$.

15. $F(x) = e^{4x}(x - 2 \sin 5x) + 3x - x^2 e^{-2x} \cos x$.

For problems 16–23, determine the general solution to the given DE. Derive your trial solution using the annihilator technique.

16. $y'' + y = 6e^x$.

17. $y'' + 4y' + 4y = 5xe^{-2x}$.

18. $y'' + 16y = 4 \cos x$.

19. $y'' - y' - 2y = 5e^{2x}$.

20. $y'' + 2y' + 5y = 3 \sin 2x$.

21. $y''' + 2y'' - 5y' - 6y = 4x^2$.

22. $y''' - y'' + y' - y = 9e^{-x}$.

23. $y''' + 3y'' + 3y' + y = 2e^{-x} + 3e^{2x}$.

24. Solve the given IVP.

$$(D - 1)(D - 2)(D - 3) y = 6e^{4x},$$
$$y(0) = 4, \quad y'(0) = 10, \quad y''(0) = 30.$$

For problems 25–31, determine an appropriate trial solution for the given DE. Do *not* solve for the constants that arise in your trial solution.

25. $(D - 2)(D - 3)y = 7e^{2x}$.

26. $(D + 1)(D^2 + 1)y = 4xe^x$.

27. $(D^2 + 4D + 13)^2 y = 5e^{-2x} \cos 3x$.

28. $(D^2 + 4)(D - 2)^3 y = 4x + 9xe^{2x}$.

29. $D^2(D - 1)(D^2 + 4)^2 y = 11e^x - \sin 2x$.

30. $(D^2 - 2D + 2)^3(D - 2)^2(D + 4)y = e^x \cos x - 3e^{2x}$.

31. $D(D^2 - 9)(D^2 - 4D + 5)y = 2e^{3x} + e^{2x} \sin x$.

32. Derive an appropriate trial solution for the DE

$$P(D)y = cx^k e^{ax} \cos bx.$$

7.4   THE VARIATION-OF-PARAMETERS METHOD

We now consider the generalization of the variation-of-parameters method to linear nonhomogeneous DE of arbitrary order n. In this case, the basic equation is

$$y^{(n)} + a_1(x)y^{(n-1)} + \cdots + a_{n-1}(x)y' + a_n(x)y = F(x), \qquad (7.4.1)$$

where we assume that the functions $a_1, a_2, \ldots, a_n$ and F are at least continuous on the interval I. Let $\{y_1(x), y_2(x), \ldots, y_n(x)\}$ be a LI set of solutions to the associated homogeneous equation

$$y^{(n)} + a_1(x)y^{(n-1)} + \cdots + a_{n-1}(x)y' + a_n(x)y = 0 \qquad (7.4.2)$$

on I, so that the general solution to equation (7.4.2) on I is

$$y_c(x) = c_1 y_1(x) + c_2 y_2(x) + \cdots + c_n y_n(x).$$

We now look for a particular solution to equation (7.4.1) of the form

$$y_p(x) = u_1(x)y_1(x) + u_2(x)y_2(x) + \cdots + u_n(x)y_n(x). \qquad (7.4.3)$$

The idea is to substitute for y_p into equation (7.4.1) and choose the functions $u_1, u_2, \ldots, u_n$, so that the resulting y_p is indeed a solution. However, equation (7.4.1) will only give one constraint on the functions $u_1, u_2, \ldots, u_n$ and their derivatives. Since we have n functions, we might expect that we can impose $n - 1$ further constraints on these functions. Following the steps taken in the second-order case, we differentiate y_p n times, while imposing the constraint that the sum of the terms involving derivatives of the $u_1, u_2, \ldots, u_n$ that arise at each stage (except the last) should equal zero. For example, at the first stage, we obtain

$$y_p' = u_1 y_1' + u_1' y_1 + u_2 y_2' + u_2' y_2 + \cdots + u_n y_n' + u_n' y_n,$$

and so, we impose the constraint

$$u_1' y_1 + u_2' y_2 + \cdots + u_n' y_n = 0$$

in which case the foregoing expression for y_p' reduces to

$$y_p' = u_1 y_1' + u_2 y_2' + \cdots + u_n y_n'.$$

Continuing in this manner leads to the following expressions for y_p and its derivatives

$$\begin{aligned}
y_p &= u_1 y_1 &+ u_2 y_2 &+ \cdots &+ u_n y_n, \\
y_p' &= u_1 y_1' &+ u_2 y_2' &+ \cdots &+ u_n y_n',
\end{aligned}$$

$$\vdots \tag{7.4.4}$$

$$y_p^{(n)} = u_1 y_1^{(n)} + u_2 y_2^{(n)} + \cdots + u_n y_n^{(n)} + [u_1' y_1^{(n-1)} + u_2' y_2^{(n-1)}$$

$$+ \cdots + u_n' y_n^{(n-1)}],$$

together with the corresponding constraint conditions

$$\begin{aligned}
u_1' y_1 &+ u_2' y_2 &+ \cdots + u_n' y_n &= 0, \\
u_1' y_1' &+ u_2' y_2' &+ \cdots + u_n' y_n' &= 0,
\end{aligned}$$

$$\vdots \tag{7.4.5}$$

$$u_1' y_1^{(n-2)} + u_2' y_2^{(n-2)} + \cdots + u_n' y_n^{(n-2)} = 0.$$

Substitution from (7.4.4) into (7.4.1) yields the following condition in order for y_p to be a solution [simply multiply each equation in (7.4.4) by the appropriate a_i and add the elements in each column]:

$$u_1 [y_1^{(n)} + a_1 y_1^{(n-1)} + \cdots + a_{n-1} y_1' + a_n y_1]$$

$$+ u_2 [y_2^{(n)} + a_1 y_2^{(n-1)} + \cdots + a_{n-1} y_2' + a_n y_2]$$

$$+ \cdots + u_n [y_n^{(n)} + a_1 y_n^{(n-1)} + \cdots + a_{n-1} y_n' + a_n y_n]$$

$$+ [u_1' y_1^{(n-1)} + u_2' y_2^{(n-1)} + \cdots + u_n' y_n^{(n-1)}] = F(x).$$

The terms in each of the brackets except the last vanish, since $y_1, y_2, \ldots, y_n$ are solutions of equation (7.4.2). We are therefore left with the condition

$$u_1' y_1^{(n-1)} + u_2' y_2^{(n-1)} + \cdots + u_n' y_n^{(n-1)} = F(x).$$

Combining this with the constraints given in (7.4.5) leads to the following linear system of equations for determining $u_1', u_2', \ldots, u_n'$:

$$\begin{aligned}
y_1 u_1' &+ y_2 u_2' &+ \cdots + y_n u_n' &= 0, \\
y_1' u_1' &+ y_2' u_2' &+ \cdots + y_n' u_n' &= 0,
\end{aligned}$$

$$\vdots \tag{7.4.6}$$

$$y_1^{(n-2)} u_1' + y_2^{(n-2)} u_2' + \cdots + y_n^{(n-2)} u_n' = 0,$$

$$y_1^{(n-1)} u_1' + y_2^{(n-1)} u_2' + \cdots + y_n^{(n-1)} u_n' = F(x).$$

The determinant of the matrix of coefficients of this system is the Wronskian of the functions $y_1, y_2, \ldots, y_n$, which is necessarily nonzero on I since $y_1, y_2, \ldots, y_n$ are LI on I.

Consequently, the system (7.4.6) has a unique solution for u_1', u_2', ..., u_n', from which we can determine u_1, u_2, ..., u_n by integration. Having found u_1, u_2, ..., u_n, we can obtain y_p by substitution into equation (7.4.3).

Our results are summarized in the next theorem.

Theorem 7.4.1 (Variation-of-Parameters):   Consider

$$y^{(n)} + a_1(x)y^{(n-1)} + \cdots + a_{n-1}(x)y' + a_n(x)y = F(x), \tag{7.4.7}$$

where a_1, a_2, ..., a_n, F are assumed to be (at least) continuous on the interval I. Let $\{y_1, y_2, ..., y_n\}$ be a LI set of solutions to the associated homogeneous equation

$$y^{(n)} + a_1(x)y^{(n-1)} + \cdots + a_{n-1}(x)y' + a_n(x)y = 0$$

on I. Then a particular solution to equation (7.4.7) is

$$y_p = u_1 y_1 + u_2 y_2 + \cdots + u_n y_n,$$

where the functions u_1, u_2, ..., u_n satisfy (7.4.6).

Example 7.4.1   Find the general solution to

$$y''' - 3y'' + 3y' - y = 36 e^x \ln x. \tag{7.4.8}$$

Solution   In this case, the auxiliary polynomial of the associated homogeneous equation is

$$P(r) = r^3 - 3r^2 + 3r - 1 = (r-1)^3,$$

so that three LI solutions are

$$y_1(x) = e^x, \quad y_2(x) = x e^x, \quad y_3(x) = x^2 e^x.$$

According to the variation-of-parameters method, there is a particular solution to equation (7.4.8) of the form

$$y_p(x) = e^x u_1(x) + x e^x u_2(x) + x^2 e^x u_3(x), \tag{7.4.9}$$

where u_1, u_2, and u_3 satisfy (7.4.6), which, in this case (after division by e^x) assumes the form

$$u_1' + \qquad x u_2' \qquad\qquad x^2 u_3' \ = 0,$$
$$u_1' + (x+1)u_2' + (x^2 + 2x)u_3' \ = 0,$$
$$u_1' + (x+2)u_2' + (x^2 + 4x + 2)u_3' = 36 \ln x.$$

To solve this system, we reduce its augmented matrix to row-echelon form:

$$\begin{bmatrix} 1 & x & x^2 & 0 \\ 1 & x+1 & x^2+2x & 0 \\ 1 & x+2 & x^2+4x+2 & 36 \ln x \end{bmatrix} \sim \begin{bmatrix} 1 & x & x^2 & 0 \\ 0 & 1 & 2x & 0 \\ 0 & 2 & 4x+2 & 36 \ln x \end{bmatrix}$$

$$\sim \begin{bmatrix} 1 & x & x^2 & 0 \\ 0 & 1 & 2x & 0 \\ 0 & 0 & 2 & 36 \ln x \end{bmatrix} \sim \begin{bmatrix} 1 & x & x^2 & 0 \\ 0 & 1 & 2x & 0 \\ 0 & 0 & 1 & 18 \ln x \end{bmatrix}.$$

Consequently,

$$u_1{}' = 18x^2 \ln x, \quad u_2{}' = -36x \ln x, \quad u_3{}' = 18 \ln x.$$

By integrating, we obtain

$$u_1(x) \ = 18 \int x^2 \ln x \, dx = 2x^3(3 \ln x - 1),$$

$$u_2(x) \ = -36 \int x \ln x \, dx = 9x^2(1 - 2 \ln x),$$

$$u_3(x) \ = 18 \int \ln x \, dx = 18x(\ln x - 1),$$

where we have set the integration constants to zero without loss of generality. Substituting these expressions for u_1, u_2, and u_3 into equation (7.4.9) yields the particular solution

$$y_p(x) = x^3 e^x(6 \ln x - 11).$$

The general solution to the given DE is therefore

$$y(x) = e^x[c_1 + c_2 x + c_3 x^2 + x^3(6 \ln x - 11)].$$

GREEN'S FUNCTIONS

If we let $W_k(x)$ denote the determinant that is obtained when the kth column of

$$W[y_1, y_2, \ldots, y_n](x) \text{ is replaced by } \begin{bmatrix} 0 \\ 0 \\ \vdots \\ F(x) \end{bmatrix}, \text{ then, using Cramer's rule, the solution to}$$

the system (7.4.6) can be written in the form

$$u_k{}' = \frac{W_k(x)}{W(x)}, \quad k = 1, 2, \ldots, n,$$

where, for simplicity, we have denoted $W[y_1, y_2, \ldots, y_n](x)$ by $W(x)$. A particular solution to the DE (7.4.7) is therefore given by

$$y_p(x) = y_1(x) \int_{x_0}^{x} \frac{W_1(t)}{W(t)} \, dt + y_2(x) \int_{x_0}^{x} \frac{W_2(t)}{W(t)} \, dt + \cdots + y_n(x) \int_{x_0}^{x} \frac{W_n(t)}{W(t)} \, dt,$$

where x_0 is any convenient point in the interval of interest. Although this is an elegant mathematical formula, it is usually computationally more efficient to solve the system (7.4.6) using Gaussian elimination, rather than Cramer's rule, when dealing with DE of order higher than two.

As shown in Exercise 11, the preceding formula can be written in the equivalent form

$$y_p(x) = \int_{x_0}^{x} K(x, t)F(t)\, dt, \tag{7.4.10}$$

where K is given by

$$K(x, t) = \frac{y_1(x)\tilde{W}_1(t) + y_2(x)\tilde{W}_2(t) + \cdots + y_n(x)\tilde{W}_n(t)}{W(t)}, \tag{7.4.11}$$

with $\tilde{W}_k(t)$ being obtained by replacing the kth column vector in $W(t)$ with

$$\begin{bmatrix} 0 \\ 0 \\ \vdots \\ 1 \end{bmatrix}.$$

The function $K(x, t)$ is called a **Green's function** associated with the problem (see Chapter 2 for the second order case).

Example 7.4.2   Use a Green's function to determine a particular solution for

$$(D - 1)(D - 2)(D + 4)y = F(x).$$

Solution   Three LI solutions to the associated homogeneous problem are

$$y_1(x) = e^x, \quad y_2(x) = e^{2x}, \quad y_3(x) = e^{-4x}.$$

The Wronskian of these solutions is

$$W[y_1, y_2, y_3](x) = 30e^{-x}.$$

Furthermore,

$$\tilde{W}_1(t) = \begin{vmatrix} 0 & e^{2t} & e^{-4t} \\ 0 & 2e^{2t} & -4e^{-4t} \\ 1 & 4e^{2t} & 16e^{-4t} \end{vmatrix} = -6e^{-2t},$$

$$\tilde{W}_2(t) = \begin{vmatrix} e^t & 0 & e^{-4t} \\ e^t & 0 & -4e^{-4t} \\ e^t & 1 & 16e^{-4t} \end{vmatrix} = 5e^{-3t},$$

$$\tilde{W}_3(t) = \begin{vmatrix} e^t & e^{2t} & 0 \\ e^t & 2e^{2t} & 0 \\ e^t & 4e^{2t} & 1 \end{vmatrix} = e^{3t}.$$

Consequently,

$$K(x, t) = \frac{1}{30e^{-t}}[e^x(-6e^{-2t}) + e^{2x}(5e^{-3t}) + e^{-4x}(e^{3t})]$$

$$= \frac{1}{30}[e^{-4(x-t)} + 5e^{2(x-t)} - 6e^{(x-t)}],$$

and a particular solution to the given DE is

$$y_p(x) = \frac{1}{30}\int_{x_0}^{x} [e^{-4(x-t)} + 5e^{2(x-t)} - 6e^{(x-t)}]\, F(t)\, dt.$$

EXERCISES 7.4

For problems 1–4, use the variation-of-parameters technique to determine a particular solution to the given DE.

1 . $y''' - 6y'' + 12y' - 8y = 36e^{2x} \ln x$.

2 . $y''' - 3y'' + 3y' - y = 2x^{-2}e^x, \; x > 0$.

3 . $y''' + 3y'' + 3y' + y = \dfrac{2e^{-x}}{1 + x^2}$

4 . $y''' - 6y'' + 9y' = 12e^{3x}$. Suggest a better method for solving this problem.

For problems 5–8, use a Green's function to determine a particular solution to the given DE.

5 . $(D + 3)(D - 3)(D + 5)y = F(x)$.

6 . $(D + 1)(D^2 + 9)y = F(x)$.

7 . $(D^2 + 8D + 16)(D - 2)y = F(x)$.

8 . $(D^2 - 4D + 13)(D - 3)y = F(x)$.

9 . If F is continuous on the interval $[a, b]$, use the variation-of-parameters technique to show that a particular solution to

$$(D - r)^3 y = F(x), \quad r \text{ constant}$$

is

$$y_p(x) = \frac{1}{2} \int_a^x F(t)\,(x - t)^2\, e^{r(x-t)}\, dt.$$

10. Determine the general solution to the nonhomogeneous Cauchy–Euler equation

$$x^3\, y''' + x\, y' - y = 24x \ln x, \quad x > 0.$$

(Hint: Try for solutions to the associated homogeneous equation of the form $y(x) = x^r$.)

11. Establish formula (7.4.10).

8

Systems of Differential Equations

8.1   INTRODUCTION

So far, our discussion of DE has centered around solving a single DE for a single unknown function $y(x)$. However, in practice, most applied problems involve more than one unknown function for their formulation and hence require the solution to a system of DE. Perhaps the simplest way to see how systems naturally arise is to consider the motion of an object in space. If this object has mass m and is moving under the influence of a force $F = (F_1, F_2, F_3)$, then, according to Newton's second law of motion, the position of the object at time t, $(x(t), y(t), z(t))$, is obtained by solving the system

$$m\frac{d^2x}{dt^2} = F_1, \qquad m\frac{d^2y}{dt^2} = F_2, \qquad m\frac{d^2z}{dt^2} = F_3.$$

In this chapter, we consider the formulation and solution of systems of DE. The majority of the chapter is concerned with *linear* systems of DE. In this case, the following familiar questions need addressing:

Question 1:   How can we formulate our problems in a way suitable for solution?

Question 2:   How many solutions, if any, does our differential system possess?

Question 3:   How do we find the solutions that arise in Question 2?

As in the case of the linear algebraic systems discussed in Chapter 3, we will find that the answer to Question 1 lies in the use of matrices. Answering Question 2 will once more require the vector space techniques from Chapters 5 and 6, whereas, in the case when our linear systems have constant coefficients, we will find an elegant answer to Question 3 using eigenvalues and eigenvectors of appropriate matrices. Towards the end of the chapter, we will give a brief introduction to more general techniques for exploring the qualitative behavior of solutions to nonlinear systems of DE.

Before beginning the general development of the theory for systems of DE, we consider two physical problems that can be formulated mathematically in terms of such systems.

Consider the coupled spring-mass system that consists of two masses m_1, m_2 connected by two springs whose spring constants are k_1 and k_2, respectively. (See Figure 8.1.1.) Let $x(t)$ and $y(t)$ denote the displacement of m_1 and m_2, respectively, from their

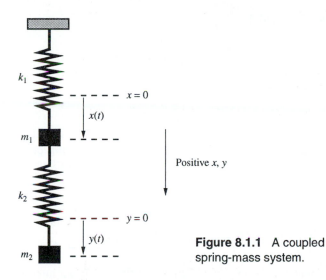

Figure 8.1.1 A coupled spring-mass system.

positions when the system is in the static equilibrium position. Then, using Hooke's law and Newton's second law, it follows that the motion of the masses is governed by the system of DE

$$m_1 \frac{d^2x}{dt^2} = -k_1x + k_2(y - x), \tag{8.1.1}$$

$$m_2 \frac{d^2y}{dt^2} = -k_2(y - x). \tag{8.1.2}$$

We would expect the problem to have a unique solution once we have specified the initial positions and velocities of the masses.

As a second example, consider the mixing problem depicted in Figure 8.1.2. Two tanks contain a solution consisting of chemical dissolved in water. A solution containing c grams/liter of the chemical flows into tank 1 at a rate of r liters/minute, and the solution in tank 2 flows out at the same rate. In addition, the solution flows into tank 1 from tank 2 at a rate of r_{12} liters/minute and into tank 2 from tank 1 at a rate of r_{21} liters/minute. We wish to determine the amounts of chemical $A_1(t)$ and $A_2(t)$, in tanks 1 and 2 at any time t. A similar analysis to that used in Section 1.7, yields the following system of DE

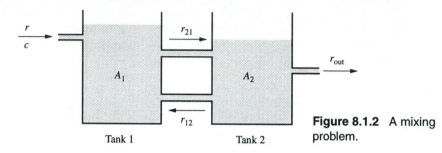

Figure 8.1.2 A mixing problem.

the behavior of A_1, and A_2:

$$\frac{dA_1}{dt} = -\frac{r_{21}}{V_1} A_1 + \frac{r_{12}}{V_2} A_2 + cr, \tag{8.1.3}$$

$$\frac{dA_2}{dt} = \frac{r_{21}}{V_1} A_1 - \frac{(r_{12} + r)}{V_2} A_2, \tag{8.1.4}$$

where V_1 and V_2 denote the volume of solution in each tank at time t.

We will give a full discussion of both of the foregoing problems in Section 8.8 once we have developed the theory and solution techniques for linear differential systems.

8.2   FIRST-ORDER LINEAR SYSTEMS

As mentioned in the previous section, we first focus our attention on *linear* systems of DE. Once such a system has been appropriately formulated, vector space methods can be applied to derive the complete theory regarding their solution properties.

Definition 8.2.1: A system of DE of the form

$$\frac{dx_1}{dt} = a_{11}(t)x_1(t) + a_{12}(t)x_2(t) + \cdots + a_{1n}(t)x_n(t) + b_1(t),$$

$$\frac{dx_2}{dt} = a_{21}(t)x_1(t) + a_{22}(t)x_2(t) + \cdots + a_{2n}(t)x_n(t) + b_2(t),$$

$$\vdots \tag{8.2.1}$$

$$\frac{dx_n}{dt} = a_{n1}(t)x_1(t) + a_{n2}(t)x_2(t) + \cdots + a_{nn}(t)x_n(t) + b_n(t),$$

where the $a_{ij}(t)$ and $b_i(t)$ are specified functions on an interval I, is called a **first-order linear system**. If $b_1 = b_2 = \cdots = b_n = 0$, then the system is called **homogeneous**. Otherwise, it is called **nonhomogeneous**.

REMARKS

1. It is important to notice the structure of a first-order linear system. The highest derivative occurring in such a system is a first derivative. Further, there is precisely one equation involving the derivative of each separate unknown function. Finally, the terms that appear on the right-hand side of the equations do *not* involve any derivatives and are linear in the unknown functions $x_1, x_2, ..., x_n$.

2. We will usually denote $\dfrac{dx_i}{dt}$ by $x_i{}'$.

Example 8.2.1   An example of a nonhomogeneous first-order linear system is

$$x_1{}' = e^t x_1 + t^2 x_2 + \sin t,$$
$$x_2{}' = \ tx_1 + 3x_2 - \cos t.$$

The associated homogeneous system is

$$x_1{}' = e^t x_1 + t^2 x_2,$$
$$x_2{}' = \ tx_1 + 3x_2.$$ □

Definition 8.2.2:   By a **solution** to the system (8.2.1) on an interval I we mean an ordered n-tuple of functions $x_1(t), x_2(t), ..., x_n(t)$, which, when substituted into the left-hand side of the system, yield the right-hand side for all t in I.

Example 8.2.2   Verify that

$$x_1(t) = -2e^{5t} + 4e^{-t}, \quad x_2(t) = e^{5t} + e^{-t} \tag{8.2.2}$$

is a solution to

$$x_1{}' = \ x_1 - 8x_2 \tag{8.2.3}$$
$$x_2{}' = -x_1 + 3x_2 \tag{8.2.4}$$

on $(-\infty, \infty)$.

Solution   From (8.2.2) it follows that the left-hand side of equation (8.2.3) is

$$x_1{}'(t) = -10e^{5t} - 4e^{-t},$$

whereas the right-hand side is

$$x_1(t) - 8x_2(t) = (-2e^{5t} + 4e^{-t}) - 8(e^{5t} + e^{-t}) = -10e^{5t} - 4e^{-t}.$$

Consequently,

$$x_1{}' = x_1 - 8x_2,$$

so that equation (8.2.3) is satisfied by the given functions for all $t \in (-\infty, \infty)$. Similarly, it is easily shown that, for all $t \in (-\infty, \infty)$,

$$x_2{}' = -x_1 + 3x_2,$$

so that equation (8.2.4) is also satisfied. It follows that x_1 and x_2 do define a solution to the given system on $(-\infty, \infty)$.   □

We now derive a simple technique for solving the system (8.2.1) that can be used when the coefficients $a_{ij}(t)$ in the system are constants. Although we will develop a superior technique for such systems in the later sections, the method introduced here does have importance and will be useful in motivating some of the subsequent results. For simplicity, we will only consider $n = 2$. Under the assumption that all a_{ij} are constants, the system (8.2.1) reduces to

$$x_1{'} = a_{11}x_1 + a_{12}x_2 + b_1(t),$$
$$x_2{'} = a_{21}x_1 + a_{22}x_2 + b_2(t).$$

This system can be written in the equivalent form

$$(D - a_{11})x_1 - \qquad a_{12}x_2 = b_1(t), \qquad (8.2.5)$$
$$-a_{21}x_1 + (D - a_{22})x_2 = b_2(t), \qquad (8.2.6)$$

where D is the differential operator d/dt. The idea behind the solution technique is that we can now easily eliminate x_2 between these two equations by operating on equation (8.2.5) with $D - a_{22}$, multiplying equation (8.2.6) by a_{12}, and adding the resulting equations. This yields a second-order constant coefficient linear DE for x_1 only, which can be solved using the techniques of Chapter 2. Substituting the expression thereby obtained for x_1 into equation (8.2.5) will then yield x_2.[1] We illustrate the technique with an example.

Example 8.2.3   Solve the system

$$x_1{'} = \quad x_1 + 2x_2, \qquad (8.2.7)$$
$$x_2{'} = 2x_1 - 2x_2. \qquad (8.2.8)$$

Solution   We begin by rewriting the system in operator form as

$$(D - 1)x_1 - \qquad 2x_2 = 0, \qquad (8.2.9)$$
$$-2x_1 + (D + 2)x_2 = 0. \qquad (8.2.10)$$

To eliminate x_2 between these two equations, we first operate on equation (8.2.9) with $D + 2$ to obtain

$$(D + 2)(D - 1)x_1 - 2(D + 2)x_2 = 0.$$

Adding twice equation (8.2.10) to this equation eliminates x_2 and yields

$$(D + 2)(D - 1)x_1 - 4x_1 = 0.$$

That is,

$$(D^2 + D - 6)x_1 = 0.$$

This constant coefficient DE has auxiliary polynomial

$$P(r) = r^2 + r - 6 = (r + 3)(r - 2).$$

Consequently,

$$x_1(t) = c_1 e^{-3t} + c_2 e^{2t}. \qquad (8.2.11)$$

We now determine x_2. From equation (8.2.9), we have

$$x_2(t) = \frac{1}{2}(D - 1)x_1.$$

---

[1] If $a_{12} = 0$, we can determine x_1 directly from equation (8.2.5), and then x_2 can be determined from equation (8.2.6).

Inserting the expression for x_1 from (8.2.11) into the previous equation yields

$$x_2(t) = \frac{1}{2}(Dx_1 - x_1) = \frac{1}{2}(-4c_1 e^{-3t} + c_2 e^{2t}).$$

Hence, the solution to the system of DE (8.2.7) and (8.2.8) is

$$x_1(t) = c_1 e^{-3t} + c_2 e^{2t}, \quad x_2(t) = \frac{1}{2}(-4c_1 e^{-3t} + c_2 e^{2t}),$$

where c_1 and c_2 are arbitrary constants.   ❏

   In solving an applied problem that is governed by a system of DE, we usually require the particular solution to the system that corresponds to the specific problem of interest. Such a particular solution is obtained by specifying appropriate auxiliary conditions. This leads to the idea of an initial-value problem for linear systems.

Definition 8.2.3:   Solving the system (8.2.1) subject to n auxiliary conditions imposed at the *same* value of the independent variable is called an **initial-value problem (IVP)**. Thus, the general form of the auxiliary conditions for an IVP is:

$$x_1(t_0) = \alpha_1, x_2(t_0) = \alpha_2, \ldots, x_n(t_0) = \alpha_n,$$

where $\alpha_1, \alpha_2, \ldots, \alpha_n$ are constants.

Example 8.2.4   Solve the IVP

$$x_1{'} = x_1 + 2x_2, \quad x_2{'} = 2x_1 - 2x_2,$$

$$x_1(0) = 1, \quad x_2(0) = 0.$$

Solution   We have already seen in the previous example that the solution to the given system of DE is

$$x_1(t) = c_1 e^{-3t} + c_2 e^{2t}, \quad x_2(t) = \frac{1}{2}(-4c_1 e^{-3t} + c_2 e^{2t}), \tag{8.2.12}$$

where c_1 and c_2 are arbitrary constants. Imposing the two initial conditions yields the following equations for determining c_1 and c_2:

$$c_1 + c_2 = 1,$$

$$-4c_1 + c_2 = 0.$$

Consequently,

$$c_1 = \frac{1}{5}, \quad c_2 = \frac{4}{5}.$$

Substituting for c_1 and c_2 into (8.2.12) yields the unique solution

$$x_1(t) = \frac{1}{5}(e^{-3t} + 4e^{2t}), \quad x_2(t) = \frac{2}{5}(e^{2t} - e^{-3t}).$$ ❏

   It might appear that restricting to *first-order* linear systems means that we are only considering very special types of linear differential systems. In fact this is incorrect, since most systems of k DE that are linear in k unknown functions and their derivatives can be rewritten as an equivalent first-order system by redefining the dependent variables. We illustrate with an example.

Example 8.2.5   Rewrite the linear system

$$\frac{d^2x}{dt^2} - 4y = e^t, \tag{8.2.13}$$

$$\frac{d^2y}{dt^2} + t^2\frac{dx}{dt} = \sin t, \tag{8.2.14}$$

as an equivalent first-order system.

Solution   We introduce new dependent variables relative to which equations (8.2.13) and (8.2.14) reduce to first-order DE. Let

$$x_1 = x, \quad x_2 = \frac{dx}{dt}, \quad x_3 = y, \quad x_4 = \frac{dy}{dt}. \tag{8.2.15}$$

Then equations (8.2.13) and (8.2.14) can be replaced by

$$\frac{dx_2}{dt} - 4x_3 = e^t, \quad \frac{dx_4}{dt} + t^2x_2 = \sin t.$$

These equations must also be supplemented with equations for x_1 and x_3. From (8.2.15), we see that

$$\frac{dx_1}{dt} = x_2, \quad \frac{dx_3}{dt} = x_4.$$

Consequently, the given system of DE is equivalent to the first-order linear system

$$\frac{dx_1}{dt} = x_2, \quad \frac{dx_2}{dt} = 4x_3 + e^t, \quad \frac{dx_3}{dt} = x_4, \quad \frac{dx_4}{dt} = -t^2x_2 + \sin t. \qquad \square$$

Finally, consider the general nth-order linear DE

$$x^{(n)} + a_1(t)\, x^{(n-1)} + \cdots + a_{n-1}(t)x' + a_n(t)x = F(t). \tag{8.2.16}$$

If we introduce the new variables $x_1, x_2, \ldots, x_n$ defined by

$$x_1 = x, \quad x_2 = x', \quad \ldots, \quad x_n = x^{(n-1)},$$

then equation (8.2.16) can be replaced by the equivalent first-order linear system

$$x_1' = x_2, \quad x_2' = x_3, \quad \ldots, \quad x_{n-1}' = x_n,$$

$$x_n' = -a_n(t)x_1 - a_{n-1}(t)x_2 - \cdots - a_1(t)\, x_n + F(t).$$

Consequently, any nth-order linear DE can be replaced by an equivalent system of DE.

Example 8.2.6   Write the following DE as an equivalent first-order system

$$\frac{d^2x}{dt^2} + 4e^t\frac{dx}{dt} - 9t^2x = 7t^2.$$

Solution   We introduce new variables x_1 and x_2 defined by

$$x_1 = x, \quad x_2 = \frac{dx}{dt}.$$

Then the given DE can be replaced by the first-order system

$$\frac{dx_1}{dt} = x_2, \quad \frac{dx_2}{dt} = 9t^2x_1 - 4e^tx_2 + 7t^2.$$

EXERCISES 8.2

For problems 1–7, solve the given system of DE.

1. $x_1' = 2x_1 - 3x_2,\ x_2' = x_1 - 2x_2$.

2. $x_1' = 4x_1 + 2x_2,\ x_2' = -x_1 + x_2$.

3. $x_1' = 2x_1 + 4x_2,\ x_2' = -4x_1 - 6x_2$.

4. $x_1' = 2x_2,\ x_2' = -2x_1$.

5. $x_1' = x_1 - 3x_2,\ x_2' = 3x_1 + x_2$.

6. $x_1' = 2x_1,\ x_2' = x_2 - x_3,\ x_3' = x_2 + x_3$.

7. $x_1' = -2x_1 + x_2 + x_3,\ x_2' = x_1 - x_2 + 3x_3,$
$x_3' = -x_2 - 3x_3$.

For problems 8–10, solve the given IVP.

8. $x_1' = 2x_2,\ x_2' = x_1 + x_2,\ x_1(0) = 3,\ x_2(0) = 0$.

9. $x_1' = 2x_1 + 5x_2,\ x_2' = -x_1 - 2x_2,\ x_1(0) = 0,$
$x_2(0) = 1$.

10. $x_1' = 2x_1 + x_2,\ x_2' = -x_1 + 4x_2,\ x_1(0) = 1,$
$x_2(0) = 3$.

For problems 11–13, solve the given nonhomogeneous system.

11. $x_1' = x_1 + 2x_2 + 5e^{4t},\ x_2' = 2x_1 + x_2$.

12. $x_1' = -2x_1 + x_2 + t,\ x_2' = -2x_1 + x_2 - 1$.

13. $x_1' = x_1 + x_2 + e^{2t},\ x_2' = 3x_1 - x_2 + 5e^{2t}$.

For problems 14 and 15, convert the given system of DE to a first-order linear system.

14. $\dfrac{dx}{dt} - ty = \cos t,\ \dfrac{d^2y}{dt^2} - \dfrac{dx}{dt} + x = e^t$.

15. $\dfrac{d^2x}{dt^2} - 3\dfrac{dy}{dt} + x = \sin t,\ \dfrac{d^2y}{dt^2} - t\dfrac{dx}{dt} - e^ty = t^2$.

For problems 16–18, convert the given linear DE to a first-order linear system.

16. $x'' + 2tx' + x = \cos t$.

17. $x'' + ax' + bx = F(t),\ a, b,$ constants.

18. $x''' + t^2x' - e^tx = t$.

19. The IVP that governs the behavior of a coupled spring-mass system is (see Section 8.1)

$$m_1 \frac{d^2x}{dt^2} = -k_1 x + k_2(y - x)$$

$$m_2 \frac{d^2y}{dt^2} = -k_2(y - x).$$

$$x(0) = \alpha_1,\ x'(0) = \alpha_2,\ y(0) = \alpha_3,\ y'(0) = \alpha_4,$$

where $\alpha_1, \alpha_2, \alpha_3,$ and α_4 are constants. Convert this problem into an IVP for an equivalent first-order linear system. (You must give the appropriate initial conditions in the new variables.)

20. Solve the IVP:

$$x_1' = -\tan t\, x_1 + 3\cos^2 t$$
$$x_2' = x_1 + \tan t\, x_2 + 2\sin t$$

$$x_1(0) = 4,\ x_2(0) = 0.$$

8.3   VECTOR FORMULATION

The first step in developing the general theory for first-order linear systems is to formulate the problem of solving such a system as an appropriate vector space problem. The key to this formulation is the realization that the scalar system

$$x_1' = a_{11}(t)x_1(t) + a_{12}(t)x_2(t) + \cdots + a_{1n}(t)x_n(t) + b_1(t),$$

$$x_2' = a_{21}(t)x_1(t) + a_{22}(t)x_2(t) + \cdots + a_{2n}(t)x_n(t) + b_2(t),$$

$$\vdots \tag{8.3.1}$$

$$x_n' = a_{n1}(t)x_1(t) + a_{n2}(t)x_2(t) + \cdots + a_{nn}(t)x_n(t) + b_n(t),$$

can be written as the equivalent vector equation

$$\boxed{\mathbf{x}'(t) = A(t)\mathbf{x}(t) + \mathbf{b}(t),} \tag{8.3.2}$$

where

$$\mathbf{x}(t) = \begin{bmatrix} x_1(t) \\ x_2(t) \\ \vdots \\ x_n(t) \end{bmatrix}, \quad \mathbf{x}'(t) = \begin{bmatrix} x_1'(t) \\ x_2'(t) \\ \vdots \\ x_n'(t) \end{bmatrix}$$

and

$$A(t) = \begin{bmatrix} a_{11}(t) & a_{12}(t) & \cdots & a_{1n}(t) \\ a_{21}(t) & a_{22}(t) & \cdots & a_{2n}(t) \\ \vdots & \vdots & & \vdots \\ a_{n1}(t) & a_{n2}(t) & \cdots & a_{nn}(t) \end{bmatrix}, \quad \mathbf{b}(t) = \begin{bmatrix} b_1(t) \\ b_2(t) \\ \vdots \\ b_n(t) \end{bmatrix}.$$

Example 8.3.1   The system of equations

$$x_1' = 3x_1 + (\sin t)x_2 + e^t,$$
$$x_2' = 7tx_1 + \quad t^2 x_2 - 4e^{-t}.$$

can be written as

$$\begin{bmatrix} x_1' \\ x_2' \end{bmatrix} = \begin{bmatrix} 3 & \sin(t) \\ 7t & t^2 \end{bmatrix} \begin{bmatrix} x_1 \\ x_2 \end{bmatrix} + \begin{bmatrix} e^t \\ -4e^{-t} \end{bmatrix}.$$

That is,

$$\mathbf{x}'(t) = A(t)\mathbf{x}(t) + \mathbf{b}(t)$$

where

$$\mathbf{x}(t) = \begin{bmatrix} x_1(t) \\ x_2(t) \end{bmatrix}, \quad A(t) = \begin{bmatrix} 3 & \sin t \\ 7t & t^2 \end{bmatrix}, \quad \mathbf{b}(t) = \begin{bmatrix} e^t \\ -4e^{-t} \end{bmatrix}. \qquad \square$$

We see from (8.3.2) that matrices provide a natural framework for studying first-order *linear* systems. Notice, however, that in this formulation, we are, in fact, dealing with matrix *functions*, that is, matrices whose elements are themselves functions. The following properties of matrix functions will be required in the remainder of this chapter:

1. The algebra of matrix functions is the same as that for matrices of constants.

2. The calculus of matrix functions is defined elementwise. In particular,

    (a) The derivative of a matrix function is obtained by differentiating *every* element of the matrix. Thus, if $A(t) = [a_{ik}(t)]$, then $\dfrac{dA}{dt} = [da_{ik}/dt]$, provided that each of the a_{ik} are differentiable.

    (b) It follows from (a), and the index form of the matrix product, that if A and B are both differentiable and the product AB is defined, then

$$\frac{d}{dt}(AB) = A\,\frac{dB}{dt} + \frac{dA}{dt}\,B.$$

The key point to notice is that the order of the multiplication must be preserved.

(c)   If $A(t) = [a_{ik}(t)]$, where each $a_{ik}(t)$ is integrable on the interval $[a, b]$, then

$$\int_a^b A(t)\,dt = \left[\int_a^b a_{ik}(t)\,dt\right].$$

Example 8.3.2   If $A(t) = \begin{bmatrix} 2t & 1 \\ 6t^2 & 4e^{2t} \end{bmatrix}$ determine $\dfrac{dA}{dt}$ and $\displaystyle\int_0^1 A(t)\,dt$.

Solution:

$$\frac{dA}{dt} = \begin{bmatrix} 2 & 0 \\ 12t & 8e^{2t} \end{bmatrix},$$

whereas

$$\int_0^1 A(t)\,dt = \begin{bmatrix} \displaystyle\int_0^1 2t\,dt & \displaystyle\int_0^1 1\,dt \\[2ex] \displaystyle\int_0^1 6t^2\,dt & \displaystyle\int_0^1 4e^{2t}\,dt \end{bmatrix} = \begin{bmatrix} 1 & 1 \\ 2 & 2(e^2 - 1) \end{bmatrix}.$$

COLUMN VECTOR FUNCTIONS

An $n \times 1$ matrix function is called a **column n-vector function** (or just a column vector function if the number of elements is clear). Thus, $\mathbf{x}$, $\mathbf{x}'$ and $\mathbf{b}$ in (8.3.2) are column n-vector functions. We let $V_n(I)$ denote the set of all column n-vector functions defined on an interval I, and define addition and scalar multiplication within this set in the same manner as for column vectors. The following results concerning $V_n(I)$ will be needed in the remaining sections:

Theorem 8.3.1: $V_n(I)$ is a vector space.

PROOF   Verifying that $V_n(I)$ together with the operations of addition and scalar multiplication just defined satisfies Definition 5.2.1 is left as an exercise.   ■

Since $V_n(I)$ is a vector space, we can discuss linear dependence and linear indep–endence of column vector functions. We first need a definition.

Definition 8.3.1:   Let $\mathbf{x}_1(t)$, $\mathbf{x}_2(t)$, ..., $\mathbf{x}_n(t)$ be vectors in $V_n(I)$. Then the **Wronskian** of these vector functions, denoted $W[\mathbf{x}_1, \mathbf{x}_2, ..., \mathbf{x}_n](t)$, is defined by

$$W[\mathbf{x}_1, \mathbf{x}_2, ..., \mathbf{x}_n](t) = \det([\mathbf{x}_1(t), \mathbf{x}_2(t), ..., \mathbf{x}_n(t)]).$$

REMARK   Notice that the Wronskian introduced in this definition refers to column vector functions in the vector space $V_n(I)$, whereas the Wronskian defined previously in the text refers to functions in $C^n(I)$. The relationship between these two Wronskians is investigated in Problem 26.

Example 8.3.3   Determine the Wronskian of the column vector functions

$$\mathbf{x}_1(t) = \begin{bmatrix} e^t \\ 2e^t \end{bmatrix}, \quad \mathbf{x}_2(t) = \begin{bmatrix} 3\sin t \\ \cos t \end{bmatrix}.$$

Solution   From Definition 8.3.1, we have

$$W[\mathbf{x}_1, \mathbf{x}_2](t) = \begin{vmatrix} e^t & 3\sin t \\ 2e^t & \cos t \end{vmatrix} = e^t(\cos t - 6\sin t).$$ ☐

Our next theorem indicates that the Wronskian plays a familiar role in determining the linear independence of a set of vectors in $V_n(I)$.

Theorem 8.3.2:   Let $\mathbf{x}_1(t)$, $\mathbf{x}_2(t)$, ..., $\mathbf{x}_n(t)$ be vectors in $V_n(I)$. If $W[\mathbf{x}_1(t), \mathbf{x}_2(t), ..., \mathbf{x}_n(t)](t_0)$ is *nonzero* at any point t_0 in I, then $\{\mathbf{x}_1(t), \mathbf{x}_2(t), ..., \mathbf{x}_n(t)\}$ is LI on I.

PROOF   Consider

$$c_1\mathbf{x}_1(t) + c_2\mathbf{x}_2(t) + \cdots + c_n\mathbf{x}_n(t) = \mathbf{0},$$

where $c_1, c_2, ..., c_n$ are scalars. We can write this as the vector equation

$$X(t)\mathbf{c} = \mathbf{0},$$

where $\mathbf{c} = [c_1 \ c_2 \ ... \ c_n]^T$, and $X(t) = [\mathbf{x}_1(t), \mathbf{x}_2(t), ..., \mathbf{x}_n(t)]$. Let t_0 be in I. If $\det([X(t_0)]) \neq 0$, Corollary 4.2.2 implies that the only solution to this $n \times n$ system of linear equations is $\mathbf{c} = \mathbf{0}$. Consequently, $\{\mathbf{x}_1(t), \mathbf{x}_2(t), ..., \mathbf{x}_n(t)\}$ is LI on I. But $\det([X(t_0)]) = W[\mathbf{x}_1, \mathbf{x}_2, ..., \mathbf{x}_n](t_0)$, and hence, the result follows.   ■

Example 8.3.4   The vector functions

$$\mathbf{x}_1(t) = \begin{bmatrix} e^t \\ 2e^t \end{bmatrix}, \quad \mathbf{x}_2(t) = \begin{bmatrix} 3\sin t \\ \cos t \end{bmatrix}$$

are LI on $(-\infty, \infty)$ since, for example, $W[\mathbf{x}_1, \mathbf{x}_2](0) = 1 \neq 0$.   ☐

VECTOR DIFFERENTIAL EQUATIONS

A system of linear DE written in the vector form

$$\mathbf{x}'(t) = A(t)\mathbf{x}(t) + \mathbf{b}(t)$$

will be called a **vector differential equation** (VDE). We emphasize that within this formulation the primary unknown is the column vector function $\mathbf{x}(t)$ whose components, $x_1(t), x_2(t), \ldots, x_n(t)$ are the unknowns in the corresponding scalar differential system (8.3.1). The problem of determining all solutions to the general first-order linear system of DE (8.3.1) can now be formulated as the vector space problem

Find all column vector functions $\mathbf{x}(t) \in V_n(I)$ satisfying the VDE $\mathbf{x}'(t) = A(t)\mathbf{x}(t) + \mathbf{b}(t)$.

The vector space $V_n(I)$ is not finite–dimensional, since there is no finite set of LI vectors that spans $V_n(I)$. The key to solving linear differential systems comes from the realization that, if $A(t)$ is an $n \times n$ matrix function, then the set of all solutions to the homogeneous VDE

$$\mathbf{x}'(t) = A(t)\mathbf{x}(t)$$

is an n-dimensional subspace of $V_n(I)$. This is illustrated in the next example and established in general in the next section.

Example 8.3.5   Consider the homogeneous linear system of DE

$$\begin{aligned} x_1' &= x_1 + 2x_2, \\ x_2' &= 2x_1 - 2x_2. \end{aligned} \tag{8.3.3}$$

This can be formulated as the equivalent VDE

$$\mathbf{x}' = A\mathbf{x}, \quad A = \begin{bmatrix} 1 & 2 \\ 2 & -2 \end{bmatrix}. \tag{8.3.4}$$

In the previous section, we derived the following solution to the scalar system (8.3.3)

$$x_1(t) = c_1 e^{-3t} + c_2 e^{2t}, \quad x_2(t) = \frac{1}{2}(-4c_1 e^{-3t} + c_2 e^{2t}).$$

Therefore, the solution vector to the VDE (8.3.4) is

$$\mathbf{x}(t) = \begin{bmatrix} c_1 e^{-3t} + c_2 e^{2t} \\ \frac{1}{2}(-4c_1 e^{-3t} + c_2 e^{2t}) \end{bmatrix},$$

which can be written in the equivalent form

$$\mathbf{x}(t) = c_1 \begin{bmatrix} e^{-3t} \\ -e^{-3t} \end{bmatrix} + c_2 \begin{bmatrix} e^{2t} \\ \frac{1}{2} e^{2t} \end{bmatrix}.$$

Consequently, in this particular example, the set of all solutions to the VDE is the two–dimensional subspace of $V_2(I)$ spanned by the LI column vector functions

$$\mathbf{x}_1(t) = \begin{bmatrix} e^{-3t} \\ -e^{-3t} \end{bmatrix}, \quad \mathbf{x}_2(t) = \begin{bmatrix} e^{2t} \\ \frac{1}{2} e^{2t} \end{bmatrix}.$$

EXERCISES 8.3

For problems 1–6, convert the given equation or system to a first-order linear system and write the resulting system in matrix form.

1. $x_1' + 4x_1 - 3x_2 = 4t$, $x_2' - 6x_1 + 4x_2 = t^2$.

2. $x_1' + tx_2 - t^2x_1 = 0$, $x_2' + (\sin t)x_1 - x_2 = 0$.

3. $x_1' + (\sin t)x_2 - x_3 = t$, $x_2' + e^tx_1 - t^2x_3 = t^3$, $x_3' + t x_1 - t^2x_2 = 1$.

4. $x'' + t^2x' - e^tx = \sin t$.

5. $x''' - (\sin t) x'' + (t^2 - a^2)x = e^t$, a constant .

6. $x'' - 4 (\sin t) x' + \cos t\, y' - y = \tan t$, $y'' + 5t\, y' + 4\, t^2y - x = t$.

For problems 7–10, determine the derivative of the given matrix function.

7. $A(t) = \begin{bmatrix} e^{-2t} \\ \sin t \end{bmatrix}$.

8. $A(t) = \begin{bmatrix} t & \sin t \\ \cos t & 4t \end{bmatrix}$.

9. $A(t) = \begin{bmatrix} e^t & e^{2t} & t^2 \\ 2e^t & 4e^{2t} & 5t^2 \end{bmatrix}$.

10. $A(t) = \begin{bmatrix} \sin t & \cos t & 0 \\ -\cos t & \sin t & t \\ 0 & 3t & 1 \end{bmatrix}$.

11. Let $A = [a_{ik}(t)]$ be an $m \times n$ matrix function and let $B = [b_{ik}(t)]$ be an $n \times p$ matrix function. Use the definition of matrix multiplication to prove that

$$\frac{d}{dt}(AB) = A\frac{dB}{dt} + \frac{dA}{dt}B .$$

For problems 12 and 13, determine $\int_a^b A(t)\, dt$ for the given matrix function.

12. $A(t) = \begin{bmatrix} \cos t \\ \sin t \end{bmatrix}$, $a = 0$, $b = \pi/2$.

13. $A(t) = \begin{bmatrix} e^t & e^{-t} \\ 2e^t & 5e^{-t} \end{bmatrix}$, $a = 0$, $b = 1$.

14. Evaluate $\int A(t)\, dt$ if $A(t) = \begin{bmatrix} 2t \\ 3t^2 \end{bmatrix}$.

For problems 15 and 16, verify that the given vector function defines a solution to $\mathbf{x}' = A\mathbf{x} + \mathbf{b}$ for the given A and $\mathbf{b}$.

15. $\mathbf{x}(t) = \begin{bmatrix} e^{4t} \\ -2e^{4t} \end{bmatrix}$, $A = \begin{bmatrix} 2 & -1 \\ -2 & 3 \end{bmatrix}$,

$\mathbf{b}(t) = \begin{bmatrix} 0 \\ 0 \end{bmatrix}$.

16. $\mathbf{x}(t) = \begin{bmatrix} 4e^{-2t} + 2\sin t \\ 3e^{-2t} - \cos t \end{bmatrix}$, $A = \begin{bmatrix} 1 & -4 \\ -3 & 2 \end{bmatrix}$,

$\mathbf{b}(t) = \begin{bmatrix} -2(\cos t + \sin t) \\ 7\sin t + 2\cos t \end{bmatrix}$.

For problems 17–20, show that the given vector functions are LI on $(-\infty, \infty)$.

17. $\mathbf{x}_1(t) = \begin{bmatrix} e^t \\ -e^t \end{bmatrix}$, $\mathbf{x}_2(t) = \begin{bmatrix} e^t \\ e^t \end{bmatrix}$.

18. $\mathbf{x}_1(t) = \begin{bmatrix} t \\ t \end{bmatrix}$, $\mathbf{x}_2(t) = \begin{bmatrix} t \\ t^2 \end{bmatrix}$.

19. $\mathbf{x}_1(t) = \begin{bmatrix} t + 1 \\ t - 1 \\ 2t \end{bmatrix}$, $\mathbf{x}_2(t) = \begin{bmatrix} e^t \\ e^{2t} \\ e^{3t} \end{bmatrix}$,

$\mathbf{x}_3(t) = \begin{bmatrix} 1 \\ \sin t \\ \cos t \end{bmatrix}$.

20. $\mathbf{x}_1(t) = \begin{bmatrix} \sin t \\ \cos t \\ 1 \end{bmatrix}$, $\mathbf{x}_2(t) = \begin{bmatrix} t \\ 1 - t \\ 1 \end{bmatrix}$,

$\mathbf{x}_3(t) = \begin{bmatrix} \sinh t \\ \cosh t \\ 1 \end{bmatrix}$.

For problems 21 and 22, show that the given vector functions are linearly dependent on $(-\infty, \infty)$.

21. $\mathbf{x}_1(t) = \begin{bmatrix} e^t \\ 2e^{2t} \end{bmatrix}$, $\mathbf{x}_2(t) = \begin{bmatrix} 4e^t \\ 8e^{2t} \end{bmatrix}$.

22. $\mathbf{x}_1(t) = \begin{bmatrix} \sin^2t \\ \cos^2t \\ 2 \end{bmatrix}$, $\mathbf{x}_2(t) = \begin{bmatrix} 2\cos^2t \\ 2\sin^2t \\ 1 \end{bmatrix}$.

$$\mathbf{x}_3(t) = \begin{bmatrix} 2 \\ 2 \\ 5 \end{bmatrix}.$$

23. Prove that $V_n(I)$ is a vector space.

24. Let $A(t)$ be an $n \times n$ matrix function. Prove that the set of all solutions to $\mathbf{x}' = A(t)\,\mathbf{x}$ is a subspace of $V_n(I)$.

25. If $A = \begin{bmatrix} 2 & -4 \\ 1 & -3 \end{bmatrix}$, determine two LI solutions to $\mathbf{x}' = A\mathbf{x}$ on $(-\infty, \infty)$.

Problem 26 investigates the relationship between the Wronskian defined in this section for column vector functions in $V_n(I)$ and the Wronskian defined previously for functions in $C^n(I)$.

26. Consider the differential equation

$$\frac{d^2y}{dt^2} + a\frac{dy}{dt} + by = 0, \qquad (26.1)$$

where a and b are arbitrary functions of t.

(a) Show that equation (26.1) can be replaced by the equivalent linear system

$$\mathbf{x}' = A\mathbf{x}, \qquad (26.2)$$

where

$$A = \begin{bmatrix} 0 & 1 \\ -b & -a \end{bmatrix} \text{ and } x_1 = y,\, x_2 = y'.$$

(b) If $y_1 = f_1(t)$ and $y_2 = f_2(t)$ are solutions to equation (26.1) on an interval I, show that the corresponding solutions to the system (26.2) are

$$\mathbf{x}_1(t) = \begin{bmatrix} f_1(t) \\ f_1'(t) \end{bmatrix}, \qquad \mathbf{x}_2(t) = \begin{bmatrix} f_2(t) \\ f_2'(t) \end{bmatrix}.$$

(c) Show that

$$W[\mathbf{x}_1, \mathbf{x}_2](t) = W[y_1, y_2](t).$$

8.4 GENERAL RESULTS FOR FIRST-ORDER LINEAR DIFFERENTIAL SYSTEMS

We now show how the formulation of a linear system of DE as a single VDE enables us to derive the underlying theory for linear differential systems as an application of the vector space results from Chapter 5. We emphasize that although the derivation of the results are based on the VDE formulation, the results themselves apply to any first-order linear differential system, since such a system can always be formulated as a VDE.

The fundamental theoretical result that will be used in deriving the underlying theory for the solution of VDE is the following existence and uniqueness theorem:

Theorem 8.4.1:   The IVP

$$\mathbf{x}'(t) = A(t)\mathbf{x} + \mathbf{b}(t), \quad \mathbf{x}(t_0) = \mathbf{x}_0,$$

where $A(t)$ and $\mathbf{b}(t)$ are continuous on an interval I, has a unique solution on I.

PROOF   The proof is omitted. (See, for example, F.J. Murray and K.S. Miller, *Existence Theorems*, New York University Press, 1954.)   ∎

HOMOGENEOUS VDE

Just as for single nth order linear equations, the solution to a nonhomogeneous linear differential system can, in theory, be obtained once we have solved the associated homogeneous differential system. Consequently, we begin by developing the theory for the homogeneous VDE:

$$\mathbf{x}'(t) = A(t)\mathbf{x}(t), \qquad (8.4.1)$$

where A is an $n \times n$ matrix function. This is where the vector space techniques are required. We first show that the set of all solutions to (8.4.1) is an n-dimensional subspace of the vector space of all column n-vector functions.

Theorem 8.4.2:   The set of all solutions to $\mathbf{x}'(t) = A(t)\mathbf{x}(t)$, where $A(t)$ is an $n \times n$ matrix function that is continuous on an interval I, is a vector space of dimension n.

PROOF   Let S denote the set of all solutions to $\mathbf{x}' = A\mathbf{x}$. S is certainly nonempty, since (8.4.1) always has the trivial solution $\mathbf{x}(t) = \mathbf{0}$. We establish that S is a subspace of $V_n(I)$ by verifying that S is closed under addition and scalar multiplication. Let $\mathbf{u}(t)$ and $\mathbf{v}(t)$ be in S. Then $\mathbf{u}' = A\mathbf{u}$ and $\mathbf{v}' = A\mathbf{v}$, so that

$$(\mathbf{u} + \mathbf{v})' = \mathbf{u}' + \mathbf{v}' = A\mathbf{u} + A\mathbf{v} = A(\mathbf{u} + \mathbf{v}),$$

which implies that $\mathbf{u} + \mathbf{v}$ is in S. Furthermore, for any scalar c,

$$(c\mathbf{u})' = c\mathbf{u}' = cA\mathbf{u} = A(c\mathbf{u}),$$

which implies that $c\mathbf{u}$ is also in S. Consequently, S is indeed a subspace of $V_n(I)$.

We now prove that the dimension of S is n by constructing a basis for S containing n vectors. We first show that there exist n LI solutions to $\mathbf{x}' = A\mathbf{x}$. Let e_i denote the ith column vector of the identity matrix I_n. Then, from Theorem 8.4.1, for each i the IVP

$$\begin{cases} x_i'(t) = A(t)x_i(t), \\ x_i(t_0) = e_i, \end{cases} \qquad i = 1, 2, \ldots, n$$

has a unique solution. Further, $W[\mathbf{x}_1, \mathbf{x}_2, \ldots, \mathbf{x}_n](t_0) = \det(I_n) = 1 \neq 0$, so that $\{\mathbf{x}_1(t), \mathbf{x}_2(t), \ldots, \mathbf{x}_n(t)\}$ is LI on I. Next we establish that these solutions span the solution space. Let $\mathbf{x}(t)$ be any real solution to $\mathbf{x}' = A\mathbf{x}$ on I. Then, since $\{\mathbf{x}_1(t_0), \mathbf{x}_2(t_0), \ldots, \mathbf{x}_n(t_0)\}$ is LI in $\mathbf{R}^n$ it is a basis for $\mathbf{R}^n$. Consequently

$$\mathbf{x}(t_0) = c_1\mathbf{x}_1(t_0) + c_2\mathbf{x}_2(t_0) + \cdots + c_n\mathbf{x}_n(t_0)$$

for some scalars $c_1, c_2, \ldots, c_n$. It follows that $\mathbf{x}(t)$ is the *unique* (by Theorem 8.4.1) solution to the IVP

$$\begin{cases} x'(t) = A(t)x(t), \\ x(t_0) = c_1 x_1(t_0) + c_2 x_2(t_0) + \cdots + c_n x_n(t_0). \end{cases} \qquad (8.4.2)$$

But

$$\mathbf{u}(t) = c_1\mathbf{x}_1(t) + c_2\mathbf{x}_2(t) + \cdots + c_n\mathbf{x}_n(t)$$

also satisfies the IVP (8.4.2), and so, by uniqueness, we must have

$$\mathbf{x}(t) = \mathbf{u}(t) = c_1\mathbf{x}_1(t) + c_2\mathbf{x}_2(t) + \cdots + c_n\mathbf{x}_n(t).$$

We have therefore shown that any solution to $\mathbf{x}' = A\mathbf{x}$ on I can be written as a linear combination of the n LI solutions $\mathbf{x}_1, \mathbf{x}_2, \ldots, \mathbf{x}_n$, and hence, $\{\mathbf{x}_1, \mathbf{x}_2, \ldots, \mathbf{x}_n\}$ is a basis for the solution space. Consequently, the dimension of the solution space is n. ∎

It follows from Theorem 8.4.2 that if $\{\mathbf{x}_1(t), \mathbf{x}_2(t), \ldots, \mathbf{x}_n(t)\}$ is *any* set of n LI solutions to (8.4.1), then every solution to the system can be written as

$$\mathbf{x}(t) = c_1\mathbf{x}_1(t) + c_2\mathbf{x}_2(t) + \cdots + c_n\mathbf{x}_n(t), \qquad (8.4.3)$$

for appropriate constants $c_1, c_2, \ldots, c_n$. In keeping with the terminology that we have used throughout the text, we will refer to (8.4.3) as the **general solution** to the VDE (8.4.1).

The following definition introduces some important terminology for homogeneous VDE.

Definition 8.4.1:   Let $A(t)$ be an $n \times n$ matrix function that is continuous on an interval I. Any set of n solutions, $\{\mathbf{x}_1(t), \mathbf{x}_2(t), \ldots, \mathbf{x}_n(t)\}$, to $\mathbf{x}' = A\mathbf{x}$ that is LI on I is called a **fundamental solution set** on I. The corresponding matrix $X(t)$ defined by

$$X(t) = [\mathbf{x}_1(t), \mathbf{x}_2(t), \ldots, \mathbf{x}_n(t)]$$

is called a **fundamental matrix** for the VDE $\mathbf{x}' = A\mathbf{x}$.

REMARK   If $X(t)$ is a fundamental matrix for (8.4.1) then, applying Theorem 3.2.1, the general solution (8.4.3) can be written in vector form as $\mathbf{x}(t) = X(t)\mathbf{c}$, where $\mathbf{c} = [c_1\ c_2\ \ldots\ c_n]^T$. This will be our starting point when the variation-of-parameters method is derived in Section 8.7.

Now suppose that $\mathbf{x}_1(t), \mathbf{x}_2(t), \ldots, \mathbf{x}_n(t)$ are solutions to $\mathbf{x}' = A\mathbf{x}$ on an interval I. We have shown in the previous section that if $W[\mathbf{x}_1, \mathbf{x}_2, \ldots, \mathbf{x}_n](t) \neq 0$ at any point in I then the solutions are LI on I. We now prove the converse.

Theorem 8.4.3:   Let $A(t)$ be an $n \times n$ matrix function that is continuous on an interval I. If $\{\mathbf{x}_1(t), \mathbf{x}_2(t), \ldots, \mathbf{x}_n(t)\}$ is a *LI* set of *solutions* to $\mathbf{x}' = A\mathbf{x}$ on I, then

$$W[\mathbf{x}_1, \mathbf{x}_2, \ldots, \mathbf{x}_n](t) \neq 0$$

at every point in I.

PROOF   It is easier to prove the equivalent statement that if $W[\mathbf{x}_1, \mathbf{x}_2, \ldots, \mathbf{x}_n](t_0) = 0$ at some point t_0 in I then $\{\mathbf{x}_1, \mathbf{x}_2, \ldots, \mathbf{x}_n\}$ is LD on I. We proceed as follows. If $W[\mathbf{x}_1, \mathbf{x}_2, \ldots, \mathbf{x}_n](t_0) = 0$ then, from Corollary 5.5.2, the set of vectors $\{\mathbf{x}_1(t_0), \mathbf{x}_2(t_0), \ldots, \mathbf{x}_n(t_0)\}$ is LD in R^n. Thus there exist scalars $c_1, c_2, \ldots, c_n$, not all zero, such that

$$c_1\mathbf{x}_1(t_0) + c_2\mathbf{x}_2(t_0) + \cdots + c_n\mathbf{x}_n(t_0) = \mathbf{0}. \tag{8.4.4}$$

Now let

$$\mathbf{x}(t) = c_1\mathbf{x}_1(t) + c_2\mathbf{x}_2(t) + \cdots + c_n\mathbf{x}_n(t). \tag{8.4.5}$$

It follows from equations (8.4.4) and (8.4.5) and Theorem 8.4.1 that $\mathbf{x}(t)$ is the unique solution to the IVP

$$\mathbf{x}'(t) = A(t)\mathbf{x}(t), \ \mathbf{x}(t_0) = \mathbf{0}.$$

However, this IVP has the solution $\mathbf{x}(t) = \mathbf{0}$, and so, by uniqueness, we must have

$$c_1\mathbf{x}_1(t) + c_2\mathbf{x}_2(t) + \cdots + c_n\mathbf{x}_n(t) = \mathbf{0}.$$

Since not all of the c_i are zero, it follows that the set of vector functions $\{\mathbf{x}_1, \mathbf{x}_2, \ldots, \mathbf{x}_n\}$ is indeed LD on I. ∎

Thus, to determine whether $\{x_1(t), x_2(t), \ldots, x_n(t)\}$ is a fundamental solution set for $x' = Ax$ on an interval I, we can compute the Wronskian of $x_1(t), x_2(t), \ldots, x_n(t)$ at *any* convenient point t_0 in I. If $W[x_1, x_2, \ldots, x_n](t_0) \neq 0$ then the solutions are LI on I, whereas if $W[x_1, x_2, \ldots, x_n](t_0) = 0$, then the solutions are LD on I.

Example 8.4.1   Consider the VDE $x' = Ax$, where $A = \begin{bmatrix} 1 & 2 \\ -2 & 1 \end{bmatrix}$, and let

$$x_1(t) = \begin{bmatrix} -e^t \cos 2t \\ e^t \sin 2t \end{bmatrix}, \quad x_2(t) = \begin{bmatrix} e^t \sin 2t \\ e^t \cos 2t \end{bmatrix}.$$

(a)   Verify that $\{x_1, x_2\}$ is a fundamental set of solutions for the VDE on any interval, and write the general solution to the VDE.

(b)   Solve the IVP $x' = Ax$, $x(0) = \begin{bmatrix} 3 \\ 2 \end{bmatrix}$, and write the corresponding scalar solutions.

Solution

(a)   Differentiating the given vector functions with respect to t yields, respectively,

$$x_1'(t) = \begin{bmatrix} e^t(-\cos 2t + 2\sin 2t) \\ e^t(\sin 2t + 2\cos 2t) \end{bmatrix}, \quad x_2'(t) = \begin{bmatrix} e^t(\sin 2t + 2\cos 2t) \\ e^t(\cos 2t - 2\sin 2t) \end{bmatrix},$$

whereas

$$Ax_1 = \begin{bmatrix} 1 & 2 \\ -2 & 1 \end{bmatrix} \begin{bmatrix} -e^t \cos 2t \\ e^t \sin 2t \end{bmatrix} = \begin{bmatrix} e^t(-\cos 2t + 2\sin 2t) \\ e^t(\sin 2t + 2\cos 2t) \end{bmatrix},$$

and

$$Ax_2 = \begin{bmatrix} 1 & 2 \\ -2 & 1 \end{bmatrix} \begin{bmatrix} e^t \sin 2t \\ e^t \cos 2t \end{bmatrix} = \begin{bmatrix} e^t(\sin 2t + 2\cos 2t) \\ e^t(\cos 2t - 2\sin 2t) \end{bmatrix}.$$

Hence,

$$x_1' = Ax_1 \quad \text{and} \quad x_2' = Ax_2$$

so that x_1 and x_2 are indeed solutions to the given VDE. Furthermore, the Wronskian of these solutions is

$$W[x_1, x_2](t) = \begin{vmatrix} -e^t \cos 2t & e^t \sin 2t \\ e^t \sin 2t & e^t \cos 2t \end{vmatrix} = -e^{2t}.$$

Since the Wronskian is never zero, it follows that $\{x_1, x_2\}$ is LI on any interval and so is a fundamental set of solutions for the given VDE. Therefore, the general solution to the system is

$$x(t) = c_1 x_1(t) + c_2 x_2(t) = c_1 \begin{bmatrix} -e^t \cos 2t \\ e^t \sin 2t \end{bmatrix} + c_2 \begin{bmatrix} e^t \sin 2t \\ e^t \cos 2t \end{bmatrix}.$$

Combining the two column vector functions on the right-hand side yields

$$\mathbf{x}(t) = \left[\begin{array}{c} e^t(-c_1\cos 2t + c_2\sin 2t) \\ e^t(c_1\sin 2t + c_2\cos 2t) \end{array}\right].$$

(b) Imposing the given initial condition $\mathbf{x}(0) = \left[\begin{array}{c} 3 \\ 2 \end{array}\right]$ requires that

$$\left[\begin{array}{c} -c_1 \\ c_2 \end{array}\right] = \left[\begin{array}{c} 3 \\ 2 \end{array}\right],$$

so that $c_1 = -3$, $c_2 = 2$. Hence,

$$\mathbf{x}(t) = \left[\begin{array}{c} e^t(3\cos 2t + 2\sin 2t) \\ e^t(-3\sin 2t + 2\cos 2t) \end{array}\right].$$

The corresponding scalar solutions are

$$x_1(t) = e^t(3\cos 2t + 2\sin 2t), \quad x_2(t) = e^t(-3\sin 2t + 2\cos 2t).$$

NONHOMOGENEOUS VDE

The preceding results have dealt with the case of a homogeneous VDE. We end this section with the main theoretical result that will be needed for nonhomogeneous VDE. In view of our previous experience with nonhomogeneous linear problems, the following theorem should not be too surprising:

Theorem 8.4.4:   Let $A(t)$ be a matrix function that is continuous on the interval I, and let $\{\mathbf{x}_1, \mathbf{x}_2, ..., \mathbf{x}_n\}$ be a fundamental solution set for the VDE $\mathbf{x}'(t) = A(t)\mathbf{x}(t)$ on I. If $\mathbf{x}_p(t)$ is any particular solution to the nonhomogeneous VDE

$$\mathbf{x}'(t) = A(t)\mathbf{x}(t) + \mathbf{b}(t) \tag{8.4.6}$$

on I, then every solution to (8.4.6) on I is of the form

$$\mathbf{x}(t) = c_1\mathbf{x}_1 + c_2\mathbf{x}_2 + \cdots + c_n\mathbf{x}_n + \mathbf{x}_p.$$

PROOF   Since $\mathbf{x} = \mathbf{x}_p(t)$ is a solution to $\mathbf{x}'(t) = A(t)\mathbf{x}(t) + \mathbf{b}(t)$ on I, we have

$$\mathbf{x}_p{}'(t) = A(t)\mathbf{x}_p(t) + \mathbf{b}(t). \tag{8.4.7}$$

Now let $\mathbf{x} = \mathbf{u}(t)$ be any other solution to $\mathbf{x}'(t) = A(t)\mathbf{x}(t) + \mathbf{b}(t)$ on I. We then also have

$$\mathbf{u}'(t) = A(t)\mathbf{u}(t) + \mathbf{b}(t). \tag{8.4.8}$$

Subtracting (8.4.7) from (8.4.8) yields

$$(\mathbf{u} - \mathbf{x}_p)' = A(\mathbf{u} - \mathbf{x}_p).$$

Thus, the vector function $\mathbf{x} = \mathbf{u} - \mathbf{x}_p$ is a solution to the associated homogeneous system $\mathbf{x}' = A\mathbf{x}$ on I. Since $\{\mathbf{x}_1, \mathbf{x}_2, ..., \mathbf{x}_n\}$ spans the solution space of this system, it follows that

$$\mathbf{u} - \mathbf{x}_p = c_1\mathbf{x}_1 + c_2\mathbf{x}_2 + \cdots + c_n\mathbf{x}_n,$$

for some scalars $c_1, c_2, ..., c_n$. Consequently,

$$\mathbf{u} = c_1\mathbf{x}_1 + c_2\mathbf{x}_2 + \cdots + c_n\mathbf{x}_n + \mathbf{x}_p,$$

and the result is proved.   ∎

Theorem 8.4.4 implies that in order to solve a nonhomogeneous VDE, we must first find the general solution to the associated homogeneous system. In the next two sections, we will concentrate on homogeneous VDE, and then, in Section 8.7, we will see how the variation-of-parameters technique can be used to determine a particular solution to a nonhomogeneous VDE.

EXERCISES 8.4

For problems 1–3, show that the given functions are solutions of the system $\mathbf{x}' = A\mathbf{x}$ for the given matrix A, and hence, find the general solution to the system (remember to check linear independence). If auxiliary conditions are given, find the particular solution that satisfies these conditions.

1. $\mathbf{x}_1(t) = \begin{bmatrix} e^{4t} \\ 2e^{4t} \end{bmatrix}$, $\mathbf{x}_2(t) = \begin{bmatrix} 3e^{-t} \\ e^{-t} \end{bmatrix}$,

$A = \begin{bmatrix} -2 & 3 \\ -2 & 5 \end{bmatrix}$, $\mathbf{x}(0) = \begin{bmatrix} -2 \\ 1 \end{bmatrix}$.

2. $\mathbf{x}_1(t) = \begin{bmatrix} e^{2t} \\ -e^{2t} \end{bmatrix}$, $\mathbf{x}_2(t) = \begin{bmatrix} e^{2t}(1+t) \\ -te^{2t} \end{bmatrix}$,

$A = \begin{bmatrix} 3 & 1 \\ -1 & 1 \end{bmatrix}$.

3. $\mathbf{x}_1(t) = \begin{bmatrix} -3 \\ 9 \\ 5 \end{bmatrix}$, $\mathbf{x}_2(t) = \begin{bmatrix} e^{2t} \\ 3e^{2t} \\ e^{2t} \end{bmatrix}$,

$\mathbf{x}_3(t) = \begin{bmatrix} e^{4t} \\ e^{4t} \\ e^{4t} \end{bmatrix}$, $A = \begin{bmatrix} 2 & -1 & 3 \\ 3 & 1 & 0 \\ 2 & -1 & 3 \end{bmatrix}$.

For problems 4–7, determine two LI solutions to the given system.

4. $\mathbf{x}' = A\mathbf{x}$, where $A = \begin{bmatrix} -1 & 2 \\ 2 & 2 \end{bmatrix}$.

5. $\mathbf{x}' = A\mathbf{x}$, where $A = \begin{bmatrix} 0 & 3 \\ -3 & 0 \end{bmatrix}$.

6. $\mathbf{x}' = A\mathbf{x}$, where $A = \begin{bmatrix} 1 & -2 \\ 2 & 1 \end{bmatrix}$.

7. $\mathbf{x}' = A\mathbf{x}$, where $A = \begin{bmatrix} -3 & -1 \\ 4 & 1 \end{bmatrix}$.

8. If $\mathbf{x}_1, \mathbf{x}_2, \ldots, \mathbf{x}_n$ are solutions to $\mathbf{x}' = A(t)\mathbf{x}$ and $X = [\mathbf{x}_1, \mathbf{x}_2, \ldots, \mathbf{x}_n]$, prove that

$$X' = A(t)X.$$

9. Let $X(t)$ be a fundamental matrix for $\mathbf{x}' = A(t)\mathbf{x}$ on the interval I.

(a) Show that the general solution to the linear system can be written as

$$\mathbf{x} = X(t)\mathbf{c},$$

where $\mathbf{c}$ is a vector of constants.

(b) If $t_0 \in I$, show that the solution to the IVP

$$\mathbf{x}' = A\mathbf{x}, \mathbf{x}(t_0) = \mathbf{x}_0,$$

can be written as

$$\mathbf{x} = X(t)\,X^{-1}(t_0)\,\mathbf{x}_0.$$

8.5   HOMOGENEOUS CONSTANT COEFFICIENT VDE: NONDEFECTIVE COEFFICIENT MATRIX

The theory that we have developed in the previous section is valid for any first-order linear system. However, in practice, these are too difficult to solve in general, and so, we must make a simplifying assumption in order to develop solution techniques applicable to a broad class of linear systems. The assumption that we will make is that the coefficient matrix is a constant matrix.[1] In the next two sections, we will consider only homogeneous linear systems

$$\mathbf{x}' = A\mathbf{x},$$

where A is an $n \times n$ matrix of real *constants*. For example,

$$\left.\begin{array}{l} x_1' = 2x_1 - 3x_2 \\ x_2' = -x_1 + 4x_2 \end{array}\right\} \quad \Leftrightarrow \quad \mathbf{x}' = A\mathbf{x}, \text{ where } A = \left[\begin{array}{cc} 2 & -3 \\ -1 & 4 \end{array}\right].$$

To motivate the new solution technique to be developed, we recall from Example 8.3.5 that two LI solutions to the VDE

$$\mathbf{x}' = A\mathbf{x}, \quad A = \left[\begin{array}{cc} 1 & 2 \\ 2 & -2 \end{array}\right]$$

are

$$\mathbf{x}_1(t) = \left[\begin{array}{c} e^{-3t} \\ -2e^{-3t} \end{array}\right], \quad \mathbf{x}_2(t) = \left[\begin{array}{c} e^{2t} \\ \frac{1}{2} e^{2t} \end{array}\right],$$

which we write as

$$\mathbf{x}_1(t) = e^{-3t}\left[\begin{array}{c} 1 \\ -2 \end{array}\right], \quad \mathbf{x}_2(t) = e^{2t}\left[\begin{array}{c} 1 \\ \frac{1}{2} \end{array}\right].$$

The key point to notice is that both of these solutions are of the form

$$\mathbf{x}(t) = e^{\lambda t}\,\mathbf{v} \tag{8.5.1}$$

where λ is a scalar and $\mathbf{v}$ is a constant vector. This suggests that the general VDE

$$\mathbf{x}' = A\mathbf{x} \tag{8.5.2}$$

may also have solutions of the form (8.5.1). We now investigate this possibility. Differentiating (8.5.1) with respect to t yields

$$\mathbf{x}' = \lambda e^{\lambda t}\mathbf{v}.$$

Thus, $\mathbf{x}(t) = e^{\lambda t}\mathbf{v}$ is a solution to (8.5.2) if and only if

$$\lambda e^{\lambda t}\mathbf{v} = e^{\lambda t}A\mathbf{v},$$

that is, if and only if λ and $\mathbf{v}$ satisfy

$$A\mathbf{v} = \lambda\mathbf{v}.$$

But this is the statement that λ and $\mathbf{v}$ must be an eigenvalue/eigenvector pair for A. Consequently, we have established the following fundamental result:

Theorem 8.5.1: Let A be an $n \times n$ matrix of real constants, and let λ be an eigenvalue of A with corresponding eigenvector $\mathbf{v}$. Then

$$\mathbf{x}(t) = e^{\lambda t}\mathbf{v}$$

is a solution to the constant coefficient VDE $\mathbf{x}' = A\mathbf{x}$ on any interval. ∎

---

[1]This assumption should not be too surprising in view of the discussion of linear nth-order equations in Chapters 2 and 7.

REMARK   Notice that we have not assumed that the eigenvalues and eigenvectors of A are real; the preceding result holds in the complex case also.

We now illustrate how Theorem 8.5.1 can be used to find the general solution to constant coefficient VDE.

Example 8.5.1   Find the general solution to

$$
\begin{aligned}
x_1' &= 2x_1 + x_2 \\
x_2' &= -3x_1 - 2x_2.
\end{aligned}
\tag{8.5.3}
$$

Solution   The given system can be written as the VDE

$$
\mathbf{x}' = A\mathbf{x}, \text{ where } A = \begin{bmatrix} 2 & 1 \\ -3 & -2 \end{bmatrix}.
\tag{8.5.4}
$$

We first find the eigenvalues and eigenvectors of A. A straightforward calculation yields

$$
\det(A - \lambda I) = \begin{vmatrix} 2 - \lambda & 1 \\ -3 & -2 - \lambda \end{vmatrix} = \lambda^2 - 1,
$$

so that A has eigenvalues $\lambda = \pm 1$.

Eigenvectors:

$\lambda = 1$:   In this case the system $(A - \lambda I)\mathbf{v} = \mathbf{0}$ is

$$
\begin{aligned}
v_1 + v_2 &= 0, \\
-3v_1 - 3v_2 &= 0,
\end{aligned}
$$

with solution $\mathbf{v} = r(1, -1)$. It follows from Theorem 8.5.1 that

$$
\mathbf{x}_1(t) = e^t \begin{bmatrix} 1 \\ -1 \end{bmatrix}
$$

is a solution to the VDE (8.5.4).

$\lambda = -1$:   In this case the system $(A - \lambda I)\mathbf{v} = \mathbf{0}$ is

$$
\begin{aligned}
3v_1 + v_2 &= 0, \\
-3v_1 - v_2 &= 0,
\end{aligned}
$$

with solution $\mathbf{v} = s(1, -3)$. Consequently,

$$
\mathbf{x}_2(t) = e^{-t} \begin{bmatrix} 1 \\ -3 \end{bmatrix}
$$

is also a solution to the VDE (8.5.4).

The Wronskian of these solutions is

$$
W[\mathbf{x}_1, \mathbf{x}_2](t) = \begin{vmatrix} e^t & e^{-t} \\ -e^t & -3e^{-t} \end{vmatrix} = -2 \neq 0,
$$

so that $\{x_1, x_2\}$ is LI on any interval. Hence the general solution to (8.5.4) is

$$\mathbf{x}(t) = c_1 \mathbf{x}_1 + c_2 \mathbf{x}_2 = c_1 e^t \begin{bmatrix} 1 \\ -1 \end{bmatrix} + c_2 e^{-t} \begin{bmatrix} 1 \\ -3 \end{bmatrix}.$$

Combining the column vectors on the right-hand side yields the solution vector

$$\mathbf{x}(t) = \begin{bmatrix} c_1 e^t + c_2 e^{-t} \\ -(c_1 e^t + c_2 e^{-t}) \end{bmatrix}.$$

Therefore, the solution to the corresponding scalar differential system (8.5.3) is

$$x_1(t) = c_1 e^t + c_2 e^{-t}, \quad x_2(t) = -(c_1 e^t + 3c_2 e^{-t}).$$ □

To find the general solution to an $n \times n$ constant coefficient VDE, we need to find n LI solutions. The preceding example together with our experience with eigenvalues and eigenvectors suggests that we will be able to find n such LI solutions provided the matrix A has n LI eigenvectors. (That is, A is nondefective). This is indeed the case, although if the eigenvalues and eigenvectors are complex, we must do some work to obtain real-valued solutions to the system. We first give the result for the case of real eigenvalues.

Theorem 8.5.2:   Let A be an $n \times n$ matrix of real constants. If A has n real LI eigenvectors $\mathbf{v}_1, \mathbf{v}_2, \ldots, \mathbf{v}_n$, with corresponding eigenvalues $\lambda_1, \lambda_2, \ldots, \lambda_n$ (not necessarily distinct), then the vector functions $\{\mathbf{x}_1, \mathbf{x}_2, \ldots, \mathbf{x}_n\}$ defined by

$$\mathbf{x}_k(t) = e^{\lambda_k t} \mathbf{v}_k, \quad k = 1, 2, \ldots, n,$$

for all t, are LI solutions to $\mathbf{x}' = A\mathbf{x}$ on any interval. The general solution to this VDE is

$$\mathbf{x}(t) = c_1 \mathbf{x}_1 + c_2 \mathbf{x}_2 + \cdots + c_n \mathbf{x}_n.$$

PROOF   We have already shown (Theorem 8.5.1) that each $\mathbf{x}_k(t)$ satisfies $\mathbf{x}' = A\mathbf{x}$ for all t. Further, using properties of determinants,

$$W[\mathbf{x}_1, \mathbf{x}_2, \ldots, \mathbf{x}_n] = \det([e^{\lambda_1 t} \mathbf{v}_1, e^{\lambda_2 t} \mathbf{v}_2, \ldots, e^{\lambda_n t} \mathbf{v}_n])$$

$$= e^{(\lambda_1 + \lambda_2 + \cdots + \lambda_n)t} \det([\mathbf{v}_1, \mathbf{v}_2, \ldots, \mathbf{v}_n])$$

$$\neq 0,$$

since the eigenvectors are LI by assumption, and hence the solutions are LI on any interval.   ∎

Example 8.5.2   Find the general solution to $\mathbf{x}' = A\mathbf{x}$ if $A = \begin{bmatrix} 0 & 2 & -3 \\ -2 & 4 & -3 \\ 2 & 2 & -1 \end{bmatrix}$.

Solution   We first determine the eigenvalues and eigenvectors of A. For the given matrix, we have

$$\det(A - \lambda I) = \begin{vmatrix} -\lambda & 2 & -3 \\ -2 & 4-\lambda & -3 \\ -2 & 2 & -1-\lambda \end{vmatrix} = -(\lambda + 1)(\lambda - 2)^2,$$

so that the eigenvalues are $\lambda = -1, 2$ (multiplicity 2).

Eigenvectors

$\lambda = -1$: It is easily shown that all eigenvectors corresponding to this eigenvalue are of the form $\mathbf{v} = r(1, 1, 1)$, so that we can take

$$\mathbf{v}_1 = (1, 1, 1).$$

$\lambda = 2$: In this case, the system for the eigenvectors reduces to the single equation

$$2v_1 - 2v_2 + 3v_3 = 0,$$

which has solution $\mathbf{v} = r(1, 1, 0) + s(-3, 0, 2)$. Therefore, two LI eigenvectors corresponding to $\lambda = 2$ are

$$\mathbf{v}_2 = (1, 1, 0), \quad \mathbf{v}_3 = (-3, 0, 2).$$

It follows from Theorem 8.5.2 that three LI solutions to the given VDE are

$$\mathbf{x}_1(t) = e^{-t}\begin{bmatrix} 1 \\ 1 \\ 1 \end{bmatrix}, \quad \mathbf{x}_2(t) = e^{2t}\begin{bmatrix} 1 \\ 1 \\ 0 \end{bmatrix}, \quad \mathbf{x}_3(t) = e^{2t}\begin{bmatrix} -3 \\ 0 \\ 2 \end{bmatrix}.$$

Consequently, the general solution to the given system is

$$\mathbf{x}(t) = c_1 e^{-t}\begin{bmatrix} 1 \\ 1 \\ 1 \end{bmatrix} + c_2 e^{2t}\begin{bmatrix} 1 \\ 1 \\ 0 \end{bmatrix} + c_3 e^{2t}\begin{bmatrix} -3 \\ 0 \\ 2 \end{bmatrix},$$

which can be combined to obtain the solution vector

$$\mathbf{x}(t) = \begin{bmatrix} c_1 e^{-t} + c_2 e^{2t} - 3c_3 e^{2t} \\ c_1 e^{-t} + c_2 e^{2t} \\ c_1 e^{-t} + 2c_3 e^{2t} \end{bmatrix}. \qquad \square$$

We now consider the case when some (or all) of the eigenvalues are complex. Since we are restricting attention to systems of equations with *real* constant coefficients, it follows that the matrix of the system will have real entries, and hence, from Theorem 6.5.1, the eigenvalues *and* eigenvectors will occur in conjugate pairs. The corresponding solutions to $\mathbf{x}' = A\mathbf{x}$ guaranteed by Theorem 8.5.1 will also be complex conjugate. However, as we now show, each conjugate pair gives rise to two real-valued solutions.

Theorem 8.5.3:   Let $\mathbf{u}(t)$ and $\mathbf{v}(t)$ be real-valued vector functions. If

$$\mathbf{w}_1(t) = \mathbf{u}(t) + i\mathbf{v}(t) \text{ and } \mathbf{w}_2(t) = \mathbf{u}(t) - i\mathbf{v}(t)$$

are complex conjugate solutions to $\mathbf{x}' = A\mathbf{x}$, then

$$\mathbf{x}_1(t) = \mathbf{u}(t) \text{ and } \mathbf{x}_2(t) = \mathbf{v}(t)$$

are themselves *real-valued* solutions of $\mathbf{x}' = A\mathbf{x}$.

PROOF   Since $\mathbf{w}_1$ and $\mathbf{w}_2$ are solutions to the VDE, so is any linear combination of them. In particular,

$$\mathbf{x}_1(t) = \frac{1}{2}[\mathbf{w}_1(t) + \mathbf{w}_2(t)] = \mathbf{u}(t)$$

and

$$\mathbf{x}_2(t) = \frac{1}{2i}[\mathbf{w}_1(t) - \mathbf{w}_2(t)] = \mathbf{v}(t)$$

are solutions to the VDE.

∎

We now explicitly derive two appropriate real-valued solutions corresponding to a complex conjugate pair of eigenvalues. Suppose that $\lambda = a + ib$ ($b \neq 0$) is an eigenvalue of A with corresponding eigenvector $\mathbf{v} = \mathbf{r} + i\mathbf{s}$. Then, applying Theorem 8.5.1, a complex-valued solution to $\mathbf{x}' = A\mathbf{x}$ is

$$\mathbf{w}(t) = e^{(a + ib)t}(\mathbf{r} + i\mathbf{s}) = e^{at}(\cos bt + i\sin bt)(\mathbf{r} + i\mathbf{s}),$$

which can be written as

$$\mathbf{w}(t) = e^{at}(\cos bt\ \mathbf{r} - \sin bt\ \mathbf{s}) + i\,e^{at}(\sin bt\ \mathbf{r} + \cos bt\ \mathbf{s}).$$

Theorem 8.5.3 implies that two real-valued solutions to $\mathbf{x}' = A\mathbf{x}$ are

$$\mathbf{x}_1(t) = e^{at}(\cos bt\ \mathbf{r} - \sin bt\ \mathbf{s}), \quad \mathbf{x}_2(t) = e^{at}(\sin bt\ \mathbf{r} + \cos bt\ \mathbf{s}).$$

It can further be shown that the set of all real-valued solutions obtained in this manner is LI on any interval.

REMARK   Notice that we do not have to derive the solution corresponding to the conjugate eigenvalue $\bar{\lambda} = a - ib$, since it does not yield any new LI solutions to $\mathbf{x}' = A\mathbf{x}$.

Example 8.5.3   Find the general solution to the VDE $\mathbf{x}' = A\mathbf{x}$ if $A = \begin{bmatrix} 0 & 2 \\ -2 & 0 \end{bmatrix}$.

Solution The characteristic polynomial of A is

$$p(\lambda) = \begin{vmatrix} -\lambda & 2 \\ -2 & -\lambda \end{vmatrix} = \lambda^2 + 4.$$

Consequently, A has complex conjugate eigenvalues $\lambda = \pm 2i$. The eigenvectors corresponding to the eigenvalue $\lambda = 2i$ are obtained by solving

$$v_1 + iv_2 = 0$$

and are therefore of the form $\mathbf{v} = r(-i, 1)$. Hence, a complex-valued solution to the given DE is

$$\mathbf{w}(t) = e^{2it}\begin{bmatrix} -i \\ 1 \end{bmatrix}.$$

We must now do some algebra to obtain the corresponding real-valued solutions. Using Euler's formula, we can write

$$\mathbf{w}(t) = (\cos 2t + i\sin 2t)\begin{bmatrix} -i \\ 1 \end{bmatrix} = \begin{bmatrix} \sin 2t - i\cos 2t \\ \cos 2t + i\sin 2t \end{bmatrix}$$

$$= \begin{bmatrix} \sin 2t \\ \cos 2t \end{bmatrix} + i\begin{bmatrix} -\cos 2t \\ \sin 2t \end{bmatrix}.$$

Applying Theorem 8.5.3, we directly obtain the two real-valued solutions

$$\mathbf{x}_1(t) = \begin{bmatrix} \sin 2t \\ \cos 2t \end{bmatrix}, \quad \mathbf{x}_2(t) = \begin{bmatrix} -\cos 2t \\ \sin 2t \end{bmatrix}.$$

Consequently, the general solution to the given VDE is

$$\mathbf{x}(t) = c_1\begin{bmatrix} \sin 2t \\ \cos 2t \end{bmatrix} + c_2\begin{bmatrix} -\cos 2t \\ \sin 2t \end{bmatrix}.$$

$$= \begin{bmatrix} c_1\sin 2t - c_2\cos 2t \\ c_1\cos 2t + c_2\sin 2t \end{bmatrix}.$$

We note that, in this case, the eigenvalues of A were pure imaginary and that the components of the corresponding solution vector are oscillatory. Once more, this illustrates the importance of complex scalars when modeling oscillatory physical behavior. ❏

As illustrated in the next example, when the real part of a complex eigenvalue is nonzero, the algebra can become a little bit more tedious.

Example 8.5.4   Find the general solution to the VDE $\mathbf{x}' = A\mathbf{x}$ if $A = \begin{bmatrix} 2 & -1 \\ 2 & 4 \end{bmatrix}$.

Solution A has characteristic polynomial

$$\det(A - \lambda I) = \begin{vmatrix} 2 - \lambda & -1 \\ 2 & 4 - \lambda \end{vmatrix} = \lambda^2 - 6\lambda + 10,$$

so that the eigenvalues are $\lambda = 3 \pm i$. We need only to find the eigenvectors corres–ponding to one of these conjugate eigenvalues. When $\lambda = 3 + i$, the eigenvectors are obtained by solving

$$-(1 + i)v_1 - \quad v_2 = 0,$$
$$2v_1 + (1 - i)v_2 = 0,$$

which yield the complex eigenvectors $\mathbf{v} = r(1, -(1 + i))$. Hence a complex-valued solution to the given system is

$$\mathbf{w}(t) = e^{3t}(\cos t + i\sin t)\begin{bmatrix} 1 \\ -(1 + i) \end{bmatrix} = e^{3t}\begin{bmatrix} \cos t + i\sin t \\ -(1 + i)(\cos t + i\sin t) \end{bmatrix}$$

$$= e^{3t}\begin{bmatrix} \cos t + i\sin t \\ (\sin t - \cos t) - i(\sin t + \cos t) \end{bmatrix}$$

$$= e^{3t} \left\{ \begin{bmatrix} \cos t \\ \sin t - \cos t \end{bmatrix} + i \begin{bmatrix} \sin t \\ -(\sin t + \cos t) \end{bmatrix} \right\}.$$

From Theorem 8.5.3, the real and imaginary parts of this complex-valued solution yield the following two *real-valued* LI solutions

$$\mathbf{x}_1(t) = e^{3t} \begin{bmatrix} \cos t \\ \sin t - \cos t \end{bmatrix}, \quad \mathbf{x}_2(t) = e^{3t} \begin{bmatrix} \sin t \\ -(\sin t + \cos t) \end{bmatrix}.$$

Hence, the general solution to the given system is

$$\mathbf{x}(t) = c_1 e^{3t} \begin{bmatrix} \cos t \\ \sin t - \cos t \end{bmatrix} + c_2 e^{3t} \begin{bmatrix} \sin t \\ -(\sin t + \cos t) \end{bmatrix}$$

$$= e^{3t} \left\{ c_1 \begin{bmatrix} \cos t \\ \sin t - \cos t \end{bmatrix} + c_2 \begin{bmatrix} \sin t \\ -(\sin t + \cos t) \end{bmatrix} \right\}$$

$$= \begin{bmatrix} e^{3t}(c_1 \cos t + c_2 \sin t) \\ e^{3t}[c_1 (\sin t - \cos t) - c_2 (\sin t + \cos t)] \end{bmatrix}. \qquad \square$$

The results of this section are summarized in the next theorem.

Theorem 8.5.4:   Let A be an $n \times n$ matrix of real constants.

1. Suppose λ is a real eigenvalue of A of multiplicity m with corresponding LI eigenvectors $\mathbf{v}_1, \mathbf{v}_2, \ldots, \mathbf{v}_k$ $(1 \le k \le m)$. Then k LI solutions to $\mathbf{x}' = A\mathbf{x}$ are

$$\mathbf{x}_j(t) = e^{\lambda t} \mathbf{v}_j, \quad j = 1, 2, \ldots, k.$$

2. Suppose $\lambda = a + ib$ is a complex eigenvalue of multiplicity m with corresponding LI eigenvectors $\mathbf{v}_1, \mathbf{v}_2, \ldots, \mathbf{v}_k$ $(1 \le k \le m)$, where $\mathbf{v}_j = \mathbf{r}_j + i\mathbf{s}_j$. Then k complex-valued solutions to $\mathbf{x}' = A\mathbf{x}$ are

$$\mathbf{u}_j(t) = e^{\lambda t} \mathbf{v}_j, \quad j = 1, 2, \ldots, k.$$

and $2k$ *real-valued* LI solutions to $\mathbf{x}' = A\mathbf{x}$ are

$$\mathbf{x}_1(t) = e^{at}(\cos bt\, \mathbf{r}_1 - \sin bt\, \mathbf{s}_1), \mathbf{x}_2(t) = e^{at}(\cos bt\, \mathbf{r}_2 - \sin bt\, \mathbf{s}_2), \ldots,$$

$$\mathbf{x}_k(t) = e^{at}(\cos bt\, \mathbf{r}_k - \sin bt\, \mathbf{s}_k), \mathbf{x}_{k+1} = e^{at}(\sin bt\, \mathbf{r}_1 + \cos bt\, \mathbf{s}_1), \ldots,$$

$$\mathbf{x}_{2k}(t) = e^{at}(\sin bt\, \mathbf{r}_k + \cos bt\, \mathbf{s}_k).$$

Further, the set of all solutions to $\mathbf{x}' = A\mathbf{x}$ obtained in this manner is LI on any interval.

Corollary 8.5.1:   If A is nondefective, then the solutions obtained from parts (1) and (2) of the previous theorem yield a fundamental set of solutions to $\mathbf{x}' = A\mathbf{x}$, and the general solution to this VDE is

$$\mathbf{x}(t) = c_1 \mathbf{x}_1 + c_2 \mathbf{x}_2 + \cdots + c_n \mathbf{x}_n.$$

EXERCISES 8.5

For problems 1–15, determine the general solution to the system $\mathbf{x}' = A\mathbf{x}$ for the given matrix A.

1. $\begin{bmatrix} -2 & -7 \\ -1 & 4 \end{bmatrix}$.

2. $\begin{bmatrix} 0 & -4 \\ 4 & 0 \end{bmatrix}$.

3. $\begin{bmatrix} 1 & -2 \\ 5 & -5 \end{bmatrix}$.

4. $\begin{bmatrix} -1 & 2 \\ -2 & -1 \end{bmatrix}$.

5. $\begin{bmatrix} 2 & 0 & 0 \\ 0 & 5 & -7 \\ 0 & 2 & -4 \end{bmatrix}$.

6. $\begin{bmatrix} -1 & 0 & 0 \\ 1 & 5 & -1 \\ 1 & 6 & -2 \end{bmatrix}$.

7. $\begin{bmatrix} 0 & 1 & 0 \\ -1 & 0 & 0 \\ 0 & 0 & 5 \end{bmatrix}$.

8. $\begin{bmatrix} 2 & 0 & 3 \\ 0 & -4 & 0 \\ -3 & 0 & 2 \end{bmatrix}$.

9. $\begin{bmatrix} 3 & 2 & 6 \\ -2 & 1 & -2 \\ -1 & -2 & -4 \end{bmatrix}$.

10. $\begin{bmatrix} 0 & -3 & 1 \\ -2 & -1 & 1 \\ 0 & 0 & 2 \end{bmatrix}$.

11. $\begin{bmatrix} 3 & 0 & -1 \\ 0 & -3 & -1 \\ 0 & 2 & -1 \end{bmatrix}$.

12. $\begin{bmatrix} 1 & 1 & -1 \\ 1 & 1 & 1 \\ -1 & 1 & 1 \end{bmatrix}$.

13. $\begin{bmatrix} 2 & -1 & 3 \\ 2 & -1 & 3 \\ 2 & -1 & 3 \end{bmatrix}$.

14. $\begin{bmatrix} 1 & 2 & 3 & 4 \\ 4 & 3 & 2 & 1 \\ 4 & 5 & 6 & 7 \\ 7 & 6 & 5 & 4 \end{bmatrix}$.

15. $\begin{bmatrix} 0 & 1 & 0 & 0 \\ -1 & 0 & 0 & 0 \\ 0 & 0 & 0 & -1 \\ 0 & 0 & 1 & 0 \end{bmatrix}$.

For problems 16–18, solve the IVP $\mathbf{x}' = A\mathbf{x}$, $\mathbf{x}(0) = \mathbf{x}_0$.

16. $A = \begin{bmatrix} -1 & 4 \\ 2 & -3 \end{bmatrix}$, $\mathbf{x}_0 = \begin{bmatrix} 3 \\ 0 \end{bmatrix}$.

17. $A = \begin{bmatrix} -1 & -6 \\ 3 & 5 \end{bmatrix}$, $\mathbf{x}_0 = \begin{bmatrix} 2 \\ 2 \end{bmatrix}$.

18. $A = \begin{bmatrix} 2 & -1 & 3 \\ 3 & 1 & 0 \\ 2 & -1 & 3 \end{bmatrix}$, $\mathbf{x}_0 = \begin{bmatrix} -4 \\ 4 \\ 4 \end{bmatrix}$.

19. Solve the IVP

$$\mathbf{x}' = A\mathbf{x}, \; \mathbf{x}(0) = \begin{bmatrix} 1 \\ 1 \end{bmatrix},$$

when $A = \begin{bmatrix} 0 & 4 \\ -4 & 0 \end{bmatrix}$. Sketch the solution in the $x_1 x_2$-plane.

20. Consider the differential equation

$$\frac{d^2 x}{dt^2} + b\frac{dx}{dt} + cx = 0, \qquad (20.1)$$

where b and c are constants.

(a) Show that equation (20.1) can be replaced by the equivalent first-order linear system $\mathbf{x}' = A\mathbf{x}$,

where $A = \begin{bmatrix} 0 & 1 \\ -c & -b \end{bmatrix}$.

(b) Show that the characteristic polynomial of A coincides with the auxiliary polynomial of equation (20.1).

21. Let $\lambda = a + ib$, $b \neq 0$, be an eigenvalue of the $n \times n$ (real) matrix A with corresponding eigenvector $\mathbf{v} = \mathbf{r} + i\mathbf{s}$. Then we have shown in the text that two real-valued solutions to $\mathbf{x}' = A\mathbf{x}$ are

$$\mathbf{x}_1(t) = e^{at}[\cos bt \, \mathbf{r} - \sin bt \, \mathbf{s}],$$
$$\mathbf{x}_2(t) = e^{at}[\sin bt \, \mathbf{r} + \cos bt \, \mathbf{s}].$$

Prove that $\mathbf{x}_1$, $\mathbf{x}_2$ are LI on any interval. (You may assume that $\mathbf{r}$ and $\mathbf{s}$ are LI in $\mathbf{R}^n$.)

The remaining problems in this section investigate general properties of solutions to $\mathbf{x}' = A\mathbf{x}$, where A is a nondefective matrix.

22. Let A be a 2×2 nondefective matrix. If all eigenvalues of A have negative real part, prove that every solution to $\mathbf{x}' = A\mathbf{x}$ satisfies

$$\lim_{t \to \infty} \mathbf{x}(t) = \mathbf{0}. \qquad (22.1)$$

23. Let A be a 2×2 nondefective matrix. If *every* solution to $\mathbf{x}' = A\mathbf{x}$ satisfies (22.1), prove that all eigenvalues of A have negative real part.

24. Determine the general solution to $\mathbf{x}' = A\mathbf{x}$ if

$$A = \begin{bmatrix} 0 & b \\ -b & 0 \end{bmatrix}, \text{ where } b > 0. \text{ Describe the}$$

behavior of the solutions.

25. Describe the behavior of the solutions to $\mathbf{x}' = A\mathbf{x}$, if $A = \begin{bmatrix} a & b \\ -b & a \end{bmatrix}$, where $a < 0$, $b > 0$.

26. What conditions on the eigenvalues of an $n \times n$ matrix A would guarantee that the system $\mathbf{x}' = A\mathbf{x}$ has at least one solution satisfying

$$\mathbf{x}(t) = \mathbf{x}_0,$$

for all t, where $\mathbf{x}_0$ is a constant vector?

27. The motion of a certain physical system is described by the system of differential equations

$$x_1' = x_2, \quad x_2' = -bx_1 - ax_2,$$

where a and b are positive constants and $a \neq 2b$. Show that the motion of the system dies out as $t \to +\infty$.

8.6   HOMOGENEOUS CONSTANT COEFFICIENT VDE: DEFECTIVE COEFFICIENT MATRIX

The results of the previous section enable us to solve any constant coefficient linear differential system $\mathbf{x}' = A\mathbf{x}$, provided that A is nondefective. We recall from Chapter 6 that if m denotes the multiplicity of an eigenvalue of A, then the dimension k of the corresponding eigenspace satisfies the inequality

$$1 \leq k \leq m$$

and the condition for A to be nondefective is that the dimension of each eigenspace equals the multiplicity of the corresponding eigenvalue. (See Theorem 6.6.3.) We now turn our attention to the case when A is defective. Then, for at least one eigenvalue, the dimension k of the corresponding eigenspace is strictly less than the multiplicity m of the eigenvalue. In this case, there are only k LI eigenvectors corresponding to λ, and so Theorem 8.5.4 will only yield k LI solutions to the VDE $\mathbf{x}' = A\mathbf{x}$. We must therefore find an additional $m - k$ LI solutions. In order to motivate the main result of this section (which is rather difficult to prove), we consider a particular example.

Example 8.6.1   Find the general solution to

$$\mathbf{x}' = A\mathbf{x}, \quad A = \begin{bmatrix} 0 & 1 \\ -9 & 6 \end{bmatrix}. \qquad (8.6.1)$$

Solution   We try using the technique from the previous section. The coefficient matrix A has the single eigenvalue $\lambda = 3$ of multiplicity 2, and it is straightforward to show that there is just one corresponding LI eigenvector which we may take to be

$\mathbf{v}_0 = (1, 3)$. Consequently, A is defective, and we obtain only one LI solution to the VDE, namely,

$$\mathbf{x}_1(t) = e^{3t} \begin{bmatrix} 1 \\ 3 \end{bmatrix}. \tag{8.6.2}$$

In order to obtain a second LI solution to the VDE (8.6.1), we consider the corresponding scalar differential system

$$x_1' = \qquad x_2$$
$$x_2' = -9x_1 + 6x_2.$$

Applying the solution technique introduced in Section 8.2 yields the scalar solutions

$$x_1(t) = c_1 e^{3t} + c_2 t e^{3t}, \quad x_2(t) = 3c_1 e^{3t} + c_2 e^{3t}(3t + 1).$$

Consequently, the general solution to (8.6.1) is

$$\mathbf{x}(t) = \begin{bmatrix} c_1 e^{3t} + c_2 t e^{3t} \\ 3c_1 e^{3t} + c_2 e^{3t}(3t + 1) \end{bmatrix},$$

which can be written in the equivalent form

$$\mathbf{x}(t) = c_1 e^{3t} \begin{bmatrix} 1 \\ 3 \end{bmatrix} + c_2 e^{3t} \left\{ \begin{bmatrix} 0 \\ 1 \end{bmatrix} + t \begin{bmatrix} 1 \\ 3 \end{bmatrix} \right\}.$$

We see that two LI solutions to the given system are

$$\mathbf{x}_1(t) = e^{3t} \begin{bmatrix} 1 \\ 3 \end{bmatrix}, \quad \mathbf{x}_2(t) = e^{3t} \left\{ \begin{bmatrix} 0 \\ 1 \end{bmatrix} + t \begin{bmatrix} 1 \\ 3 \end{bmatrix} \right\}.$$

The first of these solutions coincides with the solution (8.6.2), which was derived using the eigenvalue/eigenvector technique of the previous section. The key point to notice is that, in this particular example, there is a second LI solution of the form

$$\mathbf{x}_2(t) = e^{\lambda t}(\mathbf{v}_1 + t\mathbf{v}_2). \qquad \square$$

It can be shown that the basic *form* of the second LI solution obtained in the preceding example holds in the general case when the coefficient matrix has an eigenvalue of multiplicity 2 with a corresponding one-dimensional eigenspace. Further, it can be generalized to include the case of arbitrary multiplicity. Indeed, in view of the previous example, we might suspect that if an eigenvalue has multiplicity m, but only k corresponding LI eigenvectors, then we should be able to find $m - k$ solutions of the form

$$\mathbf{x}_i(t) = e^{\lambda t}(\mathbf{v}_0 + t\mathbf{v}_1 + \cdots + t^i\mathbf{v}_i), \quad i = 1, 2, \ldots, m - k.$$

This is indeed the case, although a rigorous proof requires some deeper results from linear algebra (the Jordan canonical form of a matrix) than we have considered in this text. In the next theorem we state, without proof, the different possibilities that can arise for multiplicities of either two or three. These are the only possibilities for 2×2 or 3×3 matrices.

Theorem 8.6.1:   Consider the VDE $\mathbf{x}' = A\mathbf{x}$, where A is a constant matrix. Let m denote the multiplicity of the eigenvalue λ and let k denote the number of corresponding LI eigenvectors.

1 . $m = 2, k = 1$: There exist two LI solutions to $\mathbf{x}' = A\mathbf{x}$ of the form

$$\mathbf{x}_1(t) = e^{\lambda t}\mathbf{v}_0, \quad \mathbf{x}_2(t) = e^{\lambda t}(\mathbf{v}_1 + t\mathbf{v}_2).$$

2 . $m = 3, k = 1$: There exist three LI solutions to $\mathbf{x}' = A\mathbf{x}$ of the form[1]

$$\mathbf{x}_1(t) = e^{\lambda t}\mathbf{v}_0, \quad \mathbf{x}_2(t) = e^{\lambda t}(\mathbf{v}_1 + t\mathbf{v}_2), \quad \mathbf{x}_3(t) = e^{\lambda t}\left(\mathbf{v}_3 + t\mathbf{v}_4 + \frac{1}{2}t^2\mathbf{v}_5\right).$$

3 . $m = 3, k = 2$: There exist three LI solutions to $\mathbf{x}' = A\mathbf{x}$ of the form

$$\mathbf{x}_1(t) = e^{\lambda t}\mathbf{v}_0, \quad \mathbf{x}_2(t) = e^{\lambda t}\mathbf{v}_1, \quad \mathbf{x}_3(t) = e^{\lambda t}(\mathbf{v}_2 + t\mathbf{v}_3),$$

where $\mathbf{v}_0$ and $\mathbf{v}_1$ are LI eigenvectors of A corresponding to the eigenvalue λ.

The preceding theorem only tells us the *form* of the appropriate LI solutions. We can consider them as trial solutions. To obtain the solutions themselves, we must substitute the trial solutions into the given VDE and explicitly determine the appropriate $\mathbf{v}_k$. It is important to notice that there could be more than one LI solution corresponding to each consecutive trial solution. Also note that if λ is a complex eigenvalue, then the resulting solutions will themselves be complex-valued. In such a situation, we can obtain real-valued LI solutions in the same manner as in the previous section. We illustrate the use of the Theorem 8.6.1 with several examples.

Example 8.6.2   Solve the IVP $\mathbf{x}' = A\mathbf{x}$, $\mathbf{x}(0) = \begin{bmatrix} -1 \\ 1 \end{bmatrix}$, if $A = \begin{bmatrix} 6 & -8 \\ 2 & -2 \end{bmatrix}$.

Solution   We first determine the eigenvalues and eigenvectors of A. We have

$$\det(A - \lambda I) = \begin{vmatrix} 6-\lambda & -8 \\ 2 & -2-\lambda \end{vmatrix} = (\lambda - 2)^2, \tag{8.6.3}$$

so that there is only one eigenvalue $\lambda = 2$, with multiplicity 2. The eigenvectors corresponding to this eigenvalue are of the form

$$\mathbf{v}_0 = r(2, 1). \tag{8.6.4}$$

We therefore obtain only one LI solution to the given system, namely,

$$\mathbf{x}_1(t) = e^{2t}\begin{bmatrix} 2 \\ 1 \end{bmatrix}.$$

From Theorem 8.6.1, it follows that there is another solution to the system of the form

$$\mathbf{x}_2(t) = e^{2t}(\mathbf{v}_1 + t\mathbf{v}_2). \tag{8.6.5}$$

The vectors $\mathbf{v}_1$ and $\mathbf{v}_2$ must be chosen so that $\mathbf{x}_2$ is a solution to the given VDE, that is, so that $\mathbf{x}_2' = A\mathbf{x}_2$. Differentiating (8.6.5) with respect to t yields

$$\mathbf{x}_2' = e^{2t}[(2\mathbf{v}_1 + \mathbf{v}_2) + 2t\mathbf{v}_2].$$

---

[1] The reason for including the factor 1/2 in x_3 is that it simplifies the computation of v_5.

Consequently, $x_2' = Ax_2$, provided that v_1 and v_2 satisfy

$$e^{2t}[(2v_1 + v_2) + 2tv_2] = e^{2t}(Av_1 + tAv_2),$$

that is, they satisfy

$$(2v_1 + v_2) + 2tv_2 = Av_1 + tAv_2$$

which holds for all t if and only if

$$Av_1 = 2v_1 + v_2, \qquad (8.6.6)$$

$$Av_2 = 2v_2. \qquad (8.6.7)$$

This last equation implies that v_2 must be an eigenvector of A corresponding to the eigenvalue $\lambda = 2$ and hence, from (8.6.4), must be of the form

$$v_2 = r(2, 1). \qquad (8.6.8)$$

We now rewrite (8.6.6) in the equivalent form

$$(A - 2I)v_1 = v_2, \qquad (8.6.9)$$

which is a nonhomogeneous linear algebraic system for v_1. If we let $v_1 = (a, b)$ and substitute into (8.6.9) for $A - 2I$ and v_2, we obtain the following system for a and b:[1]

$$4a - 8b = 2r, \qquad (8.6.10)$$

$$2a - 4b = r. \qquad (8.6.11)$$

These equations are consistent for all values of r, and so, we can choose r to be any convenient *nonzero* value. We set $r = 2$, in which case the system of equations (8.6.10) and (8.6.11) reduces to the single equation

$$a - 2b = 1.$$

This has solution $a = 1 + 2s$, $b = s$, where s is a free variable. Since we require only one solution, we set $s = 0$, in which case $a = 1$, $b = 0$, so that

$$v_1 = (1, 0).$$

Substituting $r = 2$ into equation (8.6.8) yields

$$v_2 = (4, 2).$$

Equation (8.6.5) therefore implies that a second LI solution to the given VDE is

$$x_2(t) = e^{2t}\left\{ \begin{bmatrix} 1 \\ 0 \end{bmatrix} + t\begin{bmatrix} 4 \\ 2 \end{bmatrix} \right\} = e^{2t}\begin{bmatrix} 1 + 4t \\ 2t \end{bmatrix}.$$

Consequently, the general solution to the given VDE is

$$x(t) = c_1 e^{2t}\begin{bmatrix} 2 \\ 1 \end{bmatrix} + c_2 e^{2t}\begin{bmatrix} 1 + 4t \\ 2t \end{bmatrix} = \begin{bmatrix} e^{2t}[2c_1 + c_2(1 + 4t)] \\ e^{2t}(c_1 + 2c_2 t) \end{bmatrix}.$$

We now impose the given initial condition. Setting $t = 0$ in the general solution yields

---

[1] Note that the elements in the coefficient matrix $A - 2I$ can be obtained directly by setting $\lambda = 2$ in the determinant appearing in equation (8.6.3).

$$\mathbf{x}(0) = \begin{bmatrix} 2c_1 + c_2 \\ c_1 \end{bmatrix},$$

so that $\mathbf{x}(0) = \begin{bmatrix} -1 \\ 1 \end{bmatrix}$ if and only if

$$2c_1 + c_2 = -1,$$
$$c_1 \quad = 1.$$

Thus, $c_1 = 1$, $c_2 = -3$, and so, the solution to the given IVP is

$$\mathbf{x}(t) = e^{2t} \begin{bmatrix} -(1 + 12t) \\ 1 - 6t \end{bmatrix}.$$

⊓

In the previous example, the matrix A had an eigenvalue of multiplicity 2, but only a one-dimensional eigenspace. By rearranging equations (8.6.6) and (8.6.7) we see that the vectors $\mathbf{v}_1$ and $\mathbf{v}_2$ arising in the solution $\mathbf{x}_2(t) = e^{2t}(\mathbf{v}_1 + t\mathbf{v}_2)$ were determined from the two systems

$$(A - \lambda I)\mathbf{v}_2 = \mathbf{0},$$

$$(A - \lambda I)\mathbf{v}_1 = \mathbf{v}_2,$$

where $\lambda = 2$, the eigenvalue of A. The first of these equations tells us that $\mathbf{v}_2$ must be an eigenvector of A corresponding to the eigenvalue λ. Any vector satisfying the second equation is called a **generalized eigenvector** of A. We now show that these equations hold more generally.

Theorem 8.6.2:   If the VDE $\mathbf{x}' = A\mathbf{x}$ has a solution of the form

$$\mathbf{x}(t) = e^{\lambda t}(\mathbf{v}_1 + t\mathbf{v}_2),$$

then $\mathbf{v}_1$ and $\mathbf{v}_2$ must satisfy

$$(A - \lambda I)\mathbf{v}_2 = \mathbf{0},$$
$$(A - \lambda I)\mathbf{v}_1 = \mathbf{v}_2.$$

PROOF   If $\mathbf{x}(t) = e^{\lambda t}(\mathbf{v}_1 + t\mathbf{v}_2)$, then, differentiating with respect to t, we obtain

$$\mathbf{x}'(t) = e^{\lambda t}[(\lambda \mathbf{v}_1 + \mathbf{v}_2) + \lambda t\mathbf{v}_2].$$

Thus, $\mathbf{x}' = A\mathbf{x}$ if and only if

$$e^{\lambda t}[(\lambda \mathbf{v}_1 + \mathbf{v}_2) + \lambda t\mathbf{v}_2] = e^{\lambda t}(A\mathbf{v}_1 + tA\mathbf{v}_2),$$

that is, if and only if

$$(\lambda \mathbf{v}_1 + \mathbf{v}_2) + \lambda t\mathbf{v}_2 = A\mathbf{v}_1 + tA\mathbf{v}_2.$$

For this equation to hold for all values of t, we must have

$$A\mathbf{v}_1 = \lambda \mathbf{v}_1 + \mathbf{v}_2,$$
$$A\mathbf{v}_2 = \lambda \mathbf{v}_2.$$

Rearranging these equations yields

$$(A - \lambda I)\mathbf{v}_2 = \mathbf{0},$$
$$(A - \lambda I)\mathbf{v}_1 = \mathbf{v}_2,$$

as required. ∎

Example 8.6.3   Find the general solution to $\mathbf{x}' = A\mathbf{x}$ if $A = \begin{bmatrix} 6 & 3 & 6 \\ 1 & 4 & 2 \\ -2 & -2 & -1 \end{bmatrix}$.

Solution   In this case,

$$A - \lambda I = \begin{bmatrix} 6-\lambda & 3 & 6 \\ 1 & 4-\lambda & 2 \\ -2 & -2 & -1-\lambda \end{bmatrix}, \tag{8.6.12}$$

and a short computation yields the single eigenvalue $\lambda = 3$ (multiplicity 3). The associ–ated eigenvectors are of the form

$$\mathbf{v} = r(-1, 1, 0) + s(-2, 0, 1). \tag{8.6.13}$$

Thus, two LI solutions to the given VDE are

$$\mathbf{x}_1(t) = e^{3t}\begin{bmatrix} -1 \\ 1 \\ 0 \end{bmatrix}, \quad \mathbf{x}_2(t) = e^{3t}\begin{bmatrix} -2 \\ 0 \\ 1 \end{bmatrix}.$$

According to Theorem 8.6.1, there exists a third solution to the VDE of the form

$$\mathbf{x}_3(t) = e^{3t}(\mathbf{v}_1 + t\mathbf{v}_2). \tag{8.6.14}$$

Substituting from equation (8.6.14) into the given system, it follows that $\mathbf{x}_3$ is a solution ($\mathbf{x}_3' = A\mathbf{x}_3$) if and only if

$$e^{3t}(3\mathbf{v}_1 + \mathbf{v}_2 + 3t\mathbf{v}_2) = e^{3t}(A\mathbf{v}_1 + tA\mathbf{v}_2).$$

Equating coefficients of e^t and te^t on either side of the equation yields the two conditions

$$A\mathbf{v}_1 = 3\mathbf{v}_1 + \mathbf{v}_2,$$
$$A\mathbf{v}_2 = 3\mathbf{v}_2,$$

which we can write as

$$(A - 3I)\mathbf{v}_2 = \mathbf{0}, \tag{8.6.15}$$
$$(A - 3I)\mathbf{v}_1 = \mathbf{v}_2. \tag{8.6.16}$$

(These equations could have been written down directly as an application of Theorem 8.6.2.)   It follows from equation (8.6.15) that $\mathbf{v}_2$ must be an eigenvector of A corresponding to the eigenvalue $\lambda = 3$, and hence, from equation (8.6.13) is of the form

$$\mathbf{v}_2 = r(-1,1,0) + s(-2, 0, 1). \tag{8.6.17}$$

We now solve equation (8.6.16) for $\mathbf{v}_1$. If we let (a, b, c) denote the components of $\mathbf{v}_1$, then (8.6.16) can be written as [use equation (8.6.12) with $\lambda = 3$ to get the coefficient matrix]

$$3a + 3b + 6c = -r - 2s,$$
$$a + b + 2c = r, \qquad\qquad (8.6.18)$$
$$-2a - 2b - 4c = s.$$

This is a nonhomogeneous linear system, and so, we must check for consistency. The augmented matrix of the system is

$$\begin{bmatrix} 3 & 3 & 6 & -r - 2s \\ 1 & 1 & 2 & r \\ -2 & -2 & -4 & s \end{bmatrix}$$

with RREF

$$\begin{bmatrix} 1 & 1 & 2 & r \\ 0 & 0 & 0 & -2(2r + s) \\ 0 & 0 & 0 & 2r + s \end{bmatrix}.$$

It follows that the system (8.6.18) is consistent if and only if $2r + s = 0$. For convenience, we take $r = 1$ and $s = -2$. It then follows from equation (8.6.17) that

$$\mathbf{v}_2 = (-1, 1, 0) - 2(-2, 0, 1) = (3, 1, -2). \qquad\qquad (8.6.19)$$

Setting $r = 1$ and $s = -2$ in (8.6.18) yields the single equation

$$a + b + 2c = 1.$$

Although we can solve this equation in general, we only need one solution to determine $\mathbf{v}_1$. A simple choice is $a = 1$ and $b = c = 0$. Then

$$\mathbf{v}_1 = (1, 0, 0). \qquad\qquad (8.6.20)$$

Substituting from equations (8.6.19) and (8.6.20) into equation (8.6.14) yields the third LI solution

$$\mathbf{x}_3(t) = e^{3t} \left\{ \begin{bmatrix} 1 \\ 0 \\ 0 \end{bmatrix} + t \begin{bmatrix} 3 \\ 1 \\ -2 \end{bmatrix} \right\} = e^{3t} \begin{bmatrix} 1 + 3t \\ t \\ -2t \end{bmatrix}.$$

Consequently, the general solution to the given VDE is

$$\mathbf{x}(t) = c_1 e^{3t} \begin{bmatrix} -1 \\ 1 \\ 0 \end{bmatrix} + c_2 e^{3t} \begin{bmatrix} -2 \\ 0 \\ 1 \end{bmatrix} + c_3 e^{3t} \begin{bmatrix} 1 + 3t \\ t \\ -2t \end{bmatrix},$$

or, equivalently,

$$\mathbf{x}(t) = \begin{bmatrix} -e^{3t}[c_1 + 2c_2 - c_3(1 + 3t)] \\ e^{3t}(c_1 + c_3 t) \\ e^{3t}(c_2 - 2c_3 t) \end{bmatrix}. \qquad \square$$

We give one final example to illustrate the case when the eigenvectors are complex.

Example 8.6.4   Find the general solution to $\mathbf{x}' = A\mathbf{x}$ if

$$A = \begin{bmatrix} 0 & 2 & 1 & 0 \\ -2 & 0 & 0 & 1 \\ 0 & 0 & 0 & 2 \\ 0 & 0 & -2 & 0 \end{bmatrix}.$$

Solution A has characteristic polynomial

$$\det(A - \lambda I) = \begin{vmatrix} -\lambda & 2 & 1 & 0 \\ -2 & -\lambda & 0 & 1 \\ 0 & 0 & -\lambda & 2 \\ 0 & 0 & -2 & -\lambda \end{vmatrix} = (\lambda^2 + 4)^2 \qquad (8.6.21)$$

so that we obtain the complex conjugate eigenvalues $\lambda = \pm 2i$, each of multiplicity 2. If we consider the eigenvalue $\lambda = 2i$, then a short computation yields the complex eigenvectors

$$\mathbf{v} = r(-i, 1, 0, 0).$$

This gives the complex-valued solution

$$\mathbf{w}_1(t) = e^{2it} \begin{bmatrix} -i \\ 1 \\ 0 \\ 0 \end{bmatrix} = (\cos 2t + i \sin 2t) \begin{bmatrix} -i \\ 1 \\ 0 \\ 0 \end{bmatrix}.$$

Taking the real and imaginary parts of $\mathbf{w}_1$ yields the following two LI real-valued solutions

$$\mathbf{x}_1(t) = \begin{bmatrix} \sin 2t \\ \cos 2t \\ 0 \\ 0 \end{bmatrix}, \quad \mathbf{x}_2(t) = \begin{bmatrix} -\cos 2t \\ \sin 2t \\ 0 \\ 0 \end{bmatrix}.$$

We therefore require two more LI real-valued solutions or, equivalently, one more LI complex-valued solution. According to Theorem 8.6.1, there is another complex-valued solution of the form

$$\mathbf{w}_2(t) = e^{2it}(\mathbf{v}_1 + t\mathbf{v}_2). \qquad (8.6.22)$$

Further, from Theorem 8.6.2, $\mathbf{v}_1$ and $\mathbf{v}_2$ satisfy

$$(A - 2iI)\mathbf{v}_2 = \mathbf{0}, \qquad (8.6.23)$$

$$(A - 2iI)\mathbf{v}_1 = \mathbf{v}_2. \qquad (8.6.24)$$

It follows from equation (8.6.23) that $\mathbf{v}_2$ must be an eigenvector of A corresponding to the eigenvalue $\lambda = 2i$ and hence, from equation (8.6.21), is of the form

$$\mathbf{v}_2 = r(-i, 1, 0, 0). \qquad (8.6.25)$$

To determine $\mathbf{v}_1$, we must solve the linear system (8.6.24). The RREF of the augmented matrix of this system is

$$\begin{bmatrix} 1 & i & 0 & 0 & 0 \\ 0 & 0 & 1 & 0 & -ir \\ 0 & 0 & 0 & 1 & r \\ 0 & 0 & 0 & 0 & 0 \end{bmatrix},$$

so that equation (8.6.24) is consistent for all values of r. We choose $r = 1$, in which case

$$\mathbf{v}_1 = (-is,\ s,\ -i,\ 1),$$

where s is a free variable. Since we only require one solution and there are no further consistency requirements to be satisfied, we choose, for simplicity, $s = 0$. It then follows that

$$\mathbf{v}_1 = (0,\ 0,\ -i,\ 1),$$

and, from equation (8.6.25) with $r = 1$,

$$\mathbf{v}_2 = (-i,\ 1,\ 0,\ 0).$$

Substituting into equation (8.6.22) yields the complex-valued solution

$$\mathbf{w}_2(t) = e^{2it}\left\{ \begin{bmatrix} 0 \\ 0 \\ -i \\ 1 \end{bmatrix} + t\begin{bmatrix} -i \\ 1 \\ 0 \\ 0 \end{bmatrix} \right\},$$

which can be written as

$$\mathbf{w}_2(t) = (\cos 2t + i\sin 2t)\begin{bmatrix} -it \\ t \\ -i \\ 1 \end{bmatrix}.$$

Taking the real and imaginary parts of this complex-valued solution, we obtain the two LI real-valued solutions

$$\mathbf{x}_3(t) = \begin{bmatrix} t\sin 2t \\ t\cos 2t \\ \sin 2t \\ \cos 2t \end{bmatrix}, \quad \mathbf{x}_4(t) = \begin{bmatrix} -t\cos 2t \\ t\sin 2t \\ -\cos 2t \\ \sin 2t \end{bmatrix}.$$

Consequently the general solution to the given VDE is

$$\mathbf{x}(t) = c_1\begin{bmatrix} \sin 2t \\ \cos 2t \\ 0 \\ 0 \end{bmatrix} + c_2\begin{bmatrix} -\cos 2t \\ \sin 2t \\ 0 \\ 0 \end{bmatrix} + c_3\begin{bmatrix} t\sin 2t \\ t\cos 2t \\ \sin 2t \\ \cos 2t \end{bmatrix} + c_4\begin{bmatrix} -t\cos 2t \\ t\sin 2t \\ -\cos 2t \\ \sin 2t \end{bmatrix},$$

which can be written as

$$\mathbf{x}(t) = \begin{bmatrix} c_1\sin 2t - c_2\cos 2t + c_3 t\sin 2t - c_4 t\cos 2t \\ c_1\cos 2t + c_2\sin 2t + c_3 t\cos 2t + c_4 t\sin 2t \\ c_3\sin 2t - c_4\cos 2t \\ c_3\cos 2t + c_4\sin 2t \end{bmatrix}.$$

□

Finally, we state the generalization of the results of this section to higher order multiplicities.

Theorem 8.6.3:   Let λ be an eigenvalue of the $n \times n$ matrix A of multiplicity m, and suppose that the dimension of the corresponding eigenspace is k. Then there exist m LI solutions to the VDE $\mathbf{x}' = A\mathbf{x}$ of the form

$$\mathbf{x}_{i+1}(t) = e^{\lambda t}\left[\mathbf{v}_{i(i+1)/2} + t\mathbf{v}_{i(i+1)/2+1} + \cdots + \frac{1}{i!}t^i\mathbf{v}_{i(i+3)/2}\right], \quad i = 0, 1, \ldots, m-k.$$

Further, the set of all solutions to $\mathbf{x}' = A\mathbf{x}$ corresponding to different eigenvalues obtained in this manner is LI on any interval.

PROOF   See, for example, M.W. Hirsch and S. Smale, *Differential Equations, Dynamical Systems, and Linear Algebra*, Academic Press, 1974.   ■

Theorem 8.6.4:   If the VDE $\mathbf{x}' = A\mathbf{x}$ has a solution of the form

$$\mathbf{x}(t) = e^{\lambda t}\left(\mathbf{v}_0 + t\mathbf{v}_1 + \frac{1}{2!}t^2\mathbf{v}_2 + \cdots + \frac{1}{n!}t^n\mathbf{v}_n\right),$$

then $\mathbf{v}_0, \mathbf{v}_1, \ldots, \mathbf{v}_n$ must satisfy

$$(A - \lambda I)\mathbf{v}_n = \mathbf{0}$$
$$(A - \lambda I)\mathbf{v}_{n-1} = \mathbf{v}_n$$
$$(A - \lambda I)\mathbf{v}_{n-2} = \mathbf{v}_{n-1}$$
$$\vdots$$
$$(A - \lambda I)\mathbf{v}_0 = \mathbf{v}_1.$$

PROOF   We leave the proof of this theorem as an exercise.   ■

EXERCISES 8.6

For problems 1–13, determine the general solution to the system $\mathbf{x}' = A\mathbf{x}$ for the given matrix A.

1. $\begin{bmatrix} 0 & -2 \\ 2 & 4 \end{bmatrix}$.

2. $\begin{bmatrix} -3 & -2 \\ 2 & 1 \end{bmatrix}$.

3. $\begin{bmatrix} 0 & 1 & 0 \\ 0 & 0 & 1 \\ 1 & 1 & -1 \end{bmatrix}$.

4. $\begin{bmatrix} 2 & 2 & -1 \\ 2 & 1 & -1 \\ 2 & 3 & -1 \end{bmatrix}$.

5. $\begin{bmatrix} -2 & 0 & 0 \\ 1 & -3 & -1 \\ -1 & 1 & -1 \end{bmatrix}$.

6. $\begin{bmatrix} 15 & -32 & 12 \\ 8 & -17 & 6 \\ 0 & 0 & -1 \end{bmatrix}$.

7. $\begin{bmatrix} 4 & 0 & 0 \\ 1 & 4 & 0 \\ 0 & 1 & 4 \end{bmatrix}$.

8. $\begin{bmatrix} 1 & 0 & 0 \\ 0 & 3 & 2 \\ 2 & -2 & -1 \end{bmatrix}$.

9. $\begin{bmatrix} 3 & 1 & 0 \\ -1 & 5 & 0 \\ 0 & 0 & 4 \end{bmatrix}$.

10. $\begin{bmatrix} -1 & 1 & 0 \\ -2 & -3 & 1 \\ 1 & 1 & -2 \end{bmatrix}$.

11. $\begin{bmatrix} 0 & -1 & 0 & 0 \\ 1 & 0 & 0 & 0 \\ 1 & 0 & 2 & 1 \\ 0 & 1 & 0 & 2 \end{bmatrix}$.

12. $\begin{bmatrix} -2 & 3 & 0 & 0 \\ 3 & -2 & 0 & 0 \\ 1 & 0 & 1 & -1 \\ 0 & 1 & 0 & 1 \end{bmatrix}$.

13. $\begin{bmatrix} 0 & -1 & 0 & 0 \\ 1 & 0 & 0 & 0 \\ 1 & 0 & 0 & -1 \\ 0 & 1 & 1 & 0 \end{bmatrix}$.

For problems 14 and 15, solve the IVP.

14. $\mathbf{x}' = A\mathbf{x}$, $\mathbf{x}(0) = \mathbf{x}_0$, where $A = \begin{bmatrix} -2 & -1 \\ 1 & -4 \end{bmatrix}$,

$\mathbf{x}_0 = \begin{bmatrix} 0 \\ -1 \end{bmatrix}$.

15. $\mathbf{x}' = A\mathbf{x}$, $\mathbf{x}(0) = \mathbf{x}_0$, where

$$A = \begin{bmatrix} -2 & -1 & 4 \\ 0 & -1 & 0 \\ -1 & -3 & 2 \end{bmatrix}, \ \mathbf{x}_0 = \begin{bmatrix} -2 \\ 1 \\ 1 \end{bmatrix}.$$

16. (a) Show that if the differential system $\mathbf{x}' = A\mathbf{x}$ has a solution of the form

$$\mathbf{x}(t) = e^{\lambda t}[\mathbf{v}_0 + t\mathbf{v}_1 + \frac{1}{2!}t^2\mathbf{v}_2],$$

then the constant vectors $\mathbf{v}_0$, $\mathbf{v}_1$, and $\mathbf{v}_2$ must satisfy:

$(A - \lambda I)\mathbf{v}_2 = \mathbf{0}$, $(A - \lambda I)\mathbf{v}_1 = \mathbf{v}_2$, $(A - \lambda I)\mathbf{v}_0 = \mathbf{v}_1$.

(b) Prove Theorem 8.6.4.

17. Let A be a 2×2 real matrix. Prove that all solutions to $\mathbf{x}' = A\mathbf{x}$ satisfy

$$\lim_{t \to +\infty} \mathbf{x}(t) = 0$$

if and only if all eigenvalues of A have negative real part.

18. Extend the result of the previous exercise to the system $\mathbf{x}' = A\mathbf{x}$, where A is an arbitrary (real) $n \times n$ matrix.

8.7 VARIATION-OF-PARAMETERS FOR LINEAR SYSTEMS

We now consider solving the nonhomogeneous VDE

$$\mathbf{x}'(t) = A(t)\mathbf{x}(t) + \mathbf{b}(t), \tag{8.7.1}$$

where A is an $n \times n$ matrix function and $\mathbf{b}$ is a column n-vector function. The homogeneous equation associated with equation (8.7.1) is

$$\mathbf{x}'(t) = A(t)\mathbf{x}(t). \tag{8.7.2}$$

According to Theorem 8.4.4, every solution to the system (8.7.1) is of the form

$$\mathbf{x}(t) = c_1\mathbf{x}_1(t) + c_2\mathbf{x}_2(t) + \cdots + c_n\mathbf{x}_n(t) + \mathbf{x}_p(t),$$

where $\mathbf{x}_1, \mathbf{x}_2, \ldots, \mathbf{x}_n$ are n LI solutions to the associated homogeneous system (8.7.2), and $\mathbf{x}_p$ is a *particular* solution to (8.7.1). In this section, we derive the variation-of-parameters method for determining $\mathbf{x}_p$, assuming that we know n LI solutions to (8.7.2).

    Theorem 8.7.1 (Variation-of-Parameters Method):   Let $A(t)$ be an $n \times n$ matrix function and let $\mathbf{b}(t)$ be a column n-vector function both of which are continuous on an interval I. If $\{\mathbf{x}_1, \mathbf{x}_2, \ldots, \mathbf{x}_n\}$ is any LI set of solutions to $\mathbf{x}'(t) = A(t)\mathbf{x}(t)$ and $X(t) = [\mathbf{x}_1(t), \mathbf{x}_2(t), \ldots, \mathbf{x}_n(t)]$, then a particular solution to

$$\mathbf{x}'(t) = A(t)\mathbf{x}(t) + \mathbf{b}(t) \tag{8.7.3}$$

is

$$\mathbf{x}_p(t) = X(t)\mathbf{u}(t),$$

where $\mathbf{u}(t)$ satisfies

$$X(t)\mathbf{u}'(t) = \mathbf{b}(t).$$

Explicitly,

$$\mathbf{x}_p(t) = X(t) \int^t X^{-1}(s)\,\mathbf{b}(s)\,ds.$$

PROOF   The general solution to $\mathbf{x}'(t) = A(t)\mathbf{x}(t)$ is

$$\mathbf{x}_c(t) = c_1\mathbf{x}_1(t) + c_2\mathbf{x}_2(t) + \cdots + c_n\mathbf{x}_n(t),$$

which can be written in the form

$$\mathbf{x}_c(t) = X(t)\mathbf{c},$$

where $X(t) = [\mathbf{x}_1(t), \mathbf{x}_2(t), \ldots, \mathbf{x}_n(t)]$ and $\mathbf{c} = [c_1 \ c_2 \ \ldots \ c_n]^T$. We try for a particular solution to equation (8.7.3) of the form[1]

$$\mathbf{x}_p(t) = X(t)\mathbf{u}(t) \tag{8.7.4}$$

where $\mathbf{u}(t) = [u_1(t) \ u_2(t) \ \ldots \ u_n(t)]^T$. Substituting (8.7.4) into (8.7.3), it follows that $\mathbf{x}_p$ is a solution to (8.7.3) provided that $\mathbf{u}$ satisfies

$$(X\mathbf{u})' = A(X\mathbf{u}) + \mathbf{b}. \tag{8.7.5}$$

Applying the product rule for differentiation to the left-hand side of this equation, we obtain

$$X'\mathbf{u} + X\mathbf{u}' = A(X\mathbf{u}) + \mathbf{b}. \tag{8.7.6}$$

By definition we have

$$X = [\mathbf{x}_1, \mathbf{x}_2, \ldots, \mathbf{x}_n],$$

so that

$$X' = [\mathbf{x}_1', \mathbf{x}_2', \ldots, \mathbf{x}_n']. \tag{8.7.7}$$

Since each of the vector functions $\mathbf{x} = \mathbf{x}_i$ is a solution to the associated homogeneous equation $\mathbf{x}' = A\mathbf{x}$, we can write (8.7.7) in the form

$$X' = [A\mathbf{x}_1, A\mathbf{x}_2, \ldots, A\mathbf{x}_n].$$

That is,

$$X' = AX.$$

Substituting this expression for X' into (8.7.6) yields

$$(AX)\mathbf{u} + X\mathbf{u}' = A(X\mathbf{u}) + \mathbf{b},$$

---

[1] That is we replace the constants in $\mathbf{x}_c$ by arbitrary functions.

so that

$$Xu' = b.$$

This implies that[1]

$$u' = X^{-1}b.$$

Consequently,

$$u(t) = \int^t X^{-1}(s)\,b(s)\ ds$$

(we have set the integration constant to zero without loss of generality) and hence, from (8.7.4), a particular solution to (8.7.3) is

$$x_p(t) = X(t)\int^t X^{-1}(s)\,b(s)\ ds.$$ ■

REMARKS

1. You should not memorize the formula for x_p given in the previous theorem. Rather, you should remember that a particular solution to $x' = Ax + b$ is

$$x_p(t) = Xu,$$

where X is a fundamental matrix for the associated homogeneous VDE and u' is determined by solving the linear system

$$Xu' = b. \qquad (8.7.8)$$

2. Whereas in the proof of the previous theorem we used X^{-1} to obtain a simple formula for the solution to (8.7.8), in practice any of the methods for solving systems of linear algebraic equations that we have derived in the text can be applied. For 2×2 systems, Cramer's rule is quite effective. Alternatively, the inverse of X can be determined very quickly using the adjoint method. For systems bigger than 2×2, it is computationally more efficient to use Gaussian elimination to solve (8.7.8) for u' and then integrate the resulting vector to determine u.

Example 8.7.1   Solve the IVP $x' = Ax + b$, $x(0) = \begin{bmatrix} 3 \\ 0 \end{bmatrix}$ if $A = \begin{bmatrix} 1 & 2 \\ 4 & 3 \end{bmatrix}$ and $b(t) = \begin{bmatrix} 12\,e^{3t} \\ 18\,e^{2t} \end{bmatrix}$.

Solution   We first solve the associated homogeneous equation $x' = Ax$. For the given matrix A, we find that

$$\det(A - \lambda I) = (\lambda - 5)(\lambda + 1),$$

so that the eigenvalues of A are $\lambda = -1, 5$.

Eigenvectors

$\lambda = -1$ implies that $v_1 = r(-1, 1)$

$\lambda = 5$ implies that $v_2 = s(1, 2)$.

---

[1]Note that X^{-1} exists, since $\det(X) \neq 0$. (Why?)

Consequently, two LI solutions to $\mathbf{x}' = A\mathbf{x}$ are

$$\mathbf{x}_1(t) = e^{-t}\begin{bmatrix} -1 \\ 1 \end{bmatrix}, \quad \mathbf{x}_2(t) = e^{5t}\begin{bmatrix} 1 \\ 2 \end{bmatrix},$$

and, therefore, a fundamental matrix for $\mathbf{x}' = A\mathbf{x}$ is

$$X(t) = \begin{bmatrix} -e^{-t} & e^{5t} \\ e^{-t} & 2e^{5t} \end{bmatrix}.$$

It follows from Theorem 8.7.1 that a particular solution to $\mathbf{x}' = A\mathbf{x} + \mathbf{b}$ is

$$\mathbf{x}_p = X\mathbf{u},$$

where

$$X\mathbf{u}' = \mathbf{b}.$$

Since this is a 2×2 system, we will solve for the components of $\mathbf{u}'$ using Cramer's rule. We have

$$\det(X(t)) = -3e^{4t},$$

so that

$$u_1'(t) = \frac{\begin{vmatrix} 12e^{3t} & e^{5t} \\ 18e^{2t} & 2e^{5t} \end{vmatrix}}{-3e^{4t}} = -8e^{4t} + 6e^{3t}$$

and

$$u_2'(t) = \frac{\begin{vmatrix} -e^{-t} & 12e^{3t} \\ e^{-t} & 18e^{2t} \end{vmatrix}}{-3e^{4t}} = 6e^{-3t} + 4e^{-2t}.$$

Integrating these two expressions yields

$$u_1(t) = -2e^{4t} + 2e^{3t}, \quad u_2(t) = -2e^{-3t} - 2e^{-2t},$$

where we have set the integration constants to zero. Hence,

$$\mathbf{u}(t) = \begin{bmatrix} -2e^{4t} + 2e^{3t} \\ -2e^{-2t} - 2e^{-3t} \end{bmatrix}.$$

It follows that a particular solution to the given VDE is

$$\mathbf{x}_p(t) = X(t)\mathbf{u}(t) = \begin{bmatrix} -e^{-t} & e^{5t} \\ e^{-t} & 2e^{5t} \end{bmatrix}\begin{bmatrix} -2e^{4t} + 2e^{3t} \\ -2e^{-2t} - 2e^{-3t} \end{bmatrix} = \begin{bmatrix} -4e^{2t} \\ -2e^{2t} - 6e^{3t} \end{bmatrix}.$$

Consequently, the general solution to the given nonhomogeneous VDE is

$$\mathbf{x}(t) = c_1 e^{-t}\begin{bmatrix} -1 \\ 1 \end{bmatrix} + c_2 e^{5t}\begin{bmatrix} 1 \\ 2 \end{bmatrix} - \begin{bmatrix} 4e^{2t} \\ 2e^{2t} + 6e^{3t} \end{bmatrix},$$

or, equivalently,

$$\mathbf{x}(t) = \begin{bmatrix} -c_1 e^{-t} + c_2 e^{5t} - 4e^{2t} \\ c_1 e^{-t} + 2c_2 e^{5t} - 2(e^{2t} + 3 e^{3t}) \end{bmatrix}. \tag{8.7.9}$$

The initial condition $\mathbf{x}(0) = \begin{bmatrix} 3 \\ 0 \end{bmatrix}$ requires that

$$c_1 \begin{bmatrix} -1 \\ 1 \end{bmatrix} + c_2 \begin{bmatrix} 1 \\ 2 \end{bmatrix} - \begin{bmatrix} 4 \\ 8 \end{bmatrix} = \begin{bmatrix} 3 \\ 0 \end{bmatrix},$$

that is,

$$-c_1 + c_2 = 7,$$
$$c_1 + 2c_2 = 8.$$

Thus, $c_1 = -2$ and $c_2 = 5$. Substituting these values into (8.7.9) yields

$$\mathbf{x}(t) = \begin{bmatrix} 2e^{-t} + 5e^{5t} - 4e^{2t} \\ -2e^{-t} + 10e^{5t} - 2(e^{2t} + 3e^{3t}) \end{bmatrix}.$$

EXERCISES 8.7

For problems 1–8, use the variation-of-parameters technique to find a particular solution $\mathbf{x}_p$ to $\mathbf{x}' = A\mathbf{x} + \mathbf{b}$, for the given A and $\mathbf{b}$. Also obtain the general solution to the system of differential equations.

1. $A = \begin{bmatrix} 2 & -1 \\ -1 & 2 \end{bmatrix}$, $\mathbf{b} = \begin{bmatrix} 0 \\ 4e^t \end{bmatrix}$.

2. $A = \begin{bmatrix} 4 & -3 \\ 2 & -1 \end{bmatrix}$, $\mathbf{b} = \begin{bmatrix} e^{2t} \\ e^t \end{bmatrix}$.

3. $A = \begin{bmatrix} -1 & 1 \\ 3 & 1 \end{bmatrix}$, $\mathbf{b} = \begin{bmatrix} 20e^{3t} \\ 12e^t \end{bmatrix}$.

4. $A = \begin{bmatrix} -1 & 2 \\ -2 & 4 \end{bmatrix}$, $\mathbf{b} = \begin{bmatrix} 54te^{3t} \\ 9e^{3t} \end{bmatrix}$.

5. $A = \begin{bmatrix} 2 & 4 \\ -2 & -2 \end{bmatrix}$, $\mathbf{b} = \begin{bmatrix} 8\sin 2t \\ 8\cos 2t \end{bmatrix}$.

6. $A = \begin{bmatrix} 3 & 2 \\ -2 & -1 \end{bmatrix}$, $\mathbf{b} = \begin{bmatrix} -3e^t \\ 6te^t \end{bmatrix}$.

7. $A = \begin{bmatrix} 1 & 0 & 0 \\ 2 & -3 & 2 \\ 1 & -2 & 2 \end{bmatrix}$, $\mathbf{b} = \begin{bmatrix} -e^t \\ 6e^{-t} \\ e^t \end{bmatrix}$.

8. $A = \begin{bmatrix} -1 & -2 & 2 \\ 2 & 4 & -1 \\ 0 & 0 & 3 \end{bmatrix}$, $\mathbf{b} = \begin{bmatrix} -e^{3t} \\ 4e^{3t} \\ 3e^{3t} \end{bmatrix}$.

9. Let $X(t)$ be a fundamental matrix for the differential system $\mathbf{x}' = A(t)\mathbf{x}$, where $A(t)$ is an $n \times n$ matrix function. Show that the solution to the IVP

$$\mathbf{x}' = A(t)\mathbf{x} + \mathbf{b}(t), \quad \mathbf{x}(t_0) = \mathbf{x}_0$$

can be written as

$$\mathbf{x}(t) = X(t) X^{-1}(t_0)\mathbf{x}_0 + X(t) \int_{t_0}^{t} X^{-1}(s) \mathbf{b}(s) \, ds.$$

10. Consider the nonhomogeneous system

$$x_1' = 2x_1 - 3x_2 + 34 \sin t,$$
$$x_2' = 4x_1 - 2x_2 + 17 \cos t.$$

Find the general solution to this system by first solving the associated homogeneous system, and then using the *method of undetermined coefficients* to obtain a particular solution. (Hint:

The form of the nonhomogeneous term suggests a trial solution of the form:

$$\mathbf{x}_p = \left[\begin{array}{c} A_1 \cos t + B_1 \sin t \\ A_2 \cos t + B_2 \sin t \end{array} \right],$$

where the constants A_1, A_2, B_1, and B_2 can be determined by substituting into the given system.)

8.8 SOME APPLICATIONS OF LINEAR SYSTEMS OF DIFFERENTIAL EQUATIONS

In this section, we analyze the two problems that were briefly introduced in Section 8.1. We begin with the coupled spring-mass system that consists of two masses m_1 and m_2 connected by two springs whose spring constants are k_1 and k_2, respectively. (See Figure 8.8.1.) Let $x(t)$ and $y(t)$ denote the displacement of m_1 and m_2 from their equilibrium positions. When the system is in motion, the extension of spring 1 is

$$L_1(t) = x(t),$$

whereas the *net* extension of spring 2 is

$$L_2(t) = y(t) - x(t).$$

Consequently, using Hooke's law, the net forces acting on masses m_1 and m_2 at time t are

$$F_1(t) = -k_1 x(t) + k_2[y(t) - x(t)], \quad F_2(t) = -k_2[y(t) - x(t)],$$

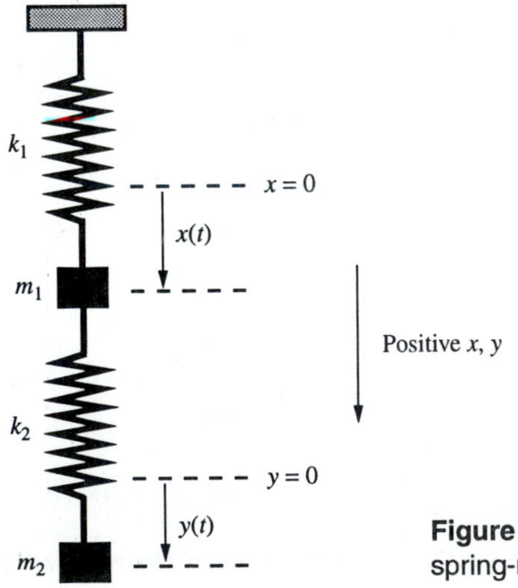

Figure 8.8.1 A coupled spring-mass system.

respectively. Thus, applying Newton's second law to each mass yields the system of DE

$$m_1 \frac{d^2x}{dt^2} = -k_1 x + k_2(y - x), \tag{8.8.1}$$

$$m_2 \frac{d^2y}{dt^2} = -k_2(y - x). \tag{8.8.2}$$

The motion of the spring-mass system will be fully determined once we have specified appropriate initial conditions of the form

$$x(t_0) = \alpha_1, \frac{dx}{dt}(t_0) = \alpha_2, \, y(t_0) = \alpha_3, \frac{dy}{dt}(t_0) = \alpha_4, \tag{8.8.3}$$

where α_1, α_2, α_3, and α_4 are constants.

To apply the techniques that we have developed in this chapter for solving systems of DE, we must convert equations (8.8.1) and (8.8.2) into a first-order system. We introduce new variables x_1, x_2, x_3, and x_4 defined by

$$x_1 = x, \qquad x_2 = x', \qquad x_3 = y, \qquad x_4 = y'. \tag{8.8.4}$$

Then equations (8.8.1) and (8.8.2) can be replaced by the equivalent system

$$x_1' = x_2, \qquad x_2' = -\frac{k_1}{m_1} x_1 + \frac{k_2}{m_1}(x_3 - x_1), \qquad x_3' = x_4, \qquad x_4' = -\frac{k_2}{m_2}(x_3 - x_1).$$

Rearranging terms yields the first-order linear system

$$x_1' = x_2, \tag{8.8.5}$$

$$x_2' = -\left(\frac{k_1}{m_1} + \frac{k_2}{m_1}\right)x_1 + \frac{k_2}{m_1}x_3, \tag{8.8.6}$$

$$x_3' = x_4, \tag{8.8.7}$$

$$x_4' = \frac{k_2}{m_2}x_1 - \frac{k_2}{m_2}x_3. \tag{8.8.8}$$

In the new variables, the initial conditions (8.8.3) are

$$x_1(t_0) = \alpha_1, \quad x_2(t_0) = \alpha_2, \quad x_3(t_0) = \alpha_3, \quad x_4(t_0) = \alpha_4. \tag{8.8.9}$$

This IVP for x_1, x_2, x_3, x_4 can be written in vector form as

$$\mathbf{x}' = A\mathbf{x}, \quad \mathbf{x}(t_0) = \mathbf{x}_0,$$

where

$$\mathbf{x} = \begin{bmatrix} x_1 \\ x_2 \\ x_3 \\ x_4 \end{bmatrix}, \, A = \begin{bmatrix} 0 & 1 & 0 & 0 \\ -\dfrac{1}{m_1}(k_1 + k_2) & 0 & \dfrac{k_2}{m_1} & 0 \\ 0 & 0 & 0 & 1 \\ \dfrac{k_2}{m_2} & 0 & -\dfrac{k_2}{m_2} & 0 \end{bmatrix}, \, \mathbf{x}_0 = \begin{bmatrix} \alpha_1 \\ \alpha_2 \\ \alpha_3 \\ \alpha_4 \end{bmatrix}.$$

We leave the analysis of the general system for the exercises and consider a particular example.

Example 8.8.1   Consider the spring-mass system with

$$k_1 = 4 \text{ Nm}^{-1}, \quad k_2 = 2 \text{ Nm}^{-1}, \quad m_1 = 2 \text{ kg}, \quad m_2 = 1 \text{ kg}.$$

At $t = 0$, both masses are pulled down a distance 1 m from equilibrium and released from rest. Determine the subsequent motion of the system.

Solution   The motion of the system is governed by the IVP

$$2\frac{d^2x}{dt^2} = -4x + 2(y - x), \tag{8.8.10}$$

$$\frac{d^2y}{dt^2} = -2(y - x), \tag{8.8.11}$$

$$x(0) = 1, \quad \frac{dx}{dt}(0) = 0, \quad y(0) = 1, \quad \frac{dy}{dt}(0) = 0.$$

Introducing new variables $x_1 = x$, $x_2 = x'$, $x_3 = y$, $x_4 = y'$ yields the equivalent IVP

$$x_1' = x_2,$$
$$x_2' = -3x_1 + x_3,$$
$$x_3' = x_4,$$
$$x_4' = 2x_1 - 2x_3,$$
$$x_1(0) = 1, \quad x_2(0) = 0, \quad x_3(0) = 1, \quad x_4(0) = 0.$$

In vector form, we have

$$\mathbf{x}' = A\mathbf{x}, \quad \mathbf{x}(0) = \mathbf{x}_0, \tag{8.8.12}$$

where

$$A = \begin{bmatrix} 0 & 1 & 0 & 0 \\ -3 & 0 & 1 & 0 \\ 0 & 0 & 0 & 1 \\ 2 & 0 & -2 & 0 \end{bmatrix} \text{ and } \mathbf{x}_0 = \begin{bmatrix} 1 \\ 0 \\ 1 \\ 0 \end{bmatrix}.$$

The characteristic polynomial of A is[1]

$$\det(A - \lambda I) = \begin{vmatrix} -\lambda & 1 & 0 & 0 \\ -3 & -\lambda & 1 & 0 \\ 0 & 0 & -\lambda & 1 \\ 2 & 0 & -2 & -\lambda \end{vmatrix}$$

$$= \lambda^4 + 5\lambda^2 + 4$$
$$= (\lambda^2 + 1)(\lambda^2 + 4).$$

Thus the eigenvalues of A are

$$\lambda = \pm i, \pm 2i.$$

We now determine the eigenvectors.

$\lambda = i$:   The system $(A - \lambda I)\mathbf{v}_1 = \mathbf{0}$ has augmented matrix

---

[1] In problem 1 the reader is asked to fill in the missing details of this computation.

$$\begin{bmatrix} -i & 1 & 0 & 0 & 0 \\ -3 & -i & 1 & 0 & 0 \\ 0 & 0 & -i & 1 & 0 \\ 2 & 0 & -2 & -i & 0 \end{bmatrix}$$

with RREF

$$\begin{bmatrix} 1 & 0 & 0 & i/2 & 0 \\ 0 & 1 & 0 & -1/2 & 0 \\ 0 & 0 & 1 & i & 0 \\ 0 & 0 & 0 & 0 & 0 \end{bmatrix}.$$

Consequently, the eigenvectors are

$$\mathbf{v}_1 = r(-i,\ 1,\ -2i,\ 2),$$

so that a complex-valued solution to $\mathbf{x}' = A\mathbf{x}$ is

$$\mathbf{u}_1(t) = e^{it} \begin{bmatrix} -i \\ 1 \\ -2i \\ 2 \end{bmatrix} = (\cos t + i \sin t) \begin{bmatrix} -i \\ 1 \\ -2i \\ 2 \end{bmatrix}.$$

Taking the real and imaginary parts of this complex-valued solution yields the two LI real-valued solutions

$$\mathbf{x}_1(t) = \begin{bmatrix} \sin t \\ \cos t \\ 2 \sin t \\ 2 \cos t \end{bmatrix}, \qquad \mathbf{x}_2(t) = \begin{bmatrix} -\cos t \\ \sin t \\ -2 \cos t \\ 2 \sin t \end{bmatrix}.$$

$\lambda = 2i$:   In this case, the augmented matrix of the system $(A - \lambda I)\mathbf{v}_2 = \mathbf{0}$ is

$$\begin{bmatrix} -2i & 1 & 0 & 0 & 0 \\ -3 & -2i & 1 & 0 & 0 \\ 0 & 0 & -2i & 1 & 0 \\ 2 & 0 & -2 & -2i & 0 \end{bmatrix}$$

with RREF

$$\begin{bmatrix} 1 & 0 & 0 & -i/2 & 0 \\ 0 & 1 & 0 & 1 & 0 \\ 0 & 0 & 1 & i/2 & 0 \\ 0 & 0 & 0 & 0 & 0 \end{bmatrix}.$$

The corresponding eigenvectors are therefore of the form

$$\mathbf{v}_2 = s(i,\ -2,\ -i,\ 2)$$

so that a complex-valued solution to the system $\mathbf{x}' = A\mathbf{x}$ is

$$
\mathbf{u}_2(t) = e^{2it} \begin{bmatrix} i \\ -2 \\ -i \\ 2 \end{bmatrix} = (\cos 2t + i \sin 2t) \begin{bmatrix} i \\ -2 \\ -i \\ 2 \end{bmatrix}.
$$

Taking the real and imaginary parts of this complex-valued solution yields the additional real-valued LI solutions

$$
\mathbf{x}_3(t) = \begin{bmatrix} -\sin 2t \\ -2\cos 2t \\ \sin 2t \\ 2\cos 2t \end{bmatrix}, \quad \mathbf{x}_4(t) = \begin{bmatrix} \cos 2t \\ -2\sin 2t \\ -\cos 2t \\ 2\sin 2t \end{bmatrix}.
$$

Consequently, the VDE in (8.8.12) has general solution

$$
\mathbf{x}(t) = c_1 \begin{bmatrix} \sin t \\ \cos t \\ 2\sin t \\ 2\cos t \end{bmatrix} + c_2 \begin{bmatrix} -\cos t \\ \sin t \\ -2\cos t \\ 2\sin t \end{bmatrix} + c_3 \begin{bmatrix} -\sin 2t \\ -2\cos 2t \\ \sin 2t \\ 2\cos 2t \end{bmatrix} + c_4 \begin{bmatrix} \cos 2t \\ -2\sin 2t \\ -\cos 2t \\ 2\sin 2t \end{bmatrix}.
$$

Combining the vector functions yields the solution vector

$$
\mathbf{x}(t) = \begin{bmatrix} c_1 \sin t - c_2 \cos t - c_3 \sin 2t + c_4 \cos 2t \\ c_1 \cos t + c_2 \sin t - 2c_3 \cos 2t - 2c_4 \sin 2t \\ 2c_1 \sin t - 2c_2 \cos t + c_3 \sin 2t - c_4 \cos 2t \\ 2(c_1 \cos t + c_2 \sin t + c_3 \cos 2t + c_4 \sin 2t) \end{bmatrix}.
$$

We now impose the initial condition $\mathbf{x}(0) = \begin{bmatrix} 1 \\ 0 \\ 1 \\ 0 \end{bmatrix}$. This requires c_1, c_2, c_3, and c_4 to satisfy

$$
\begin{aligned}
-c_2 \quad\quad + c_4 &= 1, \\
c_1 \quad\quad - 2c_3 \quad\quad &= 0, \\
-2c_2 \quad\quad - c_4 &= 1, \\
2c_1 \quad\quad + 2c_3 \quad\quad &= 0,
\end{aligned}
$$

which has solution $c_1 = 0$, $c_2 = -2/3$, $c_3 = 0$, $c_4 = 1/3$. Thus

$$\mathbf{x}(t) = \begin{bmatrix} \dfrac{1}{3}\,(2\cos t + \cos 2t) \\[2ex] -\dfrac{2}{3}\,(\sin t + \sin 2t) \\[2ex] \dfrac{1}{3}\,(4\cos t - \cos 2t) \\[2ex] \dfrac{2}{3}\,(-2\sin t + \sin 2t) \end{bmatrix}.$$

Since $x = x_1$ and $y = x_3$, it follows that the motion of the spring-mass system is given by

$$x(t) = \frac{1}{3}\,(2\cos t + \cos 2t),$$

$$y(t) = \frac{1}{3}\,(4\cos t - \cos 2t).$$

The motion of both masses is periodic, with period 2π. Looking at the second and third components of the solution vector $\mathbf{x}$ (or by differentiating the previous expressions for x and y), we see that

$$x'(t) = -\frac{2}{3}(\sin t + \sin 2t) = -\frac{2}{3}(1 + 2\cos t)\sin t,$$

$$y'(t) = \frac{2}{3}(-2\sin t + \sin 2t) = \frac{4}{3}(\cos t - 1)\sin t.$$

Consequently, on the interval $[0, 2\pi]$, x' has zeros when $t = 0$, $2\pi/3$, π, $4\pi/3$, and 2π, whereas the only zeros of y' are 0, π, 2π. Notice that both y'' and y''' vanish at $t = 0$, 2π. Hence, the graph of y is very flat in the neighborhood of these points. This motion is depicted in Figure 8.8.2.

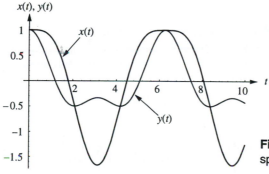

Figure 8.8.2 The solutions for the spring-mass system in Example 8.8.1.

Next consider the mixing problem depicted in Figure 8.8.3. Two tanks contain a solution consisting of chemical dissolved in water. A solution containing c_{in} grams/liter of chemical flows into tank 1 at a rate of r_{in} liters/min and solution of concentration c_{out} grams/liter flows out of tank 2 at a rate of r_{out} liters/min. In addition, solution of concentration c_{12} gm/liter flows into tank 1 from tank 2 at a rate of r_{12} liters /min and solution of concentration c_{21} grams/liter flows into tank 2 from tank 1 at a rate of r_{21} liters/minute. We wish to determine $A_1(t)$ and $A_2(t)$, the amounts of chemical in tank 1 and tank 2, respectively. The analysis is similar to that used in Section 1.7. Assuming that the solution in each tank is well mixed, it follows immediately that

$$c_{12} = c_{out} = \frac{A_2}{V_2}, \quad c_{21} = \frac{A_1}{V_1}$$

where V_i denotes the volume of solution in tank i at time t.

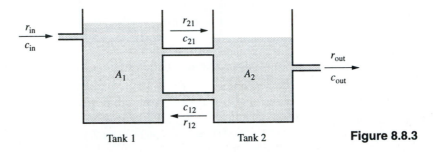

Tank 1   Tank 2   **Figure 8.8.3**

Consider a short time interval Δt. The total amount of chemical entering tank 1 in this time interval is approximately

$$(c_{in} r_{in} + c_{12} r_{12}) \, \Delta t \text{ grams},$$

whereas approximately

$$c_{21} r_{21} \, \Delta t \text{ grams}$$

of chemical leave tank 1 in the same time interval. Consequently, the change in the amount of chemical in tank 1 in the time interval Δt, denoted ΔA_1, is approximately

$$\Delta A_1 \approx [(c_{in} r_{in} + c_{12} r_{12}) - c_{21} r_{21}] \Delta t,$$

that is

$$\Delta A_1 \approx \left[c_{in} r_{in} + r_{12} \frac{A_2}{V_2} - r_{21} \frac{A_1}{V_1} \right] \Delta t. \tag{8.8.13}$$

Similarly, the change in the amount of chemical in tank 2 in the time interval Δt, denoted ΔA_2, is approximately

$$\Delta A_2 \approx [r_{21} c_{21} - (r_{12} c_{12} + r_{out} c_{out})] \Delta t$$

or, equivalently,

$$\Delta A_2 \approx \left[r_{21} \frac{A_1}{V_1} - (r_{12} + r_{out}) \frac{A_2}{V_2} \right] \Delta t. \tag{8.8.14}$$

Dividing equations (8.8.13) and (8.8.14) by Δt and taking the limit as $\Delta t \to 0^+$ yields the following system of DE for A_1 and A_2.

$$\frac{dA_1}{dt} = -r_{21} \frac{A_1}{V_1} + r_{12} \frac{A_2}{V_2} + c_{in} r_{in},$$

$$\frac{dA_2}{dt} = r_{21} \frac{A_1}{V_1} - (r_{12} + r_{out}) \frac{A_2}{V_2}.$$

We will now assume that V_1 and V_2 are constant. This imposes the conditions

$$r_{in} + r_{12} - r_{21} = 0,$$
$$r_{21} - r_{12} - r_{out} = 0.$$

(See Problem 7.) Consequently, the foregoing system of DE reduces to

$$\frac{dA_1}{dt} = -\frac{r_{21}}{V_1}A_1 + \frac{r_{12}}{V_2}A_2 + c_{in}r_{in},$$

$$\frac{dA_2}{dt} = \frac{r_{21}}{V_1}A_1 - \frac{r_{21}}{V_2}A_2.$$

This is a constant coefficient system for A_1 and A_2, and, therefore, it can be solved using the techniques that we have developed in this chapter.

Example 8.8.2   Two tanks each contain 20 L of a solution consisting of salt dissolved in water. A solution containing 4 g/L of salt flows into tank 1 at a rate of 3 L/min and the solution in tank 2 flows out at the same rate. In addition, solution flows

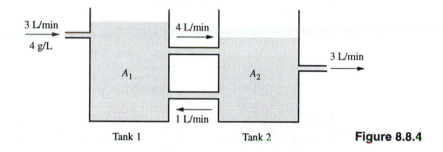

3 L/min
4 g/L

4 L/min

3 L/min

A_1

A_2

1 L/min

Tank 1   Tank 2   **Figure 8.8.4**

into tank 1 from tank 2 at a rate of 1 L/min and into tank 2 from tank 1 at a rate of 4 L/min. Initially tank 1 contained 40 g of salt and tank 2 contained 20 g of salt. Find the amount of salt in each tank at time t.

Solution   The inflow and outflow rates from each tank are indicated in Figure 8.8.4. We notice that the total amount of solution flowing into tank 1 is 4 liters/min, and the same volume of solution flows out of tank 1 per minute. Consequently, the volume of solution in tank 1 remains constant at 20 L. The same is true for tank 2. Let $A_1(t)$ and $A_2(t)$ denote the amounts of salt in tanks 1 and 2, respectively, and let c_{ij} denote the concentration of salt in the solution flowing into tank i from tank j. Now consider a short time interval Δt. The overall change in the amount of salt in tank 1 in this time interval, Δt, is approximately

$$\Delta A_1 \approx (12 + 1 \cdot c_{12})\Delta t - 4c_{21}\Delta t,$$

that is,

$$\Delta A_1 \approx \left(12 + \frac{1}{20}A_2 - \frac{1}{5}A_1\right)\Delta t. \qquad (8.8.15)$$

A similar analysis of the change in the amount of salt in tank 2 in the time interval Δt yields

$$\Delta A_2 \approx \left(\frac{1}{5} A_1 - \frac{1}{20} A_2 - \frac{3}{20} A_2 \right) \Delta t,$$

that is,

$$\Delta A_2 \approx \left(\frac{1}{5} A_1 - \frac{1}{5} A_2 \right) \Delta t. \tag{8.8.16}$$

Dividing equations (8.8.15) and (8.8.16) by Δt and taking the limit as $\Delta t \to 0^+$ yields the system of DE

$$\frac{dA_1}{dt} = -\frac{1}{5} A_1 + \frac{1}{20} A_2 + 12,$$

$$\frac{dA_2}{dt} = \frac{1}{5} A_1 - \frac{1}{5} A_2.$$

We are also given the initial conditions

$$A_1(0) = 40, \quad A_2(0) = 20.$$

In vector form, we must therefore solve the IVP

$$\mathbf{x}' = A\mathbf{x} + \mathbf{b}, \quad \mathbf{x}(0) = \mathbf{x}_0,$$

where

$$\mathbf{x} = \begin{bmatrix} A_1 \\ A_2 \end{bmatrix}, \quad A = \begin{bmatrix} -\frac{1}{5} & \frac{1}{20} \\ \frac{1}{5} & -\frac{1}{5} \end{bmatrix}, \quad \mathbf{b} = \begin{bmatrix} 12 \\ 0 \end{bmatrix}, \quad \mathbf{x}_0 = \begin{bmatrix} 40 \\ 20 \end{bmatrix}.$$

A has characteristic polynomial

$$\det(A - \lambda I) = \begin{vmatrix} -\frac{1}{5} - \lambda & \frac{1}{20} \\ \frac{1}{5} & -\frac{1}{5} - \lambda \end{vmatrix} = (\lambda + \frac{1}{5})^2 - \frac{1}{100}.$$

Consequently, the eigenvalues of A are

$$\lambda = -\frac{1}{5} \pm \frac{1}{10},$$

that is,

$$\lambda_1 = -\frac{1}{10}, \quad \lambda_2 = -\frac{3}{10}.$$

The corresponding eigenvectors are

$$\mathbf{v}_1 = (1, 2), \quad \mathbf{v}_2 = (1, -2),$$

respectively, so that two LI solutions to $\mathbf{x}' = A\mathbf{x}$ are

$$\mathbf{x}_1(t) = e^{-t/10} \begin{bmatrix} 1 \\ 2 \end{bmatrix}, \quad \mathbf{x}_2(t) = e^{-3t/10} \begin{bmatrix} 1 \\ -2 \end{bmatrix}.$$

Thus,

$$\mathbf{x}_c(t) = c_1 e^{-t/10} \begin{bmatrix} 1 \\ 2 \end{bmatrix} + c_2 e^{-3t/10} \begin{bmatrix} 1 \\ -2 \end{bmatrix}.$$

We now need a particular solution to $\mathbf{x}' = A\mathbf{x} + \mathbf{b}$. According to the variation-of-parameters technique, a particular solution is

$$\mathbf{x}_p = X\mathbf{u},$$

where

$$X\mathbf{u}' = \mathbf{b} \tag{8.8.17}$$

and

$$X(t) = \begin{bmatrix} e^{-t/10} & e^{-3t/10} \\ 2 e^{-t/10} & -2 e^{-3t/10} \end{bmatrix}.$$

The system (8.8.17) is

$$e^{-t/10} u_1' + e^{-3t/10} u_2' = 12,$$
$$e^{-t/10} u_1' - e^{-3t/10} u_2' = 0,$$

which has solution

$$u_1' = 6 e^{t/10}, \quad u_2' = 6 e^{3t/10}.$$

By integrating, we obtain

$$u_1(t) = 60 e^{t/10}, \quad u_2(t) = 20 e^{3t/10},$$

where we have set the integration constants to zero without loss of generality. Consequently,

$$\mathbf{x}_p(t) = \begin{bmatrix} e^{-t/10} & e^{-3t/10} \\ 2 e^{-t/10} & -2 e^{-3t/10} \end{bmatrix} \begin{bmatrix} 60 e^{t/10} \\ 20 e^{3t/10} \end{bmatrix} = \begin{bmatrix} 80 \\ 80 \end{bmatrix}.$$

Hence, the general solution to the system $\mathbf{x}' = A\mathbf{x} + \mathbf{b}$ is

$$\mathbf{x}(t) = c_1 e^{-t/10} \begin{bmatrix} 1 \\ 2 \end{bmatrix} + c_2 e^{-3t/10} \begin{bmatrix} 1 \\ -2 \end{bmatrix} + \begin{bmatrix} 80 \\ 80 \end{bmatrix}.$$

That is,

$$\mathbf{x}(t) = \begin{bmatrix} c_1 e^{-t/10} + c_2 e^{-3t/10} + 80 \\ 2c_1 e^{-t/10} - 2c_2 e^{-3t/10} + 80 \end{bmatrix}.$$

Imposing the initial condition $\mathbf{x}(0) = \begin{bmatrix} 40 \\ 20 \end{bmatrix}$ requires

$$\begin{bmatrix} c_1 + c_2 + 80 \\ 2c_1 - 2c_2 + 80 \end{bmatrix} = \begin{bmatrix} 40 \\ 20 \end{bmatrix}.$$

That is

$$c_1 + c_2 = -40,$$
$$c_1 - c_2 = -30.$$

Thus, $c_1 = -35$ and $c_2 = -5$, and the solution to the IVP is

$$\mathbf{x}(t) = - 35 \, e^{-t/10} \begin{bmatrix} 1 \\ 2 \end{bmatrix} - 5 e^{-3t/10} \begin{bmatrix} 1 \\ -2 \end{bmatrix} + \begin{bmatrix} 80 \\ 80 \end{bmatrix}.$$

Consequently, the amounts of salt in tanks 1 and 2 at time t are, respectively,

$$A_1(t) = 80 - 35 \, e^{-t/10} - 5 e^{-3t/10},$$

$$A_2(t) = 80 - 70 e^{-t/10} + 10 e^{-3t/10}.$$

We see that both A_1 and A_2 approach the constant value 80 g as $t \to \infty$. Why is this a reasonable result?

EXERCISES 8.8

1. Derive the eigenvalues and eigenvectors given in Example 8.8.1.

2. Determine the motion of the coupled spring-mass system which has

$$k_1 = 3 \text{ Nm}^{-1}, k_2 = \frac{1}{2} \text{ Nm}^{-1}, m_1 = \frac{1}{2} \text{ kg}, m_2 = \frac{1}{12} \text{ kg}$$

given that at $t = 0$ both masses are set in motion from their equilibrium positions with a velocity of 1 m/s.

3. Determine the general motion of the coupled spring-mass system that has

$$k_1 = 3 \text{ Nm}^{-1}, k_2 = 4 \text{ Nm}^{-1},$$

$$m_1 = 1 \text{ kg}, m_2 = 4/3 \text{ kg}.$$

4. Determine the general motion of the coupled spring-mass system which has

$$k_1 = 2 \, k_2, m_1 = 2 \, m_2.$$

[Hint: Let $\omega^2 = k_2/(2m_2)$.]

5. Consider the general coupled spring-mass system whose motion is governed by the system (8.8.5)–(8.8.8). Show that the coefficient matrix of the system has characteristic equation

$$\lambda^4 + \left[\frac{k_2}{m_2} + \frac{(k_1 + k_2)}{m_1} \right] \lambda^2 + \frac{k_1 k_2}{m_1 m_2} = 0$$

and that the corresponding eigenvalues are of the form

$$\lambda = \pm i\omega_1, \pm i\omega_2,$$

where ω_1, ω_2 are positive real numbers.

6. Two masses m_1 and m_2, rest on a horizontal frictionless plane. The masses are attached to fixed walls by springs whose spring constants are k_1 and k_3. (See Figure 8.8.5.) The masses are

also connected by a spring whose spring constant is k_2. Determine a first-order system of DE that governs the motion of the system.

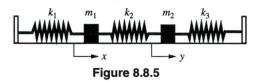

Figure 8.8.5

7. Show that the assumption that V_1 and V_2 are constant in the general mixing problem considered in the text imposes the conditions

$$r_{in} + r_{12} = r_{21},$$

$$r_{21} - r_{12} = r_{out}.$$

8. Solve the IVP arising in Example 8.8.2 using the technique derived in Section 8.2.

9. Solve the mixing problem depicted in Figure 8.8.6, given that at $t = 0$, the volume of solution in both tanks is 60 L, and tank 1 contains 60 g of chemical whereas tank 2 contains 200 g of chemical.

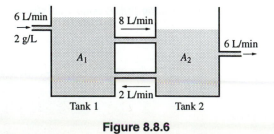

Figure 8.8.6

10. In the mixing problem shown in Figure 8.8.7, there is no inflow from or outflow to the outside. For this reason, the system is said to be *closed*. If tank 1 contains 6 L of solution and tank 2 contains 12 L of solution, determine the amount of chemical in each tank at time t, given that initially tank 1 contains 5 g of chemical and tank 2 contains 25 g of chemical.

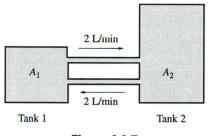

Tank 1   Tank 2

Figure 8.8.7

11. Consider the general closed system depicted in Figure 8.8.8.

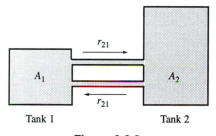

Tank 1   Tank 2

Figure 8.8.8

(a) Derive the system of DE that governs the behavior of A_1 and A_2.

(b) Define the constant β by $V_2 = \beta V_1$, where V_1 and V_2 denote the volume of solution in tank 1 and tank 2, respectively.   Show that the eigenvalues of the coefficient matrix of the system derived in (a) are

$$\lambda_1 = 0, \qquad \lambda_2 = -\frac{(1+\beta)}{\beta V_1} r_{21}.$$

(c) Determine A_1 and A_2, given that $A_1(0) = \alpha_1$, $A_2(0) = \alpha_2$, where α_1 and α_2 are positive constants.

(d) Show that

$$\lim_{t \to +\infty} \frac{A_1}{V_1} = \lim_{t \to +\infty} \frac{A_2}{V_2} = \frac{(\alpha_1 + \alpha_2)}{(1+\beta)V_1}.$$

Is this result reasonable?

8.9   AN INTRODUCTION TO THE MATRIX EXPONENTIAL FUNCTION

In the next two sections we give a brief introduction to the matrix exponential function and indicate the role that this function plays in the analysis and solution of systems of linear differential equations.

Definition 8.9.1:   Let A be an $n \times n$ matrix of constants.   We define the **matrix exponential function**, denoted e^{At}, by

$$e^{At} = I + At + \frac{1}{2!}(At)^2 + \frac{1}{3!}(At)^3 + \cdots + \frac{1}{k!}(At)^k + \cdots. \qquad (8.9.1)$$

It can be shown that for all $n \times n$ matrices A and all values of $t \in (-\infty, \infty)$, the infinite series appearing on the right-hand side of (8.9.1) converges to an $n \times n$ matrix. Consequently, e^{At} is a well-defined $n \times n$ matrix.

PROPERTIES OF THE MATRIX EXPONENTIAL FUNCTION

1. If A and B are $n \times n$ matrices satisfying $AB = BA$, then

$$\boxed{e^{(A+B)t} = e^{At}\, e^{Bt}.}$$

2. For all $n \times n$ matrices A, e^{At} is nonsingular and

$$\boxed{(e^{At})^{-1} = e^{(-A)t} = e^{-At}}$$

that is,

$$e^{At}\, e^{-At} = I_n.$$

The proof of these results requires a precise definition of convergence of an infinite series of matrices. This would take us too far astray from the main focus of this text, and hence, the proofs are omitted. (See, for example, M.W. Hirsch and S. Smale, *Differential Equations, Dynamical Systems, and Linear Algebra*, Academic Press, 1974.)

Before investigating the relationship between the matrix exponential function and systems of differential equations, we discuss some direct methods for finding e^{At}.

Example 8.9.1   Compute e^{At} if $A = \begin{bmatrix} 2 & 0 \\ 0 & -1 \end{bmatrix}$.

Solution   In this case we see that

$$At = \begin{bmatrix} 2t & 0 \\ 0 & -t \end{bmatrix}, \quad (At)^2 = \begin{bmatrix} (2t)^2 & 0 \\ 0 & (-t)^2 \end{bmatrix}, \quad (At)^3 = \begin{bmatrix} (2t)^3 & 0 \\ 0 & (-t)^3 \end{bmatrix}, \dots,$$

$$(At)^k = \begin{bmatrix} (2t)^k & 0 \\ 0 & (-t)^k \end{bmatrix}, \dots,$$

so that

$$e^{At} = \begin{bmatrix} 1 & 0 \\ 0 & 1 \end{bmatrix} + \begin{bmatrix} 2t & 0 \\ 0 & -t \end{bmatrix} + \frac{1}{2!}\begin{bmatrix} (2t)^2 & 0 \\ 0 & (-t)^2 \end{bmatrix} + \dots + \frac{1}{k!}\begin{bmatrix} (2t)^k & 0 \\ 0 & (-t)^k \end{bmatrix} + \dots$$

$$= \begin{bmatrix} \displaystyle\sum_{k=0}^{\infty} \frac{1}{k!}(2t)^k & 0 \\ 0 & \displaystyle\sum_{k=0}^{\infty} \frac{1}{k!}(-t)^k \end{bmatrix}.$$

Hence,

$$e^{At} = \begin{bmatrix} e^{2t} & 0 \\ 0 & e^{-t} \end{bmatrix}.$$

∎

More generally, it can be shown (see problem 1) that

If $A = \mathrm{diag}(d_1, d_2, ..., d_n)$, then $e^{At} = \mathrm{diag}(e^{d_1 t}, e^{d_2 t}, ..., e^{d_n t})$.

If A is not a diagonal matrix, then the computation of e^{At} is more involved. The next simplest case that can arise is when A is nondefective. In this case, as we have shown in Section 6.7, A is *similar* to a diagonal matrix, and we might suspect that this would lead to a simplification in the evaluation of e^{At}. We now show that this is indeed the case. Suppose that A has n LI eigenvectors $s_1, s_2, ..., s_n$, and define the $n \times n$ matrix S by

$$S = [s_1, s_2, ..., s_n].$$

Then, from Theorem 6.7.2,

$$S^{-1}AS = \mathrm{diag}(\lambda_1, \lambda_2, ..., \lambda_n), \qquad (8.9.2)$$

where $\lambda_1, \lambda_2, ..., \lambda_n$ are the eigenvalues of A corresponding to the eigenvectors $s_1, s_2, ..., s_n$. Premultiplying equation (8.9.2) by S and postmultiplying by S^{-1} yields

$$A = SDS^{-1},$$

where

$$D = \mathrm{diag}(\lambda_1, \lambda_2, ..., \lambda_n).$$

We now compute e^{At}. From Definition 8.9.1,

$$e^{At} = I + At + \frac{1}{2!}(At)^2 + \frac{1}{3!}(At)^3 + \cdots + \frac{1}{k!}(At)^k + \ldots$$

$$= SS^{-1} + (SDS^{-1})t + \frac{1}{2!}(SDS^{-1})^2 t^2 + \cdots + \frac{1}{k!}(SDS^{-1})^k t^k + \ldots. \qquad (8.9.3)$$

In order to proceed with this computation, we need the following lemma:

Lemma 8.9.1:   If S is an $n \times n$ matrix and D is an $n \times n$ diagonal matrix, then

$$(SDS^{-1})^k = SD^k S^{-1}. \qquad (8.9.4)$$

PROOF   The proof is by mathematical induction. The result is trivially true when $k = 1$. Now suppose that for some $m \geq 1$,

$$(SDS^{-1})^m = SD^m S^{-1}. \qquad (8.9.5)$$

We must prove that this implies the validity of equation (8.9.4) when $k = m+1$. This can be accomplished as follows.

$$(SDS^{-1})^{m+1} = (SDS^{-1})^m (SDS^{-1}).$$

That is, using equation (8.9.5),

$$(SDS^{-1})^{m+1} = (SD^mS^{-1})(SDS^{-1}) = SD^m(S^{-1}S)DS^{-1}$$
$$= SD^mIDS^{-1}$$
$$= SD^{m+1}S^{-1}$$

as required. It follows by induction that equation (8.9.4) is true for all positive integers k. ∎

We now return to (8.9.3). Using the result of the foregoing lemma, we can therefore write

$$e^{At} = S[I + Dt + \frac{1}{2!}(Dt)^2 + \cdots + \frac{1}{k!}(Dt)^k + \cdots]S^{-1},$$

that is,

$$e^{At} = Se^{Dt}S^{-1}.$$

Consequently, we have established the next theorem.

Theorem 8.9.1:   Let A be a *nondefective* $n \times n$ matrix with LI eigenvectors s_1, s_2, ..., s_n, and corresponding eigenvalues $\lambda_1, \lambda_2, ..., \lambda_n$. Then

$$e^{At} = Se^{Dt}S^{-1},$$

where $S = [s_1, s_2, ..., s_n]$ and $D = \text{diag}(\lambda_1, \lambda_2, ..., \lambda_n)$.

Example 8.9.2   Determine e^{At} if $A = \begin{bmatrix} 3 & 3 \\ 5 & 1 \end{bmatrix}$.

Solution A has eigenvalues $\lambda_1 = 6$, $\lambda_2 = -2$ and therefore is nondefective. A straightforward computation yields the following eigenvectors:

$\lambda_1 = 6$ implies that $s_1 = (1, 1)$; $\lambda_2 = -2$ implies that $s_2 = (-3, 5)$.

It follows from Theorem 8.9.1 that if we set

$$S = \begin{bmatrix} 1 & -3 \\ 1 & 5 \end{bmatrix} \text{ and } D = \text{diag}(6, -2),$$

then

$$e^{At} = Se^{Dt}S^{-1}.$$

That is,

$$e^{At} = S\begin{bmatrix} e^{6t} & 0 \\ 0 & e^{-2t} \end{bmatrix}S^{-1}. \tag{8.9.6}$$

It is easily shown that

$$S^{-1} = \begin{bmatrix} \frac{5}{8} & \frac{3}{8} \\ -\frac{1}{8} & \frac{1}{8} \end{bmatrix},$$

so that, substituting into equation (8.9.6),

$$e^{At} = \begin{bmatrix} 1 & -3 \\ 1 & 5 \end{bmatrix} \begin{bmatrix} e^{6t} & 0 \\ 0 & e^{-2t} \end{bmatrix} \begin{bmatrix} \frac{5}{8} & \frac{3}{8} \\ -\frac{1}{8} & \frac{1}{8} \end{bmatrix} = \begin{bmatrix} 1 & -3 \\ 1 & 5 \end{bmatrix} \begin{bmatrix} \frac{5}{8}e^{6t} & \frac{3}{8}e^{6t} \\ -\frac{1}{8}e^{-2t} & \frac{1}{8}e^{-2t} \end{bmatrix}$$

Consequently,

$$e^{At} = \begin{bmatrix} \frac{1}{8}(5e^{6t} + 3e^{-2t}) & \frac{3}{8}(e^{6t} - e^{-2t}) \\ \frac{5}{8}(e^{6t} - e^{-2t}) & \frac{1}{8}(3e^{6t} + 5e^{-2t}) \end{bmatrix}.$$

◻

The computation of e^{At} when A is a defective matrix is best accomplished by relating e^{At} to the solution of the corresponding VDE $\mathbf{x}' = A\mathbf{x}$. This is one of the goals of the next section.

EXERCISES 8.9

1. If $A = \text{diag}(d_1, d_2, ..., d_n)$, prove that

$$e^{At} = \text{diag}(e^{d_1 t}, e^{d_2 t}, ..., e^{d_n t}).$$

2. If $A = \begin{bmatrix} -3 & 0 \\ 0 & 5 \end{bmatrix}$, determine e^{At} and e^{-At}.

3. Prove that for all values of the constant λ,

$$e^{\lambda I_n t} = e^{\lambda t} I_n.$$

4. Consider the matrix $A = \begin{bmatrix} a & b \\ 0 & a \end{bmatrix}$. We can

write $A = B + C$, where $B = \begin{bmatrix} a & 0 \\ 0 & a \end{bmatrix}$ and

$C = \begin{bmatrix} 0 & b \\ 0 & 0 \end{bmatrix}$.

(a) Verify that $BC = CB$.

(b) Verify that $C^2 = 0_2$, and determine e^{Ct}.

(c) Use property (1) of the matrix exponential function to find e^{At}.

5. If $A = \begin{bmatrix} a & b \\ -b & a \end{bmatrix}$, use property (1) of the

matrix exponential function and Definition 8.9.1

to show that $e^{At} = e^{at} \begin{bmatrix} \cos bt & \sin bt \\ -\sin bt & \cos bt \end{bmatrix}$.

For problems 6–12, show that A is nondefective, and use Theorem 8.9.1 to find e^{At}.

6. $A = \begin{bmatrix} 1 & 2 \\ 0 & 3 \end{bmatrix}$.

7. $A = \begin{bmatrix} 3 & 1 \\ 1 & 3 \end{bmatrix}$.

8. $A = \begin{bmatrix} 0 & 2 \\ -2 & 0 \end{bmatrix}$.

9. $A = \begin{bmatrix} -1 & 3 \\ -3 & -1 \end{bmatrix}$.

10. $A = \begin{bmatrix} a & b \\ -b & a \end{bmatrix}$.

11. $A = \begin{bmatrix} 3 & -2 & -2 \\ 1 & 0 & -2 \\ 0 & 0 & 3 \end{bmatrix}$.

12. $A = \begin{bmatrix} 6 & -2 & -1 \\ 8 & -2 & -2 \\ 4 & -2 & 1 \end{bmatrix}$, and you may assume

that $p(\lambda) = -(\lambda - 2)^2(\lambda - 1)$.

An $n \times n$ matrix A that satisfies $A^k = 0_n$, for some k, is called *nilpotent*. For problems 13–17, show that the given matrix is nilpotent, and use Definition 8.9.1 to determine e^{At}.

13. $A = \begin{bmatrix} -3 & 9 \\ -1 & 3 \end{bmatrix}$.

14. $A = \begin{bmatrix} 1 & 1 \\ -1 & -1 \end{bmatrix}$.

15.
$$A = \begin{bmatrix} 0 & 0 & 0 \\ 1 & 0 & 0 \\ 0 & 1 & 0 \end{bmatrix}.$$

16. $A = \begin{bmatrix} -1 & -6 & -5 \\ 0 & -2 & -1 \\ 1 & 2 & 3 \end{bmatrix}.$

17. $A = \begin{bmatrix} 0 & 1 & 0 & 0 \\ 0 & 0 & 1 & 0 \\ 0 & 0 & 0 & 1 \\ 0 & 0 & 0 & 0 \end{bmatrix}.$

18. Let A be the $n \times n$ matrix whose only nonzero elements are

$$a_{i+1\,i} = 1, \quad i = 1, 2, \ldots, n-1.$$

Determine e^{At}. (See problem 15 for the case $n = 3$.)

8.10   THE MATRIX EXPONENTIAL FUNCTION AND SYSTEMS OF DIFFERENTIAL EQUATIONS

In this section we investigate the relationship between the matrix exponential function, e^{At}, and the solutions to the corresponding VDE

$$\mathbf{x}' = A\mathbf{x}.$$

We begin by defining the derivative of e^{At}. It can be shown that the infinite series (8.9.1) defining e^{At} can be differentiated term by term and the resulting series converges for all $t \in (-\infty, \infty)$. Thus, through differentiating (8.7.1), we have

$$\frac{d}{dt}(e^{At}) = A + A^2 t + \frac{1}{2!}A^3 t^2 + \cdots + \frac{1}{(k-1)!}A^k t^{k-1} + \cdots.$$

That is,

$$\frac{d}{dt}(e^{At}) = A\left[I + At + \frac{1}{2!}(At)^2 + \frac{1}{3!}(At)^3 + \cdots + \frac{1}{k!}(At)^k + \cdots \right].$$

Hence,

$$\boxed{\frac{d}{dt}(e^{At}) = A e^{At}.} \tag{8.10.1}$$

Now recall from Chapter 1 that for all values of the constants a and x_0, the unique solution to the IVP

$$\frac{dx}{dt} = ax, \quad x(0) = x_0$$

is

$$x(t) = x_0 e^{at}.$$

Our next theorem shows that the same formula holds for the VDE

$$\mathbf{x}' = A\mathbf{x},$$

provided we replace e^{at} by e^{At}. This is a very elegant result which has far-reaching consequences in both the computation of e^{At} and the analysis of VDE.

Theorem 8.10.1:   Let $\mathbf{x}_0$ be an arbitrary vector. Then the unique solution to the IVP

$$\mathbf{x}' = A\mathbf{x}, \ \mathbf{x}(0) = \mathbf{x}_0$$

is

$$\mathbf{x}(t) = e^{At}\mathbf{x}_0.$$

PROOF   If $\mathbf{x}(t) = e^{At}\mathbf{x}_0$, then $\mathbf{x}'(t) = Ae^{At}\mathbf{x}_0$. That is

$$\mathbf{x}' = A\mathbf{x}.$$

Further, setting $t = 0$,

$$\mathbf{x}(0) = e^{0A}\mathbf{x}_0 = I\mathbf{x}_0 = \mathbf{x}_0.$$

Consequently, $\mathbf{x}(t) = e^{At}\mathbf{x}_0$ is a solution to the given IVP. The uniqueness of the solution follows from Theorem 8.4.1.   ∎

We now investigate how the result of Theorem 8.10.1 combined with our previous techniques for solving $\mathbf{x}' = A\mathbf{x}$ can be used to determine e^{At}. To this end, let $\mathbf{x}_1, \mathbf{x}_2, \ldots, \mathbf{x}_n$ be LI solutions to the VDE

$$\mathbf{x}' = A\mathbf{x}, \tag{8.10.2}$$

where A is an $n \times n$ matrix of constants. We recall from Section 8.4 that the corresponding matrix function

$$X(t) = [\mathbf{x}_1, \mathbf{x}_2, .., \mathbf{x}_n]$$

is called a fundamental matrix for (8.10.2) and that the general solution to (8.10.2) can be written as

$$\mathbf{x}(t) = X(t)\mathbf{c},$$

where $\mathbf{c}$ is a column vector of arbitrary constants. If $X(t)$ is any fundamental matrix for (8.10.2) and B is any *nonsingular* matrix of constants, then the matrix function

$$Y(t) = X(t)B$$

is also a fundamental matrix for (8.10.2), since its column vectors are linear combinations of the column vectors of X and hence are LI solutions of (8.10.2). (The linear independence follows since $Y(t)$ is nonsingular.) We focus our attention on a particular fundamental matrix.

Definition 8.10.1:   The unique fundamental matrix for $\mathbf{x}' = A\mathbf{x}$ that satisfies

$$X(0) = I_n$$

is called the **transition matrix** for $\mathbf{x}' = A\mathbf{x}$ based at $t = 0$, and it is denoted by $X_0(t)$.

In terms of the transition matrix, the solution to the IVP

$$\mathbf{x}' = A\mathbf{x} , \ \mathbf{x}(0) = \mathbf{x}_0$$

is just

$$\mathbf{x}(t) = X_0(t)\mathbf{x}_0,$$

so that the transition matrix does indeed describe the transition of the system from its state at time $t = 0$ to its state at time t. Further, if $X(t)$ is any fundamental matrix for $\mathbf{x}' = A\mathbf{x}$, then the transition matrix can be determined from (see problem 1)

$$X_0(t) = X(t)X^{-1}(0). \tag{8.10.3}$$

We now prove that $X_0(t)$ is in fact e^{At}. From equation (8.10.1), we have

$$\frac{d}{dt}(e^{At}) = Ae^{At},$$

so that the column vectors of e^{At} are solutions to $\mathbf{x}' = A\mathbf{x}$. Further, setting $t = 0$ yields

$$e^{0A} = I_n, \tag{8.10.4}$$

which implies that

$$\det(e^{0A}) = 1 \neq 0.$$

Hence, the column vectors of e^{At} are LI on any interval. Consequently, e^{At} is a fund–amental matrix for $\mathbf{x}' = A\mathbf{x}$. Finally, combining (8.10.4) with the uniqueness of the transition matrix leads to the required conclusion, namely,

$$e^{At} = X_0(t). \tag{8.10.5}$$

Thus, if A is an $n \times n$ matrix and $X(t)$ is *any* fundamental matrix for the corresponding VDE $\mathbf{x}' = A\mathbf{x}$, then equations (8.10.3) and (8.10.5) imply that

$$\boxed{e^{At} = X(t)X^{-1}(0).} \tag{8.10.6}$$

Consequently, to determine e^{At}, we can use the techniques from Sections 8.4 and 8.5 to find a fundamental matrix for $\mathbf{x}' = A\mathbf{x}$, and then e^{At} can be obtained directly from equation (8.10.6).

Example 8.10.1   Determine e^{At} if $A = \begin{bmatrix} 6 & -8 \\ 2 & -2 \end{bmatrix}$.

Solution   We first find a fundamental matrix for

$$\mathbf{x}' = A\mathbf{x}.$$

This system has been solved in Example 8.6.2, where it was found that two LI solutions are[1]

$$\mathbf{x}_1(t) = e^{2t}\begin{bmatrix} 2 \\ 1 \end{bmatrix}, \quad \mathbf{x}_2(t) = e^{2t}\begin{bmatrix} 1 + 4t \\ 2t \end{bmatrix}.$$

Thus a fundamental matrix for $\mathbf{x}' = A\mathbf{x}$ is

$$X(t) = \begin{bmatrix} 2e^{2t} & (1 + 4t)e^{2t} \\ e^{2t} & 2te^{2t} \end{bmatrix},$$

whose inverse at $t = 0$ is

$$X^{-1}(0) = \begin{bmatrix} 0 & 1 \\ 1 & -2 \end{bmatrix}.$$

---

[1]Notice that in this example A is defective.

Consequently,

$$e^{At} = X(t)X^{-1}(0) = \begin{bmatrix} 2e^{2t} & (1+4t)e^{2t} \\ e^{2t} & 2te^{2t} \end{bmatrix} \begin{bmatrix} 0 & 1 \\ 1 & -2 \end{bmatrix}.$$

That is,

$$e^{At} = \begin{bmatrix} (1+4t)e^{2t} & -8te^{2t} \\ 2te^{2t} & (1-4t)e^{2t} \end{bmatrix},$$

which can be written as

$$e^{At} = e^{2t} \begin{bmatrix} 1+4t & -8t \\ 2t & 1-4t \end{bmatrix}. \qquad \square$$

We now indicate how the matrix exponential function can be used to directly derive LI solutions to $\mathbf{x}' = A\mathbf{x}$. Let $\mathbf{v}_1, \mathbf{v}_2, \ldots, \mathbf{v}_n$ be LI vectors in R^n (or C^n), and consider the corresponding vector functions

$$\mathbf{x}_1(t) = e^{At}\mathbf{v}_1, \quad \mathbf{x}_2(t) = e^{At}\mathbf{v}_2, \quad \ldots, \quad \mathbf{x}_n(t) = e^{At}\mathbf{v}_n.$$

Theorem 8.10.1 implies that each of these vector functions is a solution to $\mathbf{x}' = A\mathbf{x}$; further,

$$\det([\mathbf{x}_1, \mathbf{x}_2, \ldots, \mathbf{x}_n]) = \det(e^{At}[\mathbf{v}_1, \mathbf{v}_2, \ldots, \mathbf{v}_n]) = \det(e^{At}) \cdot \det([\mathbf{v}_1, \mathbf{v}_2, \ldots, \mathbf{v}_n]),$$

which is nonzero, since $\mathbf{v}_1, \mathbf{v}_2, \ldots, \mathbf{v}_n$ are LI and e^{At} is nonsingular. Consequently,

$$X(t) = [\mathbf{x}_1, \mathbf{x}_2, \ldots, \mathbf{x}_n]$$

is a fundamental matrix for $\mathbf{x}' = A\mathbf{x}$. Now, we can certainly write

$$A = \lambda I + (A - \lambda I),$$

and since the matrices $B = \lambda I$ and $C = A - \lambda I$ satisfy $BC = CB$, it follows from property (1) of the matrix exponential function that

$$e^{At}\mathbf{v} = e^{[\lambda It + (A - \lambda I)t]}\mathbf{v} = e^{\lambda It}e^{(A - \lambda I)t}\mathbf{v}.$$

Further,

$$e^{\lambda It} = \mathrm{diag}(e^{\lambda t}, e^{\lambda t}, \ldots, e^{\lambda t}) = e^{\lambda t}I,$$

so that

$$e^{At}\mathbf{v} = e^{\lambda t}\left[\mathbf{v} + (A - \lambda I)\mathbf{v} + \frac{1}{2!}(A - \lambda I)^2\mathbf{v} + \cdots \right]. \qquad (8.10.7)$$

In general, the preceding series contains an infinite number of terms and hence is intractable. However, if we can find vectors $\mathbf{v}$ such that

$$(A - \lambda I)^k\mathbf{v} = \mathbf{0},$$

for some k, then the series will terminate after a finite number of terms. For example, if $\mathbf{v}$ is an eigenvector of A [that is, $(A - \lambda I)\mathbf{v} = \mathbf{0}$], then the series in (8.10.7) has only one term, and we obtain the result of Theorem 8.5.1, namely, that

$$\mathbf{x}(t) = e^{\lambda t}\mathbf{v}$$

is a solution to $\mathbf{x}' = A\mathbf{x}$ whenever λ and $\mathbf{v}$ are an eigenvalue/eigenvector pair for A. Hence, if A is nondefective, equation (8.10.7) yields n LI solutions to $\mathbf{x}' = A\mathbf{x}$ in the usual manner. Suppose, however, that A is defective. Then there is at least one eigenvalue of A that has an eigenspace whose dimension is less than the multiplicity of the eigenvalue. In this case, Theorem 8.6.3 gives us the general form of the solutions. We now derive an equivalent technique for deriving these solutions based on the matrix exponential function. We need the following theorem from linear algebra.

Theorem 8.10.2: Let A be an $n \times n$ matrix, and suppose that λ is an eigen-value of multiplicity m. Then the system of equations

$$(A - \lambda I)^m \mathbf{v} = \mathbf{0} \tag{8.10.8}$$

has m LI solutions. Further, nontrivial solutions to equation (8.10.8) corresponding to distinct eigenvalues are LI.

PROOF   The proof of this theorem is best left for a second course in linear algebra. ∎

Suppose that λ is an eigenvalue of multiplicity m. Then according to Theorem 8.10.2, we can obtain m LI vectors, $\mathbf{v}_1, \mathbf{v}_2, \ldots, \mathbf{v}_m$ satisfying equation (8.10.8). Further, using these vectors, the infinite series (8.10.7) terminates after a finite number of terms, specifically,[1]

$$e^{At}\mathbf{v}_i = e^{\lambda t}\left[\mathbf{v}_i + t(A - \lambda I)\mathbf{v}_i + \frac{t^2}{2!}(A - \lambda I)^2\mathbf{v}_i + \cdots + \frac{t^{m-1}}{(m-1)!}(A - \lambda I)^{m-1}\mathbf{v}_i\right],$$

where $i = 1, 2, \ldots, m$. Proceeding in this manner for each eigenvalue, we can obtain n LI solutions to $\mathbf{x}' = A\mathbf{x}$. This will determine a fundamental matrix for $\mathbf{x}' = A\mathbf{x}$ from which e^{At} can be determined in the usual manner.

Example 8.10.2   Let $A = \begin{bmatrix} 6 & 8 & 1 \\ -1 & -3 & 3 \\ -1 & -1 & 1 \end{bmatrix}$.

(a) Determine a fundamental matrix for $\mathbf{x}' = A\mathbf{x}$, and thereby determine the general solution to the VDE.

(b) Determine e^{At}.

Solution

(a) The characteristic polynomial of A is

$$p(\lambda) = -(\lambda - 3)^2(\lambda + 2).$$

Hence, A has eigenvalues $\lambda_1 = 3$ (multiplicity 2), and $\lambda_2 = -2$.

$\lambda_1 = 3$: In this case, we must determine two LI solutions to

$$(A - 3I)^2 \mathbf{v} = \mathbf{0}. \tag{8.10.9}$$

The coefficient matrix of this system is

---

[1]Often the series will terminate after less than m terms.

$$(A - 3I)^2 = \begin{bmatrix} 3 & 8 & 1 \\ -1 & -6 & 3 \\ -1 & -1 & -2 \end{bmatrix} \begin{bmatrix} 3 & 8 & 1 \\ -1 & -6 & 3 \\ -1 & -1 & -2 \end{bmatrix} = \begin{bmatrix} 0 & -25 & 25 \\ 0 & 25 & -25 \\ 0 & 0 & 0 \end{bmatrix},$$

so that the system (8.10.9) reduces to the single equation

$$v_2 - v_3 = 0,$$

which has two free variables. We set $v_1 = r$, and $v_3 = s$, in which case $v_2 = s$. Hence, equation (8.10.9) has solution

$$\mathbf{v} = r(1, 0, 0) + s(0, 1, 1).$$

Consequently, two LI solutions to equation (8.10.9) are

$$\mathbf{v}_1 = (1, 0, 0), \quad \mathbf{v}_2 = (0, 1, 1).$$

Since $(A - 3I)^2 \mathbf{v}_1 = \mathbf{0}$, $e^{At}\mathbf{v}_1$ reduces to

$$e^{At}\mathbf{v}_1 = e^{3t}[\mathbf{v}_1 + t(A - 3I)\mathbf{v}_1]$$

$$= e^{3t}\left\{ \begin{bmatrix} 1 \\ 0 \\ 0 \end{bmatrix} + t \begin{bmatrix} 3 & 8 & 1 \\ -1 & -6 & 3 \\ -1 & -1 & -2 \end{bmatrix} \begin{bmatrix} 1 \\ 0 \\ 0 \end{bmatrix} \right\}$$

$$= e^{3t}\left\{ \begin{bmatrix} 1 \\ 0 \\ 0 \end{bmatrix} + t \begin{bmatrix} 3 \\ -1 \\ -1 \end{bmatrix} \right\}.$$

Hence, one solution to the given system is

$$\mathbf{x}_1(t) = e^{At}\mathbf{v}_1 = e^{3t} \begin{bmatrix} 1 + 3t \\ -t \\ -t \end{bmatrix}.$$

Similarly,

$$e^{At}\mathbf{v}_2 = e^{3t}[\mathbf{v}_2 + t(A - 3I)\mathbf{v}_2]$$

$$= e^{3t}\left\{ \begin{bmatrix} 0 \\ 1 \\ 1 \end{bmatrix} + t \begin{bmatrix} 3 & 8 & 1 \\ -1 & -6 & 3 \\ -1 & -1 & -2 \end{bmatrix} \begin{bmatrix} 0 \\ 1 \\ 1 \end{bmatrix} \right\}$$

$$= e^{3t}\left\{ \begin{bmatrix} 0 \\ 1 \\ 1 \end{bmatrix} + t \begin{bmatrix} 9 \\ -3 \\ -3 \end{bmatrix} \right\}.$$

Thus a second LI solution to the given system is

$$\mathbf{x}_2(t) = e^{At}\mathbf{v}_2 = e^{3t} \begin{bmatrix} 9t \\ 1 - 3t \\ 1 - 3t \end{bmatrix}.$$

$\lambda_2 = -2$:   It is easily shown that the eigenvectors corresponding to $\lambda = -2$ are all scalar multiples of

$$\mathbf{v}_3 = (-1, 1, 0).$$

Hence, a third LI solution to the given system is

$$\mathbf{x}_3(t) = e^{At}\mathbf{v}_3 = e^{-2t} \begin{bmatrix} -1 \\ 1 \\ 0 \end{bmatrix}.$$

Consequently, a fundamental matrix for $\mathbf{x}' = A\mathbf{x}$ is

$$X(t) = [e^{At}\mathbf{v}_1, e^{At}\mathbf{v}_2, e^{At}\mathbf{v}_2] = \begin{bmatrix} e^{3t}(1 + 3t) & 9te^{3t} & -e^{-2t} \\ -te^{3t} & e^{3t}(1 - 3t) & e^{-2t} \\ -te^{3t} & e^{3t}(1 - 3t) & 0 \end{bmatrix}, \qquad (8.10.10)$$

so that the given VDE has general solution

$$\mathbf{x}(t) = X(t)\mathbf{c} = \begin{bmatrix} e^{3t}(1 + 3t) & 9te^{3t} & -e^{-2t} \\ -te^{3t} & e^{3t}(1 - 3t) & e^{-2t} \\ -te^{3t} & e^{3t}(1 - 3t) & 0 \end{bmatrix} \begin{bmatrix} c_1 \\ c_2 \\ c_3 \end{bmatrix}.$$

(b) From equation (8.10.10), we have

$$X(0) = \begin{bmatrix} 1 & 0 & -1 \\ 0 & 1 & 1 \\ 0 & 1 & 0 \end{bmatrix},$$

and, using the Gaussian–Jordan method, we find that

$$X^{-1}(0) = \begin{bmatrix} 1 & 1 & -1 \\ 0 & 0 & 1 \\ 0 & 1 & -1 \end{bmatrix}.$$

Consequently,

$$e^{At} = X(t) X^{-1}(0) = \begin{bmatrix} e^{3t}(1 + 3t) & 9te^{3t} & -e^{-2t} \\ -te^{3t} & e^{3t}(1 - 3t) & e^{-2t} \\ -te^{3t} & e^{3t}(1 - 3t) & 0 \end{bmatrix} \begin{bmatrix} 1 & 1 & -1 \\ 0 & 0 & 1 \\ 0 & 1 & -1 \end{bmatrix}.$$

That is,

$$e^{At} = \begin{bmatrix} e^{3t}(1 + 3t) & e^{3t}(1 + 3t) - e^{-2t} & e^{3t}(6t - 1) + e^{-2t} \\ -te^{3t} & e^{-2t} - te^{3t} & e^{3t}(1 - 2t) - e^{-2t} \\ -te^{3t} & -te^{3t} & e^{3t}(1 - 2t) \end{bmatrix}. \qquad \square$$

REMARK   As the examples in this section indicate, the computation of e^{At} can be quite tedious. The main use of the matrix exponential is theoretical.

We end this section by showing that the results so far obtained in this chapter are a generalization of those from Chapter 1. Consider the IVP

$$\frac{dx}{dt} - ax = b(t), \quad x(0) = x_0,$$

where a is a constant. Using the technique developed in Section 1.6 for solving linear DE, it is easily shown that the solution to the IVP is

$$x(t) = e^{at}\left[\int_0^t e^{-as}b(s)\,ds + x_0\right]. \tag{8.10.11}$$

Now consider the corresponding IVP for VDE, namely,

$$\mathbf{x}' = A\mathbf{x} + \mathbf{b}(t), \quad \mathbf{x}(0) = \mathbf{x}_0. \tag{8.10.12}$$

According to the variation-of-parameters method, a particular solution to the system is

$$\mathbf{x}_p(t) = X(t)\int_0^t X^{-1}(s)\mathbf{b}(s)\,ds,$$

where $X(t)$ is any fundamental matrix for $\mathbf{x}' = A\mathbf{x}$. Further, the complementary function for (8.10.12) is

$$\mathbf{x}_c(t) = X(t)\mathbf{c}.$$

If we use the matrix exponential function e^{At} as the fundamental matrix, then combining $\mathbf{x}_c$ and $\mathbf{x}_p$, the general solution to the system (8.10.12) assumes the form

$$\mathbf{x}(t) = e^{At}\mathbf{c} + e^{At}\int_0^t e^{-As}\mathbf{b}(s)\,ds. \tag{8.10.13}$$

Imposing the initial condition $\mathbf{x}(0) = \mathbf{x}_0$ yields

$$\mathbf{c} = \mathbf{x}_0.$$

Substituting into (8.10.13) and simplifying, we finally obtain

$$\mathbf{x}(t) = e^{At}\left[\mathbf{x}_0 + \int_0^t e^{-As}\mathbf{b}(s)\,ds\right]$$

which is the generalization of (8.10.11) to systems.

EXERCISES 8.10

1 . If $X(t)$ is any fundamental matrix for $\mathbf{x}' = A\mathbf{x}$, show that the transition matrix based at $t = 0$ is given by

$$X_0 = X(t)X^{-1}(0).$$

For problems 2–4, use the techniques from Section 8.5 and Section 8.6 to determine a fundamental matrix for $\mathbf{x}' = A\mathbf{x}$, and hence, find e^{At}.

2 . $A = \begin{bmatrix} 1 & 2 \\ 0 & -1 \end{bmatrix}$.

3 . $A = \begin{bmatrix} 2 & 1 \\ 0 & 2 \end{bmatrix}$.

4 . $A = \begin{bmatrix} 3 & 0 & 0 \\ 0 & 3 & -1 \\ 0 & 1 & 1 \end{bmatrix}$.

For problems 5–11, find n LI solutions to $\mathbf{x}' = A\mathbf{x}$ of the form $e^{At}\mathbf{v}$, and hence find e^{At}.

5 . $A = \begin{bmatrix} 3 & -1 \\ 4 & -1 \end{bmatrix}$.

6 . $A = \begin{bmatrix} -3 & -2 \\ 2 & 1 \end{bmatrix}$.

7 . $A = \begin{bmatrix} 2 & 0 & 0 \\ 0 & 1 & -8 \\ 0 & 2 & -7 \end{bmatrix}$.

For problems 8–10, solve $\mathbf{x}' = A\mathbf{x}$ by determining n LI solutions of the form $\mathbf{x} = e^{At}\mathbf{v}$.

8 . $A = \begin{bmatrix} -8 & 6 & -3 \\ -12 & 10 & -3 \\ -12 & 12 & -2 \end{bmatrix}$. You may assume

that $p(\lambda) = -(\lambda + 2)^2(\lambda - 4)$.

9 . $A = \begin{bmatrix} 0 & 1 & 3 \\ 2 & 3 & -2 \\ 1 & 1 & 2 \end{bmatrix}$. You may assume that

$p(\lambda) = -(\lambda + 1)(\lambda - 3)^2$.

10. $A = \begin{bmatrix} 1 & 0 & 0 & 0 \\ 0 & 6 & -7 & 3 \\ 0 & 0 & 3 & -1 \\ 0 & -4 & 9 & -3 \end{bmatrix}$. You may assume

that $p(\lambda) = (\lambda - 1)(\lambda - 2)^3$.

11. The matrix $A = \begin{bmatrix} 0 & -1 & 0 & 0 \\ 1 & 0 & 0 & 0 \\ 1 & 0 & 0 & -1 \\ 0 & 1 & 1 & 0 \end{bmatrix}$ has

characteristic polynomial $p(\lambda) = (\lambda^2 + 1)^2$. Determine two *complex-valued* solutions to $\mathbf{x}' = A\mathbf{x}$ of the form $\mathbf{x} = e^{At}\mathbf{v}$, and hence, find four LI *real-valued* solutions to the differential system.

8.11   THE PHASE PLANE FOR LINEAR AUTONOMOUS SYSTEMS

So far in this chapter we have developed the general theory for linear systems of DE, and have derived particular solution techniques for solving such systems in the constant coefficient case. If we drop either the constant coefficient assumption or the linearity assumption, in general, it is not possible to explicitly solve the resulting systems. Consequently, we need to resort either to a qualitative analysis of the system or to numerical techniques. In the final two sections of this chapter, we give a brief introduction to the qualitative approach in the case of systems of the form

$$\frac{dx}{dt} = F(x, y), \tag{8.11.1}$$

$$\frac{dy}{dt} = G(x, y), \tag{8.11.2}$$

where F and G depend on x and y only. Such a system in which t does not explicitly occur in F and G is called an **autonomous system**. We can interpret the two equations in the system as determining the components of the velocity of a particle that is moving in the xy-plane. As t increases, the particle moves along a curve in the xy-plane called a **trajectory**.[1] The xy-plane itself is referred to as the **phase plane**, and the totality of all trajectories gives the **phase portrait**. Note that each trajectory has a natural direction associated with it, namely, the direction that the particle moves along a trajectory as t increases. From equation (8.11.1) and equation (8.11.2) we see that the DE determining the trajectories is

$$\frac{dy}{dx} = \frac{G(x, y)}{F(x, y)}.$$

Even if we cannot solve this DE, it is possible to obtain much qualitative information about the behavior of the trajectories by constructing, either by hand or using technology,

---

[1]Other terms used for trajectories are *phase paths* or *orbits*.

the slope field associated with it.

For the system of equations (8.11.1) and (8.11.2), any values of x and y for which *both* F and G vanish are called **equilibrium points**. If (x_0, y_0) is an equilibrium point, then $x(t) = x_0$, $y(t) = y_0$ is a solution to the system (8.11.1), (8.11.2) and is called an **equilibrium solution**. We will see that equilibrium points play a key role in the analysis of the phase plane.

Example 8.11.1   Determine all equilibrium points for the system

$$x' = x + y, \; y' = 2x - 3y.$$

Solution   To determine any equilibrium points, we must solve

$$x + y = 0, \quad 2x - 3y = 0.$$

Since the determinant of the matrix of coefficients of this homogeneous algebraic linear system is nonzero, the only solution is $(0, 0)$. Hence, this is the only equilibrium point.

Example 8.11.2   Determine the phase portrait for the system

$$\frac{dx}{dt} = x, \quad \frac{dy}{dt} = -y. \tag{8.11.3}$$

Solution   The only equilibrium point for the system is $(0, 0)$. In this example, the system has constant coefficients and has solutions

$$x(t) = c_1 e^t, \; y(t) = c_2 e^{-t}. \tag{8.11.4}$$

To determine the phase portrait, however, we need the relationship between x and y. This is easily found, since (8.11.4) implies that

$$xy = c_1 c_2, \tag{8.11.5}$$

so that the trajectories consist of rectangular hyperbola $xy = k$, together with the two asymptotes $y = 0$ and $x = 0$. The equilibrium solution $x(t) = 0$, $y(t) = 0$ corresponds to the origin in the phase plane, and therefore no other trajectories can pass through the origin. We see from the first equation in (8.11.3) that $dx/dt > 0$ when $x > 0$ and that $dx/dt < 0$ when $x < 0$. Furthermore, when $x = 0$, the second equation in (8.11.3) implies that $dy/dt > 0$ when $y < 0$ and $dy/dt < 0$ when $y > 0$. This determines the directions shown on the trajectories in Figure 8.11.1.

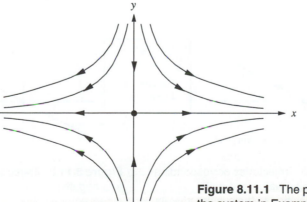

Figure 8.11.1 The phase portrait for the system in Example 8.11.2.

Before analyzing the general autonomous system of equations (8.11.1) and (8.11.2), we need to look at the simpler case when F and G are linear functions of x and y. The system then reduces to the general homogeneous constant coefficient system

$$x' = ax + by, \quad y' = cx + dy, \tag{8.11.6}$$

where a, b, c, and d are constants. Consequently, the DE for determining the trajectories (or slope field) is

$$\frac{dy}{dx} = \frac{cx + dy}{ax + by},$$

which falls into the first-order homogeneous type that we studied in Chapter 1. Whereas in many cases it is possible to solve this DE using the change of variables $y = xV$, we can more easily determine the general behavior in the phase plane by working with the equivalent VDE and using the results already obtained in this chapter. Consequently, we write (8.11.6) as the VDE

$$\mathbf{x}' = A\mathbf{x}, \quad A = \begin{bmatrix} a & b \\ c & d \end{bmatrix}. \tag{8.11.7}$$

We will assume that $\det(A) \neq 0$, so that the only equilibrium point for the system is the origin $(0, 0)$. Since this corresponds to a solution to the system, it follows that no other trajectory can pass through the origin. As we now show, the eigenvalues and eigenvectors of A play a basic role in the structure of the phase plane. First suppose that λ ($\neq 0$) and $\mathbf{v}$ are a real eigenvalue/eigenvector pair for A. Then a solution to (8.11.7) is

$$\mathbf{x}(t) = e^{\lambda t}\mathbf{v}.$$

Since $\mathbf{v}$ is a constant vector, the corresponding trajectories are two half-lines that emanate from the equilibrium point $(0, 0)$ and are parallel to the eigenvector $\mathbf{v}$. The initial conditions would determine which half-line corresponded to a particular motion. If $\lambda > 0$, then, due to the $e^{\lambda t}$ term, the direction along the trajectory is away from the origin. (See Figure 8.11.2.) Interpreting $(x(t), y(t))$ as the coordinates of a particle at time t, these trajectories correspond to a particle emitted from the equilibrium point at $t = -\infty$ and moving outwards along the appropriate half-line. If $\lambda < 0$, then the direction along the trajectory is towards the origin. (See Figure 8.11.3). In this case, we can interpret the trajectory as corresponding to a point particle moving along the half-line towards the origin, but which does not reach the origin in a finite time.

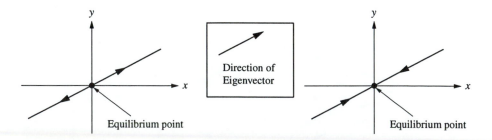

Figure 8.11.2 Trajectories corresponding to a positive eigenvalue and real eigenvector solution to the system $\mathbf{x}' = A\mathbf{x}$.

Figure 8.11.3 Trajectories corresponding to a negative eigenvalue and real eigenvector solution to the system $\mathbf{x}' = A\mathbf{x}$.

We now consider the general phase plane for the system (8.11.7). We let λ_1 and λ_2 denote the eigenvalues of A. The analysis is split up into several cases.

CASE 1: λ_1 and λ_2 real and distinct. Let v_1 and v_2 denote corresponding LI eigenvectors. Then we have the basic solutions

$$x_1(t) = e^{\lambda_1 t} v_1, \quad x_2(t) = e^{\lambda_2 t} v_2, \tag{8.11.8}$$

and the general solution to the system is

$$x(t) = c_1 e^{\lambda_1 t} v_1 + c_2 e^{\lambda_2 t} v_2. \tag{8.11.9}$$

The two solutions in (8.11.8) give rise to four half-line trajectories, as previously discussed. Since trajectories cannot intersect, the phase plane is divided into four regions. The specific behavior of the remaining trajectories depends on the relationship between λ_1 and λ_2.

(a) $\lambda_2 < \lambda_1 < 0$: The general properties of the trajectories are summarized as follows:

1. Since λ_1 and λ_2 are both negative, $\lim_{t \to \infty} x(t) = 0$, so that as $t \to \infty$ all trajectories approach the equilibrium point $(0, 0)$.

2. Writing (8.11.9) as

$$x(t) = e^{\lambda_1 t} [c_1 v_1 + c_2 v_2 e^{(\lambda_2 - \lambda_1)t}]$$

and using the fact that $\lambda_2 - \lambda_1 < 0$ in this case, we see that for large t, and $c_1 \neq 0$, the second term in the brackets is negligible compared to the first term. Consequently, apart from the trajectories corresponding to the eigenvector solution $x_2(t) = e^{\lambda_2 t} v_2$, all trajectories are parallel to v_1 as $t \to \infty$.

3. Writing (8.11.9) as

$$x(t) = e^{\lambda_2 t} [c_1 v_1 e^{(\lambda_1 - \lambda_2)t} + c_2 v_2]$$

and using the fact that $\lambda_1 - \lambda_2 > 0$, we see that apart from the trajectories corresponding to the eigenvector solution $x_1(t) = e^{\lambda_1 t} v_1$ all trajectories are parallel to v_2 as $t \to -\infty$.

A generic sketch of the phase plane in this case is given in Figure 8.11.4. The equilibrium

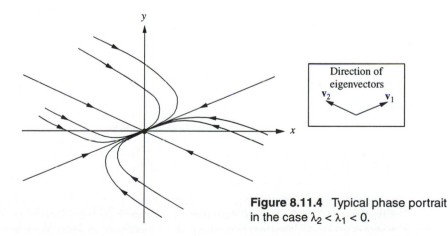

Figure 8.11.4 Typical phase portrait in the case $\lambda_2 < \lambda_1 < 0$.

point $(0, 0)$ is called a **node**. It is stable, since all solutions approach the node as $t \to \infty$.

(b) $0 < \lambda_1 < \lambda_2$: The general behavior in this case is the same as that in Case 1a, except the arrows are reversed on each trajectory. The critical point is still called a **node**, but in this case it is unstable.

(c) $\lambda_2 < 0 < \lambda_1$:   The following general behavior can be identified.

1. The only trajectories that approach the equilibrium point as $t \to \infty$ are those corresponding to the eigenvector solution $\mathbf{x}_2(t) = e^{\lambda_2 t}\mathbf{v}_2$.

2. Writing (8.11.9) as

$$\mathbf{x}(t) = e^{\lambda_1 t}\left[c_1\mathbf{v}_1 + c_2\mathbf{v}_2\,e^{(\lambda_2 - \lambda_1)t}\right]$$

and using the fact that $\lambda_2 - \lambda_1 < 0$, it follows that apart from the trajectories corresponding to the eigenvector solution $\mathbf{x}_2(t) = e^{\lambda_2 t}\mathbf{v}_2$, all trajectories are parallel to $\mathbf{v}_1$ as $t \to \infty$.

3. Writing (8.11.9) as

$$\mathbf{x}(t) = e^{\lambda_2 t}\left[c_1\mathbf{v}_1\,e^{(\lambda_1 - \lambda_2)t} + c_2\mathbf{v}_2\right]$$

and using the fact that $\lambda_1 - \lambda_2 > 0$, we see that, apart from the trajectories corresponding to the eigenvector solution $\mathbf{x}_1(t) = e^{\lambda_1 t}\mathbf{v}_1$, all trajectories are parallel to $\mathbf{v}_2$ as $t \to -\infty$.

A typical phase portrait is sketched in Figure 8.11.5. In this case, the equilibrium point is called a **saddle point**, and is unstable.

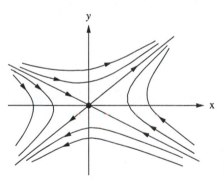

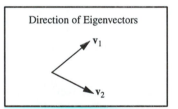

Figure 8.11.5 Typical phase portrait in the case when $\lambda_2 < 0 < \lambda_1$. The equilibrium point is a saddle point and is unstable.

CASE 2: $\lambda_1 = \lambda_2 = \lambda \neq 0$.

(a) Nondefective coefficient matrix A.   In this case, the general solution to the system is

$$\mathbf{x}(t) = e^{\lambda t}(c_1\mathbf{v}_1 + c_2\mathbf{v}_2),$$

which, for each pair of values for c_1 and c_2, is the equation of a line through the origin. Moreover, since $\mathbf{v}_1$ and $\mathbf{v}_2$ are LI, all directions in the xy-plane are obtained as c_1 and c_2 assume all possible values. Consequently, the trajectories consist of all half-lines through the origin. The equilibrium point is called a **proper node** in this case. If $\lambda < 0$, then all trajectories approach the equilibrium point as $t \to \infty$, whereas if $\lambda > 0$, the direction along the trajectories is away from the equilibrium point. A representative sketch of the phase portraits is given in Figure 8.11.6.

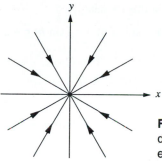

Figure 8.11.6 Typical phase portrait for a non–defective coefficient matrix with $\lambda_1 = \lambda_2 < 0$. The equilibrium point is a proper node and is stable.

(b) Defective coefficient matrix A. The general solution to the system is now

$$\mathbf{x}(t) = e^{\lambda t}[c_1\mathbf{v}_1 + c_2(\mathbf{v}_0 + t\mathbf{v}_1)],$$

where $\mathbf{v}_1$ is an eigenvector corresponding to the eigenvalue λ, and $\mathbf{v}_0$ is a generalized eigenvector. For $c_2 = 0$, we have the trajectories corresponding to the eigenvector solution. If $c_2 \neq 0$, then the dominant term in the general solution as $t \to \pm\infty$ is

$$\mathbf{x}(t) \approx c_2 t e^{\lambda t}\mathbf{v}_1.$$

Consequently, if $\lambda < 0$, we have the following results.

1. As $t \to \infty$, all trajectories approach the equilibrium point $(0, 0)$ tangent to the eigenvector $\mathbf{v}_1$.

2. As $t \to -\infty$, all trajectories are parallel to $\mathbf{v}_1$.

See Figure 8.11.7 for a typical phase portrait. If $\lambda > 0$, then the direction along the trajectories is reversed. In this case, the equilibrium point $(0, 0)$ is called a **degenerate node** and is said to be stable or unstable depending on whether λ is negative or positive respectively.

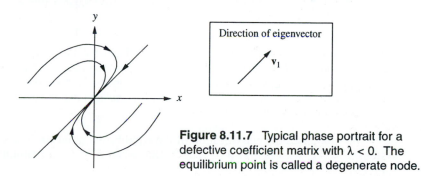

Direction of eigenvector

$\mathbf{v}_1$

Figure 8.11.7 Typical phase portrait for a defective coefficient matrix with $\lambda < 0$. The equilibrium point is called a degenerate node.

CASE 3: Complex conjugate eigenvalues $\lambda = a \pm ib$.

If we let $\mathbf{v} = \mathbf{r} + i\mathbf{s}$ denote a complex eigenvector corresponding to the eigenvalue $\lambda = a + ib$, then according to our results from Section 8.5, two LI solutions to the system of DE are of the form

$$\mathbf{x}_1(t) = e^{at}(\cos bt\, \mathbf{r} - \sin bt\, \mathbf{s}), \qquad \mathbf{x}_2(t) = e^{at}(\sin bt\, \mathbf{r} + \cos bt\, \mathbf{s}).$$

Consequently, the general solution in this case is

$$\mathbf{x}(t) = e^{at}[c_1(\cos bt\ \mathbf{r} - \sin bt\ \mathbf{s}) + c_2(\sin bt\ \mathbf{r} + \cos bt\ \mathbf{s})],$$

which we write as

$$\mathbf{x}(t) = e^{at}\mathbf{v}(t),   \qquad\qquad (8.11.10)$$

where

$$\mathbf{v}(t) = c_1(\cos bt\ \mathbf{r} - \sin bt\ \mathbf{s}) + c_2(\sin bt\ \mathbf{r} + \cos bt\ \mathbf{s}).$$

The key point to notice is that

$$\mathbf{v}(t + 2\pi/b) = \mathbf{v}(t).$$

Consequently, $\mathbf{v}(t)$ has period $T = 2\pi/b$ and we can therefore draw the following conclusions.

(a) If $a = 0$, equation (8.11.10) implies that $\mathbf{x}(t + 2\pi/b) = \mathbf{x}(t)$. All trajectories are therefore closed curves and so the corresponding solutions are periodic. In this case the equilibrium point $(0, 0)$ is called a **center** and is stable.

(b) If $a \neq 0$, then the trajectories spiral around the origin. (See Figure 8.11.9.) The equilibrium point $(0, 0)$ is called a **spiral point**. Furthermore,

1. If $a > 0$, the trajectories spiral away from the equilibrium point, and, therefore, it is called an **unstable spiral point**.

2. If $a < 0$, the trajectories spiral towards the origin, and, therefore, it is called a **stable spiral point**.

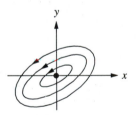

Figure 8.11.8 A typical phase portrait for the case of pure imaginary eigenvalues. The equilibrium point is called a center and is stable.

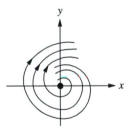

Figure 8.11.9 Typical phase portrait for the case of complex conjugate eigenvalues $\lambda = a + ib$, with $a > 0$. The equilibrium point is called a spiral point.

This completes the classification of the equilibrium point $(0, 0)$ associated with the system $\mathbf{x}' = A\mathbf{x}$, with $\det(A) \neq 0$. The results when $\lambda_1 \neq \lambda_2$ are summarized in Table 8.11.1.

Based on the preceding analysis, it is not too difficult to obtain a general sketch of the phase plane once we have determine the eigenvalues and eigenvectors of A. However, as in the case of slope fields considered in Chapter 1, this is an area where technology is a definite benefit. In the examples that follow we have provided Maple plots of the phase planes.

TABLE   8.11.1

Eigenvalues	Type of Equilibrium Point
Real and negative	Stable node
Real and positive	Unstable node
Opposite sign	Saddle
Pure imaginary	Stable center
Complex with positive real part	Unstable spiral
Complex with negative real part	Stable Spiral

Example 8.11.3   Characterize the equilibrium point for the system $\mathbf{x}' = A\mathbf{x}$ and sketch the phase portrait.

(a) $A = \begin{bmatrix} -1 & -2 \\ -2 & -1 \end{bmatrix}$ (b) $\begin{bmatrix} -1 & -2 \\ 2 & -1 \end{bmatrix}$.   (c) $\begin{bmatrix} 1 & 3 \\ -2 & -4 \end{bmatrix}$.

Solution

(a) The matrix A has eigenvalues $\lambda_1 = 1$, $\lambda_2 = -3$ with corresponding LI eigenvectors $\mathbf{v}_1 = (1, -1)$, $\mathbf{v}_2 = (1, 1)$. Since the eigenvalues have different signs, the equilibrium point $(0, 0)$ is a saddle point. A Maple sketch including the slope field is given in Figure 8.11.10. Notice that we have appended arrow heads to each line segment in the slope field to indicate the direction that trajectories are traversed. The resulting slope field is usually called a **direction field**.

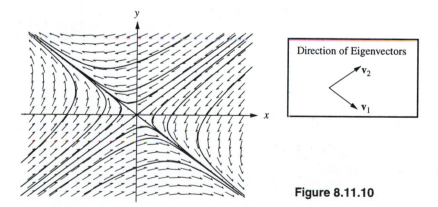

Figure 8.11.10

(b) In this case, the matrix has complex conjugate eigenvalues $\lambda = -1 \pm 2i$. Since the real part of the eigenvalues is negative, the equilibrium point is a stable spiral. To determine whether the trajectories spiral clockwise or counterclockwise towards the origin, we check the sign of dy/dt at points where the trajectories intersect the positive x-axis. From the given system when $x = 0$ and $y > 0$, we see that $dy/dt < 0$, so that the trajectories spiral counterclockwise around the origin. A Maple plot of the phase plane is given in Figure 8.11.11.

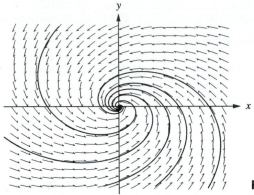

Figure 8.11.11

(c) A has eigenvalues $\lambda_1 = -1$ and $\lambda_2 = -2$, with corresponding LI eigenvectors $\mathbf{v}_1 = (3, -2)$ and $\mathbf{v}_2 = (1, -1)$. We see that the equilibrium point is a stable node. All trajectories approach $(0, 0)$ tangent to the eigenvector $\mathbf{v}_1$. The phase plane for this system is given in Figure 8.11.12.

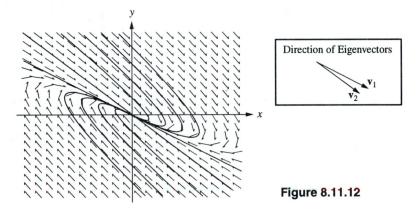

Figure 8.11.12

EXERCISES 8.11

For problems 1–3, determine all equilibrium points of the given system.

1. $x' = x(x - y + 1)$, $y' = y(y + 2x)$.

2. $x' = x(2x + y)$, $y' = y(x - 2y + 4)$.

3. $x' = x(x^2 + y^2 - 1)$, $y' = 2y(xy - 1)$.

For problems 4–19, characterize the equilibrium point for the system $\mathbf{x}' = A\mathbf{x}$ and sketch the phase portrait.

4. $A = \begin{bmatrix} 1 & 3 \\ 1 & -1 \end{bmatrix}$.

5. $A = \begin{bmatrix} 0 & 2 \\ -2 & 0 \end{bmatrix}$.

6. $A = \begin{bmatrix} 1 & 0 \\ 3 & 1 \end{bmatrix}$.

7. $A = \begin{bmatrix} 2 & 3 \\ -1 & -2 \end{bmatrix}$.

8. $A = \begin{bmatrix} -2 & 3 \\ -3 & -2 \end{bmatrix}$.

9. $A = \begin{bmatrix} -2 & 1 \\ 1 & -2 \end{bmatrix}$.

10. $A = \begin{bmatrix} 5 & 4 \\ 4 & 5 \end{bmatrix}$.

11. $A = \begin{bmatrix} 0 & -1 \\ 1 & 0 \end{bmatrix}$.

12. $A = \begin{bmatrix} 3 & -2 \\ 2 & -1 \end{bmatrix}$.

13. $A = \begin{bmatrix} 2 & -1 \\ 1 & 2 \end{bmatrix}$.

14. $A = \begin{bmatrix} 2 & -5 \\ 4 & -7 \end{bmatrix}$.

15. $A = \begin{bmatrix} 2 & 1 \\ 3 & 4 \end{bmatrix}$.

16. $A = \begin{bmatrix} 3 & 4 \\ 4 & -3 \end{bmatrix}$.

17. $A = \begin{bmatrix} 1 & 1 \\ -9 & -5 \end{bmatrix}$.

18. $A = \begin{bmatrix} 1 & -1 \\ 1 & 2 \end{bmatrix}$.

19. $A = \begin{bmatrix} 3 & 0 \\ 0 & 3 \end{bmatrix}$.

20. Characterize the equilibrium point $(0, 0)$ for the system $\mathbf{x}' = A\mathbf{x}$ if $A = \begin{bmatrix} -1 & 2 \\ -2 & -1 \end{bmatrix}$. Solve the system of DE, and show that the components of the solution vector satisfy

$$x^2 + y^2 = e^{-4t}(c_1{}^2 + c_2{}^2), \qquad (20.1)$$

where c_1 and c_2 are constants. As t varies in equation (20.1), describe the curve that is generated in the phase plane.

For problems 21–24, convert the given DE to a first-order system using the substitution $u = y$, $v = dy/dt$, and determine the phase portrait for the resulting system.

21. $\dfrac{d^2y}{dt^2} + 6\dfrac{dy}{dt} + 9y = 0$.

22. $\dfrac{d^2y}{dt^2} + 16y = 0$.

23. $\dfrac{d^2y}{dt^2} + 4\dfrac{dy}{dt} + 5y = 0$.

24. $\dfrac{d^2y}{dt^2} - 25y = 0$.

25. Consider the DE

$$\frac{d^2y}{dt^2} + 2c\frac{dy}{dt} + ky = 0,$$

where c and k are positive constants, that governs the behavior of a spring-mass system. Convert the DE to a first-order linear system and sketch the corresponding phase portraits. (You will need to distinguish the three cases $c^2 > k$, $c^2 < k$, and $c^2 = k$.) In each case, use your phase portrait to describe the behavior of y for various initial conditions.

8.12 NONLINEAR SYSTEMS

We now briefly discuss the qualitative analysis of general autonomous systems of the form

$$\frac{dx}{dt} = F(x, y), \quad \frac{dy}{dt} = G(x, y), \qquad (8.12.1)$$

where, throughout the remainder of the discussion, we will assume that F and G have continuous partial derivatives up to order at least two. We are interested in making a similar classification of the equilibrium points for this system as we were able to do in the linear case. The approach that we will take is to approximate (8.12.1) with a corresponding linear system. To see how the approximation arises, we recall from elementary calculus that if we are given a function $f(x, y)$ defined in some region of the xy-plane, then the equation $z = f(x, y)$ defines a surface in space. Further, the tangent plane to this surface at any point (x_0, y_0) has equation

$$z = f(x_0, y_0) + (x - x_0)\frac{\partial f}{\partial x}(x_0, y_0) + (y - y_0)\frac{\partial f}{\partial y}(x_0, y_0),$$

and this plane gives the best linear approximation to $f(x, y)$ at (x_0, y_0). Returning to the system (8.12.1), we define the linear approximation to this system at (x_0, y_0) by

$$\frac{dx}{dt} = F(x_0, y_0) + (x - x_0)\frac{\partial F}{\partial x}(x_0, y_0) + (y - y_0)\frac{\partial F}{\partial y}(x_0, y_0)$$

$$\frac{dy}{dt} = G(x_0, y_0) + (x - x_0)\frac{\partial G}{\partial x}(x_0, y_0) + (y - y_0)\frac{\partial G}{\partial y}(x_0, y_0).$$

In the case when (x_0, y_0) is an equilibrium point of (8.12.1), the approximate system reduces to

$$\frac{dx}{dt} = (x - x_0)\frac{\partial F}{\partial x}(x_0, y_0) + (y - y_0)\frac{\partial F}{\partial y}(x_0, y_0),$$

$$\frac{dy}{dt} = (x - x_0)\frac{\partial G}{\partial x}(x_0, y_0) + (y - y_0)\frac{\partial G}{\partial y}(x_0, y_0).$$

The **Jacobian matrix**, $J(x, y)$, is defined by

$$J(x, y) = \begin{bmatrix} \dfrac{\partial F}{\partial x} & \dfrac{\partial F}{\partial y} \\ \dfrac{\partial G}{\partial x} & \dfrac{\partial G}{\partial y} \end{bmatrix}.$$

Using this matrix, the linear approximation to (8.12.1) at an equilibrium point (x_0, y_0) can be written as the VDE

$$\frac{dx}{dt} = J(x_0, y_0)\begin{bmatrix} x - x_0 \\ y - y_0 \end{bmatrix},$$

or, equivalently, as

$$\mathbf{u}' = J(x_0, y_0)\mathbf{u},$$

where $\mathbf{u} = \begin{bmatrix} x - x_0 \\ y - y_0 \end{bmatrix}.$

Example 8.12.1   Determine the linear approximation to the system

$$x' = \cos x + 3y - 1 \ , \ y' = 2x + \sin y.$$

at the equilibrium point $(0, 0)$.

Solution   For the given system, we have

$$F(x, y) = \cos x + 3y - 1, \qquad G(x, y) = 2x + \sin y,$$

so that

$$J(x, y) = \begin{bmatrix} -\sin x & 3 \\ 2 & \cos y \end{bmatrix}.$$

Hence,

$$J(0, 0) = \begin{bmatrix} 0 & 3 \\ 2 & 1 \end{bmatrix},$$

and the linear approximation to the given system at $(0, 0)$ is

$$\mathbf{x}' = \begin{bmatrix} 0 & 3 \\ 2 & 1 \end{bmatrix} \begin{bmatrix} x \\ y \end{bmatrix},$$

or, equivalently,

$$\frac{dx}{dt} = 3y, \quad \frac{dy}{dt} = 2x + y. \qquad \Box$$

Example 8.12.2 Determine all equilibrium points for the system

$$x' = x(1 - y), \quad y' = y(2 - x),$$

and determine the linear approximation to the system at each equilibrium point.

Solution The system has the two equilibrium points $(0, 0)$ and $(2, 1)$. In this case, the Jacobian matrix is

$$J(x, y) = \begin{bmatrix} 1 - y & -x \\ -y & 2 - x \end{bmatrix}.$$

Thus,

$$J(0, 0) = \begin{bmatrix} 1 & 0 \\ 0 & 2 \end{bmatrix},$$

so that the linear approximation at the equilibrium point $(0, 0)$ is

$$\mathbf{x}' = J(0, 0)\mathbf{x}.$$

That is

$$x' = x, \quad y' = 2y.$$

Similarly, the linear approximation at the equilibrium point $(2, 1)$ is

$$\mathbf{u}' = J(2, 1)\mathbf{u} = \begin{bmatrix} 0 & -2 \\ -1 & 0 \end{bmatrix} \mathbf{u},$$

where $\mathbf{u} = \begin{bmatrix} u \\ v \end{bmatrix} = \begin{bmatrix} x - 2 \\ y - 1 \end{bmatrix}$. In scalar form, we have

$$u' = -2v, \quad v' = -u. \qquad \Box$$

It perhaps seems reasonable to expect that the behavior of the trajectories to the nonlinear system (8.12.1) is closely approximated by the trajectories of the corresponding linear system, provided that we do not move too far away from (x_0, y_0). This is indeed true in most cases. In Table 8.12.1 we summarize the relationship between the behavior at an equilibrium point of a nonlinear system and the behavior at the corresponding equilibrium point of the linear approximation in the case of distinct eigenvalues. This indicates that apart from the case of pure imaginary eigenvalues (or a repeated eigenvalue), the phase portrait for a nonlinear system looks similar to the linear approximation in the neighborhood of an equilibrium point.

TABLE 8.12.1

Eigenvalues	Linear Approximation	Nonlinear System
Real and negative	Stable node	Stable node
Real and positive	Unstable node	Unstable node
Opposite signs	Saddle	Saddle
Complex with negative real part	Stable spiral	Stable spiral
Complex with positive real part	Unstable spiral	Unstable spiral
Pure imaginary	Stable center	Center or spiral point, stability indeterminate

Example 8.12.3 Determine and classify all equilibrium points for the given system

    (a) $x' = x(x - 1)$, $y' = y(2 + xy^2)$.

    (b) $x' = x - y$, $y' = y(2x + y - 3)$.

Solution

    (a) The equilibrium points are obtained by solving

$$x(x - 1) = 0, \quad y(2 + xy^2) = 0.$$

The first of these equations implies that $x = 0$ or 1. In both cases, substitution into the second equation yields $y = 0$. Consequently, the only equilibrium points are $(0, 0)$ and $(1, 0)$. The Jacobian for the given system is

$$J(x, y) = \begin{bmatrix} 2x - 1 & 0 \\ y^3 & 2 + 3xy^2 \end{bmatrix}.$$

Hence, $J(0, 0) = \begin{bmatrix} -1 & 0 \\ 0 & 2 \end{bmatrix}$, which has eigenvalues $\lambda_1 = 2$ and $\lambda_2 = -1$. Since the eigenvalues have different signs, the equilibrium point $(0, 0)$ is a saddle point in both the linear approximation to the given system and the given system itself. At the equilibrium point $(1, 0)$, we have

$$J(1, 0) = \begin{bmatrix} 1 & 0 \\ 0 & 2 \end{bmatrix}.$$

This matrix has eigenvalues $\lambda_1 = 2$ and $\lambda_2 = 1$, which implies that the equilibrium point is an unstable node. Figure 8.12.1 gives a Maple plot of the phase plane for the given nonlinear system.

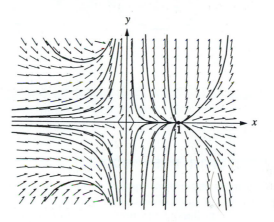

Figure 8.12.1

(b) To determine the equilibrium points, we must solve

$$x - y = 0, \quad y(2x + y - 3) = 0.$$

Substituting $y = x$ from the first equation into the second yields the condition

$$3x(x - 1) = 0.$$

Hence, the equilibrium points are $(0, 0)$ and $(1, 1)$. The Jacobian for the given system is

$$J(x, y) = \begin{bmatrix} 1 & -1 \\ 2y & 2x + 2y - 3 \end{bmatrix}.$$

At the equilibrium point $(0, 0)$, we have

$$J(0, 0) = \begin{bmatrix} 1 & -1 \\ 0 & -3 \end{bmatrix},$$

which has eigenvalues $\lambda_1 = 1$ and $\lambda_2 = -3$, with corresponding eigenvectors $\mathbf{v}_1 = (1, 0)$ and $\mathbf{v}_2 = (1, 4)$. Since the eigenvalues have different signs, the equilibrium point $(0, 0)$ is a saddle point. At the equilibrium point $(1, 1)$, we have

$$J(1, 1) = \begin{bmatrix} 1 & -1 \\ 2 & 1 \end{bmatrix},$$

which has eigenvalues $\lambda = 1 \pm i\sqrt{2}$. Hence, the equilibrium point is an unstable spiral point. A Maple plot of the phase plane is given in Figure 8.12.2.

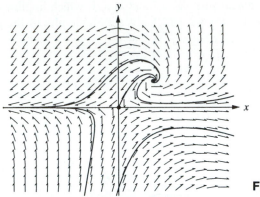

Figure 8.12.2   ❒

A PREDATOR PREY MODEL

As an example of an applied problem that is modeled by a nonlinear system of DE
we consider the interaction of two species. One species is a predator, and the other is the
prey. Let $x(t)$ denote the prey population at time t, and let $y(t)$ denote the predator
population. Then the model equations that we discuss are

$$\frac{dx}{dt} = x(a - by), \tag{8.12.2}$$

$$\frac{dy}{dt} = y(cx - d), \tag{8.12.3}$$

where a, b, c, and d are positive constants. To interpret these equations, we see that in the
absence of any predators, equation (8.12.2) reduces to the simple Malthusian exponential
growth law. The inclusion of predators is taken account of by subtracting a term
proportional to the number of predators present from the growth rate of the prey.
Similarly, from equation (8.12.3), in the absence of prey, the predator population would
decay exponentially. To account for the inclusion of prey, a term proportional to the
number of prey has been added to the growth rate of the predator. This model is called the
Lotka–Volterra system. We see that the system is nonlinear with equilibrium points at
$(0, 0)$ and $(d/c, a/b)$. Computing the Jacobian of the system yields

$$J(x, y) = \begin{bmatrix} a - by & -bx \\ cy & cx - d \end{bmatrix},$$

so that

$$J(0, 0) = \begin{bmatrix} a & 0 \\ 0 & -d \end{bmatrix}$$

with eigenvalues $\lambda_1 = a$ and $\lambda_2 = -d$. Consequently, the equilibrium point $(0, 0)$ is a
saddle point. We note that LI eigenvectors in this case are $\mathbf{v}_1 = (1, 0)$, $\mathbf{v}_2 = (0, 1)$. At
the equilibrium point $(d/c, a/b)$, we have

$$J(d/c, a/b) = \begin{bmatrix} 0 & -bd/c \\ ca/b & 0 \end{bmatrix},$$

with eigenvalues $\lambda = \pm i \sqrt{ad}$. Consequently, in the linear approximation the equilibrium point at $(d/c, a/b)$ is a center. Therefore, according to the results given in Table 8.12.1, the nonlinear system either has a center or a spiral point. In Figure 8.12.3 we give a Maple plot of the phase plane using typical values for the constants a, b, c, and d. This indicates that the equilibrium point in the nonlinear model is also a center. Consequently, the corresponding solutions for both x and y are periodic in time. The model therefore predicts that the population of both species is periodic, and hence both species would survive. The general qualitative behavior starting at small values for both the predator and prey can be seen from the trajectories. The prey initially increases, and the predator population remains approximately constant. Then, since there is plenty of food (prey) the predator population increases with a corresponding decrease in the prey population. This gives rise to a situation where there are too many predators for the prey population, and, therefore, the predator population decreases while the prey population remains approximately constant. Then the cycle repeats itself. We see from the three different trajectories in Figure 8.12.3 that the specific behavior varies quite significantly depending on the initial conditions.

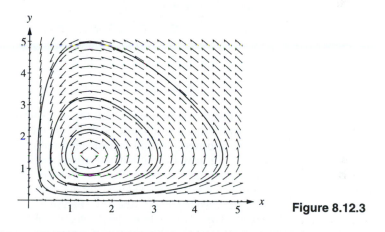

Figure 8.12.3

THE VAN DER POL EQUATION

Finally in this section, we use the nonlinear DE

$$\ddot{y} + \mu(y^2 - 1)\dot{y} + y = 0, \quad \mu > 0, \tag{8.12.4}$$

to illustrate a new type of behavior that does not arise in linear systems. The DE (8.12.4) is called the Van der Pol equation and arises in the study of nonlinear circuits. If the parameter μ is zero, then, equation (8.12.4) reduces to that of the simple harmonic oscillator, which has periodic solutions and circular trajectories. For μ small and positive and $|y| > 1$, we can presumably interpret the dy/dt term as a damping term and expect the system to behave somewhat like a damped harmonic oscillator. However, for $0 < y < 1$, the coefficient of dy/dt is negative, and, therefore, this term would tend to amplify, rather than dampen, any oscillations. This suggests that there may be an isolated periodic solution (closed trajectory) with the property that all trajectories that start within it approach the closed path as t increases, and all trajectories that start outside the closed path spiral towards it as $t \rightarrow \infty$. Such a closed path, if it exists, is called a **limit cycle**. To analyze the Van der Pol equation, we introduce the phase plane variables

$$u = y, \; v = \frac{dy}{dt},$$

thereby obtaining the equivalent first-order system

$$\frac{du}{dt} = v, \quad \frac{dv}{dt} = -u - \mu(u^2 - 1)v. \tag{8.12.5}$$

The only equilibrium point of the system (8.12.5) is (0, 0), and the Jacobian of this system is

$$J(u, v) = \begin{bmatrix} 0 & 1 \\ -1 - 2\mu v & -\mu(u^2 - 1) \end{bmatrix}.$$

Hence,

$$J(0, 0) = \begin{bmatrix} 0 & 1 \\ -1 & \mu \end{bmatrix}.$$

This matrix has characteristic polynomial

$$p(\lambda) = \lambda^2 - \mu\lambda + 1,$$

so that the eigenvalues are

$$\lambda = \frac{1}{2} (\mu \pm \sqrt{\mu^2 - 4}).$$

For $\mu > 2$, there are two positive eigenvalues, and the equilibrium point is an unstable node. For $\mu < 2$, however, the equilibrium point is an unstable spiral. Hence, the trajectories close to the equilibrium point do indeed spiral outwards. However, this local analysis does not give us information about the global behavior of the trajectories. Although we do not have the tools to prove that the Van der Pol equation does indeed have a limit cycle, further convincing evidence for its existence can be obtained by studying the phase portraits associated with the DE. Figure 8.12.4 contains a Maple plot in the case when $\mu = 0.1$. The limit cycle is clearly visible and is almost circular, as we would expect with a small μ value. Figure 8.12.5 contains a similar plot with $\mu = 1$. The limit cycle is still visible, but it no longer resembles a circle.

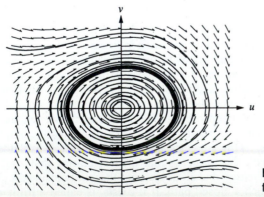

Figure 8.12.4 Maple plot of the phase plane for the Van der Pol equation with $\mu = 0.1$.

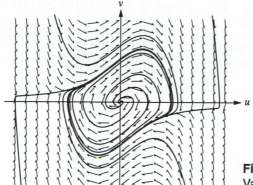

Figure 8.12.5 The phase plane for the Van der Pol equation with $\mu = 1$.

EXERCISES 8.12

For problems 1–9, determine all equilibrium points of the given system and, if possible, characterize them as centers, spirals, saddles, or nodes.

1. $x' = y(3x - 2)$, $y' = 2x + 9y^2$.

2. $x' = y(3x - 2)$, $y' = 2x - 9y^2$.

3. $x' = x - y^2$, $y' = y(9x - 4)$.

4. $x' = x + 3y^2$, $y' = y(x - 2)$.

5. $x' = 2x + 5y^2$, $y' = y(3 - 4x)$.

6. $x' = 2y + \sin x$, $y' = x(\cos y - 2)$.

7. $x' = x - 2y + 5xy$, $y' = 2x + y$.

8. $x' = x(1 - y)$, $y' = y(x + 1)$.

9. $x' = 4x - y - y\sin x$, $y = x + 2y$.

The remaining problems require the use of some form of technology to generate the phase plane for the system of DE.

◆ **10.** Sketch the phase portrait of the system in problem 1 for $-1 \le x \le 1$, $-1 \le y \le 1$, and thereby determine whether the equilibrium point $(0, 0)$ is a center or a spiral.

◆ **11.** Sketch the phase portrait of the system in problem 6 for $-2 \le x \le 2$, $-2 \le y \le 2$, and thereby determine whether the equilibrium point $(0, 0)$ is a center or a spiral.

◆ **12.** Sketch the phase portrait of the system in problem 8 for $-2 \le x \le 2$, $-2 \le y \le 2$. By

inspection, guess the equation of one particular trajectory.

For problems 13–18, sketch the phase portrait of the given system for $-2 \le x \le 2$, $-2 \le y \le 2$. Comment on the types of equilibrium points.

◆ **13.** The system in problem 2.

◆ **14.** The system in problem 3.

◆ **15.** The system in problem 4.

◆ **16.** The system in problem 5.

◆ **17.** The system in problem 7.

◆ **18.** The system in problem 9.

◆ **19.** Consider the predator-prey model

$$\frac{dx}{dt} = x(2 - y), \quad \frac{dy}{dt} = y(x - 2).$$

Sketch the phase plane for $0 \le x \le 10$, $0 \le y \le 10$. Compare the behavior of the two specific cases corresponding to the initial conditions $x(0) = 1$, $y(0) = 0.1$, and $x(0) = 1$, $y(0) = 1$.

◆ **20.** Consider the predator-prey model

$$\frac{dx}{dt} = x(3 - x - y), \quad \frac{dy}{dt} = y(x - 1).$$

Sketch the phase plane for $0 \le x \le 4$, $0 \le y \le 4$. What happens to the populations of both species as $t \to +\infty$?

◆ **21.** Consider the DE

$$\ddot{y} + 0.1(y - 4)(y + 1)\dot{y} + y = 0.$$

(a) Convert the DE to a first-order system using the substitution $u = y$, $v = dy/dt$, and characterize the equilibrium point $(0, 0)$.

(b) Sketch the phase plane for the system on the square $-2 \leq u \leq 2$, $-2 \leq v \leq 2$. Based on the resulting sketch, do you think the DE has a limit cycle?

(c) Repeat (b) using the square $-8 \leq u \leq 8$, $-8 \leq v \leq 8$, and including the trajectories corresponding to the initial conditions $u(0) = 1$, $v(0) = 0$, and $u(0) = 6$, $v(0) = 0$.

9

The Laplace Transform
and Some Elementary
Applications

9.1 THE DEFINITION OF THE LAPLACE TRANSFORM

In this chapter, we introduce another technique for solving linear, constant coefficient ordinary DE. Actually the technique has a much broader usage than this, for example, in the solution of linear systems of DE, partial DE, and also integral equations (see Section 9.9). The reader's immediate reaction is probably to question the need for introducing a new method for solving constant coefficient equations, since our results from Chapter 7 can be applied to any such equation. To answer this question, consider the DE

$$y'' + ay' + by = F,$$

where a and b are constants. We have seen how to solve this equation when F is a continuous function on some interval I. However, in many problems that arise in engineering, physics, and applied mathematics, F represents the external force that is acting on the system under investigation, and often this force acts intermittently or even instantaneously.[1] Whereas the techniques from Chapter 7 can be extended to cover these cases, the computations involved are tedious. In contrast, the approach introduced here

---

[1] For example the switch in an RLC circuit may be turned on and off several times, or the mass in a spring-mass system may be dealt an instantaneous blow at $t = t_0$.

can handle such problems quite easily.
    First we need a definition.

Definition 9.1.1:   Let f be a function defined on the interval $[0, \infty)$. The **Laplace transform** of f is the function $F(s)$ defined by

$$F(s) = \int_0^\infty e^{-st}f(t)\, dt, \qquad (9.1.1)$$

provided that the improper integral converges. We will usually denote the Laplace transform of f by $L[f]$.

Recall that the improper integral appearing in (9.1.1) is defined by

$$\int_0^\infty e^{-st}f(t)\, dt = \lim_{N\to\infty} \int_0^N e^{-st}f(t)\, dt$$

and that this improper integral converges if and only if the limit on the right-hand side exists and is finite. It follows that *not* all functions defined on $[0, \infty)$ have a Laplace transform. In the next section, we will address some of the theoretical aspects associated with determining the types of functions for which (9.1.1) converges. For the remainder of this section, we focus our attention on gaining familiarity with Definition 9.1.1 and derive some basic Laplace transforms.

Example 9.1.1 Determine the Laplace transform of the following functions:
    (a) $f(t) = 1$.
    (b) $f(t) = t$.
    (c) $f(t) = e^{at}$, where a is constant.
    (d) $f(t) = \cos bt$, where b is constant.

Solution
    (a) From the foregoing definition, we have

$$L[1] = \int_0^\infty e^{-st}dt = \lim_{N\to\infty}\left[-\frac{1}{s}e^{-st} \right]_0^N = \lim_{N\to\infty}\left[\frac{1}{s} - \frac{1}{s}e^{-sN} \right] = \frac{1}{s}, \quad s > 0.$$

Notice that the restriction $s > 0$ is required for the improper integral to converge.
    (b) In this case, we use integration by parts to obtain

$$L[t] = \int_0^\infty e^{-st}\, t\, dt = \lim_{N\to\infty}\left[-\frac{t\, e^{-st}}{s} \right]_0^N + \int_0^\infty \frac{1}{s}e^{-st}\, dt.$$

But,

$$\lim_{N\to\infty} Ne^{-sN} = 0, \quad s > 0,$$

so that

$$L[t] = \int_0^\infty \frac{1}{s}e^{-st}\, dt = \lim_{N\to\infty}\left[-\frac{e^{-st}}{s^2} \right]_0^N = \frac{1}{s^2}, \; s > 0.$$

It is left as an exercise to show that more generally, for all positive integers n,

$$L[t^n] = \frac{n!}{s^{n+1}}, \; s > 0.$$ (9.1.2)

(c) In this case, we have

$$L[e^{at}] = \int_0^\infty e^{-st}e^{at} \, dt = \int_0^\infty e^{(a-s)t} \, dt = \lim_{N \to \infty} \left[\frac{1}{a-s} e^{(a-s)t} \right]_0^N = \frac{1}{s-a},$$

provided that $s > a$. Thus,

$$L[e^{at}] = \frac{1}{s-a}, \; s > a.$$ (9.1.3)

(d) From the definition of the Laplace transform,

$$L[\cos bt] = \int_0^\infty e^{-st}\cos bt \, dt.$$

Using the standard integral

$$\int e^{at}\cos bt \, dt = \frac{e^{at}}{a^2+b^2}(a \cos bt + b \sin bt) + c,$$

it follows that

$$L[\cos bt] = \lim_{N \to \infty} \left[\frac{e^{-st}}{s^2+b^2}(b \sin bt - s \cos bt) \right]_0^N = \frac{s}{s^2+b^2}$$

provided that $s > 0$. Thus,

$$L[\cos bt] = \frac{s}{s^2+b^2}, \; s > 0.$$ (9.1.4)

Similarly, it can be shown that

$$L[\sin bt] = \frac{b}{s^2+b^2}, \; s > 0.$$ (9.1.5)

As illustrated by the preceding examples, the range of values that s can assume must often be restricted to ensure the convergence of the improper integral (9.1.1).

LINEARITY OF THE LAPLACE TRANSFORM

Suppose that the Laplace transform of both f and g exists for $s > \alpha$, where α is a constant. Then, using properties of convergent improper integrals, it follows that for $s > \alpha$,

$$L[f+g] \quad = \int_0^\infty e^{-st}[f(t)+g(t)] \, dt = \int_0^\infty e^{-st}f(t) \, dt + \int_0^\infty e^{-st}g(t) \, dt$$

$$= L[f] + L[g].$$

Further, if c is any real number, then

$$L[cf] = \int_0^\infty e^{-st}cf(t) \, dt = c \int_0^\infty e^{-st}f(t) \, dt = cL[f].$$

Consequently,

> **1.** $L[f+g] = L[f] + L[g]$,
>
> **2.** $L[cf] = L[f]$,

so that the Laplace transform satisfies the basic properties of a linear transformation. This linearity of L enables us to determine the Laplace transform of complicated functions from a knowledge of the Laplace transform of some basic functions. This will be used continually throughout the chapter.

Example 9.1.2   Determine the Laplace transform of

$$f(t) = 4e^{3t} + 2\sin 5t - 7t^3.$$

Solution   Since the Laplace transform is linear it follows that

$$L[4e^{3t} + 2\sin 5t - 7t^3] = 4L[e^{3t}] + 2L[\sin 5t] - 7L[t^3].$$

Using the results of the previous example, we therefore obtain

$$L[4e^{3t} + 2\sin 5t - 7t^3] = \frac{4}{s-3} + \frac{10}{s^2+25} - \frac{42}{s^4}, \quad s > 3.$$

PIECEWISE CONTINUOUS FUNCTIONS

The functions that we have considered in the foregoing examples have all been continuous on $[0, \infty)$. As we will see in the later sections the real power of the Laplace transform comes from the fact that piecewise continuous functions can be transformed. Before illustrating this point we recall the definition of a piecewise continuous function.

> **Definition 9.1.2:**   A function f is called **piecewise continuous** on the interval $[a, b]$ if we can divide $[a, b]$ into a finite number of subintervals in such a manner that
>
> **1.** f is continuous on each subinterval, and
>
> **2.** f approaches a finite limit as the endpoints of each subinterval are approached from within.
>
> If f is piecewise continuous on every interval of the form $[0, b]$, where b is a constant, then we say that f is piecewise continuous on $[0, \infty)$.

Example 9.1.3   The function f defined by

$$f(t) = \begin{cases} t^2 + 1, & 0 \leq t \leq 1, \\ 2 - t, & 1 < t \leq 2, \\ 1, & 2 < t \leq 3, \end{cases}$$

is piecewise continuous on $[0, 3]$, whereas

$$f(t) = \begin{cases} \dfrac{1}{1 - t}, & 0 \leq t < 1, \\ t, & 1 \leq t \leq 3, \end{cases}$$

is not piecewise continuous on $[0, 3]$. The graphs of these functions are shown in Figure 9.1.1.

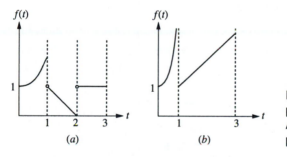

(a)   (b)

Figure 9.1.1   (a) An example of a piecewise continuous function. (b) An example of a function that is not piecewise continuous on $[0, 3]$.

Example 9.1.4   Determine the Laplace transform of the piecewise continuous function

$$f(t) = \begin{cases} t, & 0 \leq t < 1, \\ -1, & t \geq 1. \end{cases}$$

Solution   The function is sketched in Figure 9.1.2. To determine the Laplace transform of f, we use Definition 9.1.1.

$$L[f] = \int_0^\infty e^{-st} f(t)\, dt = \int_0^1 e^{-st} f(t)\, dt + \int_1^\infty e^{-st} f(t)\, dt$$

$$= \int_0^1 e^{-st} t\, dt - \int_1^\infty e^{-st}\, dt = \left[-\frac{1}{s}\, t e^{-st} - \frac{1}{s^2}\, e^{-st} \right]_0^1 + \lim_{N \to \infty} \left[\frac{1}{s}\, e^{-st} \right]_1^N$$

$$= -\frac{1}{s}\, e^{-s} - \frac{1}{s^2}\, e^{-s} + \frac{1}{s^2} - \frac{1}{s}\, e^{-s},$$

provided that $s > 0$. Thus,

$$L[f] = \frac{1}{s^2}[1 - e^{-s}(2s + 1)], \quad s > 0.$$

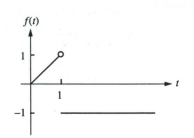

Figure 9.1.2 The piecewise continuous function in Example 9.1.4.

EXERCISES 9.1

For problems 1–12, use (9.1.1) to determine $L[f]$.

1. $f(t) = e^{2t}$.

2. $f(t) = t - 1$.

3. $f(t) = \sin bt$, where b is constant.

4. $f(t) = t\, e^{t}$.

5. $f(t) = \cosh bt$, where b is constant.

6. $f(t) = \sinh bt$, where b is constant.

7. $f(t) = 2t$.

8. $f(t) = 3e^{2t}$.

9. $f(t) = \begin{cases} 1, & 0 \le t < 2, \\ -1, & t \ge 2. \end{cases}$

10. $f(t) = \begin{cases} t^2, & 0 \le t \le 1, \\ 1, & t > 1. \end{cases}$

11. $f(t) = e^{t} \sin t$.

12. $f(t) = e^{2t} \cos 3t$.

For problems 13–22, use the linearity of L and the formulas derived in this section to determine $L[f]$.

13. $f(t) = 2t - e^{3t}$.

14. $f(t) = 2 \sin 3t + 4t^3$.

15. $f(t) = \cosh bt$, where b is a constant.

16. $f(t) = \sinh bt$, where b is a constant.

17. $f(t) = 3t^2 - 5 \cos 2t + \sin 3t$.

18. $f(t) = 7e^{-2t} + 1$.

19. $f(t) = 2e^{-3t} + 4e^{t} - 5 \sin t$.

20. $f(t) = 4\cos(t - \pi/4)$.

21. $f(t) = 4\cos^2 bt$, where b is constant.

22. $f(t) = 2\sin^2 4t - 3$.

For problems 23–30, sketch the given function and determine whether it is piecewise continuous on $[0, \infty)$.

23. $f(t) = \begin{cases} 3, & 0 \le t \le 1, \\ 0, & 1 \le t < 3, \\ -1, & t \ge 3. \end{cases}$

24. $f(t) = \begin{cases} 1, & 0 \le t \le 1, \\ 1 - t, & 1 < t \le 2, \\ 1, & t > 2. \end{cases}$

25. $f(t) = \begin{cases} 1, & 0 \le t \le 1, \\ 1/(t - 1), & t > 1. \end{cases}$

26. $f(t) = \begin{cases} t, & 0 \le t \le 1, \\ 1/t^2, & t > 1. \end{cases}$

27. $f(t) = n, \quad n \le t < n + 1, \quad n = 0, 1, 2, \dots$.

28. $f(t) = t, \ 0 \le t < 1, \ f(t+1) = f(t)$.

29. $f(t) = \dfrac{1}{t - 2}$.

30. $f(t) = \dfrac{2}{t + 1}$.

For problems 31–34, sketch the given function and determine its Laplace transform.

31. $f(t) = \begin{cases} t, & 0 \le t \le 1, \\ 0, & t \ge 1. \end{cases}$

32. $f(t) = \begin{cases} 1, & 0 \le t \le 2, \\ -1, & t > 2. \end{cases}$

33. $f(t) = \begin{cases} 0, & 0 \le t \le 1, \\ t, & 1 < t \le 2, \\ 0 & t > 2. \end{cases}$

34. $f(t) = \begin{cases} t, & 0 \le t < 1, \\ 1, & 1 \le t < 3, \\ e^{(t-3)}, & t > 3. \end{cases}$

35. Recall that according to Euler's formula,

$$e^{ibt} = \cos bt + i \sin bt.$$

Since the Laplace transformation is linear, it follows that

$$L[\cos bt] = \mathrm{Re}(L[e^{ibt}]),$$

$$L[\sin bt] = \mathrm{Im}(L[e^{ibt}]).$$

Find $L[e^{ibt}]$, and hence, derive (9.1.4) and (9.1.5).

36. Use the technique introduced in the previous problem to determine

$$L[e^{at} \cos bt] \quad \text{and} \quad L[e^{at} \sin bt],$$

where a and b are arbitrary constants.

37. Use mathematical induction to prove that for any positive integer n

$$L[t^n] = \frac{n!}{s^{n+1}}.$$

38. (a) By making the change of variables $t = \dfrac{x^2}{s}$, $s > 0$, in the integral that defines the Laplace transform, show that

$$L[t^{-1/2}] = 2s^{-1/2} \int_0^\infty e^{-x^2} \, dx.$$

(b) Use your result in (a) to show that

$$(L[t^{-1/2}])^2 = 4s^{-1} \int_0^\infty \int_0^\infty e^{-(x^2 + y^2)} \, dx \, dy.$$

(c) By changing to polar coordinates evaluate the double integral in (b), and hence, show that

$$L[t^{-1/2}] = (\pi/s)^{1/2}, \quad s > 0.$$

9.2   THE EXISTENCE OF THE LAPLACE TRANSFORM AND THE INVERSE TRANSFORM

In the previous section, we derived the Laplace transform of several elementary functions. In this section, we address some of the more theoretical aspects of the Laplace transform. The first question that we wish to answer is the following:

What types of functions have a Laplace transform?

We will not be able to answer this question completely, since it requires a deeper mathematical background than we assume of the reader. However, we can identify a very large class of functions that are Laplace transformable.

By definition, the Laplace transform of a function f is

$$L[f] = \int_0^\infty e^{-st} f(t) \, dt, \tag{9.2.1}$$

provided that the integral converges. If f is piecewise continuous on an interval $[a, b]$, then it is a standard result from calculus that f is also integrable over $[a, b]$. Thus, if we restrict attention to functions that are piecewise continuous on $[0, \infty)$, it follows that the integral

$$\int_0^b e^{-st} f(t) \, dt \tag{9.2.2}$$

exists for all positive (and finite) b. However, it does not follow that the Laplace transform of f exists, since the improper integral in (9.2.1) may still diverge. To guarantee convergence of the integral, we must ensure that the integrand in (9.2.1) approaches zero rapidly enough as $t \to \infty$. As we show next, this will be the case provided that, in addition to being piecewise continuous, f also satisfies the following definition:

> **Definition 9.2.1:**   A function f is said to be of **exponential order** if there exist constants M and α such that
>
> $$|f(t)| \le Me^{\alpha t},$$
>
> for all $t > 0$.

Example 9.2.1   The function $f(t) = 10e^{7t}\cos 5t$ is of exponential order, since

$$|f(t)| = 10e^{7t}|\cos 5t| \le 10e^{7t}. \qquad \square$$

Now let $E(0, \infty)$ denote the set of all functions that are both piecewise continuous on $[0, \infty)$ and of exponential order. If we add two functions that are in $E(0, \infty)$, the result is a new function that is also in $E(0, \infty)$. Similarly, if we multiply a function in $E(0, \infty)$ by a constant the result is once more a function in $E(0, \infty)$. It follows from Theorem 5.3.1 that $E(0, \infty)$ is a subspace of the vector space of all functions defined on $[0, \infty)$. We will show next that the functions in the vector space $E(0, \infty)$ have a Laplace transform. Before doing so, we need to state a basic theorem about the convergence of improper integrals. First a lemma.

> *Lemma 9.2.1 (**The Comparison Test for Improper Integrals**):*   Suppose that $0 \le G(t) \le H(t)$ for $0 \le t < \infty$. If $\displaystyle\int_0^\infty H(t)\,dt$ converges, then so does $\displaystyle\int_0^\infty G(t)\,dt$.

PROOF   See any textbook on Advanced Calculus.   ∎

We also recall that if $\displaystyle\int_0^\infty |F(t)|\,dt$ converges, then so does $\displaystyle\int_0^\infty F(t)\,dt$. We can now establish a key existence theorem for the Laplace transform.

> **Theorem 9.2.1:**   If f is in $E(0, \infty)$, then there exists a constant α such that
>
> $$L[f] = \int_0^\infty e^{-st}f(t)\,dt$$

exists for all $s > \alpha$.

PROOF   Since f is piecewise continuous on $[0, \infty)$, $e^{-st}f(t)$ is integrable over any finite interval. Further, since f is in $E(0, \infty)$, there exist constants M and α such that

$$|f(t)| \le Me^{\alpha t},$$

for all $t > 0$. We now use the comparison test for integrals to establish that the improper integral defining the Laplace transform converges. *Let*

$$F(t) = |e^{-st}f(t)|.$$

Then,

$$F(t) = e^{-st}|f(t)| \le Me^{(\alpha-s)t}.$$

But, for $s > \alpha$,

$$\int_0^\infty Me^{(\alpha-s)t}\, dt = \lim_{N\to\infty} \int_0^N Me^{(\alpha-s)t}\, dt = \frac{M}{s-\alpha}.$$

Applying the comparison test for improper integrals with $F(t)$ as just defined and $G(t) = Me^{(\alpha-s)t}$, it follows that

$$\int_0^\infty |e^{-st}f(t)|\, dt$$

converges for $s > \alpha$, and hence, so also does

$$\int_0^\infty e^{-st}f(t)\, dt.$$

Thus, we have shown that $L[f]$ exists for $s > \alpha$, as required.   ∎

REMARK   The preceding theorem gives only sufficient conditions that guarantee the existence of the Laplace transform. There are functions that are not in $E(0, \infty)$, but that do have a Laplace transform. For example, $f(t) = t^{-1/2}$ is certainly not in $E(0, \infty)$, but $L[t^{1/2}] = (\pi/s)^{1/2}$. (See problem 38 in the previous section.)

THE INVERSE LAPLACE TRANSFORM

Let V denote the subspace of $E(0, \infty)$ consisting of all *continuous* functions of exponential order. We have seen in the previous section that the Laplace transform satisfies

$$L[f + g] = L[f] + L[g], \quad L[cf] = cL[f].$$

Consequently, L defines a linear transformation of V onto $\text{Rng}(L)$. Further, it can be shown that L is also one-to-one, and, therefore, from the results of Section 6.4, the inverse transformation, L^{-1}, exists and is defined as follows:

Definition 9.2.2:   The linear transformation $L^{-1} : \text{Rng}(L) \to V$ defined by

$$L^{-1}[F](t) = f(t) \text{ if and only if } L[f](s) = F(s) \qquad (9.2.3)$$

is called the **inverse Laplace transform**.

REMARK   We emphasize the fact that L^{-1} is a *linear transformation*, so that

$$L^{-1}[F + G] = L^{-1}[F] + L^{-1}[G],$$

and

$$L^{-1}[cF] = cL^{-1}[F],$$

for all F and G in $\text{Rng}[L]$ and all real numbers c.

In Section 9.1, we derived the transforms

$$L[t^n] = \frac{n!}{s^{n+1}}, \quad L[e^{at}] = \frac{1}{s-a}, \quad L[\cos bt] = \frac{s}{s^2 + b^2}, \quad L[\sin bt] = \frac{b}{s^2 + b^2},$$

from which we directly obtain the inverse transforms

$$L^{-1}\left[\frac{1}{s^{n+1}}\right] = \frac{1}{n!}\,t^n, \qquad L^{-1}\left[\frac{1}{s-a}\right] = e^{at},$$

$$L^{-1}\left[\frac{s}{s^2+b^2}\right] = \cos bt, \qquad L^{-1}\left[\frac{b}{s^2+b^2}\right] = \sin bt.$$

Example 9.2.2   Find $L^{-1}[F](t)$ if

(a) $F(s) = \dfrac{2}{s^2}$.

(b) $F(s) = \dfrac{3s}{s^2+4}$.

(c) $F(s) = \dfrac{3s+2}{(s-1)(s-2)}$.

Solution

(a) $L^{-1}\left[\dfrac{2}{s^2}\right] = 2L^{-1}\left[\dfrac{1}{s^2}\right] = 2t.$

(b) $L^{-1}\left[\dfrac{3s}{s^2+4}\right] = 3L^{-1}\left[\dfrac{s}{s^2+4}\right] = 3\cos 2t.$

(c) In this case, it is not obvious at first sight what the appropriate inverse transform is. However, decomposing $F(s)$ into partial fractions yields[1]

$$F(s) = \frac{3s+2}{(s-1)(s-2)} = \frac{8}{s-2} - \frac{5}{s-1}.$$

Consequently, using the linearity of L^{-1},

$$L^{-1}\left[\frac{3s+2}{(s-1)(s-2)}\right] = L^{-1}\left[\frac{8}{s-2}\right] - L^{-1}\left[\frac{5}{s-1}\right]$$

$$= 8L^{-1}\left[\frac{1}{s-2}\right] - 5L^{-1}\left[\frac{1}{s-1}\right]$$

$$= 8\,e^{2t} - 5\,e^{t}. \qquad \square$$

If we relax the assumption that V contain only *continuous* functions of exponential order, then it is no longer true that L is one-to-one, and so, for a given $F(s)$, there will be (infinitely) many piecewise continuous functions f with the property that

$$L[f] = F(s).$$

Thus, we lose the uniqueness of $L^{-1}[F]$. However, it can be shown (see for example R.V. Churchill, *Modern Operational Mathematics in Engineering*, McGraw–Hill, 1944) that if two functions have the same Laplace transform, then they can only differ in their values at points of discontinuity. This does not affect the solution to our problems, and, therefore, we will use (9.2.3) to determine the inverse Laplace transform, even if f is piecewise continuous.

---

[1]It is very important in this chapter to be able to perform partial fraction decompositions. A review of this technique is given in Appendix 2.

Example 9.2.3   In the previous section we have shown that the Laplace transform of the piecewise continuous function

$$f(t) = \begin{cases} t, & 0 \le t < 1, \\ -1, & t \ge 1, \end{cases}$$

is

$$L[f] = \frac{1}{s^2}[1 - e^{-s}(2s + 1)], \quad s > 0.$$

Consequently,

$$L^{-1}\left\{\frac{1}{s^2}[1 - e^{-s}(2s + 1)]\right\} = f(t). \qquad \Box$$

It is possible to give a general formula for determining the inverse Laplace transform of $F(s)$ in terms of a contour integral in the complex plane. However, this is beyond the scope of the present treatment of the Laplace transform. In practice, as in the previous examples, we determine inverse Laplace transforms by recognizing $F(s)$ as being the Laplace transform of an appropriate function $f(t)$. In order for this approach to work, we need to memorize a few basic transforms and then be able to use these transforms to determine the inverse Laplace transform of more complicated functions. This is similar to the way that we learn how to integrate. The transform pairs that you will need for the remainder of the text are listed in Table 9.2.1. Several of the transforms given in this table will be derived in the following sections. More generally, very large tables of Laplace transforms have been compiled for use in applications, and most current computer algebra systems (such as Maple, Mathematica ...) have the built in capability to determine Laplace transforms.

Table 9.2.1

Function $f(t)$	Laplace Transform $F(s)$
$f(t) = t^n$, n a nonnegative integer	$F(s) = \dfrac{n!}{s^{n+1}}$, $s > 0$.
$f(t) = e^{at}$, a constant	$F(s) = \dfrac{1}{s - a}$, $s > a$.
$f(t) = \sin bt$, b constant	$F(s) = \dfrac{b}{s^2 + b^2}$, $s > 0$.
$f(t) = \cos bt$, b constant	$F(s) = \dfrac{s}{s^2 + b^2}$, $s > 0$.
$f(t) = t^{-1/2}$	$F(s) = (\pi/s)^{1/2}$, $s > 0$.
$f(t) = u_a(t)$ (see Section 9.7)	$F(s) = \dfrac{1}{s}e^{-as}$.
$f(t) = \delta(t - a)$ (see Section 9.8)	$F(s) = e^{-as}$.

Transform of Derivatives (see Section 9.4)	
f'	$L[f'] = sL[f] - f(0)$.
f''	$L[f''] = s^2L[f] - sf(0) - f'(0)$.

Shifting Theorems (see Section 9.5 and 9.7)	
$e^{at} f(t)$	$F(s - a)$.
$u_a(t) f(t - a)$	$e^{-as}F(s)$.

EXERCISES 9.2

For problems 1–5, show that the given function is of exponential order.

1 . $f(t) = \cos 2t$.

2 . $f(t) = e^{2t}$.

3 . $f(t) = e^{3t} \sin 4t$.

4 . $f(t) = t\, e^{-2t}$.

5 . $f(t) = t^n e^{at}$, where a and n are positive integers.

6 . Show that if f and g are in $E(0, \infty)$, then so are $f + g$ and cf for any scalar c.

For problems 7–21, determine the inverse Laplace transform of the given function.

7 . $F(s) = \dfrac{2}{s}$.

8 . $F(s) = \dfrac{3}{s - 2}$.

9 . $F(s) = \dfrac{5}{s + 3}$.

10. $F(s) = \dfrac{1}{s^2 + 4}$.

11. $F(s) = \dfrac{2s}{s^2 + 9}$.

12. $F(s) = \dfrac{4}{s^3}$.

13. $F(s) = \dfrac{s + 6}{s^2 + 1}$.

14. $F(s) = \dfrac{2s + 1}{s^2 + 16}$.

15. $F(s) = \dfrac{2}{s} - \dfrac{3}{s + 1}$.

16. $F(s) = \dfrac{4}{s^2} - \dfrac{s + 2}{s^2 + 9}$.

17. $F(s) = \dfrac{1}{s(s + 1)}$.

18. $F(s) = \dfrac{s - 2}{(s + 1)(s^2 + 4)}$.

19. $F(s) = \dfrac{2s + 3}{(s - 2)(s^2 + 1)}$.

10. $F(s) = \dfrac{s + 4}{(s - 1)(s + 2)(s - 3)}$.

21. $F(s) = \dfrac{2s + 3}{(s^2 + 4)(s^2 + 1)}$.

9.3   PERIODIC FUNCTIONS AND THE LAPLACE TRANSFORM

Many of the functions that arise in engineering applications are periodic on some interval. Due to the symmetry associated with a periodic function, we might suspect that the evaluation of the Laplace transform of such a function can be reduced to an integration over one period of the function. Before establishing this result, we first recall the definition of a periodic function.

Definition 9.3.1:   A function f defined on the interval $[0, \infty)$ is said to be **periodic with period T** if it satisfies

$$f(t + T) = f(t),$$

for all $t \geq 0$.

The most familiar examples of periodic functions are the trigonometric functions sine and cosine, which have period 2π.

Example 9.3.1   The function f defined by

$$f(t) = \begin{cases} 2, & 0 \le t \le 1, \\ 1, & 1 < t < 2, \end{cases} \qquad f(t + 2) = f(t),$$

is periodic on $[0, \infty)$ with period 2. (See Figure 9.3.1.)

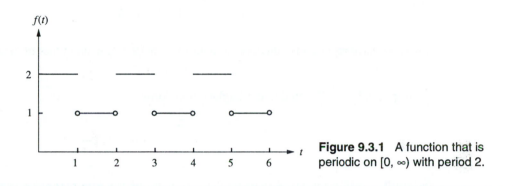

Figure 9.3.1 A function that is periodic on $[0, \infty)$ with period 2.

The following theorem can be used to simplify the evaluation of the Laplace transform of a periodic function.

   Theorem 9.3.1:   Let f be in $E(0, \infty)$. If f is periodic on $[0, \infty)$ with period T, then

$$L[f] = \frac{1}{1 - e^{-sT}} \int_0^T e^{-st} f(t) \, dt. \qquad (9.3.1)$$

PROOF   By definition of the Laplace transform we have

$$L[f] = \int_0^\infty e^{-st}f(t) \, dt = \int_0^T e^{-st}f(t) \, dt + \int_T^{2T} e^{-st}f(t) \, dt + \cdots + \int_{nT}^{(n+1)T} e^{-st}f(t) \, dt + \cdots.$$

Now consider the general integral

$$I = \int_{nT}^{(n+1)T} e^{-st}f(t) \, dt.$$

If we let $x = t - nT$, then $dx = dt$. Further, $t = nT$ corresponds to $x = 0$, whereas $t = (n + 1)T$ corresponds to $x = T$. Hence, I can be written in the equivalent form

$$I = \int_0^T e^{-s(x+nT)}f(x + nT) \, dx = e^{-nsT} \int_0^T e^{-sx}f(x) \, dx,$$

where we have used the fact that f is periodic of period T to replace $f(x + nT)$ by $f(t)$. All of the integrals that arise in the expression for $L[f]$ are of the preceding form for an appropriate value of n. It follows, therefore, that we can write

$$L[f] = (1 + e^{-sT} + e^{-2sT} + \cdots + e^{-nsT} + \cdots) \int_0^T e^{-sx} f(x)\, dx. \qquad (9.3.2)$$

However, the term multiplying the integral in (9.3.2) is just a geometric series with common ratio e^{-sT}.[1] Consequently, the sum of the geometric series is $\dfrac{1}{1 - e^{-sT}}$, so that

$$L[f] = \frac{1}{1 - e^{-sT}} \int_0^T e^{-sT} f(t)\, dt,$$

where we have replaced the dummy variable x by t in (9.3.2) without loss of generality.   ∎

Example 9.3.2   Determine the Laplace transform of

$$f(t) = \begin{cases} \sin t, & 0 \le t \le \pi, \\ 0, & \pi \le t < 2\pi, \end{cases} \qquad f(t + 2\pi) = f(t).$$

Solution   Since the given function is periodic on $[0, \infty)$ with period 2π (see Figure 9.3.2), we can use Theorem 9.3.1 to determine $L[f]$. We have

$$L[f] = \frac{1}{1 - e^{-2\pi s}} \int_0^{2\pi} e^{-st} f(t)\, dt = \frac{1}{1 - e^{-2\pi s}} \int_0^{\pi} e^{-st} \sin t\, dt.$$

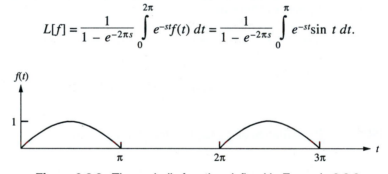

Figure 9.3.2 The periodic function defined in Example 9.3.2.

Using the standard integral:

$$\int e^{at} \sin bt\, dt = \frac{1}{(a^2 + b^2)} e^{at}(a \sin bt - b \cos bt) + c,$$

it follows that

$$L[f] = \frac{1}{1 - e^{-2\pi s}} \left\{ -\frac{1}{s^2 + 1} \left[e^{-st}(\cos t + s \sin t) \right]_0^{\pi} \right\}$$

$$= \frac{1}{1 - e^{-2\pi s}} \left[\frac{e^{-s\pi} + 1}{s^2 + 1} \right].$$

Substituting for

---

[1] Recall that an infinite series of the form $a + ar + ar^2 + ar^3 + \cdots$ is called a geometric series with common ratio r. If $|r| < 1$, then the sum of such a series is $a/(1 - r)$.

$$1 - e^{-2\pi s} = (1 - e^{-\pi s})(1 + e^{-\pi s})$$

yields

$$L[f] = \frac{1}{(s^2 + 1)(1 - e^{-\pi s})}.$$

EXERCISES 9.3

For problems 1–9, determine the Laplace transform of the given function.

1. $f(t) = t, 0 \le t < 1, f(t + 1) = f(t)$.

2. $f(t) = t^2, 0 \le t < 2, f(t + 2) = f(t)$.

3. $f(t) = \sin t, 0 \le t < \pi, f(t + \pi) = f(t)$.

4. $f(t) = \cos t, 0 \le t < \pi, f(t + \pi) = f(t)$.

5. $f(t) = e^t, 0 \le t < 1, f(t + 1) = f(t)$.

6. $f(t) = \begin{cases} 1, & 0 \le t \le 1, \\ -1, & 1 \le t < 2, \end{cases} \quad f(t + 2) = f(t)$.

7. $f(t) = \begin{cases} 2t/\pi, & 0 \le t < \pi/2, \\ \sin t, & \pi/2 \le t < \pi, \end{cases} \quad f(t + \pi) = f(t)$.

8. $f(t) = |\cos t|, 0 \le t < \pi, f(t + \pi) = f(t)$.

9. The triangular wave function (see Figure 9.3.3)

$$f(t) = \begin{cases} t/a, & 0 \le t < a, \\ (2a - t)/a, & a \le t < 2a, \end{cases} \quad f(t + 2a) = f(t),$$

where a is a positive constant.

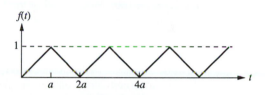

Figure 9.3.3   A triangular wave function.

10. Use Theorem 9.3.1 together with the fact that $f(t) = \sin at$ is periodic on the interval $[0, 2\pi/a]$, to determine $L[f]$.

11. Repeat the previous question for the function $f(t) = \cos at$.

9.4   THE TRANSFORM OF DERIVATIVES AND THE SOLUTION OF INITIAL-VALUE PROBLEMS

The reason that we have introduced the Laplace transform is that it provides an alternative technique for solving DE. To see how this technique arises, we must first consider how the derivative of a function transforms.

Theorem 9.4.1:   Suppose that f is of exponential order on $[0, \infty)$ and that f' exists and is piecewise continuous on $[0, \infty)$. Then $L[f']$ exists and is given by

$$L[f'] = sL[f] - f(0).$$

PROOF   For simplicity we consider the case when f' is continuous on $[0, \infty)$. The extension to the case of piecewise continuity is straightforward. Since f is differentiable and of exponential order on $[0, \infty)$, it follows that it belongs to $E(0, \infty)$, and hence its

Laplace transform exists. By definition of the Laplace transform, we have

$$L[f'] = \int_0^\infty e^{-st} f'(t)\, dt = \left[e^{-st} f(t)\right]_0^\infty + s \int_0^\infty e^{-st} f(t)\, dt.$$

That is, since f is of exponential order on $[0, \infty)$,

$$L[f'] = sL[f] - f(0). \qquad \blacksquare$$

Example 9.4.1   Solve the IVP

$$\frac{dy}{dt} = t,\ y(0) = 1.$$

Solution   This problem can be solved by a direct integration. However, we will use the Laplace transform. Taking the Laplace transform of both sides of the given DE and using the result of the previous theorem, we obtain

$$sY(s) - y(0) = \frac{1}{s^2}.$$

This is an *algebraic* equation for $Y(s)$. Substituting in the initial condition and solving algebraically for $Y(s)$ yields

$$Y(s) = \frac{1}{s^3} + \frac{1}{s}.$$

To determine the solution of the original problem, we now take the inverse Laplace transform of both sides of this equation. The result is

$$y(t) = L^{-1}\left[\frac{1}{s^3} + \frac{1}{s}\right].$$

That is, since $L^{-1}\left[\dfrac{1}{s^{n+1}}\right] = \dfrac{1}{n!}\, t^n$,

$$y(t) = \frac{1}{2} t^2 + 1. \qquad \square$$

The foregoing example illustrates the basic steps in solving an IVP using the Laplace transform. We proceed as follows:

---

1. Take the Laplace transform of the given DE, and substitute in the given initial conditions.

2. Solve the resulting equation algebraically for $Y(s)$.

3. Take the inverse Laplace transform of the resulting equation to determine the solution $y(t)$ of the given IVP.

These steps are illustrated in Figure 9.4.1.

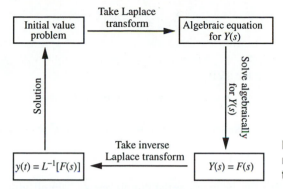

Figure 9.4.1 A schematic representation of the Laplace transform method for solving IVP.

To extend the technique introduced in the previous example to higher order DE, we need to determine how the higher order derivatives transform. This can be derived quite easily from Theorem 9.4.1. We illustrate for the case of second-order derivatives and leave the derivation of the general case as an exercise.

Assuming that f'' is sufficiently smooth, it follows from Theorem 9.4.1 that

$$L[f''] = sL[f'] - f'(0).$$

Thus, applying Theorem 9.4.1 once more yields

$$L[f''] = s^2L[f] - sf(0) - f'(0).$$

We leave it as an exercise to establish that more generally

$$L[f^{(n)}] = s^nL[f] - s^{n-1}f(0) - s^{n-2}f'(0) - \cdots - sf^{(n-2)}(0) - f^{(n-1)}(0).$$

Example 9.4.2   Use the Laplace transform to solve the IVP

$$y'' - y' - 6y = 0, \ y(0) = 1, \ y'(0) = 2.$$

Solution   We take the Laplace transform of both sides of the DE to obtain

$$[s^2Y(s) - sy(0) - y'(0)] - [sY(s) - y(0)] - 6Y(s) = 0.$$

Substituting in the given initial values and rearranging terms yields

$$(s^2 - s - 6)Y(s) = s + 1.$$

That is

$$Y(s) = \frac{s + 1}{(s - 3)(s + 2)}.$$

Thus, we have solved for the Laplace transform of $y(t)$. To find y itself, we must take the inverse Laplace transform. We first decompose the right-hand side into partial fractions to obtain

$$Y(s) = \frac{4}{5(s - 3)} + \frac{1}{5(s + 2)}.$$

We recognize the terms on the right-hand side as being the Laplace transform of appropriate exponential functions. Taking the inverse Laplace transform yields

$$y(t) = \frac{4}{5} e^{3t} + \frac{1}{5} e^{-2t},$$

and the IVP is solved.

Example 9.4.3   Solve the IVP

$$y'' + y = e^{2t}, \quad y(0) = 0, \quad y'(0) = 1.$$

Solution   Once more we take the Laplace transform of both sides of the DE to obtain

$$[s^2 Y(s) - sy(0) - y'(0)] + Y(s) = \frac{1}{s - 2}.$$

That is, upon substituting for the given initial conditions and simplifying,

$$Y(s) = \frac{s - 1}{(s - 2)(s^2 + 1)}.$$

We must now determine the partial fraction decomposition of the right-hand side. We have

$$\frac{s - 1}{(s - 2)(s^2 + 1)} = \frac{A}{s - 2} + \frac{Bs + C}{s^2 + 1},$$

for appropriate constants A, B, and C. Multiplying both sides of this equality by $(s - 2)(s^2 + 1)$ yields

$$s - 1 = A(s^2 + 1) + (Bs + C)(s - 2).$$

Equating coefficients of s^0, s^1, and s^2 results in the three conditions

$$A - 2C = -1, \quad -2B + C = 1, \quad A + B = 0.$$

Solving for A, B, and C we obtain

$$A = \frac{1}{5}, \quad B = -\frac{1}{5}, \quad C = \frac{3}{5}.$$

Thus,

$$Y(s) = \frac{1}{5(s - 2)} - \frac{s - 3}{5(s^2 + 1)}.$$

That is,

$$Y(s) = \frac{1}{5(s - 2)} - \frac{s}{5(s^2 + 1)} + \frac{3}{5(s^2 + 1)}.$$

Taking the inverse Laplace transform of both sides of this equation yields

$$y(t) = \frac{1}{5} e^{2t} - \frac{1}{5} \cos t + \frac{3}{5} \sin t.$$ ☐

The structure of the solution obtained in the previous example has a familiar form. The first term represents a particular solution to the DE that could have been obtained by the method of undetermined coefficients, whereas the last two terms come from the complementary function. There are no arbitrary constants in the solution, since we have solved an IVP. Notice the difference between solving an IVP using the Laplace transform and our previous techniques. In the Laplace transform technique, we impose the initial values at the beginning of the problem and just solve the IVP. In our previous techniques, we first found the general solution to the DE and *then* imposed the initial values to solve the IVP. We note, however, that the Laplace transform can also be used to determine the general solution of a DE (see problem 27).

It should be apparent from the results of the previous two sections that the main difficulties in applying the Laplace transform technique to the solution of IVP is in steps 1 and 3. In order for the technique to be useful, we need to know the transform and inverse transform of a large number of functions. So far, we have only determined the Laplace transform of some very basic functions, namely, t^n, e^{at}, $\sin bt$, $\cos bt$. We will show in the remaining sections how these basic transforms can be used to determine the Laplace transform of almost any function that is likely to arise in the applications. You are once more strongly advised to memorize the basic transforms.

EXERCISES 9.4

For problems 1–26, use the Laplace transform to solve the given IVP.

1. $y' + y = 8e^{3t}$, $y(0) = 2$.

2. $y' + 3y = 2e^{-t}$, $y(0) = 3$.

3. $y' + 2y = 4t$, $y(0) = 1$.

4. $y' - y = 6\cos t$, $y(0) = 2$.

5. $y' - y = 5\sin 2t$, $y(0) = -1$.

6. $y' + y = 5e^t \sin t$, $y(0) = 1$.

7. $y'' + y' - 2y = 0$, $y(0) = 1$, $y'(0) = 4$.

8. $y'' + 4y = 0$, $y(0) = 5$, $y'(0) = 1$.

9. $y'' - 3y' + 2y = 4$, $y(0) = 0$, $y'(0) = 1$.

10. $y'' - y' - 12y = 36$, $y(0) = 0$, $y'(0) = 12$.

11. $y'' + y' - 2y = 10e^{-t}$, $y(0) = 0$, $y'(0) = 1$.

12. $y'' - 3y' + 2y = 4e^{3t}$, $y(0) = 0$, $y'(0) = 0$.

13. $y'' - 2y' = 30e^{-3t}$, $y(0) = 1$, $y'(0) = 0$.

14. $y'' - y = 12e^{2t}$, $y(0) = 1$, $y'(0) = 1$.

15. $y'' + 4y = 10e^{-t}$, $y(0) = 4$, $y'(0) = 0$.

16. $y'' - y' - 6y = 6(2 - e^t)$, $y(0) = 5$, $y'(0) = -3$.

17. $y'' - y = 6\cos t$, $y(0) = 0$, $y'(0) = 4$.

18. $y'' - 9y = 13\sin 2t$, $y(0) = 3$, $y'(0) = 1$.

19. $y'' - y = 8\sin t - 6\cos t$, $y(0) = 2$, $y'(0) = -1$.

20. $y'' - y' - 2y = 10\cos t$, $y(0) = 0$, $y'(0) = -1$.

21. $y'' + 5y' + 4y = 20\sin 2t$, $y(0) = -1$, $y'(0) = 2$.

22. $y'' + 5y' + 4y = 20\sin 2t$, $y(0) = 1$, $y'(0) = -2$.

23. $y'' - 3y' + 2y = 3\cos t + \sin t$, $y(0) = 1$, $y'(0) = 1$.

24. $y'' + 4y = 9\sin t$, $y(0) = 1$, $y'(0) = -1$.

25. $y'' + y = 6\cos 2t$, $y(0) = 0$, $y'(0) = 2$.

26. $y'' + 9y = 7\sin 4t + 14\cos 4t$, $y(0) = 1$, $y'(0) = 2$.

27. Use the Laplace transform to find the general solution to $y'' - y = 0$.

28. Use the Laplace transform to solve the IVP

$$y'' + \omega^2 y = A\sin \omega_0 t + B\cos \omega_0 t$$

$$y(0) = y_0, \quad y'(0) = y_1,$$

where A, B, ω and ω_0 are positive constants and $\omega \neq \omega_0$.

29. The current, $i(t)$, in an RL circuit is governed by the DE

$$\frac{di}{dt} + \frac{R}{L}i = \frac{1}{L}E(t),$$

where R and L are constants.

(a) Use the Laplace transform to determine $i(t)$ if $E(t) = E_0$, a constant. There is no current flowing initially.

(b) Repeat (a) in the case when $E(t) = E_0 \sin \omega t$, where ω is a constant.

The Laplace transform can also be used to solve IVP for systems of linear DE. The remaining problems deal with this.

30. Consider the IVP

$$x_1' = a_{11}x_1 + a_{12}x_2 + b_1(t),$$

$$x_2' = a_{21}x_1 + a_{22}x_2 + b_2(t),$$

$$x_1(0) = \alpha_1, \quad x_2(0) = \alpha_2,$$

where the a_{ij}, α_1 and α_2 are constants. Show that the Laplace transforms of $x_1(t)$, $x_2(t)$ must satisfy the linear algebraic system

$$(s - a_{11})X_1(s) - a_{12}X_2(s) = \alpha_1 + B_1(s),$$

$$-a_{21}X_1(s) + (s - a_{22})X_2(s) = \alpha_2 + B_2(s).$$

This system can be solved quite easily (for example by Cramer's rule) to determine $X_1(s)$ and $X_2(s)$, and then $x_1(t)$, $x_2(t)$ can be obtained by taking the inverse Laplace transform.

For problems 31 and 32, solve the given IVP.

31. $x_1' = -4x_1 - 2x_2$, $x_2' = x_1 - x_2$, $x_1(0) = 0$,

$x_2(0) = 1$.

32. $x_1' = -3x_1 + 4x_2$, $x_2' = -x_1 + 2x_2$, $x_1(0) = 2$, $x_2(0) = 1$.

9.5   THE FIRST SHIFTING THEOREM

In order for the Laplace transform to be a useful tool for solving DE, we need to be able to find $L[f]$ for a large class of functions. Trying to apply the definition of the Laplace transform to determine $L[f]$ for every function we encounter is not an appropriate way to proceed. Instead, we derive some general theorems that will enable us to obtain the Laplace transform of most elementary functions from a knowledge of the transforms of the functions given in Table 9.2.1. For the remainder of the section, we will be assuming that all of the functions that we encounter do have a Laplace transform.

> **Theorem 9.5.1 (First Shifting Theorem):**   If $L[f] = F(s)$, then
>
> $$\boxed{L[e^{at}f(t)] = F(s - a).}$$

Conversely, if $L^{-1}[F(s)] = f(t)$, then

$$\boxed{L^{-1}[F(s - a)] = e^{at}f(t).}$$

PROOF   From the definition of the Laplace transform, we have

$$L[e^{at}f(t)] = \int_0^\infty e^{-st}e^{at}f(t)\,dt = \int_0^\infty e^{-(s-a)t}f(t)\,dt. \tag{9.5.1}$$

But,

$$F(s) = \int_0^\infty e^{-st}f(t)\,dt,$$

so that

$$F(s - a) = \int_0^\infty e^{-(s-a)t}f(t)\,dt. \tag{9.5.2}$$

Comparing (9.5.1) with (9.5.2), we obtain

$$L[e^{at}f(t)] = F(s - a), \tag{9.5.3}$$

as required. Taking the inverse Laplace transform of both sides of (9.5.3) yields

$$L^{-1}[F(s - a)] = e^{at}f(t). \qquad \blacksquare$$

We illustrate the use of the preceding theorem with several examples. (See also Figure 9.5.1.)

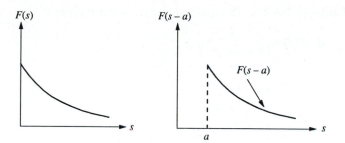

Figure 9.5.1   An illustration of the first shifting theorem. Multiplying $f(t)$ by e^{at} has the effect of shifting $F(s)$ by a units in s-space.

Example 9.5.1   Find $L[f]$ for each $f(t)$.

(a) $f(t) = e^{5t}\cos 4t$.

(b) $f(t) = e^{at}\sin bt$, a, b constants.

(c) $f(t) = e^{at}t^n$, where a is a constant and n is a positive integer.

Solution

(a) From Table 9.2.1, we have

$$L[\cos 4t] = \frac{s}{s^2 + 16},$$

so that, applying the first shifting theorem with $a = 5$ yields

$$L[e^{5t}\cos 4t] = \frac{(s - 5)}{(s - 5)^2 + 16}.$$

(b) Since $L[\sin bt] = \dfrac{b}{s^2 + b^2}$,

$$L[e^{at}\sin bt] = \frac{b}{(s - a)^2 + b^2}. \tag{9.5.4}$$

Similarly, it follows from Table 9.2.1 and the first shifting theorem that

$$L[e^{at}\cos bt] = \frac{s - a}{(s - a)^2 + b^2}. \tag{9.5.5}$$

(c) From Table 9.2.1, we have

$$L[t^n] = \frac{n!}{s^{n+1}},$$

so that

$$L[e^{at}t^n] = \frac{n!}{(s - a)^{n+1}}. \tag{9.5.6}$$

◻

The previous example dealt with the direct use of the first shifting theorem to obtain the Laplace transform of a function. Of equal importance is its use in determining inverse transforms. Once more, we illustrate this with several examples.

Example 9.5.2   Determine $L^{-1}[F(s)]$ for the given F.

(a) $F(s) = \dfrac{3}{(s - 2)^2 + 9}$.

(b) $F(s) = \dfrac{6}{(s - 4)^3}$.

(c) $F(s) = \dfrac{s + 4}{s^2 + 6s + 13}$.

(d) $F(s) = \dfrac{s - 2}{s^2 + 2s + 3}$.

Solution

(a) From Table 9.2.1,

$$L^{-1}\left[\frac{3}{s^2 + 9}\right] = \sin 3t,$$

so that, by the first shifting theorem,

$$L^{-1}\left[\frac{3}{(s - 2)^2 + 9}\right] = e^{2t}\sin 3t.$$

(b) From Table 9.2.1,

$$L^{-1}\left[\frac{6}{s^3}\right] = 3t^2.$$

Thus, applying the first shifting theorem yields

$$L^{-1}\left[\frac{6}{(s - 4)^3}\right] = 3t^2 e^{4t}.$$

(c) In this case,

$$F(s) = \frac{s + 4}{s^2 + 6s + 13},$$

which we do not recognize as being a shift of the transform of any of the functions given in Table 9.2.1. However, completing the square in the denominator of $F(s)$ yields[1]

$$F(s) = \frac{s + 4}{(s + 3)^2 + 4}. \tag{9.5.7}$$

We still cannot write down the inverse transform directly, but, by the first shifting theorem, we have

$$L^{-1}\left[\frac{s + 3}{(s + 3)^2 + 4}\right] = e^{-3t}\cos 2t, \tag{9.5.8}$$

$$L^{-1}\left[\frac{2}{(s + 3)^2 + 4}\right] = e^{-3t}\sin 2t. \tag{9.5.9}$$

---

[1]Recall that we can always write $x^2 + ax + b = (x + a/2)^2 + b - a^2/4$. This procedure is known as completing the square.

This suggests that we rewrite (9.5.7) in the equivalent form

$$F(s) = \frac{s + 3}{(s + 3)^2 + 4} + \frac{1}{(s + 3)^2 + 4},$$

so that, using the linearity of L^{-1} and equations (9.5.8) and (9.5.9),

$$L^{-1}[F(s)] = L^{-1}\left[\frac{s + 3}{(s + 3)^2 + 4}\right] + L^{-1}\left[\frac{1}{(s + 3)^2 + 4}\right]$$

$$= e^{-3t}\cos 2t + \frac{1}{2} e^{-3t}\sin 2t.$$

(d)   We proceed as in the previous example. In this case, we have

$$F(s) = \frac{s - 2}{(s + 1)^2 + 2},$$

which can be written as

$$F(s) = \frac{s + 1}{(s + 1)^2 + 2} - \frac{3}{(s + 1)^2 + 2}.$$

Then, using Table 9.2.1 and the first shifting theorem, it follows that

$$L^{-1}[F(s)] = e^{-t} \cos \sqrt{2}\, t - \frac{3}{\sqrt{2}} e^{-t} \sin \sqrt{2}\, t.$$

EXERCISES 9.5

For problems 1–10, determine $f(t - a)$ for the given function f and the given constant a.

1 . $f(t) = t$, $a = 1$.

2 . $f(t) = 1$, $a = 3$.

3 . $f(t) = t^2 - 2t$, $a = -2$.

4 . $f(t) = e^{3t}$, $a = 2$.

5 . $f(t) = e^{2t} \cos t$, $a = \pi$.

6 . $f(t) = t\, e^{2t}$, $a = -1$.

7 . $f(t) = e^{-t} \sin 2t$, $a = \pi/6$.

8 . $f(t) = \dfrac{t}{t^2 + 4}$, $a = 1$.

9 . $f(t) = \dfrac{t + 1}{t^2 - 2t + 2}$, $a = 2$.

10. $f(t) = e^{-t}(\sin 2t + \cos 2t)$, $a = \pi/4$.

For problems 11–16, determine $f(t)$.

11. $f(t - 1) = (t - 1)^2$.

12. $f(t - 1) = (t - 2)^2$.

13. $f(t - 2) = (t - 2)e^{3(t-2)}$.

14. $f(t - 1) = t\sin[3(t - 1)]$.

15. $f(t - 3) = te^{-(t-3)}$.

16. $f(t - 4) = \dfrac{t + 1}{(t - 1)^2 + 4}$.

For problems 17–26, determine the Laplace transform of f.

17. $f(t) = e^{3t}\cos 4t$.

18. $f(t) = e^{-4t}\sin 5t$.

19. $f(t) = te^{2t}$.

20. $f(t) = 3te^{-t}$.

21. $f(t) = t^3 e^{-4t}$.

22. $f(t) = e^t - te^{-2t}$.

23. $f(t) = 2e^{3t} \sin t + 4e^{-t}\cos 3t$.

24. $f(t) = e^{2t}(1 - \sin^2 t)$.

25. $f(t) = t^2(e^t - 3)$.

26. $f(t) = e^{-2t} \sin(t - \pi/4)$.

For problems 27–41, determine $L^{-1}[F]$.

27. $F(s) = \dfrac{1}{(s-3)^2}$.

28. $F(s) = \dfrac{4}{(s+2)^3}$.

29. $F(s) = \dfrac{2}{(s+3)^{1/2}}$.

30. $F(s) = \dfrac{2}{(s-1)^2 + 4}$.

31. $F(s) = \dfrac{s+2}{(s+2)^2 + 9}$.

32. $F(s) = \dfrac{s}{(s-3)^2 + 4}$.

33. $F(s) = \dfrac{5}{(s-2)^2 + 16}$.

34. $F(s) = \dfrac{6}{s^2 + 2s + 2}$.

35. $F(s) = \dfrac{s-2}{s^2 + 2s + 26}$.

36. $F(s) = \dfrac{2s}{s^2 - 4s + 13}$.

37. $F(s) = \dfrac{s}{(s+1)^2 + 4}$.

38. $F(s) = \dfrac{2s+3}{(s+5)^2 + 49}$.

39. $F(s) = \dfrac{4}{s(s+2)^2}$.

40. $F(s) = \dfrac{2s+1}{(s-1)^2(s+2)}$.

41. $F(s) = \dfrac{2s+3}{s(s^2 - 2s + 5)}$.

For problems 42–52, solve the given IVP.

42. $y'' - y = 8e^t$, $y(0) = 0$, $y'(0) = 0$.

43. $y'' - 4y = 12e^{2t}$, $y(0) = 2$, $y'(0) = 3$.

44. $y'' - y' - 2y = 6e^{-t}$, $y(0) = 0$, $y'(0) = 1$.

45. $y'' + y' - 2y = 3e^{-2t}$, $y(0) = 3$, $y'(0) = -1$.

46. $y'' - 4y' + 4y = 6e^{2t}$, $y(0) = 1$, $y'(0) = 0$.

47. $y'' + 2y' + y = 2e^{-t}$, $y(0) = 2$, $y'(0) = 1$.

48. $y'' - 4y = 2te^t$, $y(0) = 0$, $y'(0) = 0$.

49. $y'' + 3y' + 2y = 12te^{2t}$, $y(0) = 0$, $y'(0) = 1$.

50. $y'' + y = 5te^{-3t}$, $y(0) = 2$, $y'(0) = 0$.

51. $y'' - y = 8e^t \sin 2t$, $y(0) = 2$, $y'(0) = -2$.

52. $y'' + 2y' - 3y = 26e^{2t} \cos t$, $y(0) = 1$, $y'(0) = 0$.

53. Solve the IVP
$$x_1' = 2x_1 - x_2, \quad x_2' = x_1 + 2x_2,$$
$$x_1(0) = 1, \quad x_2(0) = 0.$$

9.6 THE UNIT STEP FUNCTION

In applications of DE such as

$$y'' + by' + cy = F(t)$$

to engineering problems, it often arises that the forcing term $F(t)$ is either piecewise continuous or even discontinuous. In such a situation, the Laplace transform is ideally suited for determining the solution of the DE as compared with the techniques developed in Chapters 2 and 7. To specify piecewise continuous functions in an appropriate manner, it is useful to introduce the unit step function, defined as follows:

Definition 9.6.1:   The **unit step function** or **Heaviside step function**, $u_a(t)$, is defined by

$$u_a(t) = \begin{cases} 0, & 0 \le t < a, \\ 1, & t \ge a, \end{cases}$$

where a is any positive number. (See Figure 9.6.1.)

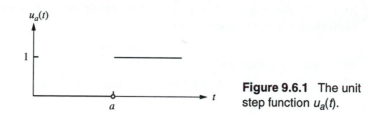

Figure 9.6.1 The unit step function $u_a(t)$.

Example 9.6.1   Sketch the function $f(t) = u_a(t) - u_b(t)$, $b > a$.

Solution   By definition of the unit step function, we have

$$f(t) = \begin{cases} 0, & 0 \le t < a, \\ 1, & a \le t < b, \\ 0, & t \ge b, \end{cases}$$

so that the graph of f is as given in Figure 9.6.2.

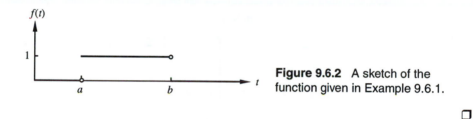

Figure 9.6.2 A sketch of the function given in Example 9.6.1.

The real power of the unit step function is that it enables us to model the situation when a force acts intermittently or in a nonsmooth manner. For example, the function f in Figure 9.6.2 can be interpreted as representing a force of unit magnitude that begins to act at $t = a$ and that stops acting at $t = b$. More generally, it is useful to regard the unit step function $u_a(t)$ as giving a mathematical description of a switch that is turned on at $t = a$.

The remaining examples in this section indicate how $u_a(t)$ can be useful for representing functions that are piecewise continuous.

Example 9.6.2   Express the following function in terms of the unit step function:

$$f(t) = \begin{cases} 0, & 0 \le t < 1, \\ t - 1, & 1 \le t < 2, \\ 1, & t \ge 2. \end{cases}$$

Solution   We view the given function in the following way. The contribution $f_1(t) = t - 1$ is "switched on" at $t = 1$ and is "switched off" again at $t = 2$. Mathematically this can be described by

$$f_1(t) = \underbrace{u_1(t)(t - 1)}_{\text{switch on at } t = 1} - \underbrace{u_2(t)(t - 1)}_{\text{switch off at } t = 2}.$$

At $t = 2$, the contribution $f_2(t) = 1$ switches on and remains on for all $t \geq 2$. Mathematically this is described by

$$f_2(t) = u_2(t).$$

The function f is then given by

$$f(t) = f_1(t) + f_2(t) = (t-1)u_1(t) - (t-1)u_2(t) + u_2(t),$$

which can be written in the equivalent form

$$f(t) = (t-1)u_1(t) - (t-2)u_2(t). \tag{9.6.1}$$

A sketch of $f(t)$ is given in Figure 9.6.3. Notice that this sketch is more easily determined from the original definition of f, rather than from (9.6.1).

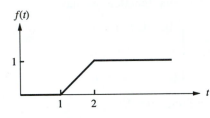

Figure 9.6.3 A sketch of the function given in Example 9.6.2.

Example 9.6.3 Make a sketch of the function $f(t)$ defined by

$$f(t) = \begin{cases} t, & 0 \leq t < 2, \\ -1, & 2 \leq t < 4, \\ t-4, & 4 \leq t < 5, \\ e^{(5-t)}, & t \geq 5, \end{cases}$$

and express f in terms of the unit step function.

Solution   The function is sketched in Figure 9.6.4. Using the unit step function, we see that f consists of the following different parts

$$f_1(t) = t[1 - u_2(t)], \quad f_2(t) = -1[u_2(t) - u_4(t)],$$
$$f_3(t) = (t-4)[u_4(t) - u_5(t)], \quad f_4(t) = e^{(5-t)}u_5(t).$$

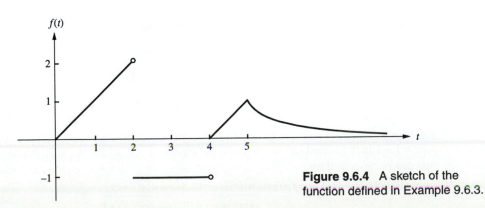

Figure 9.6.4 A sketch of the function defined in Example 9.6.3.

Thus,

$$f(t) = t[1 - u_2(t)] - [u_2(t) - u_4(t)] + (t - 4)[u_4(t) - u_5(t)] + e^{(5-t)}u_5(t).$$

EXERCISES 9.6

For problems 1–6, make a sketch of the given function on the interval $[0, \infty)$.

1 . $f(t) = 2u_1(t) - 4u_3(t)$.

2 . $f(t) = 1 + (t - 1)u_1(t)$.

3 . $f(t) = t(1 - u_1(t))$.

4 . $f(t) = u_1(t) + u_2(t) + u_3(t) + u_4(t)$.

5 . $f(t) = u_1(t) + u_2(t) + \cdots = \displaystyle\sum_{i=1}^{\infty} u_i(t)$.

6 . $f(t) = u_1(t) - u_2(t) + u_3(t) - \cdots$

$= \displaystyle\sum_{i=1}^{\infty} (-1)^{i+1} u_i(t)$.

For problems 7–14, make a sketch of the given function, and express it in terms of the unit step function.

7 . $f(t) = \begin{cases} 3, & 0 \le t < 1, \\ -1, & t \ge 1. \end{cases}$

8 . $f(t) = \begin{cases} t^2, & 0 \le t < 1, \\ 1, & t \ge 1. \end{cases}$

9 . $f(t) = \begin{cases} 2, & 0 \le t < 2, \\ 1, & 2 \le t < 4, \\ -1, & t \ge 4. \end{cases}$

10. $f(t) = \begin{cases} 2, & 0 \le t < 1, \\ 2e^{(t-1)}, & t > 1. \end{cases}$

11. $f(t) = \begin{cases} t, & 0 \le t < 3, \\ 6 - t, & 3 \le t < 6, \\ 0, & t \ge 6. \end{cases}$

12. $f(t) = \begin{cases} 0, & 0 \le t < 2, \\ 3 - t, & 2 \le t < 4, \\ -1, & t \ge 4. \end{cases}$

13. $f(t) = \begin{cases} 1, & 0 \le t < \pi/2, \\ \sin t, & \pi/2 \le t < 3\pi/2, \\ -1, & t \ge 3\pi/2. \end{cases}$

14. $f(t) = \begin{cases} \sin t, & 2n\pi \le t < (2n + 1)\pi, \\ & \qquad n = 0, 1, 2, \ldots \\ 0, & \text{otherwise.} \end{cases}$

9.7   THE SECOND SHIFTING THEOREM

In the previous section, we saw how the unit step function can be used to represent functions that are piecewise continuous. In this section, we show that the Laplace transform provides a straightforward method for solving constant coefficient linear DE that have such functions as driving terms. We first need to determine how the unit step function transforms.

Theorem 9.7.1 (Second Shifting Theorem):   Let $L[f(t)] = F(s)$.   Then,

$$\boxed{L[u_a(t)f(t-a)] = e^{-as}F(s).}$$
(9.7.1)

Conversely,

$$\boxed{L^{-1}[e^{-as}F(s)] = u_a(t)f(t-a).}$$
(9.7.2)

PROOF   Once more we must return to the definition of the Laplace transform. We have

$$L[u_a(t)f(t-a)] = \int_0^\infty e^{-st}u_a(t)f(t-a)\,dt = \int_a^\infty e^{-st}f(t-a)\,dt,$$

where we have used the definition of the unit step function. We now make a change of variable in the integral. Let $x = t - a$. Then $dx = dt$, and the lower limit of integration $t = a$ corresponds to $x = 0$, whereas the upper limit of integration is unchanged. Thus,

$$L[u_a(t)f(t-a)] = \int_0^\infty e^{-s(x+a)}f(x)\,dx = e^{-as}\int_0^\infty e^{-sx}f(x)\,dx = e^{-as}L[f]$$

as required. Taking the inverse Laplace transform of both sides of (9.7.1) yields (9.7.2).   ■

This theorem is illustrated in Figure 9.7.1.

Corollary 9.7.1:   If $L[f] = F(s)$, then

$$\boxed{L[u_a(t)f(t)] = e^{-as}L[f(t+a)].}$$

PROOF   A direct consequence of the previous theorem.   ■

We illustrate the use of the preceding theorem and corollary with several examples.

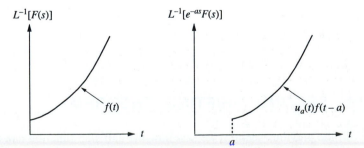

Figure 9.7.1   An illustration of the second shifting theorem. Multiplying $F(s)$ by e^{-as} has the effect of shifting $f(t)$ by a units to the right in t-space.

Example 9.7.1   Determine $L[f]$ if

$$f(t) = \begin{cases} 0, & 0 \le t < 1, \\ t - 1, & 1 \le t < 2, \\ 1, & t \ge 2. \end{cases}$$

Solution   We have already shown in Example 9.6.2 that the given function can be expressed in terms of the unit step function as

$$f(t) = (t - 1)u_1(t) - (t - 2)u_2(t).$$

If we let $g(t) = t$, then

$$f(t) = g(t - 1)u_1(t) - g(t - 2)u_2(t).$$

Using Theorem 9.7.1, it follows that

$$L[f] = e^{-s}L[g] - e^{-2s}L[g] = \frac{1}{s^2}(e^{-s} - e^{-2s}).$$

Example 9.7.2   Find $L[f]$ if

$$f(t) = \begin{cases} 1, & 0 \le t < 2 \\ e^{-(t-2)}, & t \ge 2. \end{cases}$$

Solution   To determine $L[f]$, we first express f in terms of the unit step function.   In this case, we have

$$f(t) = [1 - u_2(t)] + e^{-(t-2)}u_2(t).$$

That is,

$$f(t) = 1 + u_2(t)[e^{-(t-2)} - 1].$$

If we let $g(t) = e^{-t} - 1$, then

$$f(t) = 1 + u_2(t)g(t - 2),$$

so that, from Theorem 9.7.1,

$$L[f] = \frac{1}{s} + e^{-2s}\left(\frac{1}{s + 1} - \frac{1}{s}\right).$$

We can write this in the equivalent form

$$L[f] = \frac{1 - e^{-2s}}{s} + \frac{e^{-2s}}{s + 1}.$$

Example 9.7.3   Determine $L^{-1}\left[\dfrac{2e^{-s}}{s^2 + 4}\right]$.

Solution   From Table 9.2.1, we have

$$L[\sin 2t] = \frac{2}{s^2 + 4}.$$

Consequently,

$$L^{-1}\left[\frac{2e^{-s}}{s^2 + 4}\right] = L^{-1}\{e^{-s}L[\sin 2t]\}$$

$$= u_1(t) \sin[2(t - 1)],$$

using Theorem 9.7.1.

Example 9.7.4   Determine $L^{-1}\left[\dfrac{(s - 4)e^{-3s}}{s^2 - 4s + 5}\right]$.

Solution   Let

$$G(s) = \frac{(s - 4)e^{-3s}}{s^2 - 4s + 5}.$$

We first rewrite G in a form more suitable for determining $L^{-1}[G]$. Completing the square in the denominator yields

$$G(s) = \frac{(s - 4)\ e^{-3s}}{(s - 2)^2 + 1},$$

which can be written in the equivalent form

$$G(s) = e^{-3s}\left[\frac{(s - 2)}{(s - 2)^2 + 1} - \frac{2}{(s - 2)^2 + 1}\right].$$

Thus,

$$L^{-1}[G] = L^{-1}\{e^{-3s}L[e^{2t}\cos t - 2 e^{2t}\sin t]\}$$

$$= u_3(t)[e^{2(t-3)}\cos(t - 3) - 2 e^{2(t-3)}\sin(t - 3)].$$

$$= e^{2(t-3)}u_3(t)[\cos(t - 3) - 2\sin(t - 3)].$$ ❑

We now illustrate how the unit step function can be useful in the solution of IVP. For simplicity, we will start with a first-order DE.

Example 9.7.5   Solve the IVP

$$y' - y = 1 - (t - 1)u_1(t), \ \ y(0) = 0.$$

Solution   In this case, the forcing term on the right-hand side of the DE is sketched in Figure 9.7.2. Taking the Laplace transform of both sides of the DE yields

$$sY(s) - Y(s) - y(0) = \frac{1}{s} - \frac{e^{-s}}{s^2}.$$

By imposing the given initial condition and simplifying we obtain

$$Y(s) = \frac{1}{s(s - 1)} - e^{-s}\left[\frac{1}{s^2(s - 1)}\right].$$

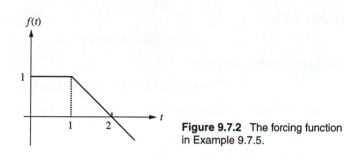

Figure 9.7.2 The forcing function in Example 9.7.5.

Decomposing the terms on the right-hand side into partial fractions yields

$$Y(s) = \frac{1}{s-1} - \frac{1}{s} - e^{-s}\left[\frac{1}{s-1} - \frac{1}{s} - \frac{1}{s^2}\right].$$

Taking the inverse Laplace transform of both sides of this equation, we obtain

$$y(t) = e^t - 1 - u_1(t)[e^{(t-1)} - 1 - (t-1)].$$

That is,

$$y(t) = e^t - 1 - u_1(t)[e^{(t-1)} - t].$$

This problem was solved in Chapter 1 (Example 1.6.4). A comparison of the two solution methods indicates the power of the Laplace transform.

Example 9.7.6   Solve the IVP

$$y'' + 2y' + 5y = f(t), \quad y(0) = 0, \quad y'(0) = 0$$

if

$$f(t) = \begin{cases} 10, & 0 \le t < 4, \\ -10, & 4 \le t < 8, \\ 0, & t \ge 8. \end{cases}$$

Solution   We first express f in terms of the unit step function.

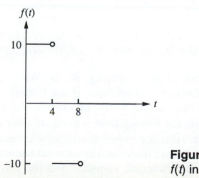

Figure 9.7.3 The forcing function $f(t)$ in Example 9.7.6.

In this case, we have

$$f(t) = 10[1 - 2u_4(t) + u_8(t)],$$

so that the DE can be written as

$$y'' + 2y' + 5y = 10[1 - 2u_4(t) + u_8(t)],$$

and we can now proceed in the usual manner. Taking the Laplace transform of both sides of the DE yields

$$[s^2Y(s) - sy(0) - y'(0)] + 2[sY(s) - y(0)] + 5Y(s) = \frac{10}{s}(1 - 2e^{-4s} + e^{-8s}).$$

That is, by imposing the given initial conditions and simplifying,

$$Y(s) = \frac{10(1 - 2e^{-4s} + e^{-8s})}{s(s^2 + 2s + 5)}.$$

The right-hand side of this equation has the following partial fraction decomposition

$$\frac{1}{s(s^2 + 2s + 5)} = \frac{1}{5s} - \frac{s + 2}{5(s^2 + 2s + 5)},$$

so that

$$Y(s) = 2(1 - 2e^{-4s} + e^{-8s})\left[\frac{1}{s} - \frac{s + 2}{(s^2 + 2s + 5)}\right]. \qquad (9.7.3)$$

Now,

$$L^{-1}\left[\frac{s + 2}{s^2 + 2s + 5}\right] = L^{-1}\left[\frac{s + 1}{(s + 1)^2 + 4} + \frac{1}{(s + 1)^2 + 4}\right]$$

$$= e^{-t}\cos 2t + \frac{1}{2}e^{-t}\sin 2t.$$

Taking the inverse Laplace transform of both sides of (9.7.3) and using Theorem 9.7.1 yields

$$y(t) = 2\Big\{1 - e^{-t}\cos 2t - \frac{1}{2}e^{-t}\sin 2t - 2u_4(t)[1 - e^{-(t-4)}\cos 2(t - 4) - \frac{1}{2}e^{-(t-4)}\sin 2(t - 4)] +$$

$$u_8(t)[1 - e^{-(t-8)}\cos 2(t - 8) - \frac{1}{2}e^{-(t-8)}\sin 2(t - 8)]\Big\}.$$

We can express this solution in the simpler form

$$y(t) = g(t) - 2u_4(t)g(t - 4) + u_8(t)g(t - 8),$$

where

$$g(t) = 2(1 - e^{-t}\cos 2t - \frac{1}{2}e^{-t}\sin 2t).$$

The given IVP can be interpreted as governing the motion of a damped spring-mass system. Due to the form of the initial conditions, if the forcing function were the same constant, F_0, for all time, the oscillations would quickly be damped out, and $y(t)$ would approach $F_0/5$. In this problem, the forcing term is constant over different time intervals. Consequently, the mass first performs damped oscillations approaching $y = 2$. Then the second part of the driving term comes into effect and the subsequent oscillations are about $y = -2$. After $t = 8$, the forcing function vanishes and the physical system performs damped oscillations about the equilibrium. These features can be seen in Figure 9.7.4.

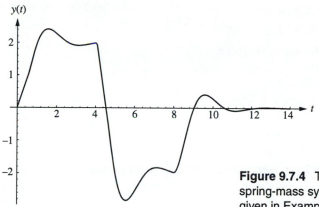

Figure 9.7.4 The response of a damped spring-mass system to the driving term given in Example 9.7.6.

EXERCISES 9.7

For problems 1–10, determine the Laplace transform of the given function f.

1. $f(t) = (t - 1)u_1(t)$.

2. $f(t) = e^{3(t - 2)}u_2(t)$.

3. $f(t) = \sin(t - \pi/4)\, u_{\pi/4}(t)$.

4. $f(t) = \cos t\, u_\pi(t)$.

5. $f(t) = (t - 2)^2\, u_2(t)$.

6. $f(t) = t\, u_3(t)$.

7. $f(t) = (t - 1)^2\, u_2(t)$.

8. $f(t) = e^{(t - 4)}(t - 4)^3 u_4(t)$.

9. $f(t) = e^{-2(t-1)}\sin 3(t - 1)\, u_1(t)$.

10. $f(t) = e^{a(t-c)}\cos b(t - c)\, u_c(t)$, where a, b, and c are positive constants.

For problems 11–25, determine the inverse Laplace transform of F.

11. $F(s) = \dfrac{e^{-2s}}{s^2}$.

12. $F(s) = \dfrac{e^{-s}}{s + 1}$.

13. $F(s) = \dfrac{e^{-3s}}{s + 4}$.

14. $F(s) = \dfrac{se^{-s}}{s^2 + 4}$.

15. $F(s) = \dfrac{e^{-3s}}{s^2 + 1}$.

16. $F(s) = \dfrac{e^{-2s}}{s + 2}$.

17. $F(s) = \dfrac{e^{-s}}{(s + 1)(s - 4)}$.

18. $F(s) = \dfrac{e^{-2s}}{s^2 + 2s + 2}$.

19. $F(s) = \dfrac{e^{-s}(s + 6)}{s^2 + 9}$.

20. $F(s) = \dfrac{e^{-5s}}{s^2 + 16}$.

21. $F(s) = \dfrac{e^{-2s}}{(s - 3)^3}$.

22. $F(s) = \dfrac{e^{-4s}(s + 3)}{s^2 - 6s + 13}$.

23. $F(s) = \dfrac{e^{-s}(2s - 1)}{s^2 + 4s + 5}$.

24. $F(s) = \dfrac{2e^{-2s}}{(s - 1)(s^2 + 1)}$.

25. $F(s) = \dfrac{50e^{-3s}}{(s + 1)^2(s^2 + 4)}$.

For problems 26–40, solve the given IVP.

26. $y' + 2y = 2u_1(t)$, $y(0) = 1$.

27. $y' - 2y = u_2(t)e^{t-2}$, $y(0) = 2$.

28. $y' - y = 4u_{\pi/4}(t)\cos(t - \pi/4)$, $y(0) = 1$.

29. $y' + 2y = u_\pi(t)\sin 2t$, $y(0) = 3$.

30. $y' + 3y = f(t)$, $y(0) = 1$, where

$$f(t) = \begin{cases} 1, & 0 \le t < 1, \\ 0, & t \ge 1. \end{cases}$$

31. $y' - 3y = f(t)$, $y(0) = 2$, where

$$f(t) = \begin{cases} \sin t, & 0 \leq t < \pi/2, \\ 1, & t \geq \pi/2. \end{cases}$$

32. $y' - 3y = 10e^{-(t-a)}\sin[2(t-a)]u_a(t)$, $y(0) = 5$, where a is a positive constant.

33. $y'' - y = u_1(t)$, $y(0) = 2$, $y'(0) = 0$.

34. $y'' - y' - 2y = 1 - 3u_2(t)$, $y(0) = 1$, $y'(0) = -2$.

35. $y'' - 4y = u_1(t) - u_2(t)$, $y(0) = 0$, $y'(0) = 4$.

36. $y'' + y = t - u_1(t)(t-1)$, $y(0) = 2$, $y'(0) = 1$.

37 $y'' + 3y' + 2y = 10u_{\pi/4}(t)\sin(t - \pi/4)$, $y(0) = 1$, $y'(0) = 0$.

38. $y'' + y' - 6y = 30u_1(t)e^{-(t-1)}$, $y(0) = 3$, $y'(0) = -4$.

39. $y'' + 4y' + 5y = 5u_3(t)$, $y(0) = 2$, $y'(0) = 1$.

40. $y'' - 2y' + 5y = 2\sin t + u_{\pi/2}(t)[1 - \sin(t - \pi/2)]$, $y(0) = 0$, $y'(0) = 0$.

For problems 41–44, solve the given IVP.

41. $y' + y = f(t)$, $y(0) = 2$, where $f(t)$ is given in Figure 9.7.5.

Figure 9.7.5

42. $y' + 2y = f(t)$, $y(0) = 0$, where $f(t)$ is given in Figure 9.7.6.

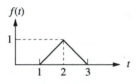

Figure 9.7.6

43. $y' - y = f(t)$, $y(0) = 2$, where $f(t)$ is given in Figure 9.7.7.

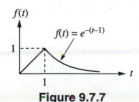

Figure 9.7.7

44. $y' - 2y = f(t)$, $y(0) = 0$, where $f(t)$ is given in Figure 9.7.8.

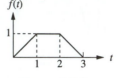

Figure 9.7.8

45. Solve the IVP

$$y' - y = f(t), \quad y(0) = 1,$$

where

$$f(t) = \begin{cases} 2, & 0 \leq t < 1, \\ -1, & t \geq 1 \end{cases}$$

in the following two ways:

(a) Directly using the Laplace transform.

(b) Using the technique for solving first-order linear equations developed in Section 1.6.

46. The current, $i(t)$, in an RL circuit is governed by the DE

$$\frac{di}{dt} + \frac{R}{L}i = \frac{1}{L}E(t)$$

where R and L are constants and $E(t)$ represents the applied EMF. At $t = 0$, the switch in the circuit is closed, and the applied EMF increases linearly from 0 V to 10 V in a time interval of 5 s. The EMF then remains constant for $t \geq 5$. Determine the current in the circuit for $t \geq 0$.

47. The DE governing the charge $q(t)$ on the capacitor in an RC circuit is

$$\frac{dq}{dt} + \frac{1}{RC}q = \frac{1}{R}E(t)$$

where R and C are constants and $E(t)$ represents the applied EMF. Over a time interval of 10 s, the applied EMF has the constant value 20 V. Thereafter the EMF decays exponentially according to $E(t) = 20e^{-(t-10)}$. If the capacitor is initially uncharged, and $RC \neq 1$, determine current in the circuit for $t > 0$. [Recall that the current $i(t)$ is related to the charge on the capacitor by $i(t) = dq/dt$.]

9.8   IMPULSIVE DRIVING TERMS: THE DIRAC DELTA FUNCTION

Consider the DE

$$y'' + by' + cy = f(t).$$

We have seen in the previous sections that the Laplace transform is useful in the case when the forcing term $f(t)$ is piecewise continuous. We now consider another type of forcing term, namely, that describing an impulsive force. Such a force arises when an object is dealt an instantaneous blow, for example, when an object is hit by a hammer. (See Figure 9.8.1.) The aim of this section is to develop a way of representing impulsive forces mathematically and then to show how the Laplace transform can be used to solve DE when the driving term is due to an impulsive force.

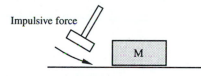

Impulsive force

M

Figure 9.8.1   An example of an impulsive force.

Suppose that a force of magnitude F acts on an object over the time interval $[t_1, t_2]$. The *impulse* of this force, I, is defined by[1]

$$I = \int_{t_1}^{t_2} F(t)\, dt.$$

Since $F(t)$ is zero for t outside the interval $[t_1, t_2]$, we can write

$$I = \int_{-\infty}^{\infty} F(t)\, dt.$$

Mathematically, I gives the area under the curve $y = F(t)$ lying over the t–axis. (See Figure 9.8.2.) We now introduce a mathematical description of a force that

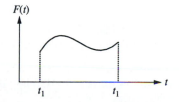

$F(t)$

t_1 t_1 t

Figure 9.8.2   When a force of magnitude F newtons acts on an object over a time interval $[t_1, t_2]$ seconds, the impulse of the force is given by the area under the curve.

instantaneously imparts an impulse of unit magnitude to an object at $t = a$. Thus the two properties that we wish to characterize are the following:

1. The force acts instantaneously.
2. The force has unit impulse.

---

[1] This represents the change in momentum of the object due to the applied force.

We proceed in the following manner. Define the function $d_\varepsilon(t-a)$ by

$$d_\varepsilon(t-a) = \frac{u_a(t) - u_{a+\varepsilon}(t)}{\varepsilon}, \qquad (9.8.1)$$

where u_a is the unit step function. (See Figure 9.8.3.) We can interpret $d_\varepsilon(t-a)$ as

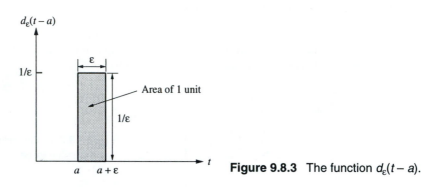

Figure 9.8.3 The function $d_\varepsilon(t-a)$.

representing a force of magnitude $1/\varepsilon$ that acts for a time interval of ε starting at $t = a$. Notice that this force does have unit impulse, since

$$I = \int_{-\infty}^{\infty} d_\varepsilon(t-a)\, dt = \int_{-\infty}^{\infty} \frac{u_a(t) - u_{a+\varepsilon}(t)}{\varepsilon}\, dt = \int_{a}^{a+\varepsilon} \frac{1}{\varepsilon}\, dt = 1.$$

To capture the idea of an instantaneous force we take the limit as $\varepsilon \to 0^+$. It follows from (9.8.1) that

$$\lim_{\varepsilon \to 0^+} d_\varepsilon(t-a) = 0 \text{ whenever } t \neq a.$$

Also, since $I = 1$ for all t,

$$\lim_{\varepsilon \to 0^+} I = 1.$$

These properties characterize mathematically the idea of a force of unit impulse acting instantaneously at $t = a$. We use them to define the *unit impulse function*.

Definition 9.8.1: The **unit impulse function**, or **Dirac delta function**, $\delta(t-a)$ is the (generalized) function that satisfies

1. $\delta(t-a) = 0, \quad t \neq a,$

2. $\displaystyle\int_{-\infty}^{\infty} \delta(t-a)\, dt = 1.$

REMARK   The unit impulse function is not a function in the usual sense. It is an example of what is called a *generalized function*. The detailed study of such functions is beyond the scope of the present text . However, all that we will require are the properties (1) and (2) of Definition 9.8.1.

Thus, to summarize,

$\delta(t - a)$ describes a force that instantaneously imparts a unit impulse to a system at $t = a$.

We now consider the possibility of determining the Laplace transform of $\delta(t - a)$. The natural way to do this is to return to the function $d_\varepsilon(t - a)$ and define the Laplace transform of $\delta(t - a)$ in the following manner.

$$L[\delta(t - a)] = \lim_{\varepsilon \to 0^+} L[d_\varepsilon(t - a)] = \lim_{\varepsilon \to 0^+} \int_0^\infty e^{-st} \left[\frac{u_a(t) - u_{a+\varepsilon}(t)}{\varepsilon} \right] dt$$

$$= \lim_{\varepsilon \to 0^+} \frac{1}{\varepsilon} \int_a^{a+\varepsilon} e^{-st} \, dt = \lim_{\varepsilon \to 0^+} \frac{1}{\varepsilon} \left\{ -\frac{1}{s} \left[e^{-s(a+\varepsilon)} - e^{-sa} \right] \right\}$$

$$= \frac{1}{s} e^{-sa} \lim_{\varepsilon \to 0^+} \left(\frac{1 - e^{-\varepsilon s}}{\varepsilon} \right).$$

Using L'Hopital's rule to evaluate the preceding limit yields

$$L[\delta(t - a)] = e^{-sa}. \tag{9.8.2}$$

In particular,

$$L[\delta(t)] = 1. \tag{9.8.3}$$

It can be shown more generally that if g is a continuous function on $(-\infty, \infty)$, then

$$\int_{-\infty}^\infty g(t) \, \delta(t - a) \, dt = g(a). \tag{9.8.4}$$

Example 9.8.1   Solve the IVP

$$y'' + 4y' + 13y = \delta(t - \pi), \quad y(0) = 2, \quad y'(0) = 1.$$

Solution   Taking the Laplace transform of both sides of the given DE and imposing the initial conditions yields

$$s^2 Y - 2s - 1 + 4(sY - 2) + 13Y = e^{-\pi s},$$

which implies that

$$Y(s) = \frac{e^{-\pi s} + 2s + 9}{s^2 + 4s + 13}$$

$$= \frac{e^{-\pi s} + 2s + 9}{(s + 2)^2 + 9}$$

$$= \frac{e^{-\pi s}}{(s+2)^2 + 9} + \frac{2(s+2)}{(s+2)^2 + 9} + \frac{5}{(s+2)^2 + 9}.$$

Taking the inverse Laplace transform of both sides gives

$$y(t) = L^{-1}\left\{\frac{1}{3} e^{-\pi s}L[e^{-2t}\sin 3t]\right\} + 2e^{-2t}\cos 3t + \frac{5}{3} e^{-2t}\sin 3t$$

$$= \frac{1}{3} u_\pi(t)\, e^{-2(t-\pi)}\sin 3(t-\pi) + 2e^{-2t}\cos 3t + \frac{5}{3} e^{-2t}\sin 3t.$$

Since $\sin 3(t - \pi) = -\sin 3t$, we finally obtain

$$y(t) = -\frac{1}{3} u_\pi(t)e^{-2(t-\pi)}\sin 3t + e^{-2t}(2\cos 3t + \frac{5}{3}\sin 3t).$$

Example 9.8.2   Consider the spring-mass system depicted in Figure 9.8.4. At $t = 0$, the mass is pulled down a distance 1 unit from equilibrium and released from rest. After 3 s, the mass is dealt an instantaneous blow that imparts five units of impulse in the upward direction. The IVP governing the motion of the mass is

$$\frac{d^2y}{dt^2} + 4y = -5\delta(t-3),\ y(0) = 1,\ \frac{dy}{dt}(0) = 0,$$

where $5\delta(t - 3)$ describes the impulsive force that acts on the mass at $t = 3$, and the positive y-direction is downward. Determine the motion of the mass for all $t > 0$.

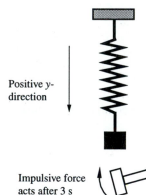

Positive y-direction

Impulsive force acts after 3 s

Figure 9.8.4 A spring-mass system in which friction is neglected and the only external force acting on the system is an impulsive force which imparts 5 units of impulse at $t = 3$.

Solution   Taking the Laplace transform of the DE and imposing the initial conditions yields

$$s^2Y(s) - s + 4\, Y(s) = -5e^{-3s},$$

so that

$$Y(s) = \frac{-5e^{-3s} + s}{s^2 + 4}$$

$$= -\frac{5e^{-3s}}{s^2 + 4} + \frac{s}{s^2 + 4}.$$

We can now take the inverse Laplace transform of both sides of this equation to obtain

$$y(t) = L^{-1}\left\{-\frac{5}{2} e^{-3s}L[\sin 2t]\right\} + \cos 2t.$$

Thus,

$$y(t) = -\frac{5}{2} u_3(t)\sin 2(t - 3) + \cos 2t.$$

The first term on the right-hand side represents the contribution from the impulsive force. Obviously this does not affect the motion of the mass until $t = 3$, but then contributes for all $t \geq 3$. More explicitly, we can write the solution as

$$y(t) = \begin{cases} \cos 2t, & 0 \leq t < 3, \\ \cos 2t - \dfrac{5}{2} \sin 2(t - 3), & t \geq 3. \end{cases}$$

The resulting motion is depicted in Figure 9.8.5, where the effect of the impulsive force is apparent. We see that $y(t)$ is continuous at $t = 3$, but that $y'(t)$ is discontinuous at $t = 3$.

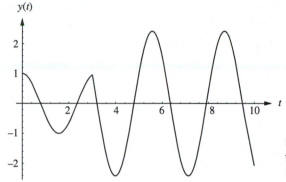

Figure 9.8.5 The motion of the spring-mass system in Example 9.8.2.

EXERCISES 9.8

For problems 1–12, solve the given IVP.

1. $y' - 2y = \delta(t - 2)$, $y(0) = 1$.

2. $y' + 4y = 3\delta(t - 1)$, $y(0) = 2$.

3. $y' - 5y = 2e^{-t} + \delta(t - 3)$, $y(0) = 0$.

4. $y'' - 3y' + 2y = \delta(t - 1)$, $y(0) = 1$, $y'(0) = 0$.

5. $y'' - 4y = \delta(t - 3)$, $y(0) = 0$, $y'(0) = 1$.

6. $y'' + 2y' + 5y = \delta(t - \pi/2)$, $y(0) = 0$, $y'(0) = 2$.

7. $y'' - 4y' + 13y = \delta(t - \pi/4)$, $y(0) = 3$, $y'(0) = 0$.

8. $y'' + 4y' + 3y = \delta(t - 2)$, $y(0) = 1$, $y'(0) = -1$.

9. $y'' + 6y' + 13y = \delta(t - \pi/4))$, $y(0) = 5$, $y'(0) = 5$.

10. $y'' + 9y = 15\sin 2t + \delta(t - \pi/6)$, $y(0) = 0$, $y'(0) = 0$.

11. $y'' + 16y = 4\cos 3t + \delta(t - \pi/3)$, $y(0) = 0$, $y'(0) = 0$.

12. $y'' + 2y' + 5y = 4\sin t + \delta(t - \pi/6)$, $y(0) = 0$, $y'(0) = 1$.

13. The motion of a spring-mass system is governed by the IVP

$$\frac{d^2y}{dt^2} + 4y = F_0\cos 3t, \quad y(0) = 0, \quad \frac{dy}{dt}(0) = 0,$$

where F_0 is a constant. At $t = 5$ s, the mass is dealt a blow in the upward direction that instantaneously imparts 4 units of impulse to the system. Determine the resulting motion of the mass.

14. The motion of a spring-mass system is governed by the IVP

$$\frac{d^2y}{dt^2} + 4\frac{dy}{dt} + 13y = 10\sin 5t, \quad y(0) = 0, \quad \frac{dy}{dt}(0) = 0.$$

At $t = 10$ s, the mass is dealt a blow in the downward direction that instantaneously imparts 2 units of impulse to the system. Determine the resulting motion of the mass.

15. Consider the spring-mass system whose motion is governed by the IVP

$$\frac{d^2y}{dt^2} + \omega_0^2 y = F_0\sin \omega t + A\,\delta(t - t_0),$$

$$y(0) = 0, \quad \frac{dy}{dt}(0) = 0,$$

where ω_0, ω, F_0, A and t_0 are positive constants and $\omega \neq \omega_0$. Solve the IVP to determine the position of the mass at time t.

9.9   THE CONVOLUTION INTEGRAL

Very often in solving a DE using the Laplace transform method we require the inverse Laplace transform of an expression of the form

$$H(s) = F(s)G(s),$$

where $F(s)$ and $G(s)$ are functions whose inverse Laplace transform is known. It is important to note that

$$L^{-1}[H(s)] \neq L^{-1}[F(s)]L^{-1}[G(s)].$$

For example,

$$L^{-1}\left[\frac{1}{(s-1)(s^2+1)}\right] = L^{-1}\left[\frac{1}{2(s-1)} - \frac{s+1}{2(s^2+1)}\right]$$

$$= \frac{1}{2}e^t - \frac{1}{2}\cos t - \frac{1}{2}\sin t,$$

whereas,

$$L^{-1}\left[\frac{1}{s-1}\right]L^{-1}\left[\frac{1}{s^2+1}\right] = e^t \sin t.$$

Consequently,

$$L^{-1}\left[\frac{1}{(s-1)(s^2+1)}\right] \neq L^{-1}\left[\frac{1}{(s-1)}\right]L^{-1}\left[\frac{1}{(s^2+1)}\right].$$

However, it is possible, at least in theory, to determine $L^{-1}[F(s)G(s)]$ directly in terms of an integral involving $f(t)$ and $g(t)$. Before showing this we require a definition.

Definition 9.9.1:   Suppose that f and g are continuous on the interval $[0, b]$. Then for t in $(0, b]$, the **convolution product**, $f * g$, of f and g is defined by

$$(f * g)(t) = \int_0^t f(t - \tau)\, g(\tau)\, d\tau.$$

Notice that $f * g$ is indeed a function of t. The integral

$$\int_0^t f(t-\tau)\,g(\tau)\,d\tau \tag{9.9.1}$$

is called a **convolution integral**.

Example 9.9.1   If $f(t) = t$ and $g(t) = \sin t$, determine $f * g$.

Solution   From Definition 9.9.1, we have

$$(f * g)(t) = \int_0^t (t-\tau)\sin \tau\,d\tau = t\int_0^t \sin \tau\,d\tau - \int_0^t \tau \sin \tau\,d\tau$$

$$= t(1 - \cos t) - (\sin t - t\cos t)$$

$$= t - \sin t. \qquad \square$$

The convolution product satisfies the three basic properties of the ordinary multiplicative product, namely,

1. $f * g = g * f.$ (Commutative)
2. $f * (g * h) = (f * g) * h.$ (Associative)
3. $f * (g + h) = f * g + f * h.$ (Distributive over addition)

The proofs of these three properties are left as exercises.

We now show how the convolution product can be useful in evaluating inverse Laplace transforms.

Theorem 9.9.1 (The Convolution Theorem):   If f and g are in $E(0, \infty)$, then

$$L[f * g] = L[f]\,L[g]. \tag{9.9.2}$$

Conversely,

$$L^{-1}[F(s)G(s)] = (f * g)(t). \tag{9.9.3}$$

PROOF   We must use the definition of the Laplace transform and the convolution product:

$$L[f * g] = \int_0^\infty e^{-st}\left\{\int_0^t f(t-\tau)g(\tau)\,d\tau\right\} dt$$

$$= \int_0^\infty \int_0^t e^{-st} f(t-\tau)g(\tau)\,d\tau\,dt.$$

It is not clear how to proceed at this point. However, when dealing with an iterated double integral, it is often worth changing the order of integration to see if any simplification arises. In this case, the limits of integration are $0 \le \tau \le t, 0 \le t < \infty$, so that the region of integration is that part of the $t\tau$-plane that lies above the t-axis and below the line $\tau = t$. This region is shown in Figure 9.9.1. Reversing the order of

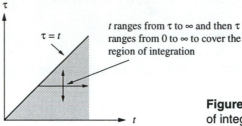

t ranges from τ to ∞ and then τ
ranges from 0 to ∞ to cover the
region of integration

Figure 9.9.1   Changing the order
of integration in Theorem 9.9.1.

integration, the new limits are $\tau \leq t < \infty$, $0 \leq \tau < \infty$. Thus, we can write

$$L[(f * g)(t)] = \int_0^\infty \int_\tau^\infty e^{-st} f(t - \tau) g(\tau) \, dt \, d\tau.$$

We now make the change of variable $u = t - \tau$ in the first iterated integral. Then $du = dt$ (remember that τ is treated as a constant when performing the first integration) and the new u-limits are $0 \leq u < \infty$. Consequently,

$$L[(f * g)(t)] = \int_0^\infty \int_0^\infty e^{-s(u+\tau)} g(\tau) f(u) \, du \, d\tau = \int_0^\infty e^{-s\tau} g(\tau) \left[\int_0^\infty e^{-su} f(u) \, du \right] d\tau$$

$$= \left[\int_0^\infty e^{-su} f(u) \, du \right] \left[\int_0^\infty e^{-s\tau} g(\tau) \, d\tau \right] = L[f] \, L[g],$$

as required. The converse, (9.9.3), is obtained in the usual manner by taking the inverse Laplace transform of (9.9.2). ∎

REMARK   It can be shown more generally that

$$L^{-1}[F_1(s) \, F_2(s) \cdots F_n(s)] = (f_1 * f_2 * \cdots * f_n)(t).$$

Example 9.9.2   Determine $L[f]$ if

$$f(t) = \int_0^t \sin(t - \tau) e^{-\tau} \, d\tau.$$

Solution   In this case, we recognize that

$$f(t) = \sin t * e^{-t},$$

so that, by the convolution theorem,

$$L[f] = L[\sin t] \, L[e^{-t}] = \frac{1}{(s^2 + 1)(s + 1)}.$$

Example 9.9.3   Find $L^{-1}\left[\dfrac{1}{s^2(s - 1)} \right]$.

Solution   We could determine the inverse Laplace transform in the usual manner by first using a partial fraction decomposition. However we will use the convolution theorem.

$$L^{-1}\left[\frac{1}{s^2(s-1)}\right] = L^{-1}\left[\frac{1}{s^2}\right] * L^{-1}\left[\frac{1}{s-1}\right] = \int_0^t (t-\tau)e^\tau \, d\tau.$$

By integrating by parts, we obtain

$$L^{-1}\left[\frac{1}{s^2(s-1)}\right] = \left[te^\tau - (\tau e^\tau - e^\tau)\right]_0^t = e^t - t - 1.$$

Example 9.9.4   Find $L^{-1}\left[\dfrac{G(s)}{(s-1)^2+1}\right]$.

Solution   Using the convolution theorem, we see that

$$L^{-1}\left[\frac{G(s)}{(s-1)^2+1}\right] = L^{-1}\left[\frac{1}{(s-1)^2+1}\right] * L^{-1}[G(s)]$$

$$= e^t \sin t * g(t).$$

That is,

$$L^{-1}\left[\frac{G(s)}{(s-1)^2+1}\right] = \int_0^t e^{t-\tau}\sin(t-\tau)g(\tau) \, d\tau.$$

Example 9.9.5   Solve the IVP

$$y'' + \omega^2 y = f(t), \quad y(0) = \alpha, \quad y'(0) = \beta,$$

where α, β, and ω (not equal to 0) are constants and $f(t)$ is an arbitrary function in $E(0, \infty)$.

Solution   Taking the Laplace transform of the DE and imposing the given initial conditions yields

$$s^2 Y(s) - \alpha s - \beta + \omega^2 Y(s) = F(s)$$

where $F(s)$ denotes the Laplace transform of f. Simplifying, we obtain

$$Y(s) = \frac{F(s)}{s^2 + \omega^2} + \frac{\alpha s}{s^2 + \omega^2} + \frac{\beta}{s^2 + \omega^2}.$$

Taking the inverse Laplace transform of both sides of this equation and using the convolution theorem yields

$$y(t) = \frac{1}{\omega}\int_0^t \sin \omega(t-\tau) f(\tau) d\tau + \alpha \cos \omega t + \frac{\beta}{\omega} \sin \omega t.$$

VOLTERRA INTEGRAL EQUATIONS

The applications of the Laplace transform that we have so far considered have been for solving DE. We now briefly discuss another type of equation whose solution can often be obtained using the Laplace transform.

An equation of the form

$$x(t) = f(t) + \int_0^t k(t - \tau) \, x(\tau) \, d\tau \tag{9.9.4}$$

is called a **Volterra integral equation**. In this equation, the unknown function is $x(t)$. The functions f and k are specified, and k is called the **kernel** of the equation. For example,

$$x(t) = 2 \sin t + \int_0^t \cos(t - \tau) \, x(\tau) \, d\tau$$

is a Volterra integral equation. We now show how the convolution theorem for Laplace transforms can be used to determine, up to the evaluation of an inverse transform, the function $x(t)$ that satisfies equation (9.9.4). The key to solving equation (9.9.4) is to notice that the integral that appears in this equation is, in fact, a convolution integral. Thus, taking the Laplace transform of both sides of equation (9.9.4) we obtain

$$X(s) = F(s) + L[k(t) * x(t)].$$

That is, using the convolution theorem,

$$X(s) = F(s) + K(s) X(s),$$

Solving algebraically for $X(s)$ yields

$$X(s) = \frac{F(s)}{1 - K(s)}$$

so that

$$x(t) = L^{-1} \left[\frac{F(s)}{1 - K(s)} \right].$$

This technique can be used to solve a wide variety of Volterra integral equations.

Example 9.9.6   Solve the Volterra integral equation

$$x(t) = 3 \cos t + 5 \int_0^t \sin(t - \tau) x(\tau) \, d\tau.$$

Solution   Taking the Laplace transform of the given integral equation and using the convolution theorem yields

$$X(s) = \frac{3s}{s^2 + 1} + \frac{5}{s^2 + 1} X(s).$$

That is,

$$X(s) \left(\frac{s^2 - 4}{s^2 + 1} \right) = \frac{3s}{s^2 + 1},$$

so that

$$X(s) = \frac{3s}{s^2 - 4}.$$

Decomposing the right-hand side into partial fractions, we obtain

$$X(s) = \frac{3}{2}\left(\frac{1}{s - 2} + \frac{1}{s + 2}\right).$$

Taking the inverse Laplace transform yields

$$x(t) = \frac{3}{2}(e^{2t} + e^{-2t}) = 3\cosh 2t.$$

EXERCISES 9.9

For problems 1–5, determine $f * g$.

1. $f(t) = t$, $g(t) = 1$.
2. $f(t) = \cos t$, $g(t) = t$.
3. $f(t) = e^t$, $g(t) = t$.
4. $f(t) = t^2$, $g(t) = e^t$.
5. $f(t) = e^t$, $g(t) = e^t \sin t$.

6. Prove that $f * g = g * f$.

7. Prove that $f * (g * h) = (f * g) * h$.

8. Prove that $f * (g + h) = f * g + f * h$.

For problems 9–13, determine $L[f * g]$.

9. $f(t) = t$, $g(t) = \sin t$.
10. $f(t) = e^{2t}$, $g(t) = 1$.
11. $f(t) = \sin t$, $g(t) = \cos 2t$.
12. $f(t) = e^t$, $g(t) = te^{2t}$.
13. $f(t) = t^2$, $g(t) = e^{3t} \sin 2t$.

For problems 14–19, determine $L^{-1}[F(s)G(s)]$ in the following two ways: **(a)** using the convolution theorem, **(b)** using partial fractions.

14. $F(s) = \frac{1}{s}$, $G(s) = \frac{1}{s - 2}$.

15. $F(s) = \frac{1}{s + 1}$, $G(s) = \frac{1}{s}$.

16. $F(s) = \frac{s}{s^2 + 4}$, $G(s) = \frac{2}{s}$.

17. $F(s) = \frac{1}{s + 2}$, $G(s) = \frac{s + 2}{s^2 + 4s + 13}$.

18. $F(s) = \frac{1}{s^2 + 9}$, $G(s) = \frac{2}{s^3}$.

19. $F(s) = \frac{1}{s^2}$, $G(s) = \frac{e^{-\pi s}}{s^2 + 1}$.

For problems 20–24, express $L^{-1}[F(s)G(s)]$ in terms of a convolution integral.

20. $F(s) = \frac{4}{s^3}$, $G(s) = \frac{s - 1}{s^2 - 2s + 5}$.

21. $F(s) = \frac{s + 1}{s^2 + 2s + 2}$, $G(s) = \frac{1}{(s + 3)^2}$.

22. $F(s) = \frac{2}{s^2 + 6s + 10}$, $G(s) = \frac{2}{s - 4}$.

23. $F(s) = \frac{s + 4}{s^2 + 8s + 25}$, $G(s) = \frac{se^{-\pi s/2}}{s^2 + 16}$.

24. $F(s) = \frac{1}{s - 4}$, $G(s)$ arbitrary.

For problems 25–31, solve the given IVP up to the evaluation of a convolution integral.

25. $y'' + y = e^{-t}$, $y(0) = 0$, $y'(0) = 1$.
26. $y'' - 2y' + 10y = \cos 2t$, $y(0) = 0$, $y'(0) = 1$.
27. $y'' + 16y = f(t)$, $y(0) = \alpha$, $y'(0) = \beta$, where α, and β are constants.
28. $y' - ay = f(t)$, $y(0) = \alpha$, where a and α are constants.
29. $y'' - a^2 y = f(t)$, $y(0) = \alpha$, $y'(0) = \beta$, where a, α, and β are constants, and $a \neq 0$.
30. $y'' - (a + b)y' + aby = f(t)$, $y(0) = \alpha$, $y'(0) = \beta$, where a, b, α, and β, are constants, and $a \neq b$.
31. $y'' - 2ay' + (a^2 + b^2)y = f(t)$, $y(0) = \alpha$, $y'(0) = \beta$, where a, b, α, and β, are constants, and $b \neq 0$.

For problems 32–37, solve the given Volterra integral equation.

32. $x(t) = e^{-t} + 4\int_0^t (t - \tau)x(\tau)\, d\tau$.

33. $x(t) = 2e^{3t} - \int_0^t e^{2(t-\tau)}x(\tau)\, d\tau$.

34. $x(t) = 4e^t + 3 \displaystyle\int_0^t e^{-(t-\tau)} x(\tau)\, d\tau.$

35. $x(t) = 1 + 2 \displaystyle\int_0^t \sin(t - \tau)\, x(\tau)\, d\tau.$

36. $x(t) = e^{2t} + 5 \displaystyle\int_0^t \cos[2(t - \tau)]\, x(\tau)\, d\tau.$

37. $x(t) = 2\left\{1 + \displaystyle\int_0^t \cos[2(t - \tau)]\, x(\tau)\, d\tau\right\}.$

38. Show that the IVP

$$y'' + y = f(t),\ y(0) = 0,\ y'(0) = 0$$

can be reformulated as the integral equation

$$x(t) = f(t) - \int_0^t (t - \tau) x(\tau)\, d\tau,$$

where $y''(t) = x(t)$.

10

Series Solutions to Linear Differential Equations

10.1 INTRODUCTION

So far, the techniques that we have developed for solving DE have involved determining a closed form solution for a given equation (or system) in terms of familiar elementary functions. Essentially, the only equations that we can derive such solutions for are as follows:

1. Constant coefficient equations.
2. Cauchy–Euler equations.

For example, we cannot at the present time determine the solution to the seemingly simple DE

$$y'' + e^x y = 0.$$

In this chapter, we consider the possibility of representing solutions to linear DE in the form of some type of infinite series. We begin in Section 10.3 with the simplest case, namely, DE whose solutions can be represented as a convergent power series,

$$y(x) = \sum_{n=0}^{\infty} a_n x^n,$$

where a_n are constants. This can be considered as a generalization of the method of undetermined coefficients to the case when we have an infinite number of constants. We will determine the appropriate values of these constants by substitution into the DE.

Not all DE have solutions that can be represented by a convergent power series. We will find that the next simplest type of series solution that is applicable to a broad class of linear DE is one of the form

$$y(x) = x^r \sum_{n=0}^{\infty} a_n x^n,$$

called a Frobenius series. Here, in addition to the coefficients a_n, we must also determine the value of the constant r (which in general will *not* be a positive integer). The analysis of this problem is quite involved, and the computations can be extremely tedious. However, the technique is an important and useful addition to the applied mathematician's tools for solving DE.

Before beginning the development of the theory, we note that for simplicity we will restrict our attention in this chapter to second-order linear homogeneous DE whose standard form is

$$y'' + p(x)y' + q(x)y = 0,$$

where p and q are functions that are specified on some interval I. The techniques can be extended easily to higher order, and also to systems of linear DE.

10.2   A REVIEW OF POWER SERIES

We begin with a very brief review of the main facts about power series which should be familiar from a previous calculus course. They will be required throughout the remainder of the chapter.

BASIC DEFINITION   An infinite series of the form

$$\sum_{n=0}^{\infty} a_n(x - x_0)^n, \tag{10.2.1}$$

where a_n and x_0 are constants, is called a power series centered at $x = x_0$. The substitution $u = x - x_0$ has the effect of transforming (10.2.1) to

$$\sum_{n=0}^{\infty} a_n u^n,$$

so that we can, without loss of generality, restrict attention to power series of the form

$$\sum_{n=0}^{\infty} a_n x^n, \tag{10.2.2}$$

whose center is $x = 0$. The series (10.2.2) is said to **converge** at $x = x_1$ if

$$\lim_{k \to \infty} \sum_{n=0}^{k} a_n x_1^n$$

exists and is finite. The set of all x for which (10.2.2) converges is called the **interval of convergence**.

BASIC CONVERGENCE THEOREM   For the power series (10.2.2), precisely one of the following is true:

1. $\displaystyle\sum_{n=0}^{\infty} a_n x^n$ converges only at $x = 0$.

2. $\displaystyle\sum_{n=0}^{\infty} a_n x^n$ converges for all real x.

3. There is a positive number R such that $\displaystyle\sum_{n=0}^{\infty} a_n x^n$ converges (absolutely) for $|x| < R$ and diverges for $|x| > R$.

REMARK   The number R occurring in possibility (3) is called the **radius of convergence**. (See Fig. 10.2.1.) The convergence or divergence of the series at the endpoints $x = \pm R$ must be treated separately. In possibility (2), we define the radius of convergence to be $R = \infty$.

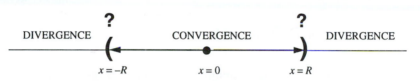

Figure 10.2.1   The radius of convergence of a power series.

RATIO TEST   The ratio test for the convergence of a power series can be stated as follows.

Ratio Test: Consider the power series $\displaystyle\sum_{n=0}^{\infty} a_n x^n$. If $\displaystyle\lim_{n\to\infty} \left| \frac{a_{n+1}}{a_n} \right| = L$, then the radius of convergence of the power series is $R = \dfrac{1}{L}$. If $L = 0$, the series converges for all x, whereas if $L = \infty$, the power series converges only at $x = 0$.

Example 10.2.1   Determine the radius of convergence of $\displaystyle\sum_{n=0}^{\infty} \frac{n^2 x^n}{3^n}$.

Solution   In this case, we have

$$\lim_{n\to\infty} \left| \frac{a_{n+1}}{a_n} \right| = \lim_{n\to\infty} \frac{(n+1)^2}{3^{n+1}} \frac{3^n}{n^2} = \lim_{n\to\infty} \frac{(n+1)^2}{3n^2} = \frac{1}{3}.$$

Thus, $L = \dfrac{1}{3}$, so that the radius of convergence is $R = 3$. It is easy to see that the series diverges at the end points $x = \pm 3$, so that the interval of convergence is $(-3, 3)$.[1]

---

[1]Recall that a necessary (but not sufficient) condition for the convergence of the infinite series $\displaystyle\sum_{n=0}^{\infty} a_n x^n$ is that $\displaystyle\lim_{n\to\infty} a_n = 0$.

The Algebra of Power Series   Two power series $\displaystyle\sum_{n=0}^{\infty} a_n x^n$ and $\displaystyle\sum_{n=0}^{\infty} b_n x^n$ are equal if and only if each corresponding coefficient is equal, that is, $a_n = b_n$ for all n. In particular,

$$\sum_{n=0}^{\infty} a_n x^n = 0 \text{ if and only if } a_n = 0 \text{ for every } n.$$

We will use this result repeatedly throughout the chapter.

Now let $\displaystyle\sum_{n=0}^{\infty} a_n x^n$ and $\displaystyle\sum_{n=0}^{\infty} b_n x^n$ be power series with radii of convergence R_1 and R_2, respectively, and let $R = \min\{R_1, R_2\}$. For $|x| < R$, define the functions f and g by

$$f(x) = \sum_{n=0}^{\infty} a_n x^n, \qquad g(x) = \sum_{n=0}^{\infty} b_n x^n.$$

Then, for $|x| < R$,

1. $f(x) + g(x) = \displaystyle\sum_{n=0}^{\infty} (a_n + b_n) x^n$ (addition of power series).

2. $cf(x) = \displaystyle\sum_{n=0}^{\infty} (ca_n) x^n$, (multiplication of a power series by a real number c).

3. $f(x)g(x) = \displaystyle\sum_{n=0}^{\infty} c_n x^n$, where $c_n = \displaystyle\sum_{k=0}^{n} a_{n-k} b_k$, (multiplication of power series).
   The coefficients c_n appearing in this formula can be written in the equivalent form

$$c_n = \sum_{k=0}^{n} a_k b_{n-k}.$$

Example 10.2.2   It is known that the coefficients in the expansion

$$f(x) = \sum_{n=0}^{\infty} a_n x^n$$

satisfy

$$\sum_{n=1}^{\infty} n a_n x^n - \sum_{n=0}^{\infty} a_n x^{n+1} = 0. \tag{10.2.3}$$

Express all a_n in terms of a_0.

Solution   We replace n by $n - 1$ in the second summation in equation (10.2.3) (and alter the range of n appropriately) to obtain a common power of x^n in both sums. The result is

$$\sum_{n=1}^{\infty} n a_n x^n - \sum_{n=1}^{\infty} a_{n-1} x^n = 0,$$

which can be written as

$$\sum_{n=1}^{\infty} (na_n - a_{n-1})x^n = 0.$$

It follows that the a_n must satisfy the **recurrence relation**

$$na_n - a_{n-1} = 0, \quad n = 1, 2, 3, \ldots,$$

that is,

$$a_n = \frac{1}{n} a_{n-1}, \quad n = 1, 2, 3, \ldots.$$

Substituting for successive values of n we obtain

$n = 1$: $a_1 = a_0$,

$n = 2$: $a_2 = \dfrac{1}{2} a_1 = \dfrac{1}{2} a_0$,

$n = 3$: $a_3 = \dfrac{1}{3} a_2 = \dfrac{1}{3 \cdot 2} a_0$,

$n = 4$: $a_4 = \dfrac{1}{4} a_3 = \dfrac{1}{4 \cdot 3 \cdot 2} a_0$.

Continuing in this manner, we see that

$$a_n = \frac{1}{n!} a_0.$$

Consequently, we can write[1]

$$f(x) = a_0 \sum_{n=0}^{\infty} \frac{1}{n!} x^n.$$

THE CALCULUS OF POWER SERIES

Suppose that $\sum_{n=0}^{\infty} a_n x^n$ has radius of convergence R, and let

$$f(x) = \sum_{n=0}^{\infty} a_n x^n, \quad |x| < R.$$

Then $f(x)$ can be differentiated an arbitrary number of times on the interval $|x| < R$; furthermore, the derivatives can be obtained by termwise differentiation. Thus,

$$f'(x) = \sum_{n=0}^{\infty} na_n x^{n-1} = \sum_{n=1}^{\infty} na_n x^{n-1},$$

$$f''(x) = \sum_{n=1}^{\infty} n(n-1)a_n x^n = \sum_{n=2}^{\infty} n(n-1)a_n x^n,$$

and so on for higher order derivatives.

Analytic Functions and Taylor Series   We now introduce one of the main definitions of the section.

---

[1]The power series is the Maclaurin expansion of e^x, so that we can write $f(x) = a_0 e^x$.

> **Definition 10.2.1:**   A function is said to be **analytic at $x = x_0$** if it can be represented by a convergent power series centered at $x = x_0$ with nonzero radius of convergence.

In a previous calculus course you should have seen that if a function is analytic at $x = x_0$, then the power series representation of that function is unique and is given by

$$f(x) = \sum_{n=0}^{\infty} \frac{f^{(n)}(x_0)}{n!} (x - x_0)^n. \tag{10.2.4}$$

This is the **Taylor series** expansion of $f(x)$ about $x = x_0$. If $x_0 = 0$, then (10.2.4) reduces to

$$f(x) = \sum_{n=0}^{\infty} \frac{f^{(n)}(0)}{n!} x^n, \tag{10.2.5}$$

which is called the **Maclaurin series** expansion of $f(x)$. Many of the familiar elementary functions are analytic at all points, in particular, the Maclaurin expansions of e^x, $\sin x$, and $\cos x$ are, respectively,

$$e^x = 1 + x + \frac{1}{2!} x^2 + \frac{1}{3!} x^3 + \cdots + \frac{1}{n!} x^n + \cdots = \sum_{n=0}^{\infty} \frac{1}{n!} x^n \tag{10.2.6}$$

$$\sin x = x - \frac{1}{3!} x^3 + \frac{1}{5!} x^5 - \cdots + \frac{(-1)^n}{(2n+1)!} x^{2n+1} + \cdots = \sum_{n=0}^{\infty} \frac{(-1)^n}{(2n+1)!} x^{2n+1} \tag{10.2.7}$$

$$\cos x = 1 - \frac{1}{2!} x^2 + \frac{1}{4!} x^4 - \cdots + \frac{(-1)^n}{(2n)!} x^{2n} + \cdots = \sum_{n=0}^{\infty} \frac{(-1)^n}{(2n)!} x^{2n}, \tag{10.2.8}$$

and each of the series on the right of these equations converges to the function on the left for all real values of x.

We can determine many other analytic functions using the next theorem.

> **Theorem 10.2.1:**   If $f(x)$ and $g(x)$ are analytic at x_0, then so also are $f(x) \pm g(x)$, $f(x)g(x)$, and $\dfrac{f(x)}{g(x)}$ [provided that $g(x_0) \neq 0$].

Of particular importance to us throughout this chapter will be polynomial functions—that is, functions of the form

$$p(x) = a_0 + a_1 x + a_2 x^2 + \cdots + a_n x^n, \tag{10.2.9}$$

where $a_0, a_1, \ldots, a_n$ are real numbers. Such a function is analytic at all points. In particular, (10.2.9) can be considered as the Maclaurin series expansion of p about $x = 0$. Since the series has only a finite number of terms, it converges for all real x. Now suppose that $p(x)$ and $q(x)$ are polynomials, and hence analytic at all points. According to Theorem 10.2.1, the rational function r, defined by $r(x) = \dfrac{p(x)}{q(x)}$, is analytic at all points $x = x_0$ such that $q(x_0) \neq 0$. However, Theorem 10.2.1 does not give us any indication of the radius of convergence of the series representation of $r(x)$. The next theorem deals with this point.

Theorem 10.2.2: If $p(x)$ and $q(x)$ are polynomials and $q(x_0) \neq 0$, then the power series representation of p/q has radius of convergence R, where R is the distance, in the complex plane, from x_0 to the nearest zero of q.

PROOF The proof of this theorem requires results from complex analysis with which we do not assume the reader is familiar. For this reason the proof is omitted. ∎

REMARK If $z = a + ib$ is a zero of q, then the distance from the center, $x = x_0$, of the power series to z is (see Fig. 10.2.2)

$$|z - x_0| = \sqrt{(a - x_0)^2 + b^2} .$$

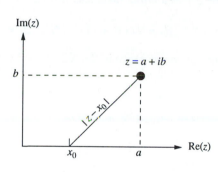

Figure 10.2.2 Determining the radius of convergence of the power series representation of a rational function centered at $x = x_0$.

Example 10.2.3 Determine the radius of convergence of the power series representation of the function

$$f(x) = \frac{1 - x}{x^2 - 4}$$

centered at **(a)** $x = 0$, **(b)** $x = 1$.

Solution Taking $p(x) = 1 - x$ and $q(x) = x^2 - 4$, we have

$$f(x) = \frac{p(x)}{q(x)},$$

and the zeros of q are $x = \pm 2$.

(a) In this case, the center of the power series is $x = 0$, so that the distance to the nearest zero of q is 2. (See Figure 10.2.3.) Consequently the radius of convergence of the power series representation of f centered at $x = 0$ is $R = 2$.

Figure 10.2.3 Determining the radius of convergence of the power series representation of the function given in Example 10.2.3.

(b) If the center of the power series is $x = 1$, then the nearest zero of q is at $x = 2$ (see

Figure 10.2.3), and hence, the radius of convergence of the power series representation of f is $R = 1$.

Example 10.2.4   Determine the radius of convergence of the power series expansion of

$$f(x) = \frac{1 - x}{(x^2 + 2x + 2)(x - 2)}$$

centered at $x = 0$.

Solution   We take $p(x) = 1 - x$ and $q(x) = (x^2 + 2x + 2)(x - 2)$. According to Theorem 10.2.2, the radius of convergence of the required power series will be given by the distance from $x = 0$ to the nearest zero of q. It is easily seen that the zeros of q are $x_1 = -1 + i$, $x_2 = -1 - i$, $x_3 = 2$. The corresponding distances from $x = 0$ are

$$d_1 = \sqrt{(1)^2 + (-1)^2} = \sqrt{2}\,,\ d_2 = \sqrt{(1)^2 + (1)^2} = \sqrt{2}\,,\ d_3 = 2.$$

Consequently, the appropriate radius of convergence is $R = \sqrt{2}$. (See Figure 10.2.4.)

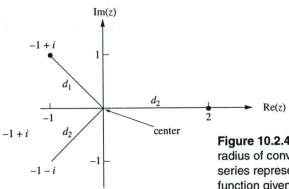

Figure 10.2.4 Determination of the radius of convergence of the power series representation of the rational function given in Example 10.2.4.

EXERCISES 10.2

For problems 1–5, determine the radius of convergence of the given power series.

1. $\displaystyle\sum_{n=0}^{\infty} \frac{x^n}{2^{2n}}.$

2. $\displaystyle\sum_{n=0}^{\infty} \frac{x^n}{n^2}.$

3. $\displaystyle\sum_{n=0}^{\infty} \frac{2^n x^n}{n}.$

4. $\displaystyle\sum_{n=0}^{\infty} n!\, x^n.$

5. $\displaystyle\sum_{n=0}^{\infty} \frac{5^n x^n}{n!}.$

For problems 6–10, determine the radius of convergence of the power series representation of the given function with center x_0.

6. $f(x) = \dfrac{x^2 - 1}{x + 2},\ x_0 = 0.$

7. $f(x) = \dfrac{x}{x^2 + 1},\ x_0 = 0.$

8. $f(x) = \dfrac{2x}{x^2 + 16},\ x_0 = 1.$

9. $f(x) = \dfrac{x^2 - 3}{x^2 - 2x + 5}$, $x_0 = 0$.

10. $f(x) = \dfrac{x}{(x^2 + 4x + 13)(x - 3)}$, $x_0 = -1$.

11. (a) Determine all values of x at which the function

$$f(x) = \frac{1}{x^2 - 1} \qquad (11.1)$$

is analytic.

(b) Determine the radius of convergence of a power series representation of the function (11.1) centered at $x = x_0$. (You will need to consider the cases $-1 < x_0 < 1$, and $|x_0| > 1$ separately.)

12. By redefining the ranges of the summations appearing on the left-hand side, show that

$$\sum_{n=2}^{\infty} n(n-1)\,a_{n-1}x^{n-2} + \sum_{n=1}^{\infty} n\,a_n x^{n-1}$$

$$= \sum_{n=0}^{\infty} (n+1)(n+3)a_{n+1}\,x^n.$$

13. If $f(x) = \displaystyle\sum_{n=0}^{\infty} a_n x^n$, where the coefficients in the expansion satisfy

$$\sum_{n=0}^{\infty} n(n+2)\,a_n\,x^n + \sum_{n=1}^{\infty} (n-3)a_{n-1}x^n = 0,$$

determine $f(x)$.

14. Suppose it is known that the coefficients in the expansion

$$f(x) = \sum_{n=0}^{\infty} a_n x^n$$

satisfy

$$\sum_{n=0}^{\infty} (n+2)\,a_{n+1}\,x^n - \sum_{n=0}^{\infty} a_n\,x^n = 0.$$

Show that

$$f(x) = \frac{a_0}{x} \sum_{n=0}^{\infty} \frac{1}{(n+1)!}\,x^{n+1},$$

and express this in terms of familiar elementary functions.

15. If

$$\sum_{n=1}^{\infty} (n+1)(n+2)a_{n+1}x^n - \sum_{n=1}^{\infty} na_{n-1}x^n = 0,$$

show that

$$a_{2k} = \frac{1 \cdot 3 \cdot 5 \cdots (2k-1)}{(2k+1)!}\,a_0, \quad a_{2k+1} = \frac{2^{k+1}k!}{(2k+2)!}\,a_1,$$

$$k = 1, 2, \ldots .$$

10.3   SERIES SOLUTIONS ABOUT AN ORDINARY POINT

We now consider the second-order linear homogeneous DE written in standard form

$$y'' + p(x)y' + q(x)y = 0.$$

Our aim is to determine a series representation of the general solution to this DE centered at $x = x_0$. We will see that the existence and form of the solution is dependent on the behavior of the functions p and q at $x = x_0$.

> **Definition 10.3.1:**   The point $x = x_0$ is called an **ordinary point** of the DE
>
> $$y'' + p(x)y' + q(x)y = 0 \qquad (10.3.1)$$
>
> if p and q are *both* analytic at $x = x_0$. Any point that is not an ordinary point of equation (10.3.1) is called a **singular point** of the DE.

Example 10.3.1   The DE

$$y'' + \frac{1}{x^2 - 4}y' + \frac{1}{x + 1}y = 0$$

has

$$p(x) = \frac{1}{x^2 - 4} \quad \text{and} \quad q(x) = \frac{1}{x + 1}.$$

We see by inspection that the only points at which p fails to be analytic are $x = \pm 2$, whereas q is analytic at all points except $x = -1$. Consequently, the only singular points of the DE are $x = \pm 2, -1$. All other points are ordinary points. $\square$

In this section, we restrict our attention to ordinary points. Since the functions p and q are analytic at an ordinary point, we might suspect that any solution to equation (10.3.1) valid at $x = x_0$ is also analytic there and hence can be represented as a convergent power series

$$y(x) = \sum_{n=0}^{\infty} a_n(x - x_0)^n,$$

for appropriate constants a_n. This is indeed the case, and the power series representation of the solution will, in general, have a nonzero radius of convergence. Before stating the general result, we consider a familiar example.

Example 10.3.2:   Determine two LI power series solutions to the DE

$$y'' + y = 0 \tag{10.3.2}$$

centered at $x = 0$. Identify the solutions in terms of familiar elementary functions.

Solution   Since $x = 0$ is an ordinary point of the DE, we try for a power series solution of the form

$$y(x) = \sum_{n=0}^{\infty} a_n x^n. \tag{10.3.3}$$

We proceed in a similar manner to the method of undetermined coefficients by substituting (10.3.3) into equation (10.3.2) and determining the values of the a_n such that (10.3.3) is indeed a solution. Differentiating (10.3.3) twice with respect to x yields

$$y'(x) = \sum_{n=1}^{\infty} n a_n x^{n-1}, \quad y''(x) = \sum_{n=2}^{\infty} n(n - 1)a_n x^{n-2},$$

where we have shifted the starting point on the summations without loss of generality. Substituting into equation (10.3.2), it follows that (10.3.3) does define a solution, provided that,

$$\sum_{n=2}^{\infty} n(n - 1)a_n x^{n-2} + \sum_{n=0}^{\infty} a_n x^n = 0. \tag{10.3.4}$$

If we replace n by $k + 2$ in the first summation, and replace n by k in the second summation the result is

$$\sum_{k=0}^{\infty} (k + 2)(k + 1)a_{k+2} x^k + \sum_{k=0}^{\infty} a_k x^k = 0.$$

Combining the summations yields

$$\sum_{k=0}^{\infty} [(k + 2)(k + 1)a_{k+2} + a_k] x^k = 0.$$

This implies that the coefficients of x^k must vanish for $k = 0, 1, 2, \ldots$. Consequently, we obtain the *recurrence relation*

$$(k + 2)(k + 1)a_{k+2} + a_k = 0, \qquad\qquad k = 0, 1, 2, \ldots.$$

Since $(k + 2)(k + 1)$ is never zero, we can write this recurrence relation in the equivalent form

$$a_{k+2} = - \frac{1}{(k + 2)(k + 1)} a_k \quad k = 0, 1, 2, \ldots. \qquad\qquad (10.3.5)$$

We now use this relation to determine the appropriate values of the coefficients. It is convenient to consider the two cases k even and k odd separately.

Even k: Substituting successively into (10.3.5), we obtain the following.

$k = 0$:

$$\boxed{a_2 = -\frac{1}{2} a_0.}$$

$k = 2$:

$$\boxed{a_4 = -\frac{1}{4 \cdot 3} a_2 = \frac{1}{4 \cdot 3 \cdot 2} a_0.}$$

That is

$$\boxed{a_4 = \frac{1}{4!} a_0.}$$

$k = 4$:

$$\boxed{a_6 = -\frac{1}{6 \cdot 5} a_4,}$$

so that

$$\boxed{a_6 = -\frac{1}{6!} a_0.}$$

Continuing in this manner, we soon recognize the pattern that is emerging, namely,

$$\boxed{a_{2n} = \frac{(-1)^n}{(2n)!} a_0.} \qquad\qquad (10.3.6)$$

Thus, all of the even coefficients are determined in terms of a_0, but a_0 itself is arbitrary.

Odd k: Now consider the recurrence relation (10.3.5) when k is an odd positive integer.

$k = 1$:

$$\boxed{a_3 = -\frac{1}{2 \cdot 3} a_1 = -\frac{1}{3!} a_1.}$$

$k = 3$:

$$\boxed{a_5 = -\frac{1}{4 \cdot 5} a_3 = \frac{1}{5!} a_1.}$$

$k = 5$:

$$\boxed{a_7 = -\frac{1}{6 \cdot 7} a_5 = -\frac{1}{7!} a_1.}$$

Continuing in this manner, we see that the general odd coefficient is

$$\boxed{a_{2n+1} = \frac{(-1)^n}{(2n+1)!}a_1.}$$
(10.3.7)

Thus, we have shown that for all values of a_0, a_1, a solution to the given DE is

$$y(x) = a_0\left(1 - \frac{1}{2!}x^2 + \frac{1}{4!}x^4 - \cdots\right) + a_1\left(x - \frac{1}{3!}x^3 + \frac{1}{5!}x^5 - \cdots\right).$$

That is,

$$y(x) = a_0\sum_{n=0}^{\infty}\frac{(-1)^n}{(2n)!}x^{2n} + a_1\sum_{n=0}^{\infty}\frac{(-1)^n}{(2n+1)!}x^{2n+1}.$$
(10.3.8)

Setting $a_1 = 0$ and $a_0 = 1$ yields the solution

$$y_1(x) = \sum_{n=0}^{\infty}\frac{(-1)^n}{(2n)!}x^{2n},$$

whereas setting $a_0 = 0$ and $a_1 = 1$ yields the solution

$$y_2(x) = \sum_{n=0}^{\infty}\frac{(-1)^n}{(2n+1)!}x^{2n+1}.$$

Applying the ratio test it is straightforward to show that both of the foregoing series converge for all real x. Finally, since y_1 and y_2 are not proportional, they are LI on any interval. It follows that (10.3.8) is the general solution of the given DE. Indeed, the power series representing y_1 is just the Maclaurin series expansion of $\cos x$, whereas the series defining y_2 is the Maclaurin series expansion of $\sin x$. Thus, we can write (10.3.8) in the more familiar form

$$y(x) = a_0\cos x + a_1\sin x. \qquad \qquad \square$$

The solution of the previous example consisted of the following four steps.

1. Assume that a power series solution of the form

$$y(x) = \sum_{n=0}^{\infty}a_nx^n$$

   exists.
2. Determine the values of the coefficients, a_n, such that y is a formal solution of the DE. This led to two distinct solutions, one determined in terms of the constant a_0, and the other in terms of the constant a_1.
3. Use the ratio test to determine the radius of convergence of the solutions and hence the interval over which the solutions are valid.
4. Check that the solutions are LI on the interval of existence.

The next theorem justifies the preceding steps and shows that the technique can be applied about any ordinary point of a DE.

Theorem 10.3.1:   Let p and q be analytic at $x = x_0$, and suppose that their power series expansions are valid for $|x - x_0| < R$. Then the general solution to the DE

$$y'' + p(x)y' + q(x)y = 0 \qquad (10.3.9)$$

can be represented as a power series centered at $x = x_0$, with radius of convergence *at least* R. The coefficients in this series solution can be determined in terms of a_0 and a_1 by

directly substituting $y(x) = \displaystyle\sum_{n=0}^{\infty} a_n(x - x_0)^n$ into (10.3.9). The resulting solution is of the form

$$y(x) = a_0 y_1(x) + a_1 y_2(x),$$

where y_1 and y_2 are LI solutions to (10.3.9) on the interval of existence. If the initial conditions $y(x_0) = \alpha$, $y'(x_0) = \beta$ are imposed, then $a_0 = \alpha$, $a_1 = \beta$.

IDEA BEHIND PROOF   We outline the steps required to prove this theorem but do not give details. The first step is to expand p and q in a power series centered at $x = x_0$. Then

we assume a solution exists of the form $y(x) = \displaystyle\sum_{n=0}^{\infty} a_n(x - x_0)^n$ and substitute this into

the DE. Upon collecting the coefficients of like powers of $x - x_0$, a recurrence relation is obtained, and it can be shown that this relation determines all of the coefficients in terms of a_0 and a_1. These steps are computationally tedious, but quite straightforward. The hard part is to show that the power series solution that has been obtained has a radius of convergence at least equal to R. This requires some ideas from advanced calculus. Having determined a power series solution, y_1 and y_2 arise as the special cases $a_0 = 1$, $a_1 = 0$, and $a_0 = 0$, $a_1 = 1$, respectively. The Wronskian of these functions satisfies $W[y_1, y_2](x_0) = 1$, so that they are LI on their interval of existence. Finally, it is easy to show that $y(x_0) = a_0$ and that $y'(x_0) = a_1$.   ∎

We now illustrate the use of the above theorem with some examples.

Example 10.3.3   Show that

$$(1 + x^2)y'' + 3xy' + y = 0 \qquad (10.3.10)$$

has two LI series solutions centered at $x = 0$, and determine a lower bound on the radius of convergence of these solutions.

Solution   We first rewrite (10.3.10) in the standard form

$$y'' + \frac{3x}{1 + x^2}y' + \frac{1}{1 + x^2}y = 0,$$

from which we can conclude that $x = 0$ is an ordinary point and hence, equation (10.3.10) does indeed have two LI series solutions centered at $x = 0$. In this case,

$$p(x) = \frac{3x}{1 + x^2} \quad \text{and} \quad q(x) = \frac{1}{1 + x^2}.$$

According to Theorem 10.3.1, the radius of convergence of the power series solutions will be *at least* equal to the smaller of the radii of convergence of the power series representations of p and q. Using Theorem 10.2.2, we see directly that the series expansions of both p and q about $x = 0$ have radius of convergence $R = 1$, so that a lower bound on the radius of convergence of the power series solutions to equation (10.3.10) is also $R = 1$.

Example 10.3.4   Determine two LI series solutions in powers of x to

$$y'' - 2xy' - 4y = 0, \qquad (10.3.11)$$

and find the radius of convergence of these solutions.

Solution   The point $x = 0$ is an ordinary point of the DE, and therefore Theorem 10.3.1 can be applied with $x_0 = 0$. In this case, we have

$$p(x) = -2x, \quad q(x) = -4.$$

Since these are both polynomials, their power series expansions about $x = 0$ are valid for all x, and hence, from the previous theorem, the general solution to equation (10.3.11) can be represented in the form

$$y(x) = \sum_{n=0}^{\infty} a_n x^n, \qquad (10.3.12)$$

and this power series solution will converge for all real x. Differentiating (10.3.12) we obtain

$$y'(x) = \sum_{n=1}^{\infty} n a_n x^{n-1}, \quad y''(x) = \sum_{n=2}^{\infty} n(n-1) a_n x^{n-2}.$$

Substitution into equation (10.3.11) yields

$$\sum_{n=2}^{\infty} n(n-1) a_n x^{n-2} - 2 \sum_{n=1}^{\infty} n a_n x^n - 4 \sum_{n=0}^{\infty} a_n x^n = 0.$$

We now redefine the ranges in the summations in order to obtain a common x^k in all terms. This is accomplished by replacing n with $k + 2$ in the first summation, and, for consistency in notation, we replace n with k in the other summations. The result is

$$\sum_{k=0}^{\infty} [(k+2)(k+1)a_{k+2} - 2ka_k - 4a_k]x^k = 0.$$

This equation requires that the coefficient of x^k vanish, and hence, we obtain the recurrence relation

$$(k+2)(k+1)a_{k+2} - 2ka_k - 4a_k = 0 \quad k = 0, 1, 2, \ldots,$$

which can be written in the equivalent form

$$a_{k+2} = \frac{2(k+2)}{(k+1)(k+2)} a_k, \quad k = 0, 1, 2, \ldots,$$

that is,

$$a_{k+2} = \frac{2}{k+1} a_k, \quad k = 0, 1, 2, \ldots. \qquad (10.3.13)$$

We see from this relation that, as in the previous example, all of the even coefficients can be expressed in terms of a_0, whereas all of the odd coefficients can be expressed in terms of a_1. We now determine the exact form of these coefficients.

Even k: From (10.3.13), we have the following:

$k = 0:$ $\qquad\qquad\qquad\qquad\qquad\qquad a_2 = 2a_0.$

$k = 2$:
$$a_4 = \frac{2}{3} a_2 = \frac{2^2}{3} a_0.$$

$k = 4$:
$$a_6 = \frac{2}{5} a_4 = \frac{2^3}{1 \cdot 3 \cdot 5} a_0.$$

$k = 6$:
$$a_8 = \frac{2}{7} a_6 = \frac{2^4}{1 \cdot 3 \cdot 5 \cdot 7} a_0.$$

The general even term is thus

$$a_{2n} = \frac{2^n}{1 \cdot 3 \cdot 5 \cdots (2n - 1)} a_0.$$

Odd k: Substituting successively into (10.3.13) yields the following:

$k = 1$:
$$a_3 = a_1.$$

$k = 3$:
$$a_5 = \frac{2}{4} a_3 = \frac{1}{1 \cdot 2} a_1.$$

$k = 5$:
$$a_7 = \frac{2}{6} a_5 = \frac{1}{1 \cdot 2 \cdot 3} a_1.$$

$k = 7$:
$$a_9 = \frac{2}{8} a_7 = \frac{1}{1 \cdot 2 \cdot 3 \cdot 4} a_1.$$

The general odd term can therefore be written as

$$a_{2n+1} = \frac{1}{n!} a_1.$$

Substituting back into (10.3.12), we obtain the solution

$$y(x) = a_0 \left[1 + 2x^2 + \frac{2^2}{1 \cdot 3} x^4 + \frac{2^3}{1 \cdot 3 \cdot 5} x^6 + \cdots + \frac{2^n}{1 \cdot 3 \cdot 5 \cdots (2n - 1)} x^{2n} + \cdots \right]$$

$$+ a_1 \left[x + x^3 + \frac{1}{2!} x^5 + \frac{1}{3!} x^7 + \cdots + \frac{1}{n!} x^{2n+1} + \cdots \right].$$

That is

$$y(x) = a_0 \left[1 + \sum_{n=1}^{\infty} \frac{2^n}{1 \cdot 3 \cdot 5 \cdots (2n - 1)} x^{2n} \right] + a_1 \left[\sum_{n=0}^{\infty} \frac{1}{n!} x^{2n+1} \right].$$

Consequently, from Theorem 10.3.1, two LI solutions to equation (10.3.11) on $(-\infty, \infty)$ are

$$y_1(x) = 1 + \sum_{n=1}^{\infty} \frac{2^n}{1 \cdot 3 \cdot 5 \cdots (2n - 1)} x^{2n}, \quad y_2(x) = \sum_{n=0}^{\infty} \frac{1}{n!} x^{2n+1}. \qquad \square$$

In Examples 10.3.2 and 10.3.4, we were able to solve the recurrence relation that arose from the power series technique. In general, this will not be possible, and hence, we must be satisfied with obtaining just a finite number of terms in each power series solution.

Example 10.3.5   Determine the terms up to x^5 in each of two LI power series solutions to

$$y'' + (2 - 4x^2)y' - 8xy = 0$$

centered at $x = 0$. Also find the radius of convergence of these solutions.

Solution   The functions $p(x) = 2 - 4x^2$, and $q(x) = -8x$ are polynomials, and hence, from Theorem 10.3.1, the power series solutions will converge for all real x. We now determine the solutions. Substituting

$$y(x) = \sum_{n=0}^{\infty} a_n x^n \tag{10.3.14}$$

into the given DE yields

$$\sum_{n=2}^{\infty} n(n-1)a_n x^{n-2} + 2\sum_{n=1}^{\infty} na_n x^{n-1} - 4\sum_{n=1}^{\infty} na_n x^{n+1} - 8\sum_{n=0}^{\infty} a_n x^{n+1} = 0.$$

Replacing n by $k + 2$ in the first summation, $k + 1$ in the second summation, and $k - 1$ in the third and fourth summations, we obtain

$$\sum_{k=0}^{\infty} (k+2)(k+1)a_{k+2}x^k + 2\sum_{k=0}^{\infty} (k+1)a_{k+1}x^k$$

$$- 4\sum_{k=2}^{\infty} (k-1)a_{k-1}x^k - 8\sum_{k=1}^{\infty} a_{k-1}x^k = 0.$$

Separating out the terms corresponding to $k = 0$ and $k = 1$, it follows that this can be written as

$$2a_2 + 2a_1 + (6a_3 + 4a_2 - 8a_0)x$$
$$\uparrow \qquad\qquad\qquad \uparrow$$
$$k = 0 \qquad\qquad k = 1$$

$$+ \sum_{k=2}^{\infty} \left\{ (k+2)(k+1)a_{k+2} + 2(k+1)a_{k+1} - [4(k-1) + 8]a_{k-1}\right\}x^k = 0.$$

Setting the coefficients of consecutive powers of x to zero yields the following:

$k = 0$: $\qquad\qquad\qquad 2a_2 + 2a_1 = 0.$ $\qquad\qquad\qquad$ (10.3.15)

$k = 1$: $\qquad\qquad\qquad 6a_3 + 4a_2 - 8a_0 = 0.$ $\qquad\qquad$ (10.3.16)

$k \geq 2$: $\quad (k+2)(k+1)a_{k+2} + 2(k+1)\,a_{k+1} - [4(k-1) + 8]a_{k-1} = 0.$ (10.3.17)

It follows from (10.3.15) and (10.3.16) that

$$a_2 = -a_1, \quad a_3 = \frac{2}{3}(2a_0 + a_1), \tag{10.3.18}$$

and (10.3.17) yields the general recurrence relation

$$a_{k+2} = \frac{4a_{k-1} - 2a_{k+1}}{k+2}, \quad k = 2, 3, 4, \dots. \tag{10.3.19}$$

In this case, the recurrence relation is quite difficult to solve. However, we were only asked to determine terms up to x^5 in the series solutions, and so we proceed to do so. We already have a_2 and a_3 expressed in terms of a_0 and a_1. Setting $k = 2$ in (10.3.19) yields

$$a_4 = \frac{1}{4}(4a_1 - 2a_3) = \frac{1}{4}[4a_1 - \frac{4}{3}(2a_0 + a_1)],$$

where we have substituted from (10.3.18) for a_3. Simplifying this expression, we obtain

$$a_4 = \frac{2}{3}(a_1 - a_0).$$

We still require one more term. Setting $k = 3$ in (10.3.19) yields

$$a_5 = \frac{1}{5}(4a_2 - 2a_4) = \frac{1}{5}[-4a_1 - \frac{4}{3}(a_1 - a_0)],$$

so that

$$a_5 = \frac{4}{15}(a_0 - 4a_1).$$

Substituting for the coefficients a_2, a_3, a_4, and a_5 into (10.3.14), we obtain

$$y(x) = a_0 + a_1 x - a_1 x^2 + \frac{2}{3}(2a_0 + a_1)x^3 + \frac{2}{3}(a_1 - a_0)x^4 + \frac{4}{15}(a_0 - 4a_1)\,x^5 + \cdots.$$

That is,

$$y(x) = a_0\left(1 + \frac{4}{3}x^3 - \frac{2}{3}x^4 + \frac{4}{15}x^5 + \cdots\right)$$

$$+ a_1\left(x - x^2 + \frac{2}{3}x^3 + \frac{2}{3}x^4 - \frac{16}{15}x^5 + \cdots\right).$$

Thus, two LI solutions to the given DE on $(-\infty, \infty)$ are

$$y_1(x) = 1 + \frac{4}{3}x^3 - \frac{2}{3}x^4 + \frac{4}{15}x^5 + \cdots,$$

$$y_2(x) = x - x^2 + \frac{2}{3}x^3 + \frac{2}{3}x^4 - \frac{16}{15}x^5 + \cdots.$$

EXERCISES 10.3

For problems 1–8, determine two LI power series solutions to the given DE centered at $x = 0$. Also determine the radius of convergence of the series solutions.

1. $y'' - y = 0$.
2. $y'' - 2xy' - 2y = 0$.
3. $y'' + 2xy' + 4y = 0$.
4. $y'' + xy = 0$.
5. $y'' - x^2 y' - 2xy = 0$.
6. $y'' - x^2 y' - 3xy = 0$.
7. $y'' + xy' + 3y = 0$.

8. $y'' + 2x^2 y' + 2xy = 0$.

For problems 9–12, determine two LI power series solutions to the given DE centered at $x = 0$. Give a lower bound on the radius of convergence of the series solutions obtained.

9. $(1 + x^2)y'' + 4xy' + 2y = 0$.
10. $(x^2 - 3)y'' - 3xy' - 5y = 0$.
11. $(x^2 - 1)y'' - 6xy' + 12y = 0$.
12. $(1 - 4x^2)y'' - 20xy' - 16y = 0$.

For problems 13–16, determine terms up to and including x^5 in two LI power series solutions of the given DE. State the radius of convergence of the series solutions.

13. $y'' + xy' + (2 + x)y = 0$.

14. $y'' + 2y' + 4xy = 0$.

15. $y'' - e^x y = 0$. (Hint: $e^x = 1 + x + \dfrac{1}{2!}x^2 + \dfrac{1}{3!}x^3 + \cdots$.)

16. $y'' + (\sin x)y' + y = 0$.

17. Consider the DE

$$xy'' - (x - 1)y' - xy = 0. \qquad (17.1)$$

(a) Is $x = 0$ an ordinary point ?

(b) Determine the first three nonzero terms in each of two LI series solutions to equation (17.1) centered at $x = 1$. (Hint: Make the change of variables $z = x - 1$ and obtain a series solution in powers of z.) Give a lower bound on the radius of convergence of each of your solutions.

18. Determine a series solution to the IVP

$$(1 + 2x^2)y'' + 7xy' + 2y = 0$$

$y(0) = 0$, $y'(0) = 1$.

19. (a) Determine a series solution to the IVP

$$4y'' + xy' + 4y = 0, \ y(0) = 1, \ y'(0) = 0. \quad (19.1)$$

(b) Find a polynomial that approximates the solution to equation (19.1) with an error less than 10^{-5} on the interval $[-1, 1]$. (Hint: The series obtained is a convergent alternating series.)

20. Consider the DE

$$(x^2 - 1)y'' + [1 - (a + b)]xy' + aby = 0, \quad (20.1)$$

where a and b are constants.

(a) Show that the coefficients in a series solution to equation (20.1) centered at $x = 0$ must satisfy the recurrence relation

$$a_{n+2} = \frac{(n - a)(n - b)}{(n + 2)(n + 1)}a_n, \ n = 0, 1, \ldots,$$

and determine two LI series solutions.

(b) Show that if either a or b is a nonnegative integer, then one of the solutions obtained in (a) is a polynomial.

(c) Show that if a is an odd positive integer and b is an even positive integer, then *both* of the solutions obtained in (a) are polynomials.

(c) If $a = 5$ and $b = 4$, determine two LI polynomial solutions to equation (20.1). Notice that in this case the radius of convergence of the solutions obtained is $R = \infty$, whereas Theorem 10.3.1 only guarantees a radius of convergence $R \geq 1$.

The power series technique can also be used to solve nonhomogeneous DE of the form

$$y'' + p(x)y' + q(x)y = r(x),$$

provided that p, q, and r are analytic at the point about which we are expanding. For problems 21 and 22, determine terms up to x^6 in the power series representation of the general solution to the given DE centered at $x = 0$. Identify those terms in your solution that correspond to the complementary function and those that correspond to a particular solution to the DE.

21. $y'' + xy' - 4y = 6e^x$.

22. $y'' + 2x^2 y' + xy = 2\cos x$.

10.4   THE LEGENDRE EQUATION

There are several linear DE that arise frequently in applied mathematics and whose solutions can only be obtained using a power series technique. Amongst the most important of these are the following:

$$(1 - x^2)y'' - 2xy' + \alpha(\alpha + 1)y = 0, \qquad \text{(Legendre Equation)}$$

$$y'' - 2xy' + 2\alpha y = 0, \qquad \text{(Hermite Equation)}$$

$$(1 - x^2)y'' - xy' + \alpha^2 y = 0, \qquad \text{(Chebyshev Equation)}$$

where α is an arbitrary constant. Since $x = 0$ is an ordinary point of these equations, we can obtain a series solution in powers of x. We will consider only the Legendre equation and leave the analysis of the remaining equations for the exercises.

The Legendre equation is

$$(1 - x^2)y'' - 2xy' + \alpha(\alpha + 1)y = 0, \tag{10.4.1}$$

where α is an arbitrary constant. To determine a lower bound on the radius of convergence of the series solutions to this equation, we divide by $1 - x^2$ to obtain

$$y'' - \frac{2x}{1 - x^2}y' + \frac{\alpha(\alpha + 1)}{1 - x^2}y = 0.$$

Since the power series expansion of $\dfrac{1}{1 - x^2}$ about $x = 0$ is valid for $|x| < 1$, it follows that a lower bound on the radius of convergence of the power series solutions to equation (10.4.1) about $x = 0$ is 1. We now determine the series solutions. Substituting

$$y(x) = \sum_{n=0}^{\infty} a_n x^n$$

into equation (10.4.1) yields

$$\sum_{n=2}^{\infty} n(n - 1)a_n x^{n-2} - \sum_{n=2}^{\infty} n(n - 1)a_n x^n - 2\sum_{n=1}^{\infty} na_n x^n + \sum_{n=0}^{\infty} \alpha(\alpha + 1)a_n x^n = 0.$$

That is, upon redefining the ranges of the summations,

$$\sum_{n=0}^{\infty} [(n + 2)(n + 1)a_{n+2} - n(n - 1)a_n - 2na_n + \alpha(\alpha + 1)a_n]x^n = 0.$$

Thus, we obtain the recurrence relation

$$a_{n+2} = \frac{[n(n + 1) - \alpha(\alpha + 1)]}{(n + 1)(n + 2)}a_n, \quad n = 0, 1, 2, \ldots, \tag{10.4.2}$$

which can be written as

$$a_{n+2} = -\frac{(\alpha - n)(\alpha + n + 1)}{(n + 1)(n + 2)}a_n, \quad n = 0, 1, 2, \ldots.$$

Even n:

$$n = 0 \implies a_2 = -\frac{\alpha(\alpha + 1)}{2}a_0,$$

$$n = 2 \implies a_4 = -\frac{(\alpha - 2)(\alpha + 3)}{3 \cdot 4}a_2 = \frac{(\alpha - 2)\alpha(\alpha + 1)(\alpha + 3)}{4!}a_0,$$

$$n = 4 \implies a_6 = -\frac{(\alpha - 4)(\alpha + 5)}{5 \cdot 6}a_4$$

$$= -\frac{(\alpha - 4)(\alpha - 2)\alpha(\alpha + 1)(\alpha + 3)(\alpha + 5)}{6!}a_0.$$

In general,

$$a_{2k} = (-1)^k \frac{(\alpha - 2k + 2)(\alpha - 2k + 4)\cdots(\alpha - 2)\alpha(\alpha + 1)(\alpha + 3)\cdots(\alpha + 2k - 1)}{(2k)!} a_0,$$

$k = 1, 2, 3, \ldots.$

Odd n:

$$n = 1 \implies a_3 = -\frac{(\alpha - 1)(\alpha + 2)}{2 \cdot 3} a_1,$$

$$n = 3 \implies a_5 = -\frac{(\alpha - 3)(\alpha + 4)}{4 \cdot 5} a_3 = \frac{(\alpha - 3)(\alpha - 1)(\alpha + 2)(\alpha + 4)}{5!} a_1,$$

$$n = 5 \implies a_7 = -\frac{(\alpha - 5)(\alpha + 6)}{6 \cdot 7} a_5$$

$$= -\frac{(\alpha - 5)(\alpha - 3)(\alpha - 1)(\alpha + 2)(\alpha + 4)(\alpha + 6)}{7!} a_1.$$

In general,

$$a_{2k+1} = (-1)^k \frac{(\alpha - 2k + 1)\cdots(\alpha - 3)(\alpha - 1)(\alpha + 2)(\alpha + 4)\cdots(\alpha + 2k)}{(2k + 1)!} a_1,$$

$k = 1, 2, 3, \ldots .$

Consequently, for $a_0 \neq 0$ and $a_1 \neq 0$, two LI solutions to the Legendre equation are

$$y_1(x) = a_0 \left[1 - \frac{\alpha(\alpha + 1)}{2} x^2 + \frac{(\alpha - 2)\alpha(\alpha + 1)(\alpha + 3)}{4!} x^4 \right.$$

$$\left. - \frac{(\alpha - 4)(\alpha - 2)\alpha(\alpha + 1)(\alpha + 3)(\alpha + 5)}{6!} x^6 + \cdots \right] \tag{10.4.3}$$

and

$$y_2(x) = a_1 \left[x - \frac{(\alpha - 1)(\alpha + 2)}{2!} x^3 \right.$$

$$\left. + \frac{(\alpha - 3)(\alpha - 1)(\alpha + 2)(\alpha + 4)}{5!} x^5 + \cdots \right], \tag{10.4.4}$$

and both of these solutions are valid for $|x| < 1$.

THE LEGENDRE POLYNOMIALS

Of particular importance in applications is the following special case of the Legendre equation

$$(1 - x^2)\, y'' - 2xy' + N(N + 1)y = 0,$$

where N is a nonnegative *integer*. In this case, the recurrence relation (10.4.2) is

$$a_{n+2} = \frac{[n(n + 1) - N(N + 1)]}{(n + 1)(n + 2)} a_n, \quad n = 0, 1, 2, \ldots,$$

which implies that

$$a_{N+2} = a_{N+4} = \cdots = 0.$$

Consequently, one of the solutions to Legendre's equation in this case is a polynomial of degree N. (Notice that such a solution converges for all x, and hence, we have a radius of convergence greater than is guaranteed by Theorem 10.3.1.)

Definition 10.4.1:   Let N be a nonnegative integer. The **Legendre poly–nomial of degree** N, denoted $P_N(x)$, is defined to be the polynomial solution to

$$(1 - x^2)y'' - 2xy' + N(N + 1)y = 0,$$

which has been normalized so that $P_N(1) = 1$.

Example 10.4.1   Determine P_0, P_1, P_2.

Solution   Substituting $\alpha = N$ into (10.4.3) and (10.4.4) yields

$N = 0$: $y_1(x) = a_0$, which implies that $P_0(x) = 1$;

$N = 1$: $y_2(x) = a_1 x$, which implies that $P_1(x) = x$;

$N = 2$: $y_1(x) = a_0(1 - 3x^2)$. Imposing the normalizing condition that $y_1(1) = 1$, we require that $a_0 = -\dfrac{1}{2}$, so that $P_2(x) = \dfrac{1}{2}(3x^2 - 1)$.   ☐

In general, it is tedious to determine $P_N(x)$ directly from (10.4.3) and (10.4.4) and various other methods have been derived. Amongst the most useful are the following:

Rodrigues' Formula

$$P_N(x) = \frac{1}{2^N N!}\frac{d^N}{dx^N}(x^2 - 1)^N, \quad N = 0, 1, 2, \ldots,$$

Recurrence Relation

$$P_{N+1}(x) = \frac{(2N + 1)xP_N(x) - NP_{N-1}(x)}{(N + 1)}, \quad N = 1, 2, \ldots.$$

We can use Rodrigues' formula to obtain P_N directly. Alternatively, starting with P_0 and P_1, we can use the recurrence relation to generate all P_N.

Example 10.4.2   According to Rodrigues' formula,

$$P_2(x) = \frac{1}{8}\frac{d^2}{dx^2}(x^2 - 1)^2 = \frac{1}{8}\frac{d}{dx}[4x(x^2 - 1)] = \frac{1}{2}(3x^2 - 1),$$

which does indeed coincide with that given in Example 10.4.1.   ☐

The first five Legendre polynomials are given in Table 10.4.1

TABLE 10.4.1   THE FIRST FIVE
LEGENDRE POLYNOMIALS

N	Legendre polynomial of degree N
0	$P_0(x) = 1$
1	$P_1(x) = x$
2	$P_2(x) = \frac{1}{2}(3x^2 - 1)$
3	$P_3(x) = \frac{1}{2}x(5x^2 - 3)$
4	$P_4(x) = \frac{1}{8}(35x^4 - 30x^2 + 3)$

ORTHGONALITY OF THE LEGENDRE POLYNOMIALS

In Section 5.9, we defined an inner product on the vector space $C^0[a, b]$ by

$$<f, g> = \int_a^b f(x)g(x)\, dx,$$

for all f and g in $C^0(a, b)$. We now show that the Legendre polynomials are orthogonal relative to the above inner product on the interval $[-1, 1]$.

Theorem 10.4.1:   The set of Legendre polynomials $\{P_0, P_1, \dots\}$ is an orthogonal set of functions on the interval $[-1, 1]$. That is,

$$\int_{-1}^1 P_M(x)P_N(x)\, dx = 0 \text{ whenever } M \neq N.$$

PROOF   It is easily seen that Legendre's equation

$$(1 - x^2)y'' - 2xy' + \alpha(\alpha + 1)y = 0$$

can be written in the form

$$[(1 - x^2)y']' + \alpha(\alpha + 1)y = 0.$$

Consequently, the Legendre polynomials $P_N(x)$ and $P_M(x)$ satisfy

$$[(1 - x^2)P_N']' + N(N + 1)P_N = 0, \tag{10.4.5}$$

$$[(1 - x^2)P_M']' + M(M + 1)P_M = 0, \tag{10.4.6}$$

respectively. Multiplying equation (10.4.5) by P_M and equation (10.4.6) by P_N and subtracting yields

$$[(1 - x^2)P_N']'P_M - [(1 - x^2)P_M']'P_N + [N(N + 1) - M(M + 1)]P_M P_N = 0,$$

which can be written as

$$\{[(1 - x^2)P_N'P_M]' - (1 - x^2)P_N'P_M'\} - \{[(1 - x^2)P_M'P_N]' - (1 - x^2)P_N'P_M'\}$$
$$+ [N(N + 1) - M(M + 1)]P_M P_N = 0.$$

That is,

$$[(1 - x^2)(P_N{'}P_M - P_M{'}\,P_N)]{'} + [N(N + 1) - M(M + 1)]P_M P_N = 0.$$

Integrating over the interval $[-1, 1]$, we obtain

$$\left[(1 - x^2)(P_N{'}P_M - P_M{'}P_N)\right]\Big|_{-1}^{1} + [N(N + 1) - M(M + 1)] \int_{-1}^{1} P_M(x)P_N(x)\,dx = 0.$$

The first term vanishes at $x = \pm 1$, and the term multiplying the integral can be factorized to yield

$$(N - M)(N + M + 1) \int_{-1}^{1} P_M(x)P_N(x)\,dx = 0.$$

Since M and N are nonnegative integers, the previous formula implies that

$$\int_{-1}^{1} P_M(x)P_N(x)\,dx = 0 \text{ whenever } M \neq N. \qquad \blacksquare$$

It can also be shown, although it is more difficult (see N. N. Lebedev, *Special Functions and their Applications*, Dover, 1972), that

$$\int_{-1}^{1} P_N{}^2(x)\,dx = \frac{2}{2N + 1}. \qquad (10.4.7)$$

Consequently, $\left\{\sqrt{\dfrac{2N + 1}{2}}\,P_N(x)\right\}$ is an orthonormal set of polynomials on $[-1, 1]$.

Since the set of Legendre polynomials $\{P_0, P_1, \ldots, P_N\}$ is LI on any interval, it is a basis for the vector space of all polynomials of degree less than or equal to N. Thus, if $p(x)$ is any such polynomial, there exist scalars $a_0, a_1, \ldots, a_N$ such that

$$p(x) = \sum_{k=0}^{N} a_k P_k(x). \qquad (10.4.8)$$

We can use the orthogonality of the Legendre polynomials to determine the coefficients a_k in this expansion as follows. Multiplying (10.4.8) by $P_j(x)$, $0 \leq j \leq N$ and integrating over the interval $[-1, 1]$ yields

$$\int_{-1}^{1} p(x)P_j(x)\,dx = \int_{-1}^{1} \sum_{k=0}^{N} a_k P_k(x)P_j(x)\,dx = \sum_{k=0}^{N} a_k \int_{-1}^{1} P_k(x)P_j(x)\,dx.$$

However, due to the orthogonality of the Legendre polynomials, all of the terms in the summation with $k \neq j$ vanish, so that

$$\int_{-1}^{1} p(x)P_j(x)\,dx = a_j \int_{-1}^{1} P_j(x)P_j(x)\,dx.$$

Consequently, using (10.4.7), we obtain

$$\int_{-1}^{1} p(x)P_j(x)\,dx = \frac{2}{2j + 1}\,a_j$$

which implies that

$$a_j = \frac{2j + 1}{2} \int_{-1}^{1} p(x)P_j(x)\, dx. \tag{10.4.9}$$

Example 10.4.3   Expand $f(x) = x^2 - x + 2$ as a series of Legendre polynomials.

Solution   Since $f(x)$ has degree 2, we can write

$$x^2 - x + 2 = a_0 P_0 + a_1 P_1 + a_2 P_2,$$

where, from (10.4.9), the coefficients are given by

$$a_j = \frac{2j + 1}{2} \int_{-1}^{1} (x^2 - x + 2)P_j(x)\, dx.$$

From Table 10.4.1,

$$P_0(x) = 1, \quad P_1(x) = x, \quad P_2(x) = \frac{1}{2}(3x^2 - 1),$$

so that

$$a_0 = \frac{1}{2} \int_{-1}^{1} (x^2 - x + 2)\, dx = \frac{7}{3},$$

$$a_1 = \frac{3}{2} \int_{-1}^{1} (x^2 - x + 2)x\, dx = -1,$$

$$a_2 = \frac{5}{2} \int_{-1}^{1} \frac{1}{2}(x^2 - x + 2)(3x^2 - 1)\, dx = \frac{2}{3}.$$

Consequently,

$$x^2 - x + 2 = \frac{7}{3}P_0 - P_1 + \frac{2}{3}P_2. \qquad \square$$

More generally, the following expansion theorem plays a fundamental role in many applications of mathematics to physics, chemistry, engineering, etc.

Theorem 10.4.2:   Let f and f' be continuous on the interval $(-1, 1)$. Then, for $-1 < x < 1$,

$$f(x) = a_0 P_0(x) + a_1 P_1(x) + \cdots + a_n P_n(x) + \cdots = \sum_{n=0}^{\infty} a_n P_n(x), \tag{10.4.10}$$

where

$$a_n = \frac{2n + 1}{2} \int_{-1}^{1} f(x)P_n(x)\, dx. \tag{10.4.11}$$

PROOF Establishing the existence of a convergent series of the form (10.4.10) is best left for a course on Fourier analysis or partial differential equations. The derivation that the coefficients in such an expansion must be given by (10.4.11) follows similar steps to those leading to equation (10.4.9) and is left as one of the exercises. ■

Example 10.4.4   Determine the terms up to and including $P_3(x)$ in the Legendre series expansion of $f(x) = \sin \pi x$, $-1 < x < 1$.

Solution   According to the previous theorem, the given function does have a Legendre series expansion with coefficients given by

$$a_n = \frac{2n + 1}{2} \int_{-1}^{1} \sin \pi x \, P_n(x) \, dx.$$

Thus,

$$a_0 = \frac{1}{2} \int_{-1}^{1} \sin \pi x \, dx = 0; \quad a_1 = \frac{3}{2} \int_{-1}^{1} x \sin \pi x \, dx = \frac{3}{\pi},$$

$$a_2 = \frac{5}{2} \int_{-1}^{1} \frac{1}{2}(3x^2 - 1) \sin \pi x \, dx = 0,$$

$$a_3 = \frac{7}{2} \int_{-1}^{1} \frac{1}{2} x(5x^2 - 3) \sin \pi x \, dx = \frac{7}{\pi^3} (\pi^2 - 15).$$

Consequently, Theorem 10.4.2 implies that, for $-1 < x < 1$,

$$\sin \pi x = \frac{3}{\pi} P_1(x) + \frac{7}{\pi^3}(\pi^2 - 15) P_3(x) + \cdots. \qquad (10.4.12)$$

To illustrate how good this approximation is, in Fig. 10.4.1 we have sketched the functions

$$f(x) = \sin \pi x \quad \text{and} \quad g(x) = \frac{3}{\pi} P_1(x) + \frac{7}{\pi^3}(\pi^2 - 15) P_3(x).$$

For comparison, in Fig. 10.4.2 we sketch $f(x)$ and the fifth-order Taylor approximation

$$h(x) = \pi x - \frac{1}{3!} (\pi x)^3 + \frac{1}{5!} (\pi x)^5.$$

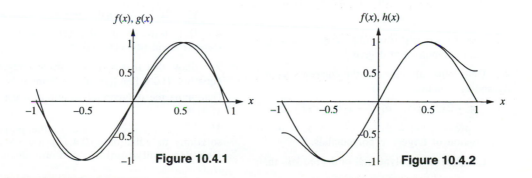

Figure 10.4.1   **Figure 10.4.2**

In Figure 10.4.3, we sketch $f(x)$ together with the Legendre polynomial approximation

$$k(x) = \frac{3}{\pi} P_1(x) + \frac{7}{\pi^3}(\pi^2 - 15) P_3(x) + \frac{11}{\pi^5} (945 - 105\pi^2 + \pi^4)P_5(x)$$

that arises when we include the next nonzero term in (10.4.12). We see that $k(x)$ gives an excellent approximation to $f(x) = \sin \pi x$ at all points in $(-1, 1)$, and in fact at the endpoints of the interval.

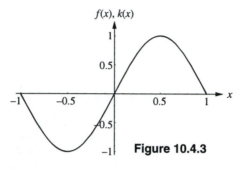

Figure 10.4.3

REMARK   Computation by hand of the coefficients in a Legendre polynomial expansion can be very tedious. However, computer algebra systems, such as Maple or Mathematica, have the Legendre polynomials as built-in functions, and therefore can be useful in computing the coefficients. For example, in Maple the command P(n, x) generates the degree-n Legendre polynomial as a function of x.

EXERCISES 10.4

1. Use (10.4.3) and (10.4.4) to determine polynomial solutions to Legendre's equation when $\alpha = 3$ and $\alpha = 4$. Hence, determine the Legendre polynomials $P_3(x)$ and $P_4(x)$.

2. Starting with $P_0(x) = 1$, $P_1(x) = x$, use the recurrence relation

$(n + 1)P_{n+1} + nP_{n-1} = (2n + 1)x\, P_n$, $n = 1, 2, 3, ...$

to determine P_2, P_3, and P_4.

3. Use Rodrigues' formula to determine the Legendre polynomial of degree 3.

4. Determine all values of the constants a_0, a_1, a_2, and a_3 such that

$$x^3 + 2x = a_0 P_0 + a_1 P_1 + a_2 P_2 + a_3 P_3.$$

5. Express $p(x) = 2x^3 + x^2 + 5$ as a linear combination of Legendre polynomials.

6. Let $Q(x)$ be a polynomial of degree less than

N. Prove that $\int_{-1}^{1} Q(x)\, P_N(x)\, dx = 0$.

7. Show that

$$\frac{d^2Y}{d\phi^2} + \cot \phi \frac{dY}{d\phi} + \alpha(\alpha + 1)Y = 0, \quad 0 < \phi < \pi,$$

is transformed into Legendre's equation by the change of variables $x = \cos \phi$.

Problems 8–10 deal with Hermite's equation

$$y'' - 2xy' + 2\alpha y = 0, \quad -\infty < x < \infty \quad (10.4.13)$$

8. Determine two LI series solutions to Hermite's equation centered at $x = 0$.

9. Show that if $\alpha = N$, a positive integer, then equation (10.4.13) has a polynomial solution. Determine the polynomial solutions when $\alpha = 0$, 1, 2, 3.

10. When suitably normalized, the polynomial solutions to equation (10.4.13) are called the **Hermite polynomials**, and are denoted by $H_N(x)$.

(a) Use equation (10.4.13) to show that $H_N(x)$ satisfies

$$(e^{-x^2}H_N{}')' + 2N\,e^{-x^2}\,H_N = 0. \qquad (10.1)$$

[Hint: Replace α with N in equation (10.4.13), and multiply the resulting equation by e^{-x^2}.]

(b) Use equation (10.1) to prove that the Hermite polynomials satisfy

$$\int_{-\infty}^{\infty} e^{-x^2}H_N(x)H_M(x)\,dx = 0,\ M \neq N. \qquad (10.2)$$

[Hint: Follow the steps taken in proving orthogonality of the Legendre polynomials. You will need to recall that

$$\lim_{x\to\pm\infty}\ e^{-x^2}p(x) = 0,$$

for any polynomial p.]

(c) Let $p(x)$ be a polynomial of degree N. Then we can write

$$p(x) = \sum_{k=1}^{N} a_k H_k(x). \qquad (10.3)$$

Given that

$$\int_{-\infty}^{\infty} e^{-x^2}H_N{}^2(x)\,dx = 2^n\,n!\sqrt{\pi},$$

use (10.2) to prove that the constants in (10.3) are given by

$$a_j = \frac{1}{2^j j!\sqrt{\pi}}\int_{-\infty}^{\infty} e^{-x^2}\,H_j(x)\,p(x)\,dx.$$

11. Consider the Chebyshev equation

$$(1 - x^2)\,y'' - xy' + \alpha^2 y = 0, \qquad (11.1)$$

where α is a constant.

(a) Show that if $\alpha = N$, a nonnegative integer, then equation (11.1) has a polynomial solution of degree N. When suitably normalized, these polynomials are called the **Chebyshev polynomials** and are denoted by $T_N(x)$.

(b) Use equation (11.1) to show that $T_N(x)$ satisfies

$$[(1 - x^2)^{1/2}T_N{}']' + \frac{N}{(1 - x^2)^{1/2}}\,T_N = 0.$$

(c) Use the result from (b) to prove that

$$\int_{-1}^{1} \frac{T_N(x)\,T_M(x)}{(1 - x^2)^{1/2}}\,dx = 0,\ M \neq N.$$

◆ **12.** Use some form of technology to determine the coefficients in the Legendre expansion of the polynomial $p(x) = 3x^3 - 1$.

For problems 13 and 14, use some form of technology to determine the first four terms in the Legendre series expansion of the given function on the interval $(-1, 1)$. Plot the given function and the approximations

$$S_0 = a_0 P_0,\ \ S_2 = a_0 P_0 + a_1 P_1 + a_2 P_2,$$

$$S_2 = a_0 P_0 + a_1 P_1 + a_2 P_2 + a_3 P_3 + a_4 P_4$$

on the interval $-1 < x < 1$. Comment on the convergence of the Legendre series to the given function for $-1 < x < 1$.

◆ **13.** $f(x) = \cos \pi x$.

◆ **14.** $f(x) = x(1 - x^2)e^x$.

10.5   SERIES SOLUTIONS ABOUT A REGULAR SINGULAR POINT

The power series technique for solving

$$y'' + P(x)\,y' + Q(x)y = 0 \qquad (10.5.1)$$

developed in the previous section is only directly applicable at ordinary points, that is, points where P and Q are both analytic. According to Definition 10.3.1, any points at which P or Q fail to be analytic are called *singular points* of equation (10.5.1), and the general analysis of the behavior of solutions to equation (10.5.1) in the neighborhood of a singular point is quite complicated. However, singular points often turn out to be the points of major interest in an applied problem, and so it is of some importance that we pursue this analysis. In the next two sections, we will show that, provided that the functions P and Q are not too badly behaved at a singular point, the power series technique can be extended to obtain solutions of the corresponding DE that are valid in the neighborhood of the singular point. We will restrict our attention to DE whose singular points satisfy the following definition.

Definition 10.5.1:   The point $x = x_0$ is called a **regular singular point** of the DE (10.5.1) if and only if the following two conditions are satisfied:

1. x_0 is a singular point of equation (10.5.1).

2. *Both* of the functions
$$p(x) = (x - x_0)P(x) \quad \text{and} \quad q(x) = (x - x_0)^2 Q(x)$$
are analytic at $x = x_0$.

A singular point of equation (10.5.1) that does not satisfy condition (2) is called an **irregular singular point**.

Example 10.5.1   Determine whether $x = 0$, $x = 1$, and $x = 2$ are ordinary points, regular singular points, or irregular singular points of the DE

$$y'' + \frac{1}{x(x-1)^2} y' + \frac{x+1}{x(x-1)^3} y = 0. \tag{10.5.2}$$

Solution   In this case, we have

$$P(x) = \frac{1}{x(x-1)^2}, \quad Q(x) = \frac{x+1}{x(x-1)^3},$$

and, by inspection, P and Q are analytic at all points except $x = 0$, and $x = 1$. Hence the only singular points of equation (10.5.2) are $x = 0$, and $x = 1$. Consequently, $x = 2$ is an ordinary point. We now determine whether the singular points are regular or irregular.

(a) Consider the singular point $x = 0$. The functions

$$p(x) = xP(x) = \frac{1}{(x-1)^2}, \quad q(x) = x^2 Q(x) = \frac{x(x+1)}{(x-1)^3}$$

are both analytic at $x = 0$, so that $x = 0$ is a *regular singular point* of equation (10.5.2).

(b) Now consider the singular point $x = 1$. Since

$$p(x) = (x-1)P(x) = \frac{1}{x(x-1)}$$

is nonanalytic at $x = 1$, it follows that $x = 1$ is an *irregular singular point* of equation (10.5.2). ☐

Now suppose that $x = x_0$ is a regular singular point of the DE

$$y'' + P(x)y' + Q(x)y = 0.$$

Multiplying this equation by $(x - x_0)^2$ yields

$$(x - x_0)^2 y'' + (x - x_0) [(x - x_0)P(x)]y' + (x - x_0)^2 Q(x)y = 0,$$

which we can write as

$$(x - x_0)^2 y'' + (x - x_0)p(x)y' + q(x)y = 0,$$

where

$$p(x) = (x - x_0)P(x), \quad q(x) = (x - x_0)^2 Q(x).$$

Since, by assumption, $x = x_0$ is a regular singular point, it follows that the functions p and q are analytic at $x = x_0$. By the change of variables $z = x - x_0$, we can always transform a regular singular point to $x = 0$, and so we will restrict attention to DE that can be written in the form

$$\boxed{x^2 y'' + x p(x) y' + q(x) y = 0,} \tag{10.5.3}$$

where p and q are analytic at $x = 0$. *This is the standard form of any equation that has a regular singular point at $x = 0$.* The simplest type of equation that falls into this category is the second order Cauchy–Euler equation

$$x^2 y'' + p_0 x y' + q_0 y = 0, \tag{10.5.4}$$

where p_0 and q_0 are constants. The solution techniques that we will develop for solving equation (10.5.3) will be motivated by the solutions to equation (10.5.4). Recall from Section 2.9 that (10.5.4) has solutions on the interval $(0, \infty)$ of the form $y(x) = x^r$, where r is a root of the indicial equation

$$r(r - 1) + p_0 r + q_0 = 0. \tag{10.5.5}$$

Now consider equation (10.5.3). Since, by assumption, p and q are analytic at $x = 0$, we can write

$$p(x) = p_0 + p_1 x + p_2 x^2 + \cdots, \qquad q(x) = q_0 + q_1 x + q_2 x^2 + \cdots, \tag{10.5.6}$$

for x in some interval of the form $(-R, R)$. It follows that equation (10.5.3) can be written as

$$x^2 y'' + x(p_0 + p_1 x + p_2 x^2 + \ldots\ )y' + (q_0 + q_1 x + q_2 x^2 + \cdots\ )y = 0.$$

For $|x| \ll 1$, this is approximately the Cauchy–Euler equation (10.5.4), and so it is reasonable to *expect* that for x in the interval $(0, R)$, equation (10.5.3) has solutions of the form

$$y(x) = \underset{\substack{\uparrow \\ \text{Cauchy–Euler} \\ \text{solution}}}{x^r} \overset{\substack{\text{Power series} \\ \downarrow}}{\sum_{n=0}^{\infty}} a_n x^n, \quad a_0 \ne 0, \tag{10.5.7}$$

where r is a root of the *indicial equation*

$$r(r - 1) + p_0 r + q_0 = 0. \tag{10.5.8}$$

A series of the form (10.5.7) is called a **Frobenius series**. We can assume without loss of generality that $a_0 \ne 0$, since if this were not the case, we could always factor the leading power of x out of the series and combine it into x^r.

The following theorem confirms our expectations:

Theorem 10.5.1:   Consider the DE

$$x^2 y'' + xp(x)y' + q(x)y = 0, \qquad x > 0, \qquad (10.5.9)$$

where p and q are analytic at $x = 0$. Suppose that

$$p(x) = \sum_{n=0}^{\infty} p_n x^n, \qquad q(x) = \sum_{n=0}^{\infty} q_n x^n,$$

for $|x| < R$. Let r_1, and r_2 denote the roots of the indicial equation

$$r(r-1) + p_0 r + q_0 = 0,$$

and assume that $r_1 \geq r_2$ if these roots are real. Then equation (10.5.9) has a solution of the form

$$y_1(x) = x^{r_1} \sum_{n=0}^{\infty} a_n x^n, \qquad a_0 \neq 0.$$

This solution is valid (at least) for $0 < x < R$. Further, *provided that r_1 and r_2 are distinct and do not differ by an integer*, there exists a second solution to equation (10.5.9) that is valid (at least) for $0 < x < R$, of the form

$$y_2(x) = x^{r_2} \sum_{n=0}^{\infty} b_n x^n, \qquad b_0 \neq 0.$$

The solutions y_1 and y_2 are LI on their interval of existence.

PROOF   The proof of this theorem, as well as its extension to the case when the roots of the indicial equation do differ by an integer, will be discussed fully in the next section. ∎

REMARK   Using the formula for the Maclaurin expansion of p and q, it follows that the constants p_0 and q_0 appearing in (10.5.6) are given by

$$p_0 = p(0), \quad q_0 = q(0).$$

Consequently, the indicial equation (10.5.8) for

$$x^2 y'' + xp(x)y' + q(x)y = 0$$

can be written directly as

$$r(r-1) + p(0)r + q(0) = 0.$$

We conclude this section with some examples that illustrate the implementation of the above theorem.

Example 10.5.2   Show that the DE

$$x^2 y'' + xe^{2x}y' - 2(\cos x)y = 0, \qquad x > 0$$

has two LI Frobenius series solutions, and determine the interval on which these solutions are valid.

Solution   Comparing the given DE with the standard form (10.5.3), we see that

$$p(x) = e^{2x}, \quad q(x) = -2 \cos x.$$

Consequently,

$$p(0) = 1, \quad q(0) = -2,$$

and so the indicial equation is

$$r(r - 1) + r - 2 = 0.$$

That is,

$$r^2 - 2 = 0.$$

Thus, the roots of the indicial equation are $r_1 = \sqrt{2}$ and $r_2 = -\sqrt{2}$. Since r_1 and r_2 are distinct and do not differ by an integer, it follows from Theorem 10.5.1 that the given DE has two Frobenius series solutions of the *form*

$$y_1(x) = x^{\sqrt{2}} \sum_{n=0}^{\infty} a_n x^n, \quad y_1(x) = x^{-\sqrt{2}} \sum_{n=0}^{\infty} b_n x^n.$$

Further, since the power series expansions of p and q about $x = 0$ are valid for all x, the preceding solutions will be defined and LI on $(0, \infty)$.

Example 10.5.3   Find the general solution to

$$2x^2 y'' + xy' - (1 + x)y = 0, \qquad x > 0. \tag{10.5.10}$$

Solution   In this case, $p(x) = \frac{1}{2}$ and $q(x) = -\frac{1}{2}(1 + x)$, both of which are analytic at $x = 0$. Thus, $x = 0$ is a regular singular point of equation (10.5.10), and so there is at least one Frobenius series solution. Furthermore, since p and q are polynomials, their power series expansions about $x = 0$ converge for all real x. Consequently, from Theorem 10.5.1, any Frobenius series solution will be valid for $0 < x < \infty$. To determine the solutions we let

$$y(x) = x^r \sum_{n=0}^{\infty} a_n x^n = \sum_{n=0}^{\infty} a_n x^{r+n}, \quad a_0 \neq 0,$$

so that

$$y'(x) = \sum_{n=0}^{\infty} (r + n)a_n x^{r+n-1}, \quad y''(x) = \sum_{n=0}^{\infty} (r + n)(r + n - 1)a_n x^{r+n-2}.$$

Substituting into equation (10.5.10) yields

$$\sum_{n=0}^{\infty} 2(r + n)(r + n - 1)a_n x^{r+n} + \sum_{n=0}^{\infty} (r + n)a_n x^{r+n} - \sum_{n=0}^{\infty} a_n x^{r+n} - \sum_{n=0}^{\infty} a_n x^{r+n+1} = 0.$$

That is, combining the first three terms and replacing n with $n - 1$ in the third sum,

$$\sum_{n=0}^{\infty} [2(r + n)(r + n - 1) + (r + n) - 1]a_n x^n - \sum_{n=1}^{\infty} a_{n-1} x^n = 0. \tag{10.5.11}$$

Thus, the coefficients of x^n must vanish for $n = 0, 1, 2, \ldots$. When $n = 0$, we obtain

$$[2r(r - 1) + r - 1] a_0 = 0.$$

Since by assumption, $a_0 \neq 0$, we must have

$$2r(r - 1) + r - 1 = 0,$$

which is just the indicial equation for equation (10.5.10). This can be written as

$$(2r + 1)(r - 1) = 0,$$

so that the roots of the indicial equation are

$$r_1 = 1, \quad r_2 = -\frac{1}{2}. \tag{10.5.12}$$

Since these roots are distinct and do not differ by an integer, there exist two LI Frobenius series solutions. From (10.5.11) when $n = 1, 2, \ldots$, we obtain the *recurrence relation*

$$(r + n - 1)(2r + 2n + 1)a_n - a_{n-1} = 0. \tag{10.5.13}$$

We now substitute the values of r obtained in (10.5.12) into this relation to determine the corresponding Frobenius series solutions.

$r = 1$:   Substitution into the recurrence relation (10.5.13) yields

$$a_n = \frac{1}{n(2n + 3)} a_{n-1}, \quad n = 1, 2, 3, \ldots. \tag{10.5.14}$$

Thus,

$n = 1$: $a_1 = \dfrac{1}{1\cdot5} a_0,$

$n = 2$: $a_2 = \dfrac{1}{2\cdot7} a_1 = \dfrac{1}{(2!)(5\cdot7)} a_0,$

$n = 3$: $a_3 = \dfrac{1}{3\cdot9} a_2 = \dfrac{1}{(3!)(5\cdot7\cdot9)} a_0,$

$n = 4$: $a_4 = \dfrac{1}{4\cdot11} a_3 = \dfrac{1}{(4!)(5\cdot7\cdot9\cdot11)} a_0.$

It follows that, in general,

$$a_n = \frac{1}{(n!)[5\cdot7\cdot9\cdots(2n + 3)]} a_0, \quad n = 1, 2, 3 \ldots,$$

so that the corresponding Frobenius series solution is

$$y_1(x) = x\left[1 + \frac{1}{5}x + \frac{1}{(2!)(5\cdot7)} x^2 + \frac{1}{(3!)(5\cdot7\cdot9)} x^3 + \cdots\right.$$
$$\left. + \frac{1}{(n!)[5\cdot7\cdot9\cdots(2n + 3)]} x^n + \cdots\right],$$

where we have set $a_0 = 1$. We can write this solution as

$$y_1(x) = x\left[1 + \sum_{n=1}^{\infty} \frac{1}{(n!)[5\cdot7\cdot9\cdots(2n + 3)]} x^n\right], \quad x > 0.$$

$r = -1/2$ In this case, the recurrence relation (10.5.13) reduces to

$$a_n = \frac{1}{n(2n - 3)} a_{n-1}, \quad n = 1, 2, \ldots. \tag{10.5.15}$$

We therefore obtain

$n = 1$: $a_1 = -a_0,$

$n = 2$: $a_2 = \dfrac{1}{2\cdot1} a_1 = -\dfrac{1}{2!} a_0,$

$n = 3$: $a_3 = \dfrac{1}{3\cdot3} a_2 = -\dfrac{1}{(3!)(1\cdot3)} a_0,$

$n = 4$: $a_4 = \dfrac{1}{4\cdot5} a_3 = -\dfrac{1}{(4!)(1\cdot3\cdot5)} a_0.$

In general, we have

$$a_n = -\frac{1}{n! \, [1 \cdot 3 \cdot 5 \cdots (2n-3)]} a_0, \quad n = 1, 2, 3, \ldots.$$

It follows that a second LI Frobenius series solution to the DE (10.5.10) on $(0, \infty)$ is

$$y_2(x) = x^{-1/2} \left[1 - x - \frac{1}{2!} x^2 - \frac{1}{(3!)(1 \cdot 3)} x^3 - \frac{1}{(4!)(1 \cdot 3 \cdot 5)} x^4 - \cdots \right.$$

$$\left. - \frac{1}{(n!)[1 \cdot 3 \cdots (2n-3)]} x^n - \cdots \right],$$

where we have once more set $a_0 = 1$. This can be written as

$$y_2(x) = x^{-1/2} \left[1 - \sum_{n=1}^{\infty} \frac{1}{(n!)[1 \cdot 3 \cdot 5 \cdots (2n-3)]} x^n \right], \quad x > 0. \qquad (10.5.16)$$

Consequently, the general solution to equation (10.5.10) on $(0, \infty)$ is

$$y(x) = c_1 y_1(x) + c_2 y_2(x). \qquad \square$$

The DE in the previous example had two LI Frobenius series solutions, and we were therefore able to determine its general solution. In the following example, there is only one LI *Frobenius series* solution:

Example 10.5.4   Determine a Frobenius series solution to

$$x^2 y'' + x(3 + x)y' + (1 + 3x)y = 0, \quad x > 0. \qquad (10.5.17)$$

Solution   By inspection we see that $x = 0$ is a regular singular point of the DE (10.5.17), and so from Theorem 10.5.1, the DE has at least one Frobenius series solution. Further, since $p(x) = 3 + x$ and $q(x) = 1 + 3x$ are both polynomials, their power series expansions about $x = 0$ are valid for all x. It follows that any Frobenius series solution will be valid for $0 < x < \infty$. To determine a solution, we let

$$y(x) = x^r \sum_{n=0}^{\infty} a_n x^n.$$

Differentiating twice with respect to x yields

$$y'(x) = \sum_{n=0}^{\infty} (r + n) a_n x^{r+n-1}, \quad y''(x) = \sum_{n=0}^{\infty} (r + n)(r + n - 1) a_n x^{r+n-2}$$

so that y is a solution to equation (10.5.17) provided that a_n and r satisfy

$$\sum_{n=0}^{\infty} (r + n)(r + n - 1) a_n x^{r+n} + 3 \sum_{n=0}^{\infty} (r + n) a_n x^{r+n} + \sum_{n=0}^{\infty} (r + n) a_n x^{r+n+1}$$

$$+ \sum_{n=0}^{\infty} a_n x^{r+n} + 3 \sum_{n=0}^{\infty} a_n x^{r+n+1} = 0.$$

Dividing by x^r and replacing n by $n - 1$ in the third and fifth sums yields

$$\sum_{n=0}^{\infty} [(r + n)(r + n - 1) + 3(r + n) + 1] a_n x^n + \sum_{n=1}^{\infty} (r + n + 2) a_{n-1} x^n = 0. \qquad (10.5.18)$$

This implies that the coefficients of x^n must vanish for $n = 0, 1, 2, \ldots$. When $n = 0$, we obtain the indicial equation

$$r(r - 1) + 3r + 1 = 0.$$

That is,

$$(r + 1)^2 = 0.$$

Thus the only value of r for which a Frobenius series solution exists is

$$r = -1.$$

For $n \geq 1$, (10.5.18) yields the recurrence relation

$$[(r + n)(r + n - 1) + 3(r + n) + 1]a_n + (r + n + 2)a_{n-1} = 0.$$

Setting $r = -1$, we obtain

$$[(n - 1)(n - 2) + 3(n - 1) + 1]a_n + (n + 1)a_{n-1} = 0,$$

which can be written as

$$a_n = -\frac{(n + 1)}{n^2} a_{n-1}, \quad n = 1, 2, \ldots.$$

Solving this recurrence relation, we have

$$n = 1: \; a_1 = -2a_0, \qquad n = 2: \; a_2 = -\frac{3}{4} a_1 = \frac{3 \cdot 2}{4} a_0,$$

$$n = 3: \; a_3 = -\frac{4}{9} a_2 = -\frac{4!}{4 \cdot 9} a_0.$$

The general term is

$$a_n = (-1)^n \frac{(n + 1)!}{2^2 \cdot 3^2 \cdots n^2} a_0, \quad n = 1, 2, 3, \ldots,$$

which can be written as

$$a_n = (-1)^n \frac{(n + 1)!}{(n!)^2} a_0.$$

That is,

$$a_n = (-1)^n \frac{(n + 1)}{n!} a_0, \quad n = 1, 2, 3, \ldots.$$

Consequently, the corresponding Frobenius series solution is

$$y(x) = x^{-1}\left[1 + \sum_{n=1}^{\infty} (-1)^n \frac{(n + 1)}{n!} x^n \right], \quad x > 0,$$

where we set $a_0 = 1$. ❏

Notice that in this problem there is only *one* LI Frobenius series solution to the given DE. To determine a second LI solution, we could, for example, use the reduction of order method introduced in Section 2.2. We will have more to say about this in the next section.

EXERCISES 10.5

For problems 1–4, determine all singular points of the given DE and classify them as regular or irregular singular points.

1 . $y'' + \dfrac{1}{1-x} y' + xy = 0.$

2 . $x^2 y'' + \dfrac{x}{(1-x^2)^2} y' + y = 0.$

3 . $(x-2)^2 y'' + (x-2)e^x y' + 4x^{-1}y = 0.$

4 . $y'' + \dfrac{2}{x(x-3)} y' - \dfrac{1}{x^3(x+3)} y = 0.$

For problems 5–8, determine the roots of the indicial equation of the given DE.

5 . $x^2 y'' + x(1-x)y' - 7y = 0.$

6 . $4x^2 y'' + xe^x y' - y = 0.$

7 . $x y'' - xy' + 2y = 0.$

8 . $x^2 y'' - x(\cos x)y' + 5e^{2x}y = 0.$

For problems 9–16, show that the indicial equation of the given DE has distinct roots that do *not* differ by an integer and find two LI Frobenius series solutions on $(0, \infty)$.

9 . $4x^2 y'' + 3xy' + xy = 0.$

10. $6x^2 y'' + x(1+18x)y' + (1+12x)y = 0.$

11. $x^2 y'' + xy' - (2+x)y = 0.$

12. $2xy'' + y' - 2xy = 0.$

13. $3x^2 y'' - x(x+8)y' + 6y = 0.$

14. $2x^2 y'' - x(1+2x)y' + 2(4x-1)y = 0.$

15. $x^2 y'' + x(1-x)y' - (5+x)y = 0.$

16. $3x^2 y'' + x(7+3x)y' + (1+6x)y = 0.$

17. Consider the DE
$$x^2 y'' + xy' + (1-x)y = 0, \quad x > 0. \quad (17.1)$$
(a) Find the indicial equation, and show that the roots are $r = \pm i$.

(b) Determine the first three terms in a complex-valued Frobenius series solution to equation (17.1).

(c) Use the solution in (b) to determine two LI real-valued solutions to equation (17.1).

18. Determine the first five nonzero terms in each of two LI Frobenius series solutions to
$$3x^2 y'' + x(1+3x^2)y' - 2xy = 0, \quad x > 0.$$

19. Consider the DE
$$4x^2 y'' - 4x^2 y' + (1+2x)y = 0.$$
(a) Show that the indicial equation has only one root, and find the corresponding Frobenius series solution.

(b) Use the *reduction of order* technique to find a second LI solution on $(0, \infty)$. (Hint: To evaluate
$$\int x^{-1}e^x\, dx,$$
expand e^x in a Maclaurin series.)

20. Find two LI solutions to
$$x^2 y'' + x(3-2x)y' + (1-2x)y = 0$$
on $(0, \infty)$.

21. Consider the DE
$$x^2 y'' + x\,(1+2N-x)y' + N^2 y = 0, \quad x > 0. \quad (21.1)$$
(a) Find the indicial equation, and show that it has only one root $r = -N$.

(b) If N is a nonnegative integer, show that the Frobenius series solution to equation (21.1) terminates after N terms.

(c) Determine the Frobenius series solutions when $N = 0, 1, 2, 3$.

(d) Show that if N is a positive integer, then the Frobenius series solution to equation (21.1) can be written as
$$y(x) =$$
$$x^{-N}\left[1 + \sum_{k=1}^{N} (-1)^k \frac{N(N-1)\cdots(N+1-k)}{1^2 \cdot 2^2 \cdots k^2} x^k\right]$$
$$= x^{-N}\left[1 + \sum_{k=1}^{N} (-1)^k \prod_{i=1}^{k} \frac{(N+1-i)}{i^2} x^k\right].$$

22. Consider the general "perturbed" Cauchy–Euler equation
$$x^2 y'' + x\,[1 - (a+b) + \beta x]y' + (ab + \gamma x)y$$
$$= 0, \quad x > 0, \quad (22.1)$$
where a, b, β, γ are constants. Assuming that a and b are distinct and do not differ by an integer, determine two LI Frobenius series solutions to equation (22.1). (Hint: Use symmetry to get the second solution.)

10.6   FROBENIUS THEORY

In the previous section, we saw how Frobenius series solutions can be obtained to the DE

$$x^2y'' + xp(x)y' + q(x)y = 0, \quad x > 0. \tag{10.6.1}$$

In this section, we give some justification for Theorem 10.5.1 and extend this theorem to the case when the roots of the indicial equation for equation (10.6.1) differ by an integer. We will first assume that $x > 0$ since our results can easily be extended to $x < 0$. We begin by establishing the existence of at least one Frobenius series solution.

Assuming that $x = 0$ is a regular singular point of equation (10.6.1), it follows that p and q are analytic at $x = 0$, and hence, we can write

$$p(x) = \sum_{n=0}^{\infty} p_n x^n, \quad q(x) = \sum_{n=0}^{\infty} q_n x^n,$$

for $|x| < R$. Consequently, equation (10.6.1) can be written as

$$x^2y'' + x\sum_{n=0}^{\infty} p_n x^n y' + \sum_{n=0}^{\infty} q_n x^n y = 0. \tag{10.6.2}$$

We try for a Frobenius series solution and therefore let

$$y(x) = x^r \sum_{n=0}^{\infty} a_n x^n, \quad a_0 \neq 0,$$

where r and a_n are constants to be determined. Differentiating y twice yields

$$y' = \sum_{n=0}^{\infty} (r+n)a_n x^{r+n-1}, \quad y'' = \sum_{n=0}^{\infty} (r+n)(r+n-1)a_n x^{r+n-2}.$$

We now substitute into equation (10.6.2) to obtain

$$\sum_{n=0}^{\infty} (r+n)(r+n-1)a_n x^n + \left[\sum_{n=0}^{\infty} p_n x^n \right]\left[\sum_{n=0}^{\infty} (r+n)a_n x^n \right]$$

$$+ \left[\sum_{n=0}^{\infty} q_n x^n \right]\left[\sum_{n=0}^{\infty} a_n x^n \right] = 0.$$

Using the formula given in Section 10.2 for the product of two infinite series gives

$$\sum_{n=0}^{\infty} (r+n)(r+n-1)a_n x^n + \sum_{n=0}^{\infty} \left[\sum_{k=0}^{n} p_{n-k} x^{n-k}(k+r)a_k x^k \right]$$

$$+ \sum_{n=0}^{\infty} \left[\sum_{k=0}^{n} q_{n-k} x^{n-k} a_k x^k \right] = 0,$$

which can be written as

$$\sum_{n=0}^{\infty} \left\{ (r+n)(r+n-1)a_n + \sum_{k=0}^{n} [p_{n-k}(k+r) + q_{n-k}]\, a_k \right\} x^n = 0.$$

Thus, a_n must satisfy the *recurrence relation*

$$(r+n)(r+n-1)a_n + \sum_{k=0}^{n} [p_{n-k}(k+r) + q_{n-k}]a_k = 0, \quad n = 1, 2, \dots. \quad (10.6.3)$$

Evaluating equation (10.6.3) when $n = 0$ yields

$$[r(r-1) + p_0 r + q_0]a_0 = 0,$$

so that, since $a_0 \neq 0$ (by assumption), r must satisfy

$$r(r-1) + p_0 r + q_0 = 0, \quad (10.6.4)$$

which we recognize as being the indicial equation for equation (10.6.1). When $n \geq 1$, we combine the coefficients of a_n in (10.6.3) to obtain

$$[(r+n)(r+n-1) + p_0(r+n) + q_0]a_n + \sum_{k=0}^{n-1} [p_{n-k}(k+r) + q_{n-k}]a_k = 0.$$

That is,

$$[(r+n)(r+n-1) + p_0(r+n) + q_0]a_n = -\sum_{k=0}^{n-1} [p_{n-k}(k+r) + q_{n-k}]a_k,$$
$$n = 1, 2, \dots. \quad (10.6.5)$$

If we define $F(r)$ by

$$F(r) = r(r-1) + p_0 r + q_0,$$

then

$$F(r+n) = (r+n)(r+n-1) + p_0(r+n) + q_0,$$

so that the indicial equation (10.6.4) is

$$F(r) = 0,$$

whereas the recurrence relation (10.6.5) can be written as

$$F(r+n)a_n = -\sum_{k=0}^{n-1} [p_{n-k}(k+r) + q_{n-k}]a_k, \quad n = 1, 2, 3, \dots. \quad (10.6.6)$$

It is tempting to divide (10.6.6) by $F(r+n)$, thereby determining a_n in terms of $a_1, a_2, \dots, a_{n-1}$. However, we can do this only if $F(r+n) \neq 0$. Let r_1, r_2 denote the roots of equation (10.6.4). The following three familiar cases arise:

1. r_1 and r_2 are real and distinct.
2. r_1 and r_2 are real and coincident.
3. r_1 and r_2 are complex conjugates.

If r_1 and r_2 are real, we assume without loss of generality that $r_1 \geq r_2$. Consider (10.6.6) when $r = r_1$. We have

$$F(r_1 + n)a_n = -\sum_{k=0}^{n-1} [p_{n-k}(k+r_1) + q_{n-k}]a_k, \quad n = 1, 2, 3, \dots. \quad (10.6.7)$$

In cases (1) and (2), it follows, since $r = r_1$ is the largest root of $F(r) = 0$, that $F(r_1 + n) \neq 0$, for any n. Also, in case (3) $F(r_1 + n) \neq 0$, and so, in all three cases we can write (10.6.7) as

$$a_n = -\frac{1}{F(r_1 + n)} \sum_{k=0}^{n-1} [p_{n-k}(k+r_1) + q_{n-k}]a_k, \quad n = 1, 2, 3, \dots. \quad (10.6.8)$$

Starting from $n = 1$, we can therefore determine all of the a_n in terms of a_0, and so we for-mally obtain the Frobenius series solution

$$y_1(x) = x^{r_1}\left[1 + \sum_{n=1}^{\infty} a_n(r_1)x^n\right],$$

where $a_n(r_1)$ denotes the coefficients obtained from (10.6.8) upon setting $a_0 = 1$. A fairly delicate analysis shows that this series solution converges for (at least) $0 < x < R$. This justifies the steps in the preceding derivation and establishes the first part of Theorem 10.5.1 stated in the previous section.

We now consider the problem of determining a second LI solution to equation (10.6.1). We must consider the three cases separately.

Case 1: r_1 and r_2 are real and distinct. Setting $r = r_2$ in (10.6.6) yields

$$F(r_2 + n)\,a_n = -\sum_{k=0}^{n-1} [p_{n-k}(k + r_2) + q_{n-k}]a_k. \tag{10.6.9}$$

Thus, provided that $F(r_2 + n) \neq 0$ for any positive integer n, the same procedure as we used when $r = r_1$ will yield a second Frobenius series solution. But, since r_1 and r_2 are the only zeros of F, it follows that $F(r_2 + n) = 0$ if and only if there exists a positive integer n such that $r_2 + n = r_1$. Consequently, provided that

$$r_1 - r_2 \neq \text{positive integer},$$

there exists a second Frobenius series solution of the form

$$y_2(x) = x^{r_2}\left[1 + \sum_{n=1}^{\infty} a_n(r_2)x^n\right],$$

where $a_n(r_2)$ denotes the values of the coefficients obtained from (10.6.6) when $r = r_2$, and once more, we have set $a_0 = 1$. Since $r_1 \neq r_2$, it follows that the Frobenius series sol-utions y_1 and y_2 are LI on (at least) $0 < x < R$.

Now suppose that $r_1 - r_2 = N$, where N is a positive integer. Then substituting for $r_2 = r_1 - N$ in (10.6.9), we obtain

$$F(r_1 + (n - N))a_n = -\sum_{k=0}^{n-1} [p_{n-k}(k + r_2) + q_{n-k}]a_k, \quad n = 1, 2, \ldots,$$

which, when $n = N$, leads to the consistency condition

$$0 \cdot a_N = -\sum_{k=0}^{N-1} [p_{N-k}(k + r_2) + q_{N-k}]a_k. \tag{10.6.10}$$

Since all of the coefficients $a_1, a_2, \ldots, a_{N-1}$ will already have been determined in terms of a_0, equation (10.6.10) will be of the form

$$0 \cdot a_N = \alpha a_0, \tag{10.6.11}$$

where α is a constant. By assumption, a_0 is nonzero, and therefore two possibilities arise. (Notice that (10.6.11) is *not* an equation for determining α, rather, it is a consistency condition for the validity of the recurrence relation.)

(a) First it may happen that $\alpha = 0$. If this occurs, then, from (10.6.11), a_N can be specified arbitrarily, and (10.6.9) determines all of the remaining Frobenius coefficients. We therefore do obtain a second LI Frobenius series solution.

(b) In the more general case, α will be nonzero. Then, since $a_0 \neq 0$ (by assumption), (10.6.11) *cannot* be satisfied, and so we cannot compute the Frobenius coefficients. Hence, there *does not exist* a Frobenius series solution corresponding to $r = r_2$. The reduction of order technique can be used, however, to prove that there exists a second LI solution to equation (10.6.1) on $(0, R)$, of the form

$$y_2(x) = A\, y_1(x)\ln x + x^{r_2} \sum_{n=0}^{\infty} b_n x^n,$$

where the constants A and b_n can be determined by direct substitution into the DE (10.6.1). The derivation is straightforward, but quite longwinded, and so the details have been relegated to Appendix 5. Notice that the foregoing form includes case (a), which arises when $A = 0$.

Case 2: $r_1 = r_2$. In this case, there certainly cannot exist a second LI *Frobenius series* solution. However, once more the reduction of order technique can be used to establish the existence of a second LI solution of the form

$$y_2(x) = y_1(x)\ln x + x^{r_1} \sum_{n=1}^{\infty} b_n x^n$$

valid for (at least) $0 < x < R$. (See Appendix 5.) The coefficients b_n can be obtained by substituting this expression for y_2 into the DE (10.6.1).

Case 3: r_1 and r_2 are complex conjugates. In this case, the solution just obtained when $r = r_1$ will be a complex-valued solution. Due to the linearity of the DE, it follows that the real and imaginary parts of this complex-valued solution will themselves be real–valued solutions. It can be shown that these real-valued solutions are LI on their interval of existence. Thus, we can always, in theory, obtain the general solution in this case.

Finally we mention that the validity of the above solutions can be extended to $-R < x < 0$ by the replacement

$$x^{r_1} \to |x|^{r_1}, \qquad x^{r_2} \to |x|^{r_2}.$$

The preceding discussion is summarized in the next theorem.

Theorem 10.6.1:   Consider the DE

$$x^2 y'' + xp(x)y' + q(x)y = 0, \tag{10.6.12}$$

where p and q are analytic at $x = 0$. Suppose that

$$p(x) = \sum_{n=0}^{\infty} p_n x^n, \qquad q(x) = \sum_{n=0}^{\infty} q_n x^n,$$

for $|x| < R$. Let r_1 and r_2 denote the roots of the indicial equation and assume that $r_1 \geq r_2$ if these roots are real. Then equation (10.6.12) has two LI solutions valid (at least) on the interval $(0, R)$. The form of the solution is determined as follows:

1. $r_1 - r_2 \neq$ integer:

$$y_1(x) = x^{r_1} \sum_{n=0}^{\infty} a_n x^n, \quad a_0 \neq 0, \tag{10.6.13}$$

$$y_2(x) = x^{r_2} \sum_{n=0}^{\infty} b_n x^n, \quad b_0 \neq 0. \tag{10.6.14}$$

2. $r_1 = r_2 = r$:

$$y_1(x) = x^r \sum_{n=0}^{\infty} a_n x^n, \quad a_0 \neq 0, \tag{10.6.15}$$

$$y_2(x) = y_1(x)\ln x + x^r \sum_{n=1}^{\infty} b_n x^n.$$

3. $r_1 - r_2 =$ positive integer:

$$y_1(x) = x^{r_1} \sum_{n=0}^{\infty} a_n x^n, \quad a_0 \neq 0, \tag{10.6.16}$$

$$y_2(x) = A y_1(x)\ln x + x^{r_2} \sum_{n=0}^{\infty} b_n x^n, \quad b_0 \neq 0. \tag{10.6.17}$$

The coefficients in each of these solutions can be determined by direct substitution into equation (10.6.12). Finally, if x^{r_1} and x^{r_2} are replaced by $|x|^{r_1}$ and $|x|^{r_2}$ respectively, we obtain LI solutions that are valid for (at least) $0 < |x| < R$.

REMARK   Since a solution of a homogeneous linear DE is only defined up to a multiplicative constant, we can use this freedom to set $a_0 = 1$ in (10.6.13), (10.6.15) and (10.6.16) and to set $b_0 = 1$ in (10.6.14) and (10.6.17). It is often convenient to make these choices in solving our problems.

We now consider several examples to illustrate the use of the preceding theorem.

Example 10.6.1   Consider the DE

$$x^2 y'' - x(3 + x)y' + (4 - x)y = 0. \tag{10.6.18}$$

Determine the general form of two LI series solutions in the neighborhood of the regular singular point $x = 0$.

Solution   By inspection we see that $x = 0$ is a regular singular point of equation (10.6.18) and that the indicial equation is

$$r(r - 1) - 3r + 4 = r^2 - 4r + 4 = (r - 2)^2 = 0.$$

Since $r = 2$ is a repeated root it follows from Theorem 10.6.1 that there exist two LI solutions to equation (10.6.18) of the form

$$y_1(x) = x^2 \sum_{n=0}^{\infty} a_n x^n, \quad y_2(x) = y_1(x)\ln |x| + x^2 \sum_{n=1}^{\infty} b_n x^n.$$

The coefficients in each of these series solutions could be obtained by direct substitution into equation (10.6.18). In this problem, we have

$$p(x) = -(3 + x), \quad q(x) = 4 - x.$$

Since these are polynomials, their power series expansions about $x = 0$ converge for all real x. It follows from Theorem 10.6.1 that the series solutions to equation (10.6.18) will be valid for all $x \neq 0$.

Example 10.6.2   Consider the DE

$$x^2 y'' + x(1 + 2x)y' - \frac{1}{4}(1 - \gamma x)y = 0, \quad x > 0, \tag{10.6.19}$$

where γ is a constant. Determine the form of two LI series solutions to this DE about the regular singular point $x = 0$.

Solution   In this case, we have $p_0 = 1$ and $q_0 = -\frac{1}{4}$, so that

$$F(r) = r(r - 1) + r - \frac{1}{4} = r^2 - \frac{1}{4}.$$

Thus, the roots of the indicial equation are $r = \pm\frac{1}{2}$. Setting $r_1 = \frac{1}{2}$ and $r_2 = -\frac{1}{2}$, it follows that $r_1 - r_2 = 1$, so that we are in Case 3 of Theorem 10.6.1. Thus, there exist two LI solutions to equation (10.6.19) on $(0, \infty)$ of the form

$$y_1(x) = x^{1/2} \sum_{n=0}^{\infty} a_n x^n, \quad y_2(x) = A y_1(x) \ln x + x^{-1/2} \sum_{n=0}^{\infty} b_n x^n. \tag{10.6.20}$$

To determine whether the constant A is zero or nonzero, we need the general recurrence relation. Substituting

$$y_1(x) = x^r \sum_{n=0}^{\infty} a_n x^n, \quad a_0 \neq 0,$$

into equation (10.6.19) yields

$$\left[\frac{4(r + n)^2 - 1}{4} \right] a_n = -[2(r + n - 1) + \gamma] a_{n-1}, \quad n = 1, 2, \dots .$$

When $r = -\frac{1}{2}$, this reduces to

$$n(n - 1)a_n = -(2n - 3 + \gamma)a_{n-1}. \tag{10.6.21}$$

As predicted from our general theory, the coefficient of a_n is zero when $n = r_1 - r_2 = 1$. Thus, a second Frobenius series solution exists [$A = 0$ in (10.6.20)] if and only if the term on the right-hand side also vanishes when $n = 1$. Setting $n = 1$ in (10.6.21) yields the consistency condition

$$0 \cdot a_1 = (1 - \gamma)\, a_0. \tag{10.6.22}$$

Since $a_0 \neq 0$, it follows from (10.6.22) that when $\gamma \neq 1$, there does not exist a second LI Frobenius series solution and so the constant A in (10.6.20) is necessarily nonzero. If $\gamma = 1$, however, then (10.6.22) is identically satisfied independently of the value of a_1. In this case, we can specify a_1 arbitrarily and then the recurrence relation (10.6.21) will determine the remaining coefficients in a second LI Frobenius series solution. To summarize,

1. If $\gamma \neq 1$, there exist two LI solutions of the form (10.6.20) with A necessarily nonzero.

2. If $\gamma = 1$, then there exist two LI Frobenius series solutions

$$y_1(x) = x^{1/2} \sum_{n=0}^{\infty} a_n x^n, \qquad y_2(x) = x^{-1/2} \sum_{n=0}^{\infty} b_n x^n.$$

In both cases, the series solutions will be valid for $0 < x < \infty$.

Example 10.6.3   Determine two LI series solutions to

$$x^2 y'' + x(3 - x)y' + y = 0, \quad x > 0. \tag{10.6.23}$$

Solution   We see by inspection that $x = 0$ is a regular singular point. Furthermore, since the functions

$$p(x) = 3 - x, \quad q(x) = 1$$

are polynomials, the LI series solutions obtained from Theorem 10.6.1 will be valid on $(0, \infty)$.

We begin by obtaining a Frobenius series solution. Substituting

$$y(x) = x^r \sum_{n=0}^{\infty} a_n x^n, \quad a_0 \neq 0,$$

into equation (10.6.23) yields

$$\sum_{n=0}^{\infty} (r + n)(r + n - 1)a_n x^{r+n} + 3 \sum_{n=0}^{\infty} (r + n)a_n x^{r+n}$$

$$- \sum_{n=0}^{\infty} (r + n)a_n x^{r+n+1} + \sum_{n=0}^{\infty} a_n x^{r+n} = 0,$$

which, upon collecting coefficients of x^{r+n} and replacing n by $n - 1$ in the third sum, can be written as

$$\sum_{n=0}^{\infty} [(r + n)(r + n + 2) + 1]a_n x^{r+n} - \sum_{n=1}^{\infty} (r + n - 1)a_{n-1} x^{r+n} = 0.$$

Dividing by x^r yields

$$\sum_{n=0}^{\infty} [(r + n)(r + n + 2) + 1]a_n x^n - \sum_{n=1}^{\infty} (r + n - 1)a_{n-1} x^n = 0. \tag{10.6.24}$$

We determine the a_n in the usual manner. When $n = 0$, we obtain

$$[r(r + 2) + 1]a_0 = 0,$$

so that the indicial equation is

$$r^2 + 2r + 1 = 0$$

that is

$$(r + 1)^2 = 0.$$

Hence, there is only one root, namely,

$$r = -1.$$

From (10.6.24), the remaining coefficients must satisfy the recurrence relation

$$[(r + n)(r + n + 2) + 1]a_n - (r + n - 1)a_{n-1} = 0, \quad n = 1, 2, \dots.$$

Setting $r = -1$, this simplifies to

$$a_n = \frac{(n - 2)}{n^2} a_{n-1}, \quad n = 1, 2, 3, \dots.$$

Consequently,

$$a_1 = -a_0, \quad a_2 = a_3 = a_4 = \cdots = 0.$$

Thus, a Frobenius series solution to equation (10.6.23) is

$$y_1(x) = x^{-1}(1 - x), \tag{10.6.25}$$

where we set $a_0 = 1$.

Since the indicial equation for equation (10.6.23) has only one root, there does not exist a second LI Frobenius series solution. However, according to Theorem 10.6.1, there is a second LI solution of the form

$$y_2(x) = y_1(x)\ln x + x^{-1}\sum_{n=1}^{\infty} b_n x^n, \tag{10.6.26}$$

where the coefficients b_n can be determined by substitution into equation (10.6.23). We now determine such a solution. The computations are quite straightforward, but they are long and tedious. The student is encouraged to pay full attention and not to be overwhelmed by the formidable look of the equations. Differentiating (10.6.26) with respect to x yields

$$y_2{'} = y_1{'}\ln x + x^{-1}y_1 + \sum_{n=1}^{\infty} (n - 1)b_n x^{n-2},$$

$$y_2{''} = y_1{''}\ln x + 2x^{-1}y_1{'} - x^{-2}y_1 + \sum_{n=1}^{\infty} (n - 2)(n - 1)b_n x^{n-3}.$$

Substituting into equation (10.6.23), we obtain the following equation for the b_n:

$$x^2\left[y_1{''}\ln x + 2x^{-1}y_1{'} - x^{-2}y_1 + \sum_{n=1}^{\infty} (n - 2)(n - 1)b_n x^{n-3} \right]$$

$$+ x(3 - x)\left[y_1{'}\ln x + x^{-1}y_1 + \sum_{n=1}^{\infty} (n - 1)b_n x^{n-2} \right] + y_1\ln x + x^{-1}\sum_{n=1}^{\infty} b_n x^n = 0,$$

or, equivalently,

$$[x^2 y{''} + x(3 - x)y_1{'} + y_1]\ln x + 2xy_1{'} + 2y_1 - xy_1 + \sum_{n=1}^{\infty} (n - 1)(n - 2)b_n x^{n-1}$$

$$+ 3\sum_{n=1}^{\infty} (n - 1)b_n x^{n-1} - \sum_{n=1}^{\infty} (n - 1)b_n x^n + \sum_{n=1}^{\infty} b_n x^{n-1} = 0.$$

Since y_1 is a solution to equation (10.6.23), the combination of terms multiplying $\ln x$

vanish.[1] Combining the coefficients of x^{n-1}, we obtain

$$2xy_1' + 2y_1 - xy_1 + \sum_{n=1}^{\infty} [(n-1)(n-2) + 3(n-1) + 1]b_n x^{n-1}$$

$$- \sum_{n=1}^{\infty} (n-1)b_n x^n = 0.$$

Simplifying the terms in the first sum and replacing n by $n-1$ in the second sum yields

$$2xy_1' + 2y_1 - xy_1 + \sum_{n=1}^{\infty} n^2 b_n x^{n-1} - \sum_{n=2}^{\infty} (n-2)b_{n-1} x^{n-1} = 0. \qquad (10.6.27)$$

From (10.6.25), we have

$$y_1(x) = x^{-1}(1-x), \quad y_1'(x) = -x^{-2}.$$

Substituting these expressions into (10.6.27) gives

$$-2x^{-1} + 2x^{-1}(1-x) - (1-x) + \sum_{n=1}^{\infty} n^2 b_n x^{n-1} - \sum_{n=2}^{\infty} (n-2)b_{n-1} x^{n-1} = 0.$$

That is,

$$-3 + x + \sum_{n=1}^{\infty} n^2 b_n x^{n-1} - \sum_{n=2}^{\infty} (n-2)b_{n-1} x^{n-1} = 0.$$

Equating the corresponding coefficients of x^{n-1} to zero for $n = 1, 2, \ldots$, we obtain the b_n in a familiar manner.

$n = 1$: $-3 + b_1 = 0$, which implies that $b_1 = 3$.

$n = 2$: $1 + 4b_2 = 0$, which implies that $b_2 = -\dfrac{1}{4}$.

$n \geq 3$: $n^2 b_n - (n-2)b_{n-1} = 0$.

That is,

$$b_n = \frac{(n-2)}{n^2} b_{n-1}, \quad n = 3, 4, \ldots . \qquad (10.6.28)$$

When $n = 3$, we have

$$b_3 = \frac{1}{3^2} b_2.$$

Substituting for $b_2 = -\dfrac{1}{4} = -\dfrac{1}{2^2}$ yields

$$b_3 = -\frac{1}{2^2 \cdot 3^2}.$$

When $n = 4$, (10.6.28) implies that

$$b_4 = \frac{2}{4^2} b_3 = -\frac{1 \cdot 2}{2^2 \cdot 3^2 \cdot 4^2}.$$

We see that in general, for $n \geq 3$,

---

[1] It should be noted that this is not just a fortuitous result for this particular example. In general, the terms multiplying $\ln x$ will vanish at this stage of the computation.

$$b_n = -\frac{(n-2)!}{2^2 \cdot 3^2 \cdots n^2},$$

which can be written as

$$b_n = -\frac{(n-2)!}{(n!)^2}, \quad n = 3, 4, \ldots.$$

Finally, substituting for b_n into (10.6.26) yields the following solution to equation (10.6.23)

$$y_2(x) = x^{-1}(1-x)\ln x + x^{-1}\left[3x - \frac{1}{4}x^2 - \sum_{n=3}^{\infty} \frac{(n-2)!}{(n!)^2}x^n\right].$$

Theorem 10.6.1 guarantees that y_1 and y_2 are LI on $(0, \infty)$. ◻

We give one final example to illustrate the case when the roots of the indicial equation differ by an integer.

Example 10.6.4   Determine two LI solutions to

$$x^2 y'' + xy' - (4+x)y = 0, \quad x > 0. \tag{10.6.29}$$

Solution   Since $x = 0$ is a regular singular point, we try for Frobenius series solutions. Substituting

$$y(x) = \sum_{n=0}^{\infty} a_n x^{r+n}$$

into equation (10.6.29) and simplifying yields the indicial equation

$$r^2 - 4 = 0$$

and the recurrence relation

$$[(r+n)^2 - 4]a_n = a_{n-1}, \quad n = 1, 2, \ldots. \tag{10.6.30}$$

It follows that the roots of the indicial equation are

$$r_1 = 2 \quad \text{and} \quad r_2 = -2,$$

which differ by an integer. Substituting $r = 2$ into equation (10.6.30) yields

$$a_n = \frac{1}{n(n+4)}a_{n-1}, \quad n = 1, 2, \ldots.$$

This is easily solved to obtain

$$a_n = \frac{4!}{n!(n+4)!}a_0.$$

Consequently, one Frobenius series solution to equation (10.6.29) is

$$y_1(x) = a_0 \sum_{n=0}^{\infty} \frac{4!}{n!(n+4)!}x^{n+2}.$$

Choosing $a_0 = \frac{1}{4!}$, this solution reduces to

$$y_1(x) = \sum_{n=0}^{\infty} \frac{1}{n!(n+4)!} x^{n+2}.$$ (10.6.31)

We now determine whether there exists a second LI *Frobenius series* solution. Substituting $r = -2$ into equation (10.6.30) yields

$$n(n-4)a_n = a_{n-1}, \quad n = 1, 2, \ldots.$$ (10.6.32)

Thus, when $n = 1$, $n = 2$, or $n = 3$, we obtain

$$a_1 = -\frac{1}{3} a_0, \quad a_2 = \frac{1}{12} a_0, \quad a_3 = -\frac{1}{36} a_0.$$

However, when $n = 4$, (10.6.32) requires

$$0 \cdot a_4 = a_3 = -\frac{1}{36} a_0,$$

which is clearly impossible, since $a_0 \neq 0$. It follows that a second LI *Frobenius series* solution does not exist. However, according to Theorem 10.6.1, there is a second LI solution of the form

$$y_2(x) = Ay_1(x) \ln x + x^{-2} \sum_{n=0}^{\infty} b_n x^n,$$ (10.6.33)

where the constants A and b_n can be determined by substitution into equation (10.6.29). Differentiating y_2 yields

$$y_2' = Ay_1' \ln x + Ax^{-1}y_1 + \sum_{n=0}^{\infty} (n-2)b_n x^{n-3},$$

$$y_2'' = Ay_1'' \ln x + 2Ax^{-1}y_1' - Ax^{-2}y_1 + \sum_{n=0}^{\infty} (n-3)(n-2)b_n x^{n-4}.$$

By substituting into equation (10.6.29) and simplifying we obtain

$$A[x^2 y_1'' + xy_1' - (4+x)y] \ln x + 2Axy_1' + \sum_{n=0}^{\infty} (n-3)(n-2)b_n x^{n-2}$$

$$+ \sum_{n=0}^{\infty} (n-2)b_n x^{n-2} - 4\sum_{n=0}^{\infty} b_n x^{n-2} - \sum_{n=0}^{\infty} b_n x^{n-1} = 0.$$

The combination of terms multiplying $\ln x$ vanish, since y_1 is a solution of (10.6.29).[1] Combining the coefficients of x^{n-2} and simplifying yields

$$2Axy_1' + \sum_{n=1}^{\infty} n(n-4)b_n x^{n-2} - \sum_{n=0}^{\infty} b_n x^{n-1} = 0.$$ (10.6.34)

We must now determine y_1'. Differentiating (10.6.31), we obtain

$$y_1'(x) = \sum_{n=0}^{\infty} \frac{(n+2)}{n!(n+4)!} x^{n+1}.$$

Substituting into equation (10.6.34) yields

---

[1] Once more, we note that this will always happen at this stage of the computation.

$$2A \sum_{n=0}^{\infty} \frac{(n+2)}{n!(n+4)!} x^{n+2} + \sum_{n=1}^{\infty} n(n-4)b_n x^{n-2} - \sum_{n=0}^{\infty} b_n x^{n-1} = 0.$$

We now replace n with $n-4$ in the first sum and replace n with $n-1$ in the third sum to obtain a common power of x in all sums. The result is

$$2A \sum_{n=4}^{\infty} \frac{(n-2)}{(n-4)!n!} x^{n-2} + \sum_{n=1}^{\infty} n(n-4)b_n x^{n-2} - \sum_{n=1}^{\infty} b_{n-1} x^{n-2} = 0.$$

That is, upon multiplying through by x^2,

$$2A \sum_{n=4}^{\infty} \frac{(n-2)}{(n-4)!n!} x^{n} + \sum_{n=1}^{\infty} n(n-4)b_n x^{n} - \sum_{n=1}^{\infty} b_{n-1} x^{n} = 0.$$

We can now determine the appropriate values of the constants by setting successive coefficients of x^n to zero in the usual manner.

$n = 1$: $\qquad\qquad\qquad\qquad -3b_1 - b_0 = 0$; hence

$$b_1 = -\frac{1}{3} b_0.$$

$n = 2$: $\qquad\qquad\qquad\qquad -4b_2 - b_1 = 0$; hence

$$b_2 = -\frac{1}{4} b_1 = \frac{1}{12} b_0.$$

$n = 3$: $\qquad\qquad\qquad\qquad -3b_3 - b_2 = 0$; hence

$$b_3 = -\frac{1}{3} b_2 = -\frac{1}{36} b_0.$$

$n = 4$: $\qquad\qquad\qquad\qquad \frac{1}{6} A - b_3 = 0$; hence

$$A = 6b_3 = -\frac{1}{6} b_0.$$

$n \geq 5$:

$$b_n = \frac{1}{n(n-4)} \left[b_{n-1} - \frac{2A(n-2)}{(n-4)!n!} \right].$$

Using this recurrence relation, we could continue to determine values of the b_n. Notice that b_4 is unconstrained in this problem and so can be set equal to any convenient value. For example we could set $b_4 = 0$. Substituting the values of the coefficients so far obtained into (10.6.33), gives

$$y_2(x) = b_0 \left[-\frac{1}{6} y_1(x) \ln x + x^{-2} \left(1 - \frac{1}{3} x + \frac{1}{12} x^2 - \frac{1}{36} x^3 + \cdots \right) \right]. \qquad (10.6.35)$$

Thus, two LI solutions to the given DE on $(0, \infty)$ are

$$y_1(x) = \sum_{n=0}^{\infty} \frac{1}{n!(n+4)!} x^{n+2}$$

and

$$y_2(x) = y_1(x) \ln x - 6x^{-2} \left(1 - \frac{1}{3} x + \frac{1}{12} x^2 - \frac{1}{36} x^3 + \cdots \right)$$

where we set $b_0 = -6$ in (10.6.35).   □

REMARKS   In the previous example, we had to be content with determining only a few terms in the second LI solution, since we could not solve the recurrence relation that arose for the b_n. This is usually the case in these types of problems.

EXERCISES 10.6

For problems 1–7, determine the roots of the indicial equation of the given DE. Also obtain the general *form* of two LI solutions to the DE on an interval $(0, R)$. Finally, if $r_1 - r_2$ equals a positive integer obtain the recurrence relation and determine whether the constant A in

$$y_2(x) = Ay_1(x) \ln x + x^{r_2} \sum_{n=0}^{\infty} b_n x^n$$

is zero or nonzero.

1. $4x^2y'' + 2x^2y' + y = 0$.

2. $x^2y'' + x(\cos x)y' - 2e^x y = 0$.

3. $x^2y'' + x^2y' - (2 + x)y = 0$.

4. $x^2y'' + 2x^2y' + (x - \frac{3}{4})y = 0$.

5. $x^2y'' + xy' + (2x - 1)y = 0$.

6. $x^2y'' + x^3y' - (2 + x)y = 0$.

7. $x^2(x^2 + 1)y'' + 7xe^x y' + 9(1 + \tan x)y = 0$.

8. Determine all values of the constant α for which

$$x^2 y'' + x(1 - 2x)y' + [2(\alpha - 1)x - \alpha^2]y = 0$$

has two LI *Frobenius series* solutions on $(0, \infty)$.

9. The indicial equation and recurrence relation for the DE

$$x^2y'' + x[(2 - b) + x]y' - (b - \gamma x)y = 0 \quad (9.1)$$

are, respectively,

$$(r + 1)(r - b) = 0,$$

$$(r + n + 1)(r + n - b)a_n = -[(r + n - 1) + \gamma]a_{n-1},$$

$$n = 1, 2, 3, ...,$$

in the usual notation, where b and γ are constants. Determine the *form* of two LI series solutions to equation (9.1) on $(0, \infty)$, in the following cases:

(a) $b \neq$ integer.

(b) $b = -1$.

(c) $b = N$, a nonnegative integer. [For solutions containing a term of the form $Ay_1(x) \ln x$ you must determine whether A is zero or nonzero.]

10. Show that

$$x^2(1 + x)y'' + x^2y' - 2y = 0$$

has two LI Frobenius series solutions on $(-1, 1)$ and find them.

11. Consider the DE

$$x^2y'' + 3xy' + (1 - x)y = 0, \quad x > 0. \quad (11.1)$$

(a) Determine the indicial equation and show that it has the repeated root $r = -1$.

(b) Obtain the corresponding Frobenius series solution.

(c) It follows from Theorem 10.6.1 that equation (11.1) has a second LI solution of the form

$$y_2(x) = y_1(x) \ln x + x^{-1} \sum_{n=1}^{\infty} b_n x^n.$$

Show that $b_1 = -2$ and that in general,

$$b_n = \frac{1}{n^2} \left[b_{n-1} - \frac{2n}{n!(n-1)!} \right], \quad n = 2, 3, 4, ...$$

Use this to find the first three terms of y_2.

12. Consider the DE

$$xy'' - y = 0, \quad x > 0. \quad (12.1)$$

(a) Show that the roots of the indicial equation are $r_1 = 1$ and $r_2 = 0$, and determine the Frobenius series solution corresponding to $r_1 = 1$.

(b) Show that there does not exist a second LI Frobenius series solution.

(c) According to Theorem 10.6.1, equation (12.1) has a second LI solution of the form

$$y_2(x) = Ay_1(x) \ln x + \sum_{n=0}^{\infty} b_n x^n.$$

Show that $A = b_0$, and determine the first three terms in y_2.

For problems 13–26, determine two LI solutions to the given DE on $(0, \infty)$.

13. $x^2y'' + x(1 - x)y' - y = 0$.

14. $x^2y'' + x(6 + x^2)y' + 6y = 0$.

15. $xy'' + y' - 2y = 0$.

16. $4x^2y'' + (1 - 4x)y = 0$.

17. $x^2y'' - x(x + 3)y' + 4y = 0$.

18. $x^2y'' + xy' - (1 + x)y = 0$.

19. $x^2y'' - x^2y' - (3x + 2)y = 0$.

20. $x^2y'' - x^2y' - 2y = 0$.

21. $4x^2y'' + 4x(1 - x) y' + (2x - 9)y = 0$.

22. $x^2y'' + x(5 - x)y' + 4y = 0$.

23. $x^2y'' - x(1 - x)y' + (1 - x)y = 0$.

24. $x^2y'' + 2x(2 + x) y' + 2(1 + x) y = 0$.

25. $4x^2y'' - (3 + 4x)y = 0$.

26. $4x^2y'' + 4x(1 + 2x)y' + (4x - 1)y = 0$.

For problems 27 and 28, determine a Frobenius series solution to the given DE and use the *reduction of order* technique to find a second LI solution on $(0, \infty)$.

27. $xy'' - xy' + y = 0$.

28. $x^2y'' + x(4 + x)y' + (2 + x) y = 0$.

29. Consider the *Laguerre* DE

$$x^2y'' + x(1 - x) y' + Nxy = 0, \qquad (29.1)$$

where N is a constant. Show that in the case when N is a positive integer, equation (29.1) has a solution that is a polynomial of degree N, and find it. When properly normalized these solutions are called the **Laguerre polynomials**.

30. Consider the DE

$$x^2y'' + x(1 + 2N + x) y' + N^2y = 0, \qquad (30.1)$$

where N is a positive integer.

(a) Show that there is only one Frobenius series solution and that it terminates after $N + 1$ terms.

Find this solution.

(b) Show that the change of variables $Y = x^Ny$ transforms equation (30.1) into the Laguerre equation (29.1).

31. Consider the DE

$$x^2y'' + x(1 + \beta x) y' + [\beta(1 - N)x - N^2]y$$
$$= 0, \; x > 0, \qquad (31.1)$$

where N is a positive integer and β is a constant.

(a) Show that the roots of the indicial equation are $r = \pm N$.

(b) Show that the Frobenius series solution corresponding to $r = N$ is

$$y_1(x) = a_0 \, x^N \sum_{n=0}^{\infty} \frac{(2N)! \, (-\beta)^n}{(2N + n)!} \, x^n$$

and that by an appropriate choice of a_0, one solution to (31.1) is

$$y_1(x) = x^{-N}\left[e^{-\beta x} - \sum_{n=0}^{2N-1} \frac{(-\beta x)^n}{n!} \right].$$

(c) Show that equation (31.1) has a second LI Frobenius series solution that can be taken as

$$y_2(x) = x^{-N} \sum_{n=0}^{2N-1} \frac{(-\beta x)^n}{n!}.$$

Hence, conclude that equation (31.1) has LI solutions

$$y_1(x) = x^{-N}e^{-\beta x}, \quad y_2(x) = x^{-N} \sum_{n=0}^{2N-1} \frac{(-\beta x)^n}{n!}.$$

10.7   BESSEL'S EQUATION OF ORDER p

One of the most important DE in applied mathematics and mathematical physics is *Bessel's equation of order p* defined by

$$x^2y'' + xy' + (x^2 - p^2)y = 0 \qquad (10.7.1)$$

where p is a *nonnegative* constant. In general it is not possible to obtain closed form solutions to this equation. However, since $x = 0$ is a regular singular point we can apply the Frobenius series technique to obtain series solutions. We will assume that $x > 0$. The indicial equation for (10.7.1) is

$$r(r - 1) + r - p^2 = 0$$

with roots

$$r = \pm p.$$

Consequently, provided $2p$ is not an integer, there will exist two LI Frobenius series solutions. In order to obtain these solutions we let

$$y(x) = x^r \sum_{n=0}^{\infty} a_n x^n \tag{10.7.2}$$

so that

$$y'(x) = \sum_{n=0}^{\infty} (r + n)a_n x^{r+n-1}, \quad y''(x) = \sum_{n=0}^{\infty} (r + n)(r + n - 1)a_n x^{r+n-2}.$$

Substituting into (10.7.1) and rearranging yields

$$\sum_{n=0}^{\infty} [(r + n)^2 - p^2]a_n x^{r+n} + \sum_{n=2}^{\infty} a_{n-2} x^{r+n} = 0. \tag{10.7.3}$$

When $n = 0$ we obtain the indicial equation whose roots are, as we have seen above,

$$r = \pm p. \tag{10.7.4}$$

When $n = 1$, (10.7.3) implies that

$$[(r + 1)^2 - p^2]a_1 = 0 \tag{10.7.5}$$

and for $n \geq 2$, we obtain the general recurrence relation:

$$[(r + n)^2 - p^2]a_n = -a_{n-2}, \quad n = 2, 3, \dots . \tag{10.7.6}$$

Consider first the root $r = p$. In this case (10.7.5) reduces to

$$(2p + 1)\, a_1 = 0$$

so that, since $p \geq 0$,

$$a_1 = 0. \tag{10.7.7}$$

Setting $r = p$ in (10.7.6) yields

$$a_n = -\frac{1}{n(2p + n)} a_{n-2}, \quad n = 2, 3, \dots . \tag{10.7.8}$$

It follows from (10.7.7) and (10.7.8) that all of the odd coefficients are zero, that is,

$$a_{2k+1} = 0, \quad k = 0, 1, 2, \dots . \tag{10.7.9}$$

Now consider the even coefficients. From (10.7.8) we obtain

$$a_2 = -\frac{1}{2(2p + 2)} a_0, \quad a_4 = -\frac{1}{4(2p + 4)} a_2 = \frac{1}{2 \cdot 4(2p + 4)(2p + 2)} a_0$$

and so on. The general even coefficient is

$$a_{2k} = \frac{(-1)^k}{2 \cdot 4 \cdots (2k)(2p + 2)(2p + 4) \cdots (2p + 2k)} a_0, \quad k = 1, 2, \dots$$

which can be written as

$$a_{2k} = \frac{(-1)^k}{2^{2k}k!(p + 1)(p + 2) \cdots (p + k)} a_0, \quad k = 1, 2, 3, \dots .$$

Consequently, the corresponding Frobenius series solution to Bessel's equation is

$$y_1(x) = a_0 x^p \left[1 + \sum_{k=1}^{\infty} \frac{(-1)^k}{2^{2k} k! (p+1)(p+2)\cdots(p+k)} x^{2k} \right]. \qquad (10.7.10)$$

This solution is valid for all $x > 0$.

BESSEL FUNCTIONS OF THE FIRST KIND[1]

In order to study the solutions of Bessel's equation obtained above it is convenient to first rewrite (10.7.10) in a different, but equivalent form. The analysis splits into two cases.

***p = N*, A POSITIVE INTEGER**   In this case the solution (10.7.10) can be written as

$$y_1(x) = a_0 x^N \left[1 + \sum_{k=1}^{\infty} \frac{(-1)^k}{2^{2k} k! (N+1)(N+2)\cdots(N+k)} x^{2k} \right] \qquad (10.7.11)$$

where the constant a_0 can be chosen arbitrarily. It is convenient to make the choice

$$a_0 = \frac{1}{N! \, 2^N}.$$

The corresponding solution of Bessel's equation is denoted $J_N(x)$ and is called the **Bessel function of the first kind of integer order** N. Substituting for a_0 into (10.7.11) we obtain

$$J_N(x) = \sum_{k=0}^{\infty} \frac{(-1)^k}{k!(N+k)!} \left(\frac{x}{2}\right)^{2k+N}. \qquad (10.7.12)$$

The most important Bessel functions of integer order are $J_0(x)$ and $J_1(x)$ since, as we shall see presently, all other integer order Bessel functions of the first kind can be expressed in terms of these two. Writing out the first few terms in (10.7.12) when $N = 0, 1$ yields, respectively

$$J_0(x) = 1 - \frac{1}{4}x^2 + \frac{1}{64}x^4 - \cdots$$

$$J_1(x) = \frac{1}{2}x \left(1 - \frac{1}{8}x^2 + \frac{1}{192}x^4 - \cdots \right).$$

An analysis of these functions shows that they both oscillate with decaying amplitude. Further, each has an infinite number of nonnegative zeros. A sketch of J_0 and J_1 on the interval $(0, 10]$ is given in Fig. 10.7.1.

---

[1]The remainder of this section includes only a brief introduction to Bessel functions. For more details and the proofs of the results stated in this section, the reader is referred to N. N. Lebedev, *Special Functions and their Applications*, Dover Publications.

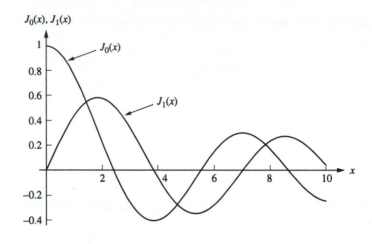

Figure 10.7.1 The Bessel functions of the first kind $J_0(x)$, $J_1(x)$.

$p > 0$, NONINTEGER   In order to obtain a formula for the Frobenius series solution (10.7.10) when p is nonintegral that is analogous to (10.7.12) we need to introduce the *gamma function*. This function can be considered as the generalization of the factorial function to the case of noninteger real numbers.

Definition 10.7.1:   The **gamma function**, $\Gamma(p)$, is defined by

$$\Gamma(p) = \int_0^\infty t^{p-1} e^{-t}\, dt, \quad p > 0.$$

It can be shown that the above improper integral converges for all $p > 0$, so that the gamma function is well–defined for all such p. To show that the gamma function is a generalization of the factorial function, we first require the following result.

Lemma 10.7.1:   For all $p > 0$,

$$\Gamma(p + 1) = p\Gamma(p). \tag{10.7.13}$$

PROOF   The proof consists of integrating the expression for $\Gamma(p + 1)$ by parts

$$\Gamma(p + 1) = \int_0^\infty t^p e^{-t}\, dt = \left[-t^p e^{-t}\right]_0^\infty + p \int_0^\infty t^{p-1} e^{-t}\, dt$$

$$= p\Gamma(p). \qquad \blacksquare$$

We also require

$$\Gamma(1) = \int_0^\infty e^{-t}\, dt = 1. \tag{10.7.14}$$

Equations (10.7.13) and (10.7.14) imply that

$$\Gamma(2) = 1\Gamma(1) = 1, \quad \Gamma(3) = 2\Gamma(2) = 2!, \quad \Gamma(4) = 3\Gamma(3) = 3!$$

and in general, for all nonnegative integers N,

$$\Gamma(N + 1) = N!.$$

This justifies the claim that the gamma function generalizes the factorial function. We now extend the definition of the gamma function to $p < 0$. From (10.7.13)

$$\Gamma(p) = \frac{\Gamma(p + 1)}{p} \tag{10.7.15}$$

for $p > 0$. We use this expression to *define* $\Gamma(p)$ for $p < 0$ as follows. If p is in the interval $(-1, 0)$ then $p + 1$ is in the interval $(0, 1)$ and so $\Gamma(p)$ given in (10.7.15) is well defined. We continue in this manner to successively define $\Gamma(p)$ in the intervals $(-2, -1)$, $(-3, -2)$, ... From (10.7.14) and (10.7.15) it follows that

$$\lim_{p \to 0^+} \Gamma(p) = +\infty, \qquad \lim_{p \to 0^-} \Gamma(p) = -\infty$$

so that the graph of the gamma function has the general form given in Figure 10.7.2. We note that the gamma function is continuous and indeed infinitely differentiable at all points of its domain. Finally, before returning to our discussion of Bessel's equation

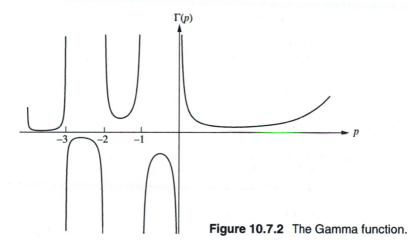

Figure 10.7.2 The Gamma function.

we require the following formula

$$\Gamma(p + 1)\left[(p + 1)(p + 2) \cdots (p + k)\right] = \Gamma(p + k + 1). \tag{10.7.16}$$

The proof of this follows by repeated application of (10.7.15), in the form

$$p\Gamma(p) = \Gamma(p + 1)$$

to the left-hand side of the above equality and is left as an exercise.

Now let us return to the solution (10.7.10) of Bessel's equation. Once more we make a specific choice for a_0. We set

$$a_0 = \frac{1}{2^p \Gamma(p + 1)}.$$

Substituting this value for a_0 into (10.7.10) and using (10.7.16) yields the **Bessel function of the first kind of order p, $J_p(x)$,** defined by

$$J_p(x) = \sum_{k=0}^{\infty} \frac{(-1)^k}{\Gamma(k + 1)\Gamma(p + k + 1)} \left(\frac{x}{2}\right)^{2k+p}. \tag{10.7.17}$$

Notice that this does reduce to $J_N(x)$ when N is a nonnegative integer.

BESSEL FUNCTIONS OF THE SECOND KIND

Now consider determining the general solution of Bessel's equation. For all $p \geq 0$ we have shown above that one solution to Bessel's equation on $(0, \infty)$ is given in (10.7.17). We therefore require a second LI solution. Since the roots of the indicial equation are $r = \pm p$, it follows from our general Frobenius theory that provided $2p$ is not equal to an integer, there will exist a second LI Frobenius series solution corresponding to the root $r = -p$. It is not too difficult to see that this solution can be obtained by replacing p by $-p$ in (10.7.17). Thus we obtain

$$J_{-p}(x) = \sum_{k=0}^{\infty} \frac{(-1)^k}{\Gamma(k + 1)\Gamma(k - p + 1)} \left(\frac{x}{2}\right)^{2k-p}. \tag{10.7.18}$$

Consequently, the general solution to Bessel's equation of order p when $2p$ is not equal to an integer is

$$y(x) = c_1 J_p(x) + c_2 J_{-p}(x). \tag{10.7.19}$$

When $2p$ is equal to an integer, two subcases arise depending on whether p is itself an integer or a half-integer (that is, $1/2$, $3/2$, ...). In the latter case a straightforward analysis of the recurrence relation (10.7.6) with $r = -p$, $p = (2j + 1)/2$, (j a nonnegative integer) shows that a second Frobenius series also exists in this case, and it is, in fact, given by (10.7.18). Thus, (10.7.19) represents the general solution of Bessel's equation provided p is not equal to an integer. In practice, rather than using J_{-p} as the second LI solution of Bessel's equation it is usual to use the following linear combination of these two solutions

$$Y_p(x) = \frac{J_p(x) \cos p\pi - J_{-p}(x)}{\sin p\pi}. \tag{10.7.20}$$

The function defined in (10.7.20) is called the **Bessel function of the second kind of order p**. Using Y_p we can therefore write the general solution of Bessel's equation, when p is not equal to a positive integer, in the form

$$y(x) = c_1 J_p(x) + c_2 Y_p(x). \tag{10.7.21}$$

The determination of a second LI solution to Bessel's equation when p is a positive integer, n, is quite a bit more complicated. From our general Frobenius theory we certainly know the form of the second LI solution, namely,

$$y_2(x) = A J_n(x) \ln x + x^{-n} \sum_{k=0}^{\infty} b_k x^k$$

and the coefficients A, b_n could be determined by direct substitution into (10.7.1). However, if we extend the definition of the Bessel function of the second kind to the case of positive integers by

$$Y_n(x) = \lim_{p \to n} \left[\frac{J_p(x)\cos p\pi - J_{-p}(x)}{\sin p\pi} \right] \qquad (10.7.22)$$

it can be shown that the above limit exists and that the resulting function is indeed a second LI solution of Bessel's equation. We could derive the series representation of (10.7.22) by evaluating the limit explicitly. The calculations are lengthy and tedious and so we omit them. The result of these calculations is

$$Y_n(x) = \frac{2}{\pi} J_n(x)\left[\ln\left(\frac{x}{2}\right) + \gamma\right] - \frac{1}{\pi}\left[\sum_{k=0}^{n-1} \frac{(n-k-1)!}{k!}\left(\frac{x}{2}\right)^{2k-n}\right.$$

$$\left. + \frac{s_n}{n!}\left(\frac{x}{2}\right)^n + \sum_{k=1}^{\infty} (-1)^k \frac{(s_k + s_{n+k})}{k!\,(n+k)!}\left(\frac{x}{2}\right)^{2k+n}\right]$$

where

$$s_k = 1 + \frac{1}{2} + \frac{1}{3} + \cdots + \frac{1}{k}$$

and

$$\gamma = \lim_{k \to \infty} (s_k - \ln k) \cong 0.577215664....$$

γ is called *Euler's constant*.

Thus (10.7.21) represents the general solution to Bessel's equation of arbitrary order p.

PROPERTIES OF BESSEL FUNCTIONS OF THE FIRST KIND

In practice we are usually interested in solutions of Bessel's equation that are bounded at $x = 0$. However, the Bessel functions of the second kind are always unbounded at $x = 0$. (When $p \neq$ integer, this follows from the fact that x^{-p} has a negative exponent, and when $p =$ integer, n, it is due to the second term in the expansion of Y_n given previously.) Thus we usually only require the Bessel functions of the first kind. In this section we list various properties of the Bessel functions of the first kind that help in either tabulating values of the functions or in working with the functions themselves.

We first recall from (10.7.17) the definition of the Bessel functions of the first kind of order p, namely

$$J_p(x) = \sum_{k=0}^{\infty} \frac{(-1)^k}{\Gamma(k+1)\Gamma(p+k+1)}\left(\frac{x}{2}\right)^{2k+p}. \qquad (10.7.23)$$

Property 1:
$$\frac{d}{dx}\left[x^p J_p(x)\right] = x^p J_{p-1}(x). \qquad (10.7.24)$$

Property 2:
$$\frac{d}{dx}\left[x^{-p} J_p(x)\right] = -x^{-p} J_{p+1}(x). \qquad (10.7.25)$$

PROOF OF PROPERTY 1   Multiplying (10.7.23) by x^p and differentiating the result with respect to x yields

$$\frac{d}{dx}[x^p J_p(x)] = \sum_{k=0}^{\infty} \frac{(-1)^k}{\Gamma(k+1)\Gamma(p+k+1)}(2k+2p)2^{p-1}\left(\frac{x}{2}\right)^{2k+2p-1}$$

$$= x^p \sum_{k=0}^{\infty} \frac{(-1)^k}{\Gamma(k+1)\Gamma(p+k+1)} (k+p) 2^{p-1} \left(\frac{x}{2}\right)^{2k+p-1}.$$

But, from (10.7.15), $\Gamma(p+k+1) = (p+k)\Gamma(p+k)$, so that,

$$\frac{d}{dx}[x^p J_p(x)] = x^p \sum_{k=0}^{\infty} \frac{(-1)^k}{\Gamma(k+1)\Gamma(p+k)} \left(\frac{x}{2}\right)^{2k+p-1}$$

$$= x^p J_{p-1}(x).$$

Property 2 is proved similarly.   ∎

We now derive two identities satisfied by the derivatives of J_p. Expanding the der–ivatives on the right-hand sides of (10.7.24) and (10.7.25) and dividing the resulting equations by x^p and x^{-p} respectively yields

$$J_p'(x) + x^{-1} p J_p(x) = J_{p-1}(x) \tag{10.7.26}$$

$$J_p'(x) - x^{-1} p J_p(x) = -J_{p+1}(x). \tag{10.7.27}$$

Subtracting (10.7.27) from (10.7.26) and rearranging terms we obtain:

Property 3: $J_{p+1}(x) = 2x^{-1} p\, J_p(x) - J_{p-1}(x). \tag{10.7.28}$

Similarly, adding (10.7.26) and (10.7.27) and rearranging yields:

Property 4: $J_p'(x) = \frac{1}{2}[J_{p-1}(x) - J_{p+1}(x)]. \tag{10.7.29}$

These formulas allow us to express high order Bessel functions and their derivatives in terms of lower order functions. For example, all integer order Bessel functions can be expressed in terms of $J_0(x)$ and $J_1(x)$.

Example 10.7.1   Express $J_2(x)$ and $J_3(x)$ in terms of $J_0(x)$ and $J_1(x)$.

Solution   Applying (10.7.28) with $p = 1$, we obtain:

$$J_2(x) = 2x^{-1} J_1(x) - J_0(x).$$

Similarly, when $p = 2$,

$$J_3(x) = 4x^{-1} J_2(x) - J_1(x) = x^{-2}(8 - x^2) J_1(x) - 4x^{-1} J_0(x).$$

A BESSEL FUNCTION EXPANSION THEOREM

It can be shown that every Bessel function of the first kind has an infinite number of positive zeros. We now show how this can be used to obtain a Bessel function expansion of an arbitrary function.

Let $\lambda_1, \lambda_2, \ldots$ denote the positive zeros of the Bessel function $J_p(x)$, where $p \geq 0$ is fixed, and consider the corresponding functions

$$u_m(x) = J_p(\lambda_m x), \quad u_n(x) = J_p(\lambda_n x)$$

for x in the interval $[0, 1]$. We begin by deriving an orthogonality relation for the functions u_m and u_n. Since $J_p(x)$ solves Bessel's equation

$$y'' + \frac{1}{x}y' + \left(1 - \frac{p^2}{x^2}\right)y = 0$$

it is not too difficult to show (see problem 15) that u_1 and u_2 satisfy

$$u_m'' + \frac{1}{x}u_m' + \left(\lambda_m{}^2 - \frac{p^2}{x^2}\right)u_m = 0 \tag{10.7.30}$$

$$u_n'' + \frac{1}{x}u_n' + \left(\lambda_n{}^2 - \frac{p^2}{x^2}\right)u_n = 0. \tag{10.7.31}$$

Multiplying (10.7.30) by u_n, (10.7.31) by u_m and subtracting yields

$$(u_m''u_n - u_n''u_m) + \frac{1}{x}(u_m'u_n - u_n'u_m) + (\lambda_m{}^2 - \lambda_n{}^2)u_mu_n = 0$$

which can be written as

$$(u_m'u_n - u_n'u_m)' + \frac{1}{x}(u_m'u_n - u_n'u_m) + (\lambda_m{}^2 - \lambda_n{}^2)u_mu_n = 0.$$

Multiplying this equation by x and combining the first two terms of the resulting equation we obtain

$$\frac{d}{dx}[x(u_m'u_n - u_n'u_m)] + x(\lambda_m{}^2 - \lambda_n{}^2)u_mu_n = 0.$$

Integrating from 0 to 1 and using the fact that $u_m(1) = u_n(1) = 0$ (since λ_m and λ_n are zeros of $J_p(x)$) yields

$$(\lambda_m{}^2 - \lambda_n{}^2) \int_0^1 x u_m u_n \, dx = 0$$

that is, since λ_m and λ_n are distinct and positive,

$$\int_0^1 x u_m u_n \, dx = 0.$$

Substituting for u_m and u_n we finally obtain:

$$\int_0^1 x J_p(\lambda_m x) J_p(\lambda_n x) \, dx = 0, \quad \text{whenever } m \neq n. \tag{10.7.32}$$

In this case we say that the set of functions $\{J_p(\lambda_n x)\}_{n=1}^{\infty}$ is *orthogonal on (0, 1) relative to the weight function* $w(x) = x$. It can further be shown (see problem 16) that when $m = n$,

$$\int_0^1 x [J_p(\lambda_n x)]^2 dx = \frac{1}{2}[J_{p+1}(\lambda_n)]^2. \tag{10.7.33}$$

We can now state a Bessel function expansion theorem.

Theorem 10.7.1:   If f and f' are continuous on the interval $[0, 1]$ then for $0 < x < 1$,

$$f(x) = a_1 J_p(\lambda_1 x) + a_2 J_p(\lambda_2 x) + \cdots + a_n J_p(\lambda_n x) + \cdots = \sum_{n=1}^{\infty} a_n J_p(\lambda_n x), \qquad (10.7.34)$$

where the coefficients can be determined from

$$a_n = \frac{2}{[J_{p+1}(\lambda_n)]^2} \int_0^1 x f(x) J_p(\lambda_n x) \, dx. \qquad (10.7.35)$$

PROOF   We show only that if a series of the form (10.7.34) exists then the coefficients are given by (10.7.35). The proof of convergence is omitted. Multiplying (10.7.34) by $x J_p(\lambda_m x)$ and integrating the resulting equation with respect to x from 0 to 1 we obtain

$$\int_0^1 x f(x) J_p(\lambda_m x) \, dx = a_m \int_0^1 x [J_p(\lambda_m x)]^2 \, dx$$

where we have used (10.7.32). Substituting from (10.7.33) for the integral on the left-hand side of the preceding equation and rearranging terms yields

$$a_m = \frac{2}{[J_{p+1}(\lambda_m)]^2} \int_0^1 x f(x) J_p(\lambda_m x) \, dx$$

which is what we wished to show.   ∎

REMARK   An expansion of the form (10.7.34) is called a **Fourier–Bessel expansion of f**.

Example 10.7.2   Determine the Fourier–Bessel expansion of $f(x) = 1$ in terms of the functions $J_0(\lambda_n x)$.

Solution   According to Theorem 10.7.1, for $0 < x < 1$ we can write

$$1 = \sum_{n=1}^{\infty} a_n J_0(\lambda_n x),$$

where

$$a_n = \frac{2}{[J_1(\lambda_n)]^2} \int_0^1 x J_0(\lambda_n x) \, dx. \qquad (10.7.36)$$

But,

$$\int_0^1 x J_0(\lambda_n x) \, dx = \frac{1}{\lambda_n^2} \int_0^{\lambda_n} u J_0(u) \, du,$$

where $u = \lambda_n x$. Applying (the integrated form of) (10.7.24) with $p = 1$ yields

$$\int_0^1 x J_0(\lambda_n x) \, dx = \frac{1}{\lambda_n^2} [u J_1(u)]_0^{\lambda_n} = \frac{1}{\lambda_n} J_1(\lambda_n).$$

Substitution into (10.7.36) gives

$$a_n = \frac{2}{\lambda_n J_1(\lambda_n)},$$

so that the appropriate Fourier–Bessel expansion is

$$1 = 2 \sum_{n=1}^{\infty} \frac{1}{\lambda_n J_1(\lambda_n)} J_0(\lambda_n x), \quad 0 < x < 1.$$

EXERCISES 10.7

1. Use the relations (10.7.5.), (10.7.6) to show that if p is a half-integer then Bessel's equation of order p has two LI Frobenius series solutions.

2. Find two LI solutions to

$$x^2 y'' + xy' + (x^2 - \frac{9}{4}) y = 0$$

on the interval $(0, \infty)$.

3. Let $\Gamma(p)$ denote the gamma function. Show that

$$\Gamma(p + 1) [(p + 1)(p + 2) \cdots (p + k)] = \Gamma(p + k + 1).$$

4. (a) By making the change of variables $t = x^2$ in the integral that defines the gamma function show that

$$\Gamma(1/2) = 2 \int_0^{\infty} e^{-x^2} \, dx.$$

(b) Use your result from (a) to show that

$$[\Gamma(1/2)]^2 = 4 \int_0^{\infty} \int_0^{\infty} e^{-(x^2+y^2)} \, dx \, dy.$$

(c) By changing to polar coordinates evaluate the double integral in (b) and hence show that

$$\Gamma(1/2) = \sqrt{\pi} \, .$$

(d) Use your result from (c) to find $\Gamma(3/2)$ and $\Gamma(-1/2)$.

5. Let $J_p(x)$ denote the Bessel function of the first kind of order p. Show that

$$\frac{d}{dx} [x^{-p} J_p(x)] = - x^{-p} J_{p+1}(x).$$

6. By manipulating the general expression for $J_{1/2}(x)$, show that it can be written in closed form as:

$$J_{1/2}(x) = \sqrt{\frac{2}{\pi x}} \sin x.$$

7. Given that

$$J_{1/2}(x) = \sqrt{\frac{2}{\pi x}} \sin x, \quad J_{-1/2}(x) = \sqrt{\frac{2}{\pi x}} \cos x,$$

express $J_{3/2}(x)$ and $J_{-3/2}(x)$ in closed form. Convince yourself that all half-integer order Bessel functions of the first kind can be expressed as a finite sum of terms involving products of $\sin x$, $\cos x$ and powers of x.

8. By integrating the recurrence relation for derivatives of the Bessel functions of the first kind show that

(a) $\displaystyle\int x^p J_{p-1}(x) \, dx = x^p J_p(x) + C.$

(b) $\displaystyle\int x^{-p} J_{p+1}(x) \, dx = - x^{-p} J_p(x) + C.$

9. Show that

(a) $J_0''(x) = -J_0(x) - x^{-1} J_0'(x).$

(b) $J_0'''(x) = x^{-1} J_0(x) + x^{-2}(2 - x^2) J_0'(x).$

10. Show that

$$J_4(x) = 8x^{-3}(6 - x^2) J_1(x) - x^{-2}(24 - x^2) J_0(x).$$

11. Show that

$$J_2'(x) = 2x^{-1} J_0(x) + x^{-2}(4 - x^2) J_0'(x).$$

12. Show that

(a) $J_2(x) = J_0(x) + 2 J_0''(x).$

(b) $J_3(x) = 3 J_1(x) + 4 J_1''(x).$

13. Determine the Fourier–Bessel expansion in the functions $J_p(\lambda_n x)$ for $f(x) = x^p$, on the interval $(0, 1)$. [Here λ_n denote the positive zeros of $J_p(x)$. Hint: You will need to use one of the results from problem 8.]

14. Determine the Fourier–Bessel expansion in the functions $J_0(\lambda_k x)$ of $f(x) = x^2$ on the interval $(0, 1)$. [Here λ_k denote the positive zeros of $J_0(x)$.]

15. Let $J_p(x)$ denote the Bessel function of the first kind of order p, and let λ be a positive real number. If $u(x) = J_p(\lambda x)$, show that u satisfies the DE

$$\frac{d^2u}{dx^2} + \frac{1}{x}\frac{du}{dx} + (\lambda^2 - \frac{p^2}{x^2})\, u = 0.$$

16. Let λ and μ be positive real numbers. Then, $J_p(\lambda x)$ and $J_p(\mu x)$ satisfy

$$\frac{d}{dx}\left\{x\frac{d}{dx}[J_p(\lambda x)]\right\} + (\lambda^2 x - \frac{p^2}{x})\, J_p(\lambda x) = 0, \quad (16.1)$$

$$\frac{d}{dx}\left\{x\frac{d}{dx}[J_p(\mu x)]\right\} + (\mu^2 x - \frac{p^2}{x})\, J_p(\mu x) = 0, \quad (16.2)$$

respectively.

(a) Show that for $\lambda \neq \mu$,

$$\int_0^1 x\, J_p(\lambda x)\, J_p(\mu x)\, dx$$

$$= \frac{\mu J_p(\lambda)\, J_p{}'(\mu) - \lambda J_p(\mu)\, J_p{}'(\lambda)}{\lambda^2 - \mu^2}. \quad (16.3)$$

[Hint: Multiply (16.1) by $J_p(\mu x)$, (16.2) by $J_p(\lambda x)$, subtract the resulting equations and integrate over $(0, 1)$.] If λ and μ are distinct zeros of $J_p(x)$, what does your result imply?

(b) In order to compute $\int_0^1 x\, [J_p(\mu x)]^2\, dx$, we take the limit as $\lambda \to \mu$ in (16.3). Use L'Hopital's rule to compute this limit and thereby show that

$$\int_0^1 x\, [J_p(\mu x)]^2\, dx$$

$$= \frac{\mu\, [J_p{}'(\mu)]^2 - J_p(\mu)\, J_p{}'(\mu) - \mu J_p(\mu)\, J_p{}''(\mu)}{2\mu}.$$

$$(16.4)$$

Substituting from Bessel's equation for $J_p{}''(\mu)$ show that (16.4) can be written as

$$\int_0^1 x\, [J_p(\mu x)]^2\, dx$$

$$= \frac{1}{2}\left\{\, [J_p{}'(\mu)]^2 + (1 - \frac{p^2}{\mu^2})\, [J_p(\mu)]^2 \,\right\}.$$

(c) In the case when μ is a zero of $J_p(x)$, use (10.7.27) to show that your result in (b) can be written as

$$\int_0^1 x\, [J_p(\mu x)]^2\, dx = \frac{1}{2}\, [J_{p+1}(\mu)]^2.$$

A Review
of Complex Numbers

Any number z of the form $z = a + ib$, where a and b are real numbers and $i = \sqrt{-1}$, is called a **complex number**. If $z = a + ib$, then we refer to a as the **real part** of z, denoted Re(z), and we refer to b as the **imaginary part** of z, denoted Im(z). Thus,

> If $z = a + ib$, then Re(z) = a and Im(z) = b.

Example A1.1 If $z = 2 - 3i$, then Re(z) = 2, Im(z) = –3.

Complex numbers can be added, subtracted, and multiplied in the usual manner, and the result is once more a complex number. Further, these operations satisfy all of the basic properties satisfied by the real numbers. All that we need to remember is that whenever we encounter the term i^2, it must be replaced by –1.

Example A1.2 If $z_1 = 3 + 4i$, $z_2 = -1 + 2i$, find $z_1 - 3z_2$, $z_1 z_2$.

Solution

$$z_1 - 3z_2 = (3 + 4i) - 3(-1 + 2i) = 6 - 2i = 2(3 - i).$$
$$z_1 z_2 = (3 + 4i)(-1 + 2i) = -3 + 6i - 4i + 8i^2 = -11 + 2i.$$

Example A1.3 If $z_1 = 4 + 3i$ and $z_2 = 4 - 3i$, determine $z_1 z_2$.

Solution In this case, $z_1 z_2 = (4 + 3i)(4 - 3i) = 16 - 12i + 12i - 9i^2 = 16 + 9 = 25.$ ❑

Notice that in the previous example the product z_1z_2 turned out to be a real number. This was not an accident. If we look at the definition of z_2, we see that it can be obtained from z_1 by replacing the imaginary part of z_1 by its negative. Complex numbers that are related in this manner are called conjugates of one another.

Definition A1.1:   If $z = a + ib$, then the complex number $\bar{z}$ defined by

$$\bar{z} = a - ib$$

is called the **conjugate** of z.

Example A1.4 If $z = 2 + 5i$, then $\bar{z} = 2 - 5i$, whereas if $z = 3 - 4i$, then $\bar{z} = 3 + 4i$.

Properties of the Conjugate

1. $\bar{\bar{z}} = z$.

2. $z\bar{z} = \bar{z}z = a^2 + b^2$.

PROOF

1. If $z = a + ib$, then $\bar{z} = a - ib$, so that $\bar{\bar{z}} = a + ib = z$.

2. $z\bar{z} = (a + ib)(a - ib) = a^2 - iab + iab - (ib)^2 = a^2 + b^2$.   ∎

If $z = a + ib$, then the real number $\sqrt{a^2 + b^2}$ is often called the **modulus** of z or the **absolute value of z** and is denoted $|z|$. It follows from property 2 that

$$|z|^2 = z\bar{z}.$$

Example A1.5   Determine $|z|$ if $z = 2 - 3i$.

Solution   By definition,

$$|z| = \sqrt{(2)^2 + (3)^2} = \sqrt{13}.$$ ◻

We now recall from elementary algebra that an expression of the form $\dfrac{1}{a + \sqrt{b}}$ can always be written with the radical in the numerator. To accomplish this, we multiply by

$$\frac{a - \sqrt{b}}{a - \sqrt{b}},$$

and the result is

$$\frac{a - \sqrt{b}}{a^2 - b^2}.$$

The reason that this works is because

$$(a + \sqrt{b})(a - \sqrt{b}) = a^2 - b^2.$$

This is similar to $(a + ib)(a - ib) = a^2 + b^2$. Now consider an expression of the form

$$\frac{1}{a + ib}.$$

As this is written, we cannot say that it is a complex number, since it is not of the form $a + ib$. However, if we multiply by

$$\frac{a - ib}{a - ib},$$

and use property 2 of the conjugate, we obtain

$$\frac{1}{a + ib} = \frac{1}{(a + ib)} \frac{(a - ib)}{(a - ib)} = \frac{a - ib}{a^2 + b^2},$$

which is a complex number.

Example A1.6   Express $z = \dfrac{1}{2 + 5i}$ in the form $a + ib$.

Solution

$$z = \frac{1}{2 + 5i} = \frac{1}{(2 + 5i)} \frac{(2 - 5i)}{(2 - 5i)} = \frac{2}{29} - \frac{5}{29} i. \qquad \square$$

More generally, if $z_1 = a + ib$ and $z_2 = x + iy$, then

$$\frac{z_1}{z_2} = \frac{a + ib}{x + iy} = \frac{(a + ib)\ (x - iy)}{(x + iy)\ (x - iy)}$$

$$= \frac{1}{x^2 + y^2} [(ax + by) + i\ (ay + bx)].$$

This illustrates that we can divide two complex numbers, and the result is once more a complex number.

Example A1.7   If $z_1 = 2 + 3i$ and $z_2 = 3 + 4i$, determine $\dfrac{z_1}{z_2}$.

Solution   In this case, we have

$$\frac{z_1}{z_2} = \frac{2 + 3i}{3 + 4i} = \frac{(2 + 3i)\ (3 - 4i)}{(3 + 4i)\ (3 - 4i)} = \frac{1}{25}(18\ +\ i).$$

COMPLEX-VALUED FUNCTIONS

A function $w(x)$ of the form

$$w(x) = u(x) + iv(x),$$

where u and v are real-valued functions of a real variable x (and $i^2 = -1$), is called a **complex-valued function** of a real variable. An example of such a function is

$$w(x) = 3 \cos 2x\ + 4i \sin 3x.$$

THE COMPLEX EXPONENTIAL FUNCTION

Recall that for all real x, the function e^x has the Maclaurin expansion

Appendix 1: A Review of Complex Numbers

$$e^x = \sum_{n=0}^{\infty} \frac{1}{n!} x^n.$$

It is also possible to discuss convergence of infinite series of complex numbers. We define e^{ib}, where b is a real number by

$$e^{ib} = \sum_{n=0}^{\infty} \frac{1}{n!} (ib)^n = 1 + ib + \frac{1}{2!} (ib)^2 + \frac{1}{3!} (ib)^3 + \cdots + \frac{1}{n!} (ib)^n + \cdots .$$

Factoring the even and odd powers of b and using the formulas

$$i^{2k} = (-1)^k, \quad i^{2k+1} = (-1)^k i$$

yields

$$e^{ib} = \left[1 - \frac{1}{2!} b^2 + \frac{1}{4!} b^4 + \cdots + \frac{(-1)^k}{(2k)!} b^{2k} + \cdots \right]$$

$$+ i \left[b - \frac{1}{3!} b^3 + \frac{1}{5!} b^5 + \cdots + \frac{(-1)^k}{(2k+1)!} b^{2k+1} + \cdots \right].$$

That is,

$$e^{ib} = \sum_{n=0}^{\infty} (-1)^n \frac{b^{2n}}{(2n)!} + i \sum_{n=0}^{\infty} (-1)^n \frac{b^{2n+1}}{(2n+1)!}.$$

The two series appearing in the foregoing equation are, respectively, the Maclaurin series expansions of $\cos b$ and $\sin b$, both of which converge for all real b. Thus, we have shown that

$$e^{ib} = \cos b + i \sin b, \tag{A1.1}$$

which is called **Euler's formula**. It is now natural to *define* e^{a+ib} by

$$e^{a+ib} = e^a \cdot e^{ib} = e^a(\cos b + i \sin b), \tag{A1.2}$$

where a and b are any real numbers.

A function of the form $f(x) = e^{rx}$ where $r = a + ib$ and x is a real variable, is called a **complex exponential function**. Replacing ib with ibx in (A1.1) and $a + ib$ with $(a + ib)x$ in (A1.2) yields the following important formulas:

$$e^{ibx} = \cos bx + i \sin bx, \quad e^{(a+ib)x} = e^{ax}(\cos bx + i \sin bx).$$

By replacing i with $-i$ in the foregoing formulas, we obtain

$$e^{-ibx} = \cos bx - i \sin bx, \quad e^{(a-ib)x} = e^{ax}(\cos bx - i \sin bx).$$

Example A1.8 Express $e^{(3-5i)x}$ in terms of trigonometric functions.

Solution

$$e^{(3-5i)x} = e^{3x}(\cos 5x - i \sin 5x). \qquad \square$$

The preceding definition of $e^{(a+ib)x}$ also enables us to attach a meaning to $x^{(a+ib)}$. We recall that for nonrational r and positive x, x^r is defined by

$$x^r = e^{r \ln x}.$$

We now extend this definition to the case when r is complex and therefore define

$$x^{a+ib} = e^{(a+ib)\ln x}.$$

Using Euler's formula, this can be written as

$$x^{a+ib} = x^a e^{ib \ln x} = x^a[\cos(b \ln x) + i \sin(b \ln x)].$$

For example,

$$x^{2+3i} = x^2[\cos(3 \ln x) + i \sin(3 \ln x)].$$

DIFFERENTIATION OF COMPLEX-VALUED FUNCTIONS

We now return to the general complex-valued function $w(x) = u(x) + iv(x)$. If $u'(x)$ and $v'(x)$ exist, then we define the derivative of w by

$$w'(x) = u'(x) + iv'(x).$$

Higher order derivatives are defined similarly. In particular, we have the following important result.

$$\frac{d}{dx}(e^{rx}) = re^{rx} \text{ when } r \text{ is complex.}$$

This coincides with the usual formula for the derivative of e^{rx} when r is a real number. To establish the above formula, we proceed as follows. If $r = a + ib$, then

$$e^{rx} = e^{(a+ib)x} = e^{ax}(\cos bx + i \sin bx).$$

Differentiating with respect to x using the product rule yields

$$\frac{d}{dx}(e^{rx}) = ae^{ax}(\cos bx + i \sin bx) + be^{ax}(-\sin bx + i\cos bx)$$
$$= ae^{ax}(\cos bx + i \sin bx) + ibe^{ax}(\cos bx + i \sin bx)$$
$$= (a + ib)e^{ax}(\cos bx + i \sin bx) = re^{rx},$$

as required.
Similarly, it can be shown that

$$\frac{d}{dx}(x^r) = rx^{r-1} \text{ when } r \text{ is complex.}$$

EXERCISES A1

For problems 1–5, determine $\bar{z}$ and $|z|$ for the given complex number.

1. $z = 2 + 5i$.
2. $z = 3 - 4i$.
3. $z = 5 - 2i$.
4. $z = 7 + i$.
5. $z = 1 + 2i$.

For problems 6–10, express $z_1 z_2$ and z_1/z_2 in the form $a + ib$.

6. $z_1 = 1 + i, z_2 = 3 + 2i$.
7. $z_1 = -1 + 3i, z_2 = 2 - i$.
8. $z_1 = 2 + 3i, z_2 = 1 - i$.
9. $z_1 = 4 - i, z_2 = 1 + 3i$.
10. $z_1 = 1 - 2i, z_2 = 3 + 4i$.

11. Show that if z_1 and z_2 are complex numbers, then

$$\overline{z_1 + z_2} = \overline{z}_1 + \overline{z}_2.$$

12. Generalize the previous example to the case when $z_1, z_2, ..., z_n$ are all complex numbers.

13. Show that if z_1 and z_2 are complex numbers then

$$\overline{z_1 z_2} = \overline{z}_1 \overline{z}_2.$$

14. Show that if z_1 and z_2 are complex numbers then

$$\overline{(z_1/z_2)} = \overline{z}_1/\overline{z}_2.$$

For problems 15–22, express the given complex-valued function in the form $u(x) + iv(x)$ for appropriate real-valued functions u and v.

15. e^{2ix}.

16. $e^{(3+4i)x}$.

17. e^{-5ix}.

18. $e^{-(2+i)x}$.

19. x^{2-i}.

20. x^{3i}.

21. x^{-1+2i}.

22. $x^{2i}e^{(3+4i)x}$.

23. Derive the famous mathematical formula

$$e^{i\pi} + 1 = 0.$$

24. Show that

$$\cos bx = \frac{1}{2}(e^{ibx} + e^{-ibx})$$

and

$$\sin bx = \frac{1}{2i}(e^{ibx} - e^{-ibx}).$$

(A comparison of these formulas with the corresponding formulas

$$\cosh bx = \frac{1}{2}(e^{bx} + e^{-bx}),$$

$$\sinh bx = \frac{1}{2}(e^{bx} - e^{-bx}),$$

indicates why the trigonometric and hyperbolic functions satisfy similar identities.)

For problems 25–27, use the result of problem 24 to express the given functions in terms of complex exponential functions.

25. $\sin 4x$.

26. $\cos 8x$.

27. $\tan x$.

28. Use the result of problem 24 to verify the identity $\sin^2 x + \cos^2 x = 1$.

A Review
of Partial Fractions

In this appendix we review the partial fraction decomposition of rational functions. No proofs are given since the reader is assumed to have seen the results in a previous calculus course.

Recall that a function of the form

$$p(x) = a_n x^n + a_{n-1} x^{n-1} + \cdots + a_1 x + a_0 \qquad \text{(A2.1)}$$

with $a_n \neq 0$ is called a *polynomial of degree n*. According to the fundamental theorem of algebra, the equation $p(x) = 0$ has precisely n roots (not all necessarily distinct). If we let $x_1, x_2, \ldots x_n$ denote these roots then $p(x)$ can be factored as

$$p(x) = K(x - x_1)(x - x_2) \cdots (x - x_n) \qquad \text{(A2.2)}$$

where K is a constant. Some of the roots may be complex. We will assume that the coefficients in (A2.1) are real numbers in which case any complex roots must occur in conjugate pairs.

A quadratic factor of the form

$$ax^2 + bx + c$$

which has no *real* linear factors is said to be **irreducible**.

Theorem A2.1   Any real polynomial[1] can be factored into linear and irreducible quadratic terms with real coefficients.

PROOF   Let $p(x)$ be a real polynomial and suppose that $x = \alpha$ is a complex root of

---

[1]By a "real polynomial" we mean a polynomial with real coefficients.

$p(x) = 0$. Then $x = \bar{\alpha}$ is also a root. Thus (A2.2) will contain the terms $(x - \alpha)(x - \bar{\alpha})$. These linear terms have complex coefficients. However, if we expand the product the result is

$$(x - \alpha)(x - \bar{\alpha}) = x^2 - (\alpha + \bar{\alpha})x + \alpha\bar{\alpha}.$$

But,

$$\alpha + \bar{\alpha} = 2\,\mathrm{Re}(\alpha), \quad \alpha\bar{\alpha} = |\,\alpha\,|^2$$

which are both real, so that the irreducible quadratic term does indeed have real coefficients.   ∎

If $p(x)$ and $q(x)$ are two polynomials (not necessarily of the same degree) then a function of the form

$$R(x) = \frac{p(x)}{q(x)}$$

is called a rational function. Suppose that $q(x)$ has been factored into linear and irreducible quadratic terms. Then $q(x)$ will consist of a product of terms of the form

$$(ax - b)^k \quad \text{or} \quad (ax^2 + bx + c)^k \tag{A2.3}$$

where a, b, c and k are constants. For example,

$$\frac{x^2 - 1}{(x + 2)(x^2 + 3)}.$$

The idea behind a partial fraction decomposition is to express a rational function as a sum of terms whose denominators are of the form (A2.3). The following rules tell us the form that such a decomposition must take.

1. Each factor of the form $(ax - b)^k$ in $q(x)$ contributes the following terms to the partial fraction decomposition of $p(x)/q(x)$

$$\frac{A_1}{(ax - b)} + \frac{A_2}{(ax - b)^2} + \cdots + \frac{A_k}{(ax - b)^k}$$

where $A_1, A_2, \ldots A_k$ are constants.

2. Each irreducible quadratic factor of the form $(ax^2 + bx + c)^k$ contributes the following terms to the partial fraction decomposition of $p(x)/q(x)$

$$\frac{A_1 x + B_1}{ax^2 + bx + c} + \frac{A_2 x + B_2}{(ax^2 + bx + c)^2} + \cdots + \frac{A_k x + B_k}{(ax^2 + bx + c)^k}.$$

Thus, for example,

$$\frac{x^2 + 1}{x(x - 1)(x^2 + 4)} = \frac{A}{x} + \frac{B}{x - 1} + \frac{Cx + D}{x^2 + 4}$$

for appropriate values of the constants A, B, C, D. Similarly,

$$\frac{x - 2}{(x + 2)^2(x^2 + 2x + 2)} = \frac{A}{x + 2} + \frac{B}{(x + 2)^2} + \frac{Cx + D}{(x^2 + 2x + 2)}$$

for appropriate A, B, C, D.

The preceding rules only give the form of a partial fraction decomposition. The next question that needs answering is the following: How do we determine the constants that arise in the partial fraction decomposition? A standard way to proceed is as follows:

---

1. Determine the general form of the partial fraction decomposition of $p(x)/q(x)$.

2. Multiply both sides of the resulting decomposition by $q(x)$.

3. Equate the coefficients of like powers of x on both sides of the resulting equation in order to determine the constants in the partial fraction decomposition.

We illustrate the procedure with several examples.

Example A2.1   Determine the partial fraction decomposition of

$$\frac{2x}{(x-1)(x+3)}.$$

Solution   The general form of the partial fraction decomposition is

$$\frac{2x}{(x-1)(x+3)} = \frac{A}{x-1} + \frac{B}{x+3}.$$

Multiplying both sides of this equation by $(x-1)(x+3)$ yields

$$2x = A(x+3) + B(x-1).$$

We now equate coefficients of like powers of x on both sides of this equation to obtain

$$A + B = 2, \quad 3A - B = 0.$$

Consequently

$$A = \frac{1}{2}, \quad B = \frac{3}{2}$$

so that

$$\frac{2x}{(x-1)(x+3)} = \frac{1}{2(x-1)} + \frac{3}{2(x+3)}.$$

Example A2.2   Determine the partial fraction decomposition of

$$\frac{x^2 + 1}{(x+1)(x^2+4)}.$$

Solution   In this case the general form of the partial fraction decomposition is

$$\frac{x^2 + 1}{(x+1)(x^2+4)} = \frac{A}{x+1} + \frac{Bx + C}{x^2 + 4}.$$

Multiplying both sides by $(x+1)(x^2+4)$ we obtain:

$$x^2 + 1 = A(x^2 + 4) + (Bx + C)(x + 1).$$

Equating coefficients of like powers of x on both sides of this equality yields

$$A + B = 1, \quad B + C = 0, \quad 4A + C = 1.$$

Solving this system of equations we obtain:

$$A = \frac{2}{5}, \quad B = \frac{3}{5}, \quad C = -\frac{3}{5}$$

so that

$$\frac{x^2 + 1}{(x + 1)(x^2 + 4)} = \frac{2}{5(x + 1)} + \frac{3(x - 1)}{5(x^2 + 4)}.$$

Example A2.3   Determine the partial fraction decomposition of

$$\frac{2x - 1}{(x + 2)^2(x^2 + 2x + 2)}.$$

Solution   The term $x^2 + 2x + 2$ is irreducible.   Thus, the partial fraction decomposition has the general form:

$$\frac{2x - 1}{(x + 2)^2(x^2 + 2x + 2)} = \frac{A}{x + 2} + \frac{B}{(x + 2)^2} + \frac{Cx + D}{x^2 + 2x + 2}.$$

Clearing the fractions yields

$$2x - 1 = A(x + 2)(x^2 + 2x + 2) + B(x^2 + 2x + 2) + (Cx + D)(x + 2)^2.$$

Equating the coefficients of like powers of x we obtain

$$A + \qquad C \qquad = 0,$$
$$4A + \quad B + 4C + \quad D = 0,$$
$$6A + 2B + 4C + 4D = 2,$$
$$4A + 2B \qquad + 4D = -1.$$

Solving this system yields

$$A = -\frac{3}{2}, \quad B = -\frac{5}{2}, \quad C = \frac{3}{2}, \quad D = \frac{5}{2}.$$

Consequently

$$\frac{2x - 1}{(x + 2)^2(x^2 + 2x + 2)} = \frac{3x + 5}{2(x^2 + 2x + 2)} - \frac{3}{2(x + 2)} - \frac{5}{2(x + 2)^2}.$$

SOME SHORT CUTS

The preceding technique for determining the constants that arise in the partial fraction decomposition of a rational function will always work. However, in practice it is often tedious to apply. We now present, without justification, some short cuts that can circumvent many of the computations.

1. *Linear Factors*: The Cover-up Rule

If $q(x)$ contains a linear factor $x - a$, then this factor contributes a term of the form

$$\frac{A}{x - a}$$

to the partial fraction decomposition of $p(x)/q(x)$. Let $P(x)$ denote the expression obtained by omitting the $x - a$ term in $p(x)/q(x)$. Then the constant A is given by

$$\boxed{A = P(a).}$$

Example A2.4   Determine the partial fraction decomposition of $\dfrac{3x - 1}{(x - 3)(x + 2)}.$

Solution   The general form of the decomposition is

$$\frac{3x - 1}{(x - 3)(x + 2)} = \frac{A}{x - 3} + \frac{B}{x + 2}.$$

To determine A, we neglect the $x - 3$ term in the given rational function and set

$$P(x) = \frac{3x - 1}{x + 2}.$$

Then, according to the preceding rule,

$$A = P(3) = \frac{8}{5}.$$

Similarly, to determine B we neglect the $x + 2$ term in the given function, and set

$$P(x) = \frac{3x - 1}{x - 3}.$$

Using the cover-up rule it then follows that

$$B = P(-2) = \frac{-7}{-5} = \frac{7}{5}.$$

Consequently

$$\frac{3x - 1}{(x - 3)(x + 2)} = \frac{8}{5(x - 3)} + \frac{7}{5(x + 2)}.$$

The idea behind the technique is to cover-up the linear factor $x - a$ in the given rational function and set $x = a$ in the remaining part of the function. The result will be the constant A in the contribution $A/(x - a)$ to the partial fraction decomposition of the rational function.

2 . Repeated Linear Factors

The cover-up rule can be extended to the case of repeated linear factors also. Suppose that $q(x)$ contains a factor of the form $(x - a)^k$. Then this contributes the terms

$$\frac{A_1}{(ax - b)} + \frac{A_2}{(ax - b)^2} + \cdots + \frac{A_k}{(ax - b)^k}$$

to the partial fraction decomposition of $p(x)/q(x)$. Let $P(x)$ be the expression obtained when the $(x - a)^k$ term is neglected in $p(x)/q(x)$. Then the constants $A_1, A_2, ..., A_k$ are given by

$$A_k = P(a), \quad A_{k-1} = P'(a), \quad A_{k-2} = \frac{1}{2!} P''(a), \quad ..., \quad A_1 = \frac{1}{(k - 1)!} P^{(k-1)}(a)$$

where a ´ denotes differentiation with respect to x.

REMARKS

1 . The above formulae look rather formidable to begin with. However, they are easy to apply in practice.

2. Notice that in the case $k = 1$ we are back to the cover-up rule.

Example A2.5   Determine the partial fraction decomposition of $\dfrac{x}{(x-1)(x+2)^2}$.

Solution   The general form of the partial fraction decomposition is

$$\frac{x}{(x-1)(x+2)^2} = \frac{A_1}{x+2} + \frac{A_2}{(x+2)^2} + \frac{A_3}{x-1} .$$

To determine A_1 and A_2 we omit the term $(x+2)^2$ in the given function to obtain

$$P(x) = \frac{x}{x-1}.$$

Applying the above rule with $k = 2$ yields

$$A_2 = P(-2) = \frac{2}{3}, \quad A_1 = P'(-2) = \left. \frac{-1}{(x-1)^2} \right|_{x=-2} = -\frac{1}{9}.$$

We now use the cover-up rule to determine A_3. Neglecting the $x - 1$ term in the given function and setting $x = 1$ in the resulting expression yields:

$$A_3 = \frac{1}{9} .$$

Thus,

$$\frac{x}{(x-1)(x+2)^2} = -\frac{1}{9(x+2)} + \frac{2}{3(x+2)^2} + \frac{1}{9(x-1)} .$$

3. *Irreducible Quadratic factors of the form $x^2 + a^2$*

The final case that we will consider is when $q(x)$ contains a factor of the form $x^2 + a^2$. This will contribute a term

$$\frac{Ax + B}{x^2 + a^2}$$

to the partial fraction decomposition of $p(x)/q(x)$. Let $P(x)$ be the expression obtained by deleting the term $x^2 + a^2$ in $p(x)/q(x)$. Then the constants A and B are given by

$$\boxed{A = \frac{1}{a} \operatorname{Im}[\, P(ia)\,], \quad B = \operatorname{Re}[\, P(ia)\,].}$$

Example A2.6   Determine the partial fraction decomposition of $\dfrac{x-1}{(x+2)(x^2+4)}$.

Solution   In this case the general form of the partial fraction decomposition is

$$\frac{x-1}{(x+2)(x^2+4)} = \frac{Ax+B}{x^2+4} + \frac{C}{x+2}.$$

In order to determine A and B we delete the $x^2 + 4$ term from the given function to obtain

$$P(x) = \frac{x-1}{x+2} .$$

Since in this case $a = 2i$, we first compute

$$P(2i) = \frac{2i - 1}{2i + 2} = \frac{1}{4}(1 + 3i).$$

Thus,

$$A = \frac{1}{2}\text{Im}[\,\frac{1}{4}(1 + 3i)\,] = \frac{3}{8}, \quad B = \text{Re}[\,\frac{1}{4}(1 + 3i)\,] = \frac{1}{4}.$$

In order to determine C we use the cover-up rule. Neglecting the $x + 2$ factor in the given function and setting $x = -2$ in the result yields

$$C = -\frac{3}{8}.$$

Thus,

$$\frac{x - 1}{(x + 2)(x^2 + 4)} = \frac{3x + 2}{8(x^2 + 4)} - \frac{3}{8(x + 2)}.$$ □

REMARK   These techniques can be extended to the case of irreducible factors of the form $(ax^2 + bx + c)^k$.

EXERCISES A2

In problems 1–18 determine the partial fraction decomposition of the given rational function.

1. $\dfrac{2x - 1}{(x + 1)(x + 2)}$.

2. $\dfrac{x - 2}{(x - 1)(x + 4)}$.

3. $\dfrac{x + 1}{(x - 3)(x + 2)}$.

4. $\dfrac{x^2 - x + 4}{(x + 3)(x - 1)(x + 2)}$.

5. $\dfrac{2x - 1}{(x + 4)(x - 2)(x + 1)}$.

6. $\dfrac{3x^2 - 2x + 14}{(2x - 1)(x + 5)(x + 2)}$.

7. $\dfrac{2x + 1}{(x + 2)(x + 1)^2}$.

8. $\dfrac{5x^2 + 3}{(x + 1)(x - 1)^2}$.

9. $\dfrac{3x + 4}{x^2(x^2 + 4)}$.

10. $\dfrac{3x - 2}{(x - 5)(x^2 + 1)}$.

11. $\dfrac{x^2 + 6}{(x - 2)(x^2 + 16)}$.

12. $\dfrac{10}{(x - 1)(x^2 + 9)}$.

13. $\dfrac{7x + 2}{(x - 2)(x + 2)^2}$.

14. $\dfrac{7x^2 - 20}{(x - 2)(x^2 + 4)}$.

15. $\dfrac{7x + 4}{(x + 1)^3(x - 2)}$.

16. $\dfrac{x(2x + 3)}{(x + 1)(x^2 + 2x + 2)}$.

17. $\dfrac{3x + 4}{(x - 3)(x^2 + 4x + 5)}$.

18. $\dfrac{7 - 2x^2}{(x - 1)(x^2 + 4)}$.

APPENDIX

3

A Review
of Integration
Techniques

In this appendix we review some of the basic integration techniques that are required throughout the text. This is a very brief refresher, and should *not* be considered as a substitute for a calculus text.

1. INTEGRATION BY PARTS The basic formula for integration by parts can be written in the form

$$\int u \, dv = uv - \int v \, du.$$

To derive this, we start with the product rule for differentiation, namely,

$$\frac{d}{dx} [u(x) \, v(x)] = u \frac{dv}{dx} + v \frac{du}{dx}.$$

Integrating both sides of this equation with respect to x yields

$$u(x)v(x) = \int \left(u \frac{dv}{dx} + v \frac{du}{dx} \right) dx$$

or, upon rearranging terms,

$$\int u \frac{dv}{dx} dx = uv - \int v \frac{du}{dx} dx.$$

Consequently,

$$\int u\ dv\ = uv - \int v\ du.$$

Example A3.1   Evaluate $\int xe^{2x}dx$.

Solution   Choosing

$$u = x, \quad dv = e^{2x}\ dx$$

it follows that

$$\frac{du}{dx} = 1, \quad v = \frac{1}{2}e^{2x}$$

so that,

$$\int xe^{2x}\ dx\ = \frac{1}{2}xe^{2x} - \frac{1}{2}\int e^{2x}\ dx$$

$$= \frac{1}{2}xe^{2x} - \frac{1}{4}e^{2x} + c$$

where c is an integration constant.

Example A3.2   Evaluate $\int x^2 \sin x\ dx$.

Solution   In this case we take

$$u = x^2, \quad dv = \sin x\ dx$$

so that

$$\frac{du}{dx} = 2x, \quad v = -\cos x.$$

Thus,

$$\int x^2 \sin x\ dx\ = -x^2 \cos x\ + 2\int x \cos x\ dx.$$

We must now evaluate the second integral. Once more integration by parts is appropriate. This time we take

$$u = x, \quad dv = \cos x\ dx.$$

Then,

$$\frac{du}{dx} = 1, \quad v = \sin x$$

so that

$$\int x^2 \sin x\ dx = -x^2 \cos x + 2(x \sin x - \int \sin x\ dx)$$

$$= -x^2 \cos x\ + 2(x \sin x + \cos x) + c. \qquad \square$$

As the previous two examples illustrate, the integration by parts technique is extremely useful for evaluating integrals of the form

$$\int x^k f(x)\ dx$$

when k is a positive integer. Such an integral can often be evaluated by successively applying the integration by parts formula until the power of x is reduced to zero. However, this will not always work.

Example A3.3   Evaluate $\int x \ln x \, dx$.

Solution   In this case if we set $u = x$ and $dv = \ln x \, dx$, then, we require the integral of $\ln x$. Instead we choose

$$u = \ln x, \quad dv = x \, dx$$

then,

$$\frac{du}{dx} = \frac{1}{x}, \quad v = \frac{1}{2}x^2.$$

Applying the integration by parts formula we obtain

$$\int x \ln x \, dx = \frac{1}{2}x^2 \ln x - \frac{1}{2}\int x \, dx$$

$$= \frac{1}{2}x^2 \ln x - \frac{1}{4}x^2 + c$$

$$= \frac{1}{4}x^2(2\ln x - 1) + c.$$

2. Integration by Substitution

This is one of the most important integration techniques. Many of the standard integrals can be derived using a substitution. We illustrate with some examples.

Example A3.4   Evaluate the following integrals:

(a) $\int xe^{x^2} \, dx$.   (b) $\int \frac{1}{\sqrt{1-x^2}} \, dx$.   (c) $\int \frac{1}{x}(\ln x)^2 \, dx$.

Solution

(a) If we let $u = x^2$, then $du = 2x \, dx$, so that

$$\int xe^{x^2} \, dx = \frac{1}{2}\int e^u \, du = \frac{1}{2}e^u + c = \frac{1}{2}e^{x^2} + c.$$

(b) Recalling that $1 - \sin^2\theta = \cos^2\theta$ the form of the integrand suggests that we let $x = \sin\theta$. Then $dx = \cos\theta \, d\theta$ and the given integral can be written as

$$\int \frac{1}{\sqrt{1-x^2}} \, dx = \int \frac{\cos\theta}{\sqrt{1-\sin^2\theta}} \, d\theta = \int d\theta = \theta + c.$$

Substituting back for $\theta = \sin^{-1}x$, we obtain

$$\int \frac{1}{\sqrt{1-x^2}} \, dx = \sin^{-1}x + c.$$

(c) In this case we recognize that the derivative of $\ln x$ is $1/x$. This suggests that we make the substitution

$$u = \ln x$$

so that

$$du = \frac{1}{x} dx.$$

Then the given integral can be written in the form:

$$\int \frac{1}{x} (\ln x)^2 \, dx = \int u^2 du = \frac{1}{3} u^3 + c.$$

Substituting back for $u = \ln x$ yields

$$\int \frac{1}{x} (\ln x)^2 \, dx = \frac{1}{3} (\ln x)^3 + c.$$

Now consider the general integral

$$\int \frac{f'(x)}{f(x)} \, dx.$$

If we let $u = f(x)$, then $du = f'(x) \, dx$, so that

$$\int \frac{f'(x)}{f(x)} \, dx = \int \frac{1}{u} \, du = \ln |u| + c.$$

Substituting back for $u = f(x)$ yields the important formula:

$$\boxed{\int \frac{f'(x)}{f(x)} \, dx = \ln |f(x)| + c.}$$

Example A3.5   Evaluate $\displaystyle\int \frac{x - 2}{x^2 - 4x + 3} \, dx.$

Solution   If we rewrite the integral in the equivalent form

$$\int \frac{x - 2}{x^2 - 4x + 3} \, dx = \frac{1}{2} \int \frac{2(x - 2)}{x^2 - 4x + 3} \, dx$$

then we see that the numerator in the second integral is the derivative of the denominator. Thus,

$$\int \frac{x - 2}{x^2 - 4x + 3} \, dx = \frac{1}{2} \ln |x^2 - 4x + 3| + c.$$

3. Integration by Partial Fractions

We can always evaluate an integral of the form

$$\int \frac{p(x)}{q(x)} \, dx \tag{A3.1}$$

when $p(x)$ and $q(x)$ are polynomials in x. Consider first the case when the degree of $p(x)$ is *less than* the degree of $q(x)$. To evaluate (A3.1) we first determine the partial fraction decomposition of the integrand (a review of partial fractions is given in Appendix 2). The result will always be integrable although we might need a substitution to carry out this integration.

Example A3.6   Evaluate $\int \dfrac{3x + 2}{(x + 1)(x + 2)}\,dx.$

Solution   In this case we require the partial fraction decomposition of the integrand. Using the rules for partial fractions it follows that

$$\frac{3x + 2}{(x + 1)(x + 2)} = \frac{A}{x + 1} + \frac{B}{x + 2}.$$

Multiplying both sides of this equality by $(x + 1)(x + 2)$ yields:

$$3x + 2 = A(x + 2) + B(x + 1).$$

Equating coefficients of like powers of x on both sides of this equality we obtain:

$$A + B = 3, \quad 2A + B = 2.$$

Solving for A and B yields

$$A = -1, \quad B = 4.$$

Thus,

$$\frac{3x + 2}{(x + 1)(x + 2)} = -\frac{1}{x + 1} + \frac{4}{x + 2}$$

so that

$$\int \frac{3x + 2}{(x + 1)(x + 2)}\,dx = -\ln |x + 1| + 4\ln |x + 2| + c.$$

Example A3.7   Evaluate $\int \dfrac{2x - 3}{(x - 2)(x^2 + 1)}\,dx.$

Solution   We first determine the partial fraction decomposition of the integrand. From the general rules of partial fractions it follows that there are constants A, B, C such that

$$\frac{2x - 3}{(x - 2)(x^2 + 1)} = \frac{A}{x - 2} + \frac{Bx + C}{x^2 + 1}. \tag{A3.2}$$

In order to determine the values of A, B, and C we multiply both sides of (A3.2) by $(x - 2)(x^2 + 1)$. This yields

$$2x - 3 = A(x^2 + 1) + (Bx + C)(x - 2).$$

Equating coefficients of like powers of x on both sides of this equality we obtain

$$A + B = 0, \quad -2B + C = 2, \quad A - 2C = -3.$$

Solving for A, B, and C yields

$$A = \frac{1}{5}, \quad B = -\frac{1}{5}, \quad C = \frac{8}{5}$$

so that

$$\frac{2x - 3}{(x - 2)(x^2 + 1)} = \frac{1}{5(x - 2)} + \frac{8 - x}{5(x^2 + 1)}.$$

Thus,

$$\int \frac{2x - 3}{(x - 2)(x^2 + 1)}\,dx = \frac{1}{5}\int \frac{1}{x - 2}\,dx + \frac{8}{5}\int \frac{1}{x^2 + 1}\,dx - \frac{1}{5}\int \frac{x}{x^2 + 1}\,dx$$

$$= \frac{1}{5} \ln | x - 2 | + \frac{8}{5} \tan^{-1}x - \frac{1}{10} \ln (x^2 + 1) + c. \qquad \square$$

Now return to the integral (A3.1). If the degree of $p(x)$ is greater than or equal to the degree of $q(x)$ then we first divide $q(x)$ into $p(x)$. The resulting expression will be integrable, although in general we will need to perform a partial fraction decomposition.

Example A3.8   Evaluate $\displaystyle\int \frac{2x - 1}{x + 3} \, dx.$

Solution   We first divide the denominator into the numerator to obtain

$$\frac{2x - 1}{x + 3} = 2 - \frac{7}{x + 3}.$$

Thus,

$$\int \frac{2x - 1}{x + 3} \, dx = \int 2 \, dx - 7 \int \frac{1}{x + 3} \, dx$$

$$= 2x - 7 \ln | x + 3| + c.$$

Example A3.9   Evaluate $\displaystyle\int \frac{3x^2 + 5}{x^2 - 3x + 2} \, dx.$

Solution   Once more we must first divide the denominator into the numerator. It is easily shown that

$$\frac{3x^2 + 5}{x^2 - 3x + 2} = 3 + \frac{9x - 1}{x^2 - 3x + 2}. \qquad (A3.3)$$

The next step is to determine the partial fraction decomposition of the second term on the right-hand side. We first notice that the denominator can be factored as $(x - 2)(x - 1)$. Using the rules for partial fraction decomposition it follows that

$$\frac{9x - 1}{(x - 2)(x - 1)} = \frac{A}{x - 2} + \frac{B}{x - 1}.$$

Clearing the fractions yields

$$9x - 1 = A(x - 1) + B(x - 2).$$

Equating coefficients of like powers of x on both sides of this equation we obtain

$$A + B = 9, \quad A + 2B = 1.$$

Solving for A and B yields

$$A = 17, \quad B = -8.$$

Substitution into (A3.3) gives

$$\frac{3x^2 + 5}{x^2 - 3x + 2} = 3 + \frac{17}{x - 2} - \frac{8}{x - 1}$$

so that

$$\int \frac{3x^2 + 5}{x^2 - 3x + 2} \, dx = 3x + 17 \ln| x - 2 | - 8 \ln| x - 1 | + c. \qquad \square$$

Table A3.1 lists some of the more important integrals. Notice that we have omitted the integration constant.

TABLE A3.1   SOME BASIC INTEGRALS

Function $F(x)$	$\int F(x)\, dx$
$x^n, n \neq -1$	$\dfrac{1}{n+1} x^{n+1}$
x^{-1}	$\ln \mid x \mid$
$e^{ax}, a \neq 0$	$\dfrac{1}{a} e^{ax}$
$\sin x$	$-\cos x$
$\cos x$	$\sin x$
$\tan x$	$\ln \mid \sec x \mid$
$\sec x$	$\ln \mid \sec x + \tan x \mid$
$\csc x$	$\ln \mid \csc x - \cot x \mid$
$e^{ax}\sin bx$	$\dfrac{1}{a^2 + b^2} e^{ax}(a \sin bx - b \cos bx)$
$e^{ax}\cos bx$	$\dfrac{1}{a^2 + b^2} e^{ax}(a \cos bx + b \sin bx)$
$\ln x$	$x \ln x - x$
$\dfrac{1}{a^2 + x^2}$	$\dfrac{1}{a} \tan^{-1}(x/a)$
$\dfrac{1}{\sqrt{a^2 - x^2}}, a > 0$	$\sin^{-1}(x/a)$
$\dfrac{1}{\sqrt{a^2 + x^2}}$	$\ln \mid x + \sqrt{a^2 + x^2} \mid$
$\dfrac{f'(x)}{f(x)}$	$\ln \mid f(x) \mid$
$e^{u(x)} \dfrac{du}{dx}$	$e^{u(x)}$

EXERCISES A3

Evaluate the given integral.

1. $\displaystyle\int x \cos x\ dx.$

2. $\displaystyle\int x^2 e^{-x}\ dx.$

3. $\displaystyle\int \ln x\ dx.$

4. $\displaystyle\int \tan^{-1}x\ dx.$

5. $\displaystyle\int x^3 e^{x^2}\ dx.$

6. $\displaystyle\int \dfrac{x}{x^2 + 1}\ dx.$

7 . $\displaystyle\int \frac{x - 1}{x + 2}\,dx.$

8 . $\displaystyle\int \frac{x + 2}{(x - 1)(x + 3)}\,dx.$

9 . $\displaystyle\int \frac{2x + 1}{x(x^2 + 4)}\,dx.$

10. $\displaystyle\int \frac{x^2 + 5}{(x - 1)(x + 4)}\,dx.$

11. $\displaystyle\int \frac{x + 3}{2x - 1}\,dx.$

12. $\displaystyle\int \frac{2x + 3}{x^2 + 3x + 4}\,dx.$

13. $\displaystyle\int \frac{3x + 2}{x(x + 1)^2}\,dx.$

14. $\displaystyle\int \frac{1}{\sqrt{4 - x^2}}\,dx.$

15. $\displaystyle\int \frac{1}{x^2 + 2x + 2}\,dx.$

16. $\displaystyle\int \frac{1}{x \ln x}\,dx.$

17. $\displaystyle\int \tan x\,dx.$

18. $\displaystyle\int \frac{x + 1}{x^2 - x - 6}\,dx.$

19. $\displaystyle\int \cos^2 x\,dx.$

20. $\displaystyle\int \sqrt{1 - x^2}\,dx.$

21. $\displaystyle\int e^{3x} \sin 2x\,dx.$

22. $\displaystyle\int e^x \sin^2 x\,dx.$

APPENDIX
4

An Existence and Uniqueness Theorem for First-Order DE

In this appendix we discuss an existence and uniqueness theorem for first order ordinary differential equations. This discussion will require the introduction of an important theoretical process known as Picard iteration. First we state the basic theorem.

Theorem A4.1: Let $f(x, y)$ be a function that is defined and continuous on the rectangle

$$R = \{(x, y) : |x - x_0| \leq a, |y - y_0| \leq b\}$$

where a and b are constants. Suppose further that $\partial f/\partial y$ is also continuous in R. Then there exists an interval I containing x_0 such that the IVP

$$\begin{cases} \dfrac{dy}{dx} = f(x, y) \\ y(x_0) = y_0 \end{cases}$$

has a unique solution for all x in I.

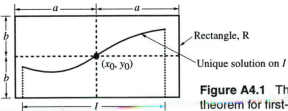

Figure A4.1 The existence and uniqueness theorem for first-order DE.

The idea behind the proof of Theorem A4.1 is to construct a sequence of functions that converges to a solution of the IVP, and then show that there can be no other solution.

The proof is quite involved and will be completed in several steps.

REFORMULATION OF AN IVP AS AN EQUIVALENT INTEGRAL EQUATION

The first step in the proof is to show that an IVP can be reformulated in terms of an equivalent integral equation. This is established in the following Lemma

> ***Lemma A4.1:***   Consider the IVP

$$\begin{cases} \dfrac{dy}{dx} = f(x, y) & \text{(A4.1)} \\[2mm] y(x_0) = y_0 & \text{(A4.2)} \end{cases}$$

where we assume that f is continuous on the rectangle

$$R = \{(x, y) : |x - x_0| \le a, |y - y_0| \le b\}.$$

Then $y(x)$ is a solution of (A4.1) and (A4.2) on an interval I if and only if it is a *continuous* solution of the integral equation

$$y(x) = y_0 + \int_{x_0}^{x} f(t, y(t))\, dt \qquad\qquad \text{(A4.3)}$$

on I.

PROOF   If y is a solution of (A4.1) and (A4.2) on an interval I then $y(x)$ and $f(x, y(x))$ are continuous on I so that we can formally integrate both sides of (A4.1) to obtain

$$y(x) - y(x_0) = \int_{x_0}^{x} f(t, y(t))\, dt$$

that is,

$$y(x) = y_0 + \int_{x_0}^{x} f(t, y(t))\, dt.$$

Conversely, suppose that $y(x)$ is a continuous solution of the integral equation (A4.3). Then $f(t, y(t))$ is continuous on I which implies that

$$\int_{x_0}^{x} f(t, y(t))\, dt$$

is differentiable on I. We can therefore differentiate (A4.3) to obtain

$$\frac{dy}{dx} = f(x, y).$$

Further, setting $x = x_0$ in (A4.3) yields

$$y(x_0) = y_0.$$

∎

Example A4.1   Reformulate the following IVP as an equivalent integral equation:

$$y' = yx^2 - \sin x, \quad y(\pi) = 1.$$

Solution   In this case we have

$$f(x, y) = yx^2 - \sin x, \quad x_0 = \pi, \quad y_0 = 1.$$

Thus the equivalent integral equation is

$$y(x) = 1 + \int_{\pi}^{x} (yt^2 - \sin t)\, dt. \qquad\qquad \Box$$

It follows from the previous Lemma that we can replace Theorem A4.1 by the following equivalent existence and uniqueness theorem.

Theorem A4.2:   Let $f(x, y)$ be a function that is defined and continuous on the rectangle

$$R = \{(x, y) : |x - x_0| \le a, |y - y_0| \le b\}$$

where a and b are constants. Suppose further that $\partial f/\partial y$ is also continuous in R. Then there exists an interval I containing x_0 such that the integral equation

$$y(x) = y_0 + \int_{x_0}^{x} f(t, y(t))\, dt.$$

has a unique *continuous* solution for all x in I.

We will concentrate on this formulation of the existence and uniqueness theorem.

THE PICARD ITERATES

As mentioned previously, the idea behind proving the existence part of Theorem A4.1 (or equivalently Theorem A4.2) is to determine a sequence of functions that converges to a solution of the IVP. Consider the integral equation (A4.3), that is

$$y(x) = y_0 + \int_{x_0}^{x} f(t, y(t))\, dt.$$

We associate with this integral equation the sequence of functions $\{y_1, y_2, \dots\}$ defined by

$$y_1(x) \;=\; y_0 + \int_{x_0}^{x} f(t, y_0)\, dt$$

$$y_2(x) \;=\; y_0 + \int_{x_0}^{x} f(t, y_1(t))\, dt$$

$$\vdots$$

$$y_n(x) \;=\; y_0 + \int_{x_0}^{x} f(t, y_{n-1}(t))\, dt$$

$$\vdots$$

The functions in this sequence are called the **Picard iterates** for the integral equation (A4.3) or, equivalently, for the IVP (A4.1) and (A4.2). Notice that each of the Picard iterates satisfies the initial condition $y_i(x_0) = y_0$.

Example A4.2   Determine the Picard iterates for the IVP

$$\frac{dy}{dx} = x - y, \quad y(0) = 0. \tag{A4.4}$$

Solution   In this case we have:

$$f(x, y) = x - y, \quad x_0 = 0, \quad y_0 = 0$$

so that

$$y_n(x) = \int_0^x f(t, y_{n-1}(t)) \, dt = \int_0^x [t - y_{n-1}(t)] \, dt.$$

Consequently

$$y_1(x) \ = \int_0^x (t - 0) \, dt = \frac{1}{2} x^2$$

$$y_2(x) \ = \int_0^x \left(t - \frac{1}{2} t^2\right) dt = \frac{1}{2} x^2 - \frac{1}{3 \cdot 2} x^3$$

$$y_3(x) \ = \int_0^x \left(t - \frac{1}{2} t^2 + \frac{1}{3 \cdot 2} t^3\right) dt$$

$$= \frac{1}{2} x^2 - \frac{1}{3 \cdot 2} x^3 + \frac{1}{4 \cdot 3 \cdot 2} x^4$$

and so on. We fairly soon recognize the pattern that is emerging, namely

$$y_n(x) = \frac{1}{2} x^2 - \frac{1}{3!} x^3 + \frac{1}{4!} x^4 - \cdots + \frac{(-1)^{n+1}}{(n+1)!} x^{n+1}.$$

In order to motivate the proof of the Existence–Uniqueness Theorem, we write this expression for y_n in the following form

$$y_n(x) = (x - 1) + 1 - x + \frac{1}{2!} x^2 - \frac{1}{3!} x^3 + \cdots + \frac{(-1)^{n+1}}{(n+1)!} x^{n+1},$$

that is,

$$y_n(x) = (x - 1) + \sum_{k=0}^{n+1} \frac{(-1)^k}{k!} x^k.$$

Taking the limit as $n \to \infty$ and recalling the Maclaurin expansion for e^{-x}, namely

$$e^{-x} = \sum_{k=0}^{\infty} \frac{(-1)^k}{k!} x^k$$

we obtain

$$\lim_{n \to \infty} y_n = (x - 1) + e^{-x}.$$

Thus if we define $y(x)$ by

$$y(x) = (x - 1) + e^{-x}$$

we have shown that the Picard iterates for the given IVP converge to $y(x)$. The important point that we wish to make is the following: a direct substitution into the DE (A4.4) shows that $y(x)$ is, in fact, a solution (for all x), and further, it satisfies the given initial condition. Consequently, *in this particular case*, the Picard iterates converge to a solution of the IVP, or, equivalently, to a continuous solution of the corresponding integral equation. We now show that this is true in general.

PROOF OF THE EXISTENCE PART OF THEOREM A4.2   We wish to prove that the Picard iterates for the integral equation (A4.3) converge to a continuous solution of the integral equation for x in some interval I. It might be thought that the appropriate interval would be $[x_0 - a, x_0 + a]$. However this is not necessarily the case. The reason why is that we must ensure that the Picard iterates are well defined, that is, that each of the y_i are themselves contained within the rectangle R. (If one of the iterates failed to remain in R then the succeeding iterates would not necessarily be defined or continuous since they would require the evaluation of f at points outside of R.) In Lemma A4.2 below we will derive an appropriate interval. However, before proceeding with the proof we need to recall some basic results from calculus.

1.   The triangle inequality for integrals:

$$\left| \int_{x_0}^{x} f(t, y(t))\, dt \right| \leq \left| \int_{x_0}^{x} |f(t, y(t))|\, dt \right|.$$

2.   The mean-value theorem which, when applied to f, states that for fixed t,

$$f(t, y_2(t)) - f(t, y_1(t)) = f_y(t, \xi)\, [y_2(t) - y_1(t)]$$

where $\xi \in (y_1(t), y_2(t))$.

3.   A function that is continuous on a closed region in the xy-plane is necessarily bounded in that region. Thus, since f and f_y in Theorem A4.1 are both continuous on the closed rectangle R, it follows that there are positive constants M and N such that

$$|f(x, y)| \leq M, \quad |f_y(x, y)| \leq N, \quad \text{for all } (x, y) \text{ in R.}$$

We will use these results repeatedly throughout the proof of Theorem A4.2.

Lemma A4.2:   Let f be continuous on the rectangle

$$R = \{(x, y) : |x - x_0| \leq a, \ |y - y_0| \leq b\}$$

and let M be a bound for f on R. If $I = [x_0 - \alpha, x_0 + \alpha]$, where $\alpha = \min(a, b/M)$, then the Picard iterates for the integral equation (A4.3) are well defined for all $x \in I$.

PROOF   The proof is by induction. It is certainly true that y_0 is well defined for $x \in I$. Now suppose that $y_0, y_1, ..., y_k$ lie in R for all $x \in I$. We must show that this implies that y_{k+1} also lies in R for all $x \in I$. Since y_k lies in R for all $x \in I$ it follows, from the continuity of f in R, that $|f(t, y_k(t))| \leq M$ for all $t \in I$. Further, by definition of the Picard iterates, we have

$$|y_{k+1}(x) - y_0| = \left| \int_{x_0}^{x} f(t, y_k(t))\, dt \right| \leq \left| \int_{x_0}^{x} |f(t, y_k(t))|\, dt \right|$$

$$\leq M\, |x - x_0| \leq M\alpha \leq b$$

so that $y_{k+1}(x)$ also lies in R for $x \in I$. Hence, by induction, all of the Picard iterates are well defined for $x \in I$. ∎

From now on we will restrict our attention to the interval $I = [x_0 - \alpha, x_0 + \alpha]$. The next step in the proof of Theorem A4.2 is to show that the Picard iterates converge to a function $y(x)$ for all $x \in I$. This is the content of the next Lemma.

Lemma A4.3: The Picard iterates for the integral equation (A4.3) converge to a function $y(x)$ for all $x \in I$.

PROOF The key behind the proof is to recognize that we can write $y_n(x)$ in the form

$$y_n(x) = y_0(x) + [y_1(x) - y_0(x)] + [y_2(x) - y_1(x)] + \cdots + [y_n(x) - y_{n-1}(x)]$$

that is,

$$y_n(x) = y_0 + \sum_{k=0}^{n-1} [y_{k+1}(x) - y_k(x)]. \tag{A4.5}$$

Convergence of the sequence of Picard iterates is equivalent to the convergence of the infinite series

$$y_0(x) + \sum_{k=0}^{\infty} [y_{k+1}(x) - y_k(x)].$$

The convergence of this series can be established by the comparison test. Consider the kth term in the series. By definition of the Picard iterates

$$| y_{k+1}(x) - y_k(x) | = \left| \int_{x_0}^{x} [f(t, y_k(t)) - f(t, y_{k-1}(t))] \, dt \right|$$

$$\leq \left| \int_{x_0}^{x} | f(t, y_k(t)) - f(t, y_{k-1}(t)) | \, dt \right|.$$

But, by the mean-value theorem, we have

$$| f(t, y_k(t)) - f(t, y_{k-1}(t)) | \leq | f_y(t, \xi) | \, | y_k(t) - y_{k-1}(t) | \leq N | y_k(t) - y_{k-1}(t) |$$

where N is a bound for f_y on R. Thus,

$$| y_{k+1}(x) - y_k(x) | \leq N \left| \int_{x_0}^{x} | y_k(t) - y_{k-1}(t) | \, dt \right|. \tag{A4.6}$$

Also,

$$| y_1(x) - y_0 | = \left| \int_{x_0}^{x} f(t, y_0) \, dt \right| \leq M | x - x_0 |. \tag{A4.7}$$

It follows from (A4.6) and (A4.7), that

$$| y_2(x) - y_1(x) | \leq N \left| \int_{x_0}^{x} M | t - x_0 | \, dt \right| \leq NM \frac{\alpha^2}{2!}$$

and it is easily shown, by induction, that

$$| y_{k+1}(x) - y_k(x) | \le \frac{NM^k \alpha^{k+1}}{(k+1)!} \qquad (A4.8)$$

for all x in I. But

$$\sum_{k=0}^{\infty} \frac{(M\alpha)^{k+1}}{(k+1)!} = e^{M\alpha} - 1$$

so that, by the comparison test,

$$\sum_{k=0}^{\infty} [y_{k+1}(x) - y_k(x)]$$

converges (absolutely and uniformly) for all x in I, and hence so also does

$$y_0 + \sum_{k=0}^{\infty} [y_{k+1}(x) - y_k(x)].$$

It follows from (A4.5) that the sequence of Picard iterates $\{y_1, y_2, \dots \}$ also converges for all x in I. We let $y(x)$ denote the limit function, so that we have proved that

$$\lim_{n \to \infty} y_n(x) = y(x)$$

for all x in I.   ∎

It is now possible to prove that the limit function $y(x)$ is continuous on I and also satisfies the integral equation on I. These results can be elegantly derived from a closer analysis of the manner in which the sequence of Picard iterates approach the limit function $y(x)$.

POINTWISE AND UNIFORM CONVERGENCE   When dealing with convergence of functions on an interval, two different types of convergence can be distinguished:

1. *Pointwise Convergence:*   Given $\varepsilon > 0$, for each x in I there exists a positive integer $N_0(\varepsilon, x)$, such that

$$| y(x) - y_n(x) | < \varepsilon \text{ whenever } n > N_0(\varepsilon, x).$$

2. *Uniform Convergence:*   Given $\varepsilon > 0$, there exists a positive integer $N_0(\varepsilon)$, such that *for all x in I,*

$$| y(x) - y_n(x) | < \varepsilon \text{ whenever } n > N_0(\varepsilon).$$

The difference between these two types of convergence is that in the case of uniform convergence for a given ε, there is one integer $N_0(\varepsilon)$ that works for *all* x in I, whereas in the case of pointwise convergence we, in general, require different integers $N_0(\varepsilon, x)$ for different points in I. Consequently, uniform convergence is a stronger condition than pointwise convergence.

We now state two important results for uniformly convergent sequences of functions.

UC1. If the sequence of functions $\{y_1, y_2, \dots\}$ converges *uniformly* to the function $y(x)$ on the interval I, and if each $y_k(x)$ is continuous on I, then the limit function $y(x)$ is also continuous on I.

UC2. If the sequence of functions $\{y_1, y_2, ...\}$ converges *uniformly* to the function $y(x)$ on the interval I, then for all x and x_0 in I,

$$\lim_{n \to \infty} \int_{x_0}^{x} y_n(x)\, dx = \int_{x_0}^{x} \lim_{n \to \infty} y_n(x)\, dx.$$

A proof of both of these results can be found, for example, in T. M. Apostol, *Mathematical Analysis*, Addison–Wesley, 1974.

We next prove that the convergence of the Picard iterates is uniform.

Lemma A4.4: The Picard iterates converge uniformly to a limit function $y(x)$ for all x in I.

PROOF   We have already established in the previous lemma that the Picard iterates converge to the function

$$y(x) = y_0 + \sum_{k=0}^{\infty} [y_{k+1}(x) - y_k(x)]$$

for all x in I. We must show that the convergence is uniform. Since

$$y_n(x) = y_0(x) + \sum_{k=0}^{n-1} [y_{k+1}(x) - y_k(x)]$$

it follows that

$$|y(x) - y_n(x)| \le \left| \sum_{k=n}^{\infty} [y_{k+1}(x) - y_k(x)] \right| \le \sum_{k=n}^{\infty} |y_{k+1}(x) - y_k(x)|.$$

Substituting from (A4.8) we therefore have

$$|y(x) - y_n(x)| \le N \sum_{k=n}^{\infty} \frac{M^k \alpha^{k+1}}{(k+1)!} = \frac{N}{M} (M\alpha)^{n+1} \sum_{j=0}^{\infty} \frac{(M\alpha)^j}{(n+j+1)!}$$

$$\le \frac{N}{M} \frac{(M\alpha)^{n+1}}{n!} \sum_{j=0}^{\infty} \frac{(M\alpha)^j}{(n+1)\cdots(n+1+j)} \le \frac{N}{M} \frac{(M\alpha)^{n+1}}{n!} \sum_{j=0}^{\infty} \frac{(M\alpha)^j}{(j+1)!}$$

$$\le \frac{N}{M} \frac{(M\alpha)^{n+1}}{n!} \sum_{j=0}^{\infty} \frac{(M\alpha)^j}{j!}$$

that is,

$$|y(x) - y_n(x)| \le \frac{N}{M} \frac{(M\alpha)^{n+1}}{n!} e^{M\alpha}. \tag{A4.9}$$

But,

$$\lim_{n \to \infty} \frac{(M\alpha)^{n+1}}{n!} = 0$$

so that, for any $\varepsilon > 0$, there exists an $N_0(\varepsilon)$ such that

$$\frac{N}{M} \frac{(M\alpha)^{n+1}}{n!} e^{M\alpha} < \varepsilon \text{ for all } n > N_0(\varepsilon).$$

Thus, from (A4.9), for any $\varepsilon > 0$ and all $x \in I$,

$$|y(x) - y_n(x)| < \varepsilon \text{ for all } n > N_0(\varepsilon)$$

so that the convergence of the Picard iterates is indeed uniform on I.   ∎

We now use the result of the above lemma together with UC1 and UC2 to establish that the limit function $y(x)$ is a continuous solution of the integral equation (A4.3).

Lemma A4.5:   The limit function $y(x)$ is a continuous solution of the integral equation (A4.3).

PROOF   Since each of the Picard iterates are continuous on I, it follows directly from UC1 that the limit function $y(x)$ is also continuous on I. Now consider the nth Picard iterate, namely,

$$y_n(x) = y_0 + \int_{x_0}^{x} f(t, y_{n-1}(t))\, dt$$

Taking the limit as $n \to \infty$ yields

$$y(x) = \lim_{n \to \infty} y_n(x) = y_0 + \lim_{n \to \infty} \int_{x_0}^{x} f(t, y_{n-1}(t))\, dt.$$

From UC2, the uniform convergence of $\{y_n\}$ allows us to take the limit inside the integral to obtain

$$y(x) = y_0 + \int_{x_0}^{x} \lim_{n \to \infty} f(t, y_{n-1}(t))\, dt.$$

Further, we can use the assumed continuity of f on I to take the limit inside the function, thereby obtaining

$$y(x) = y_0 + \int_{x_0}^{x} f(t, \lim_{n \to \infty} y_{n-1}(t))\, dt$$

that is, since $\{y_{n-1}\}$ converges to y,

$$y(x) = y_0 + \int_{x_0}^{x} f(t, y(t))\, dt.$$

Thus $y(x)$ *is a continuous* solution of the given integral equation.   ∎

This completes the proof of the existence part of Theorem A4.2.

PROOF OF UNIQUENESS   The previous lemmas have established the existence of a continuous solution to the integral equation (A4.3), and hence to the IVP (A4.1) and (A4.2), on the interval I. We now establish uniqueness. Once more it is convenient to consider the integral equation formulation.

Lemma A4.6:   The integral equation

$$y(x) = y_0 + \int_{x_0}^{x} f(t, y(t))\, dt \qquad \text{(A4.10)}$$

where f and f_y are continuous on the rectangle

$$R = \{(x, y) : |x - x_0| \leq a, |y - y_0| \leq b\}$$

has a unique continuous solution on $I = [x_0 - \alpha, x_0 + \alpha]$.

PROOF   We have already shown in the previous lemmas that the integral equation admits a continuous solution on I. We must now show that there is only one solution. In order to do so we assume the existence of two solutions to the integral equation and show that they are the same.

Let Y_1 and Y_2 be two continuous solutions to the integral equation (A4.10) on I. Then,

$$Y_2(x) - Y_1(x) = \int_{x_0}^{x} [f(t, Y_2(t)) - f(t, Y_1(t))]\, dt$$

so that,

$$|Y_2(x) - Y_1(x)| \leq \left| \int_{x_0}^{x} |[f(t, Y_2(t)) - f(t, Y_1(t))]|\, dt \right|. \qquad \text{(A4.11)}$$

But, using the mean-value theorem,

$$f(t, Y_2(t)) - f(t, Y_1(t)) = f_y(t, \xi)\, [Y_2(t) - Y_1(t)]$$

so that, since f_y is continuous on R,

$$|f(t, Y_2(t)) - f(t, Y_1(t))| \leq N\, |[Y_2(t) - Y_1(t)]|$$

where N is a bound for f_y on I. Substituting into (A4.11) yields

$$|Y_2(x) - Y_1(x)| \leq N \left| \int_{x_0}^{x} |Y_2(t) - Y_2(t)|\, dt \right|. \qquad \text{(A4.12)}$$

Now define the function $F(x)$ by

$$F(x) = \int_{x_0}^{x} |Y_2(t) - Y_1(t)|\, dt. \qquad \text{(A4.13)}$$

The remainder of the proof depends on whether $x \geq x_0$, or $x < x_0$. Suppose that $x \geq x_0$. Then, (A4.12) can be written as

$$F'(x) - NF(x) \leq 0 \qquad \text{(A4.14)}$$

for $x \geq x_0$, whereas (A4.13) implies that $F(x)$ satisfies the following conditions

$$F(x) \geq 0 \text{ for } x \geq x_0, \qquad \text{(A4.15)}$$

$$F(x_0) = 0. \qquad \text{(A4.16)}$$

Multiplying (A4.14) by e^{-Nx} yields

$$(e^{-Nx}F)' \leq 0$$

for $x \geq x_0$, which, upon integrating from x_0 to x and imposing the condition (A4.16), yields

$$e^{-Nx}F(x) \leq 0$$

for $x \geq x_0$, that is,

$$F(x) \leq 0 \qquad\qquad (A4.17)$$

for $x \geq x_0$. It follows from (A4.15) and (A4.17) that we must have

$$F(x) = 0$$

for $x \geq x_0$. We leave it as an exercise to show that $F(x) = 0$ for $x < x_0$ also, so that $F(x)$ is identically zero on I. It now follows from (A4.13) that

$$Y_1(x) = Y_2(x)$$

for all x in I, and hence we have uniqueness.   ∎

In summary, we have the following result.

Theorem A4.3:   Let $f(x, y)$ be a function that is defined and continuous on the rectangle

$$R = \{(x, y) : |x - x_0| \leq a, |y - y_0| \leq b\}$$

where a and b are constants. Suppose further that $\partial f/\partial y$ is also continuous in R. Then the Picard iterates converge to the unique solution of the IVP

$$\begin{cases} \dfrac{dy}{dx} = f(x, y) \\[2mm] y(x_0) = y_0 \end{cases}$$

on the interval $[x_0 - \alpha, x_0 + \alpha]$, where $\alpha = \min(a, b/M))$, and M is a bound for f on R.

The existence and uniqueness theorem can be generalized to include differential equations of arbitrary order and also systems of differential equations (see, for example, F. J. Murray and K. S. Miller, *Existence Theorems*, New York University Press, 1954). It forms the theoretical basis for differential equation theory.

EXERCISES A4

1. Convert the given IVP into an equivalent integral equation and determine the first two Picard iterates y_1, y_2 :

$$\frac{dy}{dx} = x(y - x), \ y(0) = 1.$$

2. Consider the IVP

$$\frac{dy}{dx} = x(x + y), \ y(0) = 0. \qquad (2.1)$$

(a) Show that the nth Picard iterate is

$$y_n(x) = \sum_{k=1}^{n} \frac{1}{1 \cdot 3 \cdots (2k + 1)} x^{2k+1}.$$

(b) Prove that

$$\lim_{n \to \infty} y_n(x) = \sum_{k=1}^{\infty} \frac{1}{1 \cdot 3 \cdots (2k + 1)} x^{2k+1}$$

by showing that the infinite series on the right–hand side converges for all real x.

(c) Show by direct substitution that

$$y(x) = \sum_{k=1}^{\infty} \frac{1}{1 \cdot 3 \cdots (2k+1)} x^{2k+1}$$

is a solution to the IVP (2.1).

3. Convert the given IVP into an equivalent integral equation and find the solution by Picard iteration:

$$\frac{dy}{dx} = 2xy, \; y(0) = 1.$$

4. Use the existence and uniqueness theorem to prove that $y(x) = 3$ is the only solution to the IVP

$$\frac{dy}{dx} = \frac{x}{x^2+1}(y^2-9), \; y(0) = 3.$$

5. One solution to the IVP

$$\frac{dy}{dx} = \frac{2}{3}(y-1)^{1/2}, \; y(1) = 1.$$

is $y(x) = 1$. Determine another solution to this IVP. Does this contradict the Existence–Uniqueness Theorem? Explain.

6. Consider the IVP

$$\frac{dy}{dx} = y^2 + 4 \sin^2 x, \; y(0) = 0. \qquad (6.1)$$

Since f and f_y are continuous for all (x, y) it follows that the existence and uniqueness theorem can be applied on any rectangle

$$R = \{(x, y) : |x| \le a, |y| \le b\}$$

where a and b are constants.

(a) Show that the unique solution guaranteed by the existence and uniqueness theorem is valid at least on the interval $I = [-\alpha, \alpha]$, where

$$\alpha = \min\left(a, \frac{b}{b^2+4}\right).$$

(b) Determine the maximum interval of existence of a solution to (6.1) guaranteed by the existence and uniqueness theorem.

7. Fill in the missing details in the proof of Lemma A4.6 when $x < x_0$.

APPENDIX
5

Linearly Independent Solutions to
$x^2y'' + xp(x)y' + q(x)y = 0$

Consider the DE

$$x^2y'' + xp(x)y' + q(x)y = 0 \qquad (A5.1)$$

where p and q are analytic at $x = 0$. Writing

$$p(x) = \sum_{n=0}^{\infty} p_n x^n, \quad q(x) = \sum_{n=0}^{\infty} q_n x^n \qquad (A5.2)$$

on $(0, R)$, it follows that the indicial equation for (A5.1) is

$$r(r - 1) + p_0 r + q_0 = 0. \qquad (A5.3)$$

If the roots of the indicial equation are distinct and do not differ by an integer, then there exist two LI Frobenius series solutions in the neighborhood of $x = 0$. In this appendix we derive the general form for two LI solutions to (A5.1) in the case that the roots of the indicial equation differ by an integer. The analysis of the case when the roots of the indicial equation coincide is left as an exercise. Let r_1 and r_2 denote the roots of the indicial equation and suppose that

$$r_1 - r_2 = N \qquad (A5.4)$$

where N is a positive integer. Then, we know that one solution to (A5.1) on $(0, R)$ is given by the Frobenius series

$$y_1(x) = x^{r_1}(1 + a_1 x + a_2 x^2 + \cdots). \qquad (A5.5)$$

According to the reduction of order technique, a second LI solution to (A5.1) on $(0, R)$ is given by

$$y_2(x) = u(x) y_1(x) \tag{A5.6}$$

where the function u can be determined by substitution into (A5.1). Differentiating (A5.6) twice and substituting into (A5.1) yields

$$x^2(y_1 u'' + 2 y_1' u' + y_1'' u) + x p(x)(y_1 u' + y_1' u) + q(x) u y_1 = 0$$

that is, since y_1 is a solution to (A5.1),

$$x^2(y_1 u'' + 2 y_1' u') + x p(x)(y_1 u') = 0.$$

Consequently u can be determined by solving

$$\frac{u''}{u'} = -\left(\frac{p(x)}{x} + 2 \frac{y_1'}{y_1} \right)$$

which, upon integrating yields

$$u' = y^{-2} e^{-\int [p(x)/x] dx}. \tag{A5.7}$$

We now determine the two terms that appear on the right-hand side. From (A5.2) we can write

$$\frac{p(x)}{x} = \frac{p_0}{x} + p_1 + p_2 x + \cdots$$

so that

$$\int \frac{p(x)}{x} dx = p_0 \ln x + P(x)$$

where

$$P(x) = p_1 x + \frac{1}{2} p_2 x^2 + \frac{1}{3} p_3 x^3 + \cdots.$$

Consequently

$$e^{-\int [p(x)/x] dx} = x^{-p_0} e^{P(x)}. \tag{A5.8}$$

Since $P(x)$ is analytic at $x = 0$, it follows that

$$e^{P(x)} = \alpha_0 + \alpha_1 x + \alpha_2 x^2 + \cdots$$

for appropriate constants $\alpha_0, \alpha_1, \alpha_2, \ldots$. Hence, from (A5.8),

$$e^{-\int [p(x)/x] dx} = x^{-p_0}(\alpha_0 + \alpha_1 x + \alpha_2 x^2 + \cdots). \tag{A5.9}$$

Now consider y_1^{-2}. From (A5.5),

$$y_1^{-2} = \frac{x^{-2 r_1}}{(1 + a_1 x + a_2 x^2 + \cdots)^2}. \tag{A5.10}$$

Further, since the series

$$1 + a_1 x + a_2 x^2 + \cdots$$

converges for $0 \le x < R$, and is nonzero at $x = 0$, it follows that $(1 + a_1 x + a_2 x^2 + \cdots)^{-2}$ is analytic at $x = 0$ and hence there exist constants $\beta_1, \beta_2, \ldots$, such that

$$(1 + a_1 x + a_2 x^2 + \cdots)^{-2} = 1 + \beta_1 x + \beta_2 x^2 + \cdots.$$

We can therefore write (A5.10) in the form

$$y_1^{-2} = x^{-2r_1}(1 + \beta_1 x + \beta_2 x + \cdots). \qquad (A5.11)$$

Substituting from (A5.9) and (A5.11) into (A5.7) yields

$$u' = x^{-(p_0+2r_1)}(\alpha_0 + \alpha_1 x + \alpha_2 x^2 + \cdots)(1 + \beta_1 x + \beta_2 x^2 + \cdots)$$

which can be written as

$$u' = x^{-(p_0+2r_1)}(A_0 + A_1 x + A_2 x^2 + \cdots) \qquad (A5.12)$$

for appropriate constants $A_0, A_1, A_2, \ldots$. Now, since the roots of the indicial equation are r_1 and $r_2 = r_1 - N$, it follows that the indicial equation has factored form

$$(r - r_1)(r - r_1 + N) = 0,$$

which upon expansion gives

$$r^2 - (2r_1 - N)r + r_1(r_1 - N) = 0.$$

Comparison with (A5.3) reveals that

$$p_0 - 1 = -(2r_1 - N),$$

so that

$$p_0 + 2r_1 = N + 1.$$

Consequently (A5.12) can be written as

$$u' = x^{-(N+1)}(A_0 + A_1 x + A_2 x^2 + \cdots)$$

that is,

$$u' = A_0 x^{-(N+1)} + A_1 x^{-N} + \cdots + A_N x^{-1} + A_{N+1} + A_{N+2} x + \cdots$$

which can be integrated directly to yield

$$u(x) = -\frac{A_0}{N}x^{-N} + \frac{A_1}{1 - N} x^{1-N} + \cdots - A_{N-1}x^{-1} + A_N \ln x + A_{N+1}x + \cdots .$$

Rearranging terms we can write this as

$$u(x) = A \ln x + x^{-N}(B_0 + B_1 x + B_2 x^2 + \cdots)$$

where we have redefined the coefficients. Substituting this expression for u into (A5.6) gives

$$y_2(x) = [A \ln x + x^{-N}(B_0 + B_1 x + B_2 x^2 + \cdots)]y_1(x)$$

that is,

$$y_2(x) = A y_1(x) \ln x + x^{-N}y_1(x)(B_0 + B_1 x + B_2 x^2 + \cdots).$$

Substituting for y_1 from (A5.5) into the second term on the right-hand side yields

$$y_2(x) = A y_1(x) \ln x + x^{r_1-N} (1 + a_1 x + a_2 x^2 + \cdots)(B_0 + B_1 x + B_2 x^2 + \cdots).$$

Finally, multiplying the two power series together and substituting $r_2 = r_1 - N$ from (A5.4), we obtain

$$y_2(x) = A y_1(x) \ln x + x^{r_2}(b_0 + b_1 x + b_2 x^2 + \cdots)$$

for appropriate constants $b_0, b_1, b_2, \ldots$, that is,

$$y_2(x) = A\,y_1(x)\ln x + x^{r_2}\sum_{n=0}^{\infty} b_n x^n.$$

The derivation of a second solution to the DE (A5.1) in the case when the indicial equation has two equal roots follows exactly the same lines as that above and is left as an exercise.

Answers
to Odd-Numbered
Problems

SECTION 1.1

1. $t = 10\sqrt{2g/g}$. **7.** $y = cx^4$. **9.** $x^2 + 2y^2 = c$. **11.** $y^2 = -2x + c$. **13.** $y^2 = -\dfrac{1}{m}x^2 + c_1$.
15. $y^2 = ce^{-2x/m}$.

SECTION 1.2

1. 2, nonlinear. **3.** 2, nonlinear. **5.** 4, linear. **7.** $(-\infty, 4)$ or $(4, \infty)$. **9.** $(-\infty, \infty)$.
11. $(-\infty, 0)$ or $(0, \infty)$. **13.** $(-\infty, \infty)$. **15.** $(-\infty, \infty)$. **17.** $r = -3, 1$. **19.** $r = \pm 1$.
25. $y(x) = x^{-1}\ln x$, $x > 0$. **27.** $y(x) = x^{-1}\sqrt{1 + \sin x}$, $x > 0$.
31. $y(x) = \dfrac{1}{(n + 1)(n + 2)} x^{n+2} + c_1 x + c_2$, if $n \neq -2, -1$; $y(x) = c_1 x + c_2 - \ln|x|$, if $n = -2$;
$y(x) = x\ln|x| + c_1 x + c_2$, if $n = -1$. Interval: $(-\infty, \infty)$ if $n \geq 0$; $(-\infty, 0)$ or $(0, \infty)$ if $n < 0$.
33. $y(x) = 3 + x - \cos x$. **35.** $y(x) = e^{-x} - e^{-1}x$.

SECTION 1.3

1. $y' = -y/x$. **3.** $y' = (y^2 - x^2)/(2xy)$. **5.** $y' = x/[-y \pm (x^2 + y^2)^{1/2}]$.
7. $y' = -(x^2 + 2xy - y^2)/(y^2 + 2xy - x^2)$. **15.** (c) Intervals: (i) $(-\infty, \infty)$;
(ii) $(-\infty, 0)$ and $(0, \infty)$; (iii) $(-\infty, -1)$, $(-1, 1)$ and $(1, \infty)$.

SECTION 1.4

1. $y(x) = ce^{x^2}$. **3.** $y(x) = \ln(c - e^{-x})$. **5.** $y(x) = c(x - 2)$.

7. $y(x) = \dfrac{cx - 3}{2x - 1}$. **9.** $y(x) = \dfrac{(x - 1) + c(x - 2)^2}{(x - 1) - c(x - 2)^2}$ and $y(x) = -1$.

11. $y(x) = c + c_1\left(\dfrac{x - a}{x - b}\right)^{1/(a-b)}$. **13.** $y(x) = a(1 - \sqrt{1 - x^2}\,)$. **15.** $y(x) = 0$.

17. (a) $v(t) = a\left(\dfrac{e^{gt/a} - e^{-gt/a}}{e^{gt/a} + e^{-gt/a}}\right) = a\tanh(gt/a)$, where $a = (mg/k)^{1/2}$. **(b)** No.

(c) $y(t) = \dfrac{a^2}{g}\ln[\cosh(gt/a)]$. **19. (b)** $x(t) = \dfrac{1}{2}\ln(1 + v_0{}^2)$. **23.** $t \approx 96.4$ minutes.
25. (a) $500°F$; **(b)** $\approx 6{:}07$ P.M.

SECTION 1.5

1. 2560. **3.** $t \approx 35.86$ h. **5.** 1091. **7. (a)** $P_1 > \dfrac{2P_0P_2}{P_0 + P_2}$, $P_1{}^2 > P_0P_2$. **(b)** No.

15. $r \approx 0.046$, $C \approx 263.95$.

SECTION 1.6

1. $y(x) = e^x(e^x + c)$. **3.** $y(x) = x^2 - 1 + ce^{-x^2}$. **5.** $y(x) = \dfrac{1}{1 + x^2}(4\tan^{-1}x + c)$.

7. $y(x) = \dfrac{x^3(3\ln x - 1) + c}{\ln x}$. **9.** $x(t) = \dfrac{4e^t(t - 1) + c}{t^2}$. **11.** $y(x) = (\tan x + c)\cos x$.

13. $y(x) = \begin{cases} \dfrac{1}{\alpha + \beta}e^{\beta x} + ce^{-\alpha x}, & \alpha + \beta \neq 0, \\ e^{-\alpha x}(x + c), & \alpha + \beta = 0. \end{cases}$ **15.** $y(x) = x^{-2}(x^4 + 1)$.

17. $x(t) = (4 - t)(1 + t)$. **19.** $y(x) = \begin{cases} 2e^{-x} + 1, & \text{if } x \leq 1, \\ e^{-x}(e + 2), & \text{if } x > 1. \end{cases}$

25. (c) $t_{\max} = 40\ln 2$. $T(t_{\max}) = 20 = T_m(t_{\max})$. **29.** $y(x) = x^{-1}(\cos x + x\sin x + c)$.
31. $y(x) = \sin x + c\csc x$.

SECTION 1.7

1. 196 g. **3.** 300 g. **5. (a)** 6.75 g. **(b)** $15(2)^{1/3}$L. **9.** $i(t) = 5(1 - e^{-40t})$.

11. $i(t) = \dfrac{3}{5}(3\sin 4t - 4\cos 4t + 4e^{-3t})$.

15. $i(t) = \dfrac{E_0}{R^2 + L^2\omega^2}(R\sin \omega t - \omega L\cos \omega t) + Ae^{-Rt/L}$,

$i_S = \dfrac{E_0}{R^2 + L^2\omega^2}(R\sin \omega t - \omega L\cos \omega t)$, $i_T = Ae^{-Rt/L}$. **17.** $i(t) = \dfrac{E_0 C}{1 - aRC}\left(\dfrac{1}{RC}e^{-t/RC} - ae^{-at}\right)$.

SECTION 1.8

1. $F(V) = (1 - V^2)/V$. **3.** $F(V) = (\sin\dfrac{1}{V} - V\sin V)/V$. **5.** Not homogeneous.

7. $F(V) = -\sqrt{1 + V^2}$. **9.** $y^2 = ce^{-3x/y}$. **11.** $y(x) = x\cos^{-1}(c/x)$. **13.** $y + \sqrt{9x^2 + y^2} = c_1x^2$.

15. $y(x) = xe^{(cx + 1)}$.　**17.** $y^2 = x^2\ln(\ln cx)$.　**19.** $y^2 = c^2 - 2cx$.　**21.** $y(x) = x\,\sin^{-1}cx$.
25. $2y^2 + xy - x^2 = 2$.　**27.** $\sin^{-1}(y/2x) = \ln x + c$.　**29.** $x^2 + y^2 = 2kx$.
31. (b) $(x + cm)^2 + (y - c)^2 = c^2(1 + m^2)$.　**33.** $(3y - x)^3 = k(2y - x)^4$.
35. (a) $(y^2 - x^2)\tan(\alpha_0) - 2xy = k$.　**37.** $y^2 = x^2(8 \sin x + c)$.

39. $y^{2/3} = x[x^2(2 \ln x - 1) + c]$.　**41.** $y(x) = \dfrac{1}{x^2(c - 2x^3)}$.

43. $y(x) = \dfrac{1}{4}\left(\dfrac{x - b}{x - a}\right)^2 [x + (b - a)\ln | x - b | + c]^2$.

45. $y(x) = [(x^2 - 1) + ce^{-x^2}]^2$.　**47.** $y(x) = \left(\dfrac{x^3 + c}{x}\right)^{1/(1 - \pi)}$.

49. $y(x) = \left(1 + \dfrac{c}{\sec x + \tan x}\right)^{1/(1 - \sqrt{3})}$.　**51.** $y^2 = \dfrac{1}{\sin^2 x(2 \cos x + 1)}$.

53. $y(x) = 3(3x - \tanh 3x)$.　**55.** $y(x) = \dfrac{1}{3}[3x - \tan^{-1}(3x + c) + 1]$.　**57.** $y(x) = x^{-1}e^{cx}$.

59. $y(x) = x^{-1}\left[\dfrac{1}{c - \ln x} - 1\right]$.　**61.** (b) $y(x) = x^{-1}\left[1 + \dfrac{1}{c - 3 \ln x}\right]$.　**63.** $y(x) = xe^{x^2}$.
65. $y(x) = \tan^{-1}(1 + ce^{-\sqrt{1 + x}})$.

SECTION 1.9

1. Exact.　**3.** Not exact.　**5.** $xy^2 - \cos y + \sin x = c$.　**7.** $6e^{2x} + 3x^2y - 3xy^2 + y^3 = c$.
9. $\tan^{-1}(y/x) + \ln x = c$.　**11.** $\sin xy + \cos x = c$.　**13.** $y(x) = x^{-1}(x^3\ln x + 5)$.
15. $y(x) = x^{-1}\ln(2 - \sin x)$.　**17.** Yes.　**19.** Yes.　**21.** $2x - y^4 = cy^2$.　**23.** $y(x) = \dfrac{c + 2x^{5/2}}{10x^{1/2}}$.

25. $y(x) = \dfrac{c + \tan^{-1}x}{1 + x^2}$.　**27.** $r = 2, s = 4$.　**29.** $r = 1, s = 2$.

SECTION 1.10

1. $y^2 = 2[(\ln x)^2 + c]$.　**3.** $y^2 + x^2y + c = 0$.　**5.** $y(x) = \dfrac{2 \cos x}{\sin^2x + c}$.　**7.** $y(x) = x\,\sin(\ln cx)$.
9. $y^2 = x^{-2}[x^5(5 \ln x - 1) + c]$.　**11.** $y(x) = \csc x \ln(c \sec x)$.　**13.** $y(x) = xe^{cx}$.
15. $y(x) = 1 + ce^{\cos x}$.　**17.** $y^2 = \dfrac{\ln x}{c + x^3(3 \ln x - 1)}$.

19. $y(x) = \left(\dfrac{x + 1}{x - 1}\right)(x - 2 \ln| x + 1 | + c)$.

SECTION 1.11

1. $y(0.5) \approx 4.8938$.　**3.** $y(0.5) \approx 1.0477$.　**5.** $y(1) \approx 0.8564$.　**7.** $y(1) \approx 0.5012$.
9. $y(1) \approx 0.7115$.　**11.** $y(0.5) \approx 5.79167$.　**13.** $y(0.5) \approx 1.0878$.　**15.** $y(1) \approx 0.9999$.

SECTION 1.12

1. $y(x) = c_1x^3 + x^4 + c_2$.　**3.** $y(x) = (c_1 + c_2e^x)^{1/3}$.　**5.** $y(x) = c_2 - \ln| c_1 - \sin x |$.
7. $y(x) = \dfrac{1}{3}x^6 + c_1x^3 + c_2$.　**9.** $y(x) = -\dfrac{1}{\alpha}\ln| c_1 + c_2e^{\beta x} |$.　**11.** $y(x) = c_1\tan^{-1}x + c_2$.

13. $y(x) = \ln(\sec x) + c_1 \ln(\sec x + \tan x) + c_2$.   **15.** $y(x) = a \cosh \omega x$.

17. $y(x) = \frac{1}{2}x^2 + c_1 x^3 + c_2 x + c_3$.

SECTION 2.1

3. General solution: $y(x) = c_1 \cos 2x + c_2 \sin 2x$.

5. $r = 2, -4$; general solution: $y(x) = c_1 x^2 + c_2 x^{-4}$.

7. (a) $y_c(x) = c_1 e^{-2x} + c_2 e^x$. **(b)** $y_p(x) = -(3 + 2x + 2x^2)$.

(c) $y(x) = c_1 e^{-2x} + c_2 e^x - (3 + 2x + 2x^2)$.

SECTION 2.2

1. $y_2(x) = x^2 \ln x$.   **3.** $y_2(x) = e^x \ln x$.   **5.** $y_2(x) = \frac{1}{2} x \ln\left(\frac{1+x}{1-x}\right) - 1$.   **7. (a)** $y_1(x) = x^m$.

(b) $y_2(x) = x^m \ln x$.   **11.** $y(x) = e^{2x}[c_1 + c_2 x + x^2(2 \ln x - 3)]$.

13. $y(x) = c_1 \cos x + c_2 \sin x + \sin x[\ln(\sin x) - x \cot x]$.

15. $y(x) = x^2(c_1 + c_2 \ln x + 2x^2)$.

SECTION 2.3

1. $y(x) = c_1 e^{2x} + c_2 e^{-x}$.   **3.** $y(x) = e^{-3x}(c_1 \cos 4x + c_2 \sin 4x)$.   **5.** $y(x) = e^{-2x}(c_1 + c_2 x)$.

7. $y(x) = e^{-5x}(c_1 + c_2 x)$.   **9.** $y(x) = e^{-4x}(c_1 \cos 2x + c_2 \sin 2x)$.   **11.** $y(x) = c_1 e^{4x} + c_2 e^{-2x}$.

13. $y(x) = 2e^{2x} + e^{-3x}$.   **15.** $y(x) = e^{2x}(3 \cos x - \sin x)$.

19. (b) $u(x, y) = e^{x/\alpha}[e^{-p\xi}(A \sin q\xi + B \cos q\xi]$.

SECTION 2.4

1. $y(x) = c_1 e^{2x} + c_2 e^x + 2e^{3x}$.   **3.** $y(x) = e^x(c_1 + c_2 x) + 3x^2 - 6$.

5. $y(x) = c_1 e^{-x} + e^{3x}(c_2 + 2x)$.   **7.** $y(x) = c_1 \cos 4x + c_2 \sin 4x + \frac{1}{17}x(24 \cos 4x + 96 \sin 4x)$.

9. $y(x) = e^{-x}(c_1 \cos x + c_2 \sin x) + 2 - 4x + 2x^2$.   **11.** $y(x) = (c_1 + 2x^2)\sin 2x + (c_2 + x)\cos 2x$.

13. $y(x) = c_1 e^{-x} + c_2 e^{2x} - 10 + 3 \cos 2x + \sin 2x$.   **15.** $y(x) = c_1 e^{2x} + c_2 e^{-2x} - \frac{2}{5} \cos x + \frac{1}{2} xe^{2x}$.

17. $y(x) = 8e^x - 4e^{-x} + e^{2x}(3x - 4)$.   **19.** $y(x) = e^{-2x} + \cos x + 3 \sin x$.

21. $q(t) = e^{-10t}(A_0 \cos 20\sqrt{2}t + B_0 \sin 20\sqrt{2}t) + 2 \sin 30t$.

$i(t) = e^{-10t}[(-10A_0 + 20\sqrt{2}B_0)\cos 20\sqrt{2}t - (20\sqrt{2}A_0 + 10B_0)\sin 20\sqrt{2}t] + 60 \cos 30t$.

SECTION 2.5

1. $y_p(x) = -(3 \cos 3x + 4 \sin 3x)$.   **3.** $y_p(x) = -(12 \cos 3x + 5 \sin 3x)$.

5. $y_p(x) = \frac{3}{10} e^x(2 \sin 2x - \cos 2x)$.   **7.** $y_p(x) = e^x[2 \sin x - 14 \cos x - 10x(2 \sin x + \cos x)]$.

9. $y_p(x) = 4xe^x \sin 3x$.   **11.** $y_p(t) = \dfrac{F_0}{\omega_0^2 - \omega^2} \cos \omega t$, $\omega_0 \neq \omega$; $y_p(t) = \dfrac{F_0}{2\omega_0} t \sin \omega_0 t$, if $\omega = \omega_0$.

SECTION 2.6

1. $\omega_0 = 2$, $A_0 = 2\sqrt{2}$, $\phi = \pi/4$, $T = \pi$.   **3. (a)** $k = 3$. **(b)** $\omega_0 = \sqrt{3}/2$, $A_0 = 2\sqrt{3}/3$, $\phi = 5\pi/6$,

$T = 4\pi\sqrt{3}/3$.   **5.** Overdamped, $y(t) = e^{-2t}(2e^t - 1)$.   **7.** Critically damped, $y(t) = e^{-t}(t - 1)$.

9. Underdamped, $y(t) = \frac{2}{3}e^{-2t}(3 \cos \sqrt{3}t + 5\sqrt{3} \sin \sqrt{3}t)$.

11. (a) $0 < \alpha < 1$. **(b)** $\alpha = 1$. **(c)** $\alpha > 1$. In (c), $y(t) = e^{-\alpha t}(c_1 e^{\mu t} + c_2 e^{-\mu t})$, $\mu = \sqrt{\alpha^2 - 1}$. The system does pass through equilibrium.

15. The IVP governing the motion is $\dfrac{d^2y}{dt^2} + \dfrac{4g}{L} y = 0$, $y(0) = \dfrac{L}{2}$, $\dfrac{dy}{dt}(0) = 0$.

Circular frequency: $\omega_0 = 2\sqrt{\dfrac{g}{L}}$. Period: $T = \pi\sqrt{\dfrac{L}{g}}$. **17.** $A_0 = \sqrt{\dfrac{\alpha^2 g + \beta^2 L}{g}}$; the phase, ϕ, is

defined by $\cos\phi = \dfrac{\alpha}{A_0}$, $\sin\phi = \dfrac{\beta}{A_0}\sqrt{\dfrac{L}{g}}$. Period: $T = 2\pi\sqrt{\dfrac{L}{g}}$. **19.** ≈ 63. **21.** $T = \dfrac{2\pi}{\omega}$, where

$\omega^2 = \dfrac{g(L_0^2 + L_0 L + L^2)}{L^2\sqrt{L^2 - L_0^2}}$. **23.** $y(t) = \dfrac{F_0}{2\omega_0}\sin(\omega_0 t) - \dfrac{F_0}{2\omega_0}t\cos(\omega_0 t)$. **27.** $T = 8\pi$.

31. $y(t) = e^{-t}(A \cos 2t + B\sin 2t) + te^{-t}\sin 2t$. As $t \to \infty$, $y(t) \to 0$.

SECTION 2.7

1. $i_S(t) = 2(3 \cos 3t + 2 \sin 3t)$. **5.** $i(t) = -A_0 e^{-3t}[3 \cos(t - \phi) + \sin(t - \phi)] - \dfrac{2\omega}{H}\sin(\omega t - \eta)$,

where $H = \dfrac{1}{2}\sqrt{(10 - \omega^2)^2 + 36\omega^2}$, $\cos\eta = \dfrac{10 - \omega^2}{2H}$, $\sin\eta = \dfrac{3\omega}{H}$. The maximum value of the

amplitude occurs when $\omega = \sqrt{10}$. **7.** $i(t) = A_0 e^{-Rt/(2L)}\cos(\mu t - \phi) - \dfrac{aCE_0}{(a^2LC - acR + 1)}e^{-at}$, where

$\mu = \dfrac{\sqrt{(4L/C) - R^2}}{2L}$.

SECTION 2.8

1. $y(x) = e^{-3x}[c_1 + c_2 x + 2x \tan^{-1}x - \ln(x^2 + 1)]$.
3. $y(x) = e^{2x}[\cos x\,(c_1 - \ln|\sec x + \tan x|) + c_2\sin x]$. **5.** $y(x) = e^{-2x}(c_1 + c_2 x - \ln x)$.
7. $y(x) = c_1 e^x + c_2 e^{-x} + 4 \tan^{-1}(e^x)\cosh x$. **9.** $y(x) = e^x[c_1 + c_2 x + x^{-1}(2 \ln x + 3)]$.

11. $y(x) = e^{-x}\left[c_1\cos 4x + c_2\sin 4x + \cos 4x \ln\left(\dfrac{\cos 4x + 2}{\cos 4x - 2}\right) + \dfrac{4}{\sqrt{3}}\sin 4x \tan^{-1}\left(\dfrac{\sin 4x}{\sqrt{3}}\right)\right]$.

13. $y(x) = e^{5x}[c_1 + c_2 x - \ln(4 + x^2) + x \tan^{-1}(x/2)]$.
15. $y(x) = c_1\cos x + c_2\sin x + 2e^x + \cos x \ln(\cos x) + x \sin x$.
17. $y(x) = c_1 e^{-2x} + c_2 x e^{-2x} + 3 \cos x + 4 \sin x + \dfrac{15}{4}x^2 e^{-2x}(2\ln x - 3)$.

19. $y_p(x) = \displaystyle\int_{x_0}^{x} \sinh(x - t)\, F(t)\, dt$. **21.** $y_p(x) = \dfrac{1}{3}\displaystyle\int_{x_0}^{x}[e^{(t-x)} - e^{4(t-x)}]F(t)\, dt$.
23. $y(x) = e^{2x}(1 - 2x + \frac{5}{6}x^3)$.

SECTION 2.9

1. $y(x) = x[c_1\sin(2 \ln x) + c_2\cos(2 \ln x)]$. **3.** $y(x) = x^2(c_1 + c_2\ln x)$. **5.** $y(x) = x^{-1}(c_1 + c_2\ln x)$.
7. $y(x) = c_1 x^7 + c_2 x^{-5}$. **9.** $y(x) = c_1 x^m + c_2 x^{-m}$. **11.** $y(x) = x^m[c_1\cos(k \ln x) + c_2\sin(k\ln x)]$.
15. $y(x) = c_1 x^3 + x^2(c_2 - \sin x)$. **17.** $y(x) = x^2[c_1 + c_2\ln x + \ln x(\ln|\ln x| - 1)]$.
19. $y(x) = c_1\cos(3 \ln x) + c_2\sin(3 \ln x) + \ln x$.

21. $y(x) = y_c(x) + y_p(x)$ where $y_c(x) = c_1 x^m + c_2 x^m \ln x$, and

$$y_p(x) = \begin{cases} \dfrac{x^m (\ln x)^{k+2}}{(k+1)(k+2)}, & \text{if } k \neq -1, -2, \\[2mm] (\ln|\ln x| - 1)x^m \ln x, & \text{if } k = -1, \\[2mm] -x^m(1 + \ln|\ln x|), & \text{if } k = -2. \end{cases}$$

SECTION 3.1

1. $a_{31} = 0$, $a_{24} = -1$, $a_{14} = 2$, $a_{32} = 2$, $a_{21} = 7$, $a_{34} = -4$.   **3.** $a = \begin{bmatrix} 2 & 1 & -1 \\ 0 & 4 & -2 \end{bmatrix}$.

5. $\begin{bmatrix} 1 & -3 & -2 \\ 3 & 6 & 0 \\ 2 & 7 & 4 \\ -4 & -1 & 5 \end{bmatrix}$.   **7.** 4.   **9.** -1.   **11.** Column vectors: $\begin{bmatrix} 1 \\ -1 \\ 2 \end{bmatrix}$, $\begin{bmatrix} 3 \\ -2 \\ 6 \end{bmatrix}$, $\begin{bmatrix} -4 \\ 5 \\ 7 \end{bmatrix}$.

Row vectors: $[1 \quad 3 \quad -4]$, $[-1 \quad -2 \quad 5]$, $[2 \quad 6 \quad 7]$.   **13.** $A = \begin{bmatrix} 1 & 2 \\ 3 & 4 \\ 5 & 1 \end{bmatrix}$.   **15.** $q \times p$.

SECTION 3.2

1. $2A = \begin{bmatrix} 2 & 4 & -2 \\ 6 & 10 & 4 \end{bmatrix}$, $-3B = \begin{bmatrix} -6 & 3 & -9 \\ -3 & -12 & -15 \end{bmatrix}$, $A - 2B = \begin{bmatrix} -3 & 4 & -7 \\ 1 & -3 & -8 \end{bmatrix}$.

3. $AB = \begin{bmatrix} 5 & 10 & -3 \\ 27 & 22 & 3 \end{bmatrix}$, CA not possible, $DB = [6 \quad 14 \quad -4]$, $CD = \begin{bmatrix} 2 & -2 & 3 \\ -2 & 2 & -3 \\ 4 & -4 & 6 \end{bmatrix}$.

5. $AB = \begin{bmatrix} -9 - 21i & 11 + 10i \\ -1 + 19i & 9 + 15i \end{bmatrix}$.   **7.** $ABC = \begin{bmatrix} -12 & -22 \\ 14 & -126 \end{bmatrix}$.

9. $\begin{bmatrix} 0 \\ -38 \end{bmatrix}$.   **11.** $\begin{bmatrix} -7 \\ 13 \\ 29 \end{bmatrix}$.   **13. (a)** $A^2 = \begin{bmatrix} -1 & -4 \\ 8 & 7 \end{bmatrix}$, $A^3 = \begin{bmatrix} -9 & -11 \\ 22 & 13 \end{bmatrix}$, $A^4 = \begin{bmatrix} -31 & -24 \\ 48 & 17 \end{bmatrix}$.

(b) $A^2 = \begin{bmatrix} -2 & 0 & 1 \\ 4 & -3 & 0 \\ 2 & 4 & -1 \end{bmatrix}$, $A^3 = \begin{bmatrix} 4 & -3 & 0 \\ 6 & 4 & -3 \\ -12 & 3 & 4 \end{bmatrix}$, $A^4 = \begin{bmatrix} 6 & 4 & -3 \\ -20 & 9 & 4 \\ 10 & -16 & 3 \end{bmatrix}$.

17. $x = 2$, $y = -1$, or $x = -1$, $y = 2$.   **19.** $\begin{bmatrix} -6 & 1 \\ 2 & 6 \end{bmatrix}$.   **27.** $x = \pm\dfrac{\sqrt{2}}{2}$, $y = \pm\dfrac{\sqrt{3}}{3}$, $z = \pm\dfrac{\sqrt{6}}{6}$.

SECTION 3.3

7. $A = \begin{bmatrix} 1 & 1 & 1 & -1 \\ 2 & 4 & -3 & 7 \end{bmatrix}$, $b = \begin{bmatrix} 3 \\ 2 \end{bmatrix}$, $A^\# = \begin{bmatrix} 1 & 1 & 1 & -1 & 3 \\ 2 & 4 & -3 & 7 & 2 \end{bmatrix}$.   **9.** $\begin{aligned} x_1 - x_2 + 2x_3 + 3x_4 &= 1, \\ x_1 + x_2 - 2x_3 + 6x_4 &= -1, \\ 3x_1 + x_2 + 4x_3 + 2x_4 &= 2. \end{aligned}$

SECTION 3.4

1. Row-echelon. **3.** Reduced row-echelon. **5.** Reduced row-echelon.

7. Reduced row-echelon. **9.** $\begin{bmatrix} 1 & -3 \\ 0 & 1 \end{bmatrix}$, rank$(A) = 2$. **11.** $\begin{bmatrix} 1 & 1 & 2 \\ 0 & 1 & 0 \\ 0 & 0 & 0 \end{bmatrix}$, rank$(A) = 2$.

13. $\begin{bmatrix} 1 & 3 \\ 0 & 1 \\ 0 & 0 \end{bmatrix}$, rank$(A) = 2$. **15.** $\begin{bmatrix} 1 & -2 & 1 & 3 \\ 0 & 1 & \frac{1}{3} & -\frac{2}{3} \\ 0 & 0 & 0 & 0 \end{bmatrix}$, rank$(A) = 2$. **17.** $\begin{bmatrix} 1 & 2 & 1 & 2 \\ 0 & 1 & 0 & 1 \\ 0 & 0 & 0 & 0 \\ 0 & 0 & 0 & 0 \end{bmatrix}$,

rank$(A) = 2$. **19.** I_2, rank$(A) = 2$. **21.** $\begin{bmatrix} 1 & -1 & 2 \\ 0 & 0 & 0 \\ 0 & 0 & 0 \end{bmatrix}$, rank$(A) = 1$. **23.** I_4, rank$(A) = 4$.

25. $\begin{bmatrix} 0 & 1 & 0 & 0 \\ 0 & 0 & 1 & 0 \\ 0 & 0 & 0 & 1 \end{bmatrix}$, rank$(A) = 3$.

SECTION 3.5

1. $(-2, 2, -1)$. **3.** No solution. **5.** $(2, -1, 3)$. **7.** $\{(1 - 2r + s - t, r, s, t) : r, s, t \in \mathbf{R}\}$.
9. $\{(1 - 2r + 3s - t, r, 2 - 4s + 3t, s, t) : r, s, t \in \mathbf{R}\}$. **11.** No solution. **13.** $\{(-s + t, -3 + s + t, s, t) : s, t \in \mathbf{R}\}$. **15.** $(1, -3, 4, -4, 2)$. **17.** $\{(-5t, -2t - 1, t), : t \in \mathbf{R}\}$. **19.** No solution.
21. (a) $k = 2$. (b) $k = -2$. (c) $k \neq \pm 2$. **31.** Trivial solution. **33.** $\{(-t, -3t, t) : t \in \mathbf{R}\}$.
35. $\{(t, 0, -3t) : t \in \mathbf{R}\}$. **37.** $\{(2t(-1 + i), t, (2 - i)t : t \in \mathbf{C}\}$. **39.** $\{(2r - 3s, r, s) : r, s \in \mathbf{R}\}$.
41. Trivial solution. **43.** $\{(t(1 - i), t) : t \in \mathbf{C}\}$. **45.** Trivial solution.
47. $\{(r(1 - 5i) + s(5 + i), 13r, 13s) : r, s \in \mathbf{C}\}$. **49.** $\{(-3t, -2t, t) : t \in \mathbf{R}\}$.
51. $\{(3s, r, s, t) : r, s, t \in \mathbf{R}\}$. **53.** (a) $k \neq -1$.

SECTION 3.6

5. $A^{-1} = \begin{bmatrix} -1 & 1 + i \\ 1 - i & -1 \end{bmatrix}$. **7.** A is singular. **9.** $A^{-1} = \begin{bmatrix} 8 & -29 & 3 \\ -5 & 19 & -2 \\ 2 & -8 & 1 \end{bmatrix}$.

11. $A^{-1} = \begin{bmatrix} 18 & -34 & -1 \\ -29 & 55 & 2 \\ 1 & -2 & 0 \end{bmatrix}$. **13.** $A^{-1} = \begin{bmatrix} -i & 1 & 0 \\ 1 - 5i & i & 2i \\ -2 & 0 & 1 \end{bmatrix}$.

15. $A^{-1} = \begin{bmatrix} 27 & 10 & -27 & 35 \\ 7 & 3 & -8 & 11 \\ -14 & -5 & 14 & -18 \\ 3 & 1 & -3 & 4 \end{bmatrix}$. **17.** $(4, -1)$. **19.** $(-2, 2, 1)$. **21.** $(-6, 1, 3)$.

33. (b) All right inverses are of the form $\begin{bmatrix} 7 + 5r & -3 + 5s \\ -2 - 2r & 1 - 2s \\ r & s \end{bmatrix}$, where r and s are arbitrary real

numbers.

SECTION 3.7

3. $M_2(-\frac{1}{7}) A_{12}(-5) P_{12}$.

5. $A_{23}(2) M_2(-1) A_{13}(-3) A_{12}(-2) = \begin{bmatrix} 1 & 0 & 0 \\ 0 & 1 & 0 \\ 0 & 2 & 1 \end{bmatrix} \begin{bmatrix} 1 & 0 & 0 \\ 0 & -1 & 0 \\ 0 & 0 & 1 \end{bmatrix} \begin{bmatrix} 1 & 0 & 0 \\ 0 & 1 & 0 \\ -3 & 0 & 1 \end{bmatrix} \begin{bmatrix} 1 & 0 & 0 \\ -2 & 1 & 0 \\ 0 & 0 & 1 \end{bmatrix}$.

7. $E_1 = \begin{bmatrix} 1 & 0 \\ \frac{1}{3} & 1 \end{bmatrix}$. $L = \begin{bmatrix} 1 & 0 \\ -\frac{1}{3} & 1 \end{bmatrix}$. **9.** $L = \begin{bmatrix} 1 & 0 \\ \frac{5}{3} & 1 \end{bmatrix}$, $U = \begin{bmatrix} 3 & 1 \\ 0 & \frac{1}{3} \end{bmatrix}$.

11. $L = \begin{bmatrix} 1 & 0 & 0 \\ -2 & 1 & 0 \\ 3 & -2 & 1 \end{bmatrix}$, $U = \begin{bmatrix} 5 & 2 & 1 \\ 0 & 2 & 5 \\ 0 & 0 & 4 \end{bmatrix}$. **13.** $L = \begin{bmatrix} 1 & 0 & 0 & 0 \\ 2 & 1 & 0 & 0 \\ -4 & -2 & 1 & 0 \\ 3 & 2 & 1 & 1 \end{bmatrix}$, $U = \begin{bmatrix} 2 & -3 & 1 & 2 \\ 0 & 5 & -1 & -3 \\ 0 & 0 & 4 & - \\ & & & \end{bmatrix}$

3,0, 0, 0, 5)).

15. $\mathbf{x} = (1, 5, -1)$. **17.** $\mathbf{x} = (\frac{677}{1300}, -\frac{9}{325}, -\frac{37}{65}, \frac{4}{13})$. **19.** $A^{-1} = \begin{bmatrix} -\frac{29}{13} & \frac{18}{13} & -\frac{14}{13} \\ -\frac{17}{13} & \frac{11}{13} & -\frac{10}{13} \\ 2 & -1 & 1 \end{bmatrix}$.

SECTION 4.1

1. (a) odd. **(b)** odd. **(c)** even. **(d)** even. **3.** 5. **5.** -17. **7.** $(2 - \sqrt{2})\pi^2$. **9.** 19.
15. (b) 70. **17. (a)** Is a term; plus sign. **(b)** Not a term. **(c)** Is a term; plus sign.
19. $(-1)^{n(n-1)/2}$.

SECTION 4.2

1. -36. **3.** 45. **5.** -103. **7.** 624. **9.** 21. **11.** 84. **13.** Nonsingular. **15.** Nonsingular.
17. Nonsingular. **19.** Singular. **21.** $k = -4, k = 1/3$. **23.** $\det(A) = 14 = \det(A^T)$,
$\det(-2A) = -112$. **25.** 1. **29.** $0, -1, 2$. **31.** -1.

SECTION 4.3

1. $M_{11} = 4, M_{21} = -3, M_{12} = 2, M_{22} = 1, C_{11} = 4, C_{21} = 3, C_{12} = -2, C_{22} = 1$. **3.** Cofactors:
$C_{11} = -5, C_{12} = 0, C_{13} = 4, C_{21} = -47, C_{22} = -2, C_{23} = 38, C_{31} = 3, C_{32} = 0, C_{33} = -2$. **5.** 5.
7. -153. **9.** 0. **11.** 9. **13.** 3. **15.** -4. **17.** 11997. **19.** -170.

23. $A^{-1} = \begin{bmatrix} \frac{1}{7} & \frac{2}{7} \\ \frac{4}{-7} & \frac{1}{-7} \end{bmatrix}$. **25.** $A^{-1} = \begin{bmatrix} \frac{7}{26} & \frac{3}{13} & -\frac{15}{26} \\ -\frac{2}{13} & \frac{2}{13} & -\frac{5}{13} \\ \frac{1}{13} & \frac{1}{13} & \frac{4}{13} \end{bmatrix}$. **27.** $A^{-1} = \begin{bmatrix} -\frac{11}{6} & \frac{3}{2} & -\frac{1}{3} \\ -\frac{1}{6} & -\frac{1}{2} & \frac{1}{3} \\ \frac{4}{3} & -1 & \frac{1}{3} \end{bmatrix}$.

29. $A^{-1} = \begin{bmatrix} -\frac{9}{14} & \frac{16}{7} & -\frac{13}{14} \\ \frac{1}{14} & -\frac{1}{7} & \frac{3}{14} \\ \frac{1}{2} & -1 & \frac{1}{2} \end{bmatrix}$. **31.** $A^{-1} = \begin{bmatrix} \frac{14}{67} & -\frac{27}{67} & \frac{3}{67} & -\frac{5}{67} \\ -\frac{23}{201} & \frac{10}{67} & \frac{19}{201} & \frac{13}{201} \\ -\frac{29}{402} & \frac{33}{134} & -\frac{11}{402} & \frac{65}{201} \\ \frac{27}{134} & -\frac{9}{134} & \frac{1}{134} & -\frac{12}{67} \end{bmatrix}$. **33.** $\frac{9}{16}$.

35. $\begin{bmatrix} e^{-t}\sin 2t & e^{-t}\cos 2t \\ -e^t\cos 2t & e^t\sin 2t \end{bmatrix}$. **39.** $(\frac{9}{7}, -\frac{66}{7}, -\frac{11}{7})$. **41.** $(\frac{11}{3}, -\frac{17}{3}, -\frac{16}{3}, -2)$. **43.** $\frac{31}{19}$.

SECTION 4.4

1. 38.   **3.** −3.   **5.** $3abc - a^3 - b^3 - c^3$.   **7.** −2196.

9. $\det(A) = -18$, $A^{-1} = \begin{bmatrix} -\frac{5}{18} & \frac{1}{18} & \frac{7}{18} \\ \frac{1}{18} & \frac{7}{18} & -\frac{5}{18} \\ \frac{7}{18} & -\frac{5}{18} & \frac{1}{18} \end{bmatrix}$.   **11.** $\det(A) = 116$, $A^{-1} = \begin{bmatrix} -\frac{51}{116} & \frac{2}{29} & \frac{31}{116} \\ -\frac{8}{29} & -\frac{5}{29} & \frac{6}{29} \\ \frac{27}{58} & \frac{3}{29} & -\frac{13}{58} \end{bmatrix}$.

SECTION 5.1

1. $\mathbf{v}_1 = (6, 2)$, $\mathbf{v}_2 = (-3, 6)$, $\mathbf{v}_3 = (3, 8)$.

SECTION 5.2

1. Not closed.   **3.** Not closed   **5.** Closed.   **11.** It is a vector space.   **13.** A5: $\mathbf{0} = (1, 1)$. No additive inverse for $(0, 0)$ so A6 does not hold.   **15.** The only axioms to hold in the definition of vector space for the given operations are A1–A3, and A9.

SECTION 5.3

3. Subspace.   **5.** Not a subspace.   **7.** Not a subspace.   **9.** Subspace.   **11.** Subspace.
13. Subspace.   **15.** Subspace.   **17.** Not a subspace.   **19.** Not a subspace.
21. Nullspace$(A) = \{(8r + 8s, -2r - 2s, r, s) : r, s \in \mathbf{R}\}$.

SECTION 5.4

1. Spans $\mathbf{R}^2$.   **3.** Does not span $\mathbf{R}^2$.   **5.** Spans $\mathbf{R}^3$.
9. $(x_1, x_2, x_3) = \frac{1}{16}(-3x_1 + x_2 + 5x_3)\mathbf{v}_1 + \frac{1}{16}(-x_1 - 5x_2 + 7x_3)\mathbf{v}_2 + \frac{1}{16}(7x_1 + 3x_2 - x_3)\mathbf{v}_3$.
13. A spanning set is $\{(2, 1, 0), (1, 0, 1)\}$.   **15.** A spanning set is $\{(1, 1, -1, 0)\}$.
17. $\begin{bmatrix} 0 & 1 \\ -1 & 0 \end{bmatrix}$.   **19.** Span$\{\mathbf{v}_1, \mathbf{v}_2\} = \{\mathbf{v} \in \mathbf{R}^3 : \mathbf{v} = (a + 2b, -a - b, 2a + 3b), a, b \in \mathbf{R}\}$.

Geometrically, span$\{\mathbf{v}_1, \mathbf{v}_2\}$ is the plane through the origin determined by the two given vectors.
23. $(5, 3, -6) \in$ span$\{\mathbf{v}_1, \mathbf{v}_2\}$.   **25.** $p \in$ span$\{p_1, p_2\}$.
27. span$\{A_1, A_2\} = \left\{A \in M_2(\mathbf{R}) : A = \begin{bmatrix} a - 2b & 2a + b \\ -a + b & 3a - b \end{bmatrix}\right\}$. $B \in$ span$\{A_1, A_2\}$.

SECTION 5.5

1. LI.   **3.** LD.   **5.** LD.   **7.** LD.   **9.** LI.   **11. (b)** No.   **13.** $k \neq -1, 2$.   **15.** LD.
17. LD.   **21.** $\{(1, 2, 3), (-3, 4, 5)\}$.   **23.** $\{(1, -1, 1), (1, -3, 1), (3, 1, 2)\}$.
25. $\left\{\begin{bmatrix} 1 & 2 \\ 3 & 4 \end{bmatrix}, \begin{bmatrix} -1 & 2 \\ 5 & 7 \end{bmatrix}\right\}$.   **27.** $\{2 + x^2, 1 + x\}$.   **33.** LD.   **35.** LI. **37.** LI.
41. $\alpha \neq \pm 1$.

SECTION 5.6

1. Basis.   **3.** Not a basis.   **5.** Basis.   **7.** Basis: $\{1, x, x^2, x^3\}$.   **9.** Basis: $\{(-2, 2, 1)\}$;
dim[nullspace(A)] = 1.   **11.** Basis: $\{(3, 1, 0), (-1, 0, 1)\}$; dim$[S] = 2$.

13. Basis: $\left\{ \begin{bmatrix} 1 & 0 \\ 0 & 0 \end{bmatrix}, \begin{bmatrix} 0 & 1 \\ 0 & 0 \end{bmatrix}, \begin{bmatrix} 0 & 0 \\ 0 & 1 \end{bmatrix} \right\}$; dim[$S$] = 3.

15. Basis: {(1, 0, 1), (0, 1, 1)}; dim[S] = 2. **17.** Basis: $\left\{ \begin{bmatrix} 1 & 3 \\ -1 & 2 \end{bmatrix}, \begin{bmatrix} -1 & 4 \\ 1 & 1 \end{bmatrix} \right\}$.

19. Components of e_1: (−1, 1). Components of e_2: (3, −2). **21.** $\alpha \neq -1$.

23. Components: $(\frac{7}{3}, -1, \frac{2}{3})$. **27.** dim[$Sym_n(\mathbf{R})$] = $\dfrac{n(n+1)}{2}$, dim[$Skew_n(\mathbf{R})$] = $\dfrac{n(n-1)}{2}$.

29. Basis: $\left\{ \begin{bmatrix} 1 & 0 \\ 0 & 1 \end{bmatrix}, \begin{bmatrix} 0 & 1 \\ 1 & 0 \end{bmatrix} \right\}$. An example of an extended basis for $M_2(\mathbf{R})$:

$\left\{ \begin{bmatrix} 1 & 0 \\ 0 & 1 \end{bmatrix}, \begin{bmatrix} 0 & 1 \\ 1 & 0 \end{bmatrix}, \begin{bmatrix} 1 & 0 \\ 0 & 0 \end{bmatrix}, \begin{bmatrix} 0 & 1 \\ 0 & 0 \end{bmatrix} \right\}$. **31.** Basis: {$e^x, e^{-3x}$}.

33. Basis: {$e^{3x}\cos 4x, e^{3x}\sin 4x$}.

SECTION 5.7

1. Basis for rowspace(A): {(1, −2)}. Basis for colspace(A): {(1, −3)}.
3. Basis for rowspace(A): {(1, 2, 3), (0, 1, 2)}. Basis for colspace(A): {(1, 5, 9), (2, 6, 10)}.
5. Basis for rowspace(A): {(1, 2, −1, 3), (0, 0, 0, 1)}. Basis for colspace(A): {(1, 3, 1, 5), (3, 5, −1, 7)}. **7.** {(1, 3, 3), (0, 1, −2)}. **9.** {(1, 4, 1, 3), (0, 0, 1, −1)}.
11. Basis for rowspace(A): {(1, 2, 4), (0, 1, 1)}. Basis for colspace(A): {(1, 5, 3), 2, 11, 7)}.

SECTION 5.8

1. Rank(A) = 1, nullity(A) = 1, n = 2. **3.** Rank(A) = 2, nullity(A) = 2, n = 4.
5. x = (−1, −7, 0, 0) + c_1(1, 3, 1, 0) + c_2(−1, 2, 0, 1).

SECTION 5.9

1. $\theta \approx 0.95$ rad. **3.** <u, v> = 19 + 11i, ‖ u ‖ = $\sqrt{35}$, ‖ v ‖ = $\sqrt{22}$, $\theta \approx 0.66$ rad.
5. <A, B> = 12, ‖ A ‖ = $\sqrt{39}$, ‖ B ‖ = $\sqrt{15}$. **9.** Using (5.9.9): <x, y> = 0; using the standard inner product in $\mathbf{R}^2$, <x, y> = −1. **11.** Using (5.9.9): <x, y> = −5; using the standard inner product in $\mathbf{R}^2$, <x, y> = 0. **13.** x = r(1, 1), or r(1, −1), where $r \in \mathbf{R}$.

SECTION 5.10

1. Orthonormal set: $\left\{ \dfrac{1}{\sqrt{6}} (2, -1, 1), \dfrac{1}{\sqrt{3}} (1, 1, -1), \dfrac{1}{\sqrt{2}} (0, 1, 1) \right\}$. **3.** Not orthogonal.

5. Orthonormal set: $\left\{ \dfrac{1}{\sqrt{14}} (1, 2, 3), \dfrac{1}{\sqrt{3}} (1, 1, -1), \dfrac{1}{\sqrt{42}} (5, -4, 1) \right\}$.

7. Orthonormal set: $\left\{ \dfrac{1}{\sqrt{5}} (1 - i, 1 + i, i), \dfrac{1}{\sqrt{3}} (0, i, 1 - i), \dfrac{1}{30} (-3 + 3i, 2 + 2i, 2i) \right\}$.

9. Orthonormal set: $\left\{ \dfrac{1}{\sqrt{2}}, \sin \pi x, \cos \pi x \right\}$. **13.** $A_4 = k \begin{bmatrix} 3 & -1 \\ 2 & 0 \end{bmatrix}$.

15. Orthonormal basis: $\left\{ \frac{1}{3}(2, 1, -2), \frac{1}{3\sqrt{2}}(-1, 4, 1) \right\}.$

17. Orthonormal basis: $\left\{ \frac{1}{\sqrt{2}}(1, 0, -1, 0), (0, 1, 0, 0), \frac{1}{\sqrt{6}}(-1, 0, -1, 2) \right\}.$

19. Orthonormal basis: $\left\{ \frac{1}{\sqrt{3}}(1, 1, -1, 0), \frac{1}{\sqrt{15}}(-1, 2, 1, 3), \frac{1}{\sqrt{15}}(3, -1, 2, 1) \right\}.$

21. Orthonormal basis: $\left\{ \frac{1}{\sqrt{3}}(1, -i, 0, i), \frac{1}{\sqrt{21}}(1, 3 + 3i, 1 - i) \right\}.$

23. Orthogonal basis: $\left\{ 1, \frac{1}{2}(2x - 1), \frac{1}{6}(6x^2 - 6x + 1) \right\}.$

25. Orthogonal basis: $\left\{ 1, \sin x, \frac{1}{\pi}(\pi \cos x - 2) \right\}.$

27. Orthogonal basis: $\left\{ \begin{bmatrix} 0 & 1 \\ 1 & 0 \end{bmatrix}, \begin{bmatrix} 1 & 0 \\ 0 & 0 \end{bmatrix}, \begin{bmatrix} 0 & 0 \\ 0 & 1 \end{bmatrix} \right\}$, the subspace of all symmetric matrices in
$M_2(\mathbf{R})$. **29.** Orthogonal basis: $\{ 1 + x^2, 1 - x - x^2 + x^3, -3 - 5x + 3x^2 + x^3 \}.$

SECTION 6.1
21. $T(\mathbf{v}_1) = 8\mathbf{v}_1 - 4\mathbf{v}_2, T(\mathbf{v}_2) = -5\mathbf{v}_1 + 3\mathbf{v}_2.$ **23.** $T(\mathbf{v}) = (3a + b)\mathbf{v}_1 + (2b - a)\mathbf{v}_2.$

SECTION 6.2
5. Shear parallel to the x-axis. **7.** Shear parallel to the y-axis. **9.** Shear parallel to the x-axis, followed by a linear stretch in the y-direction followed by a shear parallel to the y-axis.
11. Reflection in the x-axis followed by a linear stretch in the y-direction.

SECTION 6.3
3. $\text{Ker}(T) = \{\mathbf{0}\}$, $\dim[\text{Ker}(T)] = 0$; $\text{Rng}(T) = \mathbf{R}^3$, $\dim[\text{Rng}(T)] = 3.$
5. $\text{Ker}(T) = \{\mathbf{x} \in \mathbf{R}^3 : \mathbf{x} = r(-2, 0, 1) + s(1, 1, 0), r, s \in \mathbf{R}\}$, $\dim[\text{Ker}(T)] = 2$;
$\text{Rng}(T) = \{\mathbf{y} \in \mathbf{R}^2 : \mathbf{y} = t(1, -3), t \in \mathbf{R}\}$, $\dim[\text{Rng}(T)] = 1.$ **7.** Basis: $\{e^x, e^{-x}\}.$
9. Basis: $\{\cos 3x, \sin 3x\}.$ **11. (a)** $\dim[\text{Ker}(T)] = 2.$ **(b)** $\dim[\text{Rng}(T)] = 1.$
13. $\text{Ker}(T)$ is the set of all matrices that commute with B.
15. $\text{Ker}(T) = \{r(-x^2 + x + 1) : r \in \mathbf{R}\}$, $\dim[\text{Ker}(T)] = 1$; $\text{Rng}(T) = P_2$, $\dim[\text{Rng}(T)] = 2.$
17. $\text{Ker}(T) = \{\mathbf{v} \in V : \mathbf{v} = r(-\mathbf{v}_1 + \mathbf{v}_2 + \mathbf{v}_3), r \in \mathbf{R}\}$, $\dim[\text{Ker}(T)] = 1$;
$\text{Rng}(T) = W$, $\dim[\text{Rng}(T)] = 2.$

SECTION 6.4
1. $(T_1 T_2)(x_1, x_2) = (-5(x_1 + x_2), x_1 + 15x_2)$, $(T_2 T_1)(x_1, x_2) = (7(2x_1 + x_2), 2(x_1 - 2x_2)).$
3. $\text{Ker}(T_1) = \{\mathbf{x} \in \mathbf{R}^2 : \mathbf{x} = r(1, 1), r \in \mathbf{R}\}$, $\text{Ker}(T_2) = \{\mathbf{0}\}$, $\text{Ker}(T_1 T_2) = \{\mathbf{x} \in \mathbf{R}^2 : \mathbf{x} = s(2, 1)\}$,
$\text{Ker}(T_2 T_1) = \text{Ker}(T_1).$ **5.** $[T_1(f)](x) = 1 - \cos(x - a).$ **9.** Not one-to-one, not onto.

11. $T^{-1}(\mathbf{x}) = \frac{1}{\lambda}\mathbf{x}.$ **13.** T is not one-to-one, but is onto. **17.** $T(a, b, c) = \begin{bmatrix} a & b \\ 0 & c \end{bmatrix}.$

SECTION 6.5

7. $\lambda_1 = 4$, $v_1 = r(1, -1)$; $\lambda_2 = -2$, $v_2 = s(1, 5)$. **9.** $\lambda = 5$, $v = r(-2, 1)$.
11. $\lambda_1 = 1 + 2i$, $v_1 = r(1, 1, -i)$; $\lambda_2 = 1 - 2i$, $v_2 = s(1, 1 + i)$.
13. $\lambda = 2$, $v = r(3, 2, 0) + s(-1, 0, 1)$. **15.** $\lambda = 1$, $v = r(0, -1, 1)$.
17. $\lambda = -1$, $v = r(-3, 0, 4) + s(1, 1, 0)$.
19. $\lambda_1 = 1$, $v_1 = r(1, 0, 0)$; $\lambda_2 = i$, $v_2 = s(0, 1, i)$; $\lambda_3 = -i$, $v_3 = t(0, 1, -i)$.
21. $\lambda_1 = 0$, $v_1 = r(-3, 9, 5)$; $\lambda_2 = 2$, $v_2 = s(1, 3, 1)$; $\lambda_3 = 4$, $v = t(1, 1, 1)$.
23. $\lambda_1 = -2$, $v_1 = r(-1, 0, 1) + s(-1, 1, 0)$; $\lambda_2 = 4$, $v_2 = t(1, 1, 1)$.
25. $\lambda_1 = i$, $v_1 = a(0, 0, i, 1) + b(-i, 1, 0, 0)$; $\lambda_2 = -i$, $v_2 = r(0, 0, -i, 1) + s(i, 1, 0, 0)$.

27. (c) $A^{-1} = \begin{bmatrix} \frac{2}{3} & \frac{1}{6} \\ -\frac{1}{3} & \frac{1}{6} \end{bmatrix}$. **29.** $A(3v_1 - v_2) = (12, -3)$.

SECTION 6.6

1. $\lambda_1 = 5$, $m_1 = 1$, basis for E_1: $\{(1, 1)\}$, $n_1 = 1$; $\lambda_2 = -1$, $m_2 = 1$, basis for E_2: $\{(-2, 1)\}$, $n_2 = 1$.
Nondefective. **3.** $\lambda_1 = 3$, $m_1 = 2$, basis for E_1: $\{(1, 1)\}$, $n_1 = 1$. Defective. **5.** $\lambda_1 = -2$, $m_1 = 1$,
basis for E_1: $\{(1, 1, 1)\}$, $n_1 = 1$; $\lambda_2 = 3$, $m_2 = 2$, basis for E_2: $\{(1, 0, 0), (0, -1, 4)\}$, $n_2 = 2$.
Nondefective. **7.** $\lambda_1 = 4$, $m_1 = 3$, basis for E_1: $\{(0, 0, 1), (1, 1, 0)\}$, $n_1 = 2$. Defective.
9. $\lambda_1 = 2$, $m_1 = 2$, basis for E_1: $\{(-1, 2, 0)\}$, $n_1 = 1$; $\lambda_2 = -3$, $m_2 = 1$, basis for E_2: $\{(-1, 1, 1)\}$,
$n_2 = 1$. Defective. **11.** $\lambda_1 = -1$, $m_1 = 3$, basis for E_1: $\{(-3, 0, 4), (1, 1, 0)\}$, $n_1 = 2$. Defective.
13. $\lambda_1 = 0$, $m_1 = 2$, basis for E_1: $\{(-2, 0, 1), (1, 1, 0)\}$, $n_1 = 2$; $\lambda_2 = 2$, $m_2 = 1$, basis for E_2: $\{(1,$
$1, 1)\}$, $n_2 = 1$. Nondefective. **15.** $\lambda_1 = 1$, $m_1 = 2$, basis for E_1: $\{(-1, 0, 1), (-1, 1, 0)\}$, $n_1 = 2$;
$\lambda_2 = -2$, $m_2 = 1$, basis for E_2: $\{(1, 1, 1)\}$, $n_2 = 1$. Nondefective. **17.** Defective.
19. Nondefective. **21.** $\lambda_1 = 1$, basis for E_1: $\{(-1, 1)\}$; $\lambda_2 = 5$, basis for E_2: $\{(1, 3)\}$. **23.** $\lambda_1 =$
5, basis for E_1: $\{(1, 0), (0, 1)\}$. **25.** $\lambda_1 = -2$, basis for E_1: $\{(1, 1, 0)\}$. **27.** $\lambda_1 = 2$, orthogonal
basis for E_1: $\{(1, 0, 1), (-1, 2, 1)\}$; $\lambda_2 = -1$, basis for E_2: $\{(-1, -1, 1)\}$. Vectors in E_2 are
orthogonal to the vectors in E_1. **29. (d)** Sum of eigenvalues $= 24$.
Product of eigenvalues $= -607$.

SECTION 6.7

1. $S = \begin{bmatrix} 1 & 2 \\ -2 & 1 \end{bmatrix}$, $S^{-1}AS = \text{diag}(3, -2)$. **3.** Not diagonalizable. **3.** $S = \begin{bmatrix} 15 & 0 & 0 \\ -7 & 7 & 1 \\ 2 & 1 & -1 \end{bmatrix}$,

$S^{-1}AS = \text{diag}(1, 4, -4)$. **7.** $S = \begin{bmatrix} 1 & -1 & -1 \\ 1 & 0 & 1 \\ 1 & 1 & 0 \end{bmatrix}$, $S^{-1}AS = \text{diag}(-4, 2, 2)$. **9.** Not diagonalizable.

11. $S = \begin{bmatrix} -2 & 4 + 3i & 4 - 3i \\ 1 & -2 + 6i & -2 - 6i \\ 2 & 5 & 5 \end{bmatrix}$, $S^{-1}AS = \text{diag}(0, 3i, -3i)$. **13.** $S = \begin{bmatrix} 3 & -3 & 1 \\ 0 & 1 & 2 \\ 1 & 0 & 2 \end{bmatrix}$, $S^{-1}AS =$

$\text{diag}(2, 2, 1)$. **15.** $x_1(t) = 5c_1 e^{5t} - 2e^{-t}$, $x_2(t) = c_1 e^{5t} + c_2 e^{-t}$.
17. $x_1(t) = -c_1 e^{3t} + 3c_2 e^{-t}$, $x_2(t) = c_1 e^{3t} - 5c_2 e^{-t}$.
19. $x_1(t) = c_1 \sin t - c_2 \cos t$, $x_2(t) = c_1 \cos t + c_2 \sin t$.
21. $x_1(t) = e^{2t}(c_1 - c_2) + c_3 e^{-t}$, $x_2(t) = c_1 e^{2t} - c_3 e^{-t}$, $x_3(t) = c_2 e^{2t} + c_3 e^{-t}$.

SECTION 6.8

1. $S = \begin{bmatrix} -\dfrac{1}{\sqrt{5}} & \dfrac{2}{\sqrt{5}} \\ \dfrac{2}{\sqrt{5}} & \dfrac{1}{\sqrt{5}} \end{bmatrix}$, $S^T AS = \text{diag}(-2, 3)$.   **3.** $S = \begin{bmatrix} \dfrac{1}{\sqrt{2}} & -\dfrac{1}{\sqrt{2}} \\ \dfrac{1}{\sqrt{2}} & \dfrac{1}{\sqrt{2}} \end{bmatrix}$, $S^T AS = \text{diag}(3, -1)$.

5. $S = \begin{bmatrix} -\dfrac{1}{\sqrt{2}} & -\dfrac{1}{\sqrt{3}} & \dfrac{1}{\sqrt{6}} \\ 0 & \dfrac{1}{\sqrt{3}} & \dfrac{2}{\sqrt{6}} \\ \dfrac{1}{\sqrt{2}} & -\dfrac{1}{\sqrt{3}} & \dfrac{1}{\sqrt{6}} \end{bmatrix}$, $S^T AS = \text{diag}(0, 0, 6)$.   **7.** $S = \begin{bmatrix} -\dfrac{1}{\sqrt{2}} & 0 & \dfrac{1}{\sqrt{2}} \\ \dfrac{1}{\sqrt{2}} & 0 & \dfrac{1}{\sqrt{2}} \\ 0 & 1 & 0 \end{bmatrix}$,

$S^T AS = \text{diag}(-1, 1, 1)$..   **9.** $S = \begin{bmatrix} \dfrac{1}{\sqrt{6}} & \dfrac{1}{\sqrt{2}} & -\dfrac{1}{\sqrt{3}} \\ -\dfrac{1}{\sqrt{6}} & \dfrac{1}{\sqrt{2}} & \dfrac{1}{\sqrt{3}} \\ \dfrac{2}{\sqrt{6}} & 0 & \dfrac{1}{\sqrt{3}} \end{bmatrix}$, $S^T AS = \text{diag}(-1, 1, 2)$.

11. $S = \begin{bmatrix} \dfrac{1}{\sqrt{3}} & -\dfrac{1}{\sqrt{2}} & -\dfrac{1}{\sqrt{6}} \\ \dfrac{1}{\sqrt{3}} & \dfrac{1}{\sqrt{2}} & -\dfrac{1}{\sqrt{6}} \\ \dfrac{1}{\sqrt{3}} & 0 & \dfrac{2}{\sqrt{6}} \end{bmatrix}$, $S^T AS = \text{diag}(1, -5, -5)$.

13. Principal axes: $\left\{ \left(\dfrac{1}{\sqrt{2}}, \dfrac{1}{\sqrt{2}} \right), \left(\dfrac{1}{\sqrt{2}}, -\dfrac{1}{\sqrt{2}} \right) \right\}$, reduced quadratic form: $4y_1^2 - 2y_2^2$.

15. Principal axes: $\left\{ \left(\dfrac{1}{\sqrt{2}}, \dfrac{1}{\sqrt{2}}, 0 \right), \left(-\dfrac{1}{\sqrt{6}}, \dfrac{1}{\sqrt{6}}, \dfrac{2}{\sqrt{6}} \right), \left(\dfrac{1}{\sqrt{3}}, -\dfrac{1}{\sqrt{3}}, \dfrac{1}{\sqrt{3}} \right) \right\}$, reduced quadratic

form: $2y_1^2 + 2y_2^2 - y_3^2$.   **19.** $\mathbf{v}_2 = (-2, 1)$.

SECTION 7.1

1. $Lf = 2(1 - x^2) + 3e^{2x}(x - 2)$.   **3.** $Lf = 24e^{2x}(x - 1)$.
7. $L_1 L_2 = D^2 + (1 - 2x^2)D - 2x(x + 2)$, $L_2 L_1 = D^2 + (1 - 2x^2)D - 2x^2$.
9. $L_2 = D + a_1(x) + c$, where c is a constant.
11. $(D^2 + 4xD - 6x^2)y = x^2 \sin x$, $y'' + 4xy' - 6x^2 y = 0$.
15. Linearly independent solutions: $\{e^{2x}, e^{-2x}, e^{3x}\}$, general solution: $y(x) = c_1 e^{2x} + c_2 e^{-2x} + c_3 e^{3x}$.
17. Linearly independent solutions: $\{x, x^2, x^{-1}\}$, general solution: $y(x) = c_1 x + c_2 x^2 + c_3 x^{-1}$.

SECTION 7.2

1. $y(x) = c_1 e^x + c_2 \cos x + c_2 \sin x$.   **3.** $y(x) = c_1 e^{2x} + c_2 e^{4x} + c_3 e^{-4x}$.
5. $y(x) = c_1 e^{-x} + c_2 x e^{-x} + c_3 \cos 2x + c_4 \sin 2x + x(c_5 \cos 2x + c_6 \sin 2x)$.

7. $y(x) = c_1 + c_2 x + c_3 e^x$.   **9.** $y(x) = c_1 e^{2x} + c_2 e^{-2x} + c_3 \cos 2x + c_4 \sin 2x$.
11. $y(x) = c_1 e^x + c_2 e^{-5x} + e^{2x}(c_3 \cos x + c_4 \sin x)$.
13. $y(x) = c_1 e^x + c_2 e^{-x} + e^x(c_3 \cos x + c_4 \sin x)$.   **15.** $y(x) = c_1 \cos 3x + c_2 \sin 3x + x(c_3 \cos 3x + c_4 \sin 3x) + x^2(c_5 \cos 3x + c_6 \sin 3x)$.   **17.** $y(x) = e^{2x} - e^{-2x} + 2xe^{2x}$.

SECTION 7.3

1. $y(x) = c_1 e^x + c_2 e^{-x} + c_3 e^{2x} + \frac{1}{2} e^{3x}$.   **3.** $y(x) = c_1 e^{-3x} + c_2 e^{2x} + c_3 e^{-2x} - \frac{6}{25} \cos x - \frac{2}{25} \sin x$.
5. $y(x) = e^{-x}(c_1 + c_2 x + c_3 x^2 + \frac{5}{6} x^3 e^{-x}$.   **7.** $D^2(D-1)$.   **9.** $(D^2 + 16)(D-7)^4$.
11. $(D^2 - 2D + 5)(D^2 + 4)$.   **13.** $(D^2 - 10D + 26)^3$.
15. $(D-4)^2(D^2 - 8D + 41)D^2(D^2 + 4D + 5)^3$.   **17.** $y(x) = e^{-2x}(c_1 + c_2 x + \frac{5}{6} x^3)$.
19. $y(x) = c_1 e^{2x} + c_2 e^{-x} + \frac{5}{9} e^{2x}(3x - 1)$.   **21.** $y(x) = c_1 e^{2x} + c_2 e^{-x} + c_3 e^{-3x} - \frac{37}{27} + \frac{10}{9} x - \frac{2}{3} x^2$.
23. $y(x) = e^{-x}(c_1 + c_2 x + c_3 x^2 + \frac{1}{3} x^3) + \frac{1}{9} e^{2x}$.   **25.** $y_p(x) = A_0 x e^{2x}$.
27. $y_p(x) = x^2 e^{-2x}(A_0 \cos 3x + B_0 \sin 3x)$.   **29.** $y_p(x) = xe^x + x^2(A_0 \sin 2x + B_0 \cos 2x)$.
31. $y_p(x) = x(A_0 e^{3x} + A_1 \sin x + B_1 \cos x)$

SECTION 7.4

1. $y_p(x) = x^3 e^{2x}(6 \ln x - 11)$.   **3.** $y_p(x) = e^{-x}[(x^2 - 1)\tan^{-1}x - \ln(1 + x^2)]$.

5. $y_p(x) = \dfrac{1}{96} \displaystyle\int_{x_0}^{x} F(t)[2e^{3(x-t)} + 6e^{-5(x-t)} - 8e^{-3(x-t)}]\,dt$.

7. $y_p(x) = \dfrac{1}{36} \displaystyle\int_{x_0}^{x} F(t)\{e^{-4(x-t)}[6(t - x) - 1] + e^{2(x-t)}\}\,dt$.

SECTION 8.2

1. $x_1(t) = c_1 e^t + c_2 e^{-t}, \; x_2(t) = \frac{1}{3}(c_1 e^t + 3c_2 e^{-t})$.
3. $x_1(t) = e^{-2t}(c_1 + c_2 t), \; x_2(t) = \frac{1}{4} e^{-2t}[-4c_1 + c_2(1 - 4t)]$.
5. $x_1(t) = e^t(c_1 \cos 3t + c_2 \sin 3t), \; x_2(t) = e^t(c_1 \sin 3t - c_2 \cos 3t)$.
7. $x_1(t) = -e^{-2t}(2c_1 + c_2 \cos t + c_3 \sin t), \; x_2(t) = -e^{2t}[c_1 + c_2(\cos t - \sin t) + c_3(\sin t + \cos t)]$.
9. $x_1(t) = 5 \sin t, \; x_2(t) = \cos t - 2 \sin t$.
11. $x_1(t) = c_1 e^{-t} + c_2 e^{3t} + 3e^{4t}, \; x_2(t) = -c_1 e^{-t} + c_2 e^{3t} + 2e^{4t}$.
13. $x_1(t) = c_1 e^{2t} + c_2 e^{-2t} + 2te^{2t}, \; x_2(t) = c_1 e^{2t} - 3c_2 e^{-2t} + e^{2t}(1 + 2t)$.
15. $x_1' = x_2, \; x_2' = -x_1 + 3x_4 + \sin t, \; x_3' = x_4, \; x_4' = tx_2 + e^t x_3 + t^2$.
17. $x_1' = x_2, \; x_2' = -bx_1 - ax_2 + F(t)$.
19. $x_1' = x_2, \; x_2' = -\dfrac{(k_1 + k_2)}{m_1} x_1 + \dfrac{k_2}{m_1} x_3, \; x_3' = x_4, \; x_4' = \dfrac{k_2}{m_2} x_1 - \dfrac{k_2}{m_2} x_3, \; x_1(0) = \alpha_1, \; x_2(0) = \alpha_2, \; x_3(0) = \alpha_3, \; x_4(0) = \alpha_4$.

SECTION 8.3

1. $x_1' = -4x_1 + 3x_2 + 4t, \; x_2' = 6x_1 - 4x_2 + t^2$. Matrix form: $\mathbf{x}' = A\mathbf{x} + \mathbf{b}$, where

$A = \begin{bmatrix} -4 & 3 \\ 6 & -4 \end{bmatrix}$ and $\mathbf{b} = \begin{bmatrix} 4t \\ t^2 \end{bmatrix}$.   **3.** $\mathbf{x}' = A\mathbf{x} + \mathbf{b}$, where $A = \begin{bmatrix} 0 & -\sin t & 1 \\ -e^t & 0 & t^2 \\ -t & t^2 & 0 \end{bmatrix}$ and $\mathbf{b} = \begin{bmatrix} t \\ t^3 \\ 1 \end{bmatrix}$.

5. New variables: $x_1 = x$, $x_2 = x'$, $x_3 = x''$. Linear system: $\mathbf{x}' = A\mathbf{x} + \mathbf{b}$, where

$$A = \begin{bmatrix} 0 & 1 & 0 \\ 0 & 0 & 1 \\ a^2 - t^2 & 0 & \sin t \end{bmatrix} \text{ and } \mathbf{b} = \begin{bmatrix} 0 \\ 0 \\ e^t \end{bmatrix}. \quad \textbf{7.} \; \frac{dA}{dt} = \begin{bmatrix} -2e^{-2t} \\ \cos t \end{bmatrix}. \quad \textbf{9.} \; \frac{dA}{dt} = \begin{bmatrix} e^t & 2e^{2t} & 2t \\ 2e^t & 8e^{2t} & 10t \end{bmatrix}.$$

13. $\begin{bmatrix} e - 1 & 1 - e^{-1} \\ 2(e - 1) & 5(1 - e^{-1}) \end{bmatrix}.$ **17.** $W[\mathbf{x}_1, \mathbf{x}_2](0) = 2 \neq 0.$ **21.** $4\mathbf{x}_1 - \mathbf{x}_2 = \mathbf{0}.$

25. $\mathbf{x}_1 = \begin{bmatrix} 4e^t \\ e^t \end{bmatrix}, \mathbf{x}_2 = \begin{bmatrix} e^{-2t} \\ e^{-2t} \end{bmatrix}.$

SECTION 8.4

1. General solution: $\mathbf{x}(t) = c_1 \begin{bmatrix} e^{4t} \\ 2e^{4t} \end{bmatrix} + c_2 \begin{bmatrix} 3e^{-t} \\ e^{-t} \end{bmatrix}.$

Particular solution: $\mathbf{x}(t) = \begin{bmatrix} e^{4t} & 3e^{-t} \\ 2e^{4t} & e^{-t} \end{bmatrix} \begin{bmatrix} 1 \\ -1 \end{bmatrix}.$

3. General solution: $\mathbf{x}(t) = \begin{bmatrix} -3 & e^{2t} & e^{4t} \\ 9 & 3e^{2t} & e^{4t} \\ 5 & e^{2t} & e^{4t} \end{bmatrix} \begin{bmatrix} c_1 \\ c_2 \\ c_3 \end{bmatrix}.$ **5.** $\mathbf{x}_1(t) = \begin{bmatrix} \cos 3t \\ -\sin 3t \end{bmatrix}, \mathbf{x}_2(t) = \begin{bmatrix} \sin 3t \\ \cos 3t \end{bmatrix}.$

7. $\mathbf{x}_1(t) = \begin{bmatrix} e^{-t} \\ -2e^{-t} \end{bmatrix}, \mathbf{x}_2(t) = \begin{bmatrix} te^{-t} \\ -e^{-t}(1 + 2t) \end{bmatrix}.$

SECTION 8.5

1. $\mathbf{x}(t) = c_1 e^{-3t} \begin{bmatrix} 7 \\ 1 \end{bmatrix} + c_2 e^{5t} \begin{bmatrix} -1 \\ 1 \end{bmatrix}.$

3. $\mathbf{x}(t) = e^{-2t} \left\{ c_1 \begin{bmatrix} 3\cos t - \sin t \\ 5\cos t \end{bmatrix} + c_2 \begin{bmatrix} \cos t + 3\sin t \\ 5\sin t \end{bmatrix} \right\}.$

5. $\mathbf{x}(t) = c_1 e^{-2t} \begin{bmatrix} 0 \\ 1 \\ 1 \end{bmatrix} + c_2 e^{2t} \begin{bmatrix} 1 \\ 0 \\ 0 \end{bmatrix} + c_3 e^{3t} \begin{bmatrix} 0 \\ 7 \\ 2 \end{bmatrix}.$

7. $\mathbf{x}(t) = c_1 e^{5t} \begin{bmatrix} 0 \\ 0 \\ 1 \end{bmatrix} + c_2 \begin{bmatrix} \sin t \\ \cos t \\ 0 \end{bmatrix} + c_3 \begin{bmatrix} -\cos t \\ \sin t \\ 0 \end{bmatrix}.$

9. $\mathbf{x}(t) = c_1 e^{-3t} \begin{bmatrix} -1 \\ 0 \\ 1 \end{bmatrix} + c_2 e^t \begin{bmatrix} -1 \\ -2 \\ 1 \end{bmatrix} + c_3 e^{2t} \begin{bmatrix} 2 \\ -10 \\ 3 \end{bmatrix}.$

11. $\mathbf{x}(t) = c_1 e^{3t} \begin{bmatrix} 1 \\ 0 \\ 0 \end{bmatrix} + e^{-2t} \left\{ c_2 \begin{bmatrix} 5\cos t - \sin t \\ -13(\cos t + \sin t) \\ 26\cos t \end{bmatrix} + c_3 \begin{bmatrix} \cos t + 5\sin t \\ 13(\cos t - \sin t) \\ 26\sin t \end{bmatrix} \right\}.$

13. $\mathbf{x}(t) = c_1 \begin{bmatrix} -3 \\ 0 \\ 2 \end{bmatrix} + c_2 \begin{bmatrix} 1 \\ 2 \\ 0 \end{bmatrix} + c_3 e^{4t} \begin{bmatrix} 1 \\ 1 \\ 1 \end{bmatrix}.$

15. $\mathbf{x}(t) = c_1 \begin{bmatrix} 0 \\ 0 \\ -\sin t \\ \cos t \end{bmatrix} + c_2 \begin{bmatrix} 0 \\ 0 \\ \cos t \\ \sin t \end{bmatrix} + c_3 \begin{bmatrix} \sin t \\ \cos t \\ 0 \\ 0 \end{bmatrix} + c_4 \begin{bmatrix} -\cos t \\ \sin t \\ 0 \\ 0 \end{bmatrix}.$

17. $\mathbf{x}(t) = e^{2t} \begin{bmatrix} 2\cos 3t - 6\sin 3t \\ 2\cos 3t + 4\sin 3t \end{bmatrix}.$ **19.** $\mathbf{x}(t) = \begin{bmatrix} \cos 2t + \sin 2t \\ -\sin 2t + \cos 2t \end{bmatrix}.$

SECTION 8.6

1. $\mathbf{x}(t) = e^{2t} \left\{ c_1 \begin{bmatrix} -1 \\ 1 \end{bmatrix} + c_2 \begin{bmatrix} 1 - 2t \\ 2t \end{bmatrix} \right\}.$ **3.** $\mathbf{x}(t) = c_1 e^t \begin{bmatrix} 1 \\ 1 \\ 1 \end{bmatrix} + e^{-t} \left\{ c_2 \begin{bmatrix} 1 \\ -1 \\ 1 \end{bmatrix} + c_3 \begin{bmatrix} 1 + t \\ -t \\ t - 1 \end{bmatrix} \right\}.$

5. $\mathbf{x}(t) = e^{-2t} \left\{ c_1 \begin{bmatrix} 1 \\ 0 \\ 1 \end{bmatrix} + c_2 \begin{bmatrix} 1 \\ 1 \\ 0 \end{bmatrix} + c_3 \begin{bmatrix} 1 \\ t \\ -t \end{bmatrix} \right\}.$

7. $\mathbf{x}(t) = e^{4t} \left\{ c_1 \begin{bmatrix} 0 \\ 0 \\ 1 \end{bmatrix} + c_2 \begin{bmatrix} 0 \\ 1 \\ t \end{bmatrix} + c_3 \begin{bmatrix} 2 \\ 2t \\ t^2 \end{bmatrix} \right\}.$

9. $\mathbf{x}(t) = e^{4t} \left\{ c_1 \begin{bmatrix} 0 \\ 0 \\ 1 \end{bmatrix} + c_2 \begin{bmatrix} 1 \\ 1 \\ 0 \end{bmatrix} + c_3 \begin{bmatrix} t \\ 1 + t \\ 0 \end{bmatrix} \right\}.$

11. $\mathbf{x}(t) = c_1 \begin{bmatrix} 5(2\sin t - \cos t) \\ -5(2\cos t + \sin t) \\ -2(\cos t + 2\sin t) \\ -\sin t \end{bmatrix} + c_2 \begin{bmatrix} -5(2\cos t + \sin t) \\ 5(\cos t - 2\sin t) \\ 2(\cos t - \sin t) \\ \cos t \end{bmatrix} + e^{2t}$

$\left\{ c_3 \begin{bmatrix} 0 \\ 0 \\ 1 \\ 0 \end{bmatrix} + c_4 \begin{bmatrix} 0 \\ 0 \\ t \\ 1 \end{bmatrix} \right\}.$

13. $\mathbf{x}(t) = c_1 \begin{bmatrix} 0 \\ 0 \\ -\sin t \\ \cos t \end{bmatrix} + c_2 \begin{bmatrix} 0 \\ 0 \\ \cos t \\ \sin t \end{bmatrix} + c_3 \begin{bmatrix} 0 \\ \cos t \\ \cos t - t\sin t \\ \sin t + t\cos t \end{bmatrix} + c_4 \begin{bmatrix} \cos t \\ \sin t \\ \sin t + t\cos t \\ t\sin t - \cos t \end{bmatrix}.$

15. $\mathbf{x}(t) = \begin{bmatrix} 7 - 2t - 9e^{-t} \\ e^{-t} \\ 3 - t - 2e^{-t} \end{bmatrix}.$

SECTION 8.7

1. $\mathbf{x}_p(t) = \begin{bmatrix} e^t(2t+1) \\ e^t(2t-1) \end{bmatrix}$.   **3.** $\mathbf{x}_p(t) = \begin{bmatrix} 4e^t(2e^{2t}-1) \\ 4e^t(3e^{2t}-2) \end{bmatrix}$.

5. $\mathbf{x}_p(t) = \begin{bmatrix} (12t+1)\sin 2t + (1-4t)\cos 2t \\ (8t-1)\cos 2t - 4t\sin 2t \end{bmatrix}$.   **7.** $\mathbf{x}_p(t) = \begin{bmatrix} -te^t \\ 9e^{-t} \\ te^t + 6e^{-t} \end{bmatrix}$.

SECTION 8.8

3. $x(t) = 2(c_1\sin t - c_2\cos t - c_3\sin 3t + c_4\cos 3t)$, $y(t) = 3c_1\sin t - 3c_2\cos t + c_3\sin 3t - c_4\cos 3t$.

9. $A_1(t) = 120 - 50e^{-t/5} - 10e^{-t/15}$, $A_2(t) = 120 + 100e^{-t/5} - 20e^{-t/15}$.

SECTION 8.9

7. $e^{At} = \begin{bmatrix} \frac{1}{2}(e^{4t}+e^{2t}) & \frac{1}{2}(e^{4t}-e^{2t}) \\ \frac{1}{2}(e^{4t}-e^{2t}) & \frac{1}{2}(e^{4t}+e^{2t}) \end{bmatrix}$.   **9.** $e^{At} = \begin{bmatrix} e^{-t}\cos 3t & e^{-t}\sin 3t \\ -e^{-t}\sin 3t & e^{-t}\cos 3t \end{bmatrix}$.

11. $e^{At} = \begin{bmatrix} 2e^{2t}-e^t & 2(e^t-e^{2t}) & e^t-e^{3t} \\ e^{2t}-e^t & 2e^t-e^{2t} & e^t-e^{3t} \\ 0 & 0 & e^{3t} \end{bmatrix}$.   **13.** $e^{At} = \begin{bmatrix} 1-3t & 9t \\ -t & 1+3t \end{bmatrix}$.

15. $e^{At} = \begin{bmatrix} 1 & 0 & 0 \\ t & 1 & 0 \\ \frac{1}{2}t^2 & t & 1 \end{bmatrix}$.

SECTION 8.10

3. $e^{At} = e^{2t}\begin{bmatrix} 1 & t \\ 0 & 1 \end{bmatrix}$.   **5.** $e^{At} = e^t\begin{bmatrix} 1+2t & -t \\ 4t & 1-2t \end{bmatrix}$.

7. $\mathbf{x}_1(t) = e^{2t}\begin{bmatrix} 1 \\ 0 \\ 0 \end{bmatrix}$, $\mathbf{x}_2(t) = e^{-3t}\begin{bmatrix} 0 \\ 1+4t \\ 2t \end{bmatrix}$, $\mathbf{x}_3(t) = e^{-3t}\begin{bmatrix} 0 \\ -8t \\ 1-4t \end{bmatrix}$,

$e^{At} = \begin{bmatrix} e^{2t} & 0 & 0 \\ 0 & e^{-3t}(1+4t) & -8te^{-3t} \\ 0 & 2te^{-3t} & e^{-3t}(1-4t) \end{bmatrix}$.   **9.** $\mathbf{x}(t) = c_1e^{-t}\begin{bmatrix} -7 \\ 4 \\ 1 \end{bmatrix} + e^{3t}\left\{ c_2\begin{bmatrix} 1 \\ 0 \\ 1 \end{bmatrix} + c_3\begin{bmatrix} t \\ 1 \\ t \end{bmatrix} \right\}$.

11. $\mathbf{x}_1(t) = \begin{bmatrix} -\sin t \\ \cos t \\ -t\sin t \\ t\cos t \end{bmatrix}$, $\mathbf{x}_2(t) = \begin{bmatrix} \cos t \\ \sin t \\ t\cos t \\ t\sin t \end{bmatrix}$, $\mathbf{x}_3(t) = \begin{bmatrix} 0 \\ 0 \\ -\sin t \\ \cos t \end{bmatrix}$, $\mathbf{x}_4(t) = \begin{bmatrix} 0 \\ 0 \\ \cos t \\ \sin t \end{bmatrix}$.

SECTION 8.11

1. $(0, 0)$, $(-1, 0)$, $(-\frac{1}{3}, \frac{2}{3})$. **3.** $(0, 0)$, $(1, 0)$. **5.** Center. **7.** Saddle. **9.** Stable node.
11. Center. **13.** Unstable spiral. **15.** Unstable node. **17.** Stable degenerate node.
19. Unstable proper node.

SECTION 8.12

1. $(0, 0)$, center or spiral. **3.** $(0, 0)$, saddle; $(\frac{4}{9}, \frac{2}{3})$ unstable spiral; $(\frac{4}{9}, -\frac{2}{3})$, unstable spiral.
5. $(0, 0)$, unstable node. **7.** $(0, 0)$, unstable spiral; $(\frac{1}{2}, -1)$, saddle. **9.** $(0, 0)$, linearized system
has unstable degenerate node.

SECTION 9.1

1. $\dfrac{1}{s - 2}$. **3.** $\dfrac{b}{s^2 + b^2}$. **5.** $\dfrac{s}{s^2 - b^2}$. **7.** $\dfrac{2}{s^2}$. **9.** $\dfrac{1}{s}(1 - 2e^{-2s})$. **11.** $\dfrac{1}{(s - 1)^2 + 1}$.

13. $\dfrac{2}{s^2} - \dfrac{1}{s - 3}$. **15.** $\dfrac{s}{s^2 - b^2}$. **17.** $\dfrac{6}{s^3} - \dfrac{5s}{s^2 + 4} + \dfrac{3}{s^2 + 9}$. **19.** $\dfrac{2}{s + 3} + \dfrac{4}{s - 1} - \dfrac{5}{s^2 + 1}$.

21. $\dfrac{4(s^2 + 2b^2)}{s(s^2 + 4b^2)}$. **23.** Piecewise continuous. **25.** Not piecewise continuous. **27.** Piecewise

continuous. **29.** Not piecewise continuous. **31.** $\dfrac{1}{s^2}[1 - e^{-s}(s + 1)]$.

33. $\dfrac{1}{s^2} e^{-2s}[e^s(s + 1) - (2s + 1)]$.

SECTION 9.2

7. 2. **9.** $5e^{-3t}$. **11.** $2 \cos 3t$. **13.** $\cos t + 6 \sin t$. **15.** $2 - 3e^{-t}$. **17.** $1 - e^{-t}$.
19. $\frac{1}{5}(7e^{2t} - 7 \cos t - 4 \sin t)$. **21.** $\frac{1}{6}(4 \cos t + 6 \sin t - 4 \cos 2t - 3 \sin 2t)$.

SECTION 9.3

1. $F(s) = \dfrac{1}{s^2(1 - e^{-s})}[1 - e^{-s}(s + 1)]$. **3.** $F(s) = \dfrac{1 + e^{-\pi s}}{(1 - e^{-\pi s})(s^2 + 1)}$.

5. $F(s) = \dfrac{e^{(1 - s)} - 1}{(1 - s)(1 - e^{-s})}$. **7.** $F(s) = \dfrac{1}{1 - e^{-\pi s}} \left[\dfrac{2 - e^{-\pi s/2}(\pi s + 2)}{\pi s^2} + \dfrac{se^{-\pi s/2} + e^{-\pi s}}{s^2 + 1} \right]$.

9. $F(s) = \dfrac{1}{as^2} \tanh \dfrac{as}{2}$.

SECTION 9.4

1. $y(t) = 2e^{3t}$. **3.** $y(t) = 2t - 1 + 2e^{-2t}$. **5.** $y(t) = e^t - \sin 2t - 2 \cos 2t$. **7.** $y(t) = 2e^t - e^{-2t}$.
9. $y(t) = 2 + 3e^{2t} - 5e^t$. **11.** $y(t) = 2e^t + 3e^{-2t} - 5e^{-t}$. **13.** $y(t) = 3e^{2t} + 2e^{-3t} - 4$.
15. $y(t) = 2 \cos 2t + \sin 2t + 2e^{-t}$. **17.** $y(t) = \frac{7}{2} e^t - \frac{1}{2} e^{-t} - 3 \cos t$.

19. $y(t) = e^t - 2e^{-t} - 4 \sin t + 3 \cos t$. **21.** $y(t) = 2e^{-t} - e^{-4t} - 2 \cos 2t$.
23. $y(t) = \frac{1}{5}(7e^{2t} - 5e^t + 3 \cos t - 4 \sin t)$. **25.** $y(t) = 2(\cos t + \sin t - \cos 2t)$.

27. $y(t) = y(0) \cosh t + y'(0) \sinh t$. **29.** $i(t) = \dfrac{E_0}{R}[1 - e^{-(R/L)t}]$.

31. $x_1(t) = 2(e^{-3t} - e^{-2t})$, $x_2(t) = 2e^{-2t} - e^{-3t}$.

SECTION 9.5

1. $t - 1$.   **3.** $t(t + 2)$.   **5.** $-e^{2(t-\pi)}\cos t$.   **7.** $\frac{1}{2}e^{-(t-\pi/6)}(\sin 2t - \sqrt{3}\cos 2t)$.   **9.** $\dfrac{t - 1}{t^2 - 6t + 10}$.

11. t^2.   **13.** te^{3t}.   **15.** $(t + 3)e^{-t}$.   **17.** $\dfrac{s - 3}{(s - 3)^2 + 16}$.   **19.** $\dfrac{1}{(s - 2)^2}$.   **21.** $\dfrac{6}{(s + 4)^4}$.

23. $\dfrac{2(2s^3 - 9s^2 + 10s + 30)}{[(s - 3)^2 + 1][(s + 1)^2 + 9]}$.   **25.** $\dfrac{2}{(s - 1)^3} - \dfrac{6}{s^3}$.   **27.** te^{3t}.   **29.** $\dfrac{2}{\sqrt{\pi}}e^{-3t}t^{-1/2}$.

31. $e^{-2t}\cos 3t$.   **33.** $\frac{5}{4}e^{2t}\sin 4t$.   **35.** $\frac{1}{5}e^{-t}(5\cos 5t - 3\sin 5t)$.   **37.** $\frac{1}{2}e^{-t}(2\cos 2t - \sin 2t)$.

39. $1 - e^{-2t}(1 + 2t)$.   **41.** $\frac{1}{10}[6 + e^t(13\sin 2t - 6\cos 2t)]$.   **43.** $y(t) = e^{2t}(1 + 3t) + e^{-2t}$.

45. $y(t) = 2e^t + e^{-2t}(1 - t)$.   **47.** $y(t) = e^{-t}(2 + 3t + t^2)$.   **49.** $y(t) = \frac{7}{3}e^{-t} - \frac{7}{4}e^{-2t} + \frac{1}{12}e^{2t}(12t - 7)$.

51. $y(t) = 2e^t + e^{-t} - e^t(\sin 2t + \cos 2t)$.   **53.** $x_1(t) = e^{2t}\cos t, x_2(t) = e^{2t}\sin t$.

SECTION 9.6

7. $f(t) = 3 - 4u_1(t)$.   **9.** $f(t) = 2 - u_2(t) - 2u_4(t)$.   **11.** $f(t) = t + 2(3 - t)u_3(t) - (6 - t)u_6(t)$.

13. $f(t) = 1 + (\sin t - 1)u_{\pi/2}(t) - (\sin t + 1)u_{3\pi/2}(t)$.

SECTION 9.7

1. $F(s) = \dfrac{1}{s^2}e^{-s}$.   **3.** $F(s) = \dfrac{e^{-\pi s/4}}{s^2 + 1}$.   **5.** $F(s) = \dfrac{2}{s^3}e^{-2s}$.   **7.** $F(s) = \left(\dfrac{s^2 + s + 2}{s^3}\right)e^{-2s}$.

9. $F(s) = \dfrac{3}{(s + 2)^2 + 9}e^{-s}$.   **11.** $f(t) = (t - 2)u_2(t)$.   **13.** $f(t) = e^{-4(t-3)}u_3(t)$.

15. $f(t) = u_3(t)\sin(t - 3)$.   **17.** $f(t) = \frac{1}{5}u_1(t)[e^{4(t-1)} - e^{-(t-1)}]$.

19. $f(t) = [\cos 3(t - 1) + 2\sin 3(t - 1)]u_1(t)$.   **21.** $f(t) = \frac{1}{2}u_2(t)(t - 2)^2 e^{3(t-2)}$.

23. $f(t) = e^{-2(t-1)}[2\cos(t - 1) - 5\sin(t - 1)]u_1(t)$.

25. $f(t) = [2e^{-(t-3)}(5t - 13) - 4\cos 2(t - 3) - 3\sin(2t - 3)]u_3(t)$.

27. $y(t) = 2e^{2t} - u_2(t)[e^{(t-2)} - e^{2(t-2)}]$.   **29.** $y(t) = 3e^{-2t} + \frac{1}{4}u_\pi(t)[e^{-2(t-\pi)} - \cos 2t + \sin 2t]$.

31. $y(t) = \frac{1}{10}\{21e^{3t} - \cos t - 3\sin t + \frac{1}{3}u_{\pi/2}(t)[e^{3(t-\pi/2)} - 10 + 9\sin t + 3\cos t]\}$.

33. $y(t) = 2\cosh t + u_1(t)[\cosh(t - 1) - 1]$.

35. $y(t) = 2\sinh 2t + \frac{1}{4}u_1(t)[\cosh 2(t - 1) - 1] - \frac{1}{4}u_2(t)[\cosh 2(t - 2) - 1]$.

37. $y(t) = 2e^{-t} - e^{-2t} + u_{\pi/4}(t)[5e^{-(t-\pi/4)} - 2e^{-2(t-\pi/4)} - 3\cos(t - \pi/4) + \sin(t - \pi/4)]$.

39. $y(t) = e^{-2t}(2\cos t + 5\sin t) + u_3(t)\{1 - e^{-2(t-3)}[\cos(t - 3) - 2\sin(t - 3)]\}$.

41. $y(t) = 2e^{-t} + u_1(t)[e^{-(t-1)} + t - 2] - 2u_2(t)[e^{-(t-2)} + t - 3] + u_3(t)[e^{-(t-3)} + t - 4]$.

43. $y(t) = 3e^t - 1 - t - \frac{1}{2}u_1(t)[3e^{(t-1)} + e^{-(t-1)} - 2 - 2t]$.   **45.** $y(t) = 3e^t - 2 + 3u_1(t)(1 - e^{t-1})$.

47. $i(t) = \dfrac{20}{R}\left\{e^{-at} + u_{10}(t)\left[\dfrac{1}{a - 1}(ae^{-a(t-10)} - e^{-(t-10)}) - e^{-at}\right]\right\}$ where $a = 1/RC$.

SECTION 9.8

1. $y(t) = e^{2t} + u_2(t)e^{2(t-2)}$.   **3.** $y(t) = \frac{1}{3}(e^{5t} - e^{-t}) + u_3(t)e^{5(t-3)}$.

5. $y(t) = \frac{1}{2}[\sinh 2t + u_3(t)\sinh 3t]$.

7. $y(t) = e^{2t}(3\cos 3t - 2\sin 3t) - \dfrac{\sqrt{2}}{6}e^{2(t-\pi/4)}(\sin 3t + \cos 3t)u_{\pi/4}(t)$.

9. $y(t) = 5e^{-3t}(\cos 2t + 2 \sin 2t) - \frac{1}{2} e^{-3(t-\pi/4)}u_{\pi/4}(t)\cos 2t.$

11. $y(t) = \frac{4}{7} (\cos 3t - \cos 4t) + \frac{1}{4} u_{\pi/3}(t) \sin 4(t - \pi/3).$

13. $x(t) = \dfrac{F_0}{5} (\cos 2t - \cos 3t) - \dfrac{5}{2} u_1(t)\sin 2(t - 1).$

15. $y(t) = \dfrac{F_0}{\omega_0(\omega^2 - \omega_0^2)} (\omega \sin \omega_0 t - \omega_0 \sin \omega t) + \dfrac{A}{\omega_0} u_{t_0}(t)\sin \omega_0(t - t_0).$

SECTION 9.9

1. $f * g = \frac{1}{2}t^2.$ **3.** $f * f = e^t - 1 - t.$ **5.** $f * g = e^t(1 - \cos t).$ **9.** $\mathrm{L}[f * g] = \dfrac{1}{s^2(s^2 + 1)}.$

11. $\mathrm{L}[f * g] = \dfrac{s}{(s^2 + 1)(s^2 + 4)}.$ **13.** $\mathrm{L}[f * g] = \dfrac{4}{s^3[(s - 3)^2 + 4]}.$ **15.** $\mathrm{L}^{-1}[FG] = 1 - e^{-t}.$

17. $\mathrm{L}^{-1}[FG] = \dfrac{1}{3}e^{-2t}\sin 3t.$ **19.** $\mathrm{L}^{-1}[FG] = \begin{cases} 0, & \text{if } t < \pi, \\ t - \pi + \sin t, & \text{if } t \geq \pi. \end{cases}$

21. $\mathrm{L}^{-1}[FG] = \displaystyle\int_0^t e^{-(t+2\tau)}\tau \cos(t - \tau)\, d\tau.$

23. $\mathrm{L}^{-1}[FG] = \begin{cases} 0, & 0 \leq t < \pi/2, \\ \displaystyle\int_{\pi/2}^t e^{-4(t-\tau)}\cos[3(t - \tau)] \cos 4\tau\, d\tau, & t \geq \pi/2. \end{cases}$

25. $y(t) = \displaystyle\int_0^t e^{-\tau}\sin(t - \tau)\, d\tau + \sin t.$

27. $y(t) = \dfrac{1}{4} \displaystyle\int_0^t f(t - \tau)\sin 4\tau\, d\tau + \dfrac{1}{4} (4\alpha \cos 4t + \beta \sin 4t).$

29. $y(t) = \dfrac{1}{a} \displaystyle\int_0^t f(t - \tau)\sinh a\tau\, d\tau + \dfrac{1}{a} (\alpha a \cosh at + \beta \sinh at).$

31. $y(t) = \dfrac{1}{b} \displaystyle\int_0^t f(t - \tau)e^{a\tau}\sin b\tau\, d\tau + e^{at}\left[\alpha \cos bt - \dfrac{(2\alpha a - \beta)}{b} \sin bt\right].$

33. $x(t) = e^{3t} + e^t.$ **35.** $x(t) = 2 \cosh t - 1.$ **37.** $x(t) = 2 + \dfrac{4}{\sqrt{3}} \sin \sqrt{3}t.$

SECTION 10.2

1. $R = 4.$ **3.** $R = 1/2.$ **5.** $R = \infty.$ **7.** $R = 1.$ **9.** $R = \sqrt{5}.$ **11.** (a) Analytic for $x \neq \pm 1.$

(b) $R = \begin{cases} 1 - |x_0|, & \text{if } |x_0| < 1, \\ |x_0| - 1, & \text{if } |x_0| > 1. \end{cases}$ **13.** $f(x) = a_0(1 + \frac{2}{3}x + \frac{1}{12}x^2).$

SECTION 10.3

1. $y_1(x) = \displaystyle\sum_{n=0}^{\infty} \dfrac{1}{(2n)!} x^{2n},\ y_2(x) = \displaystyle\sum_{n=0}^{\infty} \dfrac{1}{(2n + 1)!} x^{2n+1},\ R = \infty.$

3. $y_1(x) = 1 + \sum\limits_{n=1}^{\infty} \dfrac{(-2)^n}{1 \cdot 3 \cdot \cdots \cdot (2n-1)} x^{2n}$, $y_2(x) = \sum\limits_{n=0}^{\infty} \dfrac{(-1)^n}{n!} x^{2n+1}$, $R = \infty$.

5. $y_1(x) = \sum\limits_{n=0}^{\infty} \dfrac{1}{3^n n!} x^{3n}$, $y_2(x) = \sum\limits_{n=0}^{\infty} \dfrac{1}{1 \cdot 4 \cdot \cdots \cdot (3n+1)} x^{3n+1}$, $R = \infty$.

7. $y_1(x) = \sum\limits_{n=0}^{\infty} (-1)^n \dfrac{1 \cdot 3 \cdot \cdots \cdot (2n+1)}{(2n)!} x^{2n}$, $y_2(x) = \sum\limits_{n=0}^{\infty} (-2)^n \dfrac{(n+1)!}{(2n+1)!} x^{2n+1}$, $R = \infty$.

9. $y_1(x) = \sum\limits_{n=0}^{\infty} (-1)^n x^{2n}$, $y_2(x) = \sum\limits_{n=0}^{\infty} (-1)^n x^{2n+1}$, $R = 1$.

11. $y_1(x) = x(1 + x^2)$, $y_2(x) = 1 + 6x^2 + x^4$, $R = \infty$.

13. $y_1(x) = 1 - x^2 - \frac{1}{6} x^3 + \frac{1}{3} x^4 + \frac{11}{120} x^5 + \cdots$, $y_2(x) = 1 - \frac{1}{2} x^3 - \frac{1}{12} x^4 + \frac{1}{8} x^5 + \cdots$, $R = \infty$.

15. $y_1(x) = 1 + \frac{1}{2} x^2 + \frac{1}{6} x^3 + \frac{1}{12} x^4 + \frac{1}{24} x^5 + \cdots$, $y_2(x) = x + \frac{1}{6} x^3 + \frac{1}{12} x^4 + \frac{1}{30} x^5 + \cdots$, $R = \infty$.

17. (a) No. **(b)** $y_1(x) = 1 + \frac{1}{2} (x-1)^2 + \frac{1}{8} (x-1)^4 + \cdots$,

$y_2(x) = (x-1) + \frac{1}{3} (x-1)^3 - \frac{1}{12} (x-1)^4 + \cdots$, R is at least 1.

19. (a) $y(x) = \sum\limits_{n=0}^{\infty} (-1)^n \dfrac{(n+1)!}{2^n (2n)!} x^{2n}$. **(b)** Polynomial approximation:

$y_8(x) = 1 - \frac{1}{2} x^2 + \frac{1}{16} x^4 - \frac{1}{240} x^6 + \frac{1}{5376} x^8$, error is less than 6.3×10^{-6} on $[-1, 1]$.

21. $y(x) = a_0(1 + 2x^2 + \frac{1}{3} x^4 + \cdots) + a_1(x + \frac{1}{2} x^3 + \frac{1}{40} x^5 + \cdots) +$

$(3x^2 + x^3 + \frac{13}{24} x^4 + \frac{1}{10} x^5 + \frac{1}{120} x^6 + \cdots)$.

SECTION 10.4

1. $\alpha = 3$: $y_2(x) = a_1 x(1 - \frac{5}{3} x^2)$; $\alpha = 4$: $y_2(x) = a_0(1 - 10x^2 + \frac{35}{3} x^4)$. $P_3(x)$ and $P_4(x)$ are given in Table 10.4.1. **5.** $2x^3 + x^2 + 5 = \frac{16}{3} P_0 + \frac{6}{5} P_1 + \frac{2}{3} P_2 + \frac{4}{5} P_3$.

9. $\alpha = 0$: $y_1(x) = 1$; $\alpha = 1$: $y_2(x) = x$; $\alpha = 2$: $y_1(x) = 1 - 2x^2$; $\alpha = 3$: $y_2(x) = x(1 - \frac{2}{3} x^2)$.

SECTION 10.5

1. $x = 1$ is a regular singular point. All other points are ordinary points.

3. $x = 0, 2$ are regu;ar singular points. All other points are ordinary points. **5.** $r = \pm\sqrt{7}$.

7. $r = 0, 1$. **9.** $y_1(x) = x^{1/4} \left\{ 1 + \sum\limits_{n=1}^{\infty} \dfrac{(-1)^n}{n! [5 \cdot 9 \cdot \cdots \cdot (4n+1)]} x^n \right\}$,

$y_2(x) = 1 + \sum\limits_{n=1}^{\infty} \dfrac{(-1)^n}{n! [3 \cdot 7 \cdot \cdots \cdot (4n-1)]} x^n$.

11. $y_1(x) = x^{\sqrt{2}} \left[1 + \sum\limits_{n=1}^{\infty} \dfrac{1}{n! (1 + 2\sqrt{2})(2 + 2\sqrt{2}) \cdot \cdots \cdot (n + 2\sqrt{2})} x^n \right]$,

$y_2(x) = x^{-\sqrt{2}} \left[1 + \sum\limits_{n=1}^{\infty} \dfrac{1}{n! (1 - 2\sqrt{2})(2 - 2\sqrt{2}) \cdot \cdots \cdot (n - 2\sqrt{2})} x^n \right]$.

13. $y_1(x) = x^3 \left[1 + \sum_{n=1}^{\infty} \frac{(n + 1)(n + 2)}{10 \cdot 13 \cdot \cdots \cdot (3n + 7)} x^n \right]$, $y_2(x) = x^{2/3} \left[1 + \sum_{n=1}^{\infty} \frac{(3n - 4)(3n - 1)}{n! 3^n} x^n \right]$.

15. $y_1(x) = x^{\sqrt{5}} \left[1 + \sum_{n=1}^{\infty} \frac{(1 + \sqrt{5})(2 + \sqrt{5}) \cdots \cdots (n + \sqrt{5})}{n!(1 + 2\sqrt{5})(2 + 2\sqrt{5}) \cdots \cdots (n + 2\sqrt{5})} x^n \right]$,

$y_2(x) = x^{\sqrt{-5}} \left[1 + \sum_{n=1}^{\infty} \frac{(1 - \sqrt{5})(2 - \sqrt{5}) \cdots \cdots (n - \sqrt{5})}{n!(1 - 2\sqrt{5})(2 - 2\sqrt{5}) \cdots \cdots (n - 2\sqrt{5})} x^n \right]$.

17. $y_1(x) = (1 + \frac{1}{5} x - \frac{3}{100} x^2 + \cdots) \cos(\ln x) + \frac{1}{25} x (10 + x + \cdots) \sin(\ln x)$,

$y_2(x) = (1 + \frac{1}{5} x - \frac{3}{100} x^2 + \cdots) \sin(\ln x) - \frac{1}{25} x (10 + x + \cdots) \cos(\ln x)$.

19. $y_1(x) = \sqrt{x}$, $y_2(x = \sqrt{x} \left(\ln x + \sum_{n=1}^{\infty} \frac{1}{n \cdot n!} x^n \right)$.

21. $N = 0$: $y(x) = 1$; $N = 1$: $y(x) = x^{-1}(1 - x)$; $N = 2$: $y(x) = x^{-2}(1 - 2x + \frac{1}{2} x^2)$;
$N = 3$: $y(x) = x^{-3}(1 - 3x + \frac{3}{2} x^2 - \frac{1}{6} x^3)$.

SECTION 10.6

1. $y_1(x) = x^{1/2} \sum_{n=0}^{\infty} a_n x^n$, $y_2(x) = y_1(x) \ln x + x^{1/2} \sum_{n=1}^{\infty} b_n x^n$. **3.** $y_1(x) = x^2 \sum_{n=0}^{\infty} a_n x^n$,

$y_2(x) = x^{-1} \sum_{n=0}^{\infty} b_n x^n$. **5.** $y_2(x) = x \sum_{n=0}^{\infty} a_n x^n$, $y_2(x) = A y_1(x) \ln x + x^{-1} \sum_{n=0}^{\infty} b_n x^n$, $A \neq 0$.

7. $y_1(x) = x^{-3} \sum_{n=1}^{\infty} a_n x^n$, $y_2(x) = y_1(x) \ln x + x^{-3} \sum_{n=1}^{\infty} b_n x^n$.

9. **(b)** $y_1(x) = x^{-1} \sum_{n=0}^{\infty} a_n x^n$, $y_2(x) = y_1(x) \ln x + x^{-1} \sum_{n=1}^{\infty} b_n x^n$.

(c) $y_1(x) = x^N \sum_{n=0}^{\infty} a_n x^n$, $y_2(x) = A y_1(x) + x^{-1} \sum_{n=0}^{\infty} b_n x^n$, where $A = 0$ if and only if
$\gamma = 1, 0, \ldots, 1 - N$.

11. $y_1(x) = x^{-1} \sum_{n=0}^{\infty} \frac{1}{(n!)^2} x^n$, $y_2(x) = y_1(x) \ln x - (2 + \frac{3}{4} x + \frac{11}{108} x^2 + \cdots)$.

13. $y_1(x) = x \sum_{n=0}^{\infty} \frac{1}{(n + 2)!} x^n = x^{-1}(e^x - x - 1)$, $y_2(x) = x^{-1}(1 + x)$.

15. $y_1(x) = \sum_{n=0}^{\infty} \frac{(2x)^n}{(n!)^2}$, $y_2(x) = y_1(x) \ln x - (4x + 3x^2 + \frac{22}{27} x^3 + \cdots)$.

17. $y_1(x) = x^2 \sum_{n=0}^{\infty} \frac{(n + 1)}{n!} x^n$, $y_2(x) = y_1(x) \ln x - x^2(3x + \frac{13}{4} x^2 + \frac{31}{18} x^3 + \cdots)$.

19. $y_1(x) = x^2 \left[1 + \sum_{n=1}^{\infty} \frac{(n + 4)}{n!} x^n \right]$, $y_2(x) = 2y_1(x) \ln x + x^{-1}(1 - x + \frac{3}{2} x^2 - 12x^4 + \cdots)$.

21. $y_1(x) = x^{3/2} \sum_{n=0}^{\infty} \frac{1}{(n + 3)!} x^n = x^{-3/2}(e^x - 1 - x - \frac{1}{2} x^2)$, $y_2(x) = x^{-3/2}(1 + x + \frac{1}{2} x^2)$.

23. $y_1(x) = x$, $y_2(x) = y_1(x) + \displaystyle\sum_{n=1}^{\infty} \frac{(-1)^n}{n \cdot n!} x^{n+1}$.

25. $y_1(x) = x^{3/2} \displaystyle\sum_{n=0}^{\infty} \frac{1}{n!(n+2)!} x^n$, $y_2(x) = 2y_1(x)\ln x - 5x^{-1/2}(1 - x + \frac{1}{5} x^3 + \frac{37}{960} x^4 + \cdots)$.

27. $y_1(x) = x$, $y_2(x) = x \ln x - 1 + \displaystyle\sum_{n=2}^{\infty} \frac{1}{n!(n-1)} x^n$.

29. $y(x) = 1 + \displaystyle\sum_{k=1}^{N} \frac{(-1)^k N(N-1)\cdots\cdot(N-k+1)}{(k!)^2} x^k$.

SECTION 10.7

7. $J_{3/2}(x) = \sqrt{\dfrac{2}{\pi}} x^{-3/2}(\sin x - x \cos x)$, $J_{-3/2}(x) = -\sqrt{\dfrac{2}{\pi}} x^{-3/2}(x \sin x + \cos x)$.

9. $x^p = \displaystyle\sum_{n=1}^{\infty} \frac{2}{\lambda_n J_{p+1}(\lambda_n)} J_p(\lambda_n x)$.

APPENDIX 1

1. $\bar{z} = 2 - 5i, |z| = \sqrt{29}$.   **3.** $\bar{z} = 5 + 2i, |z| = \sqrt{29}$.   **5.** $\bar{z} = 1 - 2i, |z| = \sqrt{5}$.

7. $z_1 z_2 = 1 + 7i, \dfrac{z_1}{z_2} = -1 + i$.   **9.** $z_1 z_2 = 7 + 11i, \dfrac{z_1}{z_2} = \dfrac{1}{10}(1 - 13i)$.

APPENDIX 2

1. $\dfrac{5}{x+2} - \dfrac{3}{x+1}$.   **3.** $\dfrac{4}{5(x-3)} + \dfrac{1}{5(x+2)}$.   **5.** $\dfrac{1}{3(x+1)} + \dfrac{1}{6(x-2)} - \dfrac{1}{2(x+4)}$.

7. $\dfrac{3}{x+1} - \dfrac{1}{(x+1)^2} - \dfrac{3}{x+2}$.   **9.** $\dfrac{3}{4x} + \dfrac{1}{x^2} - \dfrac{3x+4}{4(x^2+4)}$.   **11.** $\dfrac{1}{2(x-2)} + \dfrac{x+2}{2(x^2+16)}$.

13. $\dfrac{1}{x-2} - \dfrac{1}{x+2} + \dfrac{3}{(x+2)^2}$.   **15.** $\dfrac{2}{3(x-2)} - \dfrac{2}{3(x+1)} - \dfrac{2}{(x+1)^2} + \dfrac{1}{(x+1)^3}$.

17. $\dfrac{1}{2(x-3)} - \dfrac{x+1}{2(x^2+4x+5)}$.

APPENDIX 3

Note that we have omitted the integration constants.

1. $\cos x + x \sin x$.   **3.** $x \ln x - x$.   **5.** $\frac{1}{2} e^{x^2} (x^2 - 1)$.   **7.** $x - 3 \ln|x| + 2|$.

9. $\frac{1}{4} \ln |x| - \frac{1}{8} \ln(x^2 + 4) + \tan^{-1}\frac{x}{2}$.   **11.** $\frac{1}{2} x + \frac{7}{4} \ln| 2x - 1 |$.

13. $2 \ln |x| - \dfrac{1}{x+1} - 2 \ln|x + 1|$.   **15.** $\tan^{-1}(x + 1)$.   **17.** $-\ln| \cos x |$.   **19.** $\frac{1}{4} (2x + \sin 2x)$.

21. $\dfrac{1}{13} e^{3x} (3 \sin 2x - 2 \cos 2x)$.

APPENDIX 4

1. $y_1(x) = 1 + \dfrac{x^2}{2} - \dfrac{x^3}{3}$; $y_2(x) = 1 + \dfrac{x^2}{2} - \dfrac{x^3}{3} + \dfrac{x^2}{8} - \dfrac{x^5}{15}$.   **3.** $y(x) = \displaystyle\sum_{k=0}^{\infty} \frac{1}{k!} x^{2k} = e^{x^2}$.

Index

VECTOR SPACES

1. A set of vectors $\{\mathbf{v}_1, \mathbf{v}_2, ..., \mathbf{v}_k\}$ in a vector space V is said to:

   (a) be *linearly dependent* if there exist scalars $c_1, c_2, ..., c_k,$ *not all zero*, such that
   $$c_1\mathbf{v}_1 + c_2\mathbf{v}_2 + \cdots + c_k\mathbf{v}_k = \mathbf{0}.$$

   (b) be *linearly independent* if the *only* values of the scalars $c_1, c_2, ..., c_k$ such that
   $$c_1\mathbf{v}_1 + c_2\mathbf{v}_2 + \cdots + c_k\mathbf{v}_k = \mathbf{0} \text{ are } c_1 = c_2 = \cdots = c_k = 0.$$

2. A LI set of vectors that spans a vector space V is called a *basis* for V.

   (a) All bases in a finite-dimensional vector space V contain the same number of vectors, and this number is called the *dimension* of V, denoted by dim[V].

   (b) If dim[V] = n, then *any* LI set of n vectors in V is a basis for V.

LINEAR TRANSFORMATIONS

A mapping $T : V \rightarrow W$ from the vector space V into the vector space W is called a *linear transformation* if it satisfies
$$T(\mathbf{x} + \mathbf{y}) = T(\mathbf{x}) + T(\mathbf{y}), \text{ for all } \mathbf{x} \text{ and } \mathbf{y} \text{ in } V,$$
$$T(c\mathbf{x}) = cT(\mathbf{x}), \text{ for all } \mathbf{x} \text{ in } V \text{ and all scalars } c.$$

1. The *kernel* of T, denoted Ker(T), is the set of all vectors in V that are mapped to the zero vector in W. Thus, Ker(T) = $\{\mathbf{x} \in V : T(\mathbf{x}) = \mathbf{0}\}$. Ker($T$) is a subspace of V.

2. The *range* of T, denoted Rng(T), is the set of vectors in W that we obtain when we allow T to act on every vector in V. Equivalently, Rng(T) is the set of all transformed vectors. Thus Rng(T) = $\{T(\mathbf{v}) \in W : \mathbf{v} \in V\}$. Rng($T$) is a subspace of W.

EIGENVALUES AND EIGENVECTORS

1. For a given $n \times n$ matrix A, the eigenvalue/eigenvector problem consists of determining all scalars λ and all *nonzero* vectors $\mathbf{v}$ such that $A\mathbf{v} = \lambda\mathbf{v}$.

2. The eigenvalues of A are the roots of the characteristic polynomial
$$p(\lambda) = \det(A - \lambda I) = 0, \tag{1}$$
and the eigenvectors of A are obtained by solving the linear systems
$$(A - \lambda I)\mathbf{v} = \mathbf{0}, \tag{2}$$
when λ assumes the values obtained in (1).

3. If A is nondefective and $S = [\mathbf{v}_1, \mathbf{v}_2, ..., \mathbf{v}_n]$, where $\mathbf{v}_1, \mathbf{v}_2, ..., \mathbf{v}_n$ are LI eigenvectors of A, then
$$S^{-1}AS = \operatorname{diag}(\lambda_1, \lambda_2, ..., \lambda_n),$$
where $\lambda_1, \lambda_2, ..., \lambda_n$ are the eigenvalues of A corresponding to the eigenvectors $\mathbf{v}_1, \mathbf{v}_2, ..., \mathbf{v}_n$.

4. If A is a *real symmetric* matrix, then it has a complete orthonormal set of real eigenvectors, say $\{\mathbf{w}_1, \mathbf{w}_2, ..., \mathbf{w}_n\}$. If $S = [\mathbf{w}_1, \mathbf{w}_2, ..., \mathbf{w}_n]$, then S is an orthogonal matrix $(S^{-1} = S^T)$, and $S^TAS = \operatorname{diag}(\lambda_1, \lambda_2, ..., \lambda_n)$, where $\lambda_1, \lambda_2, ..., \lambda_n$ are the eigenvalues of A corresponding to the eigenvectors $\mathbf{w}_1, \mathbf{w}_2, ..., \mathbf{w}_n$.